Clinical Neuroanatomy:
A Neurobehavioral Approach

John E. Mendoza, Ph.D.
Anne L. Foundas, M.D.

Clinical Neuroanatomy: A Neurobehavioral Approach

John E. Mendoza
SE Louisiana Veterans Healthcare System
New Orleans, LA, USA
Tulane Medical School
Dept. Psychiatry & Neurology
LSU Medical School
Dept. of Psychiatry
john.mendoza2@med.va.gov

Anne L. Foundas
Tulane Medical School
Health Sciences Center, Dept. Psychiatry & Neurology
New Orleans, LA, USA
foundas@tulane.edu

Library of Congress Control Number: 2007923170

ISBN-13 978-0-387-36600-5   e-ISBN-13 978-0-387-36601-2

Printed on acid-free paper.

9 8 7 6 5 4 3 2 1

springer.com

DEDICATION

To the memory of Dr. James C. Young,
a brilliant clinician and a good friend.

To our students who inspire us to grow, and
from whom we often learn as much as we teach.

PREFACE

A major focus of clinical neuropsychology and cognitive-behavioral neurology is the assessment and management of cognitive and behavioral changes that result from brain injury or disease. In most instances, the task of the neuropsychologist can be divided into one of two general categories. Perhaps the most common is where patients are known to be suffering from identified neurological insults, such as completed strokes, neoplasms, major head traumas or other disease processes, and the clinician is asked to assess the impact of the resulting brain damage on behavior. The second involves differential diagnosis in cases of questionable insults to the central nervous system. Examples of the latter might be milder forms of head trauma, anoxia and dementia or suspected vascular compromise. In either instance, understanding the underlying pathology and its consequences depends in large part on an analysis of cognitive and behavioral changes, as well as obtaining a good personal and medical history. The clinical investigation will typically include assessing problems or changes in personality, social and environmental adaptations, affect, cognition, perception, as well as sensorimotor skills. Regardless of whether one approaches these questions having prior independent confirmation of the pathology versus only a suspicion of pathology, a fairly comprehensive knowledge of functional neuroanatomy is considered critical to this process.

Unfortunately as neuropsychologists we too frequently adopt a corticocentric view of neurological deficits. We recognize changes in personality, memory, or problem solving capacity as suggestive of possible cerebral compromise. We have been trained to think of motor speech problems as being correlated with the left anterior cortices, asymmetries in sensory or motor skills as a likely sign of contralateral hemispheric dysfunction, and visual perceptual deficits as being associated with the posterior lobes of the brain. At the same time there should be an awareness that multiple and diverse behavioral deficits can frequently result from strategically placed focal lesions, and that many such deficits might reflect lesions involving subcortical structures, the cerebellum, brainstem, spinal cord, or even peripheral or cranial nerves. As first noted by Hughlings Jackson in the 19th century, while the cortex is clearly central to all complex human behavior, most cortical activities begin and end with the peripheral nervous system, from sensory input to motor expression.

This current work was an outgrowth of seminars given by the principal author (JEM) at the request of neuropsychology interns and residents at the VA to broaden their clinical appreciation and application of functional neuroanatomy. In working closely with neurologists and neurosurgeons, these students also recognized the advantage of being able converse knowledgeably about patients with subtentorial deficits. While all the intricate details of the nervous system may be beyond the immediate needs of most clinicians, a general appreciation of its gross structural makeup and functional relationships is viewed as essential in working with neurological populations.

To this end, the book begins with a brief review of the gross anatomy, functional correlates, and behavioral syndromes of the spinal cord and peripheral nervous system. From there, the text carries one rostrally, looking at these same features in the cerebellum, brainstem and cranial nerves. Where this volume deviates from most textbooks of functional neuroanatomy is in its expanded treatment of supratentorial structures, particularly the cerebral cortex itself, which more directly impacts on those aspects of behavior and cognition that often represent the primary focus or interests of neuropsychologists and behavioral neurologists.

The final chapters are devoted to the vascular supply and neurochemical substrates of the brain and their clinical and pathological ramifications.

In addition to simply reviewing structural neuroanatomy and providing the classically defined behavioral correlates of the major divisions of the CNS, a major focus of this work will be to attempt to integrate functional systems and provide the reader with at least a tentative conceptual model of brain organization and how this organization is important in the understanding of behavioral syndromes. For the neuropsychologist, of equal importance to an understanding of clinical neuroanatomy is an appreciation of neuropathology, i.e., the natural history, associated signs and symptoms and physiological and/or neurological correlates underlying specific disease states. While occasional references are made to these factors throughout the text, an adequate treatment of this subject is beyond the scope of this book and the reader is advised to supplement this information with other works that specifically address these latter topics.

In going through the chapters the reader will notice some redundancy. This was purely intentional as a means of reinforcing certain key concepts and promoting their retention. Care was taken to try and resolve any discrepancies with regard to either structural or functional issues that appeared to be in dispute. However, our collective knowledge of the nervous system is still very incomplete and is often derived as much from clinical impressions and correlations as it is from definitive experimental paradigms. This is particularly true as we progress from peripheral pathways to central mechanisms in the brain. Part of the problem is the complexity of the nervous system itself and the technical (and ethical) limitations of carefully controlled studies, especially in man. Another problem is simply the startling limitations of our own knowledge. We are still far from being able to create a good working model of the brain. Thus, as we progress along the neural axis from the spinal cord to the brain, much of the data presented will be increasingly speculative. However, it is hoped that the sum total of this exercise will provide the reader not only with a broad overview of functional neuroanatomy, but will provide a beginning framework for trying to conceptualize brain-behavior relationships and the effects of focal lesions on behavior. Although this book was initially written for neuropsychologists, it provides a practical review of this subject for clinicians in other disciplines who work with the neurologically impaired, particularly neurologists and behavioral neurologists.

Finally, a number of acknowledgements are in order. Throughout the text, it will be noted that a large number of the figures use photographs of the brain and other neural structures derived from the Interactive Brain Atlas (1994). These base images were provided courtesy of the University of Washington and proved invaluable in illustrating anatomical landmarks. Two additional points should be made in this regard. First, the labeling of these images was done by one of the authors (JEM), thus any errors that might be found are not the responsibility of the University of Washington. Second, while the monochrome images used here were preferred for our text, the University of Washington has an updated version of this interactive atlas, which is highly recommended for anyone interested in an easy and entertaining way to review basic neuroanatomy. Thanks are also in order to the University of Illinois Press, Western Psychological Services, and Dr. Kenneth Heilman for permission to use published materials, as well as to Dr. Jose Suros who provided several brain images and to Dr. Enrique Palacios who was kind enough to review the radiographic images. It also seems appropriate to mention Stephen Stahl whose works on psychopharmacology provided inspiration for much of the material contained in Chapter 11.

A special acknowledgment is reserved for Mr. Eugene New, a medical illustrator from the LSU Health Sciences Center, who is responsible for all the artwork seen throughout the text.

CONTENTS

1   THE SPINAL CORD
AND DESCENDING TRACTS

OVERVIEW

The spinal cord, along with its ventral roots, represents the final common pathway for skilled, voluntary motor responses initiated in the cerebral cortex and carried out by our trunk and limbs. Beyond consciously guided or directed actions, spinal motor pathways and nuclei are important in maintaining what might be termed baseline, automatic, or supportive motor activities, such as muscle tone, balance, and various reflexes. In fact, some of the latter are apparently mediated purely at a spinal level. Descending spinal pathways apparently can influence even certain sensory phenomena, such as the perception of painful stimuli. Finally, the spinal cord plays a critical role in the homeostatic modulation of internal and external organs via the autonomic nervous system, the effects of which may be seen throughout the entire body.

Injury and disease selectively may target the spinal cord and/or its peripheral processes, often producing highly specific and very profound clinical syndromes. As similar symptoms can result from pathology at various levels of the neuroaxis, it is imperative that the clinician has some appreciation of the basic neuroanatomy of the spinal cord and its impact on these behavioral phenomena. In this chapter, we will focus on the general organization of the spinal cord, its relationship to the peripheral nerves, and the major descending (motor) pathways, including the spinal portion of the autonomic nervous system (ascending or sensory pathways will be reviewed in Chapter 2).

At the conclusion of this chapter, the reader should be able to describe the fundamental structure of the spinal cord and define the functional significance of the major descending tracts, discuss the basic mechanisms underlying spinal reflexes, and outline the general structural and functional correlates of the autonomic nervous system. While the major clinical syndromes affecting the spinal cord will be reviewed in Chapter 2, by the end of this chapter the reader should be able to discuss the basic characteristics differentiating upper and lower motor neuron lesions.

INTRODUCTION

It is easy to think of the spinal cord as simply the interface or site of synaptic connections between ascending and descending cortical pathways and the peripheral nerves. However, as the following two chapters will note, its function is indeed quite complex. Its structure and connections provide an anatomical substrate for the reciprocal excitation and inhibition of agonist and antagonist muscles crucial for effective coordinated activity. Combining both afferent feedback and motor effector mechanisms, the spinal cord plays a major role in moderating or regulating muscle tone that is the background against which all muscle activity must take place. The spinal cord serves as the initial staging area for all somatosensory input from the trunk and limbs. This sensory input is summated and dispersed to (1) the ventral posterior lateral (VPL) nuclei of the thalamus and eventually to the cortex for conscious perception; (2) to the reticular system and intralaminar nuclei of the thalamus for arousal and attention; and (3) to the cerebellum to assist with the coordination of complex, programmed activities. In addition, the interaction of both descending influences at the spinal level, as well as the interactions among various types of afferent inputs, allow for the modulation of sensory input at the spinal level. Thus, what takes place at the spinal level can impact directly on what is eventually perceived by the cortex. The spinal cord also mediates reflexes that are essential for posture and balance and withdrawal from certain types of painful stimuli and participates in other, more complex reflexes essential for executing more complex motor functions.

While damage to the spinal cord, its ascending or descending tracts, or the peripheral nerves has no impact on cognition or behavior as defined in psychological terms, it can, and typically does, result in somatosensory and both elementary and complex, coordinated motor disturbances. Such deficits not only will impact on measures of sensorimotor skills routinely used as part of a neuropsychological examination, but in a larger sense can directly impact on activities of daily living and rehabilitation efforts. Indirectly they also can drastically impact the patient's self-image and emotional or psychological adjustment. Thus, it becomes important for the neuropsychologist to have an appreciation of the functional anatomy of these and other subtentorial structures. The present chapter will focus on the general structure of the spinal cord and the descending or motor system. The ascending or somatosensory system will be reviewed in the following chapter.

GROSS ANATOMY OF THE SPINAL CORD

The spinal cord, along with the brainstem, represents the most primitive part of the central nervous system. Housed within the vertebral column, it is seen as a caudal extension of the medulla of the brainstem (Figure 1–1). Unlike the brainstem, in which discrete, more or less circumscribed nuclear groups can be identified at specific levels, the spinal cord tends to be organized in a more continuous, columnar fashion. This tends to be true not only for the descending (motor) and ascending (somatosensory) white matter pathways, but also to a large extent for its nuclear components. However, for teaching and description purposes, the

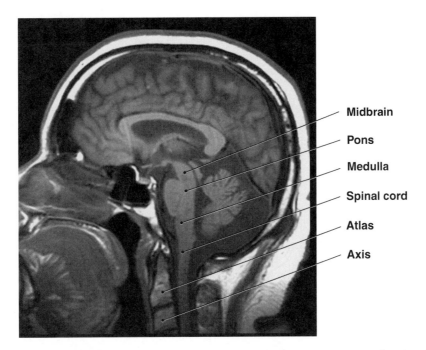

Figure 1–1. Brainstem exiting the cranial cavity through the foramen magnum forming the spinal cord as it enters the cervical vertebrae.

cord is often discussed segmentally, and, as will be seen when the internal structure of the spinal cord is reviewed, there are noticeable variations at different levels. This segmentation is based on those portions of the cord from which the 31 pairs of spinal nerves enter or exit. These 31 segments are labeled C-1 through C-8 (cervical), T-1 through T-12 (thoracic), L-1 through L-5 (lumbar), S-1 through S-5 (sacral), and coccygeal (1). While these labels refer to the intervertebral spaces through which these spinal nerves pass, it should be noted that the cord itself does not extend the entire length of the vertebral column.[1] In adults, the spinal cord terminates around L-1. The dorsal and ventral spinal nerves simply travel down the vertebral canal as the cauda equina, exiting from the intervertebral spaces for which they are named. The areas of the skin supplied by the sensory segments (dorsal roots) of the spinal nerves are referred to as dermatomes. A general map of these dermatomes can be found in Chapter 2.

In looking at a cross section of the cord (Figure 1–2), several features may be immediately obvious. The first is that, unlike the cortex, the axonal pathways (white matter) are on the outer portions of the cord with the nuclear groupings (gray matter) on the inside. On the ventral or anterior surface of cord there is a very noticeable central cleft or sulcus (the ventral median fissure). There also is a less obvious dorsal fissure known as the dorsal median sulcus. These two fissures or sulci divide the cord into a right and left half. In the central portion of the cord, the gray matter is roughly H-shaped. The small opening in the center of the "crossbar" is the central canal that is continuous with the fourth ventricle.

The white matter of each half of the cord is commonly divided into three sections or **funiculi** representing the columns of ascending and/or descending fiber tracts. These are the dorsal, the lateral, and the ventral funiculi (Figure 1–3). The most anatomically and functionally distinct is the **dorsal funiculus,** which lies between the dorsal median sulcus and the posterolateral sulcus. The latter, which represents the entry point for the posterior (sensory) roots, can be easily identified by following the dorsal horns of the central gray

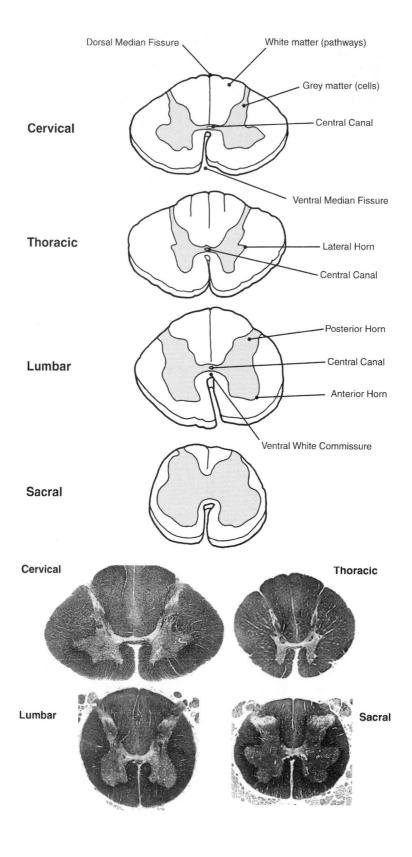

A Dorsal median sulcus
B Posterolateral (Dorsolateral) sulcus
C Ventrolateral sulcus (indistinct)
D Ventral median sulcus

A→B = Dorsal Funiculus
B→C = Lateral Funiculus
C→D = Ventral Funiculus

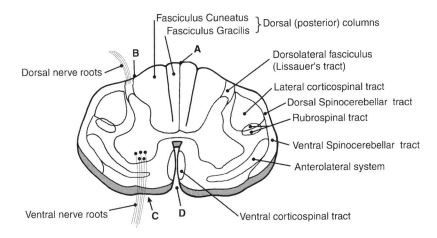

Figure 1–3. The relative positions of the dorsal, lateral and ventral funiculi, as well as a number of the more prominent ascending and descending pathways contained therein.

matter to where they intersect with the surface of the cord. In the upper levels of the cord, this dorsal funiculus is further divided into a more medial portion—the **fasciculus gracilis**—and a more lateral portion—the **fasciculus cuneatus**. As will be discussed in more detail in Chapter 2, the former includes the ascending tracts representing proprioception and stereognosis or fine tactile discrimination for the lower extremities (and lower trunk). The mnemonic device often applied here is gracilis → "graceful" → "dancer" → "legs." The more lateral fasciculus cuneatus carries the fibers for proprioception and stereognosis for the upper extremities (and upper trunk).

The remaining columns of fibers constitute the lateral and ventral funiculi. The **lateral funiculus** occupies the lateral portion of the cord, while the **ventral funiculus** is found in the ventromedial aspect of the cord. Unlike the division between the dorsal and lateral funiculi, there is no clear anatomical demarcation between the lateral and ventral funiculi. The division is arbitrarily defined by the anterolateral sulcus, which represents the point of exit for the anterior (motor) nerve roots. However, this point of departure for the motor fibers of the anterior horn cells typically cannot be seen in most cross-sections of the cord, but their location can be roughly approximated by visualizing the "5 and 7 o'clock" positions on the cord. Also, unlike the rather homogeneous nature of the dorsal funiculus, the lateral and ventral funiculi consist of a variety of ascending and descending tracts representing multiple

Figure 1–2. Axial sections of the cord at various levels are illustrated in Figure 1–2a. Note the relative differences in the proportions of gray to white matter at the various levels due to the number of ascending and descending present at a given level, and the number of synapses present in the gray matter in those areas representing innervation of the extremities. Figure 1–2b shows stained specimens at the cervical, thoracic, lumbar and sacral levels (Courtesy of Univ. Of Washington). Due to staining technique, fiber tracts (while matter) appear dark.

functional components. As a general rule, those descending motor pathways that primarily mediate discrete, fine motor skills are predominately located in the lateral funiculus. The largest of these is the lateral corticospinal tract (see below), while the tracts responsible for coordination, balance, and axial musculature tend to be located in the ventral funiculus. Fibers representing pain, temperature, and simple touch can be found both in the lateral and ventral funiculi. These tracts will be discussed in greater detail later in this chapter and in the next chapter. However, one other white matter pathway deserves mention at this point. Between the small zone of gray matter surrounding the central canal and the ventral median fissure lies a thin strip of white matter: the ventral or anterior white commissure. It is through this commissure that both descending and ascending pathways can cross from one side of the cord to the other.

The gray matter of the cord is commonly divided into the dorsal (or posterior) and ventral or (anterior) horns. The area between the two, roughly corresponding to the crossbar of the "H," is termed the intermediate zone. Approximately from T-1 through L-3 a small lateral protuberance is noted in this intermediate zone. This is known as the lateral horn and represents the preganglionic neurons of the sympathetic portion of the autonomic nervous system. The dorsal horns are the site of entry for the dorsal roots of the spinal nerves that carry somatosensory information, including the afferent feedback from the intrafusal muscle fibers. The dorsal or posterior horn contains the postsynaptic cell bodies for some of the ascending sensory pathways (see Chapter 2). However, the greatest number of spinal cord cells are **interneurons** whose processes generally remain local or, at most, within a few segments of their origin. These interneurons are important for the initial editing, integration, and synthesis of incoming sensory information; the modulation of sensory input by descending pathways; transferring information from one side of the cord to the other; and influencing motor neurons by maintaining muscle tone and triggering reflex postural adjustments.

The ventral or anterior horns contain the cells of origin for the final common motor pathways or **lower motor neurons,** which are represented by the ventral roots of the spinal nerves (see below for distinction between upper and lower motor neurons). As in the case of the dorsal horns, the ventral horns also contain a vast number of interneurons that respond both to descending influences from the cortex and brainstem (and indirectly from the cerebellum), as well as from segmental inputs from the dorsal roots. Again, these interneurons become important for maintaining optimal levels of muscle tone, the coordination of complex motor activity, postural stability, and/or postural reflexes. The relative size and shape of this central gray area will change at various levels of the cord. At levels C-5 to T-1 (cervical enlargement) and again at approximately L-2 to S-3 (lumbosacral enlargement) there are noticeable increases in the size of the cord as a whole and in the size of the ventral horns in particular. These enlargements reflect the increased number of cells serving the upper and lower extremities, respectively. At levels T-1 to L-2 one can see the lateral horn representing the cell bodies of the preganglionic fibers of the sympathetic nervous system. Also, the proportion of gray to white matter generally increases as the result of more descending fibers dropping out as one proceeds down the cord and, conversely, more ascending fibers being added as one goes rostrally in the cord.

The central gray area of the spinal cord has been further divided into ten zones or lamina that appear to have distinct anatomical and functional properties. While a detailed knowledge of these properties is probably not critical for our present purposes, for the sake of completeness a brief summary will be included. Lamina I (Figure 1–4) occupies the tip of the dorsal horn at the site of entry of the dorsal nerve roots. It contains some cells that contribute to the spinothalamic tract. Lamina II is synonymous with the substantia gelatinosa, which plays an important role in the spinal modulation of pain. Lamina III through VI are largely involved in processing somatosensory input. Lamina VII, which comprises most of the intermediate zone between the dorsal and ventral horns, is mostly made up of interneurons. In addition

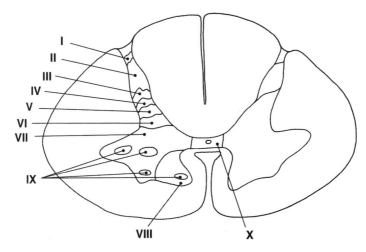

Figure 1–4. Approximate locations of Rexed's lamina in the cervical cord.

to some probable integration of sensorimotor information, the intermediate zone, as noted, serves as the site for cell bodies for preganglionic fibers of the autonomic nervous system. In the thoracic and upper lumbar regions of the cord, the nucleus dorsalis (dorsal nucleus of Clark), which is the origin of the dorsal spinocerebellar tract, also is located in the medial portion of lamina VII. Lamina VIII and IX, which are located in the ventral horn, are primarily involved with motor output. In the latter are located the cells of origin for both the large-diameter alpha neurons that innervate the striated muscles and the small-diameter gamma neurons that go to the intrafusal muscle fibers. Lamina X surrounds the central canal and its role is unclear. However, given its location, there is some speculation that it may be involved in the sharing or transfer of certain types of information between the two halves of the cord.[2]

SPINAL NERVES

At each vertebral junction, two groups of nerves enter/exit on each side of the spinal cord. As noted earlier, the dorsal (sensory) roots enter the cord in the region of the dorsal horn of the central gray. The cell bodies for these fibers lie outside the cord in the dorsal root ganglia. The ventral (motor) roots exit the cord along its ventrolateral surface and consist primarily of large alpha neurons, as well as the smaller gamma neurons (Figure 1–5). The alpha neurons

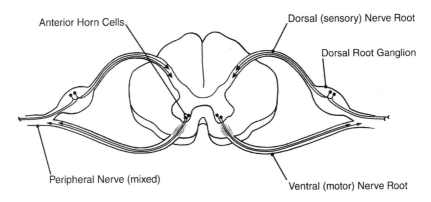

Figure 1–5. Dorsal and ventral roots, combining to form peripheral nerve.

supply the extrafusal striated muscle fibers (the ones that do the work), while the gamma neurons innervate the intrafusal muscle fibers, which are important in maintaining tone and posture and in supplying information about the state of kinetic activity in the muscles. At specific levels within the cord, these roots also contain the preganglionic fibers for the autonomic nervous system. The cell bodies for the ventral roots are contained in the ventral horns for the striated muscles and in the intermediate zone for the viscera. Distal to the dorsal root ganglia, the dorsal and ventral roots combine into the peripheral nerves that travel together to the target organs (eg, muscles, tendons, skin). The nerves that supply the upper extremities involve C-5 to T-1. The nerves that supply the lower extremities are from L-2 to S-3.

CORTICAL MOTOR TRACTS (DESCENDING PATHWAYS)

The pathway for skilled, voluntary motor activity begins in cortex and terminates in the individual fibers of the peripheral musculature. However, this system is impacted all along the way by numerous other systems that may modulate its activity. For example, any purposeful (and many reflexive) movements of the extremities occur against a background of muscle tone and antigravity forces. The smooth, successful execution of any such movement depends on the proper reciprocal excitation and inhibition of agonist and antagonist muscle groups and adaptive posturing of the trunk, head, or other extremities. To execute purposeful movement, the motor system requires extensive, ongoing feedback from the cerebellum as well as from a variety of sensory information, including touch, pressure, proprioception, kinesthetic, vestibular, visual, and visual-spatial. Constant monitoring and feedback from multiple cortical areas are required to initiate a goal-directed activity, to plan its sequence, to modify its original direction (goal or intention) as might be demanded by changing circumstances, and to simply know when to terminate the activity. In addition to being influenced by other systems, these fiber tracts also give off collaterals to other neural centers or structures, thereby creating a complex array of feedback loops. Thus, it may be useful to keep in mind that only a relatively small percentage of the fibers of the various descending pathways to be discussed actually end up synapsing on alpha (or even gamma) motor neurons (ie, directly effecting a motor response). The majority serves to modulate, in an indirect manner, various motor and sensory systems. Even the experience of painful stimuli is affected by descending pathways at the spinal level. This interdependence on other systems will become increasingly clear as we review some of the more prominent interconnecting pathways and as we later review other functional-anatomical systems.

The remaining focus of this chapter will be to outline the primary motor tracts that exit from the brain and descend through the spinal cord and to briefly describe the anatomical basis for reflex arcs and the maintenance of muscle tone. Some of the more common pathological consequences of insults to this system will be presented at the conclusion of the following chapter.

Corticospinal Tract

The **corticospinal** or **pyramidal** tract is the primary motor pathway for skilled, voluntary movements, particularly those involving the upper and lower extremities. The **corticobulbar tracts** that subserve the muscles of the face and head have similar origins, but terminate in the brainstem. These will be reviewed below and again in Chapter 5. Although the corticospinal tract is considered to be the major pathway for skilled movements, only about 55% of its fibers have their cells of origin in the primary and secondary motor cortices (Brodmann's areas 4 and 6). The somatosensory cortex (areas 3, 2, and 1) contribute approximately 35%, with other frontal and parietal areas contributing the remaining 10% (the temporal

Table 1–1. Major Descending Voluntary Motor Tracts

Tract	Origin	Termination	Primary Function
Lateral corticospinal (crossed)	Primarily sensorimotor cortices	Anterior horn cells in the spinal cord	Voluntary movement of trunk, extremities
Ventral corticospinal (uncrossed)	Same as above	Same as above	Axial, proximal muscles for posture, balance
Corticobulbar (primarily crossed)	Same as above, including frontal eye fields	Brainstem nuclei	Control of muscles supplied by cranial nerves; cerebellar inputs
Corticotectal (uncrossed)	Primarily occipital and inferior parietal cortices	Superior colliculi, which give rise to tectobulbar and tectospinal tracts	Visual tracking, visual reflexes
Corticorubral (uncrossed)	Primarily sensorimotor cortices	Red nuclei, which give rise to CTT[a] and rubrospinal tracts	May facilitate fine motor coordination
Corticoreticular (bilateral)	Primarily sensorimotor cortices?	Pontine and medullary reticular nuclei, from there giving rise to reticulospinal tracts	Posture, gross motor coordination, balance
Medial and lateral vestibulospinal	Medial and inferior vestibular nuclei	Spinal cord	Medial: same as above; Lateral: fine motor coordination?

[a] Central tegmental tract.

and occipital cortices apparently contribute little, if any, to this pathway). The large Betz (pyramidal) cells of area 4, once thought to be a major source of these fibers, contribute only about 5% of the total. Thus, from its very origin, one can see the influence of somatosensory and other higher cortical systems on this descending motor tract.

Both within the voluntary motor and the somatosensory systems, there is a clear topographical organization beginning in the cortex (for the corticospinal pathways) and continuing into the cord. At the cortical level, the ventral-most portion of the motor strip subserves the muscles used for speaking and swallowing. Proceeding dorsally, next are the cortical origins for the fibers that will eventually mediate the muscles involved with mastication and facial expression. Approximately at the midway point on the lateral surface of the motor strip is the cortical representation of the hand, followed by the arm and shoulder, then the trunk, and the hip and upper leg. Over the dorsal aspect of the motor strip and into its medial surface lies that portion that is responsible for movements of the lower leg and foot. Because this latter part of the cortex is supplied by a different arterial system (anterior cerebral artery) than the lateral surface (middle cerebral artery), patients with strokes involving the middle cerebral artery may regain better use of the lower extremity than of the arm or hand. It also should be noted that the different body parts are disproportionately represented in the motor (and somatosensory) cortex. The areas that mediate those body parts that call for finer, discrete actions, such as the tongue, the face, and the hands, have a much broader cortical representation than those body parts that, though larger in size, lack the need for such precision in their response (eg, the upper arm, shoulder, trunk, or leg).

From their cells of origin in the cortex, the corticobulbar and corticospinal fibers coalesce and descend through the genu and posterior limb of the **internal capsule,** respectively. The internal capsule, when viewed in horizontal section, looks like a boomerang with its anterior limb lying between the lenticular nuclei (putamen and globus pallidus) and the head of the caudate and its posterior limb between the lenticular nuclei and the thalamus, with its genu in the center (see Figure 1–6a).[3] As the descending corticospinal and corticobulbar fibers enter the brainstem, they constitute the central or middle portion of the cerebral

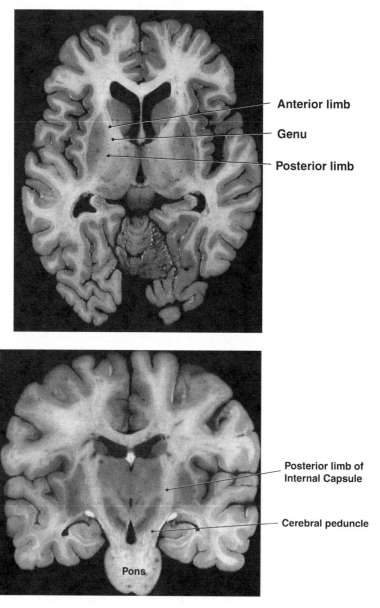

Figure 1–6. (a) Axial section showing anterior limb, genu, and posterior limb of internal capsule. Fibers mediating facial muscles are located in the genu, followed by the upper extremities, trunk, and legs in the posterior limb. (b) Coronal section showing transition of corticofugal fibers in internal capsule as they form the cerebral peduncles.

peduncles (also known as the **basis pedunculi** or **crus cerebri**) on the basal or ventral aspect of the midbrain (see Figure 1–6b). The medial and lateral portions of the cerebral peduncles are made up of descending frontopontine and parieto-occipito-temporo-pontine fibers, respectively, (ie, fiber tracts originating in the frontal, parietal, occipital, and temporal cortices and terminating within the brainstem). The corticospinal tract, largely coming from the region around the central sulcus, descends through the basilar portion of the pons. As the corticospinal tract exits from the pons, it forms the pyramids on the ventral surface of the upper medulla. It is from this localized portion of the descending corticospinal tract that the term **pyramidal tract** derives its name.[4]

In the lower medulla, approximately 90% of the corticospinal tract decussates, with about 10% of the fibers remaining ipsilateral. That portion of the corticospinal (C-S) tract that decussates descends in the cord as the **lateral corticospinal tract**. These fibers then synapse at their respective levels in the central gray (horn) cells of the cord (laminae IV through VIII), which in turn synapse with the alpha and gamma motor neurons in the ventral horn (lamina IX) (Figure 1–7). Because of this decussation, lesions above the level of the lower medulla will produce contralateral motor symptoms, while lesions below this level will result in ipsilateral deficits. The fibers of the lateral corticospinal tract are primarily responsible for mediating fine, voluntary motor movements of the extremities.

The 10% or so of the corticospinal tract that does not decussate travels down the cord in the ventral or anterior funiculus as the **ventral corticospinal tract**. These fibers then decussate at the level that they cross the cord (via the white anterior commissure) to synapse in the ventral horn (lamina IX). As opposed to the lateral corticospinal tract, which is generally associated with control over distal movements of the extremities, the ventral corticospinal tract appears to project largely to axial and proximal musculature and plays a greater role in maintaining posture, balance, and gross truncal movements. It should be noted that the fibers of the corticospinal tract serve not only to initiate and direct skilled motor movements, but also probably impact on sensory feedback loops through synapses with neurons of the ascending sensory tracts and may also affect local reflex arcs.

Corticobulbar Tract

Whereas the corticospinal tract largely subserves voluntary movement of the upper and lower extremities, the corticobulbar tract provides comparable innervation for the muscles of the head (eg, eyes, facial expression, mastication, speech) and neck supplied by the cranial nerves. The corticobulbar pathways start out with the corticospinal tract, their cells of origin being primarily the inferior portions of areas the motor (4 and 6) and somatosensory (3, 1, and 2) cortices and area 8 (frontal eye fields). They descend in the genu of the internal capsule anterior to the corticospinal tract. Most of these fibers terminate in the brainstem where they influence (both directly and via indirect, multisynaptic connections) cranial nerves III, IV, V, VI, VII, IX, X, XI, and XII. A large percentage of these fibers synapse in the basal nuclei of the pons from where they give rise to the fibers of the middle cerebellar peduncle. While the latter will be discussed in greater detail in Chapter 3, for now suffice it point out that this cortico-pontine-cerebellar pathway appears to be the primary means by which cerebellum is informed of cortical intentions. Also, as is the case with the corticospinal tract, the majority of the corticobulbar fibers likely exercise a modulating function, contributing both to indirect motor feedback loops and afferent systems.

There is one other clinically significant difference between the corticospinal tract and the corticobulbar tract. The latter's connections are often bilateral. Thus, cortical lesions involving corticobulbar pathways frequently result in transient and/or very limited weakness of the target musculature. The most notable exception involves the lower face (the muscles of

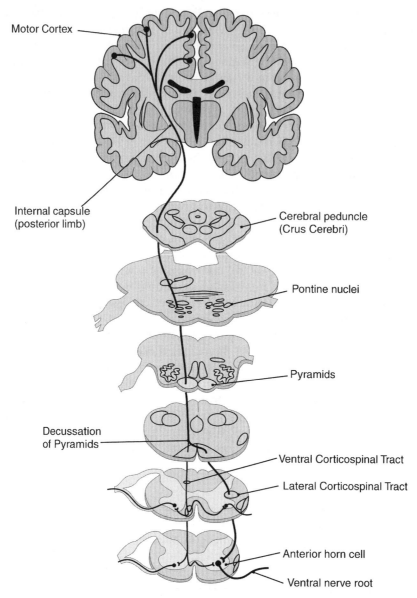

Figure 1–7. The corticospinal tracts from their origin in the cerebral cortex, through the internal capsule, brainstem, spinal cord, to their eventual synapse in the anterior horn cells are shown. The anterior horn cells, in turn, give rise to the ventral nerve roots (which themselves join with the incoming somatosensory fibers to form the peripheral nerves) that ill eventually innervate the skeletal musculature.

expression), lateral rectus and tongue in which the pattern is one of contralateral representation. Thus, a supranuclear (upper motor neuron) lesion of this tract can result in contralateral weakness of the muscles of the lower face while the muscles of the upper face (eg, those involved in frowning) appear unaffected. By contrast, both the upper and lower face will evidence weakness with lower motor neuron lesions of the seventh cranial nerve. If the lesion is located at the pontomedullary junction where the corticobulbar fibers decussate to synapse in the seventh cranial nerve nuclei, the patient may evidence ipsilateral weakness of the lower face and contralateral weakness in the extremities due to the fact that the corticospinal tracts have not yet decussated.

Corticotectal and Tectofugal Tracts

The **corticotectal tract** is one of several that is concerned with the control of eye movements.[5] The functional mechanisms and neuroanatomical connections that mediate eye movement are extremely complex and will be discussed here only in general terms (see: Chapter 5). The cells of origin for the corticotectal tract likely derive from several cortical areas, but the primary input seems to be from the occipital and possibly inferior parietal cortices. As implied by its name, the tract terminates ipsilaterally in the tectum, specifically in the superior colliculi. Fibers synapse topographically in the areas corresponding to the same portion of the visual fields as the inputs received directly from the retina. The efferent fibers from the superior colliculi do not synapse directly with the nuclei of the third, fourth, and sixth cranial nerves. Rather, through their connections with intermediary gaze centers (eg, pontine paramedian reticular formation) and other brainstem and spinal nuclei, the superior colliculi help govern tracking and other reflex movements of the eyes. Among these projection sites of the superior colliculi are the pons (**tectopontine**), the medulla (**tectobulbar**), and cervical spine (**tectospinal tracts**). In addition, the superior colliculi are a source of fibers that eventually serve afferent or arousal functions, such as the tectoreticular (to the ascending reticular activating system) and the tectothalamic tracts. The superior colliculi also receive input from the somatosensory (especially from head and neck), cerebellar, vestibular, and the auditory systems.

Let us turn momentarily to an exploration of the probable functional significance of these pathways. First, we often need or desire to actively scan our environment or to consciously shift our focus from one object to another. These voluntary eye movements appear largely mediated by corticobulbar connections originating in Brodmann's area 8 (**frontal eye fields**). Frequently, it is necessary to track moving objects or to fixate on an object while we are in motion. Depending on the circumstances, such tracking may involve volitional components, but often this is done reflexively or subconsciously. Such movements are likely accomplished, at least in part, by the corticotectal fibers coming from the occipital (and possibly inferior parietal) cortices.[6] A good example may be to read this text while slowing moving your head up and down or from side to side. In doing so you will notice that the words still remain fixed (and appear static) in your foveal vision. Feedback from the vestibular nuclei is critical for making these adjustments.

There is survival value in being able to react quickly to unexpected and/or novel peripheral visual (or auditory) stimuli, that is, to attend to and identify such stimuli as rapidly as possible. This typically involves an automatic turning of the head toward such stimuli. Tectospinal pathways mediate these functions. Rapid closure of the eyelids is likewise essential for protecting the eye from oncoming objects. Such reflex responses would appear to require rather direct connections between unfiltered visual input and the muscles of the eyelids. The tectopontine pathways to the nuclei of the seventh cranial nerve seem well-suited to this function. Of course, whenever you have muscle systems involved, you need some type of somatosensory feedback loop; thus perhaps one of the functions served by inputs to the tectal region from the somatosensory system. Direct connections from the inferior colliculi (an auditory relay center) to the superior colliculi would ensure a prompt visual response (orientation) to unexpected sounds. Finally, the presence of a sudden, unexpected (potentially dangerous) visual stimulus may be sufficient cause for all the cortex and all the senses to become more alert and attentive, at least until the stimulus is identified.

This may be one of the key functions of the tectoreticular pathways. Whereas the purpose of the tectothalamic connections is less clear, they may help maintain cortical tone along with other sensory inputs. Again, it should be noted the above explanations represent a grossly simplified and, in part, speculative explanation of a tremendously complex system.

Corticorubral and Rubrospinal Tracts

The **red nucleus** is a fairly large, well-defined structure in the midbrain at the level of the superior colliculus (see Chapter 4). In addition to ipsilateral input from the cortex, primarily from the regions surrounding the central gyrus (**corticorubral tract**), the red nucleus also receives major projections from the cerebellum via the superior cerebellar peduncle. After synapsing, the corticorubral fibers give rise to the **rubrospinal** tract, a crossed pathway that descends in the lateral funiculus of the cord and synapses in lamina V, VI, VII. In addition to its spinal projections, the red nucleus contributes significant input to the inferior olivary nucleus (via the **central tegmental tract**), which in turn projects heavily back to the cerebellum. Thus, through direct and indirect connections between the cerebellum and the red nucleus, the rubrospinal pathway may represent one of three descending pathways by which the cerebellum exerts control of motor activities in the extremities and trunk. Some authors suggest that the role of the rubrospinal tract may be to facilitate flexor and inhibit extensor alpha and gamma motor neurons, particularly those of the upper extremities.

Corticoreticular and Reticulospinal Tracts

The **reticular formation**, at least in neuropsychological literature, is most commonly associated with the ascending reticular activating system, which is intimately involved with cortical arousal. However, the reticular formation, which consists of multiple discrete nuclei, also has a descending influence on motor systems. **Corticoreticular pathways** represent bilateral descending cortical influences, probably similar in origin to that of the corticospinal tract, on the reticular formation. The pontine nuclei of the reticular formation give rise to the **medial reticulospinal tract** that descends in the medial aspect of the ventral funiculus of the cord. The **lateral reticulospinal tract**, arising from the **medullary** portion of the reticular formation, travels in the lateral funiculus of the cord. While both the medial and lateral reticulospinal tracts mainly represent ipsilateral pathways, some of the fibers cross within the cord. The lateral reticulospinal tract probably plays a significant role in mediating autonomic activities, as well as in controlling somatic musculature. The role of the reticulospinal tracts in somatic motor activity will be considered in association with the vestibulospinal tracts below.

Vestibulospinal Tracts

The vestibular nuclei, located in the upper medulla and lower pons, receive input from the utricle and saccule and the three semicircular canals or ducts of the inner ear. The utricle and the saccule are associated with orientation with respect to the pull of gravity. The semicircular canals respond to the detection of kinetic changes in space. These nuclei are intimately interconnected with the cerebellum. In addition, these nuclei are the origin of two descending fiber tracts: the **lateral** and **medial vestibulospinal tracts**. The lateral vestibular nuclei give rise to the lateral vestibulospinal tracts that descend in the ventral–lateral aspect of the cord throughout its entire length. One of the primary functions of this tract appears to be the facilitation of the extensor or antigravity musculature. A lesion that essentially transects the brainstem above the lateral vestibular nuclei can result in decerebrate rigidity as a result of the unmodulated activity of these nuclei or tracts. The medial and inferior vestibular nuclei are the source of the medial vestibulospinal tracts that descend bilaterally

Table 1–2. Upper versus Lower Motor Neuron Lesions

Upper Motor Neuron Lesion	Lower Motor Neuron Lesion
Likely affects many muscle groups	Likely affects single muscles
Increased muscle tone	Decreased muscle tone
Spastic paralysis	Flaccid paralysis
Mild, chronic atrophy due to disuse	Significant atrophy
Fasciculations absent	Fasciculations present
Hyperreflexia (hyporeflexia acutely)	Hyporeflexia or absent
Clonus possible	Clonus not seen
Babinski likely present (C-S tract)	Absent Babinski reflex
Absent abdominal/cremasteric reflexes	Absent only if those muscles are specifically involved

in the ventral medial aspect of the cord via the medial longitudinal fasciculus. These latter tracts extend only to the upper thoracic region and appear to mediate movements of the head, neck, and/or upper back in making postural adjustments or in orienting to stimuli.

In terms of general functional considerations, it would appear that those tracts that descend in the lateral funiculus, ie, the lateral corticospinal, lateral reticulospinal, and perhaps also the rubrospinal tract, primarily mediate fine, discrete skilled movements of the extremities. In animals, lesions of the pyramidal tracts at the level of the medulla or below are found to have more permanent and disabling effects on skilled movements of the extremities when accompanied by lesions of the lateral reticulospinal and rubrospinal tracts. By contrast, those pathways that are concentrated in the ventral funiculus, such as the tectospinal, interstitiospinal, medial reticulospinal, vestibulospinal and ventral corticospinal tract generally seem to serve a different function. These pathways may be more important in governing axial and peripheral muscular responses essential to the maintenance of posture, balance, and the integration or coordination of movements, including orienting the head and body to novel or threatening stimuli. As noted above, the vestibulospinal tracts in particular have extensive and fairly direct input from the cerebellum.

UPPER AND LOWER MOTOR NEURONS

The notion of distinguishing between "upper" and "lower" motor neurons has a number of very important clinical implications and is one that is commonly employed in neurology. For all practical purposes, upper motor neurons are basically all those motor neurons that originate in the cortex and/or brainstem nuclei and travel down the spinal cord to synapse either on interneurons or on the alpha (or gamma) neurons in the ventral horn of the cord. Similarly, lower motor neurons are those motor neurons whose cell bodies lie in the ventral horn and whose axons exit the cord as the ventral nerve roots (ie, the motor components of the peripheral nerves). The neural motor pathways that innervate the muscles of the head, face, and neck also are composed of upper and lower motor neuron. The corticobulbar pathways represent the upper motor neuron segment of this system, whereas the motor components of the cranial nerves or their motor nuclei constitute its lower motor neuron pathways. Table 1–2 presents a list of clinical symptoms that tend to be characteristic of upper versus lower motor neuron lesions.

Some diseases of the nervous system may tend to selectively affect upper motor neurons, such as posterolateral column syndrome; others lower motor neurons, such as anterior horn cell syndrome; or both, such as amyotrophic lateral sclerosis. Other diseases, such as tabes dorsalis (neurosyphilis), produce motor symptoms secondary to impairment of sensory feedback loops. Other lesions, such as those due to trauma or vascular disease, may produce a variety of clinical pictures, including both sensory and motor changes. Brief descriptions of a few of these more common neurological disorders affecting descending (motor) and ascending (sensory) systems will be reviewed in the next chapter.

SPINAL REFLEXES

A study of the spinal reflexes contributes to an understanding of how the motor and sensory work together as a functional unit. Neurologically, they are useful to help investigate the integrity of the central and peripheral nervous system. The clinically pathological response may be the absence of a normally present response (abdominal, corneal), the presence of a normally absent response (suck, snout, grasp, palmomental), changes in the direction of the reflex (Babinski), or in the intensity of the response (hyporeflexia or hyperreflexia).

Spinal reflexes are a stereotypic, involuntary response to specific types of stimulation to the muscles, tendons, or skin. While they can be influenced or modulated by higher centers (cerebrum, cerebellum), the basic stimulus—response arc takes place at the spinal level through either monosynaptic (muscle stretch reflex) or multisynaptic mechanisms. Functionally, reflexes help maintain the proper muscle tone to serve as a background for voluntary muscle activity, help maintain balance and posture, and serve as a protective response from noxious or threatening stimuli.

Mechanisms of Spinal Stretch Reflex

Within the striated or skeletal muscles (**extrafusal fibers**) are muscle spindles (**intrafusal muscle fibers**) that lie parallel to the extrafusal fibers. There are two major types of intrafusal fibers: nuclear bag fibers and nuclear chain fibers that stretch or contract with the stretching or contraction of the surrounding extrafusal fibers (muscles). The intrafusal fibers have both efferent and afferent connections to the spinal cord. The primary afferent connections of the intrafusal fibers are group Ia neurons, which generally synapse on central portions of intrafusal fibers and respond both to the length of the fiber (state of stretch or contraction) as well as to the rate of change. Group II neurons that synapse on the more distal portions, primarily on nuclear chain fibers, respond only to the length of the fiber. In addition to the muscles, there are sensory stretch receptors in the tendons (Golgi tendon organs) which are innervated by group Ib afferent fibers (Figure 1–8).

The primary efferent connections to the extrafusal fibers are made via the **alpha motor neurons**. **Gamma motor neurons** innervate the intrafusal (muscle spindle) fibers. The alpha motor neurons are responsible for the contraction of the muscle, while the gamma neurons are responsible for the state of contraction of the intrafusal fibers. The latter is important to maintain feedback capability on further changes in muscle tone during states of contraction and makes them more responsive to stretch. Two types of gamma neurons have been identified, one more responsive to static stretch and the other to phasic stretch.

When the quadriceps muscle is stretched by striking the patellar tendon, the quick stretching of the intrafusal fibers triggers the afferent fibers to send a signal to the cord. There they make a monosynaptic connection with the alpha motor neuron in the ventral horn of the central gray matter. The alpha motor neuron then sends a signal directly back to the extrafusal fibers of the muscle causing it to contract; hence the foot "jumps" with the knee-jerk response (Figure 1–9).

While reflex excitation of extrafusal muscle fibers, as a result of the stretching of a muscle, can serve useful functions, excessive or inappropriate reflex contractions of certain muscles in certain circumstances can be counterproductive. Hence, in addition to the facilitation, at

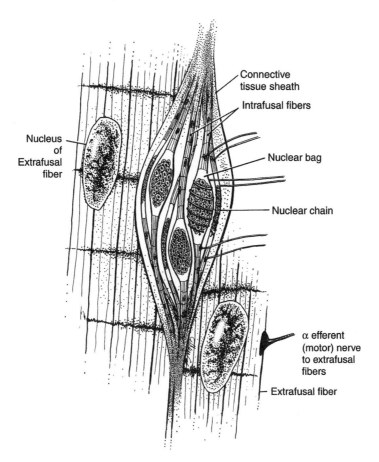

Figure 1–8. Intra and extrafusal muscle fibers.

times the inhibition of muscle groups may be needed. For example, for an extensor muscle to contract, its antagonistic flexor muscle must relax; or if one "voluntarily" stretches a muscle, as in the act of sitting, the stretch reflex of the quadriceps must be inhibited. Such inhibition can take place in a variety of ways. Golgi tendon organs, when stretched, send afferent feedback to the cord, which makes polysynaptic connections with interneurons that serve to inhibit the agonist muscle and to facilitate the antagonist muscles. The same afferent fibers that result in the excitatory connection with alpha motor neurons may send collaterals to interneurons, which in turn have an inhibitory influence on the alpha fibers going to the antagonist muscles. Even the excitatory alpha neurons themselves may send off collaterals to specialized interneurons, known as Renshaw cells, which make inhibitory connections with that same alpha neuron. In response to certain types of noxious stimuli (eg, stepping on a sharp tack), flexor muscles on that side must be facilitated and extensors inhibited, while on the contralateral side the reverse must occur (extensors facilitated and flexors inhibited). Again this all takes place by an elaborate system of excitatory and inhibitory connections involving multiple interneurons across several spinal segments. Sensory feedback from muscle activity also sends messages rostrally up the cord to higher segmental levels as well as to the cerebellum and cortex, which in turn have a downward influence on both the gamma and alpha fibers, both excitatory and inhibitory.

Muscle tone (optimal state of contractility) is maintained by this reflex arc. A state of hypotonia will occur if lesions occur either in the dorsal nerve roots, ventral nerve roots,

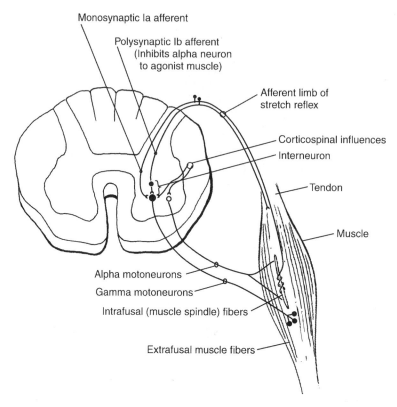

Figure 1–9. Basic afferent and efferent components mediating the stretch reflex.

anterior horn cells, or peripheral nerves (lower motor neuron lesion). It also can occur as a result of cerebellar lesions. Hypertonia results from a lesion rostral to the anterior horn cells (upper motor neuron lesion) and can take two forms: spasticity (increased resistance to movement of the "clasp knife" type) and rigidity (persistent resistance to movement throughout the range of movement of the limb).

AUTONOMIC NERVOUS SYSTEM

The autonomic nervous system (ANS) exerts control over or regulates the action of smooth and cardiac muscles and glands. It is primarily a motor or efferent system, though most if not all of its end organs have some type of afferent feedback loops built in. Following most discussions on this system, the focus here will be simply on the efferent connections.

The ANS consists of **sympathetic** and **parasympathetic** divisions that operate in an antagonistic or complementary fashion. Somewhat simplistically, the primary purpose of the ANS is to maintain homeostasis and meet demands of changes in environmental situations, both internal and external. The sympathetic system predominates in situations of physical or psychological stress and prepares the body for "flight" or "fight." Common responses might include:

(1) dilation of the bronchioles to allow more oxygen into the blood,
(2) increased heart rate,

Table 1–3. Summary of Common Autonomic Responses

Organ system	Sympathetic Response	Parasympathetic Response
Pupils	Dilation	Constriction
Lens (ciliary muscle)	Relaxes (enhances far vision)	Contracts (enhances near vision)
Lacrimal glands	No effect	Promotes tearing
Cardiac	Increases heart rate, contractility	Decreases heart rate, contractility
Pulmonary	Increases bronchiole diameter	Constricts bronchioles
Digestion	Decreases peristalsis, salivation, constriction of sphincters	Increases peristalsis, salivation, relaxation of sphincters
Pancreas	Inhibits secretion	Promotes secretion
Adrenal medulla	Promotes release of epinephrine, norepinephrine; diffuse effects	No effect
Liver	Glycogenolysis, and gluconeogensis	No effect
Skin	Sweating, piloerection	No significant effect
Blood vessels	Alpha receptors: constriction[a,b] Beta receptors – dilation	Mostly little if any effect[c]
Bladder	Relaxes bladder, contracts sphincters	Contracts bladder, relaxes sphincters
Penis/clitoris	Orgasm, ejaculation (in male)	Erection (dilates genital vessels)

[a] Predominate sympathetic response appears to be vascular constriction, but may have mixed effects. This is particularly true in heart and skeletal muscles where vasodilation may occur in response to threat. However, normally arterioles of both heart and skeletal muscles may respond primarily to nonautonomic factors, such as metabolic rate and availability of oxygen.

[b] Sympathetic response is to constrict blood vessels of the skin (eg, in response to threat, cold). Vasodilation of blood vessels in the skin in response to heat may be influenced by local factors, in addition to simply a reduced sympathetic output. For example, as will be seen in Chapter 4, Horner's syndrome, a disruption of sympathetic pathways, results in facial flushing.

[c] The one notable exception being the parasympathetic response in the genitals (see below).

(3) dilation of the arteries to heart and peripheral muscles and simultaneous constriction of the vessels to skin,

(4) secretion of glycogen by the liver

(5) release of additional red blood cells by the spleen,

(6) slowing of peristalsis,

(7) dilation of the pupils,

(8) increased sweating, and

(9) piloerection (probably a throwback to some of our hairier ancestors for whom this had more practical implications).[7]

In general, the parasympathetic division has the opposite effect on these end organs, (eg, it constricts the bronchioles and decreases heart rate, increases peristalsis, and constricts the pupils). Unlike the voluntary motor system that was discussed earlier, the efferent fibers of the ANS do not synapse directly on their target organs after leaving the central nervous system, but rather undergo one additional synapse in ganglia located outside the CNS (see Table 1–3).

In addition to the specific functional differences mentioned above, the two systems have some other very basic anatomical differences. The fibers of the parasympathetic system

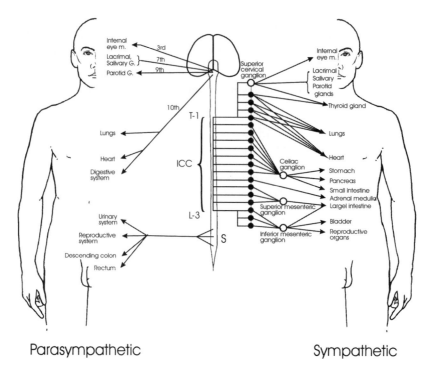

Parasympathetic   Sympathetic

Figure 1–10. Parasympathetic and sympathetic innervation of selected organ systems. Other major target organs not illustrated are the skin (e.g., sweat glands and piloerection) and blood vessels. Both tend to be responsive primarily to sympathetic innervation, although acetylcholine is used by the postganglionic sympathetic neurons innervating the sweat glands. Note: preganglionic sympathetic fibers innervate the adrenal medulla directly. Legend: ICC – intermediolateral cell column (in lateral horn).

leave the CNS at the level of the brainstem (via cranial nerves III, VII, IX, and X) and the sacral segments of the cord (S-2 to S-4). The sympathetic fibers, on the other hand, leave the CNS via the lateral horn nuclei of the spinal cord (in the intermediolateral cell column) from T-1 to L-3. Whereas the synaptic connections (ganglia) for the sympathetic system generally lie just outside the vertebral column, the ganglia or synaptic connections for the parasympathetic pathways lie in close approximation to their respective end organs (Figure 1–10). Finally, whereas the preganglionic fibers for both the sympathetic and parasympathetic, as well as the postganglionic fibers for the parasympathetic system, are **cholinergic** (release acetylcholine), the postganglionic fibers of the sympathetic system are **adrenergic** (release norepinephrine). Just for the record, one exception is the sympathetic innervation of the sweat glands, which are cholinergic.

While many of the visceral organs are able to carry out their basic functions when deprived of ANS stimulation, under such conditions they are unable to adapt to major physiological demands, ie, increased output under stress. One relatively well-known neurological condition that demonstrates disruption of the ANS is **Horner's syndrome**. It may result from injury anywhere along the pathway supplying, or supplied by, the superior cervical ganglion of the sympathetic ANS. Symptoms include a restricted (miotic) pupil, as a result of the unopposed action of the parasympathetic system; partial ptosis that can be overcome volitionally via the third cranial nerve which also innervates striated muscles of the eyelid; anhidrosis (lack of sweating); and vasodilatation on the affected side of the face (the skin feels drier, redder, and warmer). On rare occasions, both the parasympathetic and sympathetic systems may be selectively affected, producing what has been termed **dysautonomic polyneuropathy**. As might be expected, this condition is characterized by multiple signs of autonomic dysregulation, including orthostatic hypotension, gastrointestinal disturbances, sexual dysfunction, problems regulating body temperature, decreased salivation, and visual difficulties (Young et al., 1975). More common, however, is partial autonomic dysregulation (eg, orthostatic hypotension, sexual dysfunction) following more circumscribed lesions to the spinal cord or peripheral nerves, frequently accompanied by other signs of sensorimotor involvement.

Much of the time the ANS responds to the homeostatic needs of the organism, more or less independently of the person's subjective or conscious awareness. Thus, for example, the heart may pump a bit faster, the bronchioles may dilate slightly, respiratory rate may increase, and the peripheral blood vessels also may dilate and sweating is increased as one walks up an incline on a warm day. This is the body's adaptation to changes in the physical demands being placed on it. However, the ANS also is very responsive to cortical and subcortical cognitive and emotional experiences. We know, for example, that one's eyes tend to dilate if one is viewing a sufficiently pleasant or appealing picture. A more dramatic, but nonetheless common example is the increased state of autonomic (sympathetic) arousal one experiences in states of anxiety, fear, or threat. The eliciting stimulus is typically first experienced in the cortex, emotionally colored by the limbic system, and then the ANS response is likely triggered by the hypothalamus. Such response patterns are meant to be adaptive for the organism. However, as we are well aware, such responses often become triggered when the circumstances do not merit it, through faulty learning or negative conditioning. If such responses become chronic and excessive we typically label them in psychiatric terms, such as panic disorder, free-floating anxiety, or irritable bowel syndrome. On the other hand, cognitively learned techniques, such as relaxation procedures or meditation, can be used to enhance parasympathetic responses in situations that might normally tend to elicit less desirable and unnecessary sympathetic discharges. Thus, at least to some extent, many aspects of ANS function can be brought under some degree of cortical control.

Endnotes

1. The first spinal (cervical) nerve exits above the first cervical vertebra, the second cervical nerve above the second, and so forth. However, as there are only seven cervical vertebrae, the eighth cervical nerve exits between the seventh cervical vertebra and the first thoracic vertebra. From here on, each spinal nerve exits below its respective vertebra.
2. The majority of information carried by the major ascending or descending systems that cross the midline of the cord apparently does so via the anterior white commissure, which lies ventral to the central canal.
3. Greater detail regarding ascending and descending pathways within the internal capsule is provided in Chapter 8.
4. Of the approximately 20 million fibers that make up the midbrain portion of the cerebral peduncles, only about 5% remain to form the pyramids after exiting from the pons. The remaining 95% of these corticofugal fibers synapse in the brainstem (eg, red nucleus, motor cranial nerves; reticular formation; and pontine nuclei, the latter which in turn project to the cerebellum).
5. The other major cortical influence on eye movement derives from area 8 (the frontal eye fields) which are thought to be primarily responsible for directing voluntary movements via interconnections of various mesencephalic and pontine nuclei.

6. If the object is already close to us and moving closer, this requires the eyes to both move medially (convergence), as opposed to moving across the visual field, which requires the eyes to move together either vertically or horizontally (conjugate eye movements). Both would appear to require the integrity of the corticotectal pathways.

7. Under stress, sympathetic innervation of the medulla of the adrenal gland stimulates an increased secretion of epinephrine and norepinephrine, which are then released into the bloodstream. Their effect on target organs throughout the body is similar to direct sympathetic stimulation, only more prolonged.

REFERENCES AND SUGGESTED READINGS

Barr, M.L. & Kiernan, J.A. (1993) The Human Nervous System. Philadelphia: J.B. Lippincott Co., pp. 67–87; 349–376.

Brazis, P., Masdeu, J., & Biller, J. (1990) Localization in Clinical Neurology. Boston: Little, Brown & Co., pp. 70–92.

Carpenter, M.B. & Sutin, J. (1983) Human Neuroanatomy. Baltimore: Williams & Wilkins, pp. 265–314.

DeMeyer, W. (1980) Technique of the Neurologic Examination: A programmed text. New York.: McGraw-Hill, pp. 179–241.

Gilman, S. & Newman, S.W. (1992) Essentials of Clinical Neuroanatomy and Neurophysiology. Philadelphia: F.A. Davis Co., pp 29–51; 75–95.

Martin, J.H. (1996) Neuroanatomy: Text and Atlas. Stanford, CT: Appleton and Lange, pp. 249–289.

Mihialoff, G.A. & Haines, D.E. (1997) Motor system I and II. In Haines, D.E. (Ed.) Fundamental neuroscience. New York: Churchill Livingstone, pp. 335–362.

Nolte, J. (1993) The Human Brain. St. Louis: Mosby-Year Book, Inc., pp. 119–153.

Weisberg, L., Strub, R., & Garcia, C. (1989) Essentials of Clinical Neurology. Rockville, MD: Aspen Publishing.

Wiebers, D.O., Dale, A.J.D., Kokmen, E., & Swanson, J.W. (Eds) (1998) Mayo *clinic* examinations in Neurology. St. Louis: Mosby, pp. 151–253 (muscles and reflexes); 275–285 (ANS).

Young, R.R., Asbury, A.K., Corbett, J.L., & Adams, R.D. (1975) Pure pandysautonomia with recovery: Description and discussion of diagnostic criteria. Brain, *98*, 613–636.

2   THE SOMATOSENSORY SYSTEMS

CHAPTER OVERVIEW

The spinal cord encompasses three major ascending sensory pathways that provide ongoing feedback from essentially all of our body, with the exception of our face and parts of our head. Much, although not all, of this sensory input is consciously perceived and provides us with critical information not only about our immediate environment, but about the state and integrity of our bodies. This information is provided by two separate, but complementary systems; the **anterolateral** or **spinothalamic**, and what has been termed the **posterior column** or **lemniscal** system. The spinothalamic system allows us to perceive pain and temperature, as well as pressure or crude touch. In addition to giving us potentially useful information about these aspects of our environment, this system also provides a protective function by serving as the afferent limb for withdrawal reflexes in the presence of potentially harmful stimuli. The lemniscal system, on the other hand, allows us to make fine tactile discriminations (stereognosis), to be aware of the position and movement of our limbs in space, and is probably our primary means of detecting vibrations (other than sound). In addition to these "conscious" perceptions, other ascending pathways provide constant sensory feedback to the cerebellum that apparently bypasses conscious awareness. Such information is critical in maintaining posture and balance, especially while engaging in whole-body activities.

By the conclusion of this chapter the reader should be able to identify the major ascending pathways, both in terms of their general localization within the cord and general functional correlates. In addition, one should be able to describe the basic elements of a somatosensory examination, as well as the nature and clinical significance of abnormal somatosensory findings. Finally, having reviewed the basic anatomical structure of the spinal cord, as well as its major ascending and descending pathways, the reader should be able to identify and appreciate the anatomical basis for some of the more common spinal and peripheral nerve syndromes described in the chapter.

INTRODUCTION

In order to function effectively one needs constant sensory feedback from the external environment, as well as information about conditions that directly impact on one's body. Additionally, it is essential that one maintain awareness of the body in space (e.g., the position and/or movement of one's limbs through the time-space continuum). One also must modulate internal states or homeostatic mechanisms. These functions are accomplished through multiple afferent or sensory feedback systems, some of which operate at a conscious level, others at an unconscious level.

Some stimuli remain essentially detached from us. That is, we become aware of them only indirectly through what are referred to as telereceptors. These include the senses of hearing and vision. Olfaction may be considered a telereceptor as through this sense one can be aware of the presence of a sweet olive tree or a garbage dump from a sizable distance away. On the other hand, odor is perceived when airborne molecules of the object or substance come into direct contact with our nasal mucosa—not always a particularly comforting or pleasant thought.

Other external stimuli are perceived only when they come into direct contact with the body. This would include most tactile sensations such as touch, pressure, temperature, external painful stimuli, and vibration. However, under certain conditions, low-frequency, high-amplitude vibrations may be perceived at considerable distance from their source via pressure waves. Finally, there are those sensations that derive not from external sources or forces but rather from internal conditions or states. In contrast to the first two classes of stimuli for which one is either generally aware or potentially aware, stimuli that arise from within the body may or may not ever reach conscious awareness. In general, this probably has less to do with the nature of the stimulus than with the pathways that carry such information. Thus, awareness of the position of one's limbs may reach conscious awareness when such signals are conveyed through the dorsal columns of the cord through the thalamus and to the cortex. Signals from the same or similar peripheral receptors that synapse locally in the cord or travel through the spinocerebellar tracts never reach conscious levels, but still have a meaningful impact on certain aspects of motor coordination, maintaining muscle tone, and providing the afferent loops necessary for postural and other reflexes.

The special senses mediated by the cranial nerves will be discussed in Chapter 5. These include vision, hearing, taste, and olfaction, and what might be considered a sixth sense: awareness of one's orientation in or movement through space (vestibular). In the current chapter, the emphasis will be on general somatic perceptions or the somatosensory system. This includes the perception of simple touch, discriminative touch (stereognosis), pain, temperature, proprioception (awareness of the position of the limbs, head, or trunk), kinesthesia (awareness of movement), and vibration. While this chapter will focus primarily on those tracts that ascend in the cord, it will be noted that comparable mechanisms serve these same types of stimuli applied to the head or face. More detail about the neural pathways for the latter will be presented in Chapter 5 (The Cranial Nerves). Finally, in Chapters 9 (The Cerebal Cortex) and 10 (The Cerebral Vascular System), there will be a discussion of some of the more unique types of disturbances of somatic awareness that are peculiar to cortical lesions, namely, anosognosia (lack of awareness of sensorimotor deficits), Anton's syndrome (lack of awareness of blindness), finger agnosia (a specific type of autotopagnosia resulting in difficulty discriminating parts of one's body), and unilateral neglect (inattention to one half of the body or hemispace).

ASCENDING PATHWAYS

The ascending or afferent somatosensory pathways within the spinal cord, with the exception of those that primarily mediate position sense and stereognosis (the dorsal columns), are intermingled with descending motor pathways in the lateral and ventral funiculi (see Figure 1–2, Chapter 1).

The ascending pathways can be divided into three major, though not always totally distinct, systems:

1. The **anterolateral system,** which tends to be more primitive and polysynaptic and is primarily responsible for the sensations of pain, temperature, and crude (i.e., less well defined) or simple touch.
2. The **lemniscal system,** which is represented by the dorsal column pathways or dorsal funiculus, has fewer synaptic connections than the anterolateral pathways and conveys conscious information about position sense, kinesthesia, vibration, and the finer aspects of tactual discrimination.
3. The **spinocerebellar system,** which unlike the other two systems, does not project to the thalamus, and hence does not reach conscious awareness. The latter system, however, may carry much of the same type of information as the other two pathways, particularly position sense and kinesthesia, which are essential for coordinated motor activity.

These systems can be further divided into those areas that are supplied by the cranial nerves versus those innervated via the dorsal spinal roots (see Table 2–1). Again, the former will be covered in more detail in Chapter 5.

Before discussing each of these systems in greater detail, it might be useful to review features that they all have in common. Most of the fibers contained in these systems arise from specialized receptors or nerve endings within the skin, muscles, or tendons. Those that supply the trunk and limbs travel centrally in the peripheral nerves, along with the motor or efferent fibers. Just outside the cord, they separate from the motor fibers and enter the cord as the dorsal nerve roots. The cell bodies for these latter fibers reside outside the cord in the dorsal root ganglia. All dorsal root fibers synapse after entering the cord. Most synapse

Table 2–1. Major Somatosensory Tracts

Tracts	*Origin*	*Termination*	*Function*
Spinothalamic (anterolateral system)	Free nerve endings; specialized skin receptors	Spinal cord, posterior horn; secondary fibers to VPL[a] nuclei; tertiary fibers to cortex	Mediates perception of pain, temperature, pressure or simple touch; spinal reflexes
Dorsal columns (lemniscal system)	Specialized skin receptors; muscle and joint receptors	Nuclei cuneatus and gracilis; secondary fibers to VPL; tertiary fibers to cortex	Proprioception, kinesthetic feedback, vibration, fine tactile discrimination; spinal reflexes
Spinocerebellar (dorsal, ventral, rostral, and cuneocerebellar)	Similar to the above	Cerebellum, primarily in the vermal region	Coordination, balance via somatosensory and kinesthetic feedback to cerebellum
Ventral (crossed) and dorsal (uncrossed)[b] trigeminothalamic	Similar to the above, but in an area of the face	Trigeminal ganglion; secondary fibers to VPM[c]; tertiary to cortex	Provides feedback comparable to posterior columns and spinothalamic tracts, but from the facial area[d]

[a] Ventral posterior lateral nuclei of the thalamus.
[b] Appears to carry information from the oral cavity.
[c] Ventral posterior medial nuclei of the thalamus.
[d] See Chapters 4 and 5 for more detail.

locally, that is, at or within a few segments from where they entered the cord; however, some of the fibers of the posterior columns ascend to the upper cord before synapsing. These synapses within the cord would appear to serve multiple purposes. As noted in Chapter 1, some of the afferent neurons synapse either directly or indirectly (via interneurons) on alpha or gamma neurons and serve to maintain muscle tone, mediate spinal reflexes, and provide the potential for the modulation of, or modulation by, descending or efferent fibers. Thus both sensory input and motor output can be influenced at the spinal level. This branching and interconnection of the afferent fibers also allow for the summation of sensory input from more than one nerve rootlet. The commonly observed phenomenon of attempting to relieve sharp pain by rubbing the affected area seems to be accomplished in large part by the interaction of the two types of stimuli in the substantia gelatinosa.

Anterolateral System

The anterolateral system is represented primarily by the **lateral** and **ventral spinothalamic tracts** (Figure 2–1). While these two tracts were once described as carrying different and distinct types of sensory information, the current thinking is that the two have extensive

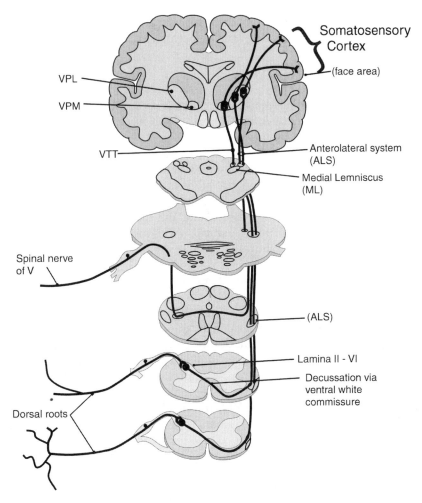

Figure 2–1. Fibers of the anterolateral (spinothalamic) system, which are important in pain and temperature perception, synapse in the dorsal horn and then cross the midline within a few segments of where they enter the cord to ascend in the lateral and ventral funiculi. Traveling in the ventrolateral and later in the dorsolateral portions of the brainstem, these second-order fibers terminate in the ventral posterior lateral (VPL) nuclei of the thalamus. From there, third-order neurons proceed to the cortex. Abbreviations: VPM, ventral posterior medial nucleus (thalamus); VPL, ventral posterior lateral nucleus; VTT, ventral trigeminothalamic tract.

functional overlap, and hence should be considered as a single system. It should be noted that the anterolateral system is most closely associated with, and probably most responsible for, the sensations of pain, temperature, and light touch, while the posterior columns (lemniscal system) are more closely associated with position sense and stereognosis. Some overlapping of functions is found within these two systems as well. In addition, some information shared by both of these systems is conveyed to the thalamus by the **lateral cervical system** (see below). One net effect of all this redundancy within the spinal cord is that unilateral lesions affecting a single tract or system unlikely will result in a total loss of the sense of touch. However, the finer aspects of tactile discrimination indeed will be generally severely compromised following lesions to the posterior columns. While loss of pain on one side of the body caudal to a lesion typically results from unilateral destruction of the anterolateral pathways (and hence, occasionally surgically produced to treat severe and otherwise intractable pain), the effect may not be permanent, again suggesting the possibility of alternate pain pathways.

Pain

The pathways for pain and temperature are difficult to separate, and hence, are usually considered together when discussed anatomically. The sensations of itching and tickling also are thought to be mediated by this system.

Unlike some of the other somatic sensations, there apparently are no specialized "pain receptors," although some may respond chemically to specific substances released during injury to tissues. The sensation of pain probably is initiated at the periphery by small-diameter myelinated (A-delta) and unmyelinated (C) fibers. Each of these appear to be responsible for different aspects or types of pain: the former or the A-delta fibers being responsible for the perception of "fast," "sharp," well-localized pain, while the C fibers carry the "slow," "dull," aching or burning chronic type of pain that tends to be less well localized.

After entering the dorsal cord, these fibers might branch into short ascending and descending axonal processes and might travel for one or more segments in posterolateral portion of the lateral funiculus (Lissauer's tract) before synapsing on interneurons in the posterior horn. While the neurons in lamina II, the substantia gelatinosa, are thought to play a significant role in the spinal regulation of pain perception, synaptic connections for the incoming pain fibers are established in lamina I through V of the spinal cord. While some of these incoming fibers make direct connections with the cells of origin for the anterolateral spinothalamic tracts, most probably make polysynaptic connections, including connections with interneurons in lamina VI through VIII, before ascending. Once these final synaptic connections have been made, the fibers then cross the midline in the ventral white commissure, which is located just anterior to the spinal canal in the central gray area of the cord. Once on the opposite side of the cord, these fibers ascend as part of the anterolateral spinothalamic tracts.

It again should be noted that because the initial peripheral nerves branched into both ascending and descending fibers as they entered the cord, any group of fibers might represent several cord segments entering the spinothalamic tracts. As a result of this arrangement, if a lesion occurs in the cord affecting the anterolateral tracts, the patient will demonstrate a loss or diminution of pain sensation on the opposite side of the body, beginning one or two segments below the level of the lesion. Conversely, irritation of a nerve root will result in pain (ipsilateral to the source of irritation) in the area served by the distribution of that nerve.

In addition to the spinothalamic tracts, the pathways just discussed also give rise to a large proportion of fibers that do not reach the thalamus, at least not directly. Rather, they synapse in the reticular system located in the brainstem, and hence, give rise to the spinoreticular tracts that travel in association with the anterolateral system. The possible function of the **spinoreticular tracts** will be discussed shortly.

As mentioned earlier, two general types of pain have been identified. The first is characterized by a sharp, acute, pricking, and generally well-localized type of pain. The second type has been described as pain that is slower in onset, more aching or burning in character, less well localized, and generally more persistent. It also was noted that the former seemed to be mediated primarily by the larger, myelinated A-delta fibers (fast pain), whereas smaller, unmyelinated C fibers were largely responsible for the latter or "slow pain." However, it also appears that these two types of pain may be conveyed centrally or rostrally in somewhat separate and distinct pathways. "Fast pain" from the contralateral hemibody appears to be related to those spinothalamic fibers originating primarily in lamina I and V and projecting to the **ventral posterolateral (VPL) nucleus** of the thalamus. Comparable fibers from the contralateral face mediated by the trigeminothalamic pathway project to the **ventral postero-medial (VPM) nucleus**. These thalamic nuclei, to which the posterior columns (medial lemniscus) and related trigeminothalamic pathways also project, are topographically well organized and project in this same topographically organized fashion to the sensorimotor cortex (Brodmann's areas 3, 1, and 2). Because of the rather direct, point-to-point representation that is retained from the periphery to the cortex, this system is capable of finer discriminations of pain (or other tactile) information. Because such finer cortical discriminations represent a more recent phylogenetic development, these pain pathways have been termed the **neospinothalamic tracts**.

On the other hand, those fibers that carry information related to chronic, "slow," dull pain seem to originate more in lamina VI through VIII. These slow or dull pain pathways project primarily to the intralaminar nuclei of the thalamus, which in turn project to the cortex in a much more diffuse manner. These latter projections then do not have the same precise topographical organization as VPL or VPM and receive considerable input from the ascending reticular system. In fact, it is thought that these intralaminar nuclei probably represent the primary thalamic projection area for the spinoreticular fibers discussed above. Because they would appear to represent a more primitive system, those spinothalamic fibers that project to the intralaminar nuclei are known as the **paleospinothalamic** tracts. Fibers within this system (perhaps neospinothalamic fibers as well) also project to the posterior nuclei of the thalamus. Both the intralaminar and posterior nuclei of the thalamus project heavily to the secondary somatosensory cortex that lies in the area of the parietal operculum (an area where the inferior parietal cortex enfolds into the lateral fissure). In fact, though relatively rare, loss of affective response to painful stimuli or **pain asymbolia** has been associated most frequently with lesions in the left parietal opercular area. Thus, pain may be "perceived" in either the primary and/or secondary somatosensory cortices. There has been some suggestion that pain also may be "perceived" in some manner even in the thalamus, although it would seem unlikely that any true discriminative perception of pain could take place at this level.

In summary, the neospinothalamic tracts, which are phylogenetically more recently developed, project to those thalamic nuclei (VPL and VPM) that are topographically organized and which, in turn, project to the sensorimotor cortex in the same organized fashion. Because of this point-to-point representation, this system is better able to discriminate and localize painful stimuli. The paleospinothalamic system, along with the spinoreticular tracts, project in a more polysynaptic, diffuse manner to the intralaminar nuclei (and posterior nuclear complex). It is this latter system that seems to be responsible for the less well-defined, more diffusely localized chronic type of pain. It also is thought that this latter system may likely be the one more responsible for an individual's affective response to pain because of its connections with the limbic system.

Before leaving the discussion of pain pathways, several additional phenomena should be reviewed. The first is the notion of **referred pain**. In general, two types of referred pain might be identified, although only the first is classically defined as referred pain. An example of the former is where afferent feedback from some internal organs are perceived as painful in the distribution more or less corresponding to the segmental portion through which these fibers enter the cord. A classic example is pain radiating down the left arm in cases of myocardial infarction. The other situation in which pain might be considered as "referred" is found, for example, in dorsal root irritation. While the problem or mechanical irritant may be located in the lower lumbar region of the spinal vertebrae, the pain may be "perceived" as radiating down the leg.

A **thalamic pain** syndrome can result from a destructive lesion affecting the posterior thalamus. Such a lesion may initially produce a loss or decrease in sensation on the contralateral side of the body.

While these deficits may persist, after a time the patient may begin to experience significant and occasionally quite distressing pain on the side contralateral to the lesion. Such pain generally takes on more of the characteristics of the kind of pain mediated by the paleospinothalamic system (i.e., poorly localized, aching or burning type of pain). It often can be set off by minimal stimuli and will tend to persist and/or expand well beyond the time and space parameters of the eliciting stimulus.

Certain surgical lesions in the brain, including frontal lobotomies, frontal leukotomies, cingulotomies, thalamotomies (especially when the lesions involved some of the intralaminar or dorsomedial nuclei), and even surgical destruction of the pituitary gland, are among the sites found capable of dramatically altering one's subjective impression of, or response to, otherwise intractable pain. The problem with most of these lesions is that they also can produce other disabling behavioral problems. It is thought that perhaps the critical element in most, if not all, of these surgeries is the disconnection of critical limbic pathways (it has been hypothesized that the effectiveness of pituitary lesions is due to secondary lesions produced in the hypothalamus). In many of these "disconnection" cases, the patient may still report "feeling the pain," but being affectively "disconnected" from it. Other surgical treatments for pain include implanting devices to produce stimulation of large-diameter fibers, which in turn suppress or inhibit the firing of small-diameter C fibers (transcutaneous stimulation) or stimulation of the periaqueductal gray matter of the midbrain to stimulate the release of endorphins. Surgery also is used to disrupt the pain pathways by sectioning single nerves, the dorsal roots (**rhizotomy**), the lateral spinothalamic tract (**cordotomy**), or the posterior and intralaminar nuclei of the thalamus. Unfortunately, many of the surgical treatments may produce only temporary relief from pain.

More conservatively, treatment for pain may simply involve a variety of procedures including drugs that change the sensitivity of the peripheral nerve endings or affect the central processing of pain.

Temperature Sense

The information that allows the brain to perceive and distinguish the temperature of objects, liquids, or air as they come into contact with the skin also is carried by the anterolateral system. Similar to pain, it appears that A-delta and C fibers are responsible for conveying this information from the periphery to the spinal cord. No specific nerve endings have been identified as responding to temperature. It appears likely that, again similar to pain, these sensations are initiated through free nerve endings in the skin. The C fibers may mediate both extremes of hot or cold sensations that are perceived as "painful." The spinothalamic pathways for the perception of temperature are virtually indistinguishable from those mediating pain. Hence, loss of pain and temperature generally occur at the same time and in the same distribution.

Pain and temperature sensations in the face are mediated by the trigeminal (V) cranial nerve, specifically through the spinal trigeminal nucleus (see Chapter 5 for a more detailed description). Before entering this nucleus, however, the fibers coming into the pons travel caudally as the spinal tract of V prior to synapsing in the spinal trigeminal nucleus. The caudal portion of this nucleus is probably the relay center for most of the face, with the exception of dental pain, which is mediated through the interpolar nucleus. From these nuclei, crossed fibers ascend via the ventral trigeminal tract to terminate in the ventral posteromedial (VPM) nucleus of the thalamus. Pain sensations probably also reach the thalamus indirectly via the reticular system and are responsible for the duller, aching, chronic pain sensations of the face.

Simple Touch (Pressure)

A number of different types of cutaneous end organs have been identified that seem to play a role in the perception of touch. These are **Meissner's** and **Pacinian corpuscles; Merkel, peritichial,** and **Ruffini nerve endings**; as well as **free nerve endings**. It is not clear whether the sensory end organs that mediate the sensations of touch and pressure that travel in the anterolateral system (spinothalamic tracts) are any different in type or distribution than the ones that subserve the lemniscal (dorsal column) system, to be discussed next.

There are, however, at least two major distinctions between the two systems other than the tracts themselves:

1. In the anterolateral system, the fibers cross the midline at or near their level of entry into the cord, while those in the lemniscal system cross at the level of the medulla.
2. While both the anterolateral system and the lemniscal system are capable of mediating the sense of touch and pressure on the skin, the anterolateral system does not appear capable of fine tactile sensory discriminations, a capacity that characterizes the dorsal column or lemniscal system.

Actually, the sensory fibers that mediate touch (and probably vibration and proprioception as well) divide into three distinct ascending pathways or systems. Some ascend in the dorsal columns (lemniscal system), while others ascend ipsilaterally in the lateral cervical tract, synapsing in the lateral cervical nucleus in the upper cord. These fibers then cross the midline of the cord and ascend with the contralateral fibers of the anterolateral spinothalamic tracts. Finally, those fibers traveling in the anterolateral spinothalamic tracts follow pathways similar to those described for pain and temperature fibers. As is true of the latter nerve fibers, those conveying touch via this system typically ascend and descend several segments in the cord before crossing and ascending to the ventral posterolateral (VPL) nucleus of the thalamus. After synapsing there, these fibers, along with other somatosensory inputs, project via the posterior limb of the internal capsule to the postcentral gyrus to relay information concerning crude touch and pressure. It should be noted that while the major cortical projection site for the VPL and VPM nuclei is the primary somatosensory cortex (Brodmann's areas 3, 1, and 2), these nuclei probably project more broadly. Other cortical projections likely include the secondary somatosensory cortex in the parietal operculum and perhaps to the precentral sensorimotor cortices. As with the pathways for pain and temperature that travel in the lateral spinothalamic tract, collaterals for touch and pressure probably also go to the reticular formation and to the intralaminar nuclei of the thalamus. The intralaminar nuclei, in turn, project to a variety of structures, including other thalamic nuclei, the basal ganglia, and the limbic system and seem to play a role in general arousal or alertness.

As previously noted, because of the diversity of these pathways, including fibers that cross shortly after entering the cord (S-T tracts) and those that ascend to the medullary junction before crossing (lateral cervical tract and posterior columns), the more elementary sensations of touch are typically preserved following unilateral trauma to the cord. However, such a lesion likely would disrupt fine tactile discrimination ipsilaterally and pain and temperature perception contralaterally. The apparent presence of small-diameter afferent fibers in the ventral roots may offer one basis for the persistence of pain that can occur following sectioning of the dorsal roots. All tactile sensation below the level of the lesion, of course, would be lost following a complete transection of the cord.

Dermatomes

As noted in Chapter 1, the area of the skin represented by the pairs of dorsal roots at each spinal level (i.e., the vertebral spaces at which the nerves enter or exit) is referred

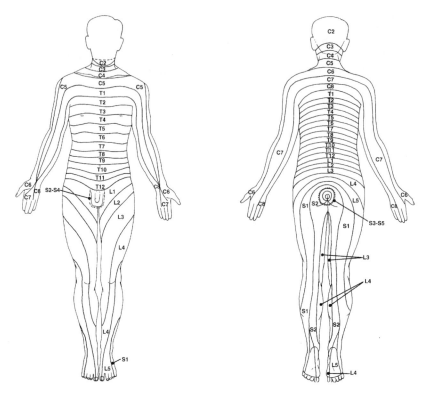

Figure 2–2. Approximate distribution of the segments of the skin represented by the various dorsal roots of the spinal cord. Considerable overlapping of adjacent segments is generally the case.

to as a **dermatome**. Figure 2–2 illustrates the distribution of the dermatomes for the spinal nerves. It should be kept in mind that the divisions shown in the illustrations are approximations and variations may be noted among various authors. Also, from a clinical standpoint, the divisions are not nearly as sharp or clear as might be implied by the diagram. There is considerable overlap between adjacent spinal segments. As a result, there may be little, if any, detectable sensory loss unless several adjacent dorsal roots are affected.

Lemniscal System

So far we have primarily focused on the anterolateral system (ventral and lateral S-T tracts), which primarily carries information regarding pain, temperature, and simple touch. In this section, the focus will be on the other major ascending system responsible for conscious perception of stimuli affecting the arms, legs and trunk, the **lemniscal** system.

The term "lemniscus" means a "ribbon" or a "band" and generically is a name that is applied to several pathways within the brainstem. The "lateral lemniscus" carries auditory fibers (see Chapter 5) and the "spinal lemniscus" results from the joining of the spinothalamic and spinotectal tracts in the medulla. The **medial lemniscus**, from which the lemniscal system derives its name, represents the ascending fibers of the posterior (dorsal) columns *after* they synapse in the nuclei cuneatus and gracilis above the spinal–medullary junction. The dorsal columns or lemniscal system represent the primary (though probably not exclusive) pathway for the perception of proprioception (position sense), kinesthesia (awareness of speed and direction of limb movement), stereognosis (fine tactual discrimination of contour, relief, texture, size, localization, etc.), and vibration sense.

Proprioception, Stereognosis, Vibratory Sense

The senses of proprioception and kinesthesia are initiated at the periphery by specialized nerve fibers. These include primary or annulospiral endings and secondary or flower-spray endings that attach to the intrafusal muscle fibers or muscles spindles, Golgi tendon organs (found where the tendon joins with the muscle), and both free and capsular (e.g., Golgi, Ruffini, and Pacinian) nerve endings in the ligaments and joints. Feedback for stereognosis is derived from the same types of specialized nerve ending in the skin (e.g., Pacinian corpuscles, Merkel's disks, Meissner's corpuscles, nerve plexus around hair follicles, and free nerve endings) that are found in the fibers which eventually ascend in the anterolateral system. Vibration probably should not be identified as a separate or special sense, but more likely represents a special condition (e.g., temporal summation) of touch. Pacinian corpuscles, because of their rapid adaptation, may play an important role in the perception of vibration.

Fibers from receptors mediating proprioception, kinesthesia, stereognosis, and vibratory sense enter directly into the dorsal columns and ascend ipsilaterally, without synapsing, to the lower medulla (Figure 2–3). The fibers from the leg ascend in the more medial portion of the dorsal columns in the fasciculus gracilis, while those fibers from the upper extremity ascend in the more lateral fasciculus cuneatus. Thus, the nerve fibers for the arm might be viewed as entering the cord after the fibers from the leg already have been laid down. These two tracts–the fasciculus gracilis and fasciculus cuneatus—terminate in the nucleus gracilis and nucleus cuneatus which are located in the lower medulla. After synapsing in these nuclei, the fibers then cross (decussate) and ascend as the medial lemniscus and terminate in the ventral posterolateral nucleus (VPL) of the thalamus. Again, comparable projections from the cranial nerves (e.g., the trigeminothalamic tract) terminate in the ventral posteromedial nucleus (VPM). From there, topographically organized projections are sent to the postcentral gyrus (BA 3, 1, 2). There probably is considerable overlap in the cortical projections of both the anterolateral and the lemniscal systems. Nevertheless, the lemniscal system's more direct and discrete topographically organized projections to the somatosensory cortex (as opposed to the more polysynaptic, diffused connections which are more characteristic of the anterolateral system) are what permit us to make fine tactile discriminations.

Facial Tactile Discrimination and Proprioception

Discrimination of tactile stimuli and proprioceptive feedback from the face is mediated by cranial nerve V. The "main sensory nucleus" of the trigeminal nucleus is the likely point of relay for this type of somatosensory information. Two tracts ascend from this nucleus. One decussates to join the medial lemniscus and terminates in the VPM nucleus of the thalamus; the other, an uncrossed tract, ascends as the dorsal trigeminal tract and terminates in a separate part of the VPM nucleus (Figure 2–3). The role of the latter is not clear, but it may relate to intraoral sensations. Proprioception for the muscles of mastication is somewhat unique. Their unipolar nerve cells are located in the mesencephalic nucleus, which is located in the upper pons and lower midbrain, rather than in a ganglion outside the CNS.

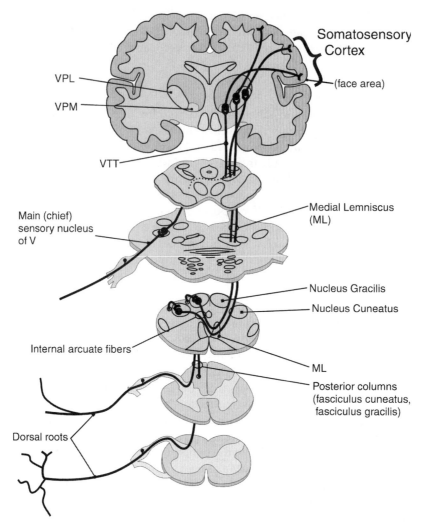

Figure 2–3. Fibers making up the lemniscal (dorsal column) system, critical for proprioception and fine tactual discrimination, enter ipsilateral dorsal funiculus without synapsing until they reach the nucleus cuneatus and nucleus gracilis in the lower medulla. Second-order neurons then cross the midline and ascend as the medical lemniscus to the VPL nuclei of the thalamus, there giving rise to third-order thalamocortical fibers.

Spinocerebellar Connections

As discussed in Chapter 1, there are many reflexes that are mediated at the spinal level. Perhaps among the most important are those that respond to noxious stimuli and those that respond to sudden changes in the antigravity muscles (necessary to maintain an upright balance).

These reflexes are executed, or at least can be executed, at the spinal level. In some instances, they may rely on a simple or monosynaptic connection between an intrafusal (proprioceptive) afferent fiber and an alpha motor neuron in the anterior horn feeding the same extrafusal muscle. At other times, although still remaining within the cord, the reflex involves several cord segments, several muscle groups, and multiple interneurons. However, for the more elaborate demands of most coordinated movements, higher-order connections must be established. Of central importance to this process are the cerebellum

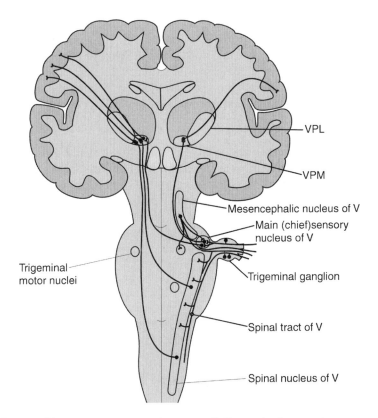

Figure 2–4. Comparable somatosensory systems mediating pain, temperature, proprioception, and fine tactual discrimination in the face are shown in this figure. Unlike the spinal pathways that synapse in the VPL, trigeminal pathways synapse in the ventral posterior medial (VPM) nuclei of the thalamus. See Chapter 5 for more detailed discussion of these pathways.

and its subsequent connections to the cortex, vestibular system, and other sensorimotor systems. Apparently there are no direct, descending connections between the cerebellum and the spinal cord. Obviously (or so it seems), what appears to be most critically needed by the cerebellum is information regarding the position of the body (both trunk and limbs) at any point in time, as well as the direction, degree, and speed of movement that is occurring. However, other types of sensory information also may be important or useful to coordinate higher-order reflex or voluntary movement. As a result of these needs, there are extensive connections between incoming somatosensory information and the cerebellum. The following is a review of these connections.

There are four major pathways by which information entering the spinal cord gets to the cerebellum. These will be discussed in greater detail in Chapter 3. For now it is important to remember that two of these pathways convey information from the lower extremities and two convey information from the upper extremities. Another point to remember is that virtually all information that goes to the cerebellum from the spinal cord remains ipsilateral, even when there is crossing of the midline by the spinocerebellar fibers (see below for the simple explanation). There are three pathways into the cerebellum: the **superior, middle,** and **inferior cerebellar peduncles**. The largest of these, the middle cerebellar peduncle, is the route by which the cortex gets information into the cerebellum. Of the pathways by which spinal information gets into the cerebellum, the majority enters via the inferior peduncle. Only one, the **ventral spinocerebellar tract**, clearly enters via the superior peduncle. The superior cerebellar peduncle, as we shall see, is also the major conduit for information leaving the cerebellum. Fibers mediating proprioception, touch, pressure, pain, and local reflex activity concerning the leg and lower torso synapse in the base of the posterior horn and then give rise to fibers that cross over to the opposite side of the cord. These then ascend to the cerebellum via the ventral spinocerebellar tract (crossed) in the lateral funiculus (Figure 2–5). However, before entering the cerebellum through the superior cerebellar peduncle (the only spinal tract that does), most of these fibers cross the midline again (i.e., double cross), so the input is still ipsilateral.

The **nucleus dorsalis of Clarke** located in the base of the dorsal horn (T-1 through L-2 or L-3) receives projections from the muscle spindles, Golgi tendon organs, and joint receptors. From there these fibers ascend, without crossing, to the cerebellum as the **dorsal spinocerebellar tract**, entering through the inferior cerebellar peduncle. This tract, which also contains information from the trunk and lower extremity, appears to be more specific in its input than the ventral spinocerebellar tract. It is primarily concerned with information regarding muscle tone, posture, and proprioceptive feedback. It enters the cerebellum through the inferior cerebellar peduncle.

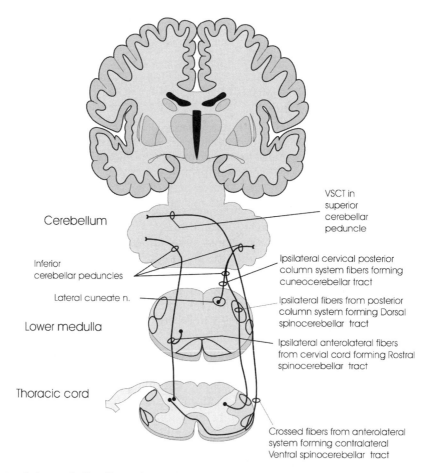

Figure 2–5. Spinocerebellar fibers shown here represent the third major group of ascending fibers in the spinal cord. Within the spinal cord itself, these consist of the dorsal and ventral spinocerebellar tract located in the lateral funiculus. Note: Spinal input to the cerebellum is primarily ipsilateral, despite the double crossing of the VSCT.

Comparable information regarding muscle tone and proprioceptive feedback from the upper extremities travels up the fasciculus cuneatus to synapse in the lateral or accessory cuneate nucleus (the upper extremity counterpart of the nucleus dorsalis). After synapsing in the accessory cuneate nucleus, the postsynaptic fibers become the **cuneocerebellar tract**. These fibers then ascend ipsilaterally and, like the dorsal spinocerebellar tract, enter the vermis of the cerebellum via the inferior cerebellar peduncle. The upper extremity counterpart of the ventral spinocerebellar tract is the **rostral spinocerebellar tract,** which originates in the nucleus centralis basalis of the cervical cord and proceeds ipsilaterally to the cerebellum.

TESTING FOR SOMATOSENSORY DEFICITS

Testing for Pain

Have the patient discriminate between the point ("sharp") and head ("dull") of a pin. You have to be careful to avoid simply tapping into the sense of touch. One should avoid using the same pin with different patients, as there is evidence that certain viruses can be transmitted in this fashion.

Testing for Proprioception

Have the patient attempt to localize his or her limb in space following passive movement by the examiner (with patients eyes closed) or indicate the state of flexion or extension of one's limb. However, perhaps the easiest and certainly one of the most sensitive and specific tests of proprioception is to passively extend or flex a digit (e.g., the great toe or a finger) while asking the patient to indicate the direction in which it is being moved. One also may check the integrity of the posterior columns by asking the patient to stand erect with the feet together. If the patient shows considerably more difficulty maintaining balance with the eyes closed than opened, this suggests posterior column compromise (Romberg's sign). If comparable difficulties are noted regardless of whether the eyes are open or closed, cerebellar disease should be suspected.

Testing for Stereognosis

Here we might ask the patient to differentiate shapes, textures, or similarly configured small objects (e.g., a paper clip versus a safety pin) by touch. The patient also may be asked to identify numbers written on the fingertip or palm of the hand (graphesthesia), to make two-point discriminations, or to localize stimuli applied to various parts of the face, limbs, or torso.

Testing for Vibration

This procedure basically calls for the application of a tuning fork (256 cps) to bony prominences of the distal upper and lower extremities. The examiner must perform trials with and without the tuning fork vibrating to assure reliability and comprehension of the test. The patient is instructed to indicate whether the tuning fork is vibrating when touching the limbs, and if it is when the vibration appears to stop. A vibratory sensory level can be determined by starting distally and working one's way proximately up the limb. The latter procedure would be important if a neurologist expects a peripheral neuropathy.

Testing for Temperature

The patient is asked to discriminate between objects of different temperatures. The examiner should ensure that the temperatures are readily discriminable, but neither is extreme. Test tubes filled with warm and cool water make reasonable testing devices.

In all cases, the examiner always should compare performances on the right versus the left side of the body. One should be alert to the possibility of cortical neglect as well as old central or peripheral injuries or disease processes that might have an effect on peripheral processes, such as peripheral neuropathy secondary to diabetes or chronic alcohol abuse.

LESIONS AND DISEASES AFFECTING THE ASCENDING AND DESCENDING TRACTS

Lesions Affecting the Posterior Limb of the Internal Capsule

Lesions restricted to the anterior portion of the posterior limb of the internal capsule normally result in a **contralateral hemiparesis** or hemiplegia (see Figure 2–6a). The distribution of weakness is distinct and characteristic of lesions of the corticospinal pathway. In the upper extremity, weakness is most pronounced in the distal extensor muscles, while in the lower extremity it is the proximal flexor muscles that are most affected. Increased deep-tendon reflexes are present on the affected side, along with pathological reflexes, such a **Babinski** or **Hoffmann's sign**. Although hypotonia may be present initially, this eventually is followed by an increase in muscle tone along with other indications of **hyperreflexia** on that side,

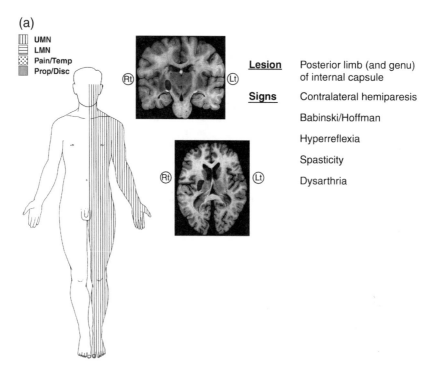

Figure 2–6. Lesions affecting sensory–motor systems. Legend: UMN, upper motor neuron deficits; LMN, lower motor neuron deficits; Pain/Temp, changes in pain and temperature; Prop/Disc, changes in proprioception, fine tactile discrimination (and vibration).

(b)

Legend:
- ⦀ UMN
- ▤ LMN
- ▨ Pain/Temp
- ▦ Prop/Disc

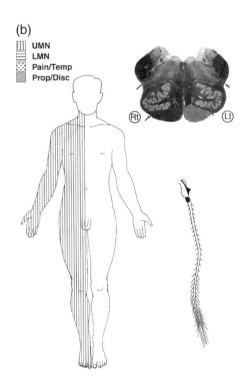

Lesion   Corticospinal tract
              (upper medulla)

Signs: Contralateral hemiparesis

              Babinski/Hoffman

              Hyperreflexia

              Spasticity

(c)

Legend:
- ⦀ UMN
- ▤ LMN
- ▨ Pain/Temp
- ▦ Prop/Disc

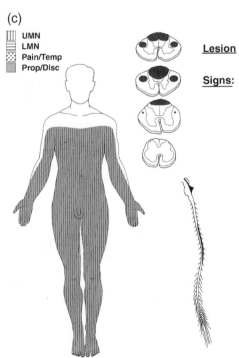

Lesion   Subacute combined degeneration

Signs: (Bilateral involvement)

              Weakness

              Positive Babinski*

              Diminished vibration/proprioception

              Positive Romberg

              Paresthesias*

              * stretch reflexes may be
              diminished and some parethesias,
              including diminished pain and temperature,
              might be present due to peripheral
              nerve involvement

(d)

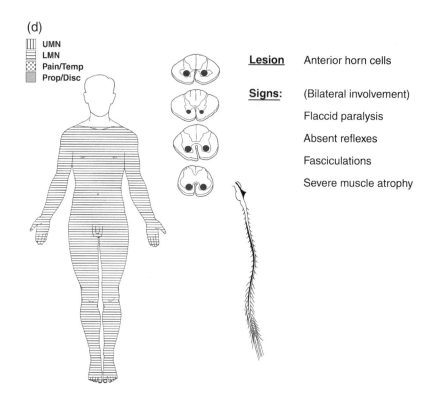

Legend:
- UMN
- LMN
- Pain/Temp
- Prop/Disc

Lesion   Anterior horn cells

Signs: (Bilateral involvement)

Flaccid paralysis

Absent reflexes

Fasciculations

Severe muscle atrophy

(e)

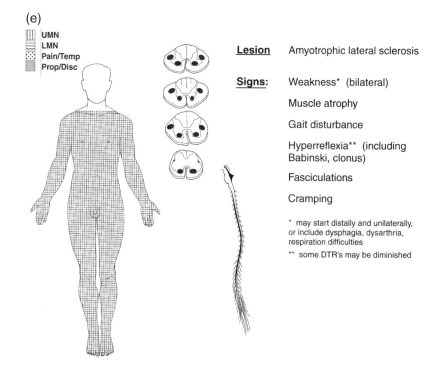

Legend:
- UMN
- LMN
- Pain/Temp
- Prop/Disc

Lesion   Amyotrophic lateral sclerosis

Signs: Weakness* (bilateral)

Muscle atrophy

Gait disturbance

Hyperreflexia** (including Babinski, clonus)

Fasciculations

Cramping

* may start distally and unilaterally, or include dysphagia, dysarthria, respiration difficulties

** some DTR's may be diminished

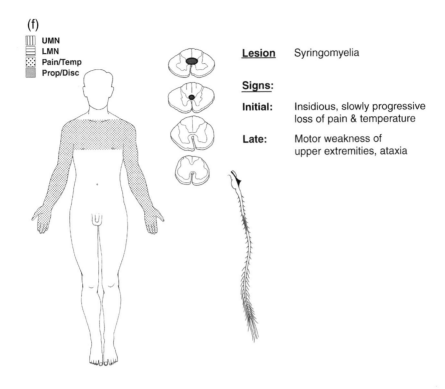

(f)

▥ UMN	
▤ LMN	
▨ Pain/Temp	
▦ Prop/Disc	

Lesion   Syringomyelia

Signs:

Initial: Insidious, slowly progressive loss of pain & temperature

Late: Motor weakness of upper extremities, ataxia

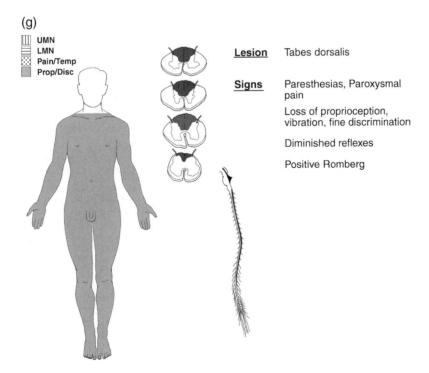

(g)

▥ UMN	
▤ LMN	
▨ Pain/Temp	
▦ Prop/Disc	

Lesion   Tabes dorsalis

Signs   Paresthesias, Paroxysmal pain

Loss of proprioception, vibration, fine discrimination

Diminished reflexes

Positive Romberg

Figure 2–6.   *(Continued)*

(h)

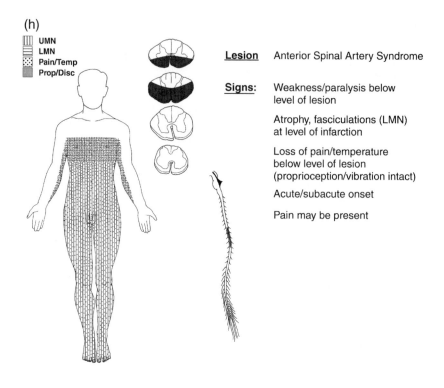

Lesion	Anterior Spinal Artery Syndrome
Signs:	Weakness/paralysis below level of lesion
	Atrophy, fasciculations (LMN) at level of infarction
	Loss of pain/temperature below level of lesion (proprioception/vibration intact)
	Acute/subacute onset
	Pain may be present

(i)

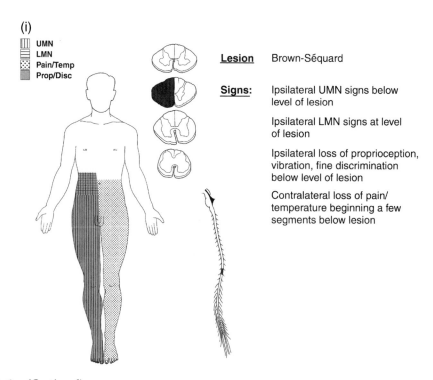

Lesion	Brown-Séquard
Signs:	Ipsilateral UMN signs below level of lesion
	Ipsilateral LMN signs at level of lesion
	Ipsilateral loss of proprioception, vibration, fine discrimination below level of lesion
	Contralateral loss of pain/ temperature beginning a few segments below lesion

Figure 2–6. *(Continued)*

which might include **clonus** and clasp-knife **spasticity**. If the lesion extends toward the genu of the internal capsule, an upper motor neuron VIIth cranial nerve deficit may be present involving the contralateral face. In upper motor neuron lesions of the VIIth nerve, only the muscles of expression in the lower face are affected (see Chapter 5).

The most common cause of such lesions are lacunar infarcts secondary to hypertension (see Chapter 10) which can be relatively small in size, and hence restricted in their sphere of influence. However, since the motor pathways converge in the internal capsule, even a small lesion can have profound and broad-reaching effects topographically. As can be seen in Figure 7–5, the thalamocortical projections from the ventral posterior nuclei of the thalamus that carry somatosensory information to the cortex are in close proximity to the descending motor pathways (generally slightly posterior to the latter, with the visual projections being even more posterior in the internal capsule). Thus, if the lacunar lesion extends slightly more posteriorly, somatosensory symptoms also may be seen. Simple touch or pain usually is not affected by such lesions, but rather the more higher-order perceptual judgments normally carried out by the postcentral cortices (3, 1, 2) such as stereognosis, two-point, weight, or texture discriminations. If there is an even more extensive disruption involving the posterior limb of the internal capsule (e.g., as a result of an intracranial hemorrhage), one might see unilateral motor, somatosensory, and visual disturbances from a single lesion, although this would be a relatively rare phenomenon.

Lesions Affecting the Corticospinal Tract

Lesions that are isolated to the C-S tract and do not involve other descending tracts are relatively rare, but may occur, for example, with focal vascular disease (strokes) in the brainstem. Lesions of the corticospinal tracts in the brainstem produce the same type of deficit described above, except the face will not be involved if the insult is below level of the nucleus of the VIIth cranial nerve (Figure 2–6b). If the brainstem lesion is above the level of the decussation, cranial nerve signs may accompany it on the side opposite the hemiplegia. However, if the damage is restricted solely to the C-S tract at the medullary level, the hemiplegia and associated reflex changes are often mild and may be limited to a pronator drift and decreased fine motor skills. The presence of other descending tracts at this point (e.g., reticulospinal, vestibulospinal, and rubrospinal) likely provides sufficient muscular innervation to prevent a more dramatic loss of function.

Subacute Combined Degeneration (Posterolateral, Dorsal Column Syndrome)

Occurring in association with **pernicious anemia** (vitamin B-12 deficiency), this syndrome tends to selectively affect the posterior columns and the lateral corticospinal tracts in the cord (Figure 2–6c). Because of the involvement of the corticospinal tracts, **upper motor neuron** signs are present (weakness, spasticity, and positive Babinski). Reflexes, however, may be diminished due to peripheral nerve involvement. Because the posterior columns are affected, there will be impairments of **proprioception, stereognosis,** and **vibration sense** and a **positive Romberg** (loss of balance with the feet together and eyes closed). As a result of the peripheral neuropathy, **paresthesias** as well as some **loss of pain and temperature** may be present. The disease typically presents with a slow, insidious onset with bilateral signs or symptoms and the legs are often affected before the upper extremities. In addition, if due to a B-12 deficiency, this syndrome may be accompanied by memory loss. Early identification and treatment are essential to prevent further progression of symptoms.

Anterior Horn Cell Disease (e.g., Poliomyelitis)

Since the disease process is typically isolated to the region of the **anterior horn cells** (lower motor neurons), the presenting clinical picture is one of flaccid paralysis, absent reflexes, fasciculations, and muscle atrophy (Figure 2–6d). As the posterior roots are unaffected, sensations remain essentially intact.

Combined Anterior Horn and Corticospinal Tract Disease (ALS)

In amyotrophic lateral sclerosis (ALS), there is a combined degeneration of the pyramidal tracts and the motor cells in the cranial nerve nuclei and in the ventral horns of the cord resulting in a **combination of upper (UMN) and lower motor neuron (LMN) symptoms** (Figure 2–6e). The LMN involvement produces weakness and atrophy in some muscle groups and the UMN lesion result in spasticity and hyperreflexia in others. The cranial nerves (motor) often are affected. Fasciculations are commonly present. As this disease process is limited to the motor system, no sensory loss is experienced.

Lesion of the Central Gray Matter (Syringomyelia)

In syringomyelia, there is an expanding cavity that forms in the central gray area of the cord, usually in the area of the cervical or lumbar enlargement. The lesion primarily affects the fibers that make up the lateral spinothalamic tract as they cross the midline of the cord. Because they affect the crossing fibers from both sides, there is a bilateral loss of pain and temperature sensation at the level of the lesion, that is, typically in both upper extremities, often described as a cape-like sensory loss (Figure 2–6f). Since the fibers that enter the cord below this defect and the lateral spinal tracts made up of those fibers are unaffected, pain and temperature in the lower extremities are unaffected. Also, since the lemniscal tracts do not decussate until the midbrain, proprioception, touch, and vibratory sense are intact in both the upper and lower extremities. Signs of upper motor neuron disease, however, may be evident in the lower extremities as a result of pressure being exerted on the corticospinal tracts by the expanding cavity. Also, as the lesion expands, it also may affect the anterior horn cells, leading to motor changes in the upper extremities. These lesions are most often congenital or associated with trauma.

Posterior Column Disease

Tabes dorsalis is a lesion associated with tertiary neurosyphilis and is anatomically limited to the posterior columns and dorsal roots of the spinal cord. Therefore, early on there may be paresthesias and sharp pains, particularly in the lower extremities (Figure 2–6g). As the disease progresses, there is a loss of sensation and loss of reflexes as a result of an interruption of the reflex arc. There is also a loss of position and vibratory sense and a positive Romberg due to the loss of proprioceptive feedback due to involvement of the posterior columns. While not directly involving the lower motor neurons, the lack of sensory feedback may produce hypotonia with little or no loss of voluntary movement or strength and no major atrophic changes or fasciculations. Gait will be affected as the patient has to visually monitor what he is doing. The gait is wide-based and clinically resembles the ataxic gait seen in midline vermal cerebellar lesions. Because the mechanism is different, the gait due to posterior column disease is called a sensory ataxia.

Thrombosis of the Anterior Spinal Artery

The anterior spinal artery supplies blood to the anterior and much of the middle sections of the cord. Atrophy, flaccid paralysis, and fasciculations (lower motor neuron syndrome) are present at the level of the lesion due to the damage to the anterior horn cells (Figure 2–6h). Spastic paralysis below the lesion (paraplegia) can result from destruction of the corticospinal tract. Loss of pain and temperature bilaterally below the lesion results from damage to the spinothalamic tracts. Because of anastomoses with radicular arteries along the course of the spinal cord, the area of infarction is often limited.

Hemisection of the Cord

A lesion that more or less locally affects one lateral half of the cord produces what is commonly referred to as a **Brown-Séquard syndrome** (Figure 2–6i). This condition usually results from a traumatic lesion, such as stab or bullet wound. Upper motor neuron signs (e.g., spastic paralysis, hyperreflexia, clonus, loss of superficial reflexes, and a positive Babinski) will be present on the side of the lesion due to disruption of the corticospinal tracts. Ipsilateral loss of position and vibratory sense and tactile discrimination (astereognosis) below the level of the lesion are observed as a result of sectioning of the dorsal column. Loss of pain and temperature on the opposite side of the lesion beginning about one or two dermatomes below the lesion are present, resulting from damage to the anterolateral spinothalamic tracts.

Peripheral Lesions

The dorsal and ventral roots, along with the combined portions of the peripheral nerves, are subject to a vast array of pathology, including trauma or mechanical injury; metabolic deficiencies; toxins, autoimmune, and infectious processes; and genetic and other disorders. Depending on the particular pathology, the symptoms may be primarily motor or sensory or a combination of the two. In addition, the effects may be unilateral and limited to or restricted to a single limb or a few dermatomes, or may be bilateral or relatively diffuse. Further confusing the picture is the fact that some disease processes directly affect the muscles rather than the nerves while producing similar symptoms. On the other hand, one disease process, **myasthenia gravis**, affects not so much the muscle or the nerve, but rather the synaptic junction between the two.[1] In addition to the history and clinical examination, nerve conduction studies, electromyography, and serum analysis are often quite helpful in making these differential diagnoses.

The array of peripheral neuropathies is quite extensive and the interested reader is referred to clinical neurology texts for a more complete listing and description of these disorders (e.g., Adams, Victor, & Ropper, 1997). There are, however, certain clinical correlates that tend to be more characteristic of peripheral nerve lesions. Because except for the nerve roots, the peripheral nerves carry both motor and sensory information, muscle weakness frequently is accompanied by sensory changes (e.g., diminished proprioception, vibration, stereognosis, and/or the presence of paresthesias, pain, or a burning sensation).[2] Frequently, although not invariably, peripheral neuropathies may tend to manifest themselves distally more than proximally and may result in a "glove and stocking" type distribution affecting several or all extremities. Because they represent lower motor neuron lesions, diminished reflexes are often prominent (see Chapter 1 for review of lower-motor neuron lesions). Muscle atrophy, a common correlate of peripheral nerve damage, may not be seen in the more acute stages of the disorder. Among the more common causes of peripheral polyneuropathy in the general population are those resulting from chronic alcohol abuse and diabetes, whereas trauma and disk herniation are among the more frequent sources of more focal neuropathies.

One particularly dramatic and potentially fatal form of peripheral neuropathy is "acute post-infectious polyneuropathy," also known as Guillain-Barré syndrome. Typically developing a week or more after a viral infection (or vaccination), this syndrome primarily attacks the peripheral nerves. Both motor (progressive weakness or paralysis, decreased tone, and markedly diminished or absent deep tendon reflexes) and sensory (paresthesias, pain, and diminished proprioception, stereognosis, and vibration) symptoms are present, although the motor are generally more severe. The symptoms are usually symmetrical and frequently start in the legs, progressing to the upper extremities and cranial nerves, often sparing the eye muscles. As the respiratory muscles are commonly involved, this may prove to be a rapidly fatal condition. However, if properly ventilated during the acute stage complete recovery is common.

Long-Tract Findings in Supraspinal Lesions

The various descending motor pathways in the spinal cord have their origins in the cerebral cortex, as well as in the brainstem. Similarly, the ascending fibers mediating conscious somatosensory perception continue either uninterrupted (spinal thalamic tracts) or following synaptic connections in the medulla (posterior columns) as discreet fiber tracts throughout the length of the brainstem before synapsing in the thalamus. As a result, brainstem lesions can result in symptoms not unlike those accompanying cord lesions. While such lesions will be reviewed in greater detail in Chapter 4, for the present it should be noted that spinal cord and brainstem lesions often might be differentiated clinically. Part of this is due to where the various fiber tracts cross the midline. For example, focal lesions of the cord are more likely to be associated with ipsilateral disturbances of proprioception and stereognosis and upper motor neuron signs below the level of the lesion. By contrast, in the brainstem, focal lesions affecting the pyramidal tract also may result in changes in pain and temperature (as well as proprioception and stereognosis) on the same side as the motor deficits. Even more critically, the brainstem is the site of most of the cranial nerves and their respective nuclei. Hence, brainstem lesions that affect the major ascending or descending pathways often also result in cranial nerve findings involving the head or face, which, of course, are not expected in lesions of the cord.[3]

Endnotes

1. Unlike peripheral neuropathy, which frequently first manifests itself in the extremities, myasthenia gravis tends to initially affect the muscles of the face and head, particularly the eyelids (levator palpebrae) and extraocular muscles, and tends to improve following rest.
2. Multiple sclerosis may similarly present with focal muscle weakness and sensory changes. However, in multiple sclerosis the locus of the motor and sensory changes may be dissociable and the symptoms tend to vary more over time and (body) space than with peripheral neuropathies.
3. One notable exception is **Horner's syndrome** (constricted pupil, partial ptosis, and ipsilateral anhidrosis) which, as noted in Chapter 1, can result from lesions of the lower cervical or upper thoracic cord, disrupting sympathetic innervation of the eye and face. As will be discussed in Chapters 4 and 5, further complicating the situation is the fact that this latter syndrome also may accompany lesions of the brainstem that interrupt descending sympathetic fibers from the hypothalamus.

REFERENCES AND SUGGESTED READINGS

Adams, R.D., Victor, M., & Ropper, A.H. (1997) Principles of Neurology (Part 5. Diseases of Spinal Cord, Peripheral Nerve, and Muscle). New York: McGraw-Hill.

Barr, M.L. & Kiernan, J.A. (1993) The Human Nervous System. Philadelphia: J.B. Lippincott Co., pp. 293–312.

Brazis, P., Masdeu, J. & Biller, J. (1990) Localization in Clinical Neurology. Boston: Little, Brown & Co., pp. 70–92.

Carpenter, M.B. & Sutin, J. (1983) Human Neuroanatomy. Baltimore: Williams & Wilkins, pp. 265–314.

Davidoff, R.A. (1989) The dorsal columns. *Neurology, 39*, 1377–1385.

DeMeyer, W. (1980) Technique of the Neurologic Examination: A programmed text. New York: McGraw-Hill, pp. 298–328.

Gilman S. & Newman, S.W. (1992) Essentials of Clinical Neuroanatomy and Neurophysiology. Philadelphia: F.A. Davis Co., pp. 55–74; 84–95.

Haines, D.E. (Ed.) Fundamental neuroscience. New York: Churchill Livingstone, pp. 219–263 (Somatosensory System I and II; Viscerosensory Pathways).

Martin, J.H. (1996) Neuroanatomy: Text and Atlas. Stamford, CT: Appleton and Lange, pp. 125–159.

Melzack, R. & Wall, P.D. (1965) Pain mechanisms: A new theory. *Science, 150*, 971–979.

Nathan, P.W., Smith, M.C., & Cook, A.W. (1986) Sensory effects in man of lesions of the posterior colums and of some other afferent pathways. *Brain, 109*, 1003–1041.

Wiebers, D.O., Dale, A.J.D., Kokmen, E., & Swanson, J.W. (Eds.) (1998) *Mayo Clinic Examinations in Neurology*. St. Louis: Mosby, pp. 255–268.

Willis, W.D. & Coggeshall, R.E. (1991) *Sensory Mechanisms of the Spinal Cord*. 2nd edition. New York: Plenum Press.

3   THE CEREBELLUM

CHAPTER OVERVIEW

Although relatively small in total volume compared to the cerebral cortex, one clue as to the behavioral significance of the cerebellum is the fact that it contains about the same number of neurons as the rest of the brain. Input to the cerebellum is extensive. As noted in the previous chapter, it obtains constant feedback from the periphery with regard to the relative position and state of activity of our limbs and muscles. In addition, it also receives extensive input from all four lobes of the cerebral cortex via the pontine nuclei, as well as directly or indirectly from various other brainstem nuclei, particularly the inferior olivary and vestibular nuclei. Its primary output is back to the sensorimotor and executive regions of the cerebral cortex via the thalamus and to the vestibular nuclei in the brainstem. Thus, the cerebellum not only has access to our plans or intentions, but also to the attitude, position, and ongoing state of activity of our physical body, as well as the eventual relative success or failure of our responses.

Although the cerebellum is intimately linked to both sensory and motor systems, we are generally not consciously aware of its activity. It is only when disease or injury compromises the normal functioning of the cerebellum that we become well aware of its critical contribution to our daily activities. When the central core of the cerebellum is affected, we may detect problems in maintaining postural equilibrium or balance when engaging in whole-body activities. By contrast, if the cerebellar hemispheres are damaged, one might be expected to experience difficulty in smoothly and accurately carrying out discreet voluntary movements. As is true of many other brain systems, some aspects of cerebellar functioning remain controversial. While its role in the execution of motor skills has been recognized for nearly a century, more recent evidence points to its contributions to learning and conditioning, timing, as well to possibly a broad range of cognitive, autonomic, and even affective behaviors.

By the end of this chapter, the reader should be able to discuss the major afferent and efferent cerebellar pathways and their likely behavioral significance. One should also be able to identify and define the more common signs and symptoms of cerebellar disease and begin to differentiate them from similar symptoms resulting from lesions in other parts of the CNS. Finally, the reader should be able to describe a number of basic clinical procedures for assessing the integrity of cerebellar functioning.

INTRODUCTION

Although the total volume of the cerebellum or "little brain" is considerably less than that of the cerebral hemispheres in man, it is more convoluted than the cerebral cortex. Consequently, its cortical surface area is much closer to that of the telencephalon or cerebral cortex than suggested by its volume alone. The cerebellum can be identified, albeit in a much more simplified form, in fishes and sharks. It shows much greater development in reptiles, especially in species such as the alligator. It is in birds, however, that we first find evidence of substantial development, not only in relative size, but also in the evolution of sulci or fissures and gyri (called "folia" in the cerebellum).

This avian expansion may provide a good introduction for trying to understand something about the functional role of the cerebellum. In the early nineteenth century in an attempt to challenge the strict localization theories of Gall and Spurzheim (see Chapter 9), Pierre Flourens systematically studied the effects of extirpations of various brain structures. He noticed that if the cerebellum of the pigeon were removed, the bird lost the capacity for coordinated movement. However, he incorrectly concluded that the cerebellum must be responsible for all motor activity. It was not until World War I and the multitude of focal brain lesions resulting from gunshot wounds that Gordon Holmes, a British neurologist, was able to articulate the basic parameters of what is now our current understanding of cerebellar function. In general, we know that although the cerebellum is believed to receive significant input from wide areas of the cerebral cortex, including sensory association areas, its major role appears to be in the maintenance of balance and equilibrium and in the coordination of complex motor activity. The latter includes the fine tuning of discrete, directed voluntary activities, such as throwing a ball; semiautomatic activities, such as walking; as well as a host of subconscious or reflex motor responses.

What does all this have to do with birds? Although an anthropomorphic account, Richard Bach's *Jonathan Livingston Seagull* provides a beautiful backdrop for trying to conceptualize the role of the cerebellum in motor activity. As much from our own observations, as from Bach's elaborate descriptions of the intricacies of flight, it becomes easy to appreciate the level of coordination that is required for such activity. It is not merely a matter of a bird flapping its wings, but the attitude of the head, the body, and, most assuredly, the tail all become as one in achieving aerial stability. When watching mallards glide in over a pond, cupping their wings at the last moment to settle gently on the water, one cannot help but being struck by the graceful symmetry of these movements. If we take this picture but one step further and visualize the hawk or the eagle diving to catch another bird in midair or a fish from the sea, it is like contrasting a man simply walking down the street to having watched Michael Jordan's acrobatics on the basketball court. It is of no coincidence that, evolutionarily speaking, birds required a highly developed cerebellum to accomplish such feats.

To continue with this analogy, what else might we learn from *Jonathan Livingston Seagull*? When we pick up his story, he has already learned the basics of flight. He can do pretty much what every other seagull can do in the air and *he seems to do it without even thinking about it.* He may "think" to himself, "I want to go search the shoals for food." However, unless something untoward occurs, the flight itself may be a largely subconscious, automatic activity. His attention may be directed to locating fish near the surface or competing with the other gulls. In a similar vein, we rarely focus on the act of walking itself or maintaining our balance while doing so (unless we are traversing extremely uneven terrain or attempting to compensate for an acute injury that might impact on those basic skills). However, when Jonathan attempts to push himself to the limits of flight, particularly mastering the power dive, he eventually becomes conscious of the effects produced by the subtle adjustment of even a single wing feather. Meanwhile, during this learning phase, other fundamentals of flight continue to operate almost reflexively. Soon too this newly learned skill of reducing the aerodynamic drag on the wings during high-speed flight will become automatic as he attempts to conquer high-velocity rolls.

The foregoing description may reveal another important aspect of cerebellar function: the cerebellum *learns*[1], thus freeing the cerebral cortex for new learning, programming, problem solving, and other higher-level cognitive and motor activities. Observe the child. Initially, the main struggle is simply to maintain one's balance, first when sitting, then while standing. Gradually, as this skill is being mastered, then walking becomes the focus of attention. Next, the child learns to run. Eventually, he or she, while running, may learn to simultaneously avoid a defender and make a behind the back pass to a teammate on the basketball court. Ponder what it would be like if, while attempting to complete such a pass or shoot a goal, one had to concentrate, as does a toddler, on maintaining their balance while putting one foot in front of the other. Thus, at each stage the cerebellum seems to learn and eventually build on patterns of rehearsed or practiced motor activities, providing a background against which conscious, volitional activities, especially new skills, can be executed. Ramnani (2006) offers a theoretical model that would appear to lend substantial support to these common observations (see endnote 4).

Some of the above phenomena may be readily demonstrated by a couple of examples. If you have ever had the misfortune to seriously injure a leg or a foot and were required to use crutches to walk, you rapidly became aware that you no longer could take vertical equilibrium for granted. As you first began to use the crutches, your first thoughts were probably not so much to ambulate, but rather not to fall. While your cerebellum was still working (in fact, probably working overtime), you became very conscious of your volitional (cerebral) motor activities, so much so that you did not wish or could not tolerate other distractions. You literally might have had trouble walking and chewing gum at the same time. However, after your body adjusted to these new demands, balance and even walking became more automatic.

We also can look at what happens when the normal function of the cerebellum is disrupted. To get some firsthand experience what this is like, you can conduct either of two experiments: one would require a fifth of liquor, the other a baseball bat. In order to be socially responsible, we will not go into further details about the former (though it is likely that a number of you first conducted that experiment on your own, usually as part of the extracurricular activities in high school or college). However, you can always try the latter, which is a common fan-participation gimmick in minor league ballparks. Simply take a baseball bat, place the barrel end of the bat on the ground (preferably in a large, open, soft, grassy area) and bend over and place your forehead on the other end. Now, while staying in that position, proceed to walk in circles around the bat about eight to ten times. Now drop the bat and try to walk, or simply stand up. The excessive vestibular stimulation has produced a functional disruption of some of the more primitive portions of your cerebellum. Although your cerebral cortex and peripheral muscles are still perfectly intact, temporarily at least, they have become relatively ineffectual without a functionally intact cerebellum.

A patient, described by Holmes (1939) as having a lesion confined to the right cerebellar hemisphere, provided an interesting account of the effects of such a lesion. Keeping in mind that the cerebellar hemispheres, unlike the cerebral hemispheres, exercise ipsilateral control over the extremities, the patient indicated that his left arm carried out activities in a normal "subconscious" manner, whereas "I have to think out each movement of the right arm" [quoted in Nolte (1993), p. 358].

The preceding was to provide a more general framework for thinking about the potential role of the cerebellum, although one should keep in mind that other systems, particularly the basal ganglia, also likely play a major role in mediating automatic or overlearned motor activities. Later in this chapter, we will return to a more detailed description of the probable functions served by the cerebellum and a more complete description of the effects of specific cerebellar lesions. Before proceeding further, however, it may be helpful to review the basic anatomy of the cerebellum, its functional and anatomical divisions, and its basic afferent and efferent connections.

GROSS ANATOMY

The cerebellum lies directly under the posterior temporal and occipital lobes in what is termed the **posterior fossa** of the cranium. It is separated from the cerebral cortex by the **tentorium**, that portion of the dura mater that covers its superior surface. The cerebellum attaches to the brainstem anteriorly at the level of the pons via three large fiber pathways called the **cerebellar peduncles** (Figure 3–1). These are the inferior (**restiform body**), the middle (**brachium pontis**), and the superior (**brachium conjunctivum**) cerebellar peduncles. When looking at the cerebellum in situ, what is first observed are its two large hemispheric masses marked by transverse or horizontal arrangements of fissures and ridges (**folia**) (Figure 3–2a). These structures constitute the cerebellar hemispheres. Approximately one third of the way from the brainstem on the dorsal surface of cerebellum

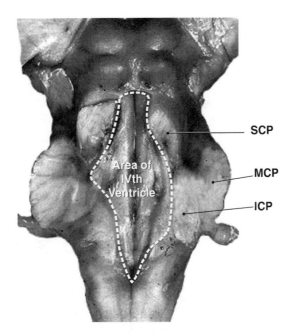

Figure 3–1. Cerebellar peduncles as seen from their points of connection to the posterior (dorsal) surface of the brainstem. Abbreviations: ICP, inferior cerebellar peduncle; MCP, middle cerebellar peduncle; SCP, superior cerebellar peduncle. (Brain images from *Interactive Brain Atlas*, courtesy of University of Washington).

is a deep fissure (the primary fissure), which separates the cerebellum into the **anterior lobe** (in front of the primary fissure) and **posterior lobe** (just about everything behind the primary fissure). Just as the cerebral cortex has undergone tremendous phylogenetic expansion, particularly in the higher apes and man, so has the cerebellum, particularly the posterior lobes of the cerebellar hemispheres. Thus perhaps as one might expect, the projections to the posterior lobe come primarily from the cerebral cortex. Between the two hemispheres lies a central portion, running along the entire anterior-posterior axis, which is called the **vermis** (Figure 3–2b), although it is not readily visible from an external view. On the ventral surface of the cerebellum, near the junction of the cerebellum and the brainstem, lie two small eminencies known as the **flocculi**. Bridging the gap between these two flocculi (normally hidden from view) is the **nodulus**. These structures taken together constitute the **flocculonodular lobe**, which, phylogenetically, is the oldest portion of the cerebellum.

Another way of structurally dividing the cerebellum, with which the reader should at least be aware, is by lobules. These lobules are essentially defined by other prominent fissures along the surface of the cerebellum and cut across the vermis and both hemispheres. Individual lobules are identified both by semantic and numerical (Roman) designations. Lobules I through V make up the anterior lobe, VI though IX constitute the posterior lobe, with the flocculonodular lobe being designated as X.

While it might be useful to be aware of these structural divisions, for clinical purposes other functional divisions are now more commonly used. These are the **vestibulocerebellum, spinocerebellum**, and **cerebrocerebellum** (or **pontocerebellum**). These will be discussed in greater detail later on.

(a)

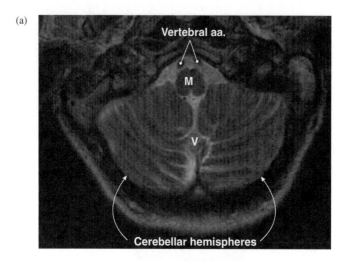

(b)

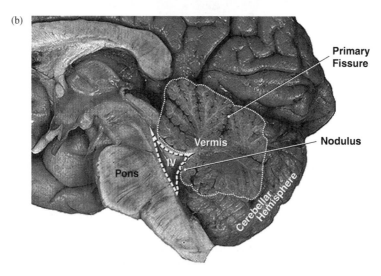

Figure 3–2. (a) Axial MRI through medulla of brainstem showing cerebellar hemispheres and vermis. (b) Midsagittal view of cerebellum and brainstem. Abbreviations: M, medulla; IV, fourth ventricle; V, vermis.

CONNECTIONS

All input to (as well as all output from) the cerebellum is by way of the three cerebellar peduncles. In fact, the arrangement is relatively simple. With one notable exception in each case, most afferents into the cerebellum are via the inferior and middle cerebellar peduncles and all output from the cerebellum is via the superior cerebellar peduncle. There are three basic areas from which the cerebellum receives input: the spinal cord (see Figure 2–4), the brainstem nuclei and the cerebral cortex. In terms of the input, again with one major exception, for the most part, all input from the spinal cord and the brainstem nuclei enter the cerebellum via the inferior cerebellar peduncle, whereas corticofugal fibers enter the cerebellum through the middle cerebral peduncle. The major afferent connections to the cerebellum are seen in Figure 3–3 and summarized in Table 3–1.[2]

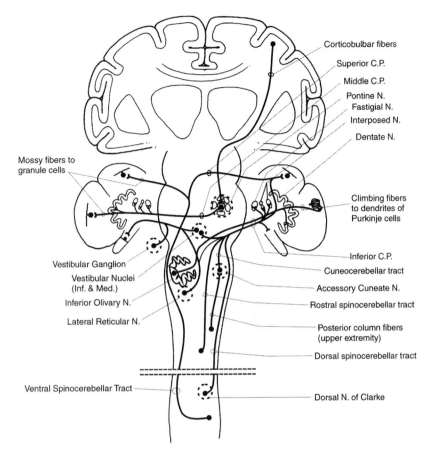

Figure 3–3. Cerebellar afferents: Cortical input (contralateral) via the middle cerebellar peduncle after synapsing in the pontine nuclei. Spinal (somatosensory, ipsilateral) input via the inferior cerebellar peduncle (ICP), by way of the dorsal spinocerebellar (lower extremities) and cuneocerebellar (upper extremities) tracts. Feedback likely representing descending motor influences and proprioceptive or kinesthetic activity appears to be carried by the rostral and ventral cerebellar tracts. The former (RCT) carries information from the upper extremities and enters the cerebellum ipsilaterally via the ICP. The latter (VCT) represents the lower extremities and enters through the superior cerebellar peduncle (both crossed and doubly crossed input). Other major input from the level of the brainstem originates in the vestibular nuclei and ganglion, the lateral reticular nuclei, and the inferior olivary nucleus.

Inferior Cerebellar Peduncles

Having provided this general schema, it may be useful to consider some of the major cerebellar inputs in somewhat greater detail. As was noted in Chapter 2, there are a number of pathways in the spinal cord that enter the cerebellum via the inferior cerebellar peduncle (restiform body), conveying proprioceptive and other somatosensory information. As seen in Figures 3–3 and 3–4, two of these pathways include the **dorsal spinocerebellar tract (DSCT)** (originating in the dorsal nucleus of Clark) and the **cuneocerebellar tract (CuCT)** (originating in the lateral cuneate nucleus of the medulla). Both of these pathways primarily convey non-conscious proprioceptive/kinesthetic information as derived from muscle spindles and tendons, with the DSCT carrying information from the lower and the CuCT from the upper extremities. The **rostral spinocerebellar tract (RSCT)**, the most obscure of the spinocerebellar tracts, also primarily enters the cerebellum through the inferior cerebellar peduncle. Carrying

Table 3–1.   Cerebellar Afferents[a]

Tract / Source	Nature of Input	Peduncle	Termination
Dorsal spinocerebellar tract (DSCT) Dorsal nucleus of Clarke (DNC)	Proprioception and cutaneous sensation from legs / lower trunk	Inferior (uncrossed)	Vermal and paravermal, (esp. anterior lobe)
Ventral spinocerebellar tract (VSCT) DNC, plus interneurons	Similar to DSCT, includes reflex information	Superior (doubly crossed)	Same as DSCT
Cuneocerebellar tract Accessory cuneate n.	Upper limb and trunk equivalent of DSCT	Inferior (uncrossed)	Same as DSCT
Rostral spinocerebellar tract Similar to VSCT	Equivalent of VSCT for arm, neck, and upper trunk	Inferior (uncrossed)	Same as DSCT
Trigeminocerebellar tract Chief sensory and spinal nuclei of V	Equivalent of DSCT for face	Inferior (uncrossed)	Same as DSCT
Pontocerebellar tracts Cerebral cortex via pontine n.	Voluntary motor, multi-sensory, association, and limbic (?)	Middle (mostly crossed)	All except flocculonodular lobe, esp. to hemispheres
Vestibulocerebellar tracts Primary: Vestibular ganglion Secondary: Inf. and medial vestibular n.	Sensation of movement and orientation in space	Juxtarestiform body Primary: uncrossed Secondary: crossed	Flocculonodular lobe and vermis
Reticulocerebellar tracts Brainstem reticular n.	Integrated sensorimotor input from cortex, cord and cerebellum?	Inferior (mostly uncrossed)	Vermal and paravermal regions
Olivocerebellar (arcuate fibers) Inferior and accessory olivary n.	Integrated sensorimotor input from red nucleus, sup. colliculi, cord, and other midbrain n.	Inferior (crossed)	All lobes

[a] This table reflects the most well-known or strongly suspected connections. This list is not meant to be comprehensive and it should be noted that most afferents give off collateral s to the deep cerebellar nuclei. Other brainstem nuclei, such as the raphe nuclei and the locus ceruleus also appear to have cerebellar projections.

information from the upper extremities, the RSCT appears to serve a role comparable to that which the **ventral spinocerebellar tract (VSCT)** serves for the lower extremities. Both likely respond to non-conscious cutaneous feedback (e.g., pressure, touch, and pain) and limb movements, as do the DSCT and CuCT. In addition, the RSCT and VSCT are thought to respond to internal activity within the spinal cord itself via interneurons. Such activity includes spinal reflexes and facilitatory and inhibitory influences both from descending cortical and bulbar-spinal tracts, as well as from peripheral afferents. The ventral spinocerebellar tract is somewhat of an oddball in that of the four spinocerebellar tracts it is the only one that crosses the midline after it enters the cord (see Figure 3–5). In addition, unlike the other three, it enters the cerebellum through the superior cerebellar peduncle, where most of its fibers once again cross the midline, resulting in mostly ipsilateral input.

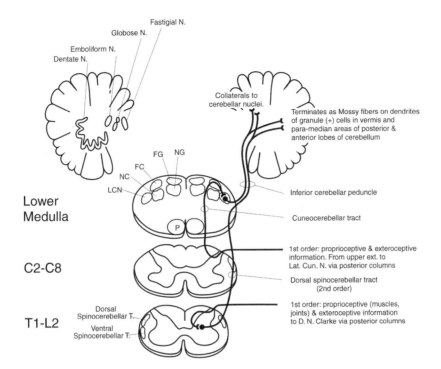

Figure 3–4. Dorsal spinocerebellar and cuneocerebellar tracts entering via the inferior cerebellar peduncle. Cerebellar nuclei are represented at upper left. Abbreviations: DN, dorsal nucleus (of Clarke); FC, fasciculus cuneatus; FG, fasciculus gracilis; LCN, lateral (accessory) cuneate nucleus; NC, nucleus cuneatus; NG, nucleus gracilis; P, pyramids.

The **trigeminocerebellar tract**, originating in the principal sensory and spinal nucleus of cranial nerve V and conveying proprioceptive and cutaneous information from the head and face, also enters via the inferior peduncle. All these pathways remain ipsilateral. Other inputs of note through the inferior cerebellar peduncle are from the inferior olivary and accessory olivary nuclei, the vestibular nuclei, and reticular nuclei (see Figure 3–6). The vestibulocerebellar fibers are worthy of note for a couple of reasons. First, as opposed to all the other inputs mentioned thus far which tend to project to portions of both the anterior and posterior lobes of the cerebellum, the fiber pathways from the vestibular nuclei project principally to the flocculonodular lobe. Second, these fibers travel in the medial portion of the inferior cerebellar peduncle (restiform body) known as the **juxtarestiform body**.

One of the exceptions noted above is that the juxtarestiform body of the inferior cerebellar peduncle also contains efferents from the flocculonodular lobe (and related fastigial nuclei) that feed back into the vestibular and reticular nuclei of the brainstem. As might be expected, the vestibular system is largely responsible for maintaining balance and equilibrium. It is believed that the descending vestibulospinal tract, which is obviously directly influenced by the cerebellum, serves to facilitate extensor motor neurons and inhibit flexor motor neurons, which in turn would have important antigravity functions.

The afferent fibers from the inferior **olivary nucleus complex** are also particularly worthy of note. These are the only fibers entering the inferior cerebellar peduncles that are completely crossed. Also, whereas all the other inferior cerebellar peduncle inputs mentioned thus far tend to project to the more central or midline portions of the cerebellum, the fibers from these nuclei tend to project to the entire contralateral hemisphere of the cerebellar

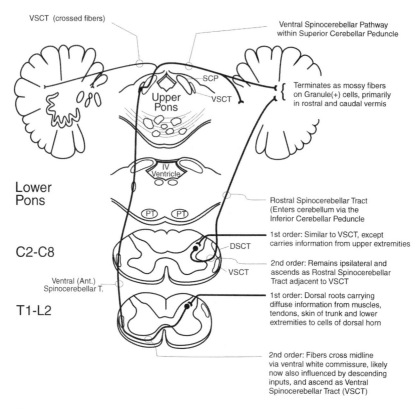

Figure 3–5. Rostral and ventral spinocerebellar pathways. Abbreviations: DSCT, dorsal spinocerebellar tract; PT, pyramidal tract; SCP, superior cerebellar peduncle; VSCT, ventral spinocerebellar tract.

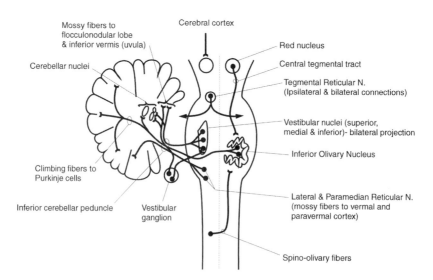

Figure 3–6. Cerebellar input from the inferior olivary nuclei and other brainstem nuclei (cerebropontocerebellar input via pontine nuclei shown in Figure 3–7).

cortex. Finally, whereas all other afferents to the cerebellar cortex are in the form of mossy fibers that synapse on the granular cells, the inferior olivary nucleus alone gives rise to the climbing fibers that synapse exclusively on the Purkinje cells (see below).

Middle Cerebellar Peduncles

The middle cerebral peduncle ("brachium pontis"), by far the largest of the three cerebellar peduncles, is the primary pathway by which the cerebral cortex influences the cerebellum. In fact, the vast majority of the corticofugal fibers descending in the cerebral peduncles of the midbrain end up in the cerebellum. These fibers synapse ipsilaterally in the pontine nuclei in the base or anterior portion of the pons. These pontine nuclei then give rise to a huge band of fibers that cross the midline (though a few may remain ipsilateral) and enter the cerebellum as the middle cerebral peduncle (Figure 3–7). It is, in fact, this band of crossed fibers that give the pons its distinctive appearance. The middle cerebellar peduncle contains fibers that project to all portions of the cerebellum except the flocculonodular lobe and, along with the climbing fibers originating in the inferior olivary nucleus, provide the major input into the cerebellar hemispheres. As might be expected, a major source of these corticopontocerebellar fibers appears to be from the regions surrounding the central sulcus (sensorimotor cortices). However, at least in humans, the prefrontal cortices in particular, as well as possibly all other cortical association areas and limbic cortices also contribute inputs (Ramnani et al., 2006; Schmahmann, 2001a). Thus it would appear that the cerebellum not only receives information about ongoing motor activity and somatosensory

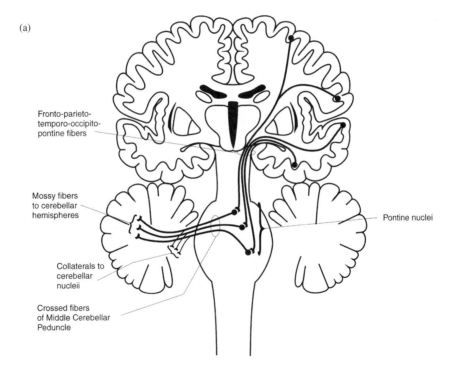

(a)

Fronto-parieto-temporo-occipito-pontine fibers

Mossy fibers to cerebellar hemispheres

Pontine nuclei

Collaterals to cerebellar nucleii

Crossed fibers of Middle Cerebellar Peduncle

Figure 3–7. (a) Cerebropontocerebellar pathways. Fibers from diverse areas of the cortex synapse in the pontine nuclei. Secondary fibers then cross the midline and enter cerebellum (primarily the cerebellar hemispheres) via the middle cerebellar peduncle. (b) Middle cerebellar peduncle as seen on axial view of MRI. Abbreviations: BA, basilar artery; CH, cerebellar hemispheres; ICA, internal carotid artery; MCP, middle cerebellar peduncle; V, vermis.

(b)

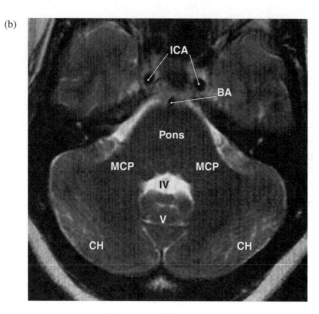

Figure 3–7. (Continued)

feedback, but also has access to information about planned or contemplated motor activity before it is actually initiated, as well as higher-order visual, auditory, and even affective information.

Superior Cerebellar Peduncles

The superior cerebellar peduncle ("brachium conjunctivum") is the major output channel for the cerebellum. The one notable exception is the efferent connections from the flocculonodular lobe and the fastigial nuclei to the vestibular and reticular systems, which exit via the inferior cerebellar peduncle. The superior cerebellar peduncle, as previously noted, contains one major afferent pathway, the anterior or ventral spinocerebellar tract (VSCT). Recall that the VSCT conveys both peripheral and spinal cord information from the lower extremities and is doubly crossed. Some authors suggest that the mesencephalic portion of the trigeminal nerve also conveys proprioceptive information regarding the muscles of mastication to the cerebellum via the superior cerebellar peduncle.

As will be discussed shortly, with the exception of some direct connections between the flocculonodular lobe and the vestibular nuclei, all efferents from the cerebellum derive from the cells of the cerebellar nuclei (fastigial, globose, emboliform, and dentate). While the fastigial nucleus transmits fibers (essentially to the vestibular and lateral reticular nuclei) via both the inferior and superior cerebellar peduncles, efferents from the other three groups of nuclei all exit via the superior cerebellar peduncle. In addition to the sites just mentioned, the major target areas for these cerebellar efferents are the contralateral inferior olivary nuclei, red nuclei, and ventral lateral nuclei of the thalamus. The latter, in turn, project to the primary motor and premotor cortices.[3] Again it might be noted that even though these pathways are crossed, the descending influences with which they interact (e.g., the corticospinal tracts) are themselves crossed, so the influence exerted by the cerebellum remains ipsilateral. As might be expected, all of these nuclei are directly involved in various motor pathways, thus allowing for the modulation of motor activity by the cerebellum at different levels and stages of programming and execution.

CEREBELLAR CORTEX

Not unlike the cerebrum, the cerebellum also has a cortex, underlying white matter consisting of axonal fibers going to and from the cerebellar cortex, and deep nuclei. However, as we shall see in Chapter 9, the cortex of the cerebellum differs in significant ways from the cerebral cortex. Unlike the cerebral cortex, the cortex of the cerebellum consists of only three layers (see Figure 3–8), is uniform throughout the cerebellum, and has distinctive and in some ways a relatively straightforward organization. The following represents a simplified picture of that organizational structure. The three layers of the cerebellar are (1) an outer, molecular layer, (2) a middle, Purkinje layer, and (3) an inner, granular layer.

Molecular Layer

The outermost or molecular layer has the least cellular density of the three layers and consists of two types of neurons, **stellate cells** and **basket cells**. However, while having relatively few cells, the molecular layer contains numerous cell processes, the most notable being the extensive dendritic trees (branches) of the Purkinje cells and the parallel fibers of the granule cells (see below). Although the branches of the Purkinje cells are quite extensive (see Figure 3–8), they branch out more or less in a single plane (at right angles to the folia in which they are found). The parallel fibers, in turn, run at right angles to the dendritic trees of the Purkinje cells upon which they have an excitatory influence. Because of their spatial arrangement, a single parallel fiber may make connections with thousands of Purkinje cells. The stellate cells have relatively short dendrites and make contact with the dendrites of a small number of Purkinje cells. Basket cells have more extensive dendritic processes and make contact (on the cell bodies) of a much large number of Purkinje cells. Both the stellate and basket cells receive excitatory input from the parallel fibers and, in turn, exert inhibitory influences on the Purkinje cells.

Purkinje Layer

The middle layer is relatively thin but consists of the densely packed cell bodies of the **Purkinje cells**. These cell bodies are the largest in the cerebellum and, with their unique dendritic trees, are probably the most distinctive cells in the entire central nervous system. The axons of the Purkinje cells descend though the granular layer and synapse on the deep cerebellar nuclei, upon which they have an inhibitory influence. Note that axons from some of the Purkinje cells in the vermis and flocculonodular lobe proceed directly to the vestibular nuclei in the brainstem. These two projections represent the only efferent connections (output) of the cerebellar cortex. Finally, the Purkinje layer contains fibers in transit to the molecular layer (climbing and aminergic fibers, processes of the granule and Golgi cells, and the dendrites of the basket cells terminating on the cell bodies of the Purkinje cells).

Granular Layer

The innermost or granular layer consists of the cell bodies of the very densely packed **granular cells** and **Golgi type II cells**. The granular cells represent an extension of the mossy fibers (see below). The axons of the granular cells extend into the molecular layer and branch out in opposite directions forming the parallel fibers (discussed above) that synapse on the dendrites of the Purkinje, stellate, and basket cells. Of all the cells that are intrinsic to the cerebellar cortex, the granule cells are the only ones that are excitatory. The other major cell type within the granular layer is the Golgi type II cells (usually simply referred to as "Golgi cells"). As previously noted, the Golgi cells send some processes into the molecular layer where they synapse with the parallel fibers; however, for the most part, they synapse on granule cell dendrites within the granular layer. The Golgi cells also exert inhibitory effects.

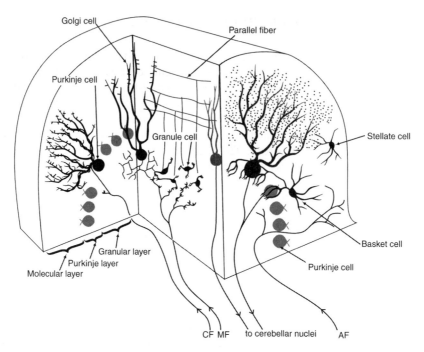

Figure 3–8. Cross section of cerebellar cortex. Abbreviations: AF, aminergic fibers; CF, climbing fibers; MF, mossy fibers. Drawing adapted from Lennart Heimer (1995).

Input and Output of Cerebellar Cerebellar

There are only three types of afferent fibers entering and very limited efferent fibers leaving the cerebellar cortex. The three types of afferent fibers are **mossy fibers, climbing fibers,** and **aminergic fibers**.

Climbing fibers emanate solely from the inferior olivary nucleus in the medulla. After leaving the inferior olivary nucleus, these fibers cross the midline (arcuate fibers) and synapse directly on the dendritic trees of the Purkinje cells. Although a climbing fiber typically makes numerous synaptic connections with the Purkinje cell, each Purkinje cell is thought to receive input from a single climbing fiber. In addition to synapsing on Purkinje cells, climbing fibers also send collaterals to the cerebellar nuclei. In either case, their influence is excitatory.

Mossy fibers represent a second and most diffuse source of input into the cerebellum, representing all other input into the cerebellum (with the exception of the aminergic fibers). Thus, all corticopontocerebellar, spinocerebellar, and vestibulocerebellar inputs are in the form of mossy fibers. Once in the cerebellum, the mossy fibers undergo extensive branching and their terminals (rosettes) synapse on the dendrites of granular cells and axons of Golgi cells in clusters called **glomeruli** within the granular layer. As noted above, the granule cells in turn give rise to the parallel fibers that influence many Purkinje cells, as well as synapsing on the dendritic trees of Golgi, stellate, and basket cells in the molecular layer. Like the climbing fibers, the mossy fibers send collaterals to the cerebellar nuclei and have excitatory effects.

Aminergic fibers are the third type of cerebellar afferents. These fibers likely originate in the raphe (serotonin) and locus ceruleus (norepinephrine) nuclei in the brainstem, as well possibly in the hypothalamus. They are thought to exert inhibitory influences on cerebellar neurons via synapses in the molecular and granular layers.

Efferent connections of the cerebellar hemispheres (as previously noted) are represented exclusively by the axons of the Purkinje cells and with the exception of some fibers that go directly to the vestibular nuclei from the flocculonodular lobe, all other Purkinje cell output is to the deep cerebellar nuclei. It also was pointed out that these efferent Purkinje fibers are inhibitory in nature. However, the nerve fibers that subsequently exit from the deep cerebellar nuclei (mostly via the superior cerebellar peduncle en route to the thalamus, red nucleus, and other brainstem nuclei) appear to be excitatory.

Summary and Implications of Input and Output

The Purkinje cell might well be considered the pivotal structure within the cerebellum. It receives major **direct** input from *climbing fibers* and **indirect** input from *mossy fibers* (via the *parallel fibers of granular cells*). Both represent excitatory input, the former from the inferior olivary nuclei, the latter from the cerebral cortex, spinal cord, and vestibular nuclei. All other inputs (*aminergic fibers*) and cells (*stellate, basket, and Golgi*) apparently serve to modulate or inhibit the Purkinje cell and/or its input. Finally, the axons of the Purkinje cell, which either synapse on the cerebellar nuclei or directly on vestibular nuclei, represent the only efferent output of the cerebellar cortex. Through this arrangement the cerebellar cortex has access to and the capacity to integrate information from multiple systems, including the:

1. Plans, intentions, commands, and expectations within the context of the total internal and external milieu (cortical input).
2. Feedback regarding the concrete, relatively immediate, unfiltered proprioceptive, kinetic, and cutaneous feedback relevant to motor activities as they are taking place (spinal cord).
3. State and orientation of the body as it moves in space (vestibular nuclei).
4. Indirect, integrated information from the cortex, spinal cord, and the cerebellum itself (via the inferior olivary nuclei).

Thus armed with all this information, the cerebellum then can help store (successful patterns of), refine (coordinate) the ongoing activity either through its own internal mechanisms or through external feedback (e.g., cerebellocorticocerebellar loops), and compare original input (intentions, directions, or commands) with subsequent integrated or processed input.

CEREBELLAR NUCLEI

Depending on the species and how they are considered, there are three or four pairs of nuclei that lie deep within the white matter of the cerebellum at its anterior end. In man, from medial to lateral, these are the **fastigial, globose, emboliform,** and **dentate nuclei**. The globose and emboliform nuclei are functionally related and together are often referred to as the **nucleus interpositus** (which also describes their relative location between the fastigial and dentate nuclei). As noted above these nuclei receive afferent collaterals from many of the sources of input to the cerebellar cortices and, with the exception of a few fibers that proceed directly to the vestibular nuclei, they are the source of all cerebellar efferents. Collectively, they are the recipient of the efferent fibers of the Purkinje cells. The most medial of these, the fastigial nuclei, lie just over the roof of the 4th ventricle and receive fibers from the flocculonodular lobe and send fibers back to the vestibular and reticular nuclei. The more centrally located globose and emboliform nuclei are only differentiated in the higher primates, including man. The primary projections to these nuclei come from those areas of the cerebellar cortex that receive heavy inputs from the spinal cord, though they also receive collateral inputs directly from these fibers as well as from other sources of cerebellar afferents, including the reticular nuclei, olivary nuclei, and probably some fibers from the red nuclei.

The largest and most recently developed of the cerebellar nuclei, the dentate nuclei are located most laterally in the white matter of the cerebellar hemispheres. Input is primarily from the cerebellar hemispheres, as well as some from the anterior lobe of the cerebellum. As with the other cerebellar nuclei, they also receive direct collateral input from outside the cerebellum, probably from the same sources as the interposed nuclei. Together with the globose and emboliform nuclei, the dentate nuclei are the primary source of projections via the superior cerebellar peduncle to the red nucleus and to the cortex by way of the ventral lateral nuclei of the thalamus. The dentate nuclei probably project primarily to the thalamus, whereas the interposed nuclei appear to have more substantial projections to the red nucleus. These nuclei also send some projections to other brainstem nuclei, notably the inferior olivary nuclei, reticular formation, and the superior colliculi (Figure 3–9).

Several features of these cerebellar outputs, which are summarized in Table 3–2, should be noted. First, the cerebellar hemispheres always end up having a predominately ipsilateral influence on appendicular (limb) movement, (i.e., if there is a lesion in the right cerebellar hemisphere, the effects will be seen on the right side of the body, whereas lesions in the central or vermal regions of the cerebellum tend to affect midline (axial) movement or motor adjustments. Second, there are a number of feedback loops within the system. For example, while there is no direct input into the spinal cord from the cerebellum, because

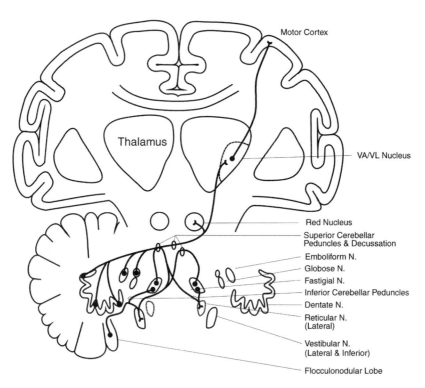

Figure 3–9. Figure illustrates several major cerebellar efferent pathways, completing feedback loops to brainstem nuclei and to the cerebral motor cortex via the ventral anterior (VA) and ventral lateral (VL) nuclei of the thalamus.

Table 3–2. Cerebellar Efferents

Lobe of Origin	Deep Nuclei	Projections	Peduncle
Flocculonodular lobe and vermis	Fastigial n.	Lateral and inferior vestibular n. (bilat.) Lateral reticular n.	Inferior
Flocculonodular lobe and vermis	(direct connections)	Vestibular n. (ipsilateral)	Inferior
Paravermal and vermal areas	Emboliform and globose n.	Red nuclei; VA/VL thalamic n. (crossed)	Superior
Cerebellar hemispheres	Dentate n.	VA/VL thalamic n. Red nuclei	Superior

Again, this represents the major or most well-established connections. The superior cerebellar peduncle also appears to have a descending limb that may project to the inferior olivary nuclei. The interposed (globose and emboliform) and dentate nuclei likely also project to the superior colliculi and oculomotor centers, thus facilitating hand-eye coordination. The fastigial nuclei appear to have some ascending thalamic connections.

of the functional loops created by efferent connections with the thalamus, red nucleus, and vestibular and reticular nuclei, the cerebellum is able to indirectly influence spinal mechanisms via the corticospinal, vestibulospinal, reticulospinal, and rubrospinal tracts. Other such loops include the dentate nuclei → thalamus → cortex → corticopontine tract → pontocerebellar fibers and the afferent and efferent interconnections between the vestibular nuclei and the flocculonodular lobe of the cerebellum.

FUNCTIONAL DIVISIONS

Earlier, it was noted that the cerebellum could be functionally divided into the **vestibulo-cerebellum, spinocerebellum**, and **cerebrocerebellum** (or **pontocerebellum**). As one might guess, these latter divisions are based primarily on the source of their primary inputs (as well as their subsequent functional significance). Before discussing their connections and suspected clinical significance it might be helpful to briefly review a few anatomical features. As previously noted, the vermis (named for its worm-like appearance) runs through the center of the cerebellum in its anterior-posterior axis. By contrast, the cerebellar hemispheres represent the most lateral portions of the cerebellum. The area of transition between the vermis (**vermal zone**) and the hemispheres (**lateral zone**) is referred to as the **paravermal zone**, whose boundaries are anatomically indistinct. The vermal zone typically has been associated with axial musculature (e.g., of head, neck, and trunk) and with such functions as balance, postural equilibrium, and related reflexes. As will be seen, more recently the vermal zone (particularly its more anterior aspects) has been tentatively associated with psychological (affective) phenomena. The paravermal zone is linked more to control over appendicular movement (hands, arms, and legs). The lateral zone, especially in the posterior lobes, is thought to largely mediate higher-order aspects of executive motor functions (e.g., learning and habit formation), as well as possibly some non-motor (cognitive) functions. Each of these zones tends to be more or less associated with specific cerebellar nuclei that serve as the primary outflow channels for these respective regions.

With this brief background, let us consider the three functional divisions of the cerebellum mentioned earlier, the (1) **vestibulocerebellum**, (2) **spinocerebellum**, and (3) **pontocere-bellum** Table 3–3).

Vestibulocerebellum

As its name implies, this is the portion of the cerebellum that receives its major input from the vestibular nuclei. All three divisions also receive substantial input from the inferior olivary nucleus. Functionally and anatomically, it overlaps to some extent with the vermal zone described above (particularly the more anterior regions of the vermis). In addition to the nodulus (included in the anterior vermis), it also encompasses the rest of the flocculonodular lobe. The Purkinje cells that lie within the vestibulocerebellum project primarily to the fastigial nuclei, although some go directly back to the vestibular nuclei. Being the most primitive of the three, it is also known as the archicerebellum. Functionally, it is apparently concerned with postural adjustments to vestibular stimulation and lesions may result in disturbances of balance, gait (ataxia), and reflex eye movements.

Spinocerebellum

That portion of the vermal and the paravermal regions that receive a large proportion of spinal afferents constitute the spinocerebellum. These spinal inputs consist not only of proprioceptive information, but all other types of cutaneous stimulation, as well as similar inputs from the trigeminal nuclei. In addition, the vermal regions receive vestibular input. Also a somewhat older anatomical division, it is also referred to as the paleocerebellum. While the vestibulocerebellum appears to be more concerned with unconscious, reflex adjustments necessary to maintain equilibrium, the vermal zones of the spinocerebellar division appears to be largely responsible for maintaining muscle tone and the coordination of synergistic muscles involved in balance, postural adaptations, and routine motor programs during consciously initiated (though not necessarily always consciously directed) motor activities such as walking, and running. The paravermal zones on the other hand are involved more with appendicular movements where, in possible conjunction with the lateral zones, they may play a role in controlling the speed, intensity, direction, sequencing, transitions, or general coordination of volitional, skilled actions.

Table 3–3. Functional Divisions of the Cerebellum

Division	Primary Function(s)[a]	Primary Input(s)	Associated Nuclei
Vestibulocerebellum (flocculonodular lobe and adjacent portions of vermis)	Balance, equilibrium, gait, reflex eye movements	Vestibular n., inf. olivary n.	Fastigial
Spinocerebellum (paravermal and vermal regions)	Muscle tone, control of axial (vermis), and limb (paravermal region) movements[b]	Spinal cord, inf. olivary, vestibular, reticular n.	Emboliform, globose, and fastigial (with vermis)
Cerebrocerebellum or pontocerebellum (cerebellar hemispheres)	Modulation, memory of skilled motor, and voluntary eye movements[c]	Cerebral cortex (via pontine n.), inf. olivary n.	Dentate

[a] For the most part, many of these represent simplified guesses as to the possible roles of the cerebellum.
[b] The anterior regions of the vermis have also been associated with affective expression.
[c] The lateral hemispheres have been associated with non-motor learning and other aspects of cognition.

Pontocerebellum

The final division within this classification is the cerebrocerebellum or pontocerebellum. Anatomically and functionally it is comparable to the lateral zone mentioned above. It derives its name from the fact that the major inputs are primarily from the cerebral cortex (especially from the prefrontal, frontal and parietal regions) via corticofugal fibers in the cerebral peduncles that synapse in pontine nuclei, from which secondary neurons enter the cerebellum via the middle cerebellar peduncle. Having greatly expanded along with the cerebral hemispheres, it is also known as the neocerebellum. As with the lateral zone, with which it is associated, its functional significance has been linked to the online modulation and learning or some type of motor memory for skilled tasks, including visual searching. Again, as will be seen below, more recent investigations have suggested a greater role in non-motor or cognitive processing for this region of the cerebellum.

FUNCTIONAL CORRELATES

Motor Behavior

While a considerable amount of information is known about the structure, anatomical connections, and the motor symptoms associated with lesions of the cerebellum, the precise mechanisms that mediate cerebellar function remain elusive. As previously noted, there is significant variation in the patterns of input and output with respect to the different regions of the cerebellar cortex. Thus, in some ways it may be considered similar to the organization of the cerebral cortex to be described in Chapter 9. However, unlike the significant variability in the cytoarchitecture of different regions within cerebral cortex, the cytoarchitecture of the cerebellum, while also extremely complex, is essentially the same throughout. This has led to the logical speculation that all areas or regions of the cerebellum likely perform the same function, the difference (e.g., in symptoms as a result of focal lesions) being the systems affected as a result of the specific pattern of inputs/outputs to that particular area, that is, the different feedback loops involved. Although it is still not yet clear what this "function" might be, it is generally believed that it serves to provide a modulating influence on behavior through its multiple feedback loops.

While there is mounting evidence that the cerebellum likely plays a crucial role in various aspects of nonverbal learning, as well as in general cognition, arousal, and emotional and autonomic responses (to be addressed below), its most commonly recognized role is in the control and/or modulation of motor activity. It is to these motor control functions that we shall now address ourselves. How does it accomplish these feats? As suggested, at this point one can only speculate.[4] To simplify things, let us consider several crucial aspects of motor activity:

1. It always takes place against a backdrop of the effects of gravity; the relative position of one's limbs, head, and trunk in space and to each other; and/or some other ongoing activity.[5]
2. For all practical purposes there is no such thing as isolated muscle movements (e.g., for each contraction of a target muscle (agonist), there is another muscle (antagonist) that must relax and typically there are a whole group of muscles (synergistic muscles) that also must contract (and relax) at the same time,
3. At least for all voluntary activity, there is an initial mental set and/or goal directing the particular action.
4. The action must be modified or changed depending on changes in external or environmental conditions, the effect and/or effectiveness of the act being executed, or changes in the mental set or intention.

If we again go back and think of the various afferent and efferent connections of the cerebellum, we can appreciate why the cerebellum is in an ideal position to contribute to this process, even if we do not fully understand how it takes place. Because of its direct interconnections with the vestibular system, the proprioceptive feedback it receives from stretch receptors via the spinocerebellar pathways, and its influence on those same antigravity muscles via the vestibulospinal (and probably the reticulospinal) pathways, the cerebellum can learn to mediate the subconscious, reflex adjustments necessary to maintain one's equilibrium.[6] As was suggested in the introductory portion of this chapter, given the cerebellum's access to diffuse proprioceptive (and other somatosensory) feedback and its capacity to influence all descending pathways both by its connections with motor nuclei in the brainstem and the motor cortex by way of the thalamus, it seems easily understandable how it could assume a key role in coordinating routine, overlearned, or quasi-automatic synergistic muscle activities.

What seems particularly intriguing is its ongoing control of skilled, voluntary motor activity. While monitoring and maintaining control of baseline functions such as equilibrium and stereotypic background activity, the cerebellum is receiving information concerning the programs or intentions for a particular motor response from the frontal and premotor cortices. It also is informed of the actual initiation of that response by the motor cortex, as well as the organism's conscious awareness of its position in space and relation to objects or stimuli relevant to carrying out that response via other corticopontocerebellar connections. At the same time, it is receiving constant feedback directly via the spinocerebellar tracts and indirectly from the somatosensory cortex as to how that movement is being executed, both absolutely and, where relevant, in juxtaposition to the external world (e.g., an object that is being moved or manipulated). All the above information is being fed to the granule cells and the cerebellar nuclei by way of mossy fiber connections. At the same time, all this can be compared to the information provided by the climbing fibers directly to the Purkinje cells from the inferior olivary nucleus which itself receives both ascending (afferent) proprioceptive information and descending input from the cortex and red nucleus. Thus the cerebellum is in an ideal position to analyze and integrate the results of the ongoing action relative to its original intention and modulate it accordingly.

When we speak of this type of behavioral analysis and subsequent modulation, it is important to differentiate the role of the cerebellum from that of the cerebral cortex and the basal ganglia, all of which play major roles in the execution of a motor program or activity. The cerebral cortex, primarily the frontal lobes, is ultimately responsible for establishing goals, initiating actions designed to accomplish those goals, determining if the goal is accomplished, and hence, whether the action should be sustained, halted, or adapted in any manner, or whether the goal itself should be changed. The specific contributions of the basal ganglia are also somewhat of an enigma and will be discussed in greater detail in Chapters 6 and 9. However, for now suffice it to say that they appear to be important in facilitating and inhibiting patterns of movements depending on the circumstances or context in which the movement occurs. Thus, depending on the nature of the lesion affecting the basal ganglia, one might witness a failure of either inhibition ("disinhibition") or facilitation resulting, respectively, in unwanted movement or bradykinesia and/or rigidity. The role of the cerebellum in such voluntary motor activities would appear to be much more specific and circumspect. It would appear that its function in these cases might be to ensure that the individual, discrete motor response as dictated and programmed by the cerebral cortex and basal ganglia is carried out in a smooth and efficient manner (e.g., control of the speed, force, direction, and termination of movement). In addition to carrying out consciously directed (voluntary), discrete motor activities, the cerebellum also is responsible for carrying out these activities while reflexively (subconsciously) making whatever postural adjustments are necessary to maintain balance and equilibrium, perhaps while engaging in a well-practiced baseline activity such as walking or engaging in other well-rehearsed activities.

While in practice it may be possible to isolate a lesion to one of these three systems (cerebral motor cortex, basal ganglia, or cerebellum) depending on the particular symptoms manifested by the patient, in analyzing normal behavior their individual contributions are often less clear. Sometimes this also can be the case in disorders of movement. For example, difficulties in making smooth, rapid transitions from one component of a motor response to the next may be found following cortical (Luria's "loss of kinetic melodies"), basal ganglia (athetosis), or cerebellar (**dysdiadochokinesia**) lesions. While each system likely contributes unique elements to the movement process, differentiating or identifying those individual components or contributions for a given action is not always easy. On the other hand, this helps to emphasize the notion that these systems are mutually dependent on and work in concert with one another. Like an orchestra, on occasion you may be able to pick out individual instruments, but for many complex movements (musical), the contributions of each are lost in the composition of the whole.

Non-Motor Functions of the Cerebellum?

In addition to its obvious control over some aspects of motor behavior, over the past 20 years in particular there has been increased interest in the role of the cerebellum in classical conditioning, timing, and other non-motor learning paradigms, as well as in cognition and affective responses. This interest has derived from numerous sources, including the appreciation of anatomical associations between the cerebellum and non-motor areas of the brain (especially the prefrontal cortex in man), physiological and functional neuroimaging studies showing activation of the cerebellum during the course of apparently non-motor, cognitive activities, and observations suggesting changes in cognitive and affective states (including psychiatric symptomatology) following lesions to the cerebellum. Glickstein and Yeo (1990), Ito (1993), Ivry, R. (1997), Leiner, Leiner, and Dow (1986, 1993), Middleton and Strick (1994, 2000), and Schmahmann (1991) were among the earlier proponents of this new way of looking at cerebellar function. Schmahmann (1991, 2001b, 2004) has proposed two key concepts in this regard: the *"dysmetria of thought hypothesis"* and the notion of a *"cerebellar cognitive affective syndrome."* With regard to the first, he suggests that just as the cerebellum "regulates the rate, force rhythm, and accuracy of movement, so it regulates the speed, capacity, consistency, and appropriateness of mental or cognitive processes" (Schmahmann, 2001b, p. 320). If this hypothesis is correct, then it would be understandable why disruptions of cerebellar function might result in a wide range of difficulties on both cognitive processes (e.g., executive and visual-spatial skills, language, memory), as well as behavioral disturbances (e.g., changes in personality, affect, or autonomic phenomena). It has been further suggested that the former result from insults to the posterior hemispheres, while affective type disturbances are associated with the anterior, vermal regions of the cerebellum. While many of the clinical cases offered in support of these hypotheses involved conditions in which more generalized involvement of the central nervous system was certainly possible (such as agenesis, tumor resections, degenerative or infectious processes), others apparently involved more restricted vascular lesions (see: Gottwald et al., 2004; Malm et al., 1998; Neau et al., 2000; Schmahmann, 2001b). Although appropriate caution is still being advised about overemphasizing the cerebellum's role in non-motoric behaviors (Glickstein, 2006), given these recent data, as clinicians we might want to pay closer attention to the potentially negative and possibly subtle impact that lesions of the cerebellum might have on cognitive and/or affective behavior.

The preceding review is not intended as a current, state-of-the-art description of the role of the cerebellum, which remains somewhat of an enigma. Rather it was provided to offer a very general overview of the possible interrelationships between the cerebellum and other neuroanatomical systems. Hopefully, this broad framework will make the following discussions more meaningful as we explore the traditional motor symptoms associated with cerebellar dysfunction and the patterns of symptoms or syndromes associated with specific types of cerebellar lesions. Table 3–4 lists some of the suspected motor feedback loops and the functions they likely serve.

CEREBELLAR SYMPTOMS

While our understanding of how the cerebellum actually functions may be somewhat limited, the motor symptoms most commonly associated with cerebellar disease or injury

Table 3–4. Feedback Loops

Balance and Equilibrium
Vestibulocerebellar pathways to flocculonodular lobe and vermis supplies information about changes in position/motion.
Ascending tracts to vermal and paravermal regions provides feedback concerning position of head and body in space and state of contraction of musculature of limbs.
Fastigiovestibular and fastigioreticular pathways give rise to vestibulospinal and medial reticulospinal tracts that appear to primarily adjust or facilitate contraction of antigravity muscles.

Whole Body Coordination
Spinocerebellar tracts to vermal and paravermal regions provides information regarding position of the body in space and state of contraction of musculature of limbs.
Olivocerebellar tracts provide cerebellum (particularly vermal and paravermal regions) with integrated, information from both cerebral cortex and spinal cord.
Interposed (and dentate) nuclei project, via the superior cerebellar peduncle, extensively to the red nuclei, and to a lesser extent, to the sensorimotor cortex (by way of the thalamus). These sites, in turn, give rise to the ventral and lateral corticospinal and the rubrospinal tracts that help control routine, whole body movements such as walking or running.

Discrete, Voluntary Limb Movements
Cerebral cortex, via the pontocerebellar pathways, provides cerebellum with information about intended action and the context (internal and external) in which this action is to take place.
Spinocerebellar tracts to vermal and paravermal regions provides information regarding the position and attitude of the limb, as well as the body in general, both at the initiation and during the course of movement.
Dentate (and interposed) nuclei project extensively to the sensorimotor cortex (by way of the VA/VL nuclei of the thalamus) and, to a lesser extent, to the red nuclei. These cerebellothalamocortical projections in particular, along with feedback from the basal ganglia, likely help modulate the output of the lateral corticospinal tracts that govern discrete movements.

Eye Movements and Hand-Eye Coordination
Information about the position and movement of the eyes comes into the cerebellum at least indirectly from the inferior olivary complex via the olivocerebellar tract and possibly through other, even more direct pathways (e.g., via pontine nuclei from the superior colliculus).
Vestibulocerebellar pathways to flocculonodular lobe and vermis supplies information about changes in position/motion, especially of the head.
The vestibulocerebellum provides input back to the vestibular nuclei which in turn send fibers via the ascending MLF to the cranial nerves III, IV, and VI which control reflex or involuntary eye movements according to changes in head movement or position.
Voluntary eye movements are likely mediated, at least in large part, by descending corticopontocerebellar pathways and returning cerebellothalamocortical projections.

have been quite well known for almost 100 years. These include loss of balance or disturbances of equilibrium, ataxia (most commonly reflected in disturbances of gait), asynergy, tremors, hypotonia, nystagmus, disturbances of motor speech, and abnormal posturing. These disturbances are not necessarily independent of one another, but may simply reflect the particular expression of one or more basic deficits. Rather than attempt to rigorously catalogue these various symptoms, it may be more practical to simply describe them.

Loss of Balance

Whether sitting or standing, we are constantly making subtle adjustments to our posture in order to remain vertically stable (recall "experiment" in Endnote 6). This is accomplished largely by the vestibulocerebellum. In patients, disturbances to this subsystem may be manifested by a wide-based gait and are tested clinically by having the patient stand with feet together. If the patient only has difficulty when the eyes are closed, this is more suggestive of posterior column disease (**Romberg's sign**). Sometimes disturbances can be elicited merely by asking the patient to sit upright or to walk.

Asynergy

Asynergy or the "decomposition" of movement is the breakdown in the rhythm and flow of a movement. Rather than a movement flowing smoothly from one component of the action to the next, it is broken down into saccades. Movements become jerky and discrete rather than smooth and continuous. Such movements have been variously described as mechanical or puppet-like. There are any of a number of ways in which disturbances of synergistic movement can be manifested. One simply observes the patient in routine activities looking for any such breakdown in the fluidity of movement. These phenomena can be tested clinically. One classic maneuver is to ask the patient to make a series of rapid alternating movements of the upper extremities, such as rapid pronation and supination movements of the outstretched hand as it rests on the thigh or in the palm of the opposite hand. With cerebellar hemispheric lesions, the movement may be excessively slow and/or irregular with a tendency to overshoot the level or desired position. Tapping out rhythms (e.g., "..._..._") with the hand (or foot) will produce irregular beats, both in terms of amplitude and frequency. Inability to perform such tasks in the absence of sufficient sensory or motor losses is called dysdiadochokinesia. Patients with cerebellar lesions also have been noted to be impaired in the discrimination of weight (barognosis). This finding has been related to the presence of asynergy in the affected limb.

Ataxia

If asynergy is a breakdown in the normal flow or the smooth sequencing of the individual components of movement, **ataxia** is the inability to properly regulate the direction, rate, extent, and intensity of movements. While the term "ataxia" is commonly used when referring to disturbances of gait, it actually can be applied to other movements of the lower extremities as well as to disturbances of movement in the upper limbs. Since gait ataxia is so common in cerebellar lesions (frequently exacerbated by problems with balance or equilibrium and/or the presence of "truncal ataxia"), special notice is usually taken of the patient's ability to walk. The patient may walk with a wide-based gait, display a certain awkwardness with excessive or exaggerated movements, perhaps with a tendency to fall or sway from one side to the other. The effect may be enhanced by asking the patient to walk a straight line, placing one foot directly in front of the other (tandem gait).

For an additional test of lower limb ataxia independent of gait, while lying or sitting, the patient may be asked to run the heel down the shin. In cerebellar disease, the patient has difficulty not only in making a smooth motion down the shin, but also has difficulty maintaining the heel in a straight line on top of the shin. Similarly, ataxia of the upper limbs can be assessed by having the patient alternately touch the tip of the nose with the finger, starting and returning to an outstretched position of the arm. The most common classical deficit seen in patients with cerebellar hemisphere lesion is that the usually straight trajectory from finger-to-nose is replaced by an irregular, sinusoidal trajectory. Actually, ataxia and asynergy may simply be different facets of the same phenomena. Where there are the components of one, it is usually possible to pick out components of the other, not only in the same patient, but also often in the same action.

In other tests for asynergy and/or ataxia of movement, the patient also can be asked to reach out (so the arm is reasonably extended) and touch the examiner's finger tip and then the patient's own nose, repeating this sequence as the examiner moves his finger from one area of space to another. Alternatively, the patient may simply be asked to reach out and pick up a small object on the desk. In addition to noting the presence or absence of tremors, attention is given as to whether the patient has trouble measuring the distance to the target. A tendency to overshoot or fall short of the intended goal is referred to as **dysmetria**.

Another sign of cerebellar involvement is manifested by the rebound phenomenon, which assesses the patient's ability to adjust to sudden displacements in the posture of the body, typically the upper extremities. The patient is asked to extend his arms and one arm is rapidly displaced downward by the examiner. In normal subjects, the arm quickly and easily returns to its former resting posture. In cerebellar disease, the arm oscillates before returning to a steady state.

A second variant of testing for the rebound phenomenon is to have the patient maintain a 90-degree flexion of the forearm. The examiner pulls against the resisting biceps muscle and then rapidly releases. Again the normal patient will quickly return to the original posturing of the arm. The cerebellar patient is not able to make this rapid adjustment and the arm will continue its trajectory or "overshooting" its mark, striking the shoulder (or, if the examiner is not careful, the patient may end up striking himself in the face). Basically, this deficit appears to reflect the loss of normal cerebellar influence in coordinating the activities of the agonist and antagonist muscles.

Tremor

During the performance of some of the above tasks, patients with cerebellar lesions may evidence certain types of tremors. Most typical is the so-called intention tremor. As opposed to diseases like parkinsonism that affect the basal ganglia, there is no "resting" tremor in cerebellar disease. If an intention tremor is present, it is likely to be observed in one of two types of situations or maneuvers. One is when the patient is asked to carry out some action, such as touching one's nose or the fingertip of the examiner (often called an "action" tremor). As the patient's finger approaches the target, a course tremor (3 to 5 cps) may become manifest. It also may be seen if the patient is asked to hold the arms extended out against the force of gravity (or sustention tremor). The presence of both an action and a sustention tremor often is referred to as a rubral tremor). Occasionally, a rhythmic bobbing of the head ("titubation") also may be present, with or without an abnormal posturing of the head.

Hypotonia

Hypotonia or loose, floppy muscles and joints ("rag doll" phenomenon) is a frequent accompaniment of cerebellar lesions, but it is most likely to be observed in the more acute stages of cerebellar lesions. As a result of this "looseness," abnormal posturing of the limbs may be possible. Deep tendon reflexes on the affected side will be hypoactive and there may be signs of decreased strength on that side. Over time, however, these particular symptoms may diminish.

Ocular Findings

Nystagmus is the most common eye finding with cerebellar disease. It is typically most pronounced on lateral gaze toward the side of the lesion, but also may be noted when fixating on a small, centrally placed object (instability of gaze). As a result of the latter, the patient may complain of diplopia (double vision) or a blurring of vision.

Motor Speech

Since cerebellar lesions can affect most aspects of motor functioning, the presence of disruptions of motor speech is not surprising. The basic disturbance is not unlike the other signs of asynergy or ataxia seen in the extremities. Similarly, there is a loss or reduction of the smooth, properly modulated flow of movement in the vocal or motor mechanisms of speech. Consequently, the speech is typically described as halting, uneven, irregular, poorly modulated, explosive, or scanning. Basically, there is a breakdown in the normal flow from one phoneme or word to the next, resulting in a breakdown in the melody of speech. Since this breakdown also can occur within single words, the speech is sometimes referred to as being **dysarthric**. Finally, as with the limbs, the amplitude of the response also is affected; thus the volume, as well as the rate and rhythm of speech can be affected, producing the "explosive" speech characteristic of cerebellar disease.

CEREBELLAR SYNDROMES

Although we have previously discussed certain motor symptoms in relation to the different parts of the cerebellum, a brief review may be useful at this point. However, before discussing these "syndromes," one should keep in mind that most lesions or diseases do not necessarily respect anatomical boundaries, not only within the cerebellum itself, but also with regard to structures outside the cerebellum (e.g., brainstem nuclei and pathways).

Vestibulocerebellar or Archicerebellar Syndrome

As we noted earlier, this primarily refers to the flocculonodular lobe and a small portion of the vermis. As this system is very closely connected with the vestibular system (balance and equilibrium), lesions involving these structures generally will produce severe disturbances in balance while sitting, standing, and walking. However, if the body is in a stable position, for example, supported in a chair or lying down in bed, the coordinated use of the limbs may be more intact. Other symptoms that may be associated with lesions of this system include titubation, abnormal posturing of the head, and nystagmus. **Note:** alcohol and certain other CNS depressants such as barbiturates selectively (but not exclusively) affect vestibulocerebellar connections; hence they are sometimes seen as a model for lesions of these systems. This is one reason why if you are stopped for suspicion of driving while intoxicated, the policeman will ask you to walk the center line (tandem gait).

Spinocerebellar or Paleocerebellar Syndrome

Lesions to this system typically are associated with the anterior lobe of the cerebellum and the vermal and paravermal regions. Again, as this area(s) of the cerebellum is heavily involved with axial mechanisms, injury to this region of the cerebellum typically will result in disturbances of stance and gait. Depending on the exact structures affected, there may be more general asynergy present that could result in nystagmus and disturbances of speech and upper limb movements, but generally the lower limbs and the trunk will evidence the greater disturbances. However, independent movements, even of the lower extremities (e.g., as in the heel-shin test) may show only relatively slight impairment or possibly none at all. Maximally disturbed are activities that require whole body movement, such as walking or running.

Pontocerebellar or Neocerebellar Syndrome

Lesions affecting the pontocerebellar system, by definition, will impact primarily on the cerebellar hemispheres or the lateral zone. Symptoms will include all the signs associated with ataxia and/or asynergy typically involving both the upper and lower extremities. Dysmetria, dysdiadochokinesia, tremor, hypotonia, as well as dysarthria and gait disturbances (secondary to limb ataxia) may be present, although balance per se likely will be intact. What is most unique about the effects of these lesions is that the behavioral (motoric) effects may affect the limbs strictly unilaterally, assuming the lesion is unilateral. Since the influence of the cerebellum is ipsilateral, the affected limbs will be on the same side as the cerebellar lesion.

Endnotes

1. As an integral part of other cortical and subcortical structures making up relevant distributed systems or feedback loops.
2. It might be noted that, unlike the cerebral cortex, there are no commissural pathways connecting the right and left cerebellar hemispheres.
3. If the cerebellum indeed receives information from all or most cortical association areas and if the expectation is that feedback loops are normally established, then one might anticipate that cerebellothalamic connections might be found in all or most of the thalamic association nuclei. However, no evidence could be found for such projections. Schmahmann (2001a) notes that the dorsomedial nucleus, as well as some of the intralaminar nuclei of the thalamus, also appears to receive input from the cerebellum. While the intralaminar nuclei do have diffuse cortical projections, these likely do not have the same intensity or specificity as the ventral lateral or dorsomedial nuclei. Also, it might be noted that the dorsomedial (also know as "medial dorsal") nucleus' primary cortical connections are with the prefrontal (and possibly cingulate) cortices. The latter would be consistent with Ramnani's (2006) suggestion that, in addition to the sensorimotor cortices, the other major corticocerebellar feedback loops likely involve the prefrontal cortex, which in turn does have corticocortical connections with the posterior association cortices.
4. But this is no different than many, if not most, of our assumptions regarding the operation of the brain. However, one particularly intriguing model is outlined by Ito (2005) and Ramnani (2006). Described as a forward control model, it essentially suggests that as learning occurs, the cortex (e.g., prefrontal motor cortices) simultaneously send out two separate "messages." One might be transmitted, for example, via the corticospinal tracts to instruct a certain action to be initiated within a particular context. The second, which follows corticopontocerebellar pathways, tells the cerebellum what is the expected outcome or consequence of the action (sensory feedback) in this particular situation (based on previous experience). These "second messages" may be stored in the cerebellum to more effectively guide or modulate well-practiced activities in specific settings, with less need for conscious cortical control or "supervision." At the

same time, online comparisons are constantly being made between expected and actual outcomes, relying heavily on corticocerebellospinoolivary loops to constantly refine (i.e., learning) outcome expectations, again within specific situations. It is thought that through multiple and discrete corticocerebellar feedback loops the cerebellum can exert its influence across a vast array of behaviors, including those in non-motor domains.

5. For example, the mental set of the football player who wishes to make a diving tackle must take precedence over the more basic instinctual response patterns of the vestibular nuclei and the antigravity reflexes that are designed to keep us vertical.

6. Assuming your cerebellum is intact, try this experiment. Without using other means of support, stand briefly on one foot (if you have decent balance, you will not fall; if you donot, you should not be trying this in the first place!). While doing this, unless you have absolutely perfect balance (extremely unlikely), you should notice subtle, movements (adjustments) of the small muscles in your foot. Again, assuming you have reasonable balance and are not about to fall over, you should also note that these movements do not seem to be under conscious control, but rather are spontaneous and involuntary. This would appear to be your cerebellum at work!

REFERENCES AND SUGGESTED READINGS

Altman, J. & Bayer, S.A. (1997) The Cerebellar System in Relation to Its Evolution, Structure, and_Functions. New York: CRC Press.

Blumenfeld, H. (2002) Neuroanatomy through Clinical Cases. Sunderland, MA: Sinauer, pp. 653–687.

Brazis, P. Masdeu, J., & Biller, J. (1990) Localization in Clinical Neurology. Boston: Little, Brown & Co., pp. 288–298.

Carpenter, M.B. & Sutin, J. (1983) *Human Neuroanatomy*. Baltimore: Williams & Wilkins, pp. 454–492.

Dow, R.S., Kramer, R.E., & Robertson, L.T. (1991) Disorders of the cerebellum. In Joynt, R. & Griggs, R.G (Eds.) Clinical Neurology. Philadelphia: J.B. Lippincott-Raven, Chapter 37, Vol 3, pp. 1–143.

Gilman, S. (1985) The cerebellum: its role in posture and movement. In. Swash, M & Kennard, C. (Eds.). Scientific Basis of Clinical Neurology. Edinburgh: Churchill Livingstone, pp. 36–55.

Gilman, S. & Newman, S.W. (1996) Essentials of Clinical Neuroanatomy and Neurophysiology. Philadelphia: F.A. Davis Co., pp. 158–168.

Glickstein, M. (2006) Thinking about the cerebellum. *Brain: A Journal of Neurology, 129*, 288–290.

Glickstein, M. & Yeo, C. (1990) The cerebellum and motor learning. *Journal of Cognitive Neuroscience. 2*, 69–80.

Gottwald, B., Wilde, B, Mihajlovic, Z., & Mehdorn, H.M. (2004) Evidence for distinct cognitive deficits after focal cerebellar lesions. *Journal of Neurology, Neurosurgery, and Psychiatry, 75*, 1524–1531.

Haines, D.E., Mihailoff, G.A., & Bloedel, J.R. (1997) The cerebellum. In Haines, D.E. (Ed.) Fundamental Neuroscience. New York: Churchill Livingstone, pp. 379–398.

Heimer, L. (1995) The human brain and spinal cord:functional neuroanatomy and dissection guide. New York: Springer-Verlag, pp. 363–376.

Holmes, G. (1939) The cerebellum of man. *Brain. 62*, 1–30.

Houk, J.C. & Wise, S.P. (1995) Distributed modular architectures linking basal ganglia, cerebellum, and cerebral cortex: their role in planning and controlling action. *Cerebral Cortex. 2*, 95–110.

Ito, M. (1993) Movement and thought: identical control mechanisms by the cerebellum. *Trends in Neurosciences, 16*, 453–454.

Ito, M. (2005) Bases and implicatioins of learning in the cerebellum adaptive control and internal model mechanism. *Progress in Brain Research, 148*, 95–109.

Ivry, R. (1997) Cerebellar timing systems. *International Review of Neurobiology, 41*, 555–573.

Keele, S.W. & Ivry, R. (1990) Does the cerebellum provide a common computation for diverse tasks? A timing hyothesis. *Annals of the New York Academy of Science. 608*, 179–211.

Leiner, H.C., Leiner, A.L., & Dow, R.S. (1986) Does the cerebellum contribute to mental skills? *Behavioral Neuroscience, 100*, 443–454.

Leiner, H.C., Leiner, A.L., & Dow, R.S. (1993) Cognitive and language functions of the human cerebellum. *Trends in Neurosciences, 16,* 444–447.

Malm, J., Kristensen, B., Karlsson, T., Carlberg, B., Fagerlund, M., & Olsson, T. (1998) Cognitive impairment in young adults with infratentorial infarcts. *Neurology, 51,* 433–440.

Martin, J.H. (1996) *Neuroanatomy Text and Atlas.* Stamford, CT: Appleton and Lange, pp. 291–322.

Middleton, F.A. & Strick, P.L. (1994) Anatomical evidence for cerebellar and basal ganglia involvement in higher cognitive function. *Science, 266,* 458–461.

Middleton, F.A. & Strick, P.L. (2000) Basal ganglia and cerebellar loops: motor and cognitive circuits. *Brain Research, Brain Research Reviews, 31,* 236–250.

Neau, J.P., Arroyo-Anllo, E., Bonnaud, V., Ingrand, P., & Gil, R. (2000) Neuropsychological disturbances in cerebellar infarcts. *Acta Neurologica Scandinavica, 102,* 363–370.

Nolte, J. (1993) The Human Brain. St. Louis: Mosby-Year Book, Inc., pp. 337–359.

Ramnani, N. (2006) The primate cortico-cerebellar system: anatomy and function. *Nature Reviews/ Neuroscience, 7,* 511–522.

Ramnani, N., Behrens, T.E., Johansen-Berg, H., Richter, M.C., Pinsk, M.A., Anderssen, J.L., Rudebeck, P., Ciccarelli, O., Richter, W., Thompson, A.L., Gross, C.G., Robson, M.D., Kastner, S., & Matthews, P.M. (2006) The evolution of prefrontal inputs to the cortico-pontine system: diffusion imaging evidence from Macaque monkeys and humans. *Cerebral Cortex, 16,* 811–818.

Schmahmann, J.D. (1991) An emerging concept. The cerebellar contribution to higher function. *Archives of Neurology, 48,* 1178–1187.

Schmahmann, J.D. (2001a) The cerebrocerebellar system: anatomic substrates of the cerebellar contribution to cognition and emotion. *International Review of Psychiatry, 13,* 247–260.

Schmahmann, J.D. (2001b) The cerebellar cognitive affective syndrome: clinical correlations of the dysmetria of thought hypothesis. *International Review of Psychiatry, 13,* 313–322.

Schmahmann, J.D. (2004) Disorders of the cerebellum: ataxia, dysmetria of thought, and the cerebellar cognitive affective syndrome. *Journal of Neuropsychiatry and Clinical Neurosciences, 16,* 367–378.

Thach, W.T. (1987) Cerebellar inputs to motor cortex. In Ciba Foundation Symposium (132), Motor areas of the cerebral cortex. New York: John Wiley & Sons, pp. 201–220.

Thach, W.T., Goodkin, H.G., & Keating, J.G. (1992) Cerebellum and the adaptive coordination of movement. *Annual Review of Neuroscience. 15,* 403–442.

4   THE BRAINSTEM

CHAPTER OVERVIEW

Among the areas covered in this book, the brainstem represents one of the smallest but perhaps the most anatomically and functionally diverse of the CNS structures. It might be described as the crossroads for the other primary divisions of the CNS, the telencephalon, diencephalon, cerebellum, and spinal cord. Not only are many if not most of the major afferent and efferent pathways serving these structures funneled through this neurological isthmus, but cell groups within the brainstem help integrate and modulate the information contained therein. Also contained within the brainstem are nuclei essential to the normal operation of all the cranial nerves, with the possible exception of olfaction. Finally, the brainstem is the source of various brain neurotransmitter substances, as well as cell groups essential for maintaining basic cortical arousal.

Given its compactness and diversity, even relatively small lesions of the brainstem can result in profound and often quite devastating consequences, not infrequently death. When and where available, the advent of magnetic resonance imaging (MRI) technology has greatly aided the clinician in localizing lesions of the brainstem. Nevertheless, an appreciation of brainstem anatomy remains essential to a fuller understanding of those neurobehavioral syndromes and their preliminary differential diagnosis.

The goals of this chapter are to provide an overview of the classic divisions of the brainstem and the more commonly defined structures contained within each. While the reader is not expected to memorize all the features discussed, hopefully one will at least be cognizant of the major sensory and motor pathways and the approximate location of some of its more prominent nuclei. Such information, along with that provided in the next chapter on the cranial nerves, should allow one, given the level of a lesion in the brainstem, to appreciate if not predict the major symptoms associated with such a lesion. To facilitate this learning, various brainstem lesions and their behavioral correlates will be reviewed and discussed at the end of the chapter. Figure 4–1 provides external views of the brainstem, while Figure 4–2 offers representative axial cuts of the midbrain, pons, and medulla. References to these figures will be made throughout the chapter.

INTRODUCTION

The preceding chapters thus far have described more or less homogeneous functional systems (e.g., movement, somatic sensations, and coordination). In the present chapter, there is no such unifying theme. The brainstem is very diverse, both structurally and functionally. This fairly primitive portion of the central nervous system packs quite a number of anatomical structures, representing a large number of functional systems into its approximate 8 to 9 centimeters of length. Among other things, it includes:

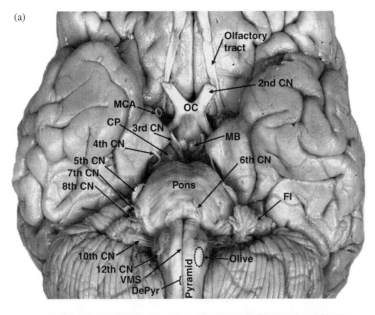

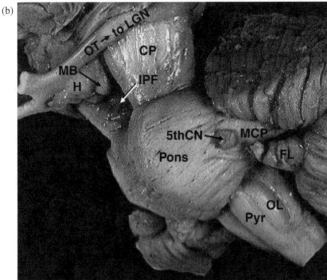

Figure 4–1. External views of the midbrain: (a) ventral view – in situ, (b) lateral view, (c) dorsal view (with cerebellum cut away). Brain images were adapted from the *Interactive Brain Atlas* (1994), courtesy of the University of Washington.

CN cranial nerve   MB mammillary body
CP cerebral peduncle   MCA middle cerebral artery
DePyr decussation of the pyramids   MCP middle cerebellar peduncle
DMS dorsal median sulcus   OC optic chiasm
Fl flocculus   OL olive
H hypothalamus (tuber cinereum)   OT optic tract
IC inferior colliculus   Pyr pyramids
ICP inferior cerebellar peduncle   SC superior colliculus
IPF interpeduncular fossa   SCP superior cerebellar peduncle
LGN lateral geniculate nucleus   VMS ventral median fissure

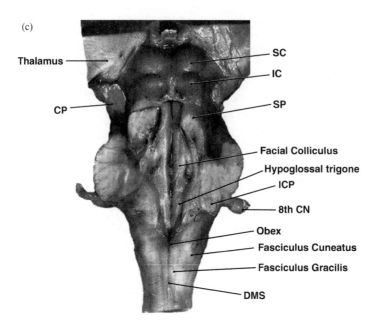

(c)

Thalamus

SC

IC

CP

SP

Facial Colliculus

Hypoglossal trigone

ICP

8th CN

Obex

Fasciculus Cuneatus

Fasciculus Gracilis

DMS

Figure 4–1. *(Continued)*

1. Fibers of passage from the cortex to the spinal cord (corticospinal tracts), from the spinal cord to the thalamus (spinothalamic tracts), and to the cerebellum (spinocerebellar tracts).
2. Nuclei that receive, integrate, and relay information to and from the cortex, cerebellum, spinal cord, and other brainstem nuclei,[1] and their respective tracts or connections.[2]
3. Ten of the twelve cranial nerves and their nuclei, as well as their interconnections with the cortex, spinal cord and cerebellum.
4. The reticular activating system.
5. Nuclei representing important sources of neurochemical transmitters.[3]

Precisely because fibers of passage and/or nuclei representing so many different functional systems are located in this relatively small neuroanatomical area, even small lesions or damage can result in quite profound and diverse behavioral effects. In fact, one of the more reliable signs of brainstem pathology is the presence of symptoms suggestive of cranial nerve involvement along with long tract findings (i.e., symptoms consistent with damage to the corticospinal and/or spinothalamic tracts). Chapter 5 will focus on the cranial nerves and the specific symptoms or syndromes associated with lesions affecting these nerves or nuclei. The present chapter will focus on the neuroanatomy of the brainstem itself, which will not only serve to complete our understanding of the functional neuroanatomy of the central nervous system but also will provide a basis for understanding the effects of lesions to this region.

GROSS ANATOMY

The brainstem is typically divided into three transverse regions (Figure 4–3). The most caudal region, which represents the rostral continuation of the spinal cord, is the **medulla** (which also is referred to as the *medulla oblongata* or less frequently the *myelencephalon*). Though

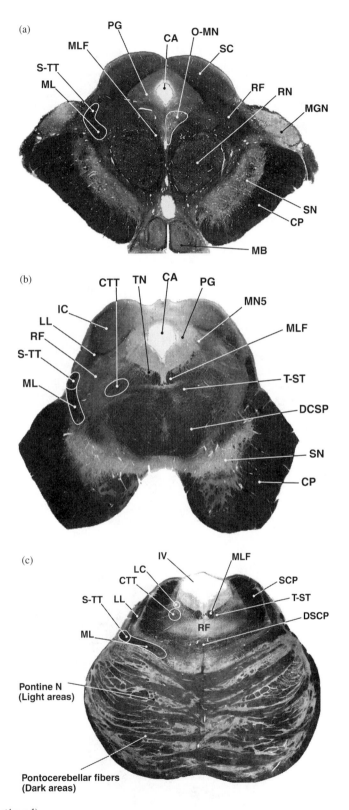

Figure 4–2. *(Continued)*

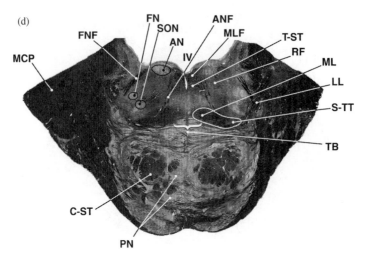

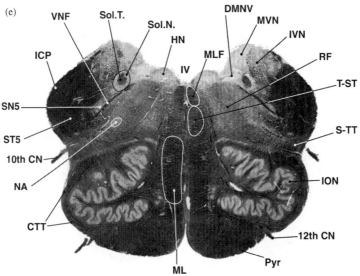

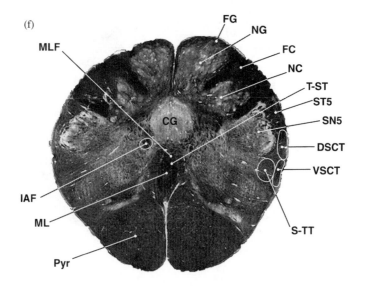

not commonly used by itself, another term applying to the medulla is "bulb." However, its adjective derivative "bulbar" is fairly common, as in the term "corticobulbar" tract. This term refers to fibers having their origin in the cortex and terminating on various nuclei in the medulla; more commonly it is used also to designate cortical tracts that terminate on any of the cranial nerve nuclei, even those above the medulla. The middle or central part is the **pons** (which, together with the cerebellum, sometimes is called the *metencephalon*). Finally, the rostralmost portion of the brainstem, which is continuous with the diencephalon (thalamus, epithalamus, hypothalamus, and subthalamic nuclei), is the **midbrain** or *mesencephalon*.

In addition to the above transverse divisions, certain longitudinal divisions also have been identified. These are the **tectum**, the **tegmentum**, and the **basis**; although not all these latter divisions typically are identified within all three transverse sections. In order to appreciate the distinction between the tectum and the tegmentum, it first is necessary to recognize the location of the ventricular system as it passes through the brainstem to become continuous with the spinal canal of the cord. From the third ventricle, lying between the two thalami, the cerebrospinal fluid flows caudally through the **cerebral aqueduct**, a small opening or

◄——

Figure 4–2. Axial sections through the brainstem: (a) upper midbrain, (b) lower midbrain, (c) upper pons, (d) lower pons, (e) upper medulla, and (f) lower medulla. Brain images were adapted from the *Interactive Brain Atlas* (1994), courtesy of the University of Washington.

10th CN vagus nerve
12th CN hypoglossal nerve
AN abducens nucleus

ANF abducens nucleus fibers
CA cerebral aqueduct
CG central gray
CP cerebral peduncle
C-ST corticospinal tract
CTT central tegmental tract
DMNV dorsal motor nucleus of vagus nerve
DSCP decussation, superior cerebellar peduncle
DSCT dorsal spinocerebellar tract
FC fasciculus cuneatus
FG fasciculus gracilis
FN facial nucleus
FNF facial nerve fibers
HN hypoglossal nucleus

IAF internal arcuate fibers
IC inferior colliculus
ICP inferior cerebellar peduncle
ION inferior olivary nucleus
IV fourth ventricle
IVN inferior vestibular nucleus
LC locus ceruleus
LL lateral lemniscus
MB mammillary body
MCP middle cerebellar peduncle
MGN medial geniculate nucleus

ML medial lemniscus
MLF medial longitudinal fasciculus
MN5 motor nucleus of trigeminal nerve
MVN medial vestibular nucleus
NA nucleus ambiguus
NC nucleus cuneatus
NG nucleus gracilis
O-MN oculomotor nucleus
PG periaqueductal gray
PN pontine nuclei
Pyr pyramidal tract (pyramids)

RF reticular formation
RN red nucleus
SC superior colliculus
SCP superior cerebellar peduncle
SN substantia nigra
SN5 spinal nucleus of the trigeminal nerve
Sol N solitary nucleus
Sol T solitary tract
SON superior olivary nucleus
S-TT spinothalamic tract
ST5 spinal tract of the trigeminal nerve
TB area of trapezoid body
TN trochlear nucleus
T-ST tectospinal tract
VSCT ventral spinocerebellar tract
VNF vagus nerve fibers

(a)

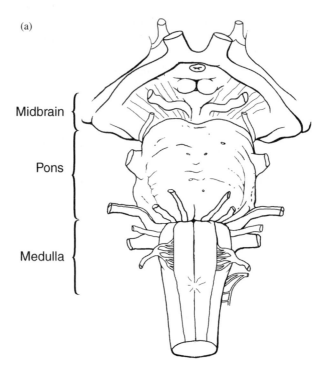

Midbrain

Pons

Medulla

(b)

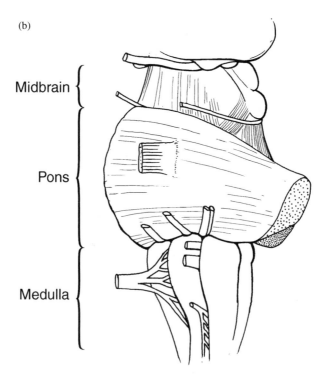

Midbrain

Pons

Medulla

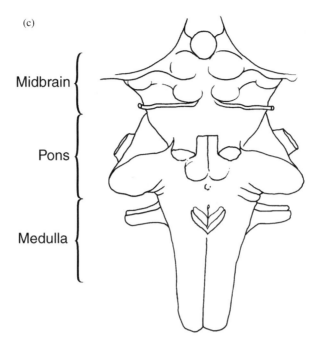

Figure 4–3.   Ventral (a), lateral (b) and dorsal (c) views of brainstem showing approximate horizontal boundaries of the midbrain, pons and medulla.

canal in the midbrain or mesencephalon. In the caudal portion of the midbrain, the cerebral aqueduct of the rostral brainstem begins to widen out at about the level of the pons and continues into the rostral medulla. This expansion constitutes the fourth ventricle, from which the cerebrospinal fluid then passes into the subarachnoid space through the **foramina of Luschka** (paired lateral openings) and the **foramen of Magendie** (a single ventral aperture).

The **tectum** (*"roof"*) is that part of the brainstem in the midbrain that lies dorsal to and forms a "roof" over the cerebral aqueduct (Figure 4–4). As can be seen, the two main structures that make up the tectum or dorsal portion of the midbrain are the **superior** and **inferior colliculi**. Collectively, these two sets of paired structures are known as the *corpora quadrigemina*; thus, the tectum is sometimes referred to as the *quadrigeminal plate*. The superior colliculi receive input from the visual system and from the spinotectal tract, while the inferior colliculi receive input from the auditory cortex, the lateral lemniscus (which conveys postsynaptic auditory information originating from the eighth cranial nerve), and the spinotectal tract. They also are the source of descending motor fibers via the tectobulbar and tectospinal tracts. These nuclei are intimately involved with a number of visual and auditory reflexes, including orientating responses of the head to visual and auditory stimuli.

The **tegmentum** (*"floor"*) is the central part of the brainstem at the level of the midbrain and forms the "floor" of the cerebral aqueduct. It represents the posterior or dorsal portion of the pons and medulla. This term, however, is most commonly used only in reference to the midbrain and pons. In the midbrain, the dorsal boundary of the tegmentum is the cerebral aqueduct and it extends ventrally to include the substantia nigra, a clearly visible dark band seen in an unstained transection of the mesencephalon (Figure 4–4b). At the level of the pons, the floor of the fourth ventricle represents the dorsal boundary of the pontine tegmentum. The pontine tegmentum extends ventrally more or less to but does not include the pontine nuclei (see below). For all practical purposes, though the term is generally not used here, the tegmentum would extend caudally into the central portion of the medulla.

(a)

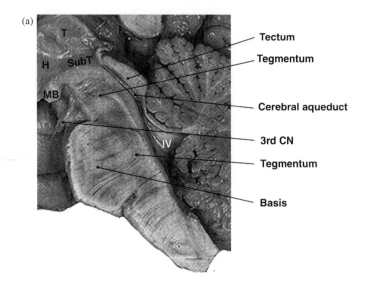

(b)

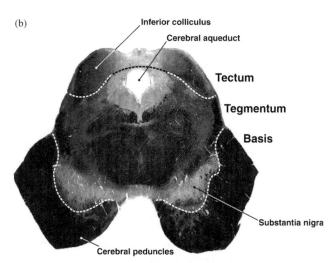

Figure 4–4. Midsagittal and axial cuts of brainstem, illustrating tectum, tegmentum, and basis. Legend: H hypothalamus, IV 4th ventricle, MB mammillary body, SubT subthalamus, T thalamus. Brain images were adapted from the *Interactive Brain Atlas* (1994), courtesy of the University of Washington.

The tegmental portion of the brainstem includes all of the cranial nerve nuclei, the reticular formation, and other supplementary motor nuclei (i.e., the red nuclei and substantia nigra in the midbrain and the inferior olivary nuclei in the medulla). The tegmentum also contains all of the ascending spinal tracts and a number of descending tracts, especially those having their origin in the nuclei of the brainstem or cerebellum.

The **basal** or ventral portion of the brainstem consists primarily of descending cortical tracts and the pontine nuclei. In the midbrain, these descending tracts (the majority of whose fibers end in the pons) are referred to as the **cerebral peduncles** or **crus cerebri** of the midbrain (Figure 4–4b). They consist of ipsilateral cortical fibers that project to the various brainstem nuclei (including cranial nerve nuclei), the cerebellum, and to the anterior horn cells of the spinal cord. In a cross section of the midbrain, the most medial portion of the crus cerebri primarily consists of frontopontine fibers, the most lateral portion representing temporal, parietal, and occipital–pontine fibers (probably more parietal than temporal or occipital), with the middle portion being made up of corticospinal and corticobulbar fibers. These latter (corticobulbar) fibers synapse in the reticular formation of the medulla and make connections with the motor nuclei of the cranial nerves in the medulla (IX–XII) and sensory medullary nuclei representing both the cranial nerves and spinal ascending pathways (e.g., the nucleus gracilis and nucleus cuneatus). The pontine nuclei are located in the basal portion of the pons (basis pontis). The majority of the fibers that constitute the cerebral peduncles synapse on these nuclei. The axons of these pontine nuclei constitute the pontocerebellar fibers which then proceed to cross the midline and enter the cerebellum via the middle cerebellar peduncle. These pontocerebellar fibers emanating from the pontine nuclei give the pons its distinctive appearance. Also included in the basal portion of the pons are corticospinal fibers that pass through this region. These latter fibers exiting from the pons make up the "basal" portion of the medulla. As these corticospinal tracts continue caudally on the ventral aspect of the medulla they are referred to as the "pyramids," hence, the term "pyramidal tract." The decussation of these pyramidal tracts can be clearly visualized on the ventral surface in the caudal half of the medulla (pyramidal decussation).

BRAINSTEM NUCLEI

Later in this chapter, we will return to a review of some of the major fiber tracts that are found within the brainstem. At this time, it may be useful to identify some of the more prominent mesencephalic, pontine and medullary brainstem nuclei. It should be noted, however, that some of these nuclear masses transverse more than one brainstem section, most notably the somatosensory portion of the nuclei of cranial nerve V and the reticular formation. For the purposes of discussion, we will also separate the cranial nerve nuclei from other types of nuclei. Finally, while some information regarding the functional significance of the individual cranial nerve nuclei will be discussed in their respective sections, more complete descriptions of the cranial nerves will be provided in the following chapter.

Nuclei of the Midbrain (Mesencephalon)

Substantia Nigra

As can be seen in Figure 4–2a, b, the substantia nigra is a very distinctive, pigmented band of cells located in the tegmentum of the midbrain, just dorsal to the cerebral peduncles. It is divided into two parts: the more medially located **pars compacta** (compact part) and the more laterally positioned **pars reticularis** (reticular part), and has been functionally related to the basal ganglia and the motor systems. The pars compacta sends efferent fibers to the putamen and caudate nucleus and is the source of dopaminergic transmission to those structures (i.e., the axons of the pars compacta release dopamine at their terminals in the basal ganglia). The reticular portion of the substantia nigra, as well as portions of the pars compacta, receive afferent input from the caudate nucleus, putamen, and globus pallidus (as well as from the cortex, amygdala, subthalamic nuclei, and a few additional structures). These afferent and efferent connections form the basis for a feedback loop with the basal ganglia, and hence, ultimately influence cortical motor activity. Another feedback loop is created via the projection of the reticular part of the substantia nigra to the thalamus (ventral anterior and ventral lateral nuclei). These nuclei in turn project back to the cortex that projects back to the thalamus and substantia nigra via the basal ganglia. Pathological changes in the substantia nigra, especially the pars compacta, occurs in Parkinson's disease and results in significantly decreased dopamine levels. An area lying adjacent to the pars compacta, the ventral tegmental area, represents another source of dopaminergic fibers to the ventral striatum and anterior limbic structures.

Red Nuclei

The red nuclei are two relatively large nuclear masses that lie in the tegmental section of the rostral portion of the midbrain, just dorsal to the substantia nigra (see Figure 4–2a). Their major input appears to be from the cerebellum (primarily from the dentate nucleus) via the superior cerebellar peduncle. The red nuclei also receive somatotopically organized projections from the cortex. Their main efferent projections are the inferior olivary nuclei[4] (via the central tegmental tract) and to the spinal cord (also somatotopically arranged) via the rubrospinal tract, as well as some direct input to cerebellar and cranial nerve nuclei.

Because of their afferent and efferent connections, the red nuclei are thought to provide an important motor feedback loop involving the cortex, cerebellum, and spinal cord. Unilateral midbrain lesions that involve the red nucleus and adjacent tracts produce contralateral motor disturbances such as tremors, ataxia, choreiform movements, and oculomotor disturbances ("Benedikt's syndrome").

Superior and Inferior Colliculi

The superior and inferior colliculi, collectively known as the corpora quadrigemina, make up the major portion of the tectum and are concerned with visual and auditory pathways respectively (see Figures 4–1c and 4–2a, b). In lower organisms, the **superior colliculi** are the primary projection site for the optic tracts, and hence would appear to play a more significant role in processing visual input. However, in higher mammals most of the visual input carried by the optic tracts eventually ends up in the lateral geniculates of the thalamus, which project to the primary visual cortex. In humans, the superior colliculi seem to be primarily concerned with eye movements, especially in response to moving, novel, or unexpected stimuli. These movements include conjugate gaze and visual tracking and certain optic reflexes, such as turning the head and neck in response to such visual (or auditory) stimuli. They receive direct visual input from the retina via collaterals from the optic tract and processed visual input from the striate cortex. In addition, the superior colliculi receive information from the frontal eye fields (which direct voluntary eye movements), somatosensory systems (via the spinotectal tracts), as well as auditory input (primarily from the inferior colliculi). These latter connections (between the superior and inferior colliculi) facilitate the reflex turning of the eyes to check out unexpected sounds, a reflex with clear survival value. Efferent output from the superior colliculi is back to the spinal level (tectospinal tract), the reticular nuclei (tectoreticular tract), the pontine nuclei (tectopontine tract), and from there to the cerebellum. While direct connections to the eye muscles have not been established, the superior colliculi appear to project to the pretectal nuclei, which in turn have connections with the ocular motor nuclei of cranial nerves III, IV, and VI, and to the Edinger–Westphal nuclei. Through these connections the superior colliculi can impact both on eye movements as well as accommodation.

The **inferior colliculi** receive most of their afferent connections via the lateral lemniscus, which carries fibers from the superior olivary nuclei and the cochlear nuclei. Both of these nuclei receive direct input from the auditory system. In addition to sending fibers to the medial geniculates (the auditory integration and relay nuclei of the thalamus) via the inferior brachium, each inferior colliculus nucleus has connections with the inferior colliculus on the opposite side and to the superior colliculus. The role of the inferior colliculi is not clearly established, but they are thought to be important in the localization of sound and the orienting response that follows (e.g., reflex turning of the eyes, head, and neck). The latter appears to be accomplished by means of connections with the superior colliculus. The superior colliculi, in turn, are connected to the cervical cord via the tectospinal tract, innervating the muscles of the head and neck.

Reticular Formation

The reticular formation is an interconnected though not always clearly demarcated collection of cellular or nuclear groups that extends throughout the central portion of the brainstem (see Figure 4–2a–e). The reticular formation has extensive afferent connections with ascending (spinal), descending (cortical and subcortical), and local brainstem systems, including the cerebellum. Many of the nuclei associated with the production of neurochemical trans-mitters fall within its boundaries (see Chapter 11). The output of its various nuclear groups appears equally diffuse, projecting caudally to the cord, rostrally to the diencephalon and cortex, as well as to other parts of the brainstem and cerebellum. Structurally, it is not uniform throughout the brainstem, but can be divided into zones or regions with differing cell structures and connections. These zones typically represent vertical columns of nuclei, some of which may run through all three divisions of the brainstem and into the spinal cord. The lateral and central (or medial) zones are the two that are most commonly defined. Both receive extensive inputs from most sensory modalities and from the cortex. In contrast, the central group appears to be the source of the majority of its efferent output. Many of the reticular nuclei that have more extensive caudal projections are located in the rostral portion in the brainstem, while nuclei with primarily ascending connections are found more caudally. The raphe nuclei (meaning, "seam"), another group of cells located along the midline and associated with the production of serotonin, represent a third reticular zone.

Given the structural diversity and complexity of the reticular formation, much of its functional significance is still largely conjectural. Nonetheless, its influence is thought to be quite varied and likely involves multiple functional systems. For example, as noted above, parts of the reticular formation receive input from the spinal cord (spinoreticular and collaterals from the spinothalamic tracts) and from the sensory components of the cranial nerves (notably the trigeminal). Reticulospinal tracts provide feedback to both sensory and motor neurons in the spinal cord. There also are reciprocal connections with the cerebellum and other motor system neurons, including the sensorimotor cortex, basal ganglia, red nuclei, and substantia nigra. While the exact function of the various descending reticular pathways is not clear, it has been suggested that they probably contribute to the regulation of axial musculature (e.g., posture and whole body activities) and muscle tone. Descending fibers from the reticular formation likely also modulate sensory input itself and may play a key role in the control of pain. These reticulospinal connections also would appear to exert some control over primitive motor response patterns (e.g., crying, sucking) and for those visceral activities necessary for maintaining homeostasis (e.g., respiration, heart rate, blood pressure).

The portion of the reticular formation that lies in the lower pons and medulla and projects rostrally into the intralaminar nuclei of the thalamus appears to be an integral part of the **reticular activating system** (RAS). This is the term that has been most closely associated with the reticular formation, although it probably represents only a fraction of its overall functional significance. The RAS plays a central role in both arousal and atten-tional mechanisms. It also is important in maintaining sleep cycles or rhythms. Arousal (and orienting responses) can be and are routinely initiated by external sensory input. Such input, at least at the reticular level, is probably diffuse, relatively nonspecific, and likely even unconscious. This "early warning (or alerting) system" has clear survival value to the organism. However, some gating of sensory input is essential or the organism would continue to respond to stimuli long after they have lost their signal or alerting value. This sensory gating is likely to occur at various levels, including the reticular formation, thalamus, and cortex. However, the cortex seems dependent upon the RAS, via the thalamus, for normal arousal (i.e., the "on" switch enabling it to function). Significant lesions of the RAS can lead to a permanent vegetative or comatose state. On the other hand, the cortex, specifically the frontal lobes, can modulate the activity of the RAS via descending or corticoreticular pathways. Thus, given at least a baseline of minimal arousal, at midnight the frontal lobes of a graduate student may invoke a greater level of arousal from the RAS when contemplating the possible consequences of failing an exam the next day.

Also considered part of the reticular formation, the **periaqueductal** (*central*) **gray area** is a region of pale-staining cells surrounding the cerebral aqueduct containing a number of smaller nuclear groupings. Diffuse connections with the limbic system and parts of the diencephalon have led to the notion that this area is concerned with visceral functions. It also has been postulated that this area may be the source of descending fibers that mediate pain control or gating mechanisms important for the perception of pain.

Cranial Nerve Nuclei of the Midbrain

Cranial nerves I (olfaction) and II (vision) do not have brainstem nuclei; only cranial nerves III through XII have paired nuclei in the brainstem. All brainstem cranial nerve nuclei lie within the tegmental portions of the brainstem. The **oculomotor (cranial nerve III) nuclei** lie in the rostral midbrain, anterior to the central gray matter at the level of the superior colliculi (Figure 4–2a). These nuclei are subdivided into smaller nuclear groupings that mediate different aspects of oculomotor nerve functioning. Control of eye movements via the superior, medial, and inferior rectus muscles, as well as the inferior oblique muscles and elevation of the eyelid, are mediated by the oculomotor nuclei. In contrast, that portion of cranial nerve III that responds to bright light (pupillary constriction) and the distance of the viewed object from the eye (accommodation of the lens) are subserved by a special nuclear grouping called the **Edinger–Westphal nuclei** that are part of the preganglionic parasympathetic autonomic nervous system.[5] Convergence of the two eyes resulting from the simultaneous stimulation of the two medial rectus muscles when focusing on a near object is under the control of the oculomotor nucleus.

The nucleus for the **trochlear nerve** (cranial nerve IV), which is responsible for activation of one of the external ocular muscles (the *contralateral* **superior oblique**), is located in the midbrain, anterior to the central gray matter at the level of the inferior colliculi (Figure 4–2b). This nucleus facilitates an inward (medial) and downward (inferior) rotation of the eye. CN IV is the only one to exit from the dorsal aspect of the brainstem (see: Chapter 5).

The **mesencephalic portion of cranial nerve V**, which mediates proprioceptive feedback from the muscles of mastication, also lies in the midbrain tegmentum (Figure 4–2b). This nucleus is unique in that it contains sensory ganglion cell bodies that migrated to be contained within the CNS. All other somatosensory ganglion cells lie outside the cord or brainstem.

Nuclei of the Pons

Locus Ceruleus

The locus ceruleus is a long, thin, pigmented[6] nuclear group that lies near the floor of the fourth ventricle (see Figure 11–13, this volume). It has connections both rostrally and caudally with cortical and subcortical structures and with the cerebellum and spinal cord. As mentioned earlier, it is a major source of norepinephrine.

Pontine Nuclei

The pontine nuclei are interspersed among both transverse and longitudinal corticofugal fibers in the ventral or anterior aspect of the pons (Figure 4–2c, d). It is upon these nuclei that the corticopontine fibers synapse and then become the source of the transverse, contralateral projections to the cerebellum via the middle cerebellar peduncle. Thus, the axons of these nuclei provide the primary means through which the cortex influences the cerebellum.

Superior Olivary Nuclei

The superior olivary nuclei lie in the lateral aspect of the tegmental portion of the pons (Figure 4–2d). They receive bilateral input from the dorsal and ventral cochlear nuclei, and hence, represent the first level at which input from both ears converges. The superior olivary nuclei are the main source of fibers for the lateral lemniscus which carries auditory information to the inferior colliculi and then to the medial geniculate nuclei. The superior olivary nuclei should not be confused with the inferior olivary nucleus of the medulla. These nuclear groups are both anatomically and functionally distinct (see below).

Cranial Nerve Nuclei of the Pons

Cranial nerve V (the *trigeminal nerve*) is very distinct, as it is the only nerve that exits through the middle of the pons (on the lateral surface) (Figure 4–1b). The nuclei for the trigeminal nerve represent the largest group of cranial nerve nuclei, collectively extending throughout the length of the brainstem. The nuclei for cranial nerve V have both motor and sensory components. The former is the smaller of the two and lies fully within the pons in the lateral portion of the tegmental region (Figure 4–5). It is responsible for the muscles of mastication.

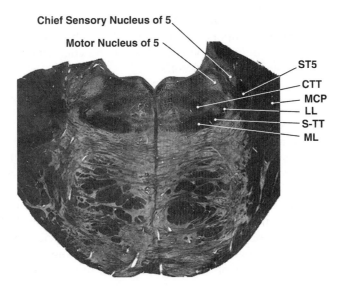

Figure 4–5. Axial section through lower pons showing the chief sensory and motor nucleus of the trigeminal nerve (5). Legend: CTT central tegmental tract, LL lateral lemniscus, MCP middle cerebellar peduncle, ML medial lemniscus, ST5 spinal tract of the trigeminal nerve, S-TT spinothalamic tract. Brain images were adapted from the *Interactive Brain Atlas* (1994), courtesy of the University of Washington.

The sensory nuclei of the fifth cranial nerve are responsible for ipsilateral sensations of touch, pain, temperature, and proprioception of the face, anterior part of the scalp, sinuses, teeth, and eyes. The sensory nucleus of the fifth cranial nerve is divided into three separate nuclei:

1. The **principle sensory nuclei of cranial nerve V** are located in the pons, just lateral to the motor nuclei of V (Figure 4–5). These nuclei appear to be responsible for discriminative touch on the face and proprioceptive feedback from the facial muscles.
2. The **mesencephalic nuclei of cranial nerve V,** which were listed under the cranial nerve nuclei of the midbrain (Figure 4–2b), are responsible for processing proprioceptive feedback from neuromuscular spindles in the muscles of mastication.
3. The **spinal nuclei of cranial nerve V,** the longest of all the cranial nerve nuclei, begins at the level of the pons and extends into the upper cervical cord, in the dorsolateral portion of the tegmental division of the medulla (see Figure 4–2e). These nuclei are responsible for sensations of light touch or pressure, pain, and temperature of the face and forehead.

It should be noted that in addition to cranial nerve V portions of other cranial nerves that have general somatic afferent functions, such as the VII, IX, and X (e.g., general somatic sensations to the external ear and to parts of the nasal and oral cavity), also have connections with the spinal trigeminal nuclei. All divisions of the sensory nuclei of cranial nerve V project to the ventral posteromedial nuclei of the thalamus via either the ventral or dorsal trigeminothalamic tract.

The **nucleus of cranial nerve VI** (the *abducens*) lies in the more caudal portion of the pons adjacent to the fourth ventricle (Figure 4–2d). It is responsible for control of the lateral rectus muscle that abducts the eye (rotates the eye horizontally in an outward or lateral direction).

The **motor nucleus of cranial nerve VII** (the facial nerve) is located in the lateral portion of the caudal pontine tegmentum (Figure 4–2d). It primarily is responsible for the muscles of facial expression. In addition to innervating the muscles of facial expression, the facial nerve also has some sensory components, primary of which is taste in the anterior two thirds of the tongue, but the nuclei responsible for this function (the nucleus solitarius or solitary nuclei) lie in the medulla (see below).

Nuclei of the Medulla

Inferior Olivary Nuclei

The most predominant nuclear masses in the medulla are the inferior olivary nuclei (Figure 4–2e). Grossly resembling the dentate nuclei of the cerebellum in appearance, the inferior olivary nuclei are prominently visible in the anterolateral portion of the upper medulla. The inferior olivary nuclei receive input from the cortex (probably from multiple areas), red nucleus, periaqueductal gray of the midbrain, vestibular nuclei, and from the deep cerebellar nuclei. Ascending fibers also reach these nuclei from the spinal cord. The major output of the inferior olivary nuclei is to the cerebellum via the inferior cerebellar peduncle in a very precise topographical fashion and is the source of the climbing fibers that synapse on the dendritic trees of the Purkinje cells of the cerebellum.

Nuclei Gracilis and Cuneatus

Recall that the fasciculi gracilis and the fasciculi cuneatus carry fibers representing stereognosis and position and vibratory sense from the lower and upper extremities, respectively. In the caudal portion of the medulla, these pathways synapse in their respective nuclei (Figure 4–2f). In keeping with the nature of these fiber tracts, there remains a sharp topographical organization within these nuclei. After leaving the nuclei gracilis and cuneatus, the postsynaptic fibers decussate as the internal arcuate fibers to form the **medial lemniscus** (Figure 4–2f), which ascends to project to the **ventral posterolateral (VPL) nuclei** of the thalamus.

Cranial Nerve Nuclei of the Medulla

The **vestibulocochlear (VIII) cranial nerve** has a dual sensory function; it serves hearing as well as vestibular senses. Hearing is mediated through the **dorsal** and **ventral cochlear nuclei** (Figure 4–6), located on the dorsolateral and ventrolateral aspects of the tegmentum (on the lateral surface of the inferior cerebellar peduncle) in the rostral medulla near the pontine–medullary junction. These nuclei receive input directly from the spiral ganglion cells that represent the first-order neurons receiving their input directly from the nerve cell endings in the ipsilateral ear (hair cells of the organ of Corti). Even at this early level, these projections to the cochlear nuclei are tonotopically organized. The output of these nuclei appears to be relatively complex, but in general they project to the inferior colliculi of the midbrain. Many of the fibers cross the midline and project directly to the **inferior colliculi** while others pass through the **superior olivary nuclei** (Figure 4–2d) (especially those from the ventral cochlear nuclei) before joining those more direct projections. The collection of auditory fibers from the cochlear nuclei and from the superior olivary nuclei that travel in the pontine tegmentum to the inferior colliculi is known as the **lateral lemniscus** (Figure 4–2b/d). This tract carries information both from the ipsilateral (less) and the contralateral (more) ear. See the section on the eighth cranial nerve in Chapter 5 for a more detailed discussion of these pathways.

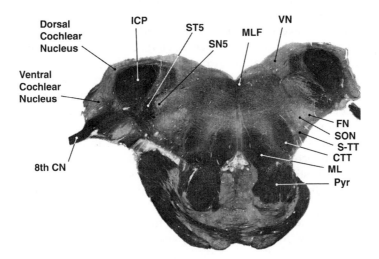

Figure 4–6. Medulla at level of the vestibulocochlear nerve. Brain images were adapted from the *Interactive Brain Atlas* (1994), courtesy of the University of Washington.

CTT central tegmental tract	SON superior olivary nucleus
FN facial nucleus	S-TT spinothalamic tract
ICP inferior cerebellar peduncle	SN5 spinal nucleus of the trigeminal nerve
ML medial lemniscus	ST5 spinal tract of the trigeminal nerve
MLF medial longitudinal fasciculus	VN vestibular nuclei (medial and inferior)
Pyr pyramidal tract	

The vestibular portion of the eighth nerve sends projections from the saccule, the utricle, and the three semicircular canals in the inner ear to the **vestibular nuclei** of the brainstem (though some may synapse directly in the flocculonodular lobe of the cerebellum). There are four vestibular nuclei: the **superior, inferior, medial, and lateral**. The superior vestibular nuclei lie just anterior to the superior cerebellar peduncle in the dorsolateral region of the caudal pons. The other three nuclei begin in the area of the pontine–medullary junction and extend caudally into the dorsolateral medulla, especially the inferior and medial vestibular nuclei (Figure 4–2e).

The peripheral receptors in the inner ear provide information about both the static orientation of the head in relation to gravitational forces (via the saccule and utricle) and movement (e.g., changes in direction and/or acceleration or deceleration). This information is relayed to the vestibular nuclei and from there both directly and indirectly (via the vestibular nuclei) to the cerebellum. In addition to inputs from cranial nerve VIII, the vestibular nuclei also receive substantial information from the spinal cord, particularly from the dorsal columns, as well as feedback from the cerebellum. From the vestibular nuclei, this information is relayed to a variety of systems that in large part mediate postural and oculomotor adjustments to movement in space or gravitational compensation. Among some of the efferent connections of the vestibular nuclei are:

1. Cerebellar, especially to the flocculonodular and vermal regions.
2. Spinal, via the lateral vestibular nuclei and the lateral vestibulospinal tract supplying the lower limbs and the medial vestibular nuclei and the medial vestibulospinal tract to the thoracic level of the cord (supplying muscles of the neck and axial muscles).[7]
3. Cortical, via the thalamus, probably both for conscious awareness of the position of the body in space (facilitating adjustments in voluntary motor activity to changes in position or motion) and to compensate for intentional changes in position as opposed to passive or unexpected changes.
4. Bulbar, with extensive connections being made with brainstem nuclei that control eye movement (II, IV, and VI nerve nuclei), certain reticular nuclei (paramedian pontine reticular formation), and the spinal nuclei of cranial nerve XI (involved in movements of the neck and head). These connections from the vestibular nuclei travel both rostrally (nuclei of cranial nerve III and IV) and caudally (nuclei of cranial nerve XI) in a tract called the **medial longitudinal fasciculus** (Figure 4–2a–f). This pathway is important for reflex adjustments in eye movement with changes of the position of one's head in space to ensure stability of the visual image despite motion of the head. These connections also form the basis of caloric tests of nystagmus to check the integrity of the vestibular system or of the brainstem in cases of coma.
5. Reticular, providing a basis for connections with the reticulospinal tract, which also seems to play a role in postural stability.
6. The nausea associated with disturbances of the vestibular system also suggests rather direct connections with the vagus nerve and its nuclei.

Afferent feedback loops from all of the above systems (cortex, cerebellum, oculomotor nerves, and spinal cord) are all essential for maintaining the operational integrity of the system.

The **solitary nuclei** (*nucleus solitarius*) that lie in the dorsolateral portion of the upper medulla (Figure 4–2e) are the primary brainstem nuclei for those special visceral afferents mediating the sense of taste [e.g., sensory fibers from the *facial* (VII), *glossopharyngeal* (IX), and *vagus* (X) nerves]. In addition to the sense of taste (which involves the more rostral portion of the nuclei), its more caudal extension subserves (1) those general visceral afferents (cranial nerves IX and X) from the walls of the larynx, pharynx, and soft palate (afferent limb of the gag reflex); (2) the carotid body and sinus (monitor for oxygen content and arterial pressure of the blood); and (3) internal organs. The feedback from the arterial system and internal organs serve autoregulatory functions of the autonomic nervous system and generally do not reach conscious awareness.

In addition to their thalamic projections carrying information about taste, the solitary nuclei send efferent fibers to the hypothalamus and other brainstem nuclei. The latter are the source of preganglionic, parasympathetic fibers that help modulate cardiovascular, respiratory, and other internal organ systems.

The **nucleus ambiguus** is a lower motor nucleus shared by the cranial nerves IX and X (glossopharyngeal and vagus) (Figure 4–2e). Its efferent output is primarily to the muscles of the larynx, pharynx, and soft palate used in swallowing and phonation. It has been suggested that the nucleus ambiguus also may supply fibers to certain visceral organs (e.g., the heart), but these pathways are not universally agreed upon at this time. The nucleus ambiguus lies just dorsal to the inferior olivary nucleus more or less in the central portion of the tegmentum of the rostral medulla. Its borders, however, are indistinct, thus its name.

The **superior** and **inferior salivatory nuclei** are two small nuclear groups in the rostral medulla that contribute to cranial nerves VII serve as the origin of fibers of cranial nerves VII and IX, which, respectively, innervate the salivary and lacrimal glands (cranial nerve VII) and the parotid glands (cranial nerve IX).

The **dorsal motor nuclei** (Figure 4–2e) are the source of the general visceral efferent fibers of the vagus nerve (cranial nerve X), which innervates the organs of the thoracic and abdominal cavities. The dorsal motor nuclei lie along the dorsal aspect of the rostral medulla, just lateral to the hypoglossal nuclei.

The **hypoglossal nuclei** (for the 12th or hypoglossal cranial nerve) are reasonably large nuclear bodies that lie on the dorsal aspect of the rostral medulla (Figure 4–2e), just on either side of the midline. Their efferent fibers form the hypoglossal nerves that innervate the muscles of the tongue.

Note: The accessory nerve (cranial nerve IX) primarily is served by the spinal nucleus of the accessory nerve that is located in the cervical spine. However, some cranial roots may emanate from nuclear groups that are a caudal extension of the nucleus ambiguus.

MAJOR FIBER TRACTS WITHIN THE BRAINSTEM

The following tracts that travel through the brainstem previously have been described but are included here for purposes of review. As these tracts may migrate or shift in their relative positions from one division of the brainstem to the next, the reader is advised to refer to the accompanying illustrations to follow their paths through the stem.

Major Afferent (Sensory) Pathways

Spinothalamic Tract

The spinothalamic tract carries information regarding pain, temperature, touch, and pressure from the lower body to the ventroposterolateral (VPL) nucleus of the thalamus. This tract starts out in the lateral aspect of the medulla (Figure 4–2e, f), just medial to the spinocerebellar tracts and just within the ventral or anterior half of the medulla. As the spinothalamic tract progresses rostrally through the brainstem, it begins to assume a more medial and dorsal position, being displaced by the cerebellar peduncles and the pontine nuclei, respectively. In the midbrain, it again resumes its more lateral position. However, with the presence of the relatively large crus cerebri and the substantia nigra, the spinothalamic tract occupies a more dorsal position relative to the other structures at this level.

Medial Lemniscus

The medial lemniscus carries information regarding proprioception, stereognosis, and vibration from the lower body to the VPL nuclei of the thalamus. In the medulla, this tract starts out along the midline, with the upper extremities represented more dorsally and the lower extremities more ventrally. As it proceeds rostrally, the medial lemniscus shifts its position so that the lower extremities are now more lateral and the upper extremities assume a more medial position in the pons. At this point, the medial lemniscus lies just above the pontine nuclei. This rotation continues in the midbrain (Figure 4–2a–f). Here, lying just on the dorsal edge of the substantia nigra, the lower extremities now occupy a more dorsal position in the tract, with the fibers representing the upper extremities lying in the ventral portion of the medial lemniscus.

Lateral Lemniscus

Beginning in the lower pons and terminating in the midbrain (Figure 4–2b–d), the lateral lemniscus carries auditory input from the cochlear and superior olivary nuclei to the inferior colliculi. The lateral lemniscus travels adjacent (dorsolateral) to the spinothalamic tracts.

Brachium of the Inferior Colliculi

Clearly visible on the dorsal surface of the midbrain (Figures 4–1c and 4–2b) (or on a axial section at the level of the superior colliculi), this pathway carries auditory information from the inferior colliculi to the medial geniculates of the thalamus.

Inferior Cerebellar Peduncle

The inferior cerebellar peduncle is a major source of input to the cerebellum from the spinal cord and brainstem nuclei. It is very prominent on the dorsolateral portion of the upper medulla (Figures 4–1c and 4–2e).

Middle Cerebellar Peduncle

The middle cerebellar peduncle is the huge band of transverse fibers that gives the external pons its characteristic appearance and coalesces to form the middle cerebellar peduncles (Figures 4–1b and 4–2d). The middle peduncles consist of corticopontocerebellar fibers representing contralateral input from the cerebral hemispheres to the cerebellum (following synapse on the pontine nuclei). Cranial nerve V can be seen to exit on the lateral surface of the pons through the middle peduncle, the only cranial nerve that does.

Trapezoid Body

The trapezoid body is the internal transverse pathway in the caudal pons, just dorsal to the pontine nuclei (Figure 4–2d), which carries decussating fibers from the ventral cochlear nuclei on one side of the brainstem to the superior olivary nuclei on the opposite side.

Dorsal and Ventral Trigeminal Lemnisci

Comparable to the spinothalamic pathways, these tracts convey somatosensory information from the face to the ventral posteromedial (VPM) nuclei of the thalamus. These pathways (not shown) lie slightly dorsal to (DTL) and contiguous with the dorsal aspect of the medial lemnisci (VTL) in the pons.

Juxtarestiform Body

The juxtarestiform body are fibers (both afferent and efferent) of the inferior cerebellar peduncle that interconnect the vestibular nuclei with the cerebellum. On gross sections, the juxtarestiform bodies essentially merge with the inferior cerebellar peduncle from which they are difficult to differentiate.

Major Efferent (Motor) Pathways

Basis Pedunculi

The basis pedunculi, also known as the cerebral peduncles or crus cerebri, consist of corticospinal, corticobulbar, and corticopontine fibers from ipsilateral cortex. These tracts occupy the ventral portion of the midbrain and lie just anterior to the substantia nigra (Figures 4–1b and 4–2a, b). In the pons, these massive peduncles break up into smaller fasciculi interspersed among the pontine nuclei where the majority of the fibers making up the basis pedunculi terminate.

Pyramids

The pyramids are located in the ventral medulla (Figures 4–1a, b and 4–2e, f) and represent the corticospinal fibers remaining from those constituting the basis pedunculi after most of the axons of the latter have synapsed in the brainstem. The majority of these fibers will decussate and descend as the lateral corticospinal tracts. The others (about 5–10%) will remain ipsilateral as the ventral corticospinal tracts.

Corticobulbar and Corticopontine Tracts

The corticobular and corticopontine tracts are fiber tracts descending in the basis pedunculi (cerebral peduncles) that terminate on various nuclei in the brainstem, including the red nuclei, inferior olivary nuclei, cranial nerve, and pontine nuclei. The latter, as was noted above, in turn project to the cerebellum as the middle cerebellar peduncles.

Central Tegmental Tracts

These tracts consist of motor fibers descending from the red nuclei to the inferior olivary nuclei, along with fibers to and from the reticular formation. As implied by their name, they generally occupy the central portion of the pontine tegmentum (Figure 4–2b, c, e).

Superior Cerebellar Peduncles

These pathways (Figures 4–1c and 4–2b, c) represent the major efferent output of the cerebellar nuclei to the cortex (by way of the ventral lateral nuclei of the thalamus). The superior cerebellar peduncle also sends collaterals to the red nuclei, lateral reticular nuclei, and the inferior olivary nuclei. However, while mostly consisting of efferent fibers, this pathway also includes some afferent input from the spinal cord (ventral spinocerebellar tract) and probably the mesencephalic motor nuclei of cranial nerve V mediating the muscles of mastication (trigeminocerebellar tract).

Medial Longitudinal Fasciculus (MLF)

This tract (Figure 4–2a–f) contains both ascending and descending fibers. The ascending fibers primarily consist of connections from the vestibular nuclei to the nuclei of cranial nerves III, IV, and VI. The MLF, in part, serves to coordinate eye movements with changes in the position of the head. This tract also extends through the medulla and into the cervical cord (where it occupies a position along the ventral median sulcus). Here it contains descending fibers from the medial vestibular nuclei and likely serves to allow postural adjustments of the head to assist in balance and/or to accommodate for changes in the position of the body.

Tectospinal Tract

This tract, which maintains a position just below the MLF both in the brainstem and in the cervical cord, probably also plays a role in the coordination of the movements of the head and the eyes (Figure 4–2b–f).

LESIONS OF THE BRAINSTEM

As previously noted, many ascending, descending, and cerebellar pathways, cranial nerves, and their nuclei, as well as other major motor, reticular, and other nuclei are located in the relatively small confines of the brainstem. It should not be surprising, therefore, that a single, discrete lesion can produce quite severe and mixed deficits. Brainstem lesions generally are distinguishable by signs of cranial nerve involvement, in combination with long-tract findings (sensory or motor) and/or cerebellar symptoms. The effects of lesions to the corticospinal, spinothalamic, and cerebellar pathways were reviewed in the first three chapters. Signs of damage to specific cranial nerves or their nuclei will become more obvious after the clinical anatomy of these structures is discussed in greater detail in Chapter 5.

Before reviewing a few of the more common brainstem syndromes, it might be helpful to review a few basic anatomical phenomena:

1. Motor tracts to the upper and lower extremities do not cross until they reach the lower medulla.
2. Ascending somatosensory fibers are crossed by the time they reach the middle medulla.
3. Ventral and lateral spinothalamic tracts that carry pain and temperature information occupy a more lateral position in the medulla and pons than the medial lemniscus (which mediates stereognosis and position sense).
4. The cerebellum has an ipsilateral influence on the spinal cord and most of the ascending spinocerebellar tracts are ipsilateral.
5. With one exception (cranial nerve IV), the cranial nerves and their nuclei exert an ipsilateral effect on the muscles and sensations of the head and face.
6. Certain patterns of deficits are more consistent with cranial nerve (or cranial nerve nucleus) involvement rather than supranuclear lesions (i.e., damage to the sensorimotor cortex or corticobulbar tracts). Examples of LMN lesions include isolated involvement of the intrinsic or extraocular muscles of the eye, unilateral weakness or atrophy of the tongue; unilateral facial weakness that includes the muscles of the forehead (see Chapter 5 for explanations).

At this point, we can review some of the more common syndromes associated with lesions of certain parts of the brainstem. However, it should be kept in mind that lesions may not respect the anatomical boundaries described below.

Lesions of the Midbrain (Mesencephalon)

Basal or Anterior Lesions

A unilateral lesion that affects the ventral portion of the mesencephalon (Figure 4–7a) will likely involve the cerebral peduncles (including the corticospinal and corticobulbar tracts), and thus may result in a complete or partial contralateral hemiparesis or hemiplegia without accompanying sensory disturbances. Because the oculomotor nerve (III) exits the midbrain anteriorly, it too is likely to be affected and will result in a third nerve palsy on the same side as the lesion. Signs of third nerve involvement may include dilated pupil, ptosis (weakness and partial closure of the eyelid), and difficulty looking up, down, or toward the midline in the affected eye. This pattern of deficits is referred to as **Weber's syndrome**.

Tegmental Lesions

If a unilateral lesion is confined to the middle or tegmental region of the midbrain (Figure 4–7b), a different syndrome will result, although the exact pattern of deficits depends on the particular placement and extent of the lesion. Critical structures in this area include the red nucleus, the medial lemniscus (and the more dorsally located spinothalamic tracts), the oculomotor (cranial nerve III) nerve complex, and crossing fibers from the superior cerebellar peduncle. Thus, in addition to signs of oculomotor nerve involvement (e.g., ipsilateral dilated pupil, ptosis, and restricted eye movement), one may find contralateral face and hemibody sensory symptoms. The latter might include decreased stereognosis, position, and vibratory sense (medial lemniscus), along with diminished perception of light touch, pain, and temperature (spinothalamic and trigeminothalamic tracts). Involvement of the red nucleus and ascending, crossing fibers from the superior cerebellar peduncle may result in ataxia and tremors on the contralateral side. Recall that the descending efferent pathways from the red nucleus also cross the midline immediately after exiting the nucleus. Likewise, the now

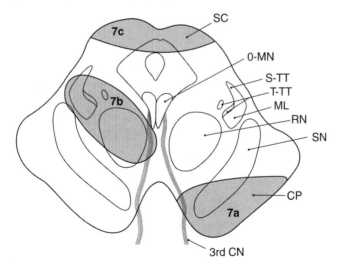

Figure 4–7. Lesions of the (a) ventral (basis), (b) central (tegmentum), and (c) dorsal (tectum) portions of the midbrain. (Adapted from Gilman and Newman, 1992).

3rd CN oculomotor nerve   SC superior colliculus
CP cerebral peduncle   SN substantia nigra
ML medial lemniscus   S-TT spinothalamic tract
O-MN oculomotor nucleus   T-TT trigeminothalamic tract
RN red nucleus

crossed ascending cerebellar fibers proceed to the ipsilateral cortex, which in turn influences the contralateral side of the body. This combination of unilateral third nerve palsy with contralateral sensory loss, ataxia, and tremor is referred to as **Benedikt's syndrome**.

Posterior Lesions

Lesions that affect the rostral portion of the dorsal mesencephalon (Figure 4–7c) basically affect the superior colliculi and connecting fibers to the pretectal area and to the Edinger–Westphal nuclei (rostral part of the oculomotor nuclear complex that is responsible for pupillary constriction). Such lesions typically will result in disturbances of conjugate upward gaze and pupillary changes (the pupils will react sluggishly to light but will constrict on accommodation). This disorder is known as **Parinaud's syndrome**. Massive or bilateral lesions severely damaging the mesencephalic reticular formation can result in a state of deep and often irreversible coma.

Lesions of the Pons

Basal Lesions

Lesions affecting the ventral portion of the pons likely will affect the descending corticospinal and corticobulbar tracts resulting in a contralateral hemiparesis or hemiplegia, including the muscles of expression of the lower face (Figure 4–8a). A more expansive lesion in the pons may affect the trigeminal nerve, resulting in ipsilateral sensory losses in the face, including an absent or diminished corneal reflex and ipsilateral weakness of the lower jaw (muscles of mastication). More caudal lesions in this region of the pons may affect the sixth cranial nerve that emerges from the anterior or ventral pontomedullary junction, producing lateral gaze palsy (lateral rectus muscle) in the eye ipsilateral to the lesion (Figure 4–8b). If this lesion extends laterally, it may include the seventh cranial nerve, resulting in ipsilateral facial weakness, which would include both the upper and lower face (i.e., a lower motor neuron syndrome).

Tegmental Lesions

If the basal lesion extends into the central tegmental area of the pons (Figure 4–8c), it may affect the medial lemniscus, producing contralateral loss of stereognosis and position sense in the extremities in addition to the contralateral weakness. As the spinothalamic tracts are more laterally localized, the perception of pain and temperature may be spared in the otherwise affected extremities. These functions also can be affected if the lesion extends more laterally in the pontine tegmentum.

In the pontine tegmentum, there is an important center for horizontal gaze. If one desires to look to the left, for example, two actions must take place: (1) the left eye turns outward (abducts), and (2) the right eye turns toward the midline (adducts). As will be discussed in greater detail in Chapter 5, the first action is mediated by the sixth cranial nerve (the abducens: abduction) through the innervation of the lateral rectus muscle of the left eye, while the adduction (looking medially) of the right eye occurs as a result of the simultaneous contraction of the medial rectus muscle via the third cranial nerve. The coordination of these two muscles necessary for lateral conjugate gaze is carried out at the pontine level by the **paramedian pontine reticular formation (PPRF)**.[8] This latter nucleus, which also has ipsilateral connections to the abducens (VI) nucleus, sends fibers across the midline (along with those internuclear fibers from the abducens nucleus). These PPRF fibers then ascend in the MLF to the contralateral oculomotor (cranial nerve III) nucleus. Thus, if a lesion extends into the dorsal–medial aspect of the upper pons, it may interrupt these ascending fibers in the MLF, resulting in difficulties with conjugate gaze to the side opposite the lesion. In

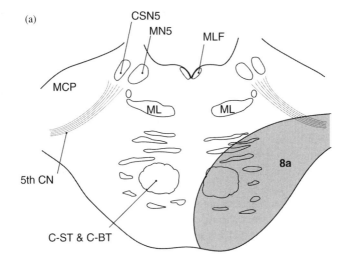

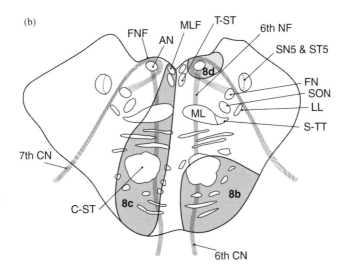

Figure 4–8. Lesions of (a)&(b) ventral, (c) central, and (d) dorsal pons. (Adapted from Gilman and Newman, 1992).

5th NF trigeminal nerve fibers
6th CN abducens nerve
6th NF abducens nerve fibers
7th CN facial nerve
AN abducens nucleus
C-BT corticobulbar tract
C-ST corticospinal tract
CSN5 chief sensory nucleus of the
   trigeminal nerve
FN facial nucleus
FNF facial nerve fibers

LL lateral lemniscus
MCP middle cerebellar peduncle
ML medial lemniscus
MLF medial longitudinal fasciculus
MN5 motor nucleus of the trigeminal nerve
SN5 spinal nucleus trigeminal nerve
ST5 spinal tract of the trigeminal nerve
SON superior olivary nucleus

S-TT spinothalamic tract
T-ST tectospinal tract

this case, the contralateral eye will abduct (but also will evidence nystagmus), while the ipsilateral eye will fail to adduct. However, if the lesion is lower in the pons and affects the PPRF and/or the abducens nucleus directly, the patient will have difficulties with lateral conjugate gaze to the same side as the lesion, involving both eyes. In this case, neither eye will be able to cross midline when attempting to look to the side of the lesion, a condition known as **lateral gaze palsy**.[9] As opposed to most other types of eye movement, difficulties with lateral conjugate gaze does not necessarily result from lesions to the brainstem or cranial nerves. Destructive cortical lesions that affect the frontal eye fields (area 8) may result in deficits of lateral conjugate gaze to the contralateral side. Thus, a destructive dorsolateral lesion in the left frontal may produce a preferential gaze to the left ("left gaze preference"), whereas a seizure focus in the same area would cause the eyes to shift to the right during the ictal event.

As illustrated in Figure 8d, a relatively small lesion of the dorsal–medial region of the caudal pons selectively may affect the nucleus of the sixth cranial nerve (abducens) and the fibers of cranial nerve VII (facial) as they loop around it before exiting from the ventrolateral pons. Such a patient would exhibit an ipsilateral sixth and seventh nerve palsy. A lesion to this region in the left pons would result in the inability to look laterally with the left eye and would produce a lower motor neuron facial syndrome (weakness of the muscles of facial expression, both upper and lower face, on the left side).

Acoustic Neuromas

These are tumors that develop on cranial nerve VIII (vestibulocochlear) as it leaves the pontomedullary junction at the so-called cerebellopontine angle. As this slowly growing tumor progresses, it first begins to compromise the functions of cranial nerve VIII, resulting in tinnitus (ringing in the ear), progressive deafness in that ear, and horizontal nystagmus due to vestibular involvement. As the lesion grows, it may affect cranial nerves V and VII, resulting in diminished pain and temperature and weakness on the face ipsilateral to the lesion. If it grows to sufficient size and impinges on the cerebellum, ipsilateral ataxia may be present.

Bilateral Lesions

A bilateral lesion in the basal portion of the lower pons can produce what is known as the "locked-in" syndrome. Because of the bilateral involvement of the corticospinal and remaining corticobulbar tracts, the patient is unable to move or speak, although remaining conscious. The only movements that remain by which the patient can communicate is blinking of the eyelids and vertical movements of the eyes (i.e., voluntary motor input to cranial nerves III and IV remain intact).

Lesions of the Medulla

Anterior Lesions

The anterior or ventral portion of the medulla contains the pyramids; thus, lesions involving this area of the medulla could result in contralateral weakness or paralysis of the extremities (Figure 4–9a).[10] The face should not be involved if the lesion is confined to the medulla as the facial fibers already have left the corticobulbar tracts. However, since the hypoglossal nerve (XII) exits from the ventrolateral sulcus, just lateral to the pyramidal tracts, lesions in this area often affect this nerve, resulting in ipsilateral weakness and atrophy of the tongue. If the lesion extends dorsally, it also will impact on the medial lemniscus, thus producing deficits of position sense, stereognosis, and vibratory perception in the contralateral extremities, while pain and temperature may be preserved.

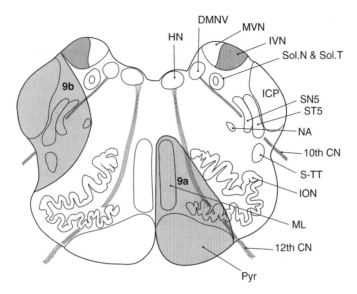

Figure 4–9. Lesions (a) ventral and (b) dorsolateral medulla. (Adapted from Gilman and Newman, 1992).

10th CN vagus nerve
12th CN hypoglossal nerve
DMNV dorsal motor nucleus of the vagus
HN hypoglossal nucleus
ICP inferior cerebellar peduncle
ION inferior olivary nucleus
IVN inferior vestibular nucleus
ML medial lemniscus

MVN medial vestibular nucleus
NA nucleus ambiguus
Pyr pyramids
SN5 spinal nucleus of the trigeminal nerve
Sol.N solitary nucleus
Sol.T. solitary tract
ST5 spinal tract of the trigeminal nerve
S-TT spinothalamic tract

Dorsolateral Lesions

Lesions that predominately affect the more lateral or dorsolateral portions of the medulla may produce what is known as a **Wallenberg's syndrome** (Figure 4–9b). Since the anterior, medial portions of the medulla are unaffected, motor strength (pyramidal tracts) position, stereognosis, and vibratory senses (medial lemniscus) are intact. However, because of the encroachment on the spinothalamic tracts, there is a contralateral loss of pain and temperature in the contralateral extremities. At the same time, there can be an ipsilateral decrease of pain and temperature in the face as a result of damage to the nucleus and tract of cranial nerve V. Since the nucleus ambiguus typically is involved in such a lesion, there is difficulty in swallowing and phonating (nucleus ambiguus supplies motor impulses to the soft palate, pharynx, and larynx). Damage to the inferior cerebellar peduncle results in ipsilateral ataxia and hypotonia. Involvement of the sympathetic pathways produces a **Horner's syndrome,** which includes a small (constricted) pupil, ptosis,[11] and a dry face (anhidrosis) ipsilateral to the lesion. Finally, there may be nystagmus if the vestibular nuclei are involved.

Endnotes

1. For example, the red nuclei, inferior olivary nuclei, lateral reticular nuclei, vestibular nuclei, and superior and inferior colliculi.

2. Such as the central tegmental tract, internal arcuate fibers, trapezoid body, the medial and lateral lemniscus, the medial and dorsal longitudinal fasciculus).

3. Substantia nigra (dopamine), raphe nuclei (serotonin), locus ceruleus (norepinephrine).

4. Which, in turn, project primarily to the cerebellum.

5. Dilation of the pupil is controlled by sympathetic fibers from the superior cervical ganglion, which also assists in elevating the eyelid. Thus ptosis (drooping of the eyelid) can result from damage to either parasympathetic (cranial nerve III) or sympathetic fibers. If the result of an oculomotor lesion, the ptosis will be present on the side of the dilated pupil (resulting from the unopposed action of the sympathetic system). If the ptosis is on the side in which the pupil remains constricted, the lesion likely involves the sympathetic system (see Horner's syndrome).

6. In addition to the locus ceruleus, there are only two other nuclear groups in the brainstem that are pigmented: the substantia nigra and the dorsal nucleus of cranial nerve X.

7. These two systems, along with the reticulospinal tract, among other functions, appear to mediate reflex changes in the extensor (antigravity) muscles of the lower extremity and help maintain postural integrity via adjustments of the muscles of the trunk, neck, and upper extremities.

8. This nucleus does not act independently. For example, it receives input from the frontal eye fields (Brodmann's area 8) for voluntary movements, from the superior colliculi that mediate reflex eye movements to external stimuli and from the vestibular nuclei to adjust to movements of the head and body. In addition, the adjacent, abducens nuclei themselves appear to be the source of crossing fibers to the oculomotor nuclei (cranial nerve III).

9. As will be seen in Chapter 5, if the lesion is confined to the sixth nerve itself (sparing the PPRF and the abducens nucleus), there will be an inability to abduct the eye on the side of the lesion, but the contralateral eye will adduct normally.

10. Lesions that are limited to the pyramids, drastically affecting motor abilities, typically do not result in total paralysis below the level of the lesion. The preserved motor functions are probably the result of other descending pathways such as the rubrospinal and vestibulospinal tracts.

11. Both cranial nerve III and the cervical segment of the sympathetic nervous system affect the muscles that elevate the eyelid. However, the main innervation of the levator palpebrae superioris is via the third cranial nerve. Hence, in Horner's syndrome, there is a partial ptosis, which can be overcome by voluntary action (of the third cranial nerve). With lesions involving the third cranial nerve directly, there is a more severe ptosis that cannot voluntarily be overcome.

REFERENCES AND SUGGESTED READINGS

Barr, M.L. & Kiernan, J.A. (1993) The Human Nervous System. Philadelphia: J.B. Lippincott Co., pp. 96–121; 148–162.

Brazis, P., Masdeu, J. & Biller, J. (1990) Localization in Clinical Neurology. Bostom: Little, Brown & Co., pp. 270–285.

Carpenter, M.B. & Sutin, J. (1983) Human Neuroanatomy. Baltimore: Williams & Wilkins., pp. 315–453.

Gilman, S. & Newman, S.W. (1992) Essentials of Clinical Neuroanatomy and Neurophysiology. Philadelphia: F.A. Davis Co., pp. 99–114; 143–152.

Haines, D.E. (Ed.) (1997) Fundamental Neuroscience. New York: Churchill Livingstone, pp. 143–188.

Martin, J.H. (1996) Neuroanatomy: Text and atlas. Stanford, CT: Appleton and Lange, pp. 353–416.

Nolte, J. (1993) The Human Brain. St. Louis: Mosby-Year Book, Inc., pp. 154–175.

5   THE CRANIAL NERVES

CHAPTER OVERVIEW

Of all the systems that a clinician might review, examination of the cranial nerves potentially can be the most accurate in terms of identifying the locus of a lesion in the CNS. As noted in Chapter 4, because of the relative compactness of the nuclei and various nerves and pathways within the brainstem, under certain circumstances it might be possible to localize a brainstem lesion to within 1 cm. Beyond this, the presence or absence of certain cranial nerve deficits might help the clinician differentiate whether certain sensory or motor deficits have their origin above or below the level of the brainstem. The two cranial nerves that lie totally outside the brainstem may help identify lesions lying within any of the four lobes of the brain.

Beyond and perhaps more important than their localizing potential the behavioral deficits associated with lesions of the cranial nerves can have a profound impact on patients' functional capacity. Whether as a direct result of the cranial nerve deficits themselves or as a function of presumed disruption of critical adjacent structures, such deficits may have vocational, social, emotional, or psychiatric implications.

Consistent with the goal of combining anatomical structure with functional relevance, a review of specific behavioral syndromes and their clinical assessment will accompany a review of the gross anatomical and functional correlates of each of the twelve cranial nerves. At the conclusion of this chapter, the reader should be able to identify each of the cranial nerves, their major functions, and how each might be tested as part of the overall clinical examination of a patient. Utilizing the information presented in this chapter, along with that reviewed in Chapters 1, 2 and 4, one hopefully should be able to derive tentative hypotheses regarding the probable level of a lesion producing sensory and/or motor deficits.

INTRODUCTION

A substantial part of the neurological examination is an assessment of the cranial nerves. In addition to providing valuable information regarding the integrity of the brainstem and/or the localization of lesions therein, the discovery of functional disturbances in one or more of the cranial nerves may herald the presence of other intracranial or extracranial pathology. To more fully understand the significance of this portion of the neurological examination, it is necessary to describe the major functional and anatomical features of each of these cranial nerves. However, before doing this a few general facts and characteristics of these nerves and their nuclei will be reviewed.

The cranial nerves are those paired sets of nerves whose constituent fibers enter (or exit) the central nervous system above the level of the foramen magnum, that is, at the level of the brainstem or above (see Figure 5–1). In fact, of the 12 pairs of cranial nerves (CN), all but two—the olfactory (CN I) and optic (CN II)—have their nuclei in the brainstem. The oculomotor (CN III) and trochlear (CN IV) exit from the midbrain. The trigeminal (CN V) enters and leaves at the pontine level. Three of the nerves—the abducens (CN VI), facial (CN VII), and vestibulocochlear (CN VIII)—are found at the pontomedullary junction. The glossopharyngeal (CN IX), vagus (CN X), and hypoglossal (CN XII) are located farther down the medulla. Finally, the spinal accessory (CN XI) has cells of origin both in the medulla and cervical cord. Of the cranial nerves whose nuclei are in the brainstem, all except one (CN IV) are found on the ventral or ventrolateral aspect of the stem (the fourth cranial nerve exits from the dorsal portion of the midbrain). It might be noted that CN IV also is unique in another respect: it is the only nerve to fully cross the midline after leaving its nucleus.

Unlike the spinal nerves that are either sensory or motor, some of the cranial nerves are purely motor (CN III, IV, VI, XI, XII), others are purely sensory (CN I, II, VIII), while others

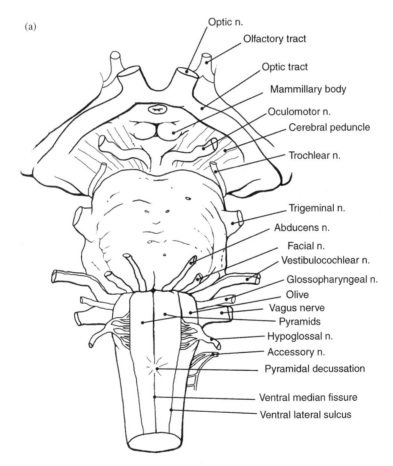

Figure 5–1. (a) Anterior, (b) lateral, and (c) posterior views of brainstem and hypothalamus showing relative location of cranial nerves. Optic nerve exits from the eye, forming optic tract after partial decussation in the optic chiasm. Olfactory nerves synapse in the olfactory bulb (not shown) and second-order fibers form the olfactory tract (see Figure 5–3).

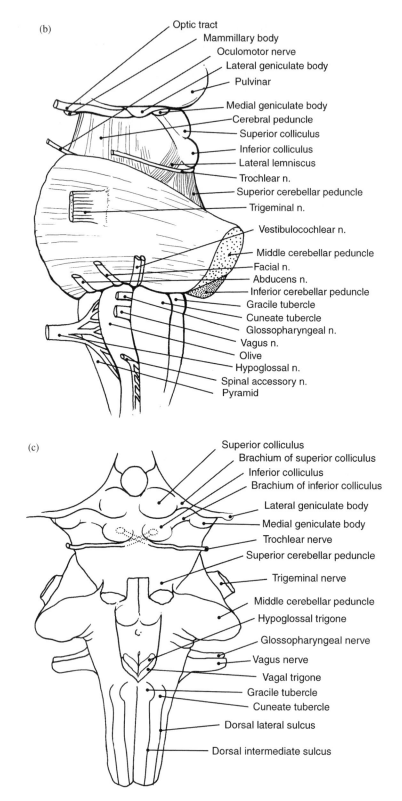

Figure 5–2. *(Continued)*

contain both motor and sensory components (CN V, VII, IX, X). Several of the brainstem nuclei are shared by more than one cranial nerve. For example, the spinal nucleus of the trigeminal nerve is shared by CN V, VI, IX, and X. Another conventional although somewhat more complex way to distinguish individual nerves or their components is based on whether they serve **general somatic functions** or **general visceral functions**. Since these can have either afferent (sensory) or efferent (motor) components, four classifications can be readily described:

1. General somatic efferents (GSE): (CN III, IV, VI, XII)
2. General visceral efferents (GVE): (CN III, VII, IX, X)
3. General somatic afferents (GSA): (CN V, VII, IX, X)
4. General visceral afferents (GVA): (CN IX, X)

The above classifications would apply equally to the spinal nerves. However, the cranial nerves are also responsible for certain special functions that are not found in the spinal nerves. Hence, we have certain "special" cranial nerves or components of those nerves. These are:

5. Special somatic afferents (SSA): (CN II, VIII)
6. Special visceral Aafferents (SVA): (CN I, VII, IX, X)
7. Special visceral efferents (SVE): (CN V, VII, IX, X, XI)[1]

Translating these classifications into more clinically meaningful or descriptive terms, we would have the following:

1. Special somatic afferents (SSA): sensory functions unique to the cranial nerves: vision (CN II), hearing and balance/equilibrium (CN VIII).
2. Special visceral afferents (SVA): taste (CN VII, IX, X) and smell (CN I).
3. General somatic afferent (GSA): perception of touch, pain, temperature, vibration, proprioception, and stereognosis (CN V, VII, IX, X).
4. General visceral afferent (GVA): feedback from such structures as the mucous membranes of the nasal and oral cavities, thoracic and abdominal organs, and the carotid sinus (CN IX, X).
5. General somatic efferents (GSE): control over the extrinsic muscles (movement) of the eyes (CN III, IV, VI), and movement of the tongue (CN XII).
6. General visceral efferents (GVE): control the constriction of the pupils and changes in the shape of the lens of the eye (CN III), lacrimal (CN VII), and salivary (CN VII, IX) glands, and the muscles related to visceral organs such as the heart and diaphragm (CN X).
7. Special visceral efferents (SVE): control over "special" visceral or "brachial" motor functions such as chewing (CN V), facial expressions and tension on the stapedius muscle (CN VII), and tympanic membrane (CN V); the muscles of the larynx and pharynx used in speaking (CN IX, X); and turning of the head (sternocleidomastoid) and shrugging of the shoulders (trapezius) (CN XI).

Following the embryological development of the central nervous system, the somatic and visceral motor nuclei (which develop from the basal plate) tend to be more medially placed in the brainstem tegmentum. In contrast, the brainstem nuclei that represent somatic and visceral afferents (evolving from the alar plate) tend to be more laterally situated. As can be seen in Figure 5–2, the somatic motor nuclei tend to be placed most medially and the somatic sensory nuclei most laterally of the four, with the visceral motor and sensory

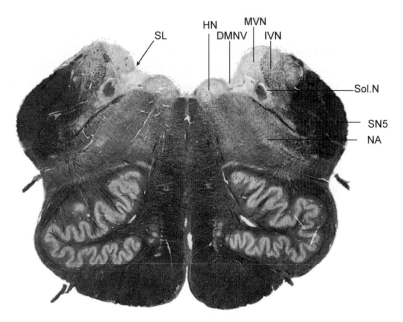

Figure 5–2. Section through the upper medulla representative of the relative arrangement of sensory and motor nuclei in the brainstem. As can be seen in the figure, the sensory nuclei are located lateral and the motor nuclei medial to the sulcus limitans. This reflects a continuation of the dorsal–ventral arrangement seen in the spinal cord as the cord "opens up," forming the brainstem. Abbreviations: DMNV, dorsal motor nucleus of the trigeminal nerve; HN, hypoglossal nucleus; IVN, inferior vestibular nucleus; NA, nucleus ambiguus; MVN, medial vestibular nucleus; SL, sulcus limitans; SN5, spinal nucleus of the trigeminal nerve; Sol.N, solitary nucleus. Brain image was adapted from the *Interactive Brain Atlas* (1994), courtesy of the University of Washington.

occupying the more intermediate positions. There is one notable exception: the special visceral efferent nuclei, for example, the facial nucleus and the nucleus ambiguus have "migrated" to occupy a more ventral and slightly lateral position in the tegmentum of the medulla.

Finally, before proceeding to a more detailed discussion of the individual cranial nerves, it may be useful to commit the names of cranial nerves to memory. One classic mnemonic phrase for recalling the names of the cranial nerves is as follows:

"On Old Olympus's Towering Tops A Fair Virtuous Girl Vends Snowy Hops"
  1   2   3   4   5   6   7   8   9   10   11   12

1. **Olfactory**	5. **Trigeminal**	9. **Glossopharyngeal**
2. **Optic**	6. **Abducens**	10. **Vagus**
3. **Oculomotor**	7. **Facial**	11. **Spinal accessory**
4. **Trochlear**	8. **Vestibulocochlear**	12. **Hypoglossal**

Table 5–1 summarizes the major functions mediated by these nerves. **Note:** The spinal accessory nerve is now more commonly known simply as the accessory nerve. Unfortunately, I am not aware of any mnemonic for remembering all the functions of each, that is simply going to require some good old memorization.

Table 5–1. Summary of Cranial Nerves

Nerve	Nuclei of Origin or Termination	Function	Deficit
I Olfactory	SVA Medial temporal, medial frontal cortical areas	Smell	Unilat. anosmia
II Optic	SSA Lat geniculates	Vision	Visual loss (nerve) Field cut (tract) ↓ Optic reflexes
III Oculomotor	GSE Oculomotor n.	Eye movements (Sup Inf & Med rectus, and Inf oblique); Elevate lid	Diff. looking up, down, medially Diplopia Ptosis
	GVE Edinger–Westphal	Constrict pupil; Adjust lens.	Dilated pupil Blurred near vision
IV Trochlear	GSE Trochlear n.	Eye movement (Sup oblique)	Diff. looking down when adducted
V Trigeminal	GSA Main sensory n.	Discrimination Proprioception	↓ Discrim. and prop. of facial m.
	GSA Spinal sensory	Touch, pain & temperature; face, cornea, teeth, tongue (ant. 2/3rds)	↓ Touch, pain & temperature on face & tongue; ↓ Corneal reflex; Neuralgia
	GSA Mesencephalic n.	Proprioception: (jaw muscles, eye movements)	↓ Jaw jerk; loss of sensory limb of reflex arc
	SVE Motor n. of V	Muscles of mastication, tensor tympani	Jaw weak, (deviates to side of lesion) ↓ Jaw jerk; (motor limb)
VI Abducens	GSE Abducens n.	Abducts the eye (Lateral rectus)	Nerve: ↓ Abduction; Diplopia Nucl: Lateral gaze palsy; Diplopia
VII Facial	SVE Facial n.	Facial expression; Closes eyelid Stapedius m.	Facial weakness; Diff. closing eyelid Hyperacousis
	GVE Sup salivatory	Salivary and lacrimal glands	Dry eye
	SVA Solitary n.	Taste (anterior 2/3 of tongue)	↓ Taste (anterior tongue)
	GSA Spinal n. of V	Tactile (ear)	
VIII Vestibulo-cochlear	SSA Vestibular	Equilibrium	Vertigo, dizziness
	SSA Cochlear	Hearing	Loss of hearing, Tinnitus

(Continued)

Table 5–1. *(Continued)*

Nerve	Nuclei of Origin or Termination	Function	Deficit
IX Glossopha-ryngeal	SVE N. ambiguus	M. of pharynx, stylopharyngeus	(see CN X)
	GVE Inf Salivatory	Parotid gland	
	SVA Solitary n. (rostral)	Taste, post. 1/3 of tongue	Taste on (back of tongue)
	GVA Solitary n. (caudal)	Visceral sensation: Carotid body, carotid sinus	
	GSA Spinal n. of V	Tactile sensation: Posterior tongue, upper pharynx, middle ear	Afferent side of gag reflex
X Vagus	SVE N. ambiguus	Muscles of soft palate, pharynx & larynx	Hoarseness, Dysphagia (mild) ↓ Gag reflex ↓ Elev. of palate
	GVE Dorsal Motor n.	Abdominal & thoracic viscera	Parasympathetic control
	GSA Spinal n. of V	Tactile: ear, pharynx	
	SVA Solitary n.	Taste: epiglottis	
	GVA Solitary n.	Sensations from viscera, larynx, trachea	
XI Accessory	SVE Accessory n.	Sternocleidomastoid, trapezius	Turning head against resistance Shrug shoulder
XII Hypog-lossal	GSE Hypoglossal n.	M. of tongue	Atrophy, weakness of tongue

Notes: GSA connections are presumed to be present in III, IV, VI, XI, and XII for proprioceptive feedback, but pathways not established.

Not all authors agree on all designations and assignments presented above, but the data presented in the table appear most representative.

Clinically, IX and X are difficult to separate; hence are often combined, e.g., IX mediating the sensory & X the motor side of the gag reflex.

CRANIAL NERVE I (OLFACTORY)

Major Function: Sense of Smell
Classification: Special Visceral Afferent (SVA)

The olfactory nerve (CN I) is somewhat unique in several respects. First, like the optic nerve (CN II), it does not have a nucleus within the brainstem. In fact, for this reason it has been suggested that perhaps it is not truly a "cranial nerve," but simply represents a fiber tract of the brain. Although other sensory systems may respond to chemical stimulation or biochemical changes within the body, olfaction along with taste represent the two major sensory systems through which we are consciously informed about our environment via chemoreceptors. Unlike the other sensory systems that project first to the thalamus and then to the cortex, the nerve fibers carrying olfactory information project initially and primarily to some of the more primitive parts of the limbic system in the basal frontal and mesial temporal cortex.

It is these rather direct connections with the limbic system that would appear to account for another somewhat unique feature of the olfactory system, namely, that *emotional or affective valences are often associated with the sense of smell.* Any stimulus, regardless of the modality of input, can elicit certain affectively charged responses through learning or experience. However, the senses of smell and to some extent taste have a more direct innate association with emotion and affective responses. Furthermore, these associations very commonly involve behaviors that are basic to survival, either of the individual or the species, for example, feeding, procreation, or defense. Examples range from the feeding frenzy that can be elicited by the smell of blood in the water by the shark, to the panic of herding animals elicited by the smell of a mountain lion, to the marking of territorial boundaries with urine by the wolf. Probably more common to our experience is the "sexual frenzy" witnessed in a male dog when there is a female dog in heat anywhere in the neighborhood. In comparison to most of our mammalian ancestors, humans have a poorly developed sense of smell, although we too evidence a close association between smell and strong, primitive drive states. Regardless of past personal experiences, almost everyone is repelled by putrid odors and attracted to the fragrance of the rose or a freshly baked apple pie. Certain odors, both manufactured and natural, manifest their ability to arouse sexual feelings. In order to have maximal survival value, it is important that such affective valences, be they positive or negative, be well retained in memory. Most of us likely have had the experience of encountering an odor that immediately triggers a host of long-forgotten memories, typically affectively tinged and usually in a manner that is quite distinct from the type of memory elicited by other stimulus modalities. Again, it probably is no coincidence that the limbic system is very closely and intimately connected with the mesial temporal and other areas so critical for memory. We will revisit these topics in Chapter 8, which covers the limbic system, but this brief preview may help explain the connections of the olfactory system to which we will now proceed.

Anatomy

The end organs of the olfactory system are the epithelial cells in the upper part of the nasal mucosa. The primary neurons of these epithelial cells pass through the cribriform plate of the ethmoid bone to the ipsilateral olfactory bulb where they make synaptic connections with second-order neurons (Figure 5–3). The two olfactory bulbs and the anterior portion of the olfactory tracts lie parallel to each other on the basal–medial aspect of the frontal lobes. The nerve cells in the bulbs (primarily mitral cells) give rise to the olfactory tracts. The positioning of the bulbs (and their tracts) and their connection with the cells of the nasal cavity makes this system particularly sensitive to certain types of tumors and trauma as will be discussed below.

As the olfactory tracts proceed posteriorly, one smaller pathway branches off the main olfactory tract, sending fibers via interneurons to the opposite bulb. The main part of the tract then proceeds as the **lateral olfactory stria** (Figure 5–4a).[2] The fibers of the lateral olfactory stria terminate in the anteromedial portion of the temporal lobe, an area that in general is referred to as the primary olfactory cortex. While there are some inconsistencies among different authors as to the exact region(s) that receives these direct inputs, the basal frontal cortex in the region of the anterior perforated substance, corticomedial portions of the amygdaloid complex, the prepyriform and pyriform cortex, and the more rostral portions of the entorhinal cortex most commonly are cited (Figure 5–4b).

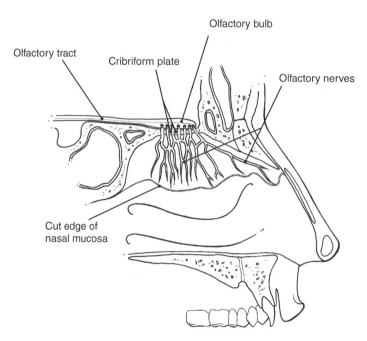

Figure 5–3. "Primary" (nerves) and "secondary" (tract) olfactory neurons synapsing in the bulb.

What is important to remember is that from these sites of primary projection substantial secondary connections are made with other limbic structures in the immediate area, including the lateral portions of the amygdala and hypothalamus (via the medial forebrain bundle). Additional indirect connections also are made with the thalamus, brainstem, and hippocampal formation. Thus, as was suggested above, the olfactory system is intimately connected with some of the more primitive brain structures, which in turn are involved in a variety of emotionally charged or affectively driven behaviors that are often critical for survival.

Because of the location of the olfactory bulbs at the base of the orbital frontal cortex the sense of smell is particularly vulnerable to lesions affecting the base of the frontal lobes. One type of lesion in particular deserves special mention: meningiomas. These very slow-growing tumors often originate from the falx cerebri (the dura lying between the two cerebral hemispheres) or from the base of the skull in the basal frontal regions and may reach considerable size before being detected clinically. However, as they begin to gradually enlarge they may impinge on one of the olfactory bulbs and/or tracts, disrupting their function. Thus, unilateral loss of the sense of smell, without other peripheral explanations (e.g., clogged nasal passage), may be the first clinical manifestation of such a lesion. Aneurysms in the area of the anterior communicating artery similarly could produce such a mass effect. While bleeds in this area also might compromise the olfactory system, other symptoms typically would predominate such as headache and possible mental status changes. Closed-head injury (CHI), particularly from decelerating types of injuries, also can result in anosmia (loss of smell). In this case, the mechanics of the pathology is different. As the brain continues to move forward relative to the skull, the nasal nerve fibers penetrating the cribriform on their way to the olfactory bulb may suffer the shearing effect of this relative motion. Contusions of orbitofrontal cortex also may occur following CHI, resulting in various other clinical syndromes, including behavioral disinhibition and seizures.

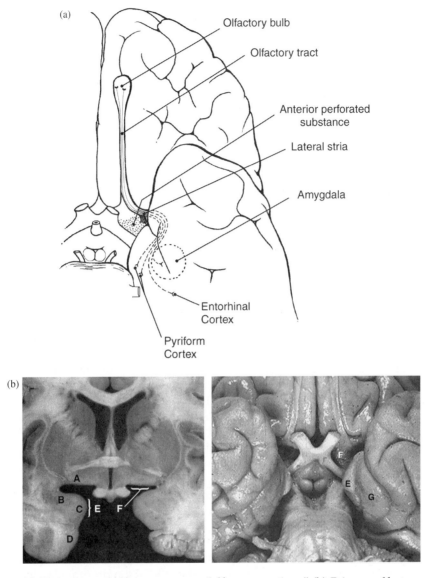

Figure 5–4. (a) Projections of olfactory tract to "olfactory cortices." (b) Primary olfactory projection areas. Abbreviations: A, basal nuclei; B, corticomedial amygdaloid nuclei; C, pyriform cortex; D, entorhinal cortex; E, uncus; F, anterior perforated substance; G, parahippocampal gyrus. **Note:** The basal nuclei, pyriform cortex, and entorhinal cortex, respectively, underlie anterior perforated substance and the uncus, which are essentially surface features. The entorhinal cortex is contained within the parahippocampal gyrus. Brain images were adapted from the *Interactive Brain Atlas* (1994), courtesy of the University of Washington.

Testing for Dysfunction

If it were not for the perception of taste, the assessment of the integrity of the olfactory nerve undoubtedly would be the most frequently neglected aspect of most routine neurological exams. The exam itself is simple enough; the problem arises (as with the assessment of taste) in the easy accessibility of adequate stimuli. A few years back when the "Scratch and Sniff" stickers were a big hit with kids, a ready and convenient supply of "odors" was easy to obtain. Some neurologists may carry small vials of aromatic substances and routinely assess the patient's sense of smell, but often this portion of the examination is omitted. Systematic examination for anosmia is more likely to occur when specific orbital frontal lesions are suspected. Testing typically entails having the patient smell a variety of easily identifiable substances or essences. Astringent substances such as ammonia, vinegar, or rubbing alcohol should be avoided as these tend to stimulate the trigeminal (CN V) nerve, rather than the olfactory (CN I). Each nostril should be tested independently, since as was noted above in the case of mass lesions only one bulb or tract may be affected. The fact that the patient may not be able to name the substance is not necessarily a pathological sign (he or she may not be quite able to place the smell or may have an aphasic or naming problem). What may be more critical to determine is whether the substance can be identified as having a distinct aroma or smell and differentiated from other substances. Before a diagnosis is made of either a unilateral or bilateral deficit, care should be taken to note whether the patient's nostrils are clear and uncongested.[3]

CRANIAL NERVE II (OPTIC)

Major Function: Sense of Vision
Classification: Special Somatic Afferent (SSA)

If to the poet, the eyes are "windows to the soul," then to the neurologist they also might be thought of as "windows to the nervous system." Through an examination of the eyes, it is possible to assess the integrity of numerous portions of the central and peripheral nervous system. This includes cortical and subcortical sites and pathways, portions of the brainstem and cerebellum, peripheral (cranial) nerves, and the autonomic nervous system, including spinal components (e.g., sympathetic fibers from the superior cervical ganglion that are responsible for dilating the pupil and assist in elevating the eyelid). While cranial nerve II relates only to vision and the afferent link of certain visual reflexes (e.g., blink, light, tracking, and accommodation), the visual pathways extend from the eyes to the thalamus (with some branching to the mesencephalon) and eventually to the posterior cortex. This extensive network creates ample opportunity for a variety of cortical or subcortical lesions to affect the visual system. In addition, movements of the eyes themselves are mediated by yet other parts of the CNS. These include such regions or structures as the frontal eye fields (Brodmann's area 8), cranial nerves III, IV, and VI, as well as other brainstem nuclei and pathways (e.g., the superior colliculi, vestibular nuclei, vestibulocerebellar pathways, the paramedian pontine reticular formation, and the medial longitudinal fasciculus). Cranial nerve V and VII likewise are involved in the tactile afferent and motoric aspects of blinking, respectively. Parasympathetic fibers of cranial nerve III (via the Edinger–Westphal nuclei) are responsible for the constriction of the pupil, while the sympathetic fibers tend to dilate it. Cranial nerve III (with contributions from the sympathetic system) elevates the eyelids, while cranial nerve VII closes the eyelids. Finally, other structures of the midbrain, including the pretectal areas and superior colliculi, also are involved in some of the above-mentioned visual reflexes. In this section, however, the primary emphasis will be on the optic nerves themselves, their pathways, and their influence on vision and certain visual reflexes. Eye movements and other related ocular phenomena (including control of the eyelids) will be covered in greater detail as the relevant cranial nerves are discussed.

Anatomy

As with the olfactory nerve, it has been suggested that the optic nerve perhaps should not be considered truly a nerve, but rather a fiber tract of the brain. However, as with the former, convention usually prevails and it will most likely continue to be referred to as cranial nerve II. The end organs for the optic nerve are the photoreceptive cells of the retina, the rods, and cones. The **rods**, which are found with relative greater frequency in the peripheral parts of the retina, are better adapted to low-intensity light, such as night vision. The **cones**, which tend to be concentrated in the **macula,** particularly in the center of the macula or **fovea**, are better adapted to the perception of color and the sharper point-to-point vision necessary for fine discriminations (foveal vision). While other cells (horizontal cells) serve as interneurons, it is basically the rods and cones that transmit visual input to the bipolar cells that lie within the retina. These in turn synapse with the ganglion cells also within the retina. The axons of these ganglion cells exit the eye through the optic disk to form the optic nerve. Visualization of the optic disk, which also contains blood vessels entering and exiting the globe, is the closest one can come to directly seeing the nervous system in the intact patient without intrusive or radiographic procedures. In fact, well before the advent of the CT scan or MRI, the presence of **papilledema** (a swelling or "choking" of the optic disk) was used as a sign of increased intracranial pressure, such as might be present with mass effect lesions or inflammatory conditions of the brain.

Before proceeding to trace the optic pathways and identify specific syndromes produced by various lesions impacting on different parts of these pathways, it is essential to first understand some properties of the retinal image. As can be seen from the accompanying diagrams, as the light rays enter the pupil and fall on the retina they cross. As shown in the two figures below, it is important to remember that the light rays cross along both the horizontal and vertical planes. Visual input representing light rays striking the left half of each eye (the temporal field of the left eye and the nasal field of the right eye) come from the right external hemispace. The reverse is true of images that lie in the observer's left visual field (Figure 5–5). In addition to the reversal of the visual field from left to right along the vertical midline as the image proceeds from the visual field to the retinal field, there is a reversal of superior–inferior along the horizontal axis. That is, input from the superior portion of the external visual field projects to the inferior parts of the retina (Figure 5–6). This organization continues in this manner to the calcarine cortex, so that the right and left as well as the superior and inferior visual fields are "reversed" as they project to the visual cortex.[4]

After converging on the bipolar and then the ganglionic cells in the retina, the visual fibers exit the eye as the **optic nerve**. The two optic nerves (each of which carries information from both the nasal and temporal halves of the eye from which it emanated) proceed to the **optic chiasm**. Upon reaching the chiasm (or "crossing"), which is located above the pituitary gland along the ventral base of the frontal lobes, those fibers in each nerve that represent the visual receptors in the nasal half of each eye decussate, while the temporal fibers of each eye remain on the same side. Thus, in the chiasm, the uncrossed temporal fibers are in the lateral or outside portions of the chiasm, with the crossed nasal fibers from both eyes making up the medial or central portion of the chiasm. This is important to keep in mind when considering the possible effects of lesions affecting the chiasm itself. Because of their position at the base of the frontal lobes, the optic nerves, the optic chiasm, and the optic tracts are particularly vulnerable to encroachment by meningiomas or pituitary tumors growing in this region.

Upon exiting the chiasm, these postganglionic visual pathways are now referred to as the **optic tracts**. However, after the partial crossing that took place in the chiasm, each optic tract now consists of fibers originating from both eyes. The left optic tract is composed of the temporal fibers of the left eye and the nasal fibers of the right eye, while the right optic tract is formed from the nerve processes emanating from the temporal half of the right eye and the nasal half of the left eye. This arrangement basically continues throughout the remainder of the visual pathways all the way to the cortex. This organization enables the

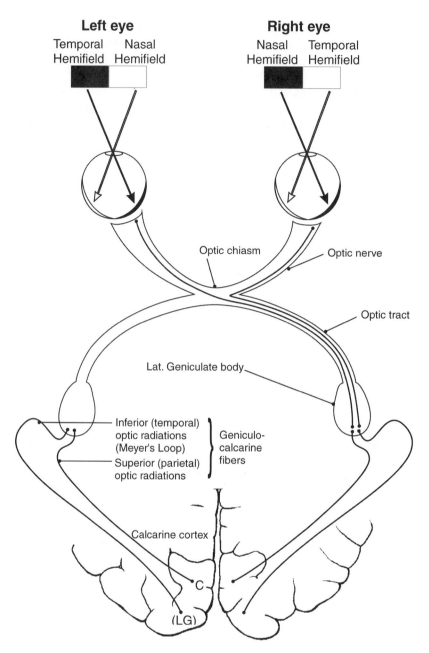

Figure 5–5. Left visual field projections.

right side of the brain to initially process information coming in from the left visual field and the left brain to respond to right visual field stimuli.

The bulk of the fibers of the optic tracts then proceed to the **lateral geniculate nuclei** of the thalamus (located just dorsal and lateral to the rostral midbrain) where they are arranged in a very precise, retinotopic manner. A few collateral fibers exit from the optic tract shortly before the lateral geniculates and proceed to the pretectal area and to the superior colliculi. The former (pretectal) connections, which synapse with the

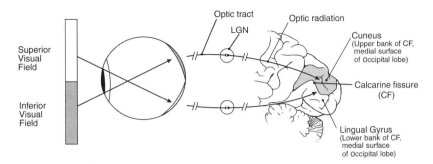

Figure 5–6. Superior–Inferior visual field projections.

Edinger–Westphal nuclei, are thought to be important for simple and consensual pupillary reflexes to light, and the latter (to the superior colliculi) for oculoskeletal reflexes (e.g., orienting responses to visual or auditory stimuli) and visual tracking, especially for novel stimuli. Additional visual reflexes such as fixating on a moving object (or stationary object if the subject is moving), convergence, accommodation, or pupillary changes to psychologically meaningful stimuli, likely are in large part mediated by pathways originating in the occipital cortex.

The lateral geniculate nuclei in turn give rise to the **optic radiations** or *geniculocalarine tracts* that project to the calcarine cortex (Brodmann's area 17) in the medial aspect of the occipital poles. It is here in the calcarine cortex that the visual stimuli begin to be processed into conscious perceptions (additional discussions on the topic of visual perception can be found in Chapter 9, this volume). As the fiber tracts that make up the optic radiations leave the lateral geniculate nuclei via the retrolenticular and sublenticular portions of the internal capsule, they become organized into superior (parietal) and inferior (temporal) pathways. The inferior fibers loop forward into the temporal lobe (Meyer's loop) before turning back toward their destination in the occipital cortex. This is particularly noticeable in those that proceed from the more lateral aspects of the nucleus. As the optic radiations fan out, these latter fibers maintain a more ventral position and synapse in the more ventral portion of the occipital cortex: the lingual gyrus that lies below the calcarine fissure. The fibers from the more medial aspects of the lateral geniculate nuclei maintain a more dorsal position, ending up in the cuneus, which is the gyrus above the calcarine fissure. These more dorsal fibers (to the cuneus in area 17) consist of third-order fibers whose antecedents originated in the dorsal halves of each retina, with the ventral fibers (and lingual gyrus) being related to the ventral parts of each retina. As a consequence of this arrangement, a lesion that might affect, for example, the more temporal–inferior (ventral) portions of the left optic radiations (or the left lingual gyrus) will produce a right superior quadrant defect (right upper quadrant field cut). Conversely, a more dorsal lesion of the superior optic radiations (e.g., deep parietal) or one involving the cuneus will result in a inferior field cut or quadrantanopia on the side opposite the lesion.

Testing for Dysfunction

Testing the integrity of the optic nerve and its central pathways involves three major areas of inquiry: (1) visual acuity or perception, (2) examination of the patient's visual fields, and (3) assessing the integrity of light reflexes. **Reduced visual acuity** can result from damage to the visual pathways, but caution must be observed as it often results from peripheral problems in the eye itself, including diseases affecting the retina, lens, cornea, or vitreous humor, or may simply reflect a refractory problem. Problems of visual perception, that is, not simply the ability to see but the ability to "recognize" what one sees (visual agnosias) typically do not result from lesions of the primary visual pathways. Rather, problems of **visual perception** generally are thought to result from lesions affecting the visual association cortices or a disconnection of the primary visual cortical areas from these association areas (see Chapter 9). In addition to their possible direct diagnostic significance, a gross assessment of basic acuity and perception (as well as visual neglect: see below) should be an integral part of any comprehensive mental status examination, since these examinations often require attention to and interpretation of visually presented stimuli.

Assessment of the patient's visual fields is a standard and important part of both neurological and neuropsychological examinations. However, before discussing visual field exams, two additional diagnostic complications should be considered. On occasion, patients may report to be "blind" or experience "tunnel vision" when in fact their vision is neurologically intact (e.g., conversion reaction). Demonstrating the presence of optokinetic nystagmus may support a diagnosis of hysterical blindness. Under proper conditions, individuals with intact vision reflexively will respond to a vertically striped stimulus passed in front of them. In these circumstances, the eyes will slowly track in the direction that the stimulus is moving and then alternately quickly shift in the opposite direction. The ability of the patient to adequately navigate their environment suggests the absence of significant tunnel vision on a neurological basis. An opposite problem is that patients can report normal vision when in fact they are totally or nearly totally blind (**Anton's syndrome**). This syndrome is most commonly the result of vascular lesions involving the calcarine cortex bilaterally (basilar artery or both posterior cerebral arteries), although it can result from other etiologies. Since the denial of cortical blindness often is associated with other cognitive deficits, including disturbances of memory, disorientation, or mental confusion, extension of the lesion to the medial temporal cortices and/or the thalamus is suspected in most cases. The presence of cortical blindness with denial of visual loss typically is easily established by asking the patient to identify specific stimuli clearly within his or her visual field.

A gross but clinically useful and fairly reliable assessment of visual fields can be accomplished via a routine confrontation examination. Initially, this can be done with both eyes open, but eventually each eye should be tested individually. There are several methods by which this is done; however, they generally all involve having the patient fixate on the nose of the examiner while the right, left, or bilateral peripheral fields are randomly stimulated. The upper, lower, and middle fields are tested in this manner while the examiner notes any tendency to ignore, suppress, or otherwise fail to report stimuli presented to a particular field. It should be noted whether failure to report such stimuli occurs only under conditions of bilateral simultaneous stimulation (suppression) or consistently even under conditions of unilateral stimulation (suggesting a true field cut). In the latter instance, the examiner may change the instructions, telling the patient that the stimulus *will* occur in the affected field, the patient's job being to report not where but *when* the stimulus occurs. If performance improves under the latter situation (assuming that movements of the eye cannot account for the enhanced reporting), it may suggest that severe neglect rather than a complete visual field cut may be present.

The accompanying diagrams (Figure 5–7) illustrate the visual field defects associated with lesions to different sites along the visual pathways.

The localization of pathology with regard to visual field cuts as illustrated in Figure 5–7, can be broadly summarized as follows:

1. Lesions of the **optic nerve** will result in blindness in that eye. Because no visual input can reach the pretectal areas from that eye (the afferent limb of the visual reflex), there will be both a loss of direct light response to that eye as well as a consensual

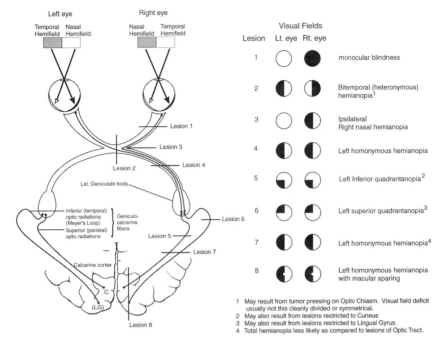

Figure 5–7. Lesions of visual system.

reflex response in the other eye. However, since the efferent limb of the reflex is unaffected, a consensual pupillary response should be present from stimulation of the intact eye and from accommodation (the convergence and pupillary constriction that takes place when attempting to focus on a near object).

2. A **bitemporal hemianopia** can result from a pituitary tumor impinging upon the medial aspect of the chiasm and/or optic tracts that more or less selectively disrupts the crossing fibers from the nasal portions of both retinas.[5]

3. A **homonymous hemianopia** (loss of comparable halves of the visual fields of both eyes) can result from either a lesion of one of the optic tracts, a complete disruption of one of the lateral geniculates or optic radiations, or a complete lesion of one of the occipital poles (area 17). In the latter case, macular sparing may be observed. If the lesion is anterior to the geniculates (e.g., optic tract), the pupillary response to direct visual stimulation (and consensual response in the opposite eye) may be absent if the light source to the eye is limited to the affected field. However, in practice it is exceedingly difficult to reliably maintain such control over the area of the retina being stimulated. In postgeniculate lesions, the pupillary responses will be intact, since the fibers mediating the affective limb of the pupillary response already have left the tract on their way to the pretectal area.

4. A **quadrantic field cut** (*inferior or superior homonymous quadrantanopia* or other partial field cuts) may result from:

    (a) An incomplete lesion of the optic radiations (e.g., deep parietal or temporal lesions producing inferior and superior field cuts, respectively).

    (b) Lesions restricted to the superior (*cuneus*) or inferior (*lingual gyrus*) bank of the calcarine fissure of the occipital pole (producing inferior and superior visual field defects respectively).

Visual Reflexes

As with all reflexes, the visual reflexes require both afferent and efferent loops. However, there are a variety of reflexes involving the eyes that utilize various sensory and motor pathways. The afferent component of the light reflex is carried by the optic nerve (CN II). The motor response—a diminution or constriction of the pupil to excessive light—is carried out by the parasympathetic fibers of cranial nerve III. The opposite response— an enlargement or dilation of the pupils in dim light—is the function of the superior cervical ganglion of the sympathetic nervous system. As noted above, fibers mediating the light reflex travel with the optic tract and detour to the pretectal area at the level of the lateral geniculate body. The pretectal area is just rostral to the superior colliculi. Connections are made with the contralateral pretectal area via the posterior commissure. From the pretectal area the Edinger–Westphal nuclei of the oculomotor nuclear complex then receives both ipsilateral and contralateral inputs (which is why shining a light in one eye will normally cause both pupils to react). These nuclei then send out parasympathetic fibers to the muscles that constrict the pupil as well as adjust the diameter of the lens. The pathway just described is active in the reflex pupillary constriction to light presented to the retina.

Opposing (and "balancing") the parasympathetic constrictor actions of cranial nerve III are the sympathetic inputs from the cord. These sympathetic inputs dilate the pupil (and also produce a tonic elevation of the eyelid). However, the latter response is more complex and can involve more than one afferent pathway. The sympathetic action typically predominates when there is minimal light striking the retina; in this instance, the afferent portion of the reflex arc is carried by the optic nerve. However, pupillary dilation also is known to occur in situations of perceived threat to the organism, in response to pain or any intense sudden and unexpected stimulation, and even in response to visual stimuli that elicit strong, positive affective responses. In the latter cases, the afferent connections mediating the reflex action are more extensive, likely involving the cortex, limbic system, and other brainstem structures. When lesions affect this latter sympathetic system, a **Horner's syndrome** may be present that includes:

1. Constricted (miotic) pupil on the affected side (resulting from the unopposed action of the parasympathetic system).
2. Mild ptosis that can be voluntarily overcome.
3. Dry (anhidrosis), warm (vasodilatation) face on the affected side.

In examining the light reflex, the basic procedure is to have the patient in a dimly lit room and to shine a light in one eye at a time. The examiner notes whether there is a direct (in the eye being stimulated) and a consensual (in the opposite eye) pupillary constriction. Normally, both should be present. If a direct response is absent but a consensual response is present, this would suggest that the afferent side of the reflex is intact but the efferent side is not functioning. A special syndrome—the **Argyll Robertson pupil**—produces dissociation between pupillary constriction from the light reflex and pupillary constriction with accommodation. In this syndrome, the pupil normally is somewhat constricted but fails to constrict further in response to light, either directly or consensually. It also typically fails to dilate in response to darkness. It also fails to respond to mydriatic (pupillodilator) drugs. However, the pupil does show the normal pattern of constriction with accommodation (see below for additional detail). The most common etiology for this syndrome is neurosyphilis and the critical lesion is thought to affect the more dorsal pathways between the pretectal area and the Edinger–Westphal nucleus.

A second afferent pathway, involving corticotectal connections (probably mostly from the occipital cortices), also is involved in pupillary constriction. This second reflex arc (for near vision) uses the same effector system for pupillary constriction to light. In this case, however, it occurs when the occipital cortex recognizes that the visual image is out of focus (e.g., when an object is presented close to the eyes). The subsequent adjustment made by the eyes is called **accommodation**. Impulses are sent again to the pretectal area. From there, connections are again made with the Edinger–Westphal nucleus of cranial nerve III. The fibers from the pretectal area occupy a slightly more ventral position than those mediating the light reflex. This difference in the position of the pretectal-Edinger–Westphal connections is what makes the Argyll Robertson pupil possible. The connections from the Edinger–Westphal nuclei to the sphincter muscles of the iris are identical for both the light reflex and the changes associated with accommodation. In accommodation, three changes take place simultaneously: (1) the lens is thickened by changes in tension of the ciliary muscles, (2) the eyes converge (via the medial rectus muscles of both eyes to prevent double vision), and (3) the pupils constrict. While this process takes place voluntarily when we choose to focus on a near object, it also occurs reflexively as an object is drawn toward the eyes.

If an object is quickly propelled toward the eye such that it might threaten the integrity of the eye itself, another reflex takes over. Instead of accommodation, the eyelids reflexively close. Here the efferent limb of the reflex arc is cranial nerve VII. This same effector response also can be brought about by other stimuli, including loud noise (CN VIII), physical contact with the cornea (CN V), or excessively bright light (CN II). The presence of recurrent stimuli passing across one's visual field with a critical frequency (CN II) can result in involuntary tracking motions of the eyes or optokinetic nystagmus (CN III, IV, and VI). Other automatic or reflex adjustments in the movement of the eyes also can take place given certain types of stimulation to the vestibular system (CN VIII). The external muscles of the eyes and cranial nerve VIII are linked in another reflex response: head (CN XI) and eye (CN III and VI) turning in reaction to a loud, unexpected sound (CN VIII). There are additional reflex changes that affect the intraocular and external muscles of the eye, but the above reflexes are the most common and provide an appreciation of the complex interactions that take place among various systems.

THE EXTRAOCULAR MUSCLES: CONTROL OF EYE MOVEMENTS

Before discussing cranial nerves III, IV, and VI, a brief understanding of the anatomy and functions of the extraocular muscles as a group might be helpful. There are six muscles controlling the movements of each eye. These are the **lateral rectus**, **medial rectus**, **superior rectus**, **inferior rectus**, **superior oblique**, and the **inferior oblique**. The lateral rectus is controlled by cranial nerve VI, the superior oblique by cranial nerve IV, and cranial nerve III controls the superior, medial, and inferior rectus and the inferior oblique. The nerves mediating the superior oblique (CN IV) and that portion of cranial nerve III that supplies the superior rectus cross the midline from their respective nuclei, whereas the inferior rectus, medial rectus, and inferior oblique (CN III) and the lateral rectus (CN VI) are all innervated by ipsilateral nuclei.

Actions of the Lateral and Medial Rectus Muscles

Two of the muscles—the lateral rectus and the medial rectus—have only a single action or effect on the eye: they rotate the eye around the vertical axis (Figure 5–8). The lateral rectus inserts on the temporal or lateral side of each eyeball and, when contracted, it rotates the eye outward or abducts the eye (abducens nerve abducts). The medial rectus inserts on the medial aspect of each eye and, when contracted, it turns the eye toward the midline or adducts the eye. The remaining muscles all have more than one potential or effective impact on eye

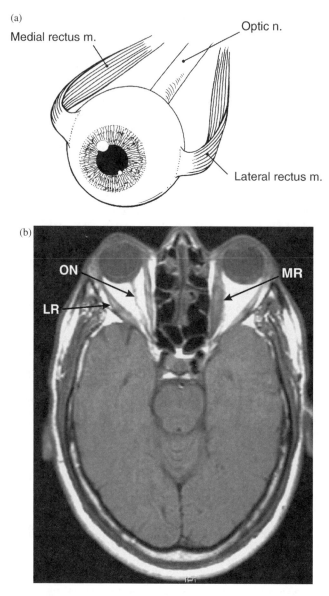

Figure 5–8. (a) Muscles for horizontal movements of the eye. (b) Lateral (LR) and medial (MR) recti muscles, and optic nerve (ON) as seen on MRI.

movement. Thus, these latter muscles have a primary as well as a secondary and a tertiary action. This is a result of the fact that the point of insertion of the remaining muscles are all slightly off center in relation to the vertical axis of the eye (Figure 5–9). The tertiary actions all involve the rotation of the eye around its anterior–posterior axis (intorsion or extorsion). But to try and keep it simple, only the primary and secondary actions will be considered here.

Actions of the Superior and Inferior Rectus Muscles

The superior and inferior rectus, which both attach toward the front of the globe from behind, are actually displaced slightly medial to the vertical axis of the eye (see accompanying

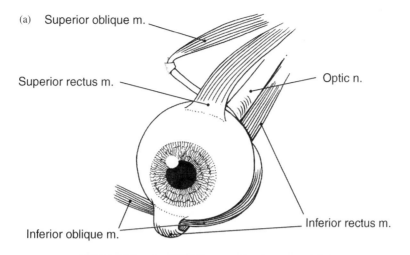

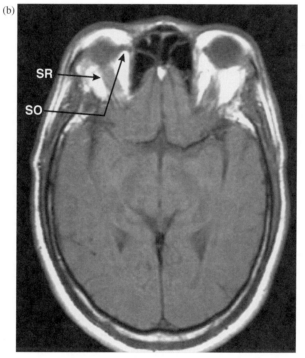

Figure 5–9. (a) Muscles for vertical and rotational movements of the eye. (b) Superior rectus (SR) and superior oblique (SO) as seen on MRI.

figures). This means that, when contracted, they tend to not only respectively elevate and depress the eye, but also tend to rotate the eye (medially) on its vertical axis. Because of this off-center attachment, the strongest elevation effect for the superior rectus will be when the eye is abducted, or looking to the outside or laterally. The same holds true for the inferior rectus.

Actions of the Superior and Inferior Oblique Muscles

Unlike the four recti muscles that pull the front of the globe up, down, or to the side from attachments behind the eye socket, the two obliques (superior and inferior) are unique in that they exert their pulling action from the back of the globe toward the front of the eye. This results from the fact that the base attachment for the obliques are either in the ventral anteromedial part of the eye socket itself (inferior oblique) or run through a pulley (*trochea*) located in the dorsal anteromedial portion of the socket. In both instances, the muscles proceed around the globe to attach more posteriorly. As is the case with the superior and inferior recti, the obliques run medially to the vertical axis of the globe. Thus when looking straight ahead, contraction of the superior oblique will tend not only to depress the eye (primary action), but also to rotate the eye outward on its vertical axis (secondary action). Thus the strongest depressing force of the superior oblique will be when the eye is turned inward or medially (adducted) when the line of force and the line of muscular attachment tend to be parallel. The same is true for the inferior obliques, except that they tend to elevate rather than depress the eye.

Summary of Actions of Extraocular Muscles

Muscle	Primary Action	Secondary Action	Maximal Effect
Medial R.	Turn eye in	(none)	
Lateral R.	Turn eye out	(none)	
Superior R.	Elevation	Adduction	Elevation when abducted
Inferior R.	Depression	Adduction	Depression when abducted
Superior O.	Depression	Abduction	Depression when adducted
Inferior O.	Elevation	Abduction	Elevation when adducted

Note: It is only when looking medially or laterally on a horizontal plane that a single muscle is involved. In looking either up or down, at least two muscles contribute to the action. Hence, from an analysis of the above it can be demonstrated that the maximal strength of each muscle (the direction patient should be told to look when testing each muscle) is as follows:

- **Lateral rectus:** looking to the outside
- **Medial rectus:** looking toward the midline
- **Superior rectus:** looking up and out
- **Inferior rectus:** looking down and out
- **Superior oblique:** looking down and toward the midline
- **Inferior oblique:** looking up and toward the midline

Conjugate Gaze and Convergence

In man, unlike certain lizards, the eyes do not move independently, but jointly. In most instances, the two eyes mirror one another: if one moves up and to the right, the other will do the same. This working together of the eyes in a complementary fashion is called **conjugate eye movement**. Thus, if patient is told to:

	Right eye uses	Left eye uses
"Look to the left"	Medial rectus	Lateral rectus
"Look up and to the left"	Inferior oblique	Superior rectus
"Look down and to the left	Superior oblique	Inferior rectus

There is one instance to which we already have alluded where the two eyes do not move in the same direction at the same time. This is when the individual needs to bring both eyes in toward the center of the visual fields to look at an object that is relatively close to the face. This is known as **convergence** (the two eyes "converge" toward the center). This action involves the simultaneous contraction of both medial recti muscles. Actually, several things take place simultaneously in this process. In addition to the convergence of the eyes, the ciliary muscles of the eyes change the shape of the lens and the pupils constrict; the two latter actions help to bring the image into sharper focus. This entire process, as we have noted earlier, is known as **accommodation**.

Although briefly reviewed in the previous chapter, it may be useful to review the basic neuroanatomical substrates of conjugate eye movements, focusing on lateral conjugate gaze, which requires the coordination of the lateral rectus muscle of one eye in concert with the

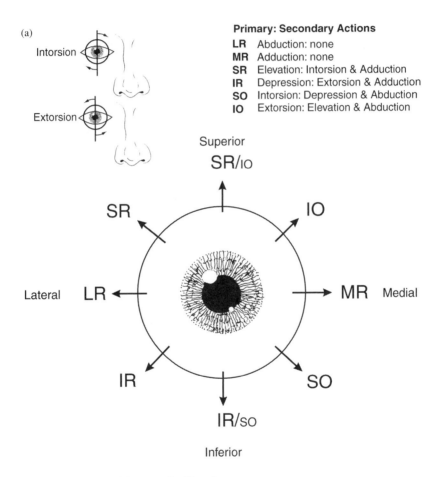

Primary: Secondary Actions

LR Abduction: none
MR Adduction: none
SR Elevation: Intorsion & Adduction
IR Depression: Extorsion & Adduction
SO Intorsion: Depression & Abduction
IO Extorsion: Elevation & Abduction

Maximum Effects of Extraocular Muscles

LR - Looking laterally
MR - Looking medially
SR - Looking laterally & up
IR - Looking laterally & down

Figure 5–10. (a) Compound (primary and secondary) eye movements. (b) Directional movements of individual muscles of the eye. Abbreviations: IO, inferior oblique; IR, inferior rectus; LR, lateral rectus; MR, medial rectus; SO, superior oblique; SR, superior rectus.

(b)

Rt. eye

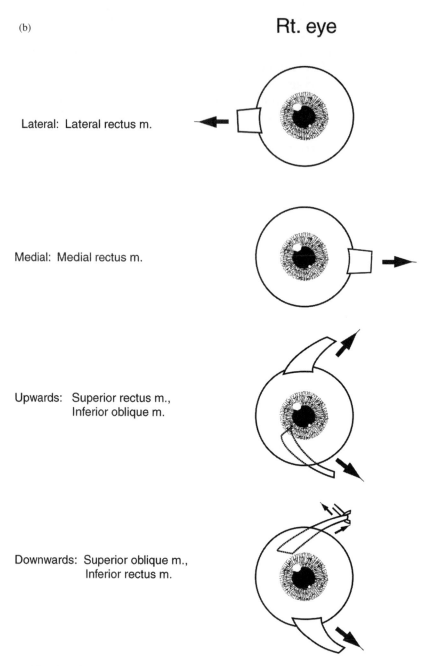

Lateral: Lateral rectus m.

Medial: Medial rectus m.

Upwards: Superior rectus m.,
Inferior oblique m.

Downwards: Superior oblique m.,
Inferior rectus m.

Figure 5–10. *(Continued)*

medial rectus of the other. The cortical center mediating lateral conjugate gaze is located in the posterior part of the frontal cortex (Brodmann's area 8 or the frontal eye fields). From there, fibers pass directly to the pontine brainstem where they decussate and synapse in an area adjacent to the nucleus of cranial nerve VI, which serves as the center for coordinating lateral conjugate gaze in the brainstem [the paramedian pontine reticular formation (PPRF)].

From the PPRF, connections are made between the abducens nucleus (CN VI) and the contralateral oculomotor nucleus via the medial longitudinal fasciculus (MLF).

Thus, if one wants to look to the left, the command is likely initiated in the right frontal eye fields (the right hemisphere is primarily concerned with the left hemispace). The message then is sent to the left PPRF nuclei and then to the left sixth cranial nerve nucleus. The left abducens nerve (CN VI) pulls the left eye laterally (to look to the left). From internuclear cells in the abducens nucleus, a signal also is sent to the right oculomotor nucleus via the MLF commanding the medial rectus of the right eye to contract, thus pulling the right eye toward the left for conjugate left horizontal gaze.

Before looking at the effects of lesions to this system, one must consider that the muscles of the eye act reciprocally in an agonist–antagonist fashion. When, for example, we stare straight ahead, this is accomplished in part by the equal pull or influence of the lateral and medial recti in each eye. What then might be expected to happen if one of those opposing influences were negated? Specifically what would happen, for example, if the lateral rectus muscle in the left eye were to lose its capacity to contract as a result of a lesion to cranial nerve VI or nucleus? In this case, the now unopposed pull of the medial rectus in the left eye would deviate the pupil of the left eye slightly toward the nose, while the pupil of the right eye would remain in its normal midline resting position. This deviation of one eye as a result of weakness or paralysis of one or more of the extrinsic muscles of the eye is referred to as **strabismus**, in this case medial strabismus of the left eye. As might be expected, this would tend to result in double vision (side by side), which then would be exacerbated if the patient were asked to look to his or her left.

As can be seen in Figure 5–11, if cranial nerve VI alone is affected, the patient would

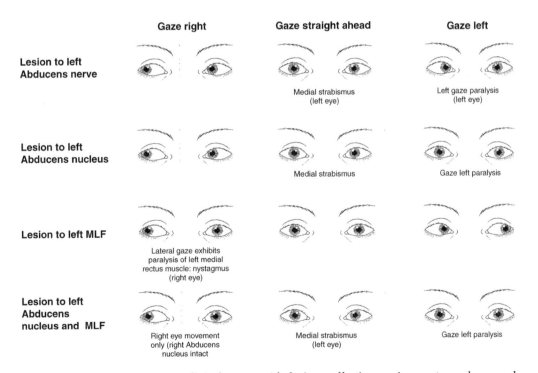

Figure 5–11. Eye deviations on clinical exam with lesions affecting various extraocular muscles. Adapted from Nolte, J (1993). *The human brain: An Introduction to its functional anatomy.* St Louis: Mosby Year Book.

be unable to abduct the eye on the side of the lesion (i.e., unable to look toward the side of the lesion with the eye on the side of the lesion). When attempting to look straight ahead, the eye on the side of the lesion would be slightly deviated toward the midline (as a result of the unopposed action of the medial rectus). However, if there is damage to the nucleus of cranial nerve VI, a slightly different pattern of deficits emerges. While strabismus is still present when the eyes are at rest (i.e., looking straight ahead), when attempting to look toward the side of the lesion not only will there be a failure of abduction, but in addition the eye opposite the lesion will not adduct. This phenomenon is referred to as **lateral gaze palsy** (to the side of the lesion). This results from the fact that not only is the lateral rectus muscle ipsilateral to the lesion affected by the lesion to the nucleus of cranial nerve VI, but the fibers from the nucleus of cranial nerve VI that cross over to supply the medial rectus of cranial nerve III via the MLF also are affected. If the lesion were to spare the nucleus of cranial nerve VI but disrupt the MLF, yet another pattern of deficit known as **internuclear ophthalmoplegia** would be found. In this case, both eyes could abduct on lateral gaze since both the nerves and nuclei of cranial nerve VI are intact. However, since the MLF represents crossed ascending fibers from the nuclei of cranial nerve VI to the oculomotor nuclei (CN III), when attempting lateral gaze, the eye on the side of lesion will not adduct (medial rectus) when attempting to look to the side opposite the lesion. Additionally, nystagmus typically will be present in the abducting (CN VI) eye. In the two latter cases there is a disruption of the function of the medial rectus muscle of one eye when attempting lateral gaze. However, during accommodation the medial rectus, which did not function on attempted lateral gaze, now performs normally, indicating the problem lies in the nuclei responsible for lateral gaze or in the MLF rather than in the oculomotor nerve or its nuclei.

Role of the Vestibular Nuclei in Vision

The vestibular nuclei also are closely associated with eye movements. When the position of the head changes or rotates in space, the eyes typically compensate by conjugate movements in the opposite direction, thus permitting the fixation of gaze to be maintained. Of course, these are reflex responses and can be overcome volitionally. Neuropathologically, the presence of these connections can offer some clinically important information. For example, if the vestibular nuclei and their connections with the centers for lateral gaze are intact, the sharp movement of the head of a coma patient to one side should elicit a conjugate movement of the eyes in the opposite direction (i.e., the eyes should keep "looking" in the same direction they were before the passive movement of the head). This is known as the **oculocephalic** (*doll's head*) **response**. If the eyes stay in a fixed position relative to the head upon passive movement of the head, this suggests a dysfunction at the pontine level (absent doll's head). Another test of related phenomenon could be accomplished by putting cold water in the patient's ear, stimulating the vestibular nerve [**oculovestibular** (*caloric*) **reflex**]. If the vestibular system and its connections with the PPRF and cranial nerve VI nucleus are intact, it should produce nystagmus with the slow component being toward the side being stimulated (see below). Finally, other disruptions of the vestibular system and its connections with the cerebellum or certain other brainstem nuclei can produce abnormal patterns of nystagmus in the conscious patient.

Cortical Control of Eye Movements

All voluntary eye muscles are ultimately under the control of the cortex, particularly Brodmann's area 8 (frontal eye fields) in the frontal lobes that work together to produce smooth pursuit eye movements. Thus lateral gaze palsy also can result from lesions to these regions of the cortex. With unilateral destructive lesions of area 8, not only may the patient be unable to voluntarily gaze to the contralateral field, but as a result of the unopposed action of the unaffected hemisphere the eyes may tend to be slightly deviated toward the side of the lesion. Conversely, if there is an irritative focus to the lesion (e.g., seizure activity), during the phases of electrical stimulation (seizure) the eyes will be "driven" to look to the side away from the lesion.

Nystagmus

Nystagmus, a rhythmic, involuntary oscillation of the eyes, is yet another phenomenon that can be of diagnostic value. Nystagmus is normal under certain circumstances, for example, watching the fence posts go by when riding in a car at a certain rate of speed. It also can be demonstrated to a slight degree in normal persons by asking the individual to deviate the eyes markedly to one side. A more pronounced effect will be seen when caloric testing is carried out (putting cold water in the ear). Pathological nystagmus can take various forms with multiple etiologies. Perhaps most commonly it is evident as excessive nystagmus on lateral gaze or any evidence of nystagmus on upward gaze. Nystagmus is typically described as either **pendular** (where the to-and-fro motions are equivalent) or **jerk** (where they are unequal, consisting of a fast component in one direction followed by a slow component in the opposite direction). In jerk nystagmus, the fast component classifies its directionality. Pendular nystagmus suggests cerebellar involvement, but it can occur in more benign forms. Jerk nystagmus, when pathological, suggests disturbances in the cerebellum, vestibular system, MLF, or other oculomotor pathways. On the other hand, a diminished, or worse, an absent response (nystagmus) to caloric stimulation implies impaired brainstem function. As noted above the latter finding makes this a valuable test to assess the integrity of the nervous system of the patient in a coma.

CRANIAL NERVE III (OCULOMOTOR)

Major Functions: Move the Eyes
Constrict the Pupils
Adjust the Lens
Elevate the Eyelid
Classification: General Somatic Efferent
General Visceral Efferent
Nuclei: Oculomotor (GSE)
Edinger–Westphal (GVE)

The oculomotor nerve is primarily a motor nerve with both somatic and visceral components. Along with cranial nerves IV and VI, it is responsible for movements of the eye (somatic efferents). Of the six extraocular muscles, the oculomotor nerve controls four. These include the:

1. **Medial rectus,** which adducts the eye (pulls it toward the midline).
2. **Superior rectus,** which, along with the inferior oblique, elevates the eye.
3. **Inferior rectus,** which, along with the superior oblique (CN IV), "depresses" the eye (when one is depressed they might say, "things are looking down").
4. **Inferior oblique,** which assists in elevating the eye, particularly when the eye is turned inward toward the midline (i.e., the right inferior rectus comes maximally into play when one looks up and to the left).

These functions are controlled by the oculomotor nuclei, which are located along the midline in the upper mesencephalon. These nuclei (and the oculomotor nerve) also innervate the levator palpebrae, which are the muscles that retract or open the eyelids (the eyelids are closed by CN VII). It should be noted that the elevation of the eyelid also is partially accomplished by the sympathetic system. Thus partial ptosis (drooping of the eyelid) may not always reflect damage to cranial nerve III (see Horner's syndrome).

In addition to these extraocular muscles, cranial nerve III also exerts parasympathetic action (visceral efferents) on the intraocular muscles. These are the **ciliary muscles,** which affect the shape of the lens, and the sphincter muscles of the iris, which affect the size of the pupil (how much light gets into the eye). Normally the ligaments that are attached to the lens are under tension, keeping the lens more or less flat. When near vision is required, the ciliary muscles contract, reducing the tension on these ligaments, which results in a thickening of the lens, which brings the near object into clear focus.[6] As was seen earlier, the pupils act reflexively to various situations, including the amount of ambient light and accommodation. The third cranial nerve is responsible for the **constriction of the pupil** (pupillary dilation is a function of sympathetic activity carried by nerves derived from the superior cervical ganglion). These latter activities are under the control of the Edinger–Westphal nuclei that lie adjacent to the oculomotor nuclei. The Edinger–Westphal nuclei are the autonomic components of the oculomotor nuclear complex.

There are no identified sensory components to the oculomotor nerves. However, it appears likely that some type of proprioceptive feedback loop exists for the extraocular muscles. Cranial nerve V is the only one of the cranial nerves for which proprioceptive feedback is typically defined. It is well known that the trigeminal nerve (CN V) mediates cutaneous feedback from the face as well as proprioceptive feedback from the muscles of mastication (the latter via the mesencephalic nucleus of V). However, it is quite likely that cranial nerve V also mediates proprioceptive feedback for the muscular activities mediated by other cranial nerves, such as the muscles of facial expression (CN VII) and tongue movements (CN XII). Except for the muscles of mastication, these proprioceptive feedback loops are likely mediated by the chief sensory nucleus of CN V.

Lesions of the Oculomotor Nerve

The classic symptoms of a complete oculomotor nerve palsy include diplopia with strabismus, failure of adduction, inability to elevate the eye, ptosis, dilated pupil, and difficulty focusing on near objects. Because of the unopposed action of the lateral rectus and superior oblique, the affected eye will be shifted slightly laterally downward at rest (strabismus), and as a result of the lack of congruent fixation the patient will experience diplopia (double vision). The diplopia is exacerbated when the patient is asked to focus on near objects or look to the side opposite the lesion. The patient will not be able to adduct (move toward midline) the affected eye because of the involvement of the medial rectus, either on command or with convergence (focusing on near object). Involvement of the superior rectus and inferior oblique interferes with upward gaze in the affected eye [remember that the superior obliques (CN IV) assist downward movements]. Because the fibers of the superior rectus cross the midline in the vicinity of the nuclei of cranial nerve III, and indeed appear to pass through the nuclei, a nuclear lesion may result in bilateral weakness of the superior rectus muscles.[7] Other lesions that are thought to affect the connections between the pretectal region and the oculomotor nuclei also may produce upward gaze palsy (without the other symptoms of third nerve involvement. **Ptosis** or drooping of the eyelid will be present on the affected side because the levator palpebrae superioris muscle that elevates the eyelid is supplied by cranial nerve III.

The sphincter muscles that control the pupil are supplied by both parasympathetic and sympathetic fibers, working in an agonist and antagonist manner. Since cranial nerve III carries the parasympathetic fibers responsible for pupillary constriction, complete lesions of the oculomotor nerve will leave the sympathetic influences unopposed. This results in a dilated pupil that will respond neither to direct or consensual light reflex nor to accommodation. Because the ciliary muscle controlling the shape of the lens is affected, the patient will experience difficulty focusing on near objects with that eye.

Before leaving the symptoms of the oculomotor nerve, it may be well to recall **Horner's syndrome**, which previously was discussed. Recall that this syndrome results from a disruption of the sympathetic pathways and produces unequal pupils (*anisocoria*) and a mild ptosis. However, in this case, the affected pupil will be constricted as a result of the compromise of the dilator function of the sympathetic system. Also the ptosis, unlike that resulting from CN III, can be voluntarily overcome and will be on the side of the small, rather than the dilated pupil. Infarctions of the lateral medulla are one common cause of this syndrome.

CRANIAL NERVE IV (TROCHLEAR)

Major Function: Eye Movement (Superior oblique)
Classification: General Somatic Efferent
Nucleus: Trochlear (GSE)

The trochlear nerve is considered a pure motor nerve. It mediates a single ocular motor muscle: the superior oblique. As previously noted, it is somewhat unique in that it is the only cranial nerve to exit dorsally from the brainstem and it is the only cranial nerve that fully crosses the midline (decussates) after leaving its nucleus. Though it exits dorsally, it curves around the mesencephalon and appears anteriorly in the lower midbrain just above the pons.

Lesions of the Trochlear Nerve

The primary action of the superior oblique is to depress the eye. It is maximally effective in this capacity when the eye is adducted (turned medially). Hence, the critical test for its function is to ask the patient to adduct the eye and at the same time to look down. Since the affected eye cannot fully carry out this motion, the patient may complain of double vision in situations requiring similar movements, such as when descending stairs or buttoning one's shirt. Clinically, it may be noted that patients with cranial nerve IV palsy often exhibit a head tilt toward the unaffected side in order to compensate for the diplopia produced by the extortion (outward rotation) of the eye on the affected side.

CRANIAL NERVE VI (ABDUCENS)

Major Function: Eye Movement (lateral rectus)
Classification: General Somatic Efferent
Nuclei: Abducens (GSE)

Like the oculomotor and trochlear nerves, the abducens is considered a pure motor nerve and innervates a single eye muscle: the lateral rectus. The function of the lateral rectus is to abduct the eye or to pull the eye laterally or temporally (as opposed to medially or toward the nose). The nerve exits the brainstem near the midline at the pontomedullary junction.

Lesions of the Abducens Nerve

Since the primary action of the lateral rectus is to pull the eye temporally, a complete lesion of cranial nerve VI will result in the affected eye not moving laterally beyond the midline. Also, given the unopposed force of the medial rectus at rest, the eye will normally show a slight deviation toward the midline (medial strabismus) and the patient may complain of double vision (horizontal). This diplopia will intensify if the patient is requested to look toward the side of the lesion. As discussed earlier, if the nucleus of cranial nerve VI is involved, the patient will likely experience lateral gaze palsy in which both the ipsilateral lateral rectus and the contralateral medial rectus will demonstrate deficits when attempting to look to the side of the lesion. *(Remember that the medial rectus in the opposite eye will retain the capacity to adduct with convergence.)* Since the facial nerve (CN VII) loops around the nucleus of cranial nerve VI, a brainstem lesion that affects the abducens nucleus also may produce ipsilateral lower motor neuron seventh nerve palsy.

CRANIAL NERVE V (TRIGEMINAL)

Major Functions: Somatosensory (face and forehead)
Motor Control of the Muscles of Mastication
Classification: General Somatic Afferent
Special Visceral Efferent
Nuclei: Main (chief) sensory n. of V. (GSA)
Spinal sensory n. of V. (GSA)
Mesencephalic n. of V. (GSA)
Motor n. of V. (GVE)

Sensory Components

The trigeminal nerve supplies the somatosensory feedback for most of the face and front part of the head, including the meninges, teeth, tongue, cornea, sinuses, nasal and oral cavities, and skin (the skin of the rest of the head is supplied by the spinal nerves). The trigeminal nerve is the only cranial nerve that enters and exits at the level of the pons. As in the case of the spinal sensory nerves, most of the sensory fibers of the trigeminal nerve have their cell bodies in an external ganglion (the trigeminal ganglion). The one exception is the fibers mediating proprioception for the muscles of the jaw that have their cell bodies in the mesencephalic nucleus of cranial nerve V. Distal to the trigeminal ganglion, the trigeminal nerve divides into three branches. The **ophthalmic** (CN V1) branch or nerve that provides sensory input from the upper part of the face and scalp, the eyes (e.g., the cornea), upper portion of the nose, part of the nasal mucosa and frontal sinuses, and the meninges (all three divisions may supply the meninges). The **maxillary** (CN V2) branch transmits sensory information from the upper jaw, teeth, lip, hard palate, maxillary sinuses, and part of the nasal mucosa. The **mandibular** (CN V3) division has both sensory and motor components. It mediates the somatosensory feedback from the lower jaw, teeth, lips, buccal mucosa, anterior two-thirds of the tongue (touch, not taste), and part of the external ear and auditory meatus. Its motor component innervates the muscles of mastication (see Figure 5–12).

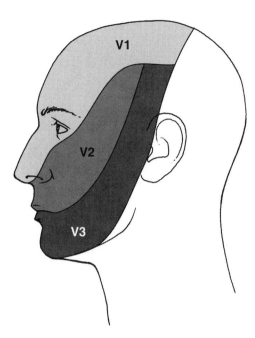

Figure 5–12. Sensory divisions of the trigeminal nerve: V1, ophthalmic; V2, maxillary; V3, mandibular.

The trigeminal nerve basically mediates all the same aspects of somatosensory sensation that are found in the dorsal roots of the spinal nerves: light touch, pain, temperature, stereognosis (fine discriminative touch), vibration, and proprioception. There are three separate nuclei that are responsible for sensory perception associated with the trigeminal nerve: (1) the **mesencephalic**, (2) **chief** (or *main*), and (3) **spinal nuclei** of cranial nerve V. The mesencephalic nucleus of cranial nerve V is the source of fibers that transmit proprioceptive information obtained primarily from stretch receptors located in the muscles of mastication. The sensory fibers synapsing in the main ("chief" or "principle") sensory nucleus of cranial nerve V, which is located in the pons, are comparable to the posterior columns of the spinal cord, mediating stereognosis, vibration, and possibly proprioception. It has been suggested that the mesencephalic nucleus largely may contribute "unconscious" proprioceptive feedback to the cerebellum, while the main sensory nucleus is associated with "conscious" proprioception. However, some authors suggest that the mesencephalic nucleus is responsible for most proprioceptive feedback, both conscious and unconscious, while some authors fail to mention proprioception at all in connection with the main sensory nucleus. Until more consistent data emerge, it seems likely that both nuclei mediate proprioception, but perhaps the mesencephalic nucleus primarily is responsible for feedback that controls mastication. While the pathways or connections have not been definitively established, one or both of these nuclei are probably also responsible for proprioceptive feedback for the muscles of the eyes, tongue, and muscles of facial expression. Lastly, the fibers derived from the spinal nucleus of cranial nerve V can be viewed as being comparable to those of the ventral and lateral spinothalamic tracts in that they appear to carry information for light touch, pain, and temperature. This latter nucleus is very long, extending from the pons into the spinal cord. As we shall see, several other cranial nerves also utilize this nucleus.

Whereas the sensory feedback from the spinal nerves travels to the contralateral ventral posterolateral (VPL) nuclei of the thalamus, the majority of the fibers from the trigeminal sensory nuclei travel to the contralateral **ventral posteromedial (VPM) nucleus** via the **ventral trigeminothalamic tracts**. The VPM nuclei, in turn, project to the more ventral portions of the sensorimotor cortices. There are a few fibers, from the main sensory nucleus in particular, that travel ipsilaterally in the dorsal trigeminothalamic tracts. Since the majority of these fibers cross, lesions affecting these tracts (as well as the VPM nucleus or somatosensory cortex) will result in contralateral sensory deficits in the face, whereas lesions of the trigeminal nuclei or nerve will produce ipsilateral deficits. There likely also are connections with the reticular nuclei, particularly from the spinal nucleus of cranial nerve V, as is the case with the spinal nerves.

Motor Components

The main motor function served by the trigeminal nerve is to innervate the muscles of mastication or the lower jaw (the masseter, temporal, internal and external pterygoids). The nerve fibers emanate from the motor nuclei of V and travel to the periphery with the mandibular branch (CN V3) of the trigeminal nerve. Similar to other muscles that act in a symmetrical fashion, the muscles of mastication receive bilateral cortical input; hence, upper motor neuron (supranuclear) lesions will have minimal effect on the patient's ability to chew. The trigeminal nerve provides sensory feedback from the ear (as do CN VII, VIII, IX, and X), but along with cranial nerve VII, it also provides motor control to the middle ear. Cranial nerve V adjusts the tension on the tympanic membrane.

Lesions of the Trigeminal Nerve

Depending on the type and location of damage to the trigeminal nerve, several different syndromes might result. Clearly, if a lesion destroyed the main branch of the trigeminal nerve, one would expect to find hemianesthesia of the face and weakness of the muscles of mastication on the same side. Commonly, however, the lesion may affect a portion of the nerve after it splits into its three divisions, thus producing reduced sensations in only one of its areas of distribution. Similarly, if nuclear lesions are involved, then only specific aspects of sensory stimulation may be affected. Even more distressing for patients is **trigeminal neuralgia** (*tic douloureux*), which although of unclear etiology, affects one or more branches of the trigeminal nerve. This latter syndrome consists of sharp, paroxysmal pain in one or more of the distributions of the trigeminal nerve. Testing for damage to the trigeminal nerve involves both motor and sensory examinations. To assess the muscles of mastication, the examiner has the patient open his or her jaw. Because of the way the muscles are attached, weakness on one side may result in a deviation of the jaw to the side of the lesion. Similarly, the patient may be asked to move the jaw laterally against resistance. There will be weakness when attempting to move the jaw opposite the side of the lesion. In lower motor neuron lesions (lesion of the nerve or motor nucleus) one might discern atrophy of the temporalis muscle on palpitation. With lesions of the mandibular branch or of the mesencephalic nucleus (afferent limb), the jaw jerk (reflex) should be diminished. The sensory exam consists of comparing light touch, pinprick (pain), two-point discrimination (stereognosis), or temperature on both sides of the face in all three trigeminal divisions. In cases of psychogenic hemianesthesia the patient may report the failure to perceive the vibrations of a tuning fork placed on the forehead of the "affected" side (in true neurological disease the sensation, which travels by bone to both sides of the forehead, is likely to be perceived bilaterally). Corneal reflexes normally are tested by touching the cornea with a wisp of cotton or tissue. Since the normal response is for both eyes to blink in response to unilateral stimulation, it may be possible to differentiate between diminished or absent sensation in the cornea (CN V) and a diminished motor response (CN VII).

CRANIAL NERVE VII (FACIAL)

Major Functions: Muscles of Facial Expression
Taste
Salivation
Classification: Special Visceral Efferent
General Visceral Efferent
Special Visceral Afferent
General Somatic Afferent
Nuclei: Facial (SVE)
Superior salivatory (GVE)
Solitary (taste) (SVA)
Spinal nucleus of V. (ear) (GSA)

Sensory Components

The facial nerve, which loops posteriorly around the nucleus of CN VI before exiting from the anterolateral portion of the brainstem at the pontomedullary junction, has both motor and sensory components. Its main afferent function is to carry **taste sensations** from the anterior two-thirds of the tongue and the palate. Each nerve supplies the ipsilateral half of the tongue. The rostral portions of the solitary nuclei in the medulla mediate the sense of taste. From the solitary nuclei, the fibers conveying information about taste travel rostrally in the central tegmental tract where they synapse in the ventral posteromedial (VPM) nuclei of the thalamus. From there the information is carried to the opercular portions of the frontal and parietal lobes and the anterior portions of the insular cortices. Cranial nerve VII also carries **somatosensory information** from the auditory meatus, tympanic membrane, and a small section of skin behind the ear. The sensations from the ear synapse in the spinal nuclei of the trigeminal nerve.

Motor Components

The major motor functions of CN VII include:

1. Innervation of the superficial muscles of the face (the muscles of facial expression, including closure of the eyelid).
2. Stimulation of the lacrimal (tearing of the eye) and submandibular and sublingual (salivation) glands and nasal mucosa.
3. Control of the stapedius muscle (regulates the stapes: one of the ossicles of the middle ear).

The seventh cranial nerve via the facial nucleus innervates all ipsilateral facial muscles (muscles of facial expression) over which we have voluntary control, with the exception of the muscles involved in chewing (CN V). Among some of the functions or expressions mediated by cranial nerve VII include furrowing the brow, closing or winking the eye, smiling, whistling, and puffing out the cheeks. There is one aspect of the neuroanatomical connections to the facial nuclei from its corresponding motor cortex (ventral portion of the precentral gyrus) that deserves special mention because of its important clinical considerations. The corticobulbar fibers, which control the muscles of the lower face, decussate and supply only the contralateral facial nucleus. However, the fibers that will eventually influence the muscles of the upper face (forehead and brow) split in the brainstem and terminate on both the ipsilateral and contralateral facial nuclei (bilateral innervation). Thus, while there is only contralateral innervation of the muscles of expression in the lower face, the muscles of the upper face receive bilateral input from the cortex.

The parasympathetic functions mediated by the salivatory nucleus of the medulla are to enhance the production of saliva (submandibular gland) and tears (lacrimal glands). As is the case with most other aspects of autonomic function, hypothalamic and limbic structures heavily influence these activities. Finally, the facial nerve exerts control of the stapedius muscle in the middle ear, which controls the action of the stapes, one of ossicles that transfers vibrations from the tympanic membrane to the oval window of the inner ear. Contraction of this muscle helps to attenuate loud sounds.

Lesions of the Facial Nerve

Lesions of the seventh cranial nerve will result in diminished or absent taste perception on the anterior two thirds of the tongue. This loss of taste should be limited to that half of the tongue that is ipsilateral to the lesion. If the branches of the nerve mediating the salivary glands are affected, the patient may complain of increased dryness of the mouth. Clinically, the most obvious effects of seventh cranial nerve involvement will center on facial motor symptoms. First, in simply observing the patient, one may note facial asymmetry. This might include a drooping of the mouth on the affected side, a smoothing of the nasal–labial fold, and an incomplete closure of the eyelid. If the patient is observed while eating, one may witness a tendency for food to collect in the cheek pouch on the affected side. There will be a decreased blink reflex to corneal stimulation, regardless of which cornea is stimulated. In fact, the eyelid may never fully close. This latter phenomenon, in conjunction with the decreased tearing in that eye from involvement of the lacrimal gland, may lead to an irritating dryness to the eye, possibly necessitating an eye patch and frequent bathing of the eye with saline solutions.

In cases where the facial weakness is less obvious, the patient may be asked to close the eyes tightly (with the examiner checking the degree of resistance to attempt to passively open them), smile, show his or her teeth, or puff out the cheeks (holding them out against resistance). The patient is then asked to furrow his/her brow (raise their eyebrows). Any significant asymmetry in the contracture of the muscles across the brow is noted. If weakness is noted in these upper facial muscles, along with signs of lower facial weakness, this implies a lesion of the facial nerve or its motor nucleus (**lower motor neuron lesion**). However, if a lower facial weakness is present, but the muscles of the forehead are intact bilaterally, an **upper motor neuron** or "*supranuclear*" **lesion** is suspected (see Figure 5–13). Remember, the upper part of the face (motor) is bilaterally innervated so only a lesion of the nerve or complete nuclear lesion will result in a total hemifacial paresis. With upper motor neuron lesions, there often will be associated weakness of the limbs, most commonly the upper extremity when the lesion involves the cerebral cortex.

An interesting phenomenon is sometimes present when a facial paresis is secondary to an upper motor neuron lesion (e.g., cortical lesion). Voluntary actions such as asking the patient to smile will reveal a unilateral facial weakness; however, if respond to a joke a spontaneous smile, no facial asymmetry is evident. This clinical feature suggests that different pathways

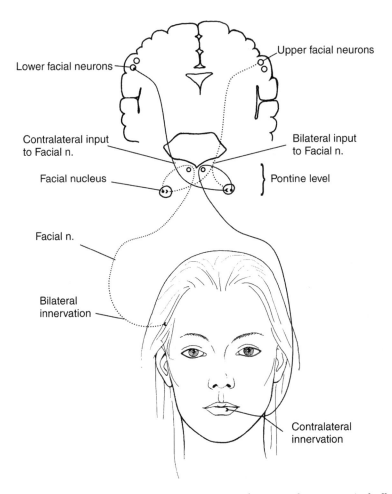

Figure 5–13. Differential innervation to upper and lower face. As shown, corticobulbar fibers that will eventually be responsible for innervation of the muscles of the lower face synapse only in the contralateral facial nucleus in the pons. Conversely, those cortical fibers destined for the upper portion of the face have inputs to both facial nuclei, from where they innervate the upper face bilaterally (not shown).

emerge for certain volitional versus emotionally generated motor responses. In addition to the deficits described above, the patient may report intolerance to high-amplitude auditory stimulation (hyperacousis) due to the impairment of the stapedius muscle.

While seventh cranial nerve paresis or paralysis may result from various conditions, including tumors and strokes, **Bell's palsy** represents perhaps the most common form of this disorder. Thought to be viral in origin, the classic symptoms of a lower motor seventh cranial nerve paralysis, as described above, typically evolve over a few hours to a few days. Fortunately, while most individuals who suffer from this condition recover within a few weeks to months, in some cases complete recovery may take much longer, if indeed it occurs at all. Finally, as in the case of the trigeminal nerve, the facial nerve has many divisions or branches after leaving the brainstem. Lesions that may affect one or more of these distal branches may produce a clinical picture of an "incomplete" seventh cranial nerve palsy.

CRANIAL NERVE VIII (VESTIBULOCOCHLEAR)

Major Function: Sense of Hearing; Balance, and Equilibrium
Classification: Special Somatic Afferent
Nuclei: Vestibular (SSA)
         Cochlear (SSA)

The vestibulocochlear nerve is the last of the three purely sensory cranial nerves. As indicated by its name, it serves two functions: auditory and vestibular. It is responsible for carrying information from the vestibular and auditory systems, whose end organs are located in the inner ear, to nuclei located in the brainstem. Both systems respond to stimuli by the mechanical displacement or bending of hair cells in a fluid medium. Like cranial nerves VI and VII, cranial nerve VIII enters the brainstem as a double nerve just posterior to cranial nerve VII and just anterior to the cerebellum at the pontomedullary junction.

Cochlear Division

The cochlear portion of cranial nerve VIII is responsible for carrying auditory information from the hair cells of the organ of Corti, which is housed in a fluid-filled, spiral bony labyrinth called the cochlea in the inner ear, to the brainstem. The energy from the airwaves picked up by the tympanic membrane (eardrum) is mechanically magnified by the ossicles of the middle ear and transferred to the fluid of the cochlea. The hair cells within the organ of Corti are tontopically organized. This tonotopic organization is maintained beginning with the first synaptic connections in the dorsal and ventral cochlear nuclei in the brainstem all the way to the primary cortical projection area (Heschl's gyrus).

Two major pathways—a dorsal and a ventral system—exit from the dorsal and ventral cochlear nuclei, respectively. All of the postsynaptic fibers from the dorsal cochlear nuclei cross the midline as the **dorsal acoustic stria** and the majority ascends directly to the contralateral inferior colliculi via the **lateral lemniscus** (auditory pathway in the brainstem). A few fibers synapse in the contralateral superior olivary nucleus before ascending in the lateral lemniscus. Postsynaptic fibers leaving the ventral cochlear nuclei take a somewhat different course, although most eventually also will make their way to the inferior colliculi. Two fiber tracts leave the ventral cochlear nucleus. One, the **intermediate acoustic stria**, like the pathway from the dorsal cochlear nucleus, curves around the dorsal aspect of the inferior cerebellar peduncle, crosses the midline, and enters the contralateral lateral lemniscus. The **ventral acoustic stria**, the largest of the three acoustic stria, follows a more ventral path around the inferior cerebellar peduncle as it exits from the ventral cochlear nucleus. As shown in Figure 5–14, fibers from the ventral acoustic stria basically take one of three courses. They may:

1. Synapse in the ipsilateral superior olivary nuclei, which in turn may send tertiary fibers to the inferior colliculi via either the ipsilateral or contralateral lateral lemniscus.
2. Send fibers to the contralateral superior olivary nucleus, which similarly projects to the inferior colliculi.
3. Send fibers directly to the contralateral inferior colliculi, bypassing the superior olivary nuclei.

These crossing fibers of the ventral acoustic stria (those either directly entering the opposite lateral lemniscus or crossing before or after synapsing in the contralateral or ipsilateral superior olivary nuclei) make up what is known as the **trapezoid body,** which is seen in the pontine tegmentum. Because of the incomplete crossing at this elementary level,

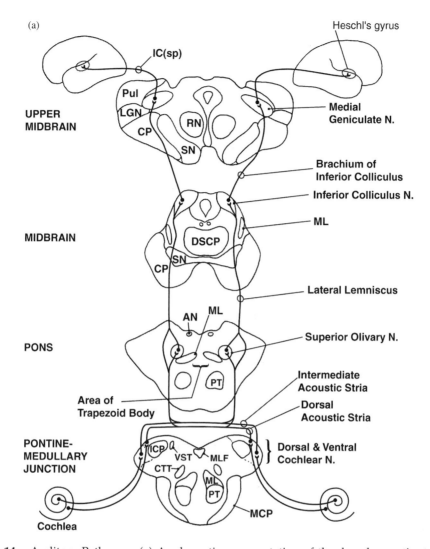

Figure 5–14. Auditory Pathways. (a) A schematic representation of the dorsal acoustic stria, which emanate from the dorsal cochlear nuclei, and the intermediate acoustic stria, which derive from the ventral cochlear nuclei. (b) In order to simplify the illustration, the ventral acoustic stria (also derived from the ventral cochlear nuclei) is shown separately. Note that the dorsal and intermediate stria both cross the midline before entering the lateral lemniscus, while the ventral stria have multiple inputs to the lateral lemniscus, both ipsilateral and contralaterally.

AN, abducens nucleus
CP, cerebral peduncle
CTT, central tegmental tract
DSCP, decussation of the superior cerebellar peduncle
IC(SP), internal capsule, sublenticular portion
ICP, inferior cerebellar peduncle
LGN lateral geniculate nucleus
MCP, middle cerebellar peduncle

ML, medial lemniscus
MLF, medial longitudinal fasciculus
PT, pyramidal tract
Pul, pulvinar
RN, red nucleus
SN, substantia nigra
VCN, vestibulocochlear nerve
VST, vestibulospinal tract

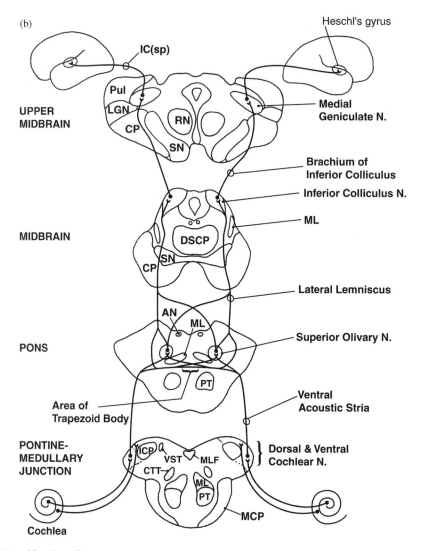

Figure 5–14. *(Continued)*

it is thought that the superior olivary nuclei may be important in the localization of sound, apparently responding to the slight time differential it takes for sound to arrive at the two ears. However, this asymmetry of input is preserved all the way to the cortex, thus also potentially allowing for some mediation of this effect at higher levels.

The above description fails to capture the full complexity of the auditory pathways at the brainstem level. A few additional considerations may help. Although the eighth cranial nerve generally is considered to be a pure sensory nerve (hearing and vestibular sense), there are several motor components that can affect hearing. The nuclear complex surrounding the superior olivary nuclei appears to be the source of an olivocochlear pathway that can modulate sound by altering the sensitivity of hair cells in the organ of Corti. The ear's sensitivity can also be altered by the respective actions of cranial nerves V and VII via differential tension on muscles controlling the tympanic membrane and stapes of the middle ear. Also, although the pathways are not well defined, some auditory input probably reaches the ascending reticular formation and those nuclei responsible for eye movements, such as CN III and VI (probably via the MLF), and movements of the head or neck (colliculi and CN XI). The former likely serves a general alerting function, while the latter probably establish the anatomical bases for various reflexes, such as orienting the eyes and head to auditory stimuli, especially when they are loud or unexpected.

As previously noted, the major proportion of fibers leaving the cochlear nuclei ends up ascending in the lateral lemniscus to the inferior colliculi. At this level there is some additional crossing of the fibers between the two colliculi. From each of the inferior colliculi, ipsilateral connections are established with the medial geniculate nuclei via the brachium of the inferior colliculi. In turn, the medial geniculate nuclei send projections (via the sublenticular portion of the internal capsule) to the primary auditory cortex (Heschl's gyrus) located in the recesses of the lateral fissure in the temporal lobe. Unlike the visual, motor, and somatosensory systems that are primarily crossed, the auditory system from the superior olives to the cortex has a more extensive bilateral representation. The left auditory cortex, however, still receives most of its input from the right ear; estimates of up to 40% of its input is from the ipsilateral ear. This redundancy in the system, or the bilateral input to the primary auditory cortex, assures that unilateral injury to **Heschl's gyrus** produces minimal clinical deficits. In order for cortical lesions to produce significant clinical deficits, bilateral lesions to Heschl's gyrus or a lesion that involves Heschl's gyrus unilaterally with extension to white matter connections to the contralateral Heschl's gyrus are necessary.

Vestibular Division

The vestibular system provides two general types of information: (1) the orientation of the head in space and (2) the movement of one's body (head) through space (both the direction of movement, as well as the sense of movement). As we have noted in earlier chapters, the integrity of the vestibular system is critical to maintain normal motor functions and for coordination, balance, or maintaining one's equilibrium. The vestibular system apparently provides highly reliable and accurate information regarding spatial orientation and direction of movement, but is still probably very dependent on collateral inputs from other sensory systems. One fairly common illustration of this fact can occur in scuba diving. In very deep dives where neither the surface nor the bottom can be visualized and there are no perceptual forces of gravity operating, it is extremely easy (and very disconcerting) to become "disoriented" and literally not know which way is up. Equally distressing is to unknowingly be diving along an incline, thinking it is a horizontal plane, then discover that one's bubbles are not "rising" but appear to be floating off in what now appears to be a horizontal direction. A diver in such situations may experience "dizziness" or other signs associated with vestibular disturbances.

The sensory organs for the vestibular system, located in the inner ear, include the three semicircular canals, the utricle, and the saccule. The semicircular canals represent three different planes or orientations in space and respond to angular acceleration and deceleration. The utricle responds to gravitational forces and to horizontal linear acceleration. The saccule responds to linear acceleration in the dorsal–ventral plane. As the fibers for the vestibular system enter the brainstem, a few course directly to the flocculonodular lobe of the cerebellum, while most synapse in the vestibular nuclei (superior, inferior, medial, and lateral).

The vestibular nuclei give rise to both the ascending and descending MLF. The former important in ensuring that changes in the position of the head will result in equal compensatory movements of the eyes in order to maintain stability of the visual image despite movement of the head in space. The connections between the vestibular nuclei, the cerebellum and the spinal cord (via the descending MLF) are reciprocal. It has been noted that the vestibular nuclei play a major role in maintaining body posture and equilibrium. While many motor responses to vestibular stimulation are reflexive in nature (e.g., excitation of antigravity muscles to maintain balance), the vestibular system also likely plays a significant role in the coordination of conscious motor activities. Finally, projections to the cortex may serve two (probably more) basic functions. Direct connections with the visual system (e.g., eye movements through the MLF) ensure that changes in the position of the head will result in equal compensatory movements of the eyes in order to maintain stability of the visual image despite movement of the head in space. Finally, projections to the cortex may serve two (probably more) basic functions. The vestibulocortical connections provide for conscious awareness of the orientation of one's body in space or movement through space, and thus allow for the conscious adjustment of skilled movements (or perhaps, the conscious inhibition of certain reflex righting movements) during the execution of particular activities.

Lesions of the Vestibulocochlear Nerve

Cochlear Division

The most common symptoms of damage or irritation to the auditory system are tinnitus and hearing loss. Tinnitus is the perception of "ringing" in one's ear, which may be characterized by either high- or low-pitch sounds. A relatively common and benign cause of tinnitus is medication (e.g., large doses of aspirin). Hearing is one of human's most acute senses, both in terms of absolute thresholds and in discriminating discrete, successive stimuli. Different forms of auditory impairment can result from lesions to the central auditory pathways (brainstem, thalamus, cerebral cortex) and peripheral systems. The peripheral auditory system can be impaired due to buildup of wax in the outer ear canal, damage to the eardrum, changes in the ossicles of the middle ear, and problems with the hair cells of the inner ear. Despite its sensitivity, the eighth cranial nerve probably is the one cranial nerve most likely to require specialty examination (by an audiologist) for more definitive diagnoses. There are, however, several routine procedures that can be carried out at bedside that can provide good preliminary hypotheses about the relative integrity of various parts of this system.

The first step is to obtain some measure of absolute thresholds of hearing sensitivity in each ear. This can be assessed in each ear by close, low-level stimulation that normally could not be picked up by the opposite ear, for example, the ticking of a watch or by rubbing two fingers together. Not uncommonly, the patient may have relative loss for certain frequencies, but this probably is best examined by an audiologist. Certain types of higher-level (e.g., cortical) hearing problems such as discrimination or perception of sounds and unilateral neglect will be covered in Chapter 9. If on routine clinical exam a hearing loss is suspected, it is important to determine whether the deficit is due to nerve damage or is the result of some type of conduction (air or bone) loss.

Two tests performed with a 512 Hz tuning fork (lower vibrations can be misinterpreted as "sound") can help make this distinction. The first test (**Weber test**) involves placing the base of a vibrating tuning fork on the forehead (centered) or on the vertex of the skull. Normally the sound is perceived as being equal in both ears or as coming from the center of the head. If the sound localizes to one ear more than the other, it will suggest either a conduction loss on the side to which the sound localizes or a neural loss in the opposite ear.

In the second (the **Rinne test**), the base of the tuning fork is placed on the mastoid process immediately behind the ear (bone conduction). When the patient reports no longer hearing the sound, the tuning fork (vibrating end) is held near the auditory canal (air conduction). Normally, the perception should increase during the second step, as air conduction is more efficient than bone conduction. If the patient hears better with the tuning fork pressed against the mastoid process than when the tuning fork is held adjacent to the ear, the patient likely suffers from an air conduction deficit. If the sound is reduced with either placement, a sensorineural hearing loss is more likely.

Vestibular Division

The most common symptoms of damage or irritation to the vestibular system (diseases of the labyrinth or neural pathways) are nystagmus and vertigo, dizziness, and nausea or vomiting. Nystagmus may be more or less constant or may be brought on by certain maneuvers (e.g., positional nystagmus). Causes of nystagmus are multiple and its vestibular origin perhaps is most clearly established when it is accompanied by vertigo. With regard to the latter, it is important to note whether the patient has any subjective symptoms of dizziness or vertigo (the room or occasionally the patient himself or herself is perceived to be spinning or the normal visual planes tend to be abnormally tilted). If present, these symptoms often are associated with nausea and vomiting. If vestibular dysfunction is suspected, caloric testing can be done, comparing the response obtained in the two ears. This test was explored earlier in our discussion of eye movements and how it could be used to test the integrity of the caudal pons or brainstem in the coma patient. In addition to providing a test of the integrity of the vestibular portion of the eighth cranial nerve and vestibular nuclei (and their connections with the centers for lateral gaze), the caloric test assesses the integrity of the peripheral mechanisms (labyrinthine complex). If the system is intact, one should produce nystagmus with the fast component to the side opposite the cold water stimulation. Although a valuable diagnostic test, cold calorics generally are not performed on a conscious patient due to the unpleasant side effects (e.g., nausea and vomiting). Although a variety of lesions can affect cranial nerve VIII and its peripheral organs, acoustic neuromas (cerebellopontine angle tumors involving the vestibulocochlear nerve) are one cause of slowly progressive hearing loss associated with tinnitus and vestibular symptoms.

CRANIAL NERVE IX (GLOSSOPHARYNGEAL)

Major Functions: Muscles of Pharynx
Sensation to Throat
Taste
Classification: Special Visceral Efferent
General Visceral Efferent
Special Visceral Afferent
General Visceral Afferent
General Somatic Afferent
Nuclei: Nucleus ambiguus (SVE)
Inferior salivatory (GVE)
Solitary (SVA)
Solitary (GVA)
Spinal n. of V. (GSA)

The glossopharyngeal nerve, as its name implies, primarily is related to functions of the tongue and pharynx. While it has both sensory and motor components, for all practical purposes it is generally regarded as a sensory nerve, since it supplies only one small muscle that has little clinical significance when lesioned (the stylopharyngeus muscle). While it is possible to isolate some specific functions of cranial nerve IX, such as touch and taste on the posterior one-third of the tongue, in practice cranial nerves IX and X typically are examined as a unit by assessing the gag reflex. Cranial nerve IX supplies the afferent limb and cranial nerve X supplies the efferent limb of this reflex response. The glossopharyngeal nerve enters (and exits) the anterolateral portions of the rostral medulla just below cranial nerve VIII.

Sensory Components

Cranial nerve IX mediates the sensation of taste from the posterior third of the tongue and the pharynx. These fibers synapse in the rostral portion of the nucleus solitarius before ascending (see cranial nerve VII). The more caudal portion of the nucleus solitarius receives afferent fibers that monitor blood oxygen levels (from carotid body) and arterial blood pressure (from carotid sinus), which are relayed to the hypothalamus. General somatic sensations from the posterior part of the oral cavity (e.g., posterior third of the tongue, upper pharynx) are mediate via glossopharyngeal connections with the spinal nucleus of the trigeminal nerve. Other GSA fibers carrying tactile information from the ear also arrive at the spinal nuclei of cranial nerve V by way of the ninth cranial nerve.

Motor Components

There are two targets for efferent motor pathway associated with cranial nerve IX. First, one group of fibers innervates one of the muscles of the pharynx (the stylopharyngeus m.), which assists in the elevation of the pharynx (although this also is accomplished to a somewhat greater degree by the vagus nerve). The nucleus ambiguus is the source of these special visceral efferents to the stylopharyngeus muscle. Second, the remaining efferent fibers of cranial nerve IX innervate the parotid gland via the inferior salivatory nuclei. Thus cranial nerves VII and IX share two common functions: (1) taste to the tongue (CN X may mediate taste to the epiglottis) and (2) control of the salivary glands.

Lesions to the Glossopharyngeal Nerve

Touch and taste to the posterior third of the tongue can be used to assess the integrity of cranial nerve IX. However, as the procedure is somewhat awkward and inconvenient, this test typically is not done as part of most neurological exams. Therefore, most commonly cranial nerve IX and X are tested together by checking the presence of the gag reflex, which assesses the afferent input necessary for this response to occur (see Vagus Nerve).

CRANIAL NERVE X (VAGUS)

Major Functions: Muscles of Pharynx and Larynx
Parasympathetic Control of Abdominal
and Thoracic Viscera
Sensation to the Throat
Classification: Special Visceral Efferent
General Visceral Efferent
Special Visceral Afferent
General Visceral Afferent
General Somatic Afferent
Nuclei: Nucleus ambiguus (SVE)
Dorsal motor nucleus of X (GVE)
Solitary nucleus (rostral) (SVA)
Solitary nucleus (caudal) (GVA)
Spinal nucleus of V (GSA)

Like the glossopharyngeal nerve, the vagus nerve is a mixed sensory and motor nerve. From a clinical viewpoint, cranial nerve IX is considered more sensory and cranial nerve X is considered more motor. Cranial nerve X is larger than cranial nerve IX and exits just below the glossopharyngeal nerve in the rostral medulla. Although cranial nerves III, VII, IX, and X all carry parasympathetic fibers, the vagus nerve is the only cranial nerve mediating parasympathetic functions that has target organs that lie outside of the cranial vault.

Sensory Components

The rostral portion of the solitary nucleus (n. solitarius) receives fibers from cranial nerve X, which carry information about taste from the epiglottis. As with cranial nerve IX, the caudal portions of the solitary nuclei also receive visceral sensory input from the trachea and larynx via the vagus nerve, as well as from some of the viscera to which it supplies efferent connections (e.g., lungs, esophagus, stomach, and intestines). Some general sensory fibers of cranial nerve X carry tactile information from the oral cavity and from the ear. This sensory input is conveyed to the spinal nucleus of cranial nerve V via the vagus nerve.

Motor Functions

The vagus nerve also supplies the muscles of the soft palate and pharynx that mediate swallowing. However, as in the case of its sensory input, the vagus nerve is the only cranial nerve that supplies motor fibers to the larynx. All these efferent fibers, which are critical for swallowing and motor speech, originate in the nucleus ambiguus. The other major efferent pathway of the vagus nerve originates from the dorsal motor nuclei of cranial nerve X that lie in the medulla. These fibers constitute the parasympathetic branches that supply the thoracic and abdominal viscera. Some of the preganglionic parasympathetic fibers that supply the heart appear to come from the nucleus ambiguus. These parasympathetic inputs reduce the heart rate, constrict the size of the bronchioles, and stimulate peristalsis as well as gastric, hepatic, and pancreatic activity.

Lesions of the Vagus Nerve

Since the vagus nerve provides most of the motor control to the muscles of the pharynx, soft palate, and larynx, the integrity of cranial nerve X can be assessed by testing and observing activities that directly require the use of those muscles. Hoarseness or dysphonia may suggest difficulties with the larynx. Dysphagia (difficulty swallowing) may suggest difficulties with the muscles of the soft palate and pharynx (in eating or drinking the patient may complain that the food or water gets up into the nasal passages). One simple direct test is to have the patient say, "Ahhh." The examiner notes whether the palate elevates symmetrically and whether the uvula deviates. With lesions affecting the tenth cranial nerve, the palate will show reduced elevation (or fail to elevate) on the affected side, causing the uvula to deviate to the unaffected side. Another direct test is to stimulate each side of the palate and observe the gag reflex. Remember, the afferent side of the "gag reflex" is mediated by the glossopharyngeal nerve, and the motor side by the vagus nerve. Thus, an absent or diminished gag reflex does not indicate whether cranial nerve IX or X is involved. If the gag reflex is intact, however, this argues for the integrity of both cranial nerves IX and X.

Cranial Nerves and Speech Production

While discussing motor functions of the larynx, pharynx, soft palate, tongue, and lips, a few general comments about motor speech function are warranted. Speech requires the integrity of various cranial nerves that mediate movement of the jaw (CN V), the lips (CN VII), the tongue (CN XII), the soft palate and the pharynx (CN IX and X), and the larynx (CN X). All of these motor systems must be intact for well-articulated speech. The afferent feedback from these muscles also must be intact. Ultimately, the motor programming for the execution of speech is derived from higher cortical centers, but these cortical commands must be executed at the level of the cranial nerves. To vocalize certain guttural sounds such as hard K or G sounds or "Ahhh," the soft palate must elevate and close off the nasal–pharyngeal passageway. Failure of the palate to elevate produces a nasal quality to the voice. Phonation is carried out by the larynx and articulation is carried out by pharynx, soft palate, tongue, lips, and to a lesser extent the mandible. Lingual sounds are those that require maximal deviation of the tongue, such as "D," "L," and "T." Labial sounds are those that emphasize the actions of the lips, such as "B," "M," and "P." Thus, to test for proper elevation of the palate, the patient can say "Ahhh"; to test for proper coordination of soft palate, tongue, and lips in speech production, the patient may be asked to repeat "kuh, tuh, buh," either singly or as a group in sequence. In general, it is important to remember is that disturbances of speech do not always translate into "cortical lesions"; disturbances at the level of the brainstem (or cerebellum) also significantly can impact motor speech production (i.e., dysarthria).

CRANIAL NERVE XI (SPINAL ACCESSORY)

Major Functions: Turn Head
Lift Shoulders
Classification: Special Visceral Efferent (SVE)
Nucleus: Accessory (SVE)

The accessory (also known as *"spinal accessory"*) nerve is considered a pure motor nerve. It supplies the sternocleidomastoid muscle, which assists in turning the head to the contralateral side, and the upper trapezius muscle, which allows for lifting or shrugging of the shoulders. Despite its apparent simplicity, this nerve has several unusual features. First, it is the only cranial nerve that has its major nucleus in the spinal cord and whose fibers arise at the spinal level. The accessory nucleus consists of cells in the lateral portion of the anterior horn of the spinal cord at the level of C-1 through C-5. The rootlets emerge from the cord at these levels, but then ascend through the foramen magnum before exiting as the 11th cranial nerve. The cerebral source of activation for the trapezius is in the contralateral precentral gyrus. The tracts for the cortical innervation of the spinal nuclei responsible for the contraction of the sternocleidomastoid muscles have not been well defined. However, it has been suggested that these connections are predominately ipsilateral. There is certain logic to this, since the activation of this muscle causes the head to turn to face the contralateral hemispace. It also is unclear whether a small group of fibers that originate in the caudal sections of the nucleus ambiguus that supply the larynx should be considered part of the accessory nerve. These fibers travel a very short distance with cranial nerve XI before joining the vagus nerve to the larynx. Some authors believe that these fibers more properly belong to the tenth cranial nerve.

Lesions of the Spinal Accessory Nerve

Sternocleidomastoid

The normal contraction of the sternocleidomastoid muscle results in the turning of the head to the opposite side. Hence, in testing the integrity of the muscle (nerve), the patient is asked to turn his or her head against resistance to each side. Weakness when attempting to turn the head to one side suggests a lesion of the contralateral 11th cranial nerve.

Trapezius

The upper trapezius serves to support the shoulders and upper back. Lesions affecting this part of the nerve may result in a slight drooping of the shoulder. In formal examination, the patient is asked to shrug the shoulders against resistance, observing for asymmetries in strength or changes in the position of the scapula (downward and outward rotation).

CRANIAL NERVE XII (HYPOGLOSSAL)

Major Function: Movement of the Tongue
Classification: General Somatic Efferent
Nucleus: Hypoglossal (GSE)

The hypoglossal nerve is considered to be a pure motor nerve. Like cranial nerve XI, the hypoglossal nerve consists of multiple rootlets. The hypoglossal nerve is found in the rostral medulla just a little below the abducens nerve (CN VI). It exits the medulla in the anterolateral sulcus (between the pyramids and the olives), thus placing the hypoglossal rootlets more anteriorly in the brainstem than cranial nerves IX, X, and XI, all of which exit lateral to the olive.

The solitary function of cranial nerve XII is to control movements of the tongue. A unilateral lesion involving the nucleus or nerve (i.e., a LMN lesion) will result in atrophy on the side of the lesion and weakness of the muscles on that side. Because of the manner in which the muscles operate the tongue when protruded, weakness on one side will result in the tip of the tongue being deviated to the same side as the lesion when the tongue is protruded. This is because as the tongue is pushed forward, the muscle on the intact side of the tongue is unopposed by the muscle on the affected side.

Lesions of the Hypoglossal Nerve

The first step in attempting to identify a possible dysfunction of the hypoglossal nerve is to simply observe the tongue for deviation (toward side of the lesion) on protrusion or for atrophy or fasciculations. One also can test the strength of the tongue when pressed against the inside wall of the cheek. Here one would look for weakness when asked to press on the cheek opposite the lesion. Consistent with the findings in the peripheral musculature, atrophy of the tongue and fasciculations are associated with lesions of the nucleus or nerve (lower motor neuron lesion). In lower motor neuron lesion the tongue will deviate toward the side of the lesions, whereas in an upper motor neuron lesion (because of the contralateral representation) the deviation will be to the side opposite the lesion.

Endnotes

1. These components are sometimes referred to as "branchial motor" since they are derived from the branchial arches: if we were fish, these would have developed into gills.

2. While a medial and (less frequently) an intermediate olfactory stria also are described, most authors seem to agree that the lateral olfactory stria represent the major cortical pathways for olfactory input. The medial and intermediate olfactory stria may have very limited olfactory input, if any, to the subcallosal or septal area and to the region of the anterior perforated substance.

3. Although perhaps one of the less frequently tested cranial nerves, detecting the presence of anosmia may offer important clinical insights, especially following closed-head injuries. Nils Varney (1988) reported that in one sample of head injury victims, the presence of anosmia was very highly correlated with failure to maintain steady employment, despite the relative absence of deficits on formal neuropsychological

measures. Of course, the lack of vocational success within this group (along with their anosmia) was likely due to lesions of the orbital frontal cortices.

4. Why is it that when you look in a mirror, the image appears reversed right to left, but not up and down?

5. A lesion that selectively affects the lateral aspect of the chiasm (i.e., the fibers deriving from the temporal portion of the retina) may produce a nasal field cut in the ipsilateral eye. An internal carotid artery aneurysm that puts pressure on the lateral portion of the chiasm might produce such a finding.

6. As an individual ages, the capacity of the lens to make this adaptation diminishes; hence, the frequent need for bifocals beginning in the fourth or fifth decades.

7. Although anatomical documentation could not be found, we suspect that this crossing of superior rectus fibers allows for synapses within the contralateral nucleus, thus establishing a mechanism for conjugate upward gaze.

REFERENCES AND SUGGESTED READINGS

Barr, M.L. & Kiernan, J.A. (1993) The Human Nervous System. J.B. Lippincott Co. Philadelphia, pp. 122–147; 313–328.

Bender, M.B., Rudolph, S.H., & Stacy, C.B. (1982) The neurology of the visual and oculomotor systems. In: Joynt, R. (Ed.) Clinical Neurology. J.B. Lippincott Co., Philadelphia, Vol 1, Chapter 12.

Brazis, P. Masdeu, J., & Biller, J. (1990) Localization in Clinical Neurology. Little, Brown & Co. Boston, pp. 94–267.

DeMeyer, W. (1980) Technique of the Neurologic Examination: A programmed text. McGraw-Hill, NY, pp 128–178; 264–297.

Gilman S., & Newman, S.W. (1992) Essentials of Clinical Neuroanatomy and Neurophysiology. F.A. Davis Co. Philadelphia, pp. 115–142; 193–206.

Martin, J.H. (1996) Neuroanatomy: Text and atlas. Stamford, CT:: Appleton and Lange, pp. 353–416.

Nolte, J. (1993) The Human Brain. Mosby-Year Book, Inc. St. Louis, pp. 176–245; 275–305.

Stern, B.J. Wityk, R.J., & Lewis, R.L. (1993) Disorders of cranial nerves and brainstem. In: Joynt, R. (Ed.) Clinical Neurology. J.B. Lippincott Co., Philadelphia, Vol 3, Chapter 40.

Varney, N. (1988) The prognostic significance of anosmia in patients with closed head trauma. *Journal of Clinical and Experimental Neuropsychology*, 10, 250–254.

Wiebers, D.O., Dale, A.J.D., Kokmen, E., & Swanson, J.W. (Eds.) (1998) Mayo Clinic_Examinations in Neurology. St. Louis: Mosby, pp. 87–150.

Wilson-Pauwek, L., Akesson, E. & Stewart, P. (1988) Cranial nerves: Anatomy and_Clinical Comments. B.C. Decker, Inc. Philadelphia.

6 THE BASAL GANGLIA

CHAPTER OVERVIEW

The basal ganglia may well come to represent the "ugly duckling" of the central nervous system. As knowledge of these structures has expanded, so too has our appreciation not only of their functional but also of their structural complexity. From a structural standpoint, new insights are constantly emerging with regard to their topographical boundaries, internal cytoarchitecture, intrinsic and extrinsic connections, and neurochemical substrates. The basal ganglia long have been viewed as being an integral part of the motor system. The identification of specific feedback loops among the cortex, basal ganglia, and thalamus involving both direct and indirect pathways, along with improved understanding of underlying neurotransmitter systems, have provided us with better insights into how these structures might contribute to both normal and abnormal movements. Equally as exciting, however, has been the relatively recent appreciation of the richness of basal ganglia connections, particularly those to prefrontal and limbic cortices. The latter, forming what might be characterized as independent neural networks subserving multiple cognitive and emotional behaviors, suggest that the basal ganglia may play an intricate role in modulating much more than simply motor expression. The concept of multiple feedback loops, especially those involving what has become recognized as the ventral striatal and ventral pallidal systems, have offered clues for a better understanding of a number of behavioral disturbances and psychiatric disorders. Knowledge of these systems along with their neurochemical substrates likely also holds the key to continued improvements in the treatment of both neurological and psychiatric disorders.

In the course of this chapter, the reader will first find a review the basic anatomy of the basal ganglia. Particular emphasis will be devoted to exploring their interrelationships with other brain structures, forming what has been characterized as multiple and at least quasi-independent neural networks, as mentioned above. Next, specific symptoms and neurobehavioral syndromes associated with lesions or functional disturbances involving various components of these networks will be explored. While in this chapter special attention will be paid to motor disturbances, the groundwork will be laid for trying to understand the role of the basal ganglia in mental and emotional disorders, a topic that will be further explored in Chapter 11. Finally, hypotheses regarding the potential role of the basal ganglia in normal behavior, both with regard to motor and non-motor activities will be reviewed. The reader is directed to Chapter 3 and Chapter 9 (Part III) for additional discussions of the probable roles of the cerebellum and various cortical regions, respectively, in motor and other behavioral expressions.

INTRODUCTION

The basal ganglia, in the general clinical connotation of the term, refers to a collection of subcortical nuclear masses that traditionally have been associated with the "extrapyramidal" motor system.[1] From a strict anatomical perspective, the term "basal ganglia" refers to those nuclear masses of the telencephalon that lie beneath the cortical mantle. These include the **caudate nucleus, putamen, globus pallidus, claustrum, amygdala,** and other basal–frontal nuclei, such as the **nucleus accumbens** and **substantia innominata.**[2] However, it is not uncommon for only the first three of these structures (the caudate nucleus, putamen, and globus pallidus) to be mentioned when clinicians discuss lesions of the basal ganglia.[3] All three structures have strong anatomical, neurochemical, and functional connections with the cerebral cortex and thalamus. However, they also have significant connections with other subcortical nuclei, particularly the subthalamus, ventral tegmental area, and the substantia nigra. These two latter areas represent the major sources of dopamine, one of the neurotransmitters most frequently associated with this system. Lesions in any of these areas can result in motor or other behavioral disturbances. Thus, when referring to the "basal ganglia" as an *anatomical entity*, the most common connotation is probably that of the caudate nuclei, putamen, and globus pallidus. However, when speaking of the basal ganglia as a *functional system*, the subthalamus, ventral tegmental area, substantia nigra, as well as parts of the thalamus proper are typically included.

There are several other terms that often are applied to portions of the basal ganglia that require definition. Unlike the term "basal ganglia," whose referents may vary, there appears to be more consensual agreement with respect to the following appellations. The term **corpus striatum** refers to the caudate nucleus, putamen, and globus pallidus.Thus it is comparable to but more precise than the term "basal ganglia" in its most common usage. In contrast, the terms **striatum** or **neostriatum** refer collectively to just the caudate nucleus and putamen. These structures share similar histology and general patterns of connectivity and in fact are joined anatomically in their most rostral aspects prior to being separated by the fibers of the internal capsule. The nucleus accumbens, which appears to represent a ventral extension of the rostral caudate and putamen (see below), is commonly referred to as the **ventral striatum**. The putamen and globus pallidus also are grouped together and referred to as the **lentiform** or **lenticular nuclei,** because of their wedge or "lens" shaped appearance, both in coronal and axial sections. Finally, the globus pallidus is sometimes known as the **pallidum** or **paleostriatum** (Table 6–1).

Just as there may be some confusion when considering the anatomy of the basal ganglia, their functional significance is also a bit of a mystery.[4] As noted above, the basal ganglia long have been considered an integral part of the motor system. Their major influence on peripheral motor systems appears to be via the corticospinal or corticobulbar tracts (i.e., the "pyramidal" system) by way of thalamocortical feedback loops or pathways. It is predominately through these thalamocortical feedback loops that the basal ganglia are believed to help "fine tune" cortically generated movements. As will also be noted, although the consequences of lesions or disease to the basal ganglia on the motor system have been well documented, there still is considerable debate when attempting to define their contribution to normal motor activities.

In addition to the more traditional association between the basal ganglia and motor functions, we also will see that a substantial portion of the afferent and efferent connections of the basal ganglia extend well beyond the boundaries of the sensorimotor cortices. The more recent discoveries of extensive connections with prefrontal and corticolimbic structures suggest that the basal ganglia likely exert significant influences on cognitive, emotional, and

Table 6–1. Various Nomenclature for Basal Ganglia Nuclei

Term	*How Used*	*Nuclei*
Basal ganglia	Broad anatomical (exceedingly rare)	Caudate, putamen, globus pallidus, claustrum, amygdala, frontobasalar n.
Basal ganglia	Common anatomical	Caudate, putamen, globus pallidus
Basal ganglia	Clinical/functional	Caudate, putamen, globus pallidus, substantia nigra, subthalamus.
Corpus striatum		Caudate, putamen, globus pallidus
Striatum (neostriatum)		Caudate and putamen
Lentiform (lenticular) nucleus		Putamen and globus pallidus
Paleostriatum (pallidum)		Globus pallidus

motivational systems. However, as with their motor functions, determining the nature and scope of these influences is often difficult. Before addressing these functional issues, it might be helpful to review the anatomy of these various nuclei and their interconnections.

ANATOMY

Caudate Nuclei

The caudate nuclei are two, deep, midline subcortical nuclear masses that developed into elongated "C"-shaped structures as the brain developed and the telencephalon expanded into its current state. The shape of the caudate nuclei resembles an elongated tadpole with an enlarged portion or "head" located anteriorly or rostrally, a tapering body and a long, slender "tail" that curves in a posterolateral and ventral direction. As can be seen in Figures 6–1 and 6–2(d–h), the head forms a distinct bulge on the lateral wall of the frontal horn of the lateral ventricle, with the body maintaining a comparable position along the lateral wall of the body of the lateral ventricle. The tail ends up lying in the dorsum or roof of the inferior (temporal) horn of the lateral ventricle, terminating near the amygdala in the antero-medial temporal lobe. When viewed in either a horizontal or coronal section, the head of the caudate nucleus protrudes into the lateral wall of the anterior horns of the lateral ventricles, creating their characteristic "boomerang" shape. The caudate is continuous with the putamen in its most rostral aspect, but the caudate and putamen soon become separated into distinct nuclei by the anterior limb of the internal capsule (Fig 6–1a). At this level, the nucleus accumbens represents the ventral extension of the conjoining of these nuclear masses. Throughout its course, the stria terminalis, a fiber pathway connecting the amygdala with the hypothalamus and septal regions of the basal forebrain, lies adjacent to the caudate.

Putamen

As previously noted, the putamen, which is histologically similar to the caudate, is continuous with the latter in their more rostral extensions (see Fig 6–2c, right side). Even as the internal capsule cleaves these bodies, strands of gray matter linking the two still may be appreciated traversing the internal capsule in Figure 6–1a. These strands are known as the *transcapsular striae* or *cell bridges*. As can be seen in the accompanying figures, as it progresses caudally, the putamen becomes clearly separated from the head of the caudate by the anterior limb of the internal capsule. The external capsule and the claustrum border the lateral aspect of the putamen. The globus pallidus lies on its medial surface, the two being separated by the **lateral** (*external*) **medullary lamina** of the globus pallidus, which may be seen in Figure 6–1c. Recall that the putamen and globus pallidus, which form a triangular or wedge-shaped mass, are collectively referred to as the "lenticular" nuclei. This perhaps can be seen most clearly on the axial sections (Figure 6–2) where the lenticular nuclei are lateral to and bounded by the anterior and posterior limbs of the internal capsule. While both the caudate and putamen receive considerable input from the cortex (see below), the putamen receives a disproportionate share from the primary sensorimotor cortices, while the caudate is more closely related to cortical association areas.

Globus Pallidus

The globus pallidus (literally, *pale sphere*) phylogenetically is an older structure (paleos-triatum). As its name implies it has a somewhat "paler" appearance than other basal ganglia structures as a result of the multitude of myelinated fibers traversing it. The globus pallidus constitutes the medial portion of lenticular nuclear complex. It is separated from the putamen

(a)

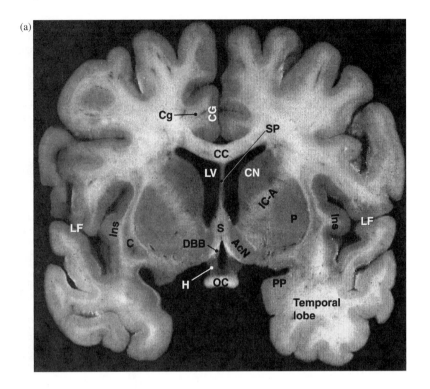

(b)

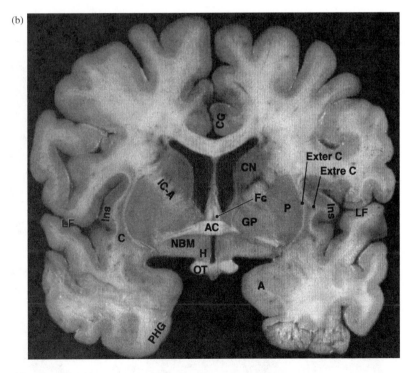

Figure 6–1. *(Continued)*

(c)

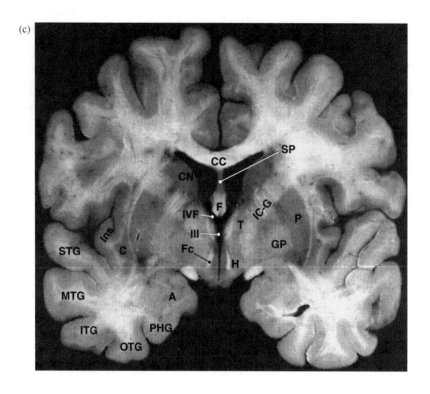

(d)

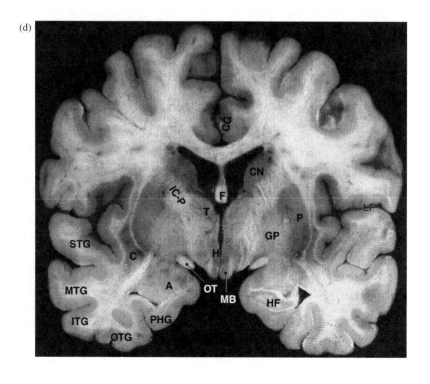

(e)

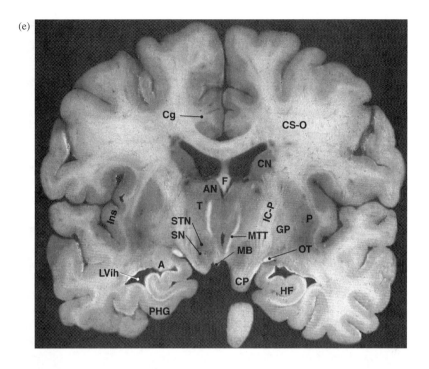

(f)

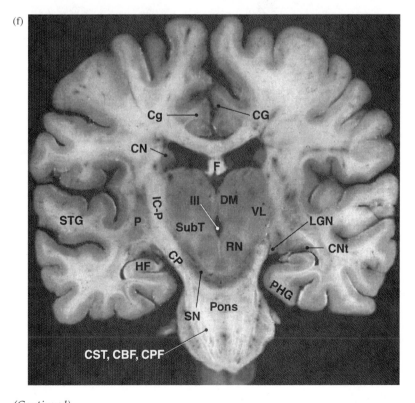

Figure 6–1. *(Continued)*

(g)

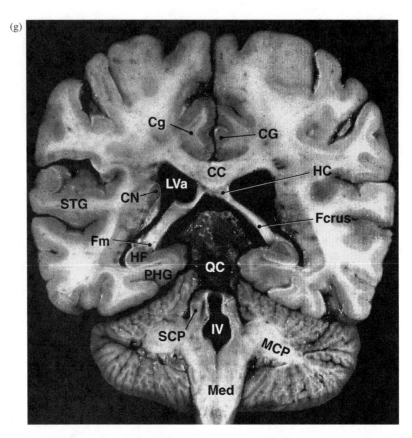

Figure 6–1. Coronal sections through the basal ganglia. Brain images were adapted from the *Interactive Brain Atlas* (1994), courtesy of the University of Washington.

III, 3rd ventricle	DBB, diagonal band of Broca
IV, 4th ventricle	DM, thalamus (dorsal medial nucleus)
A, amygdala	DN, dentate nucleus
AC, anterior commissure	Exter C, external capsule
AcN, nucleus accumbens	Extre C, extreme capsule
AN, thalamus (anterior nucleus)	F, fornix
BN, basal nuclei	F_C, fornix (columns of)
C, claustrum	F_{crus}, crus of fornix
CA, cerebral aqueduct	F_M, fimbria
CC, corpus callosum	GP, globus pallidus
CCg, corpus callosum (genu)	GR, gyrus rectus
CCs, corpus callosum (splenium)	H, hypothalamus
CG, cingulate gyrus	HC, hippocampal commissure
Cg, cingulum	HF, hippocampal formation
CN, caudate nucleus (head)	IC, inferior colliculus
CNb, caudate nucleus (body)	IC-A, internal capsule (anterior limb)
CNt, caudate nucleus (tail)	IC-G, internal capsule (genu)
CP, cerebral peduncle	IC-P, internal capsule (posterior limb)
CS-O, centrum semiovale	Ins, insular cortex
CBF, corticobulbar fibers	ITG, inferior temporal gyrus
CPF, corticopontine fibers	IVF, foramen of Monro
CST, corticospinal tract	LF, lateral fissure

by the lateral medullary lamina and from the thalamus by the internal capsule. The globus pallidus is divided into two components: a **medial** or **internal segment** (*GPi*) and a **lateral** or **external segment** (*GPe*) by a **medial** (*internal*) **medullary lamina**. Although both segments receive afferent input from the caudate and putamen, as will be seen later, the pattern and functional significance of their other connections would appear to differ significantly. For now, one might simply note that while the output of the external segment is directed primarily to the substantia nigra and subthalamic nuclei, the medial or internal segment's main output is to the thalamus. The latter represents the primary source of the lenticulothalamic fibers, which in turn constitute a major part of the cortical feedback loops via the thalamus (see below).

Ventral Striatum and Ventral Pallidum

On coronal brain sections at the level of the anterior commissure and optic chiasm (see Figure 6–1b), an area of gray matter can be found lying between the horizontal plane of the anterior commissure and the ventral surface of the brain. With the exception of the preoptic nuclei of the hypothalamus that occupy part of this region, this area was once simply referred to as the **substantia innominata**, reflecting the anatomists' uncertainty as to the specific origin or nature of these nuclei. Now it is recognized that this area includes the **basal nucleus of Meynert** (the origin of multiple cholinergic pathways), the extended amygdala,[5] as well as the ventral, posterior extension of the nucleus accumbens or the ventral striatum and the ventral extension of the globus pallidus or ventral pallidum. As we shall see, these ventral portions of the corpus striatum intimately are associated with structures that are related to emotional processing and motivation (Heimer, 2003). Furthermore, these ventral portions of the basal ganglia also contribute to cortical feedback loops through the thalamus (see discussion of cortical circuits later in this chapter).

Subthalamus

The subthalamus, which lies below or caudal to the thalamus around the lateral walls of the third ventricle (Figure 6–1e,f), contains several discrete nuclear groups. These nuclear groups include the **zona incerta**, **field H of Forel**, and the **subthalamic nucleus of Luys**.

Figure 6–1.

LGN, lateral geniculate nucleus	PHG, parahippocampal gyrus
LV, lateral ventricle	PP, prepyriform cortex
LV$_A$, Lateral ventricle (atrium)	Pul, pulvinar
LV$_{IH}$, lateral ventricle (inferior horn)	QC, quadrigeminal cistern
MB, mammillary body	RN, red nucleus
MCP, middle cerebellar peduncle	S, septal nuclei
Med, medulla	SC, superior colliculus
MGN, medial geniculate nucleus	SCP, superior cerebellar peduncle
MI, massa intermedia	SN, substantia nigra
MTG, middle temporal gyrus	SP, septum pellucidum
MTT, mammillothalamic tract	STG, superior temporal gyrus
OC, optic chiasm	STN, subthalamic nucleus
OT, optic tract	SubT, subthalamus
OTG, occipitotemporal gyrus	T, thalamus
P, putamen	V, vermis (of cerebellum)
PC, posterior commissure	VA, thalamus (ventral anterior nucleus)
	VL, thalamus (ventral lateral nucleus)

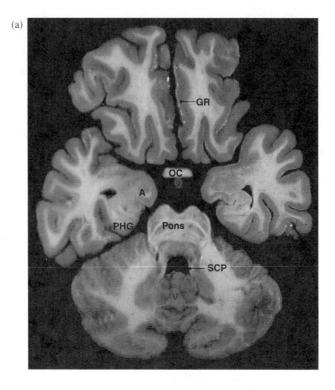

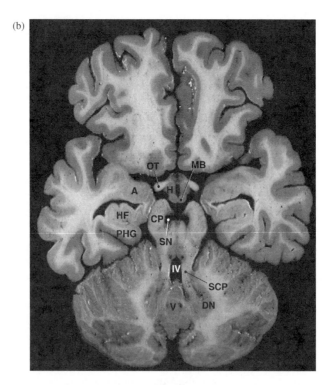

Figure 6–2. Axial sections showing structures of the basal ganglia. Brain images were adapted from the *Interactive Brain Atlas* (1994), Courtesy of the University of Washington.

(c)

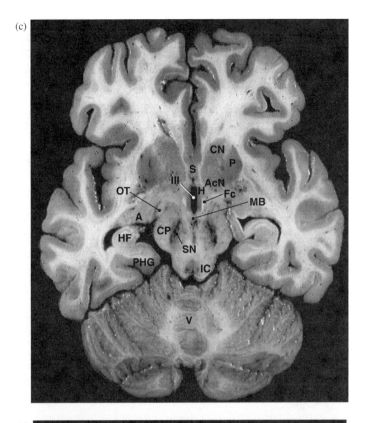

(d)

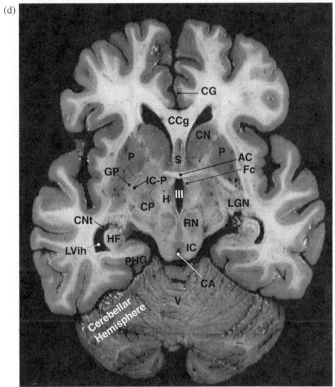

Figure 6–2. *(Continued)*

(e)

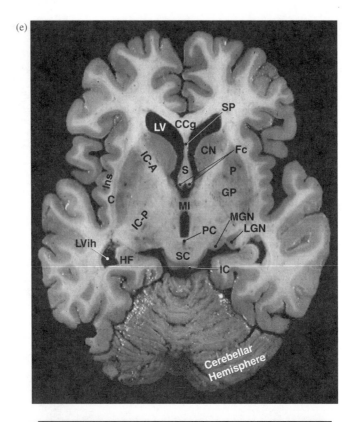

(f)

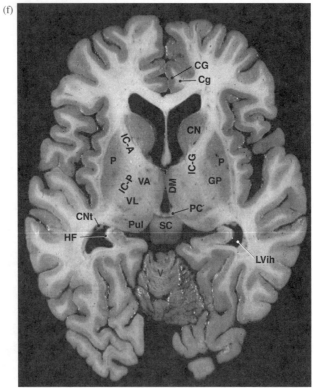

Figure 6–2. *(Continued)*

(g)

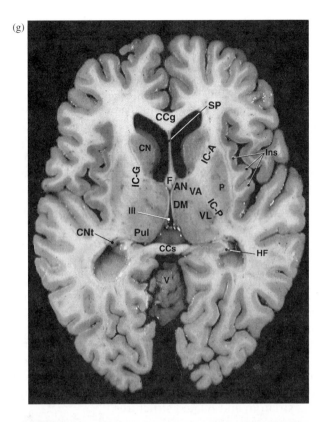

(h)

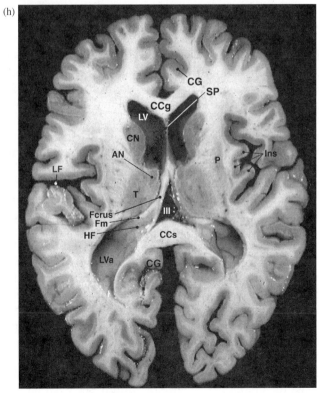

This latter nuclear group is more or less continuous with the substantia nigra in its caudal extension and plays an integral part in the corticothalamic feedback loops to be discussed later.

Substantia Nigra

As noted in Chapter 4, the substantia nigra lies in the tegmentum of the rostral midbrain just dorsal to the cerebral peduncles (Figure 6–1e,f and 6–2b,c). It is composed of two major cellular groupings: the more medially located **pars compacta** and the **pars reticularis**, which lies more laterally. The pars compacta, along with the adjacent **ventral tegmental area**, are the source of the dopaminergic pathways. The pars reticularis is structurally similar to the medial segment of the globus pallidus, and like the latter is a major source of efferent fibers to the thalamus.

CONNECTIONS

Probably the most common perception of the functions of the basal ganglia relates to their role in **feedback loops** that facilitate the smooth execution of movements or facilitate transitions between individual motor acts. These feedback loops include the cortex, corpus striatum, substantia nigra, and subthalamic and thalamic nuclei. However, as we later trace several of these major pathways or connections, it will become clear that the role of the basal ganglia is far more complex than initially thought and as noted above likely contributes to various cognitive and emotional aspects of behavior. These pathways and the effects of their disruption will be discussed below. First, it might be helpful to review the major inputs and outputs of the corpus striatum and related structures.

Striatum (Caudate and Putamen)

Afferents

The primary input to the striatum is from the cerebral cortex via both the internal and external capsules. This input is highly topographically organized and is not restricted to regions that are directly associated with motor control, but comes from diverse cortical areas, particularly association cortices. Both the caudate and putamen receive inputs from sensorimotor areas of the cortex and from frontal, parietal, and temporal association areas. These inputs, however, are differentially distributed such that the putamen receives the majority of its input from sensorimotor cortex, while the caudate has more extensive input from non-motor, association areas. The caudate receives projections throughout its length, but a disproportionate amount of fibers enter its anterior enlargement or "head." The majority of these latter fibers emanate from the tertiary or "prefrontal," dorsolateral association cortices, suggesting the caudate plays a role in the more cognitive aspects of motor or other "executive" types of behavior. Hence, lesions of the caudate can result in "frontal-type" cognitive deficits (Cummings & Benson, 1990).

In contrast, the orbital and mesial frontal cortices appear to project primarily to the ventromedial portions of the striatum, providing an anatomical basis for still further functional differentiation within the striatum. Specifically, the nucleus accumbens receives its inputs largely from structures associated with the limbic system. These sources include the anterior cingulate gyrus, the basolateral amygdala, and parts of the parahippocampal gyrus. The extended amygdala (including the bed nucleus of the stria terminalis) is structurally adjacent to the ventral striatum and may influence it either directly or through the amygdala proper via the stria terminalis. These connections may allow motivational influences to impact on the basal ganglia (e.g., the emotional valence of stimuli; see Chapter 8).

In addition to these more rostral inputs, the striatum also receives considerable input from the pars compacta of the substantia nigra and from the adjacent ventral tegmental area. Both of these areas represent dopamine projections. **Serotonergic** pathways from the raphe nuclei in the brainstem also provide input to the striatum. Finally, as we shall see, in addition to feedback loops from the cortex and substantia nigra there also is a thalamic loop. In this case, the afferent connections primarily emanate from the intralaminar (e.g., centromedian) and midline thalamic nuclei.

The cortical input to the striatum is largely if not exclusively **glutaminergic** and is facilitatory or excitatory in nature. The **dopamine pathways** from the substantia nigra (pars compacta) and ventral tegmental area appear to have either facilitatory or inhibitory influences depending on their site of action. As will be seen in the following paragraph, the striatum sends projections to both the internal and external sections of the globus pallidus. The internal globus pallidus, in turn, projects directly to the thalamus, whereas the cells of the external segment, although also eventually projecting to the internal segment and onto the thalamus, do so largely by way of the subthalamus. The dopaminergic pathways that project back to the striatal cells that in turn are destined for the external segment of the globus pallidus utilize D_2 type receptors and are inhibitory in nature. Those projecting to striatal cells destined for the internal segment of the globus pallidus appear to be excitatory, utilizing D_1 type receptors (Albin, Young, & Penny, 1989; DeLong, 1990; Mink & Thach, 1993). The possible clinical significance of these anatomical differences will be discussed later in this chapter when these connections are reviewed in light of certain disease processes affecting the basal ganglia. The serotonergic pathways from the raphe nuclei also tend to be inhibitory in nature.

Efferents

The primary outputs of the striatum are to the ipsilateral globus pallidus and to the substantia nigra. Projecting both to the internal and external segments of the pallidum, the output of the striatum, like its input, remains topographically organized. These various striatopallidal and striatonigral pathways are known to use **gamma-aminobutyric acid** (GABA) in their chemical transmission and are thought to be inhibitory in nature.[6] Just as depletions in dopamine have been associated with Parkinson's disease, reductions in GABA have been linked to Huntington's disease, which is associated with degeneration of the striatum, particularly the caudate nucleus. However, disturbances in neurochemical transmission involving some of these same substances that largely have been associated with the basal ganglia (e.g., dopamine, GABA, and serotonin) also appear to play a major role in many psychiatric disturbances (see Chapter 11). Again, this reinforces the notion that the various nuclei that make up the basal ganglia are more than just "motor" structures.

Globus Pallidus

Preliminary Considerations

Before reviewing the connections of the internal and external segments of the globus pallidus, a few preliminary observations may be useful. One, if not the major, role of the basal ganglia is thought to be to modulate the activity of the cerebral cortex. In accomplishing this, cortical information is funneled through the striatum to the thalamus via the internal and external segments of the globus pallidus. The thalamus in turn projects back to the cortex. While the pallidothalamic connections appear to be primarily inhibitory in nature, the thalamocortical projections are thought to be mostly facilitatory.

To keep it relatively simple, think for a moment about motor activity. In carrying out motor actions, it is necessary that certain motor groups (e.g., the target and agonist muscles) be facilitated, while other groups (antagonist muscles) be relaxed or inhibited. However, since most if not all thalamocortical output is excitatory, one way to modulate cortical activity would be to facilitate those thalamocortical connections to the agonist muscles, while inhibiting similar connections to the antagonist muscles. The "direct" and "indirect" pathways through the globus pallidus to the thalamus seem ideally suited to this task.

"Direct" and "Indirect" Pathways

As can be seen in Figure 6–3, in the **direct** pathway there are direct connections between the striatum (caudate nucleus and putamen) and the internal segment of the globus pallidus (GPi) before the latter sends its output to the thalamus. By contrast, in the **indirect** pathway the striatal nuclei first project to the external pallidal segment (GPe) and then detour through the subthalamus, before going on to the GPi and eventually to the thalamus. By tracing these connections it can be seen that by inhibiting the "inhibitory" pallidothalamic neurons (i.e., "disinhibition"), as is the situation in the "direct" pathways, the net effect

Direct Pathway   **Indirect Pathway**

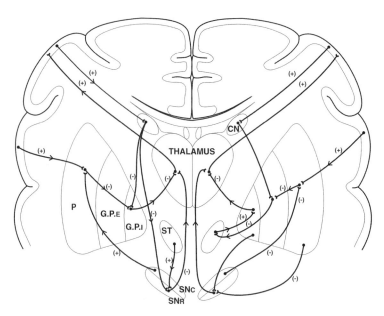

ST	Subthalamus
SNc	Substantia Nigra, Pars Compacta
SNR	Substantia Nigra, Pars Reticulata
P	Putamen
G.P.E	Globus Pallidus, Externa
G.P.I	Globus Pallidus, Interna
CN	Caudate Nucleus

Figure 6–3. Direct and indirect cortical-basal ganglia-thalamic feedback loops. Abbreviations: (+) facilitatory pathway; (−) inhibitory pathway.

will be to disinhibit the facilitatory glutaminergic thalamocortical neurons, thus creating an **excitatory** influence on those parts of the cortex. These connections then might serve to enhance the activity of agonist muscles. On the other hand, if selected (inhibitory) pallidothalamic neurons could be facilitated (as is the case in the "indirect," striatosubthalamopallidal pathways), the net effect on the targeted thalamic neurons will be **inhibitory**. Through this latter arrangement, antagonist muscle groups effectively could be toned down (i.e., normally excitatory thalamocortical feedback will be reduced).[7] Although greatly simplified, understanding these basic connections and their positive or negative influence on postsynaptic junctions helps set the stage for the discussions to follow later.

Afferents

The main source of afferent fibers into the globus pallidus is from the caudate and putamen. These fibers project to both the internal (medial) and external (lateral) segments of the pallidum from which the direct and indirect thalamic pathways are respectively derived. The other major identified source of pallidal input is from the subthalamic nucleus. Similar to the corticostriatal projections, a strong topographical organization persists, with different corticaareas ultimately projecting to distinct portions of the globus pallidus. This topographical organization is a pattern that, for the most part, appears to be maintained throughout these corpus striatal feedback loops. The input to the globus pallidus from the striatum is GABAergic and inhibitory, while the subthalamic-pallidal connections are glutaminergic and facilitatory (excitatory).

Efferents

As previously mentioned, the striatum sends its efferent projections to both the external and internal segments of the pallidum. In the more *direct route,* the internal segment sends the majority of its efferent fibers directly to the thalamus. Whereas in the *indirect pallidothalamic pathway,* the external segment establishes connections with the subthalamic nucleus before it, in turn, sends projections back to the internal segment of the globus pallidus and then on to the thalamus. Some of the more well-established pallidothalamic connections include those to the ventral–anterior (VA), ventral–lateral (VL), and dorsomedial (DM) nuclei of the thalamus. The fibers that travel to the VA and VL nuclei tend to originate in the putamen and project back to the sensorimotor cortices. Conversely, the fibers that travel to the DM nuclei generally represent caudate projections and send fibers back to the prefrontal regions (see Chapter 7). The ventral striatum, which receives input from various limbic structures, also projects to the magnocellular portion of the dorsomedial nuclei, which in turn projects to the anterior cingulate region. Finally, other probable thalamic projections include those to intralaminar nuclei (especially the centromedian and parafascicular nuclei) which project back to the striatum.

The pallidothalamic fibers, all of which emanate from the internal segment of the globus pallidus, can take one of two routes to the thalamus. The more dorsal portion of the pallidum sends fibers medially and slightly caudally, traversing the internal capsule on their way to the thalamus. This pathway is known as the **lenticular fasciculus** *(field H2 of Forel)* (Figure 6–4). By contrast, those pallidothalamic fibers leaving the more ventral portion of the internal segment take a more caudal loop into the prerubral area before joining up with the fibers of the lenticular fasciculus and with fibers ascending from the cerebellum to form the **thalamic fasciculus** *(field H1 of Forel)*. Those fibers that comprise the descending loop prior to forming the thalamic fasciculus are known as the **ansa lenticularis**. Both the fibers of the lenticular fasciculus and the ansa lenticularis are primarily **GABAergic,** and hence, are inhibitory. The globus pallidus also sends a much smaller group

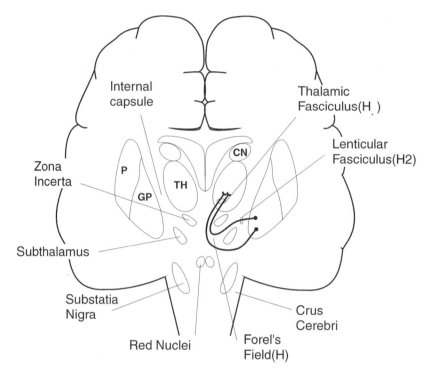

Figure 6–4. Thalamic and lenticular fasciculi.

of fibers to the habenular nuclei (epithalamus) via the **stria medullaris** and to the midbrain (pedunculopontine nuclei).

Subthalamus

Afferents

Several sources of input to this nucleus have been identified. The primary source of input, in terms of number of fibers, appears to be fibers coming from the lateral or external segment of the globus pallidus (via the **subthalamic fasciculus**). The subthalamus also receives input from the cortex, particularly the sensorimotor regions. Input also is received from the centromedian and parafascicular nuclei of the thalamus, the substantia nigra (pars compacta), and the pedunculopontine nucleus in the brainstem.[8] Like most of the other outputs of the globus pallidus, the pallidosubthalamic fibers are GABAergic and inhibitory. As in the case of the corticostriatal connections, the cortical inputs to the subthalamus are likely primarily glutaminergic (excitatory).

Efferents

The two primary outputs of the subthalamus are to the globus pallidus and to the substantia nigra (pars reticulata). Projections to the globus pallidus are to both the internal and external segments (Ma, 1997). The internal or medial segment completes the "indirect" feedback loop to the cortex (cortex → striatum → globus pallidus, external division (GPe) → subthalamus → globus pallidus, internal division (GPi) → thalamus → cortex). A smaller number of fibers from the subthalamus appear to project back to the striatum. The efferent fibers of the subthalamus are all primarily glutaminergic, and hence, have an excitatory effect on the nerve cells to which they project.

Substantia Nigra

Afferents

The most prominent inputs to the substantia nigra are those from the caudate and putamen. These inputs are largely GABAergic. Additional inputs come from the subthalamus, globus pallidus (primarily from the internal segment), and **serotonergic** fibers from the raphe nuclei. The cortex also may send a small number of fibers directly to this nuclear group. Most of the afferent connections are to the **pars reticulata** portion of the substantia nigra.

Efferents

The substantia nigra sends fibers back to the striatum, primarily from the **pars compacta**. *These nigrostriatal pathways represent the major source of dopaminergic input to the striatum.* It has been noted that these nigrostriatal connections may be either excitatory (D_1 receptors) or inhibitory (D_2 receptors), depending on the subtype of striatal dopamine receptors on which they synapse. Neurons in the caudate or putamen that project to the internal segment of the globus pallidus appear to utilize primarily D_1**-type receptors.** These synapses are **facilitatory.** On the other hand, those nigral fibers that project back to striatal cells destined for the external segment utilize D_2 **receptors** and tend to be **inhibitory**.

The **pars reticulata** portion of the substantia nigra represents the other major source of basal ganglia output to the thalamus (along with the internal segment of the globus pallidus). This (nondopaminergic) portion of the substantia nigra sends efferent fibers to the ventral anterior and ventral lateral nuclei of the thalamus. The pars reticulata also projects to the dorsomedial nucleus, which in turn projects back to prefrontal cortical areas. These nigrothalamic fibers terminate in different areas of the thalamic nuclei than do the pallidothalamic pathways, suggesting they may be mediating different functions. Additional efferent, nondopaminergic connections are sent to the superior colliculi and to the pedunculopontine nucleus (PPN). Recall that this latter nucleus (PPN) represents a confluence of cortical motor and cerebellar and basal ganglia input. Finally, the pars compacta has descending influences on the raphe nuclei (serotonergic) of the midbrain. Tables 6–2A and 6–2B provide a summary of suspected neurochemical pathways in the direct and indirect systems.

FEEDBACK LOOPS

As noted, the striatum, particularly the caudate nucleus and the ventral striatum, and nucleus accumbens receive substantial input from diverse cortical regions. Input to the putamen is primarily from the sensorimotor cortices, including the supplementary motor

Table 6–2A.   Direct Neurochemical Pathways

Pathway	Transmitter (main type)	Effect
Corticostriatal	Glutaminergic	Excitatory
Striatonigral	GABAergic	Inhibitory
Nigrostriatal	Dopaminergic(D_1)	Excitatory
Striatopallidal (internal segment)	GABAergic	Inhibitory
Pallidothalamic	GABAergic	Inhibitory
Thalamocortical	Glutaminergic (?)	Excitatory
(Net effect on cortical neurons: positive)		

Table 6–2B. Indirect Neurochemical Pathway

Pathway	Transmitter (main type)	Effect
Corticostriatal	Glutaminergic	Excitatory
Striatopallidal (external segment)	GABAergic	Inhibitory
Striatonigral	GABAergic	Inhibitory)
Nigrostriatal	Dopaminergic(D_2)	Inhibitory)
Pallidosubthalamic	GABAergic	Inhibitory
Subthalamopallidal (internal segment)	Glutaminergic	Excitatory
Pallidothalamic	GABAergic	Inhibitory
Thalamocortical	Glutaminergic (?)	Excitatory
(Net effect on cortical neurons: negative)		

area on the medial surface of the hemisphere. These latter corticostriatal pathways eventually project back to these same sensorimotor areas via the VA and VL nuclei of the thalamus. The remaining corticostriatal fibers largely come from association cortices in the frontal, parietal, and temporal lobes. The majority of these connections appear to return to the frontal association cortices. Given that these pathways originate from the cortex, project to the striatum, and then back to the cortex (via the thalamus), these pathways are referred to as *corticostriatocortical loops*. Five such corticostriatocortical pathways or loops have been identified (Alexander, DeLong, & Strick, 1986) and are described below.

Motor Circuit

This loop likely originates from cortical neurons in the **primary motor, supplementary motor, primary somatosensory**, and possibly in adjacent association **cortices** (Figure 6–5). Axonal fibers project primarily to the putamen. As is probably true of all corticostriatal pathways, there are both "direct" and "indirect" routes to the thalamus via the internal and external segments of the globus pallidus, substantia nigra, and subthalamus. In the case of the "motor" circuit, the primary thalamic projection areas are the **ventral anterior** (VA) and **ventral lateral** (VL) **nuclei**, with the topographical organization being well maintained. These nuclei, in turn, project back to the primary motor, premotor, and supplementary motor cortices. While this system primarily projects back up to the cortex, some fibers exert an independent descending influence on the spinal motor pathways via the pedunculopontine nuclei in the midbrain.

Oculomotor Circuit

It might be tempting to view this loop simply as a special subset of the motor circuit that begins and ends in the **primary** and **supplementary eye fields** were it not for a few important differences. First, some of its cells of origin appear to derive from the dorsolateral prefrontal cortex. Second, while the putamen is the primary projection site in the neostriatum for the motor circuit, the oculomotor pathway appears to converge primarily in the caudate nucleus. From there fibers (again via direct and indirect routes) are sent not only to ventral anterior, but also to the **dorsomedial** (DM) **nuclei** of the thalamus. Finally, whereas the motor circuit has connections with the pedunculopontine nucleus of the midbrain, the oculomotor loop establishes connections with the superior colliculi, which, as we have noted in Chapters 4 and 5, are important in oculomotor reflexes and eye movements. This system is apparently important in executing voluntary eye movements or in conducting visual searches of one's environment.

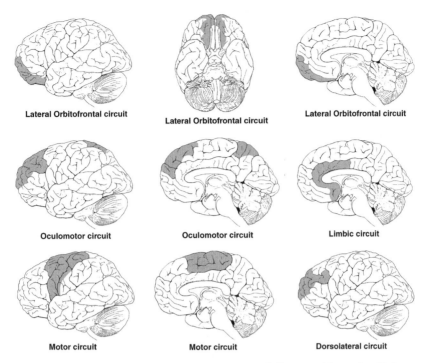

Figure 6–5. Proposed major cortical–basal ganglia circuits following Alexander, DeLong, and Strick (1986).

Dorsolateral Circuit

Again, this is where we begin to diverge from the older concepts of the basal ganglia as strictly a motor system. This "circuit" arises primarily from the **dorsolateral frontal association cortices**, although some of the parietal and temporal association areas also appear to contribute to the afferent side of this basal ganglia loop. Most of the frontal association fibers travel to the head of the caudate, while the parietal and temporal cortices likely project to the body and tail of this nucleus. After going through the usual pallidonigral loops, this system projects both to the **VA and DM nuclei** of the thalamus. From there, the majority of the thalamocortical fibers end up back in the dorsolateral or "prefrontal" association cortices. Because of its connections, this loop is thought to play a role in the higher cognitive or "executive" functions normally associated with this part of the brain (see Chapter 9).

Lateral Orbitofrontal Circuit

This circuit appears to represent a transition between the dorsolateral loop described above and the "limbic" loop to be described below. As the name implies, the origins of the corticostriatal fibers appear to be generally in the **frontal orbital regions**, although this system also may pick up some fibers from the anterior temporal cortices as well. Whereas the dorsolateral circuit tended to project most heavily to the dorsal head of the caudate, this system tends to utilize the more ventral or ventromedial portions of the head of the caudate. Similar to the dorsolateral circuit, the primary thalamic areas on which these fibers converge are also the **VA** and **DM** nuclei. However, their relative distributions within these nuclei are different. The lateral orbitofrontal system tends to concentrate on the magnocellular portions of these nuclei, in contrast to the parvicellular layer of the VA and probable more diffuse projection to the DM nuclei for the dorsolateral circuits. As is typical of these circuits, the resulting thalamocortical projections are back to the lateral orbitofrontal cortices. As we shall see later in discussing the limbic system and the frontal cortex, these orbitofrontal areas probably are important in mediating basic emotional drives such as those involved in self-defense (e.g., fight or flight) or appetitive instincts (e.g., sex attraction or hunger) and environmental contingencies or learned "social controls." Hence, disturbances of this circuit may manifest as behavioral changes (e.g., disinhibition, emotional lability). Conversely, disorders of this system that lead to excess (as opposed to deficient) activation might lead to over-control as opposed to diminished control. One example of the latter is thought to be obsessive–compulsive disorders (OCD), which have been linked with increased metabolism in this circuit (see: LaPlane et al., 1989; Insel, 1992; Mega & Cummings, 1994; Modell et al., 1989; Stahl, 1988; Baxter et al., 1992; Zald & Kim, 1996a,b).

Limbic Circuit

In the limbic circuit, striatal input primarily is from allocortex (or juxta-allocortex), including the anterior **cingulate gyrus,** medial **orbitofrontal** areas, **hippocampal gyrus,** and perhaps some portions of **temporal** neocortex. The striatal and pallidal projections also are distinct. In this case, the incoming fibers project to the region of the nucleus accumbens or ventral striatum and from there to the precommissural pallidum. These areas then project to the DM nucleus of the thalamus, which in turn projects back to the anterior cingulate and medial orbitofrontal cortex.[9] As we shall see in later chapters, disruption of these areas often leads to apathy or reduced drive or motivation and in extreme cases akinetic mutism. As was seen with the lateral orbitofrontal circuits, disturbances of this system often have been linked with psychiatric symptomatology, particularly OCD and certain features of schizophrenia. While it has been suggested for some time that schizophrenia may be associated with either an excess of dopamine or a hypersensitivity of dopaminergic receptors (see Chapter 11), it also has been hypothesized that some of the negative features of schizophrenia (such as apathy and withdrawal) may be the result of deficient neuronal activity in this medial frontal region (Pantelis et al., 1992). Finally, it is also interesting to note that selective lesions to the anterior cingulate gyrus have been found to be beneficial in relieving OCD symptoms (Ballantine, 1986).

From the foregoing discussion it is evident that while the different corticostriatocortical circuits utilize information from various sensory or posterior association areas, the thalamo-cortical radiations are invariably back to the frontal regions of the brain.[10] This organization suggests that these systems primarily are geared to impact on or modulate the execution of behavioral programs. As we shall see, this view is consistent with the current theories of the role of the basal ganglia on higher-order behavior.

Tables 6–3 and 6–4 summarize the major feedback loops that have been identified based on the different connections reviewed in the previous sections. Further specula-tions regarding the possible significance of several of these connections will be discussed following a review of some symptoms associated with lesions or dysfunction of this system. It should be kept in mind that the structures and connections listed here only offer a very broad, schematic representation of an exceedingly complex system. Each of the structures within the system has connections with structures lying outside the system, thus creating additional "open" circuits that can influence or be influenced by what takes place within these loops. Finally, refinements and/or additions to this schema can be expected in the future.

Table 6–3. Major Feedback Loops Involving Basal Ganglia Nuclei

1. **Cortical "direct" pathway**: Cerebral cortex → striatum → medial globus pallidus → thalamus → cerebral cortex[4]
2. **Cortical "indirect" pathway**: Cerebral cortex → striatum → lateral globus pallidus → subthalamus → medial globus pallidus → thalamus → cerebral cortex
3. **Cortical–nigral**: Cerebral cortex → striatum → substantia nigra → thalamus → cortex
4. **Midbrain**: Pedunculopontine nuclei → subthalamus → globus pallidus and substantia nigra → pedunculopontine nuclei
5. **Internal striatal**: Striatum → globus pallidus → thalamus → striatum
6. **Striatal–nigral**: Reciprocal, direct connections between the substantia nigra and the striatum
7. **Subthalamic**: Reciprocal, direct connections between the globus pallidus and the subthalamus

Table 6–4. Summary of Basal Ganglia – Cortical Feedback Loops

Circuit	Cortical Origins	Striatal/Thalamic Projections	Possible Functional Roles
Motor	Sensorimotor	Putamen/VA, VL	Modulate voluntary movements
Oculomotor	Visual eye fields	Caudate/VA, DM	Voluntary eye movements, visual search
Dorsolateral	Prefrontal? P-T	Caudate/VA, DM	Higher cognitive, executive functions
Lateral Orbito-frontal	Orbito-frontal, Ant. temporal	Caudate/VA, VM (ventral)	Emotional, instinctual drives and behavior
Limbic	Ant. cingulate, medial frontal, hippocampal	N.accumbens/DM	Drive, motivation

FUNCTIONAL CONSIDERATIONS

Having reviewed the better known structural associations of the basal ganglia, we might ask, "What are their clinical significance?" Before offering some speculations in this regard, it will be useful to review some of the more prominent symptoms and syndromes that have been classically linked with lesions of these nuclear masses and/or their pathways. It is to these that we now turn our attention.

Recognizing Disorders of the Basal Ganglia

Shortly after the turn of the century, S.A. Wilson described a disease that resulted from a genetic deficiency in copper metabolism. This disease, which is known as **hepatolenticular degeneration,** or *Wilson's disease,* was characterized by **muscle tremors, rigidity,** and **dystonia**. This disease was observed to first affect the liver and secondarily the brain. The lenticular nuclei, particularly the putamen, were most notably involved and Wilson was one of the first to use the term *"extrapyramidal"* in relation to the particular motor disturbances affecting the basal ganglia.

However, Wilson was not the first to identify a disease that was identified primarily with the basal ganglia. Almost 100 years earlier, James Parkinson had described a slowly progressive muscular disorder that bears his name. **Parkinson's disease**, which also is referred to as *"paralysis agitans,"* is associated with degeneration of the substantia nigra (pars compacta), which results in **depletion of dopamine**. Parkinson's disease is clinically characterized by the tetrad of (1) a coarse **resting tremor** (particularly of the hands),

(2) **muscular rigidity** and **resistance to passive movement**, (3) **bradykinesia** (slowness of movement), and (4) **diminished postural reflexes**. Parkinson patients also commonly manifest a slow shuffling gait, masked faces (a flat, expressionless face), micrographia (small, cramped handwriting), and diminished volume and prosody (emotional intonation) of speech. Memory loss and depression, although not uncommon, are less consistently present, especially during the early and middle stages of the disease.

In the latter part of the 19th century, yet another physician, George Huntington, described a disorder that bears his name, **Huntington's chorea**. It was noted that Huntington's chorea tended to run in families (an autosomal dominant disorder with 100% penetrance) and was characterized by bizarre and dramatic choreiform movements and mental deterioration. The genetic defect subsequently identified as being linked with this disorder is an excessive number of CAG (cytosine–adenine–guanine) trinucleotide sequences (>35) on chromosome 4, resulting in a mutant form of the **huntingtin gene**, which is responsible for producing the huntingtin protein. The CAG sequence, in turn, is responsible for the production of the amino acid *glutamine* and inserting it into the huntingtin protein. Thus the excessive number of CAG sequences present in the gene results in an abnormally large huntingtin protein containing an excessively large amount of glutamine, which is thought likely to have a toxic effect on neurons, particularly the GABAergic spiny neurons of the caudate and putamen. Perhaps for this reason, Huntington's disease long has been thought to be associated with relative depletions of **GABA** (Bird et al., 1973; Perry, Hanson, & Kloster, 1973) and marked degeneration of the caudate nuclei. However, as will be noted in Chapter 11, changes in other neurotransmitters are likely also involved. The degeneration in the head of the caudate results in a flattening of the convexity of the anterior horns, a change that can be identified on CT or MRI scan (Figure 6–6). **Memory disorders** and other **mental and behavioral changes**, which commonly are associated with this disorder, may either antedate or develop subsequent to the **choreiform movements**. It should be noted that although subcortical basal ganglia structures primarily are involved in both Parkinson's and Huntington's diseases, other areas of the brain, including the cerebral cortex, are likely to be affected.

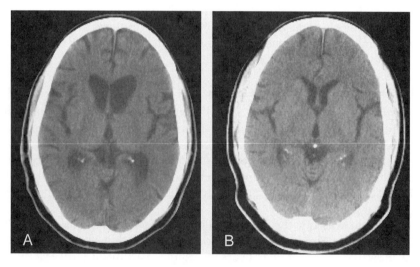

Figure 6–6. (a) Bilateral degeneration of the caudate nuclei in Huntington's disease. While some generalized cortical atrophy is present, compare the shape of the anterior horns of the lateral ventricles with that of (b) an age-matched control.

These and other acquired movement disorders (e.g., striatonigral degeneration, Sydenham's chorea, and hemiballismus) all have motor disturbances characterized by abnormal tone and/or involuntary movements due to involvement of the basal ganglia. These disorders commonly are referred to as diseases of the "extrapyramidal" motor system. This is because the basal ganglia were once thought to exert an influence on the peripheral musculature (or nervous system) that was somehow independent of the corticobulbar or corticospinal tracts or "pyramidal" system. However, as we have seen, these basal ganglia structures do not have a direct spinal pathway, but rather serve to modulate or influence cortical motor neurons that eventually give rise to the corticobulbar or corticospinal (pyramidal) tracts.

Specific Symptoms Associated with Disorders of the Basal Ganglia

Chorea (*choreiform movements*) refers to sudden, brief (although continuous over time) purposeless, unpredictable, involuntary jerks. These also have been described as "fragments of voluntary movements" and may involve the hands, limbs, trunk, or face. Although often bilateral, these abnormal movements may be restricted to one side of the body. Within these limits, the distribution of the jerking motions will be variable and apparently random. These abnormal movements are most commonly seen in Huntington's or Sydenham's chorea. In Sydenham's chorea the movements are more likely to be restricted to the limbs, whereas in **Huntington's disease**, the truncal musculature is more frequently involved.

Athetosis describes slow, irregular, writhing movements, predominately affecting the distal portions of the upper extremities, although more proximal muscles of the shoulders, hips, and trunk also can be involved. Athetosis of the facial muscles results in grimacing and abnormal movements of the tongue. Again these movements are involuntary and interfere with both active movement as well as the intention to keep a limb in a fixed position (at rest). This type of movement disorder appears to result from the simultaneous contraction of antagonistic muscles. Hypertonicity of the muscles is typically present. At times differentiation of chorea from athetosis is difficult. Elements of both might be simultaneously present and is referred to as **choreoathetosis**. Athetosis is commonly associated with **cerebral palsy** in which the basal ganglia, particularly the striatum, are affected.

Ballismus is a sudden, involuntary, "flinging" or throwing motion of an extremity. Although ballismus may involve both sides of the body, it more commonly is limited to one side of the body (*hemiballismus*) and typically is associated with lesions of the contralateral subthalamus. At times this disorder is thought to simply represent a particularly severe form of chorea.

Dystonia is characterized by a slow, sustained, or prolonged involuntary contraction of the trunk and proximal musculature, typically producing an abnormally contorted or "twisted" posture. Dystonia may result from multiple etiologies and again, either may be generalized or limited to certain, specific muscle groups. Examples of the latter include **spasmodic torticollis** or **blepharospasm** (a forceful closing of the eyelids). In some cases, the deformity may be limited to the distal portion of a single extremity. Hemidystonia may result from focal lesions (contralateral) of the striatum and/or thalamus.

Tremors may have multiple etiologies (some nonpathologic) and predominately may become manifest either at rest, in the performance of an action, or while maintaining a posture against gravity. The latter, for example, often are seen in cerebellar system lesions or chronic alcoholism. The type most commonly associated with basal ganglia disease is a **resting tremor** with a frequency of approximately 4 to 6 cps. Typically associated with **Parkinson's disease**, the tremor generally involves the fingers and wrists, often resembling a "pill-rolling" movement. This tremor typically disappears during intentional movements, although other disorders of movement might be seen (e.g., bradykinesia, rigidity, micrographia).

Table 6–5. Motor Symptoms Associated with Basal Ganglia Disease

Chorea:	Brief, purposeless, unpredictable, involuntary movements
Athetosis:	Slow, irregular, writhing movements, especially in distal upper extremities
Ballismus:	Sudden, involuntary, flinging or throwing motion of arm
Dystonia:	Slow sustained, involuntary contraction of trunk or proximal muscles
Resting tremor:	4 to 6 cps tremor while limb is at rest, most noticeable in fingers ("pill rolling") or hands
Rigidity:	Increased muscle tone, resistive to passive movement, either "lead pipe" (constant) or "cog-wheeling" (intermittent)
Bradykinesia:	Slowness in initiating or executing voluntary movement

Rigidity frequently is found in Parkinson-type syndromes and is manifested as increased muscle tone and resistance to passive movement. When the limb is passively moved, there may be evidence of rapidly alternating resistance and relaxation (*"cogwheeling"*) or steady resistance throughout the range of motion (*"lead-pipe" rigidity*), the former being the more common. When rigidity is due to basal ganglia lesions, deep tendon reflexes (DTR) may be normal, which helps to differentiate rigidity resulting from basal ganglia disease from upper motor neuron spasticity that is associated with hyperactive DTRs and "clasp-knife" rigidity on passive manipulation of the limbs.

Bradykinesia is characterized by slowness in the initiation and/or execution of voluntary movement. There also may be a reduction in the number of movements manifested (*hypokinesia*). Such patients may have difficulty in initiating walking and may walk with a slow, short, shuffling gait without a normal swing to the arms. Since patients with bradykinesia are slow to make postural adjustments, balance may be compromised. The face of such individuals may show limited emotional expressiveness (masked face). Writing tends to be slow and micrographic and speech is typically monotonic and decreased in volume (hypophonic). Table 6–5 summarizes these various symptoms.

ETIOLOGY AND EFFECTS OF DISRUPTION OF NEUROCHEMICAL PATHWAYS

Multiple syndromes involving the basal ganglia have been identified. These can result from a variety of causes, including:

1. Birth defects (cerebral palsy).
2. Genetic defects (dystonia musculorum deformans, Huntington's disease).
3. Metabolic deficiencies (Wilson's disease).
4. Infectious or inflammatory disease (Sydenham's chorea).
5. Systemic disorders (systemic lupus erythematosus).
6. Drug toxicity (e.g., phenothiazines → Parkinson-type syndrome, overdose of L-dopa → choreoathetosis).
7. Neurodegenerative disease (Parkinson's).
8. Structural lesions (stroke, tumor).

Earlier in this chapter some of the well-known neurochemical pathways that interconnect the various nuclear components of the basal ganglia and interface with the thalamus and cortex were reviewed. As was suggested previously, our knowledge and appreciation of the complexity of these pathways and their clinical significance is still quite limited.

However, while recognizing these limitations, even a highly simplified outline of some of the neurochemical mechanisms purported to underlie several of the more classic basal ganglia disorders might add to a better understanding and appreciation of the symmetry of this system. Because the chemical mechanisms and pathways that subserve the motor symptoms have been studied in greater detail, these will serve as our model. Several of the more common and well-studied syndromes and/or symptoms associated with lesions to the basal ganglia, along with their suspected primary chemical mechanisms and pathways, are reviewed below.[11]

Parkinson's Disease

It has been well established that parkinsonism is related to a degeneration of cells in the pars compacta of the substantia nigra. This cell loss and neuronal pathology in the substantia nigra results in a depletion of the neurotransmitter dopamine (Hornykiewicz & Kish, 1987; Hirsch, Graybiel, & Agid, 1988). As was seen in Figure 6–3, dopaminergic pathways normally exert a major influence on the striatum and this influence can be either facilitatory or inhibitory, depending on the particular cells to which they project and the type of receptors employed (D_1 versus D_2). Recall that the "direct" striatothalamic pathways tend to have a disinhibitory (i.e., a facilitating) effect on thalamocortical projections and the opposite is true of the "indirect" pathway, which tends to inhibit thalamic neurons. If there is a reduction in dopamine to the striatal (caudate or putamen) cells that project to the medial segment of the pallidum as a result of diminished D_1 connections ("direct" pathway), these striatal cells will show reduced rates of firing since dopamine tends to exert an excitatory influence on them. Consequently, since the striatopallidal connections (GABAergic) are inhibitory, there will be less inhibitory influences on the cells of the medial segment of the globus pallidus (GPi), which means the firing rate of these GPi cells should increase. This, in turn, should result in an increase in the normal inhibitory influence of the GPi on the VA/VL nuclei of the thalamus. Since these thalamic nuclei are thought largely to exert a facilitatory influence on the cortex, that influence will be diminished, resulting in diminished cortical excitation (reduced motor activity of targeted agonist muscles?) (see also Chapter 11).

A comparable effect would be expected with a reduction of dopaminergic input to the striatal cells projecting to the external pallidal segment ("indirect" pathway), but for different reasons. Recall that earlier it was noted that the "indirect" striatothalamic pathways are thought normally to exert an inhibitory influence on antagonist muscles as a way of inhibiting unwanted motor activity. Since dopamine is thought to exert an inhibitory influence on this "indirect" system via D_2 receptors, this means that dopamine normally would have a modulating effect on this inhibitory system. Thus, if the supply of dopamine is reduced, as in Parkinson's disease, this would be expected to result in an enhancement of the inhibitory influence of the "indirect" pathway as the normal inhibitory dopaminergic influence on the D_2 receptors in the striatum is diminished. As can be seen from Figure 6–3, this would lead to an even greater *disinhibition* of the subthalamus and an increased facilitatory influence on the medial segment (GPi). This, in turn, again means that the GPi would tend to exert a greater inhibitory influence on the VA/VL nuclei, leading to decreased thalamic output (reduced cortical excitation).[12]

However, at first glance, there would appear to be at least one major obvious flaw in this scenario. As we saw above, given the differential effect that dopamine appears to have on the striatal neurons that contribute to the direct and indirect pathways, a depletion of dopamine should have the same end result for both the direct and indirect pathways, namely, reduced cortical activation. While the majority of the motor symptoms associated with Parkinson's disease indeed involve diminished motor activation [e.g., difficulty initiating movement (akinesia), slowness of movement (bradykinesia), decreased reflexes, "masked" facial expressions, and decreased speech volume], signs suggestive of increased or overflow movements also are present (e.g., resting tremors and rigidity).[13] If the effect of dopamine depletion is to reduce cortical activation in both the direct and indirect pathways, how are we to explain the presence of tremors and the rigidity?

One potential explanation for this phenomenon hinges on the assumption that the various neurochemical transmitters contributing to the proper functioning of the cortical → basal ganglia → thalamic → cortical motor system normally are in a delicate balance. In the event of significant dopamine depletion this "balance" (whether between the direct and indirect pathways, the D_1 and D_2 receptors, or the interactions between dopamine and glutamate or between dopamine and acetylcholine) is disturbed. It is this imbalance that may contribute both to the hypokinetic and the hyperkinetic symptoms (see: Schmidt, 1995). In fact, it also has been suggested that the relative overactivity of glutamate may be as much if not more of a factor in producing symptoms in Parkinson's disease than dopamine depletion itself (Carlsson & Carlsson, 1990; Greenamyre, 1993; Starr, 1995; Lange, Kornhuber, & Riederer, 1997).[14] On the other hand, the fact that antimuscarinic (anticholinergic) drugs often are effective in reducing the tremors and rigidity in Parkinson's disease, suggests that a dopamine/cholinergic imbalance may be important in the development of these latter symptoms (recall that acetylcholine is an important interneuron neurotransmitter within the striatum).

There are additional unanswered questions with regard to Parkinson's and the model presented above (Obeso, Rodriguez, & DeLong, 1997). For example, the medial or internal segment of the globus pallidus tends to exert an inhibitory influence on the VA and VL nuclei of the thalamus. If these thalamic nuclei in turn tend to have a facilitatory influence on the motor cortex, then lesions to the internal segment of the globus pallidus, theoretically at least, should result in dyskinesias or overflow movements as a result of uninhibited thalamic output. At the same time, lesions affecting the thalamus might be expected to have a dampening effect on motor activity (i.e., produce bradykinesia or hypokinesia) as a result of diminished thalamic output. However, this is not what typically happens. While infarctions involving the basal ganglia (both the striatum and the globus pallidus) might result in a parkinson-type syndrome (Inzelberg, Bornstein, Reider, & Korczyn, 1994; Reider-Groswasser, Bornstein, & Korczyn, 1995), unilateral infarctions of the VA or VL (motor) nuclei of the thalamus rarely lead to permanent, severe motor deficits. To the contrary, therapeutically designed, stereotaxic pallidal lesions in Parkinson patients tend to relieve, rather than exacerbate some of the hyperkinetic symptoms of Parkinson's disease (e.g., tremors and rigidity). Lesions in the motor thalamus are even more effective in relieving these hyperkinetic symptoms without worsening the hypokinetic symptoms. In fact, some seem to improve in this respect, possibly as a result of reduced muscular rigidity.[15] Parenthetically, a newer, but more controversial, treatment approach to Parkinson's disease involves the transplantation of human embryonic cells from the pars compacta of the substantia nigra into the brains of Parkinson's patients. These transplanted cells provide a renewed, internal source of dopamine.

Additional Behavioral Effects of Dopamine Dysregulation

Parkinson's disease does not just affect the motor system. Affective disturbances along with other cognitive disturbances often coexist and in some cases a dementia-type syndrome may develop (Baev, 1995; Boller, 1980; Benson, 1984; Brown & Marsden, 1988, 1990; Cummings, 1988; Huber & Cummings, 1992; Gabrieli, 1995; Jacobs et al., 1997; Taylor et al, 1986). Recall that the basal ganglia have connections with most association cortices, especially the frontal lobes. Also, in addition to the more well-known nigrostriatal dopaminergic pathways, the ventral tegmental area (VTA) also provides strong dopaminergic links to the limbic and prefrontal areas. In turn, these areas also have links to the basal ganglia (Cooper, Bloom, & Roth, 1991). Disturbances of these dopaminergic pathways and/or of neurotransmitters that directly or indirectly impact on dopamine have been implicated in various psychiatric conditions, most notably schizophrenia (Cooper, Bloom, & Roth, 1991; Crow, 1980; Gray et al., 1991; Haber & Fudge, 1998; Davis, Kahn, Ko & Davidson, 1991; Tassin, 1998). These findings have led to the development of the *dopamine hypothesis of schizophrenia* (see Box 6–1; Chapter 11). While this hypothesis is now recognized to be insufficient, in and of itself, to fully explain the etiology and symptoms of schizophrenia, it does provide some insights into how the basal ganglia and its neurotransmitters may impact on a range of mental disorders.

Box 6–1. Schizophrenia and the Dopamine Hypothesis

In the early 1950s chlorpromazine (Thorazine), being used as an antihistamine, was noted to have antipsychotic effects. It subsequently was discovered that Thorazine and related neuroleptics apparently had the capacity to block dopamine receptors, particularly mesolimbic D_2 receptors. It also was discovered that chronic use of these neuroleptics often led to the development of parkinsonian-type symptoms and conversely that antiparkinsonian drugs (as well as amphetamines and cocaine, which tend to increase dopamine availability) had a tendency to exacerbate psychotic symptoms. Additionally, there were reports of increased densities of D_2 receptors in certain parts of the basal ganglia in neuroloptic-free schizophrenic patients and elevated levels of homovanillic acid (a metabolite of dopamine) in the cingulate and frontal cortices of autopsied schizophrenic patients (Cooper, Bloom, & Roth, 1991). Findings such as these led to the assumption that schizophrenia was likely related to either an excess of dopamine or a supersensitivity of D_2 (and/or D_4) receptors to the presence of dopamine. While not discarding the importance of dopamine in the manifestation of schizophrenia, most researchers currently believe that the dynamics of schizophrenia are probably much more complex and likely involve numerous other neurotransmitter interactions. For example, many of the newer, "atypical" antipsychotics, such as clozapine (Clorazil) and risperidone (Risperdal) appear to exert antipsychotic effects through their ability to block serotonin receptors (Meltzer, 1990, 1991).

   Despite the obvious complexity of these systems, it is interesting to speculate how changes in dopamine sensitivity might help account for schizophrenic symptoms. For example, if we apply the analogy of the motor systems (especially the indirect pathways discussed earlier) to the mesial frontal/limbic loops, it might be suggested that increased dopamine activity (sensitivity) at D_2 synaptic junctions may result in increased inhibition of the indirect (inhibitory) pathways. This subsequent "disinhibition" of the thalamocortical connections might then result in an overflow of sensory excitation (i.e., decreased filtering of irrelevant stimuli) and/or the inability to inhibit competing or intrusive thoughts, both of which may contribute to some of the positive symptoms of schizophrenia (hallucinations, delusions, distractibility, nonlinear thinking). Some of the "negative" symptoms of schizophrenia (apathy, flattened affect, and social withdrawal) have been associated with concomitant decreases in mesocortical dopaminergic pathways as a result of negative feedback loops (Davis et al., 1991), relative depletions of glutaminergic activity (Carlsson & Carlsson, 1990), or enhanced serotonergic activity (hence, the effectiveness of drugs like Clozaril). Finally, some of the side effects of the phenothiazines (e.g., tardive dyskinesia) are explained as probably resulting from the "supersensitivity" of D_2 receptors in the striatal motor pathways following chronic use of receptor blocking agents.

Huntington's Disease

As was noted earlier, Huntington's disease is associated with degeneration of the striatum. The pathological changes are most prominent in the head of the caudate and as noted earlier can be visualized by the loss of convexity in the anterior horns of the lateral ventricles (Figure 6–6). This degeneration leads to a dramatic loss of GABAergic fibers from the striatum to the external segment of the globus pallidus and to the substantia nigra. In many respects, the end result clinically is just the opposite of that seen in Parkinson's. If Parkinson's disease primarily can be characterized as an underarousal of the motor system or hypokinetic syndrome (e.g., bradykinesia, shuffling gait), Huntington's is marked by an overflow of movements or a *hyperkinetic* syndrome (chorea, athetosis, and ballismus). A review of the neurochemical pathways in Figure 6–3 reveals why this may be the case. If the GABAergic striatonigral pathways are generally inhibitory, the net effect of their loss is to have a disinhibitory effect on the substantia nigra. Being disinhibited, the substantia nigra thus is free to produce an excess of dopamine, which as we have just seen tends to lead to excessive motor output.

Likewise, if the normal inhibitory action of the striatopallidal fibers to the external segment (of globus pallidus) is lost or reduced, the normal inhibitory effect of the pallidosubthalamic fibers will be increased (since their cells of origin will be disinhibited). With the increased inhibition of the subthalamus, the excitatory influence it normally has on the medial portion of the globus pallidus will be reduced, resulting in turn in the diminished firing of the inhibitory pallidothalamic fibers. The now disinhibited thalamus is free to barrage the cortex with excitatory impulses, creating the potential for excessive motor output (i.e., overflow movements).

As will be discussed in greater detail in Chapter 11, increased dopamine has been associated with psychiatric disturbances, particularly schizophrenic-type illnesses. Huntington's is no exception. Frequently, although not invariably, this disorder is associated with marked behavioral disturbances early in the disease process, including blatant psychosis, paranoid delusions, and hallucinations (Dewhurst et al., 1969; Garron, 1973; Lishman, 1987; McHugh & Folstein, 1975; Shoulson, 1990). Mental or cognitive deterioration also can be encountered in the early stages of this disorder (Barr et al., 1978; Brandt, 1991; Butters et al, 1978; Caine et al., 1978). As in Parkinson's, some of these non-motor changes could be a reflection of more extensive, direct cerebral involvement, but the likelihood is that many if not most of these effects are produced via basal ganglia connections with non-motor association cortices, especially those associated with frontal and temporal–limbic systems.

Attempts to treat Huntington's commonly have involved trying to produce an increase in GABA in the CNS. However, oral administration of GABA compounds has been unsuccessful, since GABA does not readily cross the blood–brain barrier. Another way to increase CNS GABA would be to block dopamine receptors, but this means of treatment usually leads to other complications. Until more effective treatments are found, recent efforts have focused on identifying at-risk individuals before their childbearing years to reduce the expression of this disease through genetic counseling. However, the pros and cons of this approach have been debated because of the psychological impact of presymptomatic diagnosis on the at-risk individual.

Hemiballismus

Unlike Parkinson's and Huntington's, hemiballismus is a symptom rather than a disease or syndrome. Resulting primarily from a lesion affecting the subthalamus, hemiballismus can be produced by a variety of conditions, including Huntington's disease. Going back to our earlier diagram, what might we expect given a lesion affecting either the subthalamic fasciculus or the nucleus itself? In either case, the net result should be a reduction in excitatory input to the medial pallidum via the subthalamic fasciculus. As was the case with Huntington's disease where the inhibitory pathways to the external pallidal segment were disrupted, the overall effect was to reduce the excitatory influence of the subthalamus on the inhibitory pallidothalamic fibers, producing a disinhibitory influence on the VA/VL thalamic nuclei. This in turn results in increased glutaminergic (excitatory) stimulation of the motor cortex, hence possibly creating the necessary environment for overflow phenomena such as involuntary ballistic movements. These mechanisms also help explain why such symptoms can present as part of the manifestations of Huntington's disease, as well as why overdoses of L-dopa also can lead to extraneous motor activity.

ROLE OF THE BASAL GANGLIA IN NORMAL BEHAVIOR

The preceding review focused primarily on motor symptoms associated with disturbances of the basal ganglia and was intended to offer some insight into the function of these subcortical nuclei. Although we now have reason to suspect that the basal ganglia also play an important role in a wide variety of cognitive and behavioral functions, the precise role of the basal ganglia nuclei in these latter areas are not clear. Nonetheless, it may be useful to review what we know, or at least suspect, about the role of the basal ganglia in both motor and non-motor behaviors.

Motor Behavior

One of the first questions to ask is whether as a group the basal ganglia (i.e., the caudate, putamen, globus pallidus, subthalamus, and substantia nigra) have **a** role or do they subserve **multiple** roles in the overall scheme of the central nervous system? Another way of phrasing the question may be to ask whether the sum of the activity of the basal ganglia is ultimately directed to the execution of motor response programs or are they also responsible for cognitive and emotional activities independent of specific motor activities? While it might be desirable to discover a single rubric under which all the activities of the basal ganglia could be subsumed, this goal is not likely to be fulfilled in the near future. For now, the best we may be able to do is to search for pieces of the puzzle.

Perhaps the logical place to start is in the area of motor activity: *"What role do the basal ganglia play with regard to motor behavior?"* First, most authors seem to agree that the basal ganglia probably are not primarily responsible for formulating or initiating motor responses (that being primarily the role of the prefrontal and agranular frontal cortices; see Chapter 9). However, the basal ganglia would appear to have the potential to facilitate the activation of some motor responses, while at the same time having an inhibitory influence on others. Most investigators seem to agree that the basal ganglia likely play a role in the following aspects of motor behavior:

1. Preparation for specific motor responses.
2. Smooth transitioning from one motor response to another.
3. Inhibiting movements or sets of actions that might interfere with the proper execution of the primary intended action.
4. Monitoring the direction, speed and amplitude of movements as a means of preparing for the next movement.
5. Initially learning and, later, smoothly executing series or patterns of movements
6. Possibly adapting to the internal and external (environmental) demands being placed on the organism and/or comparing planned or intended actions with outcome.[16]

Table 6–6. Comparison of Symptoms of Cerebellum vs
            Basal Ganglia Disease

Cerebellum	Basal Ganglia
Loss of balance/equilibrium	Slowness of movement
Ataxia/Asynergy	Chorea/athetosis/ballismus
Intention tremors	Resting tremors
Hypotonia	Hypertonia/rigidity

So how does all this translate into everyday motor activities? A useful model might be that of driving a car. Assume you are driving along the interstate. You are in a relaxed mood, singing along with the radio, or indulging in your "fantasy de jour" as you drive. Almost unconsciously you are making occasional, minor adjustments in your steering to stay in your lane, perhaps even negotiating gentle curves in the road without giving them any thought. At this point, connections between the visual system and the frontal agranular cortices are clearly operative, as is the cerebellum (as witnessed by the smooth, delicate movements of the steering wheel). What may be less apparent is the role played by the basal ganglia under these circumstances.[17] However, suppose that suddenly the car in front of you begins to behave erratically. Your singing or fantasizing stops abruptly, your posture becomes more erect, your grip on the steering wheel tightens slightly, and you become conscious of the brake pedal, even though you may not immediately take your foot off the accelerator. You are preparing for the possibility of having to make rapid adjustments in your motor response based on a change in external events and your own internal state of apprehension. You have inhibited certain potentially distracting activities (singing or daydreaming) and are ready to facilitate others (applying the brake, making a defensive steering maneuver). It is these actions and preparations for further actions based on situational demands that likely require substantial involvement of the basal ganglia. If you become too tense (freeze up) or remain too relaxed, you may not be able to respond in an optimal manner. It is these preparations or setting the tone for possible actions that may represent contributions of the cortical–basal ganglia interactions.

In summary, the basal ganglia probably are important in the *planning* and *preparation* of an action. They facilitate responses of the organism by setting the stage for the initiation and execution of a given action.[18] The fact that overflow movements (e.g., resting tremors, chorea, athetosis, and rigidity) also are a frequent accompaniment to basal ganglia disorders suggests that one way the basal ganglia might prepare for action is by selectively inhibiting or modulating muscular tone or activity that is inappropriate to carrying out the target action. In contrast, the cerebellum may be responsible for the actual smooth execution of the action once it has been initiated. Deficits resulting from cerebellar lesions generally are observed only during the process of trying to carry out a complex motor activity (e.g., maintaining one's balance when walking or a smooth rhythm while writing). At the same time, there appears to be certain areas of functional overlap between the cerebellum and the basal ganglia. For example, both seem to play a part in maintaining what Luria (1966) described as the "kinetic melody" of a motor response (i.e., the fact that each segment or part of the action flows smoothly and orderly, without hesitation or delay, from the previous segment). Either cerebellar or basal ganglia lesions can disturb this "kinetic melody."

The key concepts to remember here are *parallel processing* and *modulation*. Both are consistent with the notions of feedback loops discussed earlier. It is useful to keep in mind that the cortex, basal ganglia, cerebellum, as well as a few other brainstem structures (such as the pedunculopontine nuclei, the red nuclei, and the inferior olivary nuclei) are all components of the motor system and act in concert with one another. Disruptions in more than one location at times may produce functional disturbances that may appear very similar. In their excellent review of some of these studies, Marsden and Obeso (1994) conclude that two of the more likely roles of the basal ganglia are to:

1. Monitor and facilitate the sequencing of cortically initiated motor responses by facilitating target or agonist actions and exercising an inhibitory influence on conflicting or "unwanted motor activity."
2. Interrupt (e.g., exert inhibitory influences and/or cease facilitating) activities that may no longer be appropriate given changes in either the internal or external milieu.

Non-Motor Behavior

The next question is, *"What role do the basal ganglia play with regard to emotional control, cognition, or other "higher cortical functions?"* Two general sources of anatomical evidence were reviewed earlier that suggest that the basal ganglia have much broader roles than simply ensuring that there is a smooth transition from one motor component to the next when executing a movement. One bit of evidence supporting the role of the basal ganglia in higher cortical functions is the fact that the parietal and temporal association cortices provide substantial input into the striatum. Other compelling evidence is the discovery of the dorsolateral, lateral orbitofrontal, and medial frontal circuits, which for the most part originate and project back to these cortical regions after converging on basal ganglia structures. These pathways, which were reviewed above, are generally associated with higher-level integrative and executive abilities, including the control of emotions, motivations, or drive states. Additional evidence touched upon earlier was the fact that in certain disease states (e.g., Parkinson's, Huntington's), various cognitive and affective disturbances frequently are reported. Some common psychiatric disorders, such as mood disturbances (Mayberg, 1994; Mayberg et al., 1988; Starkstein et al., 1987), schizophrenia, obsessive–compulsive, and other stress-related disorders, also have been linked with disturbances of the basal ganglia and related systems (Horger & Roth, 1996; Mega & Cummings, 1994; Swerdlow, 1995; Miguel, Rauch, & Leckman, 1997; Saint-Cyr et al., 1995; Zald & Kim, 1996a,b; Saxena, Brody, Schwartz, & Baxter, 1998). Discrete lesions of basal ganglia nuclei, as well as of those thalamic nuclei that constitute an integral part of the cortical–basal ganglia circuits described above, also have been associated with specific cognitive deficits such as aphasia (Robin & Schienberg, 1990; Damasio et al., 1982). Finally, in clinical practice it is not uncommon to find problems in learning or memory, visual–spatial skills, or higher-order "executive" functions in patients who have suffered lacunar infarcts involving the basal ganglia, although damage to nearby association or projection pathways also may contribute to the observed deficits (Alexander et al., 1987; Cappa et al., 1983; Naeser et al., 1982).

As described above, in addition to exerting some type of modulating influence with regard to sensorimotor activities, it would appear that the basal ganglia also perform a comparable function in relation to the cognitive and affective activities of the brain. Through the motor, oculomotor, dorsolateral, orbitofrontal, and cingulate "circuits" outlined earlier, the basal ganglia are not only in a unique position to integrate motor and sensory input with regard to motor activities, but also to integrate sensory input and its emotional valence, along with internally generated drive states to influence the executive command centers of the frontal lobes. As part of these latter "circuits," critically placed lesions within the basal ganglia may induce deficits comparable to cortical lesions involving these same circuits, that is, frontal, anterior cingulate lesions. In general, this is what frequently has been found. As

Table 6–7. Cortical versus Subcortical Dementia (Early Stages)

Dimension	Cortical dementia	Subcortical dementia
Physical appearance	Typically robust, physically active	Weak, more sedentary, stooped posture
Motor system	Generally no change	Major change: Tremors, dyskinesia, chorea, gait disturbances
Response speed	Normal, except perhaps when difficulties are encountered	More commonly slowed, even when response is correct
Speech	Good articulation, generally fluent	Dysarthric, slowed, occasionally hypophonic
Writing	Graphically intact, spelling, paraphasic errors	Dysgraphic (motorically), spelling or paraphasic errors are more rare
Language	Word finding diff., paraphasic errors, difficulty comprehending complex commands	(?)Slow to respond, but expressive and receptive language grossly intact
Memory	Difficulty learning, delayed recall poor, cuing or recognition may be of little help	Slow learning, esp. if active strategy involved, weak delayed recall, cue or recognition helpful
Cognition	Attention may be good, impaired problem solving, abstraction (not aided by extra time; perceptual, construction difficulties (due to V-S integration deficits)	Attention inconsistent, slow to problem solve, improves if given extra time, perception intact, construction impaired due to graphic problems, poor planning or impulsivity
Affect	Normal to mildly anxious or irritable	More likely to appear depressed, apathetic
Personality	Usually little change	May appear more disinhibited, inappropriate
Insight	Normal to mildly impaired (early only)	More likely to show change; indifference

Note: Not all symptoms will be present in all patients, especially in the earlier stages of either type of dementia process.

is the case with frontal cortical lesions, bilateral insults, although more rare, tend to produce more dramatic results.

Probably the blatant manifestation of these influences finds its expression in what has been termed the *subcortical dementias* (Albert et al., 1974; Cummings, 1990). Although the etiology of this general syndrome can be quite varied (for review, see Cummings, 1990) and may involve numerous subcortical structures, as well as the cortex itself, lesions affecting the basal ganglia and/or adjacent white matter pathways is common to most of these conditions. Huntington's disease, which was described earlier, is a quintessential example of this type of dementia in which there is early, bilateral involvement of the caudate nuclei, along with motor, behavioral, and cognitive symptoms. Because of their extensive connections with the frontal lobes (dorsolateral, orbitofrontal, and cingulate circuits), most of the "subcortical dementias" that typically impact these cortical → basal ganglia → thalamic → cortical pathways share many of the features of "frontal dementias" (Gustafson, 1987; Neary et al., 1988). Thus, in addition to symptoms of motor or "physical" disability, such patients commonly will manifest personality or behavioral changes (e.g., apathy, lability, and disinhibition) and cognitive changes more suggestive of "cognitive stickiness" (e.g., bradyphrenia, difficulty shifting mental sets, impaired concentration, and difficulty retrieving information). Such symptoms may be contrasted to the more frank cognitive lacunae (e.g., aphasias or agnosias) commonly observed in the more posterior cortical dementias, such as Alzheimer's.[19] Table 6–7 presents some of the more common features that may help distinguish a predominately cortical dementia from a dementia in which there is early and substantial subcortical pathology.

Even though our understanding of the role of the basal ganglia in either motor or cognitive/emotional behaviors is still incomplete, it is nonetheless important to appreciate the complexity and range of their connections (both cortical and subcortical), as well as the range of behavioral disturbances associated with lesions affecting them. Such knowledge forces us to think not only in terms of specific nuclei, lobes, or structures, but also in terms of *cerebral systems*. Recognizing that the basal ganglia (as well as other structures) are part of larger functional systems, we learn to ask, *"How does a disruption of this nucleus impact on the larger functional system of which it is a part, and how does that differ from disturbances to other parts of this functional system?"* As we begin to think in this manner, we come closer to understanding how the brain is most likely organized and how it operates.

Endnotes

1. The *extrapyramidal* nature of these motor influences is now in question, as is the exclusive "motor function" of these nuclei.
2. As will be discussed, these areas are thought to represent ventral extensions of the putamen and globus pallidus, respectively, and appear to represent important connections to limbic structures, and hence, may play an important role in affect and neuropsychiatric behaviors.
3. Although the tail of the caudate nucleus terminates in the vicinity of the amygdala, a structure more closely associated with the olfactory system and other "limbic" structures, no direct connections are known to exist between the two (however, as will be seen, indirect connections may exist via the ventral striatum). The claustrum, a narrow band of gray matter lying between the putamen and the insular cortex and separated from them by the external and extreme capsules, respectively, is known to have reciprocal connections to the cortex, particularly the posterior or sensory cortices, but its functional significance remains obscure.
4. As one joke goes, *"How are the basal ganglia and a college dean alike?"* The answer is: *"Both take up a lot of space, but no one seems to know exactly what it is that they do."*
5. This includes the bed nuclei of the stria terminalis lying ventrolateral to the anterior horns of the lateral ventricles.
6. In addition to GABA, the striatopallidal fibers destined for the internal segment contain the peptide, substance P, while those projecting to the external segment contain enkaphalin. Also, while most of the neuronal fibers exiting the striatum are GABAergic and inhibitory, those that remain as internal association fibers (striatal interneurons) are largely cholinergic and excitatory.
7. Although the "direct" and "indirect" pathways are described here in relation to motor functions, recall that the neostriatum, especially the caudate nucleus, receives projections from widespread areas of the cortex. Thus, similar patterns of facilitation or relative inhibition also may influence cognitive and perceptual activities. Likewise, as will be seen, selective disruptions of specific aspects of these systems may help account for some of the symptoms seen in certain disease states (e.g., the rigidity in Parkinson's disease; the involuntary, "overflow" movements in Huntington's disease; or possibly even the difficulty in filtering out irrelevant, extraneous thoughts or stimuli).

8. The pedunculopontine nucleus is located in the area of the decussation of the superior cerebellar peduncles, which is a particularly crucial location for involvement in a motor feedback loop. In this location, the pedunculopontine nucleus has access to input from the motor cortex, globus pallidus, substantia nigra, and the cerebellorubrothalamic pathway.

9. Given what we know about the afferent and efferent connections between these various cortical areas, particularly the anterior cingulate gyrus, one might speculate that other thalamic nuclei also might be involved in these circuits.

10. While it is possible that feedback loops project back to the posterior association cortices, so far none apparently have been identified.

11. For reviews of the chemical pathophysiology and additional syndromes associated with lesions of the basal ganglia, see: DeLong (1990); McDowell and Cedarbaum (1991); Weiner and Lang (1989); Wichmann and DeLong, (1993).

12. Given these relationships, the opposite situation (i.e., increased dopaminergic activity) should result in increased cortical activation or arousal. In fact, this is what seems to occur with the administration of dopamine agonists, such as amphetamines.

13. Miyawaki, Meah, and Koller (1997) suggest that parkinsonian tremors may be associated with changes in serotonin levels.

14. Lesions of the subthalamic nuclei in experimentally induced parkinsonism in monkeys improved the symptoms of tremor and rigidity in these animals; this was taken as an indication that these "indirect" pathways were critical to the development of these hyperkinetic-type symptoms (DeLong, 1990). What makes this finding even more interesting is that, as previously noted, lesions of this nucleus in normal persons commonly produces overflow symptoms (e.g., hemiballismus).

15. See Marsden and Obeso (1994) for a more detailed review and explanations for these findings.

16. For additional reviews of proposed functions of the basal ganglia, see Albin, Young, and Penny (1989); Alheld, Heimer, and Switzer (1990); Ciba Foundation (1984); Cote and Crutcher (1991); DeLong (1990); DeLong and Georgopoulos (1981); Denny-Brown (1962); Gunilla, Oberg, and Divac (1981); Marsden (1987); Yahr (1976).

17. The roles played by the basal ganglia versus the cerebellum in such routine movements is not always well differentiated. Certainly some of the functions normally attributed to the basal ganglia (e.g., ensuring smooth transitions from one movement to another) also would seem to describe the operation of the cerebellum. Clinically, how do they differ? While some of the differences in the symptoms manifested by the two are presented in Table 6–6, there may be a more fundamental way of viewing this problem, which has to do with learning. To stick with this same analogy, think back to when you first learned to drive. Chances are, even though the road may have been straight and unobstructed, you likely were making constant and excessive excursions of the steering wheel, a process that required your full and undivided attention. However, as you became more accustomed to driving, the steering adjustments became much more refined and you were able to relax, perhaps now more easily dividing your attention between the road and finding a better station on the radio. During the initial learning phase, a much more conscious process, the basal ganglia probably were much more important in the planning, preparation, and execution of these skills. However, with practice, normal driving rapidly became much more automatic. Under these circumstances, many of the routine movements or adjustments more easily could be controlled by the cerebellum (i.e., they became less coarse and more precise). Thus, while the basal ganglia probably continue to be important in maintaining overall muscular tone, the facilitation and inhibition of agonist and antagonist muscle groups and the

initiation and sequencing of individual muscle groups (especially when executing highly specific, fully conscious, goal-directed activities), the cerebellum can assume considerable control over those overlearned, repetitive elements that have become largely automatic (Passingham, 1993). However, both patients with cerebellar and basal ganglia disturbances can show deterioration in carrying out skilled tasks that require a fluid transition between one element and the next.

18. Increased electrical activity can be detected in the basal ganglia just prior to the initiation of an action, and as has been noted certain disease states (e.g., Parkinson's) are characterized by difficulty initiating (and occasionally stopping) movements.

19. While Alzheimer's patients also may manifest "frontal" signs, they usually occur somewhat later in the disease process.

REFERENCES AND SUGGESTED READINGS

Afifi, A.K. (1994) Basal ganglia: Functional anatomy and physiology. Parts 1 & 2. *Journal of Child Neurology, 9,* 249–260; 352–361.

Albert, M.L., Feldman, R.G., & Willis, A.L. (1974) The "subcortical dementia" of progressive supranuclear palsy. *Journal of Neurology. Neurosurgery, and Psychiatry, 37,* 121–130.

Albin, R.L., Young, A.B., & Penny, J.B. (1989) The functional anatomy of basal ganglia disorders. *Trends in Neuroscience, 12,* 366–375.

Alexander, G.E., Crutcher, M.D., & DeLong, M.R. (1990) Basal ganglia-thalamo-cortical circuits: Parallel substrates for motor, oculomotor, "prefrontal" and "limbic" functions. *Progress in Brain Research, 85,* 119–146.

Alexander, G.E., DeLong, M., & Strick, P. (1986) Parallel organization of functionally segregated circuits linking basal ganglia and cortex. *Annual Review of Neuroscience, 9,* 357–381.

Alexander, M.P., Naeser, M.A., & Palumbo, C.L. (1987) Correlations of subcortical CT lesion sites and aphasia profiles. *Brain, 110,* 961–991.

Alheld, G.F., Heimer, L., & Switzer, R.C. (1990) Basal ganglia. In: Paximos, G. (Ed.) The Human_Nervous System. San Diego: Academic Press, pp. 483–582.

Ballantine, H.T. Jr. (1986) A critical assessment of psychiatric surgery: Past, present, and future. In: Berger, P.A. & Brodie, K.H. (Eds.) *American Handbook of Psychiatry, 2nd edition,* Vol. VIII Biological Psychiatry. New York: Basic Books, pp. 1029–1047.

Barr, A.N., Heinze, W., Mendoza, J.E., & Perlik, S. (1978) Long-term treatment of Huntington's disease with L-glutamate and pyridoxine. *Neurology, 28,* 1280–1282.

Barr, M.L. & Kiernan, J.A. (1993) *The Human Nervous System,* Philadelphia, J.B. Lippincott Co. pp. 210–221.

Baev, K.V. (1995) Disturbances of learning processes in the basal ganglia in the pathogenesis of Parkinson's disease: A novel theory. *Neurological Research, 17,* 38–48.

Baxter, L.R., Schwartz, J.M., Bergman, K.S., Szuba, M.P., Guze, B.H., Mazziotta, J.C., Alazraki, A., Selin, C.E., Ferng, H-K., Munford, P., & Phelps, M.E. (1992) Caudate glucose metabolic rate changes with both drug and behavior therapy for obsessive-compulsive disorder. *Archives of General Psychiatry, 49,* 681–689.

Benson, D.F. (1984) Parkonsonian dementia: Cortical or subcortical? In: Hassler, R.G., & Christ, J.F. (Eds.). Advances in Neurology, Vol. 40. New York: Raven Press, pp. 235–240.

Benson, D.F., & Cummings, J.L. (1990) Subcortical mechanisms and human thought. In: Cummings, J.L. (Ed.) *Subcortical Dementia.* New York: Oxford University Press, pp. 251–259.

Bergman, H, Wichmann, T., & DeLong, M.R. (1990) Reversal of experimental Parkinsonism by lesions of the subthalamic nucleus. *Science, 249,* 1436.

Bird, E.D., Mackay, A.V.P., Rayner, C.N., & Iversen, L.L. (1973) Reduced glutamic-acid-decarboxylase activity of post-mortem brain in Huntington's chorea. *Lancet, 1,* 1090–1092.

Boller, F. (1980) Mental status of patients with Parkinson's disease. *Journal of Clinical Neuropsychology, 2,* 157–172.

Bradshaw, J.L. & Mattingly, J.B. (1995) Clinical *Neuropsychology: Behavioral and brain science.* New York: Academic Press. Chapters 12 & 13: Subcortical movement Disorders I & II, pp. 290–323.

Brandt, J.A. (1991) Cognitive impairments in Huntington's disease: Insights into the neuropsychology of the striatum. In: Boller, F. & Grafman, J. (Eds.) *Handbook of Neuropsychology (Vol. 5)* Amsterdam: Elsevier, pp. 241–261.

Brazis, P. Masdeu, J. & Biller, J. (1990) *Localization in Clinical Neurology,* Boston: Little, Brown & Co. pp. 346–360.

Brown, R.G. & Marsden, C.D. (1988) "Subcortical dementia": The neuropsychological evidence. *Neuroscience, 25,* 363–387.

Brown, R.G. & Marsden, C.D. (1990) Cogbnitive function in Parkinson's disease: from description to theory. *Trends in Neuroscience, 13,* 21–28.

Burchiel, K.J. (1995) Thalamotomy for movement disorders. *Neurosurgery Clinics of North America, 6,* 55–71.

Butters, N., Sax, D., Montgomery, K., & Tarlow, S. (1978) Comparison of the neuropsychological deficits associated with early and advanced Huntington's disease. *Archives of Neurology, 35,* 585–589.

Caine, E.D., Hunt, R.D., Weingartner, H., & Ebert, M.H. (1978) Huntington's dementia: Clinical and neuropsychological features. *Archives of Neurology, 35,* 377–384.

Cappa, S.F., Cavallotti, G., Guidotti, M. et al., (1983) Subcortical aphasia: Two clinical CT-scan correlations studies. *Cortex, 19,* 227–241.

Carlsson, M. & Carlsson, A. (1990) Interactions between glutamatergic and monoaminergic systems within the basal ganglia: implications for schizophrenia and Parkinson's disease. *Trends in Neurosciences, 13,* 272–276.

Carpenter, M.B. & Sutin, J. (1983) Human Neuroanatomy, Baltimore: Williams & Wilkins, pp. 579–611.

Ciba Foundation (1984) *Functions of the basal ganglia.* London: Pitman (Ciba Foundation Symposium 107).

Cooper, J.R., Bloom, F.E., & Roth, R.H. (1991) *The Biochemical Basis of Neuropharmacology.* New York: Oxford University Press, pp. 285–337.

Cote, L. & Crutcher, M.D. (1991) The basal ganglia. In: Kandell, E.R., Schwartz, J.H. & Jessell, T.M. (Eds.) *Principles of Neural Science.* New York: Elsevier, pp. 646–659.

Crow, T.J. (1980) Positive and negative schizophrenic symptoms and the role of dopamine. *British Journal of Psychiatry, 137,* 383–386.

Cummings, J.L. (Ed.) (1990) *Subcortical Dementia.* New York: Oxford University Press.

Cummings, J. (1988) Intellectual impairment in Parkinson's disease: Clinical, pathologic, and biochemical correlates. *Journal of Geriatric Psychiatry and Neurology, 1,* 24–36.

Damasio, A.R., Damasio, H., Rizzo, M., Varney, N. & Gersh, F. (1982) Aphasia with non-hemorrhagic lesions in the basal ganglia and internal capsule. *Archives of Neurology, 39,* 15–20.

Davis, K.L., Kahn, R.S., Ko, G., & Davidson, M. (1991) Dopamine in schizophrenia: A review and reconceptualization. *American Journal of Psychiatry, 148,* 1474–1486.

DeLong, M.R. (1990) Primate models of movement disorders of basal ganglia origin. *Trends in Neuroscience, 13,* 281–285.

DeLong, M.R. & Georgopoulos, A.P. (1981) Motor functions of the basal ganglia. In: Brooks, V.B. (Ed.) *Handbook of Physiology: The nervous system, Vol 2.* Bethesda, MD: American Physiological Society, pp. 1017–1061.

Denny-Brown, D. (1962) *The basal ganglia and their relation to disorders of movement.* New York: Oxford University Press.

Dewhurst, K., Oliver, J., Trick, K.L.K., & McKnight, A.L. (1969) Neuro-psychiatric aspects of Huntington's disease. *Confina Neurologica, 31,* 258–268.

Gabrieli, J. (1995) Contributions of the basal ganglia to skill learning and working memory in humans. In: Houk, J.C., Davis, J.L., & Beiser, D.G. (Eds.) Models of Information Processing in the Basal Ganglia. Cambridge, MA: MIT Press, pp. 277–294.

Garron, D.C. (1973) Huntington's chorea with schizophrenia. *Advances in Neurology, 1,* 729–734.

Gilman, S. & Newman, S.W. (1992) *Essentials of Clinical Neuroanatomy and Neurophysiology,* Philadelphia, F.A. Davis Co. pp. 183–192.

Gilroy, J. (1990) *Basic Neurology,* New York: Pergamon Press, pp. 94–117.

Gray, J.A., Feldon, J., Rawlins, J.N.P., Hemsley, D.R. & Smith, A.D. (1991) The neuropsychology of schizophrenia. *Behavioral Brain Science*, *14*, 1–20.

Greenamyre, J.T. (1993) Glutamate-dopamine interactions in the basal ganglia: Relationship to Parkinson's disease. *Journal of Neural Transmission*, *91*, 255–269.

Greybiel, A.M., Aosaki, T., Flaherty, A.W. & Kimura, M. (1994) The basal ganglia and adaptive motor control. *Science*, *265*, 1826–1831.

Groenewegen, H.J., Berendse, H.W., Wolters, J.C. & Lohman, A. (1990) The anatomical relationship of the prefrontal cortex with the striatopallidal system, the thalamus and the amygdala: evidence for a parallel organization. In: H.B.M. Uylings, C.G. Van Eden, J.C.P. DeBruin, M.A. Corner & M.G.P. Feenstra (Eds.) *Progress in Brain Research*. Elsevier Science Publishers B.V., Vol 85, pp. 95–118.

Gunilla, R., Oberg, E. & Divac, I. (1981) Commentary: The basal ganglia and the control of movement. Levels of motor planning, cognition, and control of movement. *Trends in Neuroscience*, *4*, 122–125.

Gustafson, L. (1987) Frontal lobe degeneration of the non-Alzheimer's type. II. Clinical picture and differential diagnosis. *Archives of Gerontology and Geriatrics*, *6*, 209–233.

Haber, S.N. & Fudge, J.L. (1997) The interface between dopamine neurons and the amygdala: Implications for schizophrenia. *Schizophrenia Bulletin*, *23*, 471–482.

Heimer, L. (2003) A new anatomical framework for neuropsychiatric disorders and drug abuse. *American Journal of Psychiatry*, *160 (10)*, 1726–1739.

Heimer, L. (1995) *The human brain and spinal cord: Functional Neuroanatomy and Dissection Guide*. New York: Springer-Verlag, pp. 337–360.

Hirsch, E.C., Graybiel, A.M. & Agid, Y. (1988) Melanized dopaminergic neurons are differentially affected in Parkinson's disease. *Nature*, *334*, 345–348.

Hopkins, A. (1993) *Clinical Neurology: A modern approach*. New York: Oxford. pp. 208–239.

Hornykiewicz, O. & Kish, S. (1987) Biochemical pathophysiology of Parkinson's disease. *Advances in Neurology*, *45*, 19–34.

Horger, B.A. & Roth, R.H. (1996) The role of the mesoprefrontal dopamine neurons in stress. *Critical Reviews in Neurobiology*, *10*, 395–418.

Houk, J.C., Davis, J.L. & Beiser, D.G. (1995) *Models of Information Processing in the Basal Ganglia*. Cambridge, MA: MIT Press.

Houk, J.C. & Wise, S.P. (1995) Distributed modular architectures linking basal ganglia, cerebellum and cerebral cortex: Their role in planning and controlling actions. *Cerebral Cortex*, *5*, 95–110.

Huber, S. & Cummings, J.L. (Eds.) (1992) *Parkinson's Disease: Neurobehavioral aspects*. New York: Oxford University Press.

Insel, T.R. (1992) Toward a neuroanatomy of obsessive-compulsive disorder. *Archives of General Psychiatry*, *49*, 739–744.

Inzelberg, R., Bornstein, N.M., Reider, I. & Korczyn, A.D. (1994) Basal ganglia lacunes and Parkinsonism. *Neuroepidemiology*, *13*, 108–112.

Jacobs, D.M., Stern, Y. & Mayeux, R. (1997) Dementia in Parkinson disease, Huntington disease, and other degenerative conditions. In: Feinberg, T.E. & Farah, M.J. (Eds.) *Behavioral Neurology and Neuropsychology*. New York: McGraw-Hill, pp. 579–587.

Koller, W.C., Silver, D.E. & Lieberman, A. (1994) An algorithm for the management of Parkinson's disease. *Neurology*, *44* (Suppl), 1–52.

Lange, K.W., Kornhuber, J., & Riederer, P. (1997) Dopamine/glutamate interactions in Parkinson's disease. *Neuroscience and Behavioral Reviews*, *21*, 393–400.

LaPlane, D., Levasseur, M., Pillon, B., Dubois, B., Baulac, M, Mazoyer, B., Tran Dinh, S., Sette, G., Danze, F. & Baron, J.C. (1989) Obsessive-compulsive and other behavioral changes with bilateral basal ganglia lesions. *Brain*, *112*, 699–725.

Lishman, W.A. (1987) *Organic Psychiatry*. Oxford: Blackwell Scientific Publications.

Ma, T.P. (1997) The basal ganglia. In: Haines, D.E. (Ed.) *Fundamental Neuroscience*. New York: Churchill Livingstone, pp. 363–378.

Marsden, C.D. (1985) The basal ganglia. In: M. Swash & C. Kennard (Eds.). *Scientific Basis of Clinical Neurology*. Edinburgh, Churchill Livingstone, pp. 56–73.

Marsden, C.D. (1987) What do the basal ganglia tell the premotor cortical areas? In: Ciba Foundation, *Motor areas of the cerebral cortex*. New York: John Wiley & Sons (Ciba Foundation Symposium 132), pp. 282–300.

Marsden, C.D. & Obeso, J.A. (1994) The functions of the basal ganglia and the paradox of stereotaxic surgery in Parkinson's disease. *Brain, 117,* 877–897.

Mayberg, H.S. (1994) Frontal lobe dysfunction in secondary depression. *Journal of Neuropsychiatry and Clinical Neurosciences, 6,* 428–442.

Mayberg, H.S., Robinson, R.G., Wong, D.W., Parikh, R., Bolduc, P., Starkstein, S.E., Price, T., Dannals, R.F., Links, J.M., Wilson, A.A., Ravert, H.T., & Wagner, H.N. (1988) PET imaging of cortical S2 serotonin receptors after stroke: Lateralized changes and relationship to depression. *American Journal of Psychiatry, 145,* 937–943.

McDowell, F.H. & Cedarbaum, J.M. (1991) Extrapyramidal system and disorders of movement. In: R. Joynt (Ed.) *Clinical Neurology,* Philadelphia, J.B. Lippincott-Raven, Vol 3, Chapter 38, pp. 1–120).

McGeer, P.L. & McGeer, E.G. (1993) Neurotransmitters and their receptors in the basal ganglia. *Advances in Neurology, 60,* 93–101.

McHugh, P.R., & Folstein, M.F. (1975) Psychiatric syndromes of Huntington's chorea: A clinical and phenomenologic study. In: Benson, D.F., & Blumer, D. (Eds.) Psychiatric Aspects of Neurologic Disease. New York: Grune and Stratton, pp. 267–285.

Mega, M.S. & Cummings, J.L. (1994) Frontal-subcortical circuits and neuropsychiatric disorders. *Journal of Neuropsychiatry, 6,* 358–370.

Meltzer, H.Y. (1990) The role of serotonin in the action of atypical antipsychotic drugs. *Psychiatric Annals, 20,* 571–579.

Meltzer, H.Y. (1991) The mechanism of action of novel antipsychotic drugs. *Schizophrenia Bulletin, 17,* 265–287.

Miguel, E.C., Rauch, S.L. & Leckman, J.F. (Eds.) (1997) Neuropsychiatry of the Basal Ganglia. Philadelphia: W.B. Saunders.

Mink, J.W. & Thach, W.T. (1993) Basal ganglia intrinsic circuits and their role in behavior. *Current Opinion in Neurobiology, 3,* 950–957.

Miyawaki, E., Meah, Y. & Koller, W.C. (1997) Serotonin, dopamine and motor effects in Parkinson's disease. *Clinical Neuropharmacology, 20,* 300–310.

Modell, J.G., Mountz, J.M., Curtis, G.C. & Greden, J.F. (1989) Neurophysiologic dysfunction in basal ganglia/limbic striatal and thalamocortical circuits as a pathonomic mechanism of obsessive compulsive disorder. *Journal of Neuropsychiatry and Neuroscience, 1,* 27–36.

Naser, M.A., Alexander, M.P., Helm-Estabrooks, N. et al. (1962) Aphasia with predominately subcortical lesion sites: Description of three capular/putaminal aphasia syndromes. *Archives of Neurology, 29,* 2–14.

Neary, D., Snowden, J.S., Northen, B., & Goulding, P. (1988) Dementia of the frontal lobe type. *Journal of Neurology, Neurosurgery, and Psychiatry, 51,* 353–361.

Nolte, J. (1993) *The Human Brain.* St. Louis, Mosby-Year Book, Inc. pp. 319–336.

Obeso, J.A., Rodriguez, M.C., & DeLong, M.R. (1997) Basal ganglia pathophysiology: A critical review. In: Obeso, J.A., DeLong, M.R., Ohye, C., & Marsden, C.D. (Eds.) The Basal Ganglia and New Surgical Approaches for Parkinson's Disease, *Advances in Neurology,* Vol 74. Philadelphia: Lippincott-Raven, pp. 3–18.

Pantelis, C., Barnes, T.R. & Nelson, H.E. (1992) Is the concept of frontal-subcortical dementia relevant to schizophrenia? *British Journal of Psychiatry, 160,* 442–460.

Parent, A. & Hazrati, L-N. (1995a) Functional anatomy of the basal ganglia. I. The cortico-basal ganglia-thalamo-cortical loop. *Brain Research Reviews, 20,* 91–127.

Parent, A. & Hazrati, L-N. (1995b) Functional anatomy of the basal ganglia. II. The place of subthalamic nucleus and external pallidum in basal ganglia circuitry. *Brain Research Reviews, 20,* 128–154.

Passingham, R. (1993) *The Frontal Lobes and Voluntary Action.* Oxford: Oxford Scientific Publications, Chapter 8.

Perry, T.L., Hanson, S. & Kloster, M. (1973) Huntington's chorea. Deficiency of gamma aminobutyric acid in brain. *New England Journal of Medicine, 288,* 337–342.

Reider-Groswasser, I., Bornstein, N.M. & Korczyn, A.D. (1995) Parkinsonism in patients with lacunar infarcts of the basal ganglia. *European Neurology, 35,* 46–49.

Robin, D.A. & Schienberg, S. (1990) Subcortical lesions and aphasia. *Journal of Speech and Hearing Disorders, 55,* 90–100.

Saint-Cyr, J.A., Taylor, A.E. & Nicholson, K. (1995) Behavior and the basal ganglia. In: W.J. Weiner & A.E. Lang (Eds.) Behavioral Neurology of Movement Disorders, *Advances in Neurology*, Vol 65, New York: Raven Press, pp. 1–28.

Saxena, S., Brody, A.L., Schwartz, J.M. & Baxter, L.R. (1998) Neuroimaging and frontal-subcortical circuitry in obsessive-compulsive disorder. *British Journal of Psychiatry*, 173 (Suppl 35), 26–37.

Schmidt, W.J. (1995) Balance of transmitter activities in the basal ganglia. *Journal of Neural Transmission*, 46, 57–76 (Supplementum).

Shoulson, I. (1990) Huntington's disease: Cognitive and psychiatric features. *Neuropsychiatry, Neuropsychology, and Behavioral Neurology*, 3, 15–22.

Stahl, S.M. (1988) Basal ganglia neuropharmacology amd obsessive-compulsive disorder: The obsessive complusive disorder hypothesis of basal ganglia dysfunction. *Psychopharmacology Bulletin*, 24, 370–374.

Starkstein, S.E., Robinson, R.T., & Price, T.R. (1987) Comparison of cortical and subcortical lesions in the production of post-stroke mood disorders. *Brain*, 110, 1045–1059.

Starr, M.S. (1995) Glutamate/dopamine D1/D2 balance in the basal ganglia and its relevance to Parkinson's disease. *Synapse*, 19, 264–293.

Swerdlow, N.R. (1995) Serotonin, obsessive compulsive disorder and the basal ganglia. *International Review of Psychiatry*, 7, 115–129.

Tassin, J-P. (1998) Norepinephrine–dopamine interactions in the prefrontal cortex and the ventral tegmental area: Relevance to mental disease. *Advances in Pharmacology*, Vol. 42. New York: Academic Press, pp. 712–716.

Taylor, A.E., Saint-Cyr, J.A., Lang, A.E., & Kenny, F.F. (1986) Frontal lobe dysfunction in Parkinson's disease: The cortical focus of neostriatal outflow. *Brain*, 109, 845–883.

Weiner, W.J. & Lang, A.E. (1989) *Movement Disorders: A Comprehensive Survey*. Mount Kisco, NY: Futura Publishing.

Wichmann, T. & DeLong, M.R. (1993) Pathophysiology of parkinsonian motor abnormalities. In: Narabayashi, H, Nagatsu, T., Yanagisawa, N., & Mizuno, Y. (Eds.) *Advances in Neurology*, Vol 60. New York: Raven Press, pp. 5361.

Yahr, M.D. (1976) The Basal Ganglia. New York: Raven Press.

Zald, D.H. & Kim, S.W. (1996a) Anatomy and function of the orbital frontal cortex, I: Anatomy, neurocircuitry, and obsessive-compulsive disorder. *Journal of Neuropsychiatry and Clinical Neurosciences*, 8, 125–138.

Zald, D.H. & Kim, S.W. (1996b) Anatomy and function of the orbital frontal cortex, II: Function relevance to obsessive-compulsive disorder. *Journal of Neuropsychiatry and Clinical Neurosciences*, 8, 249–261.

7   THE THALAMUS

CHAPTER OVERVIEW

Composed of multiple nuclear grouping, each with its own individual pattern of afferent and efferent connections, the thalamus both structurally and functionally occupies a central position in the brain. It essentially represents a relay station for just about all sensory information traveling from peripheral receptors to the cortex. As such, it is in an ideal position to monitor and influence cortical input. As seen in previous chapters and further explored here, specific thalamic nuclei also represent the final common pathway in most, if not all, cortical–subcortical feedback loops. These connections are thought to allow the thalamus to play an integral role in modulating ongoing behavioral programs involving motoric, cognitive, and emotional expression. Finally, the thalamus is thought to play key roles in cortical arousal and selective attention, both necessary to efficient execution of those behavioral goals or programs.

In the course of this chapter, we will review the anatomy of the thalamus, with particular emphasis on identifying its nuclei, their connections and pathways by which they communicate with others parts of the central nervous system, and their apparent individual contributions to the functions of the whole. As we shall see, these various nuclei can be divided into various groupings based both on anatomical and functional parameters. We also will review the behavioral deficits that might be expected in the events of lesions involving these nuclei, perhaps most commonly the result of hypertensive disease. Finally, the chapter will begin and conclude with an exploration of the probable functional contributions of this core structure.

DIVISIONS OF THE THALAMUS

The Diencephalon

The **diencephalon** consists of the **dorsal thalamus** or thalamus proper, the **hypothalamus,** the **subthalamus,** and the **epithalamus**. The thalamus proper is the dominant structure of the diencephalon in terms of size and lies on either side of the third ventricle (Figure 7–1). The hypothalamus, which lies on the anteroventral aspect of the thalamus proper, largely is associated with modulating internal homeostasis and primitive drive states. The hypothalamus will be discussed in greater detail in the next chapter along with other limbic structures with which the hypothalamus has close anatomical and functional associations. The subthalamus, which traditionally has been linked, along with the substantia nigra, to the basal ganglia, was discussed in the previous chapter. However, as was noted, these structures also have extensive connections with the thalamus proper as part of their cortical feedback loops.

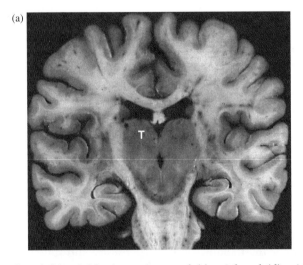

(a)

Figure 7–1.   (a) Coronal and (b) axial brain sections and (c) axial and (d) mid-sagittal MRI images showing thalamus. Abbreviations: CC, corpus callosum; F, fornix; Mid B, midbrain; T, thalamus. Figures (a) and (b) were adapted from the *Interactive Brain Atlas* (1994), courtesy of the University of Washington.

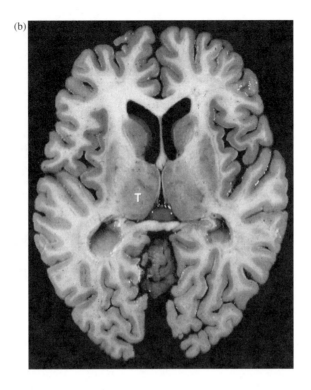

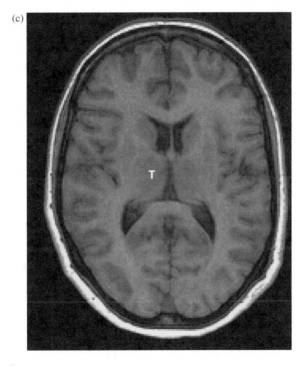

Figure 7–1. *(Continued)*

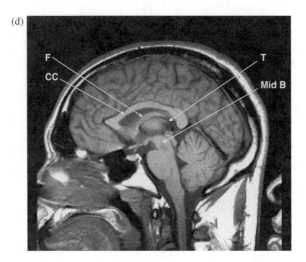

Figure 7-1. *(Continued)*

The Epithalamus

The epithalamus consists of the **pineal body** (gland) and the **habenular nuclei**. The pineal body apparently has some neuroendocrine functions and is thought to help regulate circadian rhythms, at least in some species. In man, its functional contributions are not clear. The habenular nuclei are represented by small prominences on the posteromedial surface of the thalamus proper. Each nucleus consists of a medial and lateral segment and is interconnected by the small habenular commissure that lies just dorsal to the base of the pineal body and the posterior commissure. Afferent inputs to the medial aspects of the habenular nuclei primarily come from the septal nuclei via the **stria medullaris** of the thalamus, while the globus pallidus, hypothalamus, and basal forebrain provide additional input to the more lateral portions. Both the medial and lateral segments of the habenular nuclei likely receive input from the midbrain tegmentum (e.g., serotonergic and dopaminergic fibers from the raphe nuclei and ventral tegmentum). The major output of these nuclei is to portions of the midbrain, including the interpeduncular nuclei, raphe nuclei, substantia nigra (pars compacta), reticular formation, and periaqueductal gray. These connections are made via the **fasciculus retroflexus** (*habenulointerpeduncular tract*).

While the functional significance of the habenular nuclei is still somewhat ambiguous, they have been thought to establish a major link between limbic structures, the basal ganglia, and the midbrain. Consequently, the habenular nuclei have been postulated to be involved in the regulation or modulation of various neurochemical transmitters, in the autonomic responses to olfaction, eating and mating behaviors, sleep cycles, pain, and in positive and negative reinforcements (Sandyk, 1991).

DORSAL THALAMUS

Anatomical Structure

The dorsal thalamus or thalamus proper, which is the main focus of the current chapter, is a paired oval mass that surrounds the third ventricle and lies medial to the posterior limb of the internal capsule (see Figure 7-1) Its anteroventral surface is continuous with the hypothalamus, while more posteriorly it lies above the zona incerta and the subthalamic nuclei and eventually over the red nucleus of the midbrain. The thalamus is composed of a number of discrete nuclear groups that can be differentiated on the basis of structural and functional considerations. From a structural perspective, each half of the thalamus is separated into three major divisions by a more or less vertical sheet of myelinated fibers called the **internal medullary lamina**. This lamina creates a **medial** and a **lateral** division and then, as a result of its bifurcation near the anterior pole of the thalamus, this lamina produces an **anterior division**. There also are a few other thalamic nuclei that do not fall neatly into these three divisions. Figure 7–2 schematically illustrates the general location of the major thalamic nuclei within these broad divisions. However, prior to reviewing the anatomy of the thalamus in greater detail, it may be useful to provide a very brief functional overview to serve as a conceptual framework while considering the connections of the various thalamic nuclei.

Preliminary Functional Considerations

There is some uncertainty as to the extent and/or manner in which the thalamus contributes to higher-order behaviors. Based on an analysis of its connectivity and the effects of clinical lesions, several general hypotheses are possible. First, all sensory input (with the exception of olfaction) is relayed through the thalamus prior to entering the cortex. This arrangement might allow the thalamus to influence the processing of sensory information in a number of ways, For example, depending on internal or external circumstances, including immediate behavioral goals or intentions, the thalamus might either dampen or facilitate certain sensory inputs prior to their transmission to the cortex. The thalamus also might play a role in the preliminary processing (e.g., integration or decoding) of incoming sensory information.

In addition to receiving direct sensory inputs, as has been seen, the thalamus is a critical component of the extensive feedback loops for motor, cognitive, and emotional systems that project from the cortex to the basal ganglia (and cerebellum) and back to the cortex via discrete thalamic pathways. As will be seen later in this chapter, disruptions at any point along the subcortical portions of these pathways, including the thalamus

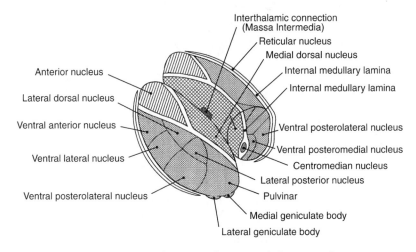

Figure 7–2. Schematic showing relative location of various thalamic nuclei.

itself, may produce effects that are qualitatively similar to deficits produced by lesions directly affecting the cortical projection sites. In addition to these extensive subcortical feedback loops, some thalamic nuclei receive input directly from the cortex and project directly back to those discrete cortical regions. In either case, it is believed that the thalamus plays a major role in activating or arousing the cortex. As has been demonstrated by PET scan studies and electrical mapping of the cortex, all cortical areas are not equally excited or aroused at the same time. Depending on task demands and environmental or contextual changes, attention to different aspects of the stimuli or different types of information processing/responses may be required. The end result can be constantly shifting sites of primary cortical activation (or *relative* cortical inhibition).[1] Thus, thalamic feedback loops or systems may be important in the relative facilitation of specific cortical zones depending upon the circumstances, needs, and intention of the organism.

Finally, while most thalamic nuclei have very specific, topographical cortical projections, others have overlapping or more diffuse projections. At least one thalamic nucleus (reticular) does not project back to the cortex at all. Rather, this latter nucleus projects back to the other nuclei of the thalamus. Thus, the thalamus has the potential to provide very specific or more diffuse modulation of the cortex or even modulation of the modulators.

At this point, it may be helpful to review, in greater detail, some of the major divisions or nuclei of the thalamus, along with their major afferent and efferent connections. The latter are schematically summarized in Figures 7–3 and 7–4. The reader should be aware that the following descriptions do not include all afferent or efferent pathways and for some nuclei, especially the smaller midline and intralaminar nuclei, the pathways have not been well established. It also should be noted that while the nomenclature used is the most common, other authors occasionally will use alternate terms and strategies in defining specific thalamic nuclei.

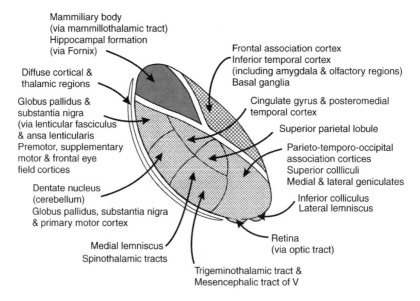

Figure 7–3. Major afferent connections of the thalamic nuclei. **Note**: Trigeminothalamic tract and mesencephalic tract of V project to the ventral posteromedial nucleus (see Figure 7–2).

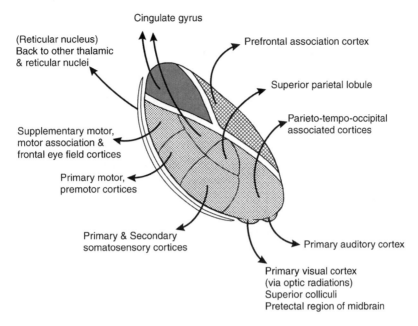

Figure 7–4. Major efferent connections of the thalamic nuclei.

Anterior Division

The anterior division's principal nucleus is the **anterior nucleus**. Its primary afferent input is from the mammillary bodies via the **mammillothalamic tract,** which in turn had received input from the hippocampus via the fornix. The hippocampus also contributes fibers to the anterior nucleus prior to entering the mammillary bodies. The output of the anterior nucleus primarily is to the cingulate gyrus. Because of these various connections, the anterior nucleus was considered to be an integral part of *Papez's circuit* and the *limbic system*, which will be reviewed in greater detail in Chapter 8. The anterior nucleus also appears to have reciprocal corticofugal or afferent connections with the cortical areas to which it projects, in this case, the cingulate gyrus. Actually, most authors assume that most thalamic nuclei have reciprocal connections with the cortical areas to which they project.

Medial Division

The medial division essentially is made up of one large nuclear mass, the **dorsomedial nucleus,** although the much smaller and less well-defined (in humans) midline nuclei sometimes are included in this division. The dorsomedial nucleus receives its input primarily from the frontal lobes, with some additional input from the inferior temporal cortex, including the amygdala, entorhinal region, and olfactory areas. As noted in Chapter 6, this nucleus also receives input from the basal ganglia and is part of the *dorsolateral, orbitofrontal,* and *limbic circuits* with much of the input from the latter two circuits coming via the ventral striatum and nucleus accumbens. Since the majority of the output of the basal ganglia is to the frontal lobes, it should not be surprising that the dorsomedial nucleus projects back to the frontal association cortices.

Lateral Division

The lateral division of the thalamus is the largest division and contains nuclei that are frequently divided into a "lateral" (more dorsal) and a "ventral" tier. The nuclei that make up the *dorsal tier* include the **lateral dorsal nucleus, lateral posterior nucleus,** and the **pulvinar.** The *ventral tier* nuclei include the **ventral anterior nucleus, ventral lateral nucleus, ventral posterior nucleus,** and the **posterior nuclear complex**. These nuclei and their known (or suspected) primary connections are discussed below.

Dorsal Tier Nuclei

Lateral Dorsal Nucleus   This nucleus is located adjacent to the posterior portion of the anterior nuclear group and appears to have somewhat similar connections. It receives input from the cingulate gyrus, as well as from the posteromedial temporal regions and projects back to the cingulate gyrus. It appears to be more extensively connected to the more posterior portions of the cingulate cortex than the anterior nucleus and to the mesial portion of the parietal lobe (precuneus), but its exact connections are uncertain.

Lateral Posterior Nucleus   This dorsal tier nucleus lies posterior to the lateral dorsal nucleus and the afferent connections of this nuclear group also are somewhat ambiguous. It appears to receive input from the superior colliculus and pretectal regions, as well as projections from the parietal lobe. The main efferent projections likely are back to the parietal lobe, especially the somatosensory association areas (eg, areas 5 and 7).

Pulvinar   The pulvinar represents the largest of the thalamic nuclei in man. It receives input from the retina, superior colliculus, both the medial and lateral geniculates, and from the parietotemporooccipital (PTO) association cortex. It projects back to the association cortices of the parietal, temporal, and occipital lobes. Although the pulvinar commonly is listed with the dorsal tier of nuclei, this huge nucleus for the most part forms the entire posterior cap of the thalamus.

Ventral Tier Nuclei

Ventral Anterior (VA) Nucleus   Whereas the thalamic nuclei thus far discussed largely project to either association cortices or limbic cortex (cingulate gyrus), the nuclei that constitute the ventral tier of the lateral division are more closely associated with primary and secondary sensory and motor areas. The most anterior of the ventral group, the ventral anterior and the ventral lateral nuclei, are intricately involved with motor systems. Input to the ventral anterior nuclei is from the substantia nigra and the globus pallidus via the **lenticular fasciculus** and the **ansa lenticularis** and from the premotor, supplementary motor, and frontal eye fields in the cortex. While the VA nucleus may receive some input from the cerebellum, the latter projects primarily to the ventral lateral nucleus. The ventral anterior nucleus, in turn, projects back to the motor association cortex (area 6), the frontal eye fields (area 8), and to the supplementary motor cortex on the medial surface of the frontal lobe.

Ventral Lateral (VL) Nucleus   Similar to the ventral anterior nucleus, the ventral lateral nucleus also receives input from the globus pallidus and the substantia nigra and also from the primary motor cortex. The ventral lateral nucleus also receives substantial input, particularly in its more caudal region, from the contralateral dentate nucleus of the cerebellum via the superior cerebellar peduncle and thalamic fasciculus. Its major output is back to the primary motor (area 4) and premotor cortices. The projections to the precentral gyrus (primary motor cortex) apparently derive primarily from that portion of the nucleus that receives cerebellar input. Both the VA and VL nuclei retain their discrete topographical arrangement as they project to the cortex.

Ventral Posterior Nucleus   The major input into this nucleus is cutaneous with somatosensory feedback from the head and body, as well as the sense of taste. The ventral posterior nucleus typically is divided into two parts: the ventral posterior lateral (VPL) and the ventral posterior medial (VPM) nuclei. The VPL nucleus receives somatosensory sensations from the body (via the medial lemniscus and spinothalamic tracts), while the VPM nucleus receives input from the head and face, including taste, via the trigeminothalamic tracts and the mesencephalic tract of V. In turn, the major cortical projection for these nuclei is to the primary somatosensory cortex (areas 3, 1, 2). In addition, there are some projections to the secondary somatosensory cortex in the parietal operculum, and probably to the more anterior sensorimotor cortices. As in the case of VA and VL, a strict topographical representation is maintained both at the thalamic and cortical projection sites.

Posterior Nuclear Complex   Lying just posterior to the ventral posterior nucleus and dorsomedial to the medial geniculate nucleus on the ventral aspect of the pulvinar, this small nuclear group inconsistently is represented in listings of thalamic nuclei. Like VPL and VPM, it also receives somatosensory inputs but primarily, if not exclusively, from the spinothalamic tracts. Unlike the somatosensory projections to the ventral posterior nuclear group, there is not the same point-to-point somatotopic organization and there even may be some degree of bilateral representation. Cells in this nucleus may be preferentially sensitive to noxious stimuli, and hence, this area not only may be important for the perception of pain but also may be a primary source of thalamic pain syndromes. The cortical projection sites for this nucleus appear to be both the primary and secondary somatosensory cortices, as well as portions of the insular cortex. In keeping with the general pattern of thalamic organization, the posterior nuclear complex likely receives reciprocal input from these same cortical areas.

   The next two thalamic nuclei to be discussed, the **medial** and **lateral geniculates,** both represent highly specific, sensory relay stations between auditory and visual inputs and their respective cortical projection areas. Collectively, these nuclear groups sometimes are referred to as the *metathalamus*. Both are seen as prominences on the ventrolateral aspect of the pulvinar; hence, the terms medial (MGB) and lateral (LGB) geniculate bodies. As its name implies, the LGB is located more laterally, lying just above and medial to the hippocampal formation on a coronal section.

Medial Geniculate Nucleus (MGN)   The major input to the MGN is from the auditory system (ie, cochlear nuclei, superior olivary nuclei, lateral lemniscus, and inferior colliculi) via the brachium of the inferior colliculi. These inputs contain bilateral auditory information, although the contralateral representation is slightly more (approximately 60% vs. 40%). Another primary source of afferent fibers is from the auditory cortex to which it projects. Efferent fibers exit the MGN via the auditory radiations and project to the ipsilateral primary auditory cortex (area 41 and probably part of 42 or, *Heschl's gyrus*) which lies in the temporal operculum. Just as the somatosensory and motor projections maintain a topographical representation on the cortex, so too do the auditory radiations, except in this case there is a tonotopic organization.

Lateral Geniculate Nucleus (LGN)   The input to the lateral geniculates is readily visualized. In dissecting the brain, it is relatively easy to follow the optic tracts as they leave the chiasm to their termination in the LGN. As reviewed in Chapter 5, the optic tracts consist of crossed (nasal half of contralateral eye) and uncrossed (temporal half of ipsilateral eye) postganglionic retinal fibers. Because of this arrangement, each LGN receives fibers that represent visual input from the contralateral visual field. The retinotopic organization of the cells, as well as the arrangement of crossed versus uncrossed fibers (for contralateral visual field perception) are maintained in the lateral geniculates. The main efferent output of the LGN is to the primary visual or **calcarine cortex** (area 17) via the **optic radiations,** which travel in the association pathways underlying the temporal and parietal cortices. Similar to other thalamic nuclei, reciprocal afferent connections are established from the striate (visual) cortex back to the LGN. Finally, the lateral geniculates also send a small segment of fibers to the superior colliculi and pretectal regions of the midbrain, as well as to other diencephalic areas (e.g., the suprachiasmatic nucleus). Unlike the medial geniculate nuclei that receive the majority of their input from the inferior colliculi, the LGN appear to receive no input from the superior colliculi. The latter project to the pulvinar nuclei, which are involved in visually guided behaviors.

Other Thalamic Nuclei and Designations

In addition to the anterior, medial, and lateral nuclear groups, there are several other smaller nuclear groups in the thalamus: the **intralaminar nuclei,** the **midline nuclei,** and the **reticular nuclei**.

Intralaminar Nuclei

These thalamic nuclei lie within the internal medullary lamina of the thalamus. The two most prominent of the intralaminar nuclei are the **centromedian** and **parafascicular nuclei**. The centromedian nucleus often clearly is visible as a central oval mass in the posterior thalamus on coronal section, with the parafascicular nucleus constituting its more medial extension. These nuclei likely receive input from ascending somatosensory fibers as well as from the reticular system and the globus pallidus. In turn, their efferent output is to diffuse areas of the cortex and back to the basal ganglia (neostriatum and subthalamic nuclei). In part because of these connections to somatosensory, reticular, and diffuse cortical regions, these nuclei are thought to play a role in generalized cortical arousal, especially in response to external stimuli.

Midline Nuclei

The midline nuclei often are included with the dorsomedial nucleus as part of the medial division of the thalamus. Lying adjacent to the lateral walls of the third ventricle, these small nuclear groups are among the smallest and least well defined of the thalamic nuclei. Although it is thought that these nuclei are connected with the basal forebrain and the limbic system, relatively little appears to be known about their functional correlates.

Reticular Nuclei (of the Thalamus)

The reticular nuclei represent a thin sheath of gray matter that lies between the posterior limb of the internal capsule and the external medullary lamina of the thalamus. The reticular nuclei receive input from diffuse cortical areas via collaterals from the multiple corticothalamic fibers entering the body of the thalamus as well as from fibers exiting the thalamus. These nuclei, unlike most others, do not project back to the cortex; rather they project to other thalamic nuclei and possibly to the reticular formation of the brainstem. Because of these connections, this nucleus is thought to perhaps exert a modulating influence on the other thalamic nuclei.

In addition to the above structural divisions of the thalamus, the various thalamic nuclei also have been classified functionally on the basis of their cortical projections, or lack thereof. The following summary represents one such functional schema.

Specific Relay Nuclei

These are nuclei that project to those areas of the cortex that have very "specific" sensory or motor functions and/or evoke relatively localized cortical responses on stimulation of the thalamic nuclei.

Specific relay nuclei	Cortical projection sites
Lateral geniculate	Visual cortex
Medial geniculate	Auditory cortex
Ventral posterior lateral	Somatosensory cortex
Ventral posterior medial	Somatosensory cortex
Ventral lateral	Motor cortex
Ventral anterior	Premotor cortex

Association Nuclei

These are nuclei that do not receive major input from the ascending (sensory) tracts and generally project to "association" cortices.

Association nuclei	Cortical projection sites
Pulvinar	PTO association cortex
Lateral posterior	Parietal association cortex
Lateral dorsal	Cingulate gyrus and precuneus
Dorsomedial	Prefrontal cortex
Anterior	Cingulate gyrus

Nonspecific Nuclei

These nuclei project to wide areas of the cortex.

Nonspecific nuclei	Projection sites
Intralaminar	Diffuse cortical areas
Midline	Probably limbic structures

Subcortical Nuclei

These nuclei do not project to the cortex.

Nucleus	Projection sites
Reticular	Other thalamic nuclei

THE INTERNAL CAPSULE

Communications between one region or structure of the brain and another are established primarily through what might be thought of as axonal "superhighways." These can be classified into three major types: **projection pathways, association pathways,** and **commissures.** The latter pathways will be discussed in greater detail in Chapter 9. Here the focus will be on the internal capsule, which represents the major projection pathway in the brain.[2] In addition, other corticofugal fibers (e.g., corticopontine, corticobulbar and corticospinal tracts), the internal capsule contains both corticothalamic and thalamocortical connections. These latter thalamocortical connections represent not only the means by which sensory information is filtered to the cortex, but also feedback loops from motor and association cortices through the basal ganglia, as well as from the cerebellum via cerebellothalamocortical pathways.

The internal capsule itself can be divided into five sections.

1. **Anterior limb,** which lies between the head of the caudate and the lenticular nuclei.
2. **Genu** or *knee,* which is the "bend" seen on axial or horizontal sections (see Figs 7–1b; 7–5).
3. **Posterior limb,** which lies between the lenticular nuclei and the thalamus.
4. **Retrolenticular portion,** which is the posterior continuation of the posterior limb.
5. **Sublenticular portion,** which represents the most caudal and ventral curvature of the capsule. Because of this curvature, it is not possible to see all parts of the internal capsule in any one brain section.

Knowledge of the general location of the above-mentioned corticofugal and corticopetal fiber tracts within the internal capsule is often clinically relevant. For example, a lesion encroaching on the posterior aspects of the internal capsule might explain the concurrence of both motor and visual symptoms. While the general schema of these pathways within the internal capsule are presented in Figure 7–5 and described below, their specific locations are not always very precise. Most do not occupy discrete sections of the internal capsule, but rather seem to overlap or interdigitate with one another, while others remain somewhat conjectural. In general, however, the various ascending and descending pathways within the internal capsule are organized as follows:[3]

1. **Anterior limb:** Dorsomedial nuclei → prefrontal cortex; anterior nuclei → cingulate gyrus; and frontopontine connections.
2. **Genu:** Corticobulbar (influencing the cranial nerve motor nuclei) and corticoreticular fibers; ventral anterior and ventral lateral nuclei → premotor and motor cortices.
3. **Posterior limb:** Additional frontopontine connections; lateral dorsal nuclei → posterior cingulate gyrus and precuneus gyrus; the corticospinal tracts with the upper extremities being located more anterior than the trunk and lower extremities; corticorubral connections; ventral posterior medial and ventral posterior lateral nuclei → somatosensory cortex; lateral posterior → somatosensory association cortex; pulvinar → PTO association cortex (the pulvinar projections probably continue into and constitute part of the retrolenticular aspect of the capsule).
4. **Retrolenticular portion:** Pulvinar → PTO cortex; corticotectal tracts; lateral geniculate body → occipital cortex (beginnings of the optic radiations, especially parietal portions).
5. **Sublenticular portion:** LGB → occipital lobes (temporal radiations); temporooccipitopontine fibers; and medial geniculate body → auditory cortex.

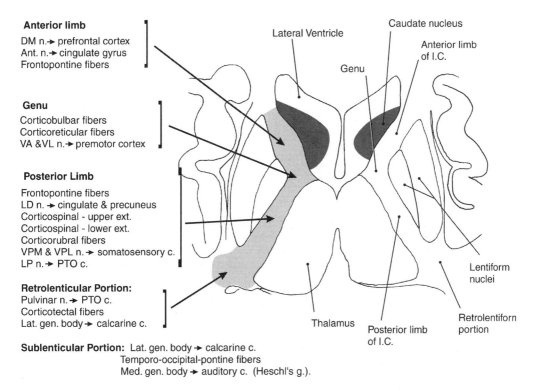

Figure 7–5. Divisions of the internal capsule (IC) containing various thalamic, cortical, and subcortical projection fibers.Abbreviations: Ant n., anterior nucleus; DM n., dorsomedial nucleus; IC, internal capsule; LD n., lateral dorsal nucleus; LP, lateral posterior nucleus; PTO c., parietal–temporal–occipital association cortex; VA & VL n., ventral anterior and ventral lateral nuclei; VPM & VPL n., ventral posterior medial and ventral posterior lateral nuclei.

FUNCTIONAL CONSIDERATIONS

As was noted in the beginning of this chapter, the role of the thalamus in the regulation or expression of behavior remains ambiguous. Perhaps stated more precisely, the functional roles of individual thalamic nuclei have not yet been completely determined. However, it is almost certain that different nuclei subserve different functions. Based on their cortical and subcortical connections and the presence of different anatomical feedback loops, it has been suggested that in general individual nuclei may be important in:

- Altering cortical (subcortical) tone or arousal
- Modulating and directing attention
- Filtering (gating), modulating, decoding (perhaps even elementary perception) of sensory input
- Facilitating relevant motor response channels
- Recruiting cortical (subcortical) circuits critical to ongoing cognitive activity

These speculations regarding the role of the thalamus in regulating and modulating behavior are derived from the fact that: (1) the ascending reticular activating system of the brainstem impacts upon the thalamus; (2) the thalamic nuclei, in turn, have direct and for the most part topographical projections to the cortex; (3) virtually all sensory input is relayed to the cortex via the thalamus, and (4) the thalamus and widespread regions of the cortex have multiple thalamic feedback loops, both directly and via other subcortical nuclei (e.g., basal ganglia, limbic structures, and hypothalamus). While an understanding of these afferent and efferent connections provide clues to the functions of the thalamus, the more convincing evidence is derived from studying the effects of lesions. However, this approach also has some limitations, especially when studying the thalamus.

As suggested above, because of the diversity of their connections, we are obliged to think in terms of individual nuclei rather than of the thalamus as a whole when considering its functional correlates. However, an infarction or hemorrhage, the more common thalamic lesions, easily may extend beyond the boundaries of a particular nucleus, as do the penetrating vessels that supply the thalamus. Therefore, thalamic lesions tend to cross nuclear boundaries and often intrude upon adjacent structures, pathways, or spaces as well (e.g., encroaching upon the internal capsule or lenticulothalamic fibers or bleeding into the ventricles). Even in small, discrete lesions, in vivo identification of the specific nucleus involved may be difficult, despite current MRI technology. Surgical lesions (e.g., for the remediation of movement abnormalities, such as in Parkinson's disease) have provided some insight into the functions of the thalamus but those surgical lesions have involved relatively few nuclei, and as noted in the previous chapter some of effects are not always easily explicable. Despite these limitations, some clinical correlations are possible.

As will be discussed in greater detail in Chapter 10 the blood supply to the thalamus primarily is supplied by four penetrating arteries derived from the posterior circulation [**tuberothalamic, paramedian, inferolateral** (or **geniculothalamic**), and the **posterior choroidal**]. The **anterior choroidal artery,** which is a branch of the internal carotid, and the **posterior choroidal artery** may supply the lateral geniculates and parts of the pulvinar. However, it should be noted that there often are significant differences in the descriptions provided by various authors of the arterial blood supply for specific nuclei. Furthermore, not only do the vessels themselves not always respect nuclear boundaries, there also may be variations in the distributions from one individual to another.

Table 7–1 lists some of the more commonly cited distributions and symptoms associated with hemorrhages or infarctions of these arteries. Again, these data should be interpreted with considerable caution for several reasons. First, as noted above, not only do clinical lesions vary considerably in terms of size and shape, but also their localization can be approximated only with current technology. Second, with the exception of those few nuclei that have been subjected to frequent surgical lesioning for the possible relief of motor disorders, there are relatively few pure clinical cases of highly discrete, well-documented, naturally occurring thalamic lesions in the literature with comprehensive clinical data. As will be discussed in greater detail below, while some thalamic lesions may produce very

Table 7–1. Thalamic Vascular Syndromes

Tuberothalamic	VA, VL, Ant?	Motor weakness;* somatosensory disturbances;* memory and speech disturbances?
Paramedian	DM, Midline	Memory deficits; dementia or mental confusion; disturbance of consciousness;* gaze palsy; contralateral neglect; visual-spatial disturbances
Inferolateral	VPL, VPM, CM?	Somatosensory disturbances; ataxia; motor weakness,*
Post. choroidal	Pulvinar, LGB?	Optokinetic disturbances; visual field defects; aphasia (with (left-sided lesions); neglect; motor weakness*
Ant. choroidal	LGB, Pulvinar?	Visual field defects; motor weakness and somatosensory deficits*

* See text.

gross, obvious symptoms, others manifestations are likely to be more subtle and may require very comprehensive and sophisticated clinical investigations to uncover their effects. Finally, as noted, thalamic lesions often extend beyond its boundaries. For example, disturbances of consciousness (and mortality) commonly are found with larger hemorrhagic lesions that bleed into the ventricles. Somatosensory disturbances, including spontaneous pain (**Dejerine-Roussy syndrome**) are among the more frequently cited symptoms following thalamic lesions. Transient motor symptoms, such as hemiparesis, can occur, but permanent deficits are more likely the result of invasion of the genu or posterior limb of the internal capsule.

Summary of Functional Correlates

Given these limitations, what can be said about the symptoms and functions of the thalamus or its nuclei? Probably the easiest place to begin is with those nuclei that have been designated as sensory relay nuclei: VPL, VPM, and the medial and lateral geniculates. There appears to be fairly strong evidence that destruction of the VPL nucleus produces contralateral somatosensory deficits in the trunk and extremities, with similar losses in the face and anterior portions of the head with lesions affecting VPM. Lesions to these nuclei may lead to deficits in all tactile modalities, including mechanical pressure (touch), pain, temperature, proprioception, and vibration. Lesions that affect the integrity of the lateral geniculate nucleus will produce a contralateral homonymous hemianopia or partial field defect, depending on the extent of the lesion. Conversely, because of the bilateral representation of hearing, unilateral lesions of the medial geniculate nucleus do not produce any dramatic sensory losses. Taste and olfaction likewise should be unaffected. Recall that only olfactory input makes its way to cortical structures without first going through the thalamus. As suggested earlier, lesions that are restricted to the thalamic "motor" nuclei (primarily VA and VL) do not typically result in either permanent loss of spontaneous movement or in abnormality of movement (dyskinesias).[4] It would appear that such lesions likely affect more subtle aspects of motor learning or programming, although the exact nature of these effects have not yet been clearly delineated in humans.

Several thalamic nuclei project to cortical association cortices (e.g., dorsomedial, lateral posterior or pulvinar) or to allocortex (anterior, lateral dorsal). As a general rule, lesions involving these nuclei produce deficits that are similar in nature, if not always in degree, to those encountered with lesions to the cortical sites to which they project. Thus lesions to these nuclei may result in such "cortical" syndromes as neglect, visual–spatial disturbances, cognitive or "intellectual" changes, memory impairments, or aphasic deficits (if the dominant thalamus is involved). As might be expected, smaller, more circumscribed lesions will more likely result in less profound and more transient deficits than larger, more destructive lesions. Similar to bilateral cortical lesions, bilateral thalamic lesions (although relatively rare) produce more extensive deficits. Discrete lesions of the midline, intralaminar, or reticular nuclei are exceedingly rare and difficult to identify; hence their clinical effects in humans are virtually unknown.

Endnotes

1. As noted in Chapter 6, the majority if not all thalamocortical connections are glutaminergic, and thus likely facilitatory. While it should be noted that "positive connections" to certain heteroreceptors or "inhibitory" interneurons may result in overall inhibition, it would appear that the primary influence of the thalamus on the cortex is to facilitate

or recruit areas or pathways important in a given behavioral context. At the same time, other cortical areas not so engaged may experience *relative inhibition* when their respective thalamic afferents in turn have been subjected to pallidal–thalamic inhibitory influences.

2. The smaller **external** and **extreme capsules** lie laterally to the putamen and claustrum, respectively, but their compositions are less clear. The external capsule contains some corticostriate (projection) fibers, while the extreme capsule likely consists mostly of association fibers between the insula and other cortical regions.

3. In a few instances, the exact location of these fiber pathways has not been confirmed.

4. While contralateral dystonia has been identified in a limited number of cases in which the lesion was apparently restricted to the thalamus, in these instances the lesion tended to be isolated to the posterior or midline nuclei rather than to VA or VL (Lee & Marsden, 1994).

REFERENCES AND SUGGESTED READINGS

Barr, M.L., & Kiernan, J.A. (1993) The Human Nervous System. Philadelphia: J.B. Lippincott Co., pp. 181–199.

Barraquer-Bordas, L., Illa, I., Escartin, A. Ruscalleda, J., & Marti-Vilalta, J.L. (1981) Thalamic hemorrhage: A study of 23 patients with diagnosis by computed tomography. *Stroke, 12*, 524–527.

Bogousslavsky, J. Regli, F., & Uske (1988) Thalamic infarcts: Clinical syndromes, etiology and prognosis. *Neurology, 38*, 837–848.

Brazis, P., Masdeu, J., & Biller, J. (1990) Localization in Clinical Neurology. Boston: Little, Brown & Co., pp. 320–344.

Bruyn, R.P.M. (1989) Thalamic aphasia: A conceptual critique. *Journal of Neurology, 236*, 21–25.

Burchiel, K.J. (1995) Thalamotomy for movement disorders. *Neurosurgery Clinics of North America, 6*, 55–71.

Caplan, L.R., DeWitt, L.D., Pessin, M.S., Gorelick, P.B., & Adelman, L.S. (1988) Lateral thalamic infarcts. *Archives of Neurology, 45*, 959–964.

Carpenter, M.B., & Sutin, J. (1983) Human Neuroanatomy. Baltimore: Williams & Wilkins. pp. 500–535.

Canavan, A.G.M., Nixon, P.D. & Passingham, R.E. (1989) Motor learning in monkeys (Macaca fascicularis) with lesions in the motor thalamus. *Experimental Brain Research, 77*, 113–126.

Gilman, S., & Newman, S. W. (1992) Essentials of Clinical Neuroanatomy and Neurophysiology. Philadelphia: F.A. Davis Co., pp. 207–215.

Graff-Radford, N.R., Damasio. H., Yamada, T., Eslinger, P.J., & Damasio, A.R. (1985) Nonhemorrhagic thalamic infarction. *Brain, 108*, 485–516.

Graff-Radford, N.R. (1997) Syndromes due to acquired thalamic damage. In: Feinberg, T.E. & Farah, M.J. (Eds.) Behavioral Neurology and Neuropsychology. New York: McGraw-Hill, pp. 433–443.

Groenwegen, H.J., Berendse, H.W., Wolters, J.G., & Lohman, A.H. (1990) The anatomical relationship of the prefrontal cortex with the striatopallidal system, the thalamus and the amygdala: Evidence for a parallel organization. *Progress in Brain Research, 85*, 95–118.

Jones, E.G. (1987) Ascending inputs to, and internal organization of, cortical motoro areas. *Ciba Foundation Symposium, 132*, 21–39.

Jones, E.G. (1985). The Thalamus. New York: Plenum Press.

Kawahara, N., Sato, K., Muraki, M., Tanaka, K., Kaneko, M., & Uemura, K. (1986) CT classification of small thalamic hemorrhages and their clinical implications. *Neurology, 36*, 165–172.

Lee, M.S. & Marsden, C.D. (1994) Movement disorders following lesions of the thalamus or subthalamic region. *Movement Disorders, 9*, 493–507.

Mair, R.G. (1994) On the role of thalamic pathology in diencephalic amnesia. *Reviews in Neuroscience, 5*, 105–140.

Marsden, C.D. & Obeso, J.A. (1994) The functions of the basal ganglia and the paradox of stereotaxic surgery in Parkinson's disease. *Brain, 117*, 877–897.

Nolte, J. (1993) The Human Brain. St. Louis: Mosby-Year Book, Inc., pp. 246–264.

Robin, D.A. & Schienberg, S. (1990) Subcortical lesions and aphasia. *Journal of Speech and Hearing Disorders*, *55*, 90–100.

Sandyk, R. (1991) Relevances of the habenular complex to neuropsychiatry: A review and hypothesis. *International Journal of Neuroscience*, *61*, 189–219.

Speelman, J.D. (Ed.). (1991) Parkinson's Disease and Stereotaxic Surgery. Amsterdam: Rodopi.

Steinke, W., Sacco, R.L., Mohr, J.P., Foulkes, M.A., Tatemichi, T.K., Wolf, P.A., Price, T.R., & Hier, D.B. (1992) Thalamic stroke: Presentation and prognosis of infarcts and hemorrhages. *Archives of Neurology*, *49*, 703–710.

Tasker, R.R. & Kiss, Z.H. (1995) The role of the thalamus in functional neurosurgery. *Neurosurgery Clinics of North America*, 73–104.

Walshe, T.M., Davis, K.R., & Fisher, C.M. (1977) Thalamic hemorrhage: A computed tomographic–clinical correlation. *Neurology*, *27*, 217–222.

Watson, R.T., Valenstein, E., & Heilman, K.M. (1981) Thalamic neglect: Possible role of the medial thalamus and nucleus reticularis in behavior. *Archives of Neurology*, *38*, 501–506.

Waxman, S.G., Ricaurte, G.A., & Tucker, S.B. (1986) Thalamic hemorrhage with neglect and memory disorder. *Journal of Neurological Science*, *75*, 105–112.

Weisberg, L.A. (1986) Thalamic hemorrhage: Clinical–CT correlations. *Neurology*, 1382–1386.

8 THE LIMBIC SYSTEM/HYPOTHALAMUS

CHAPTER OVERVIEW

If the phrase "an enigma shrouded in a mystery" ever could be used to describe a portion of the central nervous system, then it might most aptly apply to the notion of the limbic system. As will be discussed, even considering the collection of phylogenetically older cortical type tissues that largely comprise Broca's "limbic lobe" as a meaningful functional system repeatedly has been challenged. Part of the difficulty lies in the number of structures, their myriad connections, and the diversity of ascribed functions that would be represented by such a system. The problem appears twofold. First, is there sufficient functional cohesiveness to classify these limbic structures as a "system," as, for example, we might speak of a *motor* or a *visual system*? Second, there is the matter of boundaries; once we start including structures, where does one stop? Different authors have taken different stands on whether or not these structures should be construed as constituting an integrated system. Regardless, most authors will concede that this collection of allocortical tissues (along with the hypothalamus with which they are strongly connected) is critically important for a host of behaviors essential for the preservation of both the individual and the species. Listed among such behaviors typically are internally and externally stimulated drive states, emotional responsiveness, and the ability to encode into memory those life experiences relevant to those drive or emotional states.

Much of this chapter will focus on delineating those structures that are thought to play a central role in emotion, motivation (including basic biological drives), and memory. While the reader should attend to their basic anatomical features, even more critically the reader is encouraged to develop a working knowledge of their interconnections and a general understanding of their relationship to neocortical areas. Such knowledge and understanding in turn will be critical to appreciating their suspected behavioral and clinical correlates. A major goal of the authors, and presumably the readers, throughout this text.

As suggested by the opening sentence, the overall contribution of the "limbic system" or limbic structures to human behavior is still largely conjectural. Nevertheless, as clinicians and scientists it is incumbent upon us to develop and continually refine theories to help elucidate our understanding of brain–behavior relationships. Such knowledge not only is sought for its own sake, but also as a means of better understanding and appreciating pathological disease states. In trying to approach this goal, the apparent functional significance of each of these structures will be explored on an individual basis, based both on their cortical and subcortical connections, and behavioral correlates as derived from clinical and experimental studies. Having reviewed the structural and functional aspects of each of these individual structures, the chapter will conclude by summarizing a broader theoretical model of the potential role of limbic structures ("system," if you will) in the evolution of the brain and behavior.

INTRODUCTION

The first five chapters of this book focused almost exclusively on sensory and motor processes in relatively well-defined anatomical systems. Chapters 6 and 7 on the basal ganglia and the thalamus introduced systems concerned not only with sensorimotor functions, but also with what might be considered "higher-order" phenomena such as emotions and cognition. As we approach this chapter and the next on the cerebral cortex, there will be a continuing shift from more elementary sensory input and motor effector systems to an increasing focus on superordinate concepts involving emotion, motivation (drive), goals, intentions, perception, learning, memory, and cognition. From a historical perspective, it has not been that long ago that these latter processes often were ascribed to an entity (i.e., the soul) that was thought to operate independently of the central nervous system. Because of the controversies that have surrounded these more recently evolved brain structures and the speculative nature of their functional organization, it might be informative to depart from the format used in the preceding chapters. Here we will begin by looking at how the "limbic system" and its behavioral correlates have been viewed historically, even within the last century.[1]

HISTORY OF THE "LIMBIC SYSTEM"

In 1878, *Broca* defined the band of neuronal tissue that lies under the cortical mantle and more or less surrounds the upper part of the brainstem (the thalamus) as *le grande lobe limbique*. This area, which comprises a much larger percentage of the total brain in lower animals, was thought to be associated largely with the sense of smell, and hence, became known as the rhinencephalon (*smell-brain*). It included parts of the parahippocampal and cingulate gyri and the subcallosal area of frontal lobes (Figure 8–1).

In 1937, *James Papez* published a paper entitled, "A proposed mechanism of emotion." In it he describes a circuit (later referred to as **Papez's circuit**) that he felt served as the probable neurological substrate of emotional expression and, along with the contribution of the cerebral cortex, the neuronal basis for the subjective experience of emotion. This circuit was proposed based on what Papez observed to be a relatively direct series of interconnections between areas of the brain that had been implicated in emotional behaviors. Papez had noted, for example, the emotional disturbances that were commonly associated with rabies, a disease that selectively affects the hippocampal region (and cerebellum). Similarly, he noted behavioral or affective changes following lesions in the area of the cingulate gyrus, the mammillary bodies,[2] and the anterior nuclei of the thalamus. He also had noted that the precuneus (part of the medial parietal lobe that was thought to be an extension of the cingulate gyrus) was the one area of the brain that showed the greatest sex difference (being larger in males). As this also was an area that bordered on the sacral representation in the paracentral lobule, he figured this must be the area of the cortex where the sex organs were localized.[3]

Papez proposed that emotions could arise either from psychic origins (i.e., from one's cognitive or intellectual awareness of a situation) or from sensory input. With regard to the former, he suggested the following neural substrate. Cortical or psychic information first might be communicated to the hippocampal formation (e.g., via the cingulum, a pathway that connects the cingulate gyrus and other deep cortical structures with the hippocampal regions). From observations of animals and patients affected with rabies, it appeared that the hippocampal formation clearly was involved in the formation of emotions. Impulses then were thought to have been sent from the hippocampal formation to the mammillary bodies via the fornix. The earlier works of Bard (1928) and the study of "sham rage" in animals deprived of their cortex and basal ganglia had suggested that the hypothalamus was the likely source of "emotional expression." From the mammillary bodies, it was proposed

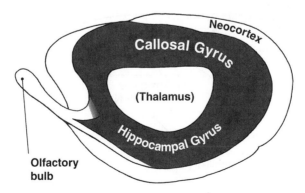

Figure 8–1. "Le Grande Lobe Limbique" as adapted from Broca's original 1878 drawing of an otter's brain. Broca's "callosal gyrus" is now termed the cingulate gyrus.

that impulses were sent to the anterior nucleus of the thalamus via the mammillothalamic tract and then from there to the cingulate gyrus. According to Papez, the cingulate gyrus was considered to be the primary receptive site for the subjective experience of emotion. From the cingulate gyrus this "emotional coloring" of the psychic experience could then be relayed to the cortex (Figure 8–2).

External sensory inputs (e.g., the sight of a snake, the roar of a lion) were thought to take on emotional tones through similar pathways. Papez recognized that most sensory input, except for olfaction, was transmitted to the cortex via the thalamus. However, he indicated that while the thalamus was relaying sensory impressions to the cortex, simultaneously the hypothalamus (again, the main source of emotional expression) also received sensory input from "primitive" sensory centers in the subthalamus. The hypothalamus, having imbued these sensory stimuli with an affective tone, conveyed this now emotionally charged information to the cingulate gyrus via the anterior nucleus of the thalamus as outlined above. Finally, these data were relayed on to the cortex, which concurrently was receiving a perceptual impression of the stimuli. These converging inputs thus attached "subjective emotional experience" (which only could be accomplished in the cortex) to the specific sensory stimulus.

This pathway (hippocampal formation → fornix → mammillary body → mammillothalamic tract → anterior thalamus → cingulate gyrus → and back to the hippocampus via the parahippocampal gyrus) was the groundwork for what later came to be known as Papez's circuit and served as the major starting point for descriptions of the "limbic system." The importance of Papez's paper was that:

1. For the first time, it attempted to lay down a systematic neurological substrate for emotional behavior.
2. It emphasized the close connection between the structures of the *limbic lobe* of Broca and the hypothalamus.
3. It served as an added impetus for research into the role of these structures or "systems" beyond olfaction.

While Papez's circuit probably is the more familiar, Yakovlev (1948) proposed a second limbic circuit that involved additional basolateral connections. This latter limbic circuit

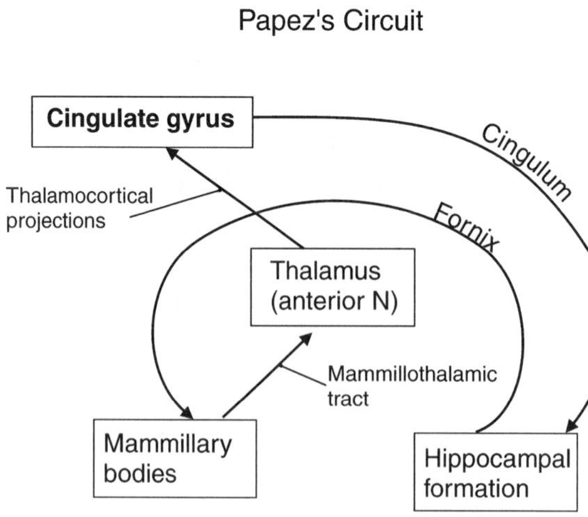

Papez's Circuit

Figure 8–2. Papez's circuit.

incorporates the dorsomedial nucleus of the thalamus, orbitofrontal cortex, temporal pole, and amygdala. Figure 8–3 illustrates the differences between the two proposed circuits. Functionally, Yakovlev's circuit would appear to be more related to the visceral aspects of emotional–affective processing. However, components of both circuits have been implicated in the neural substrates underlying learning and memory (Goldberg, 1984; Victor, Adams, & Collins, 1989; Zola-Morgan & Squire, 1993). It is important to note that both Yakovlev's and Papez's papers describe circuits consisting of closed loops; hence, lesions that disrupt any portion of these circuits, including the white matter connections, would be expected to disrupt the behavior mediated by these circuits.

In 1949, 12 years after Papez published his paper, Paul MacLean (1964) published a very interestingly written article in the journal, *Psychosomatic Medicine* in which he further developed the role of Papez's circuit in mediating the basic visceral responses of the organism, including such activities as the four "F's" (*feeding, fighting, fleeing,* and *"fooling around"*). He referred to this interconnected series of structures described by Papez as the *visceral brain.*[4] The main thrust of his paper was to establish this "visceral brain" as the source of the "unconscious" motivation or dynamics that underlie psychosomatic disorders. As he explained it, the cortex of this system (e.g., the hippocampal formation) is less well developed than that of the cerebral hemispheres. As a result, the hippocampal formation is forced to process all this visceral information (e.g., the need for food, fear, and sexual gratification, most of which began in infancy) at a very primitive, highly symbolic, nonverbal level. He also speculates that:

"[I]f the visceral brain were the kind of brain that could tie up symbolically a number of unrelated phenomena, and at the same time lack the analyzing ability of the word brain to make a nice discrimination of their differences, it is possible to conceive how it might become foolishly involved in a variety of ridiculous correlations leading to phobias, obsessive-compulsive behaviour, etc...Considered in light of Freudian psychology, the visceral brain would have many of the attributes of the unconscious id...the visceral brain [may not be] unconscious, but rather eludes the grasp of the intellect because its animalistic and primitive structure makes it impossible to communicate in verbal terms."[5]

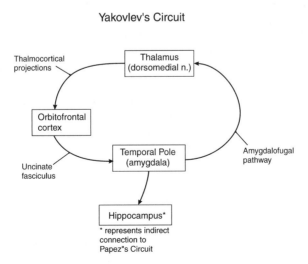

Yakovlev's Circuit

Figure 8–3.   Yakovlev's circuit.

THE CONCEPT OF A "LIMBIC SYSTEM"

Since the time of Papez (1937), Yakovlev (1948), and MacLean (1949/1964), additional clinical data have served to strengthen the concept of these limbic structures as a *system* that likely plays a major role in such basic behaviors as motivation (drive), affective arousal, emotional responsiveness, and learning and memory. For example, in some of the early studies in intracranial self-stimulation, a number of sites within this "system" have been demonstrated to have extremely positive reinforcing valences for the organism when self-stimulation procedures were made available, with some sites apparently being negatively reinforcing (Olds, 1958; Olds & Forbes, 1981). Destructive lesions and electrical stimulation involving various structures within this system were found to be capable of producing marked changes in the organism's pattern of emotional and social behaviors, including (1) aggressiveness or passivity; (2) increased or reduced sexuality; (3) fear and panic, emotional indifference (e.g., loss of previously established fear responses); or, on more rare occasions (4) positive emotional responses. (Bard, 1928; Kluver & Bucy, 1939; Terzian & Ore, 1955; Downer, 1961; Trimble, 1984, Doane, 1986, Rolls, 1986, 1990; LeDoux, 1991; Joseph, 1992). Seizure foci or tumors located in the anterior temporal and basal frontal regions frequently are associated with psychiatric symptomatology (Gloor, 1990, 1991; Gloor et al., 1982; Hermann & Chambria, 1980; Strauss et al., 1982; Sweet et al., 1969) In addition, psychosurgical procedures employed to reduce pathological anxiety (e.g., severe, intractable obsessive–compulsive disorders) or uncontrollable aggressiveness typically targeted limbic structures or their frontal connections (Diering & Bell, 1991; Valenstein, 1977).

On the other hand, other authors, most notably Brodal (1981), argued that there is no firm logical or scientific basis for maintaining the notion that these structures should be considered to represent a specific functional–anatomical "system." His argument is based, in part, on the fact that despite being composed of phylogenetically older cortical tissue, there are considerable anatomical differences among the various structures. He and more recently Kotter and Meyer (1992) and Kotter and Stephan (1997) note the anatomical and functional disparities often present in the description of the "limbic system" by various authors. Kotter and his colleagues suggest, however, that while the "limbic system" should be viewed as a hypothetical concept rather than a clear, empirically defined anatomical (or neurobehavioral) construct, the concept of a "limbic system" may have constructive "heuristic and didactic aspects," as long as its current limitations are recognized. This position is shared by Nieuwenhuys (1996), who points out that most, if not all, functional systems within the brain are characterized by some degree of anatomical uncertainty.

This chapter will not presume to resolve this debate. Rather it will focus on exploring (1) the basic anatomy and major afferent and efferent connections of certain key "limbic" structures, (2) the functional correlates most commonly associated with these structures, and (3) how their collective organization and complementary functions may contribute to the overall teleological goals of the organism. Before proceeding with an exploration of individual structures and their interrelationships, a few general principles that may facilitate our understanding of these relationships will be reviewed.

GENERAL ORGANIZATIONAL PRINCIPLES

As noted above, even among those who subscribe to the meaningfulness of the construct, there is no universal agreement as to what structures constitute the "limbic system." However, practically all anatomical definitions of the limbic system begin with the **hypothalamus** and include several other more primitive structures that have substantial direct connections with the hypothalamus. Most commonly mentioned in this regard are the **amygdala**, the **hippocampal formation**, and the **septal** nuclei. Other primitive areas that have been less frequently included are the **anterior** and **habenular** nuclei of the thalamus, the **preoptic** region, the **substantia innominata**, and the **pyriform cortex**.[6]

A second level of structural organization involves those areas defined as **paralimbic** structures (Mesulam, 1985). These areas, which are designated as *mesocortex*, include cortex within the **temporal poles**, the more caudal portions of the **orbitofrontal cortex**, the remaining portions of the **parahippocampal gyrus**, and the **cingulate gyrus**. While not having direct connections with the hypothalamus, these paralimbic regions have extensive connections with other limbic areas listed above and are thought to serve as an important link between them and the isocortex.

For the purposes of this chapter, only the following structures will be reviewed in detail: the hypothalamus, amygdala, septal nuclei, hippocampal formation, and cingulate gyrus. Although less consistently listed as part of the limbic system than the other four, the cingulate gyrus was included for both historical and practical reasons. Not only was the cingulate gyrus part of Broca's "grand limbic lobe," but it was an integral part of Papez's circuit and MacLean's original definition of the "limbic system." From a practical standpoint, as we shall see, many of the functions associated with this very prominent cortical structure appear to be directly related to the general functional constructs typically associated with the limbic system. Finally, again out of both historical and practical considerations, we also briefly will review the olfactory system and its relationship to limbic functions.

GENERAL FUNCTIONAL PRINCIPLES

As is the case with the anatomy, historically there has been less than a clear consensus regarding the primary function(s) served by the "limbic system." The "smell brain," the "visceral brain," the source of emotion and psychic energy, as well as the substrate of learning and memory are among the various designations that have been emphasized over the past century with regard to these collections of cortical and subcortical structures. Like the blind men and the elephant, each probably reflects an important element, but even collectively it is unlikely that they capture the true essence of the beast. While unraveling the mysteries of the "limbic system" is obviously still a work still in progress, there is much to be learned from what has been discovered thus far. Throughout this chapter, the probable functions of individual structures will be reviewed. But first, by way of introduction, the following represents a preliminary look at what are thought to be the major function(s) collectively subserved by these structures.

If there is one feature that is central to the notion of a "limbic system" it would have to be the *control and regulation of drive* states. As with all organisms, the most common and fundamental example of this is maintaining internal homeostasis. Without this, the organism simply does not survive. For the most part being reflexive or instinctual in nature, the drive states controlling these autonomic and visceral functions appear to reflect the most primitive core of the "limbic system." Beginning with the hypothalamus, the need for reciprocal control mechanisms readily becomes apparent. If the pupils dilate or the heart accelerates in response to a threatening stimulus, there needs to be an opposing action that results in pupillary constriction or cardiac deceleration when the threat is removed. Similarly, if eating and drinking are stimulated by low blood sugar or decreased osmotic pressure in arterial blood, there then needs to be some mechanism to inhibit these responses once these values are normalized.

However, as organisms evolved, behaviors and social structures became more complex.[7] The repertoire of animal behavior no longer was simply relegated to eating, sleeping, reproduction, and internal homeostasis. Beginning with birds, and even more evident in mammals, there was a tendency to care for the young over longer periods of time, social bonds and hierarchies became more intricate, social signaling or communication became more flexible and elaborate, and as noted by MacLean (1986) *"play behavior"* developed along with increased cortical development, particularly the cingulate gyrus. In addition, these more highly developed organisms appeared to have considerably more freedom or latitude in terms of behavioral responses to their environment. Behaviors became less ritualized and increasingly subject to modification as a result of "personal" experience (learning), a development that may be related to hippocampal expansion.

As part of this overall development, behavior not only became less "mechanical" and more individualistic (i.e., shaped by experience), but also more "emotional" (i.e., characterized by subjective arousal and drive states as positive and negative valences became experientially associated with specific stimuli). As we shall see, the amygdala apparently plays a major role in establishing these associations. This evolution of emotional responsiveness provided organisms with the means for richer and more adaptable responses to the natural environment, as well as the development of true social groups.[8] Table 8–1 provides a brief listing of several important functions served by emotions.

As the development of emotion allowed for more variable and complex drive states, the more primitive biological needs and impulses were still there and needed to be addressed. Thus, in addition to regulating internal drives, the organism needed to develop a means of regulating affective responses to external stimuli and the interactions or conflicts between the two. This was accomplished through an elaborate system of behavioral checks and balances involving diencephalic, endocrinological, limbic, and eventually cortical mechanisms. Such controls would have teleological significance not only for the individual, but also for maintaining order as social group interactions within the species became more complex. For example, while aggression and predatory behavior may be essential to meeting certain basic biological needs necessary for the survival of the individual (e.g., obtaining food) or even to maintaining the genetic viability of the species, unless controlled and directed, unrestricted aggressive or predatory behaviors might prove detrimental both for the individual and for the species.[9]

Finally, concurrent and conflicting emotions or drive states eventually would be supplemented by competition between more immediate versus long-range or even "abstract" plans or goals. As the cerebral cortex continued to develop, especially the frontal executive

Table 8–1.   Role of Emotions[a]

1. Preparation of the body for action through the initiation of an appropriate autonomic response.
2. Provides a flexible and efficient means of responding to a given stimulus that is largely based on the individual's experience.
3. Provides drive or motivation to respond.
4. Facilitates the communication of one's mood state and response propensity (e.g., anger, aggression).
5. Facilitates social bonding, as well as communication.
6. Influences the cognitive evaluation of stimuli and resulting judgments.
7. Facilitates and enhances memory storage for specific events.
8. With the assistance of memory, provides persistent motivation and direction, even in the absence of a specific stimulus.
9. Triggers the recall of specific memories.

[a] Adapted from Rolls (1995).

Table 8–2. Proposed Functions of the "Limbic System"

Maintenance of homeostasis: Primarily reflects MacLean's (1949) emphasis on the limbic system as the "visceral brain," but may be expanded to include regulatory control over a broad range of behaviors that insure that one's basic biological needs are met.

Motivated and goal-oriented behaviors: Besides basic homeostatic mechanisms, reflects any affective valences (emotions) attached to real, imagined, recalled, or anticipated situations that result in a propensity to respond in a particular manner (drive state).

Survival of the individual: Requires responding not only to homeostatic demands, but also to external (e.g., "threatening") stimuli. Depending on the organism and situation this may elicit "fight or flight" responses with accompanying sympathetic arousal.

Survival of the species: Includes sexual activities, maternal and paternal behaviors, social bonding (including affective expressions and/or vocalizations),[11] and modulation of aggressive impulses (especially between members of the same social group).

Learning and memory: Although not included in Nieuwenhuys' definition, memory and emotions are intimately linked, establishing learned emotional associations that are critical for future drive states and survival.

---

system, certain advantages were provided if the organism could modify, control, or inhibit immediate emotional tendencies or basic urges in lieu of future or alternate, higher-order goals. Conversely, the drive (motivation) to achieve more abstract or distant goals required tapping into or recruiting more primitive brain mechanisms. All these latter changes coincided with the development of what Mesulam (1985) refers to as the "paralimbic" system, which could modify the expression of these behaviors depending on external circumstances.[10]

A definition of the limbic system offered by Nieuwenhuys reflects the probable functional role(s) of the "limbic system. He states, [the limbic system] *"is concerned with specific motivated or goal-oriented behaviors, directly aimed at the maintenance of homeostasis and at the survival of the individual (organism) and of the species"* (Nieuwenhuys, 1996, p. 574). The major functional implications of this definition are presented in Table 8–2.

The preceding paragraphs provide a brief overview of the possible roles of the limbic system in behavior. At the conclusion of this chapter, some of these functions will be discussed in greater detail, along with the possible interrelationships of various limbic structures and other neural systems. First, however, we will review the major anatomical and functional aspects of the individual structures within this system, including their interconnecting pathways. Figure 8–4 provides a highly schematic perspective of some of the major structures and pathways that will be discussed below. The reader should keep in mind that this limbic network is exceedingly complex and the functional role(s) of individual structures is still incompletely understood. Also, as noted above, there still is the more basic debate as to whether "the limbic system" is a viable construct, and if it is what structures constitute it. The reader is invited to draw his or her own conclusions.

NEUROANATOMY OF THE LIMBIC SYSTEM

Hypothalamus

Location and General Anatomy

The hypothalamus represents that portion of the diencephalon that lies below and slightly anterior to the main body of the thalamus proper. It essentially surrounds the lower aspect of the third ventricle (see Figures 6–1b,c,d; 6–2c,d). The anterior boundary of the hypothalamus

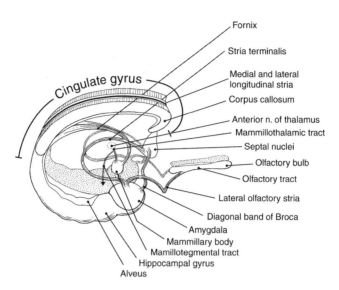

Figure 8–4. Schematic showing some of the major limbic structures and pathways.

is marked by the rostral edge of the optic chiasm ventrally and by the anterior commissure dorsally. The posterior extent of the hypothalamus generally is identified with the caudal or posterior extent of the mammillary bodies, which are part of the hypothalamus. Dorsally, the hypothalamus extends to the hypothalamic sulcus, a shallow indentation in the walls of the third ventricle. Despite its relatively small size (representing less than 0.5% of the total brain mass), the impact of the hypothalamus on behavior and normal bodily function is incalculable.

The hypothalamus is not a homogeneous structure, but rather a collection of numerous, discrete, bilateral nuclei. The hypothalamus commonly is divided into three anterior–posterior zones, a medial zone, and a lateral zone (Figure 8–5). The most **anterior zone** is represented by the nuclei above and immediately around the optic chiasm, among which are the medial and lateral preoptic nuclei and the supraoptic and suprachiasmatic nuclei. The **middle zone** is the area above the tuber cinereum and the infundibulum, which is the connection between the hypothalamus and the pituitary. Among the nuclei of note in this region are the dorsal and ventral medial nuclei and a good portion of the lateral nuclei, which will be discussed in greater detail below. The posterior **section** of the hypothalamus is the area that includes the mammillary bodies and the posterior hypothalamic nuclei. Finally, as mentioned, the hypothalamus also can be divided into **medial** and **lateral** zones. This latter division roughly is accomplished by a vertical line passing through the descending columns of the fornix. This separation will become meaningful as the effects of lesions in the ventromedial versus the lateral nuclei are discussed.

Afferent and Efferent Connections

Consistent with its role in governing multiple and diverse aspects of behavior, the hypothalamus has direct connections with the brainstem, the remaining diencephalic nuclei, many of the limbic structures, as well as the orbital frontal cortex. Structures that may not be directly connected to the hypothalamus have ample opportunity to influence, or be influenced by this center for emotional, autonomic, and endocrinological activity via indirect and neurochemical connections. For our present purposes, it will be sufficient to review only a few of the more well known pathways, but keep in mind there are additional hypothalamic

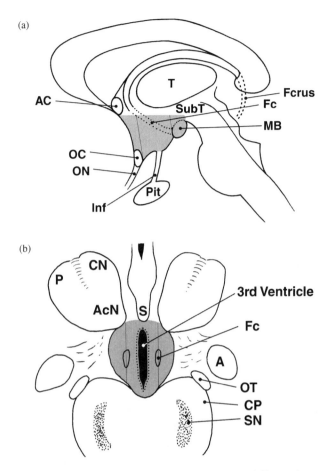

Figure 8–5. (a) The approximate boundaries of the anterior, middle, and posterior divisions, and (b) the medial and lateral zones of the hypothalamus (shaded). Hypothalamic cells immediately adjacent to the third ventricle represent a paraventricular zone. Abbreviations: A, amygdala; AC, anterior commissure; can, accumbens nucleus; CN, caudate nucleus; CP, cerebral peduncles; Fc, columns of the fornix; Fcrus, crus of fornix; Inf, infundibulum; MB, mammillary body; OC, optic chiasm; ON, optic nerve; OT, optic tract; P, putamen; Pit, pituitary gland; S, septal nuclei; SN, substantia nigra; SubT, subthalamus; T, thalamus.

connections beyond those discussed here. It is also important to remember that while some fiber tracts may be designated as primarily afferent or efferent, most probably contain reciprocal connections. This rule applies not only to hypothalamic connections, but also to most connections among the limbic structures.

As we have seen, one of the major pathways in Papez's circuit was the **fornix**. This very large and prominent subcortical fiber tract provides a major link from the hippocampal formation in the medial aspect of the temporal lobe to the mammillary bodies of the hypothalamus. As this tract loops around and over the dorsal thalamus and then recurves forward and ventrally to enter the mammillary bodies, it also gives off fibers to the anterior nucleus of the thalamus and to the septal regions. Two other major pathways into the hypothalamus from the temporal region are the **stria terminalis** and the **ventral amygdalofugal pathway**. These tracts appear to consist primarily of afferent fibers entering the hypothalamus (and preoptic areas) from the amygdala. Similar to the fornix, the stria terminalis loops around and over the dorsolateral surface of the thalamus, whereas the ventral amygdalofugal pathway is more direct. This latter pathway merges with another very large fiber system that provides input into the hypothalamus, the **medial forebrain bundle** (MFB). However, the MFB is akin to a major freeway system, carrying two-way traffic connecting the orbital frontal cortex, including olfactory and septal areas, the lateral hypothalamus, and the brainstem. Through this system the hypothalamus not only receives input from these areas, but also contributes fibers that provide efferent feedback as well.

In addition to the efferent fibers the hypothalamus contributes to the medial forebrain bundle, there are two other hypothalamic pathways that are well known for their efferent connections from this nuclear complex. One is the **mammillothalamic tract,** which figures prominently in Papez's circuit. This tract originates from the medial portion of the mammillary bodies and terminates in the anterior nuclei of the thalamus (which, in turn, projects to the cingulate gyrus). Another is the **mammillotegmental tract,** which connects the hypothalamus with the brainstem, particularly with the mesencephalic (midbrain) reticular formation (see Figure 8–4). This latter tract starts out with the mammillothalamic tract as the **fasciculus mammillaris princeps,** or the *mammillary fasciculus*, but then branches caudally to the brainstem.

Another, more indirect route from the hypothalamus to the brainstem is represented by the **stria medullaris thalami**. As noted, the hypothalamus has reciprocal connections with the septal nuclei. The septal nuclei, in turn, give rise to the stria medullaris, which adheres to the dorsomedial surface of the thalamus on its way to the habenular nuclei on the posterior dorsal aspect of the thalamus. From the habenular nuclei, a tract (the **habenulointerpeduncular tract** or the *fasciculus retroflexus*) proceeds to the interpeduncular nuclei of the midbrain. Finally, although it likely also contains reciprocal ascending fibers, the **dorsal longitudinal fasciculus** provides efferent feedback to the periventricular and periaqueductal gray of the brainstem. Other direct and/or indirect connections are established between the hypothalamus and other brainstem nuclei (e.g., dorsal motor nuclei of the vagus nerve, the nucleus ambiguus, and the nucleus solitarius), as well as with the preganglionic neurons that lie in the lateral horn of the thoracic cord. Table 8–3 outlines some of these hypothalamic connections.

Table 8–3. Major Hypothalamic Tracts[a]

Tract	Connections
Medial forebrain bundle (MFB)	Orbital and midline frontal cortex, septal nuclei, midbrain tegmentum (A&E)
Fornix	Hippocampal complex (A)
Stria terminalis	Amygdala (A)
Ventral amygdalofugal pathway	Amygdala (via MFB) (A&E)
Mammillary peduncle	Midbrain reticular formation (A)
Mammillary fasciculus	
Mammillothalamic tract	Anterior nucleus of the thalamus (E)
Mammillotegmental tract	Midbrain reticular formation (E)
Dorsal longitudinal fasciculus	Periventricular and periaqueductal gray of brainstem (A&E)
Stria medullaris thalami	Habenular nucleus of the thalamus, via septal nuclei (E)
Hypothalamospinal tracts	Sympathetic nuclei in lateral horns of the spinal cord (E)

[a] While most pathways likely contain both afferent (A) and efferent (E) connections, the designations above refer to those tracts that are typically thought to be either primarily afferent (A), primarily efferent (E), or appear to have a balance of afferent and efferent fibers (A&E).

Among other possible functions, these various pathways provide a means for interconnecting the hypothalamus with the parasympathetic and sympathetic autonomic nervous systems, orbital and medial frontal cortices, limbic structures, as well as the brainstem reticular formation. The latter includes nuclei that are responsible for the production of neurotransmitters important in homeostatic activities (e.g., the raphe nuclei, a source of serotonin, which among its other roles is involved in sleep) and sympathetic arousal (e.g., locus coeruleus, a source of norepinephrine).[12] Thus, the hypothalamus not only plays a role in regulating autonomic arousal to internal emotional states and/or external, affectively provocative stimuli, but also allows for internal changes in homeostasis to influence arousal and affective drive states via its cortical–limbic and reticular connections.

The Hypothalamic–Pituitary Connection

While the various neuronal pathways discussed above enable the hypothalamus to be informed of the internal and external environment and to effect bodily changes to meet the demands imposed by these environmental changes, these connections do not represent the only means by which the hypothalamus can effect change. In addition to neural input, the hypothalamus is bathed in a very rich supply of capillary vessels. Many of the neurons in the hypothalamus have very sensitive chemoreceptive capacities that allow them to monitor and respond to changes in the blood. For example, by monitoring an increase in the osmotic pressure in the blood, the supraoptic nuclei can signal a release of antidiuretic hormones, which increase the reabsorption of water by the kidneys. The release of this agent (vasopressin) actually is accomplished by hypothalamic neurons that axonally transport this hormone to the posterior lobe (*neurohypophysis*) of the pituitary gland where it is absorbed in the blood. Disruption of this process (e.g., by a lesion to the anterior hypothalamus) could lead to diabetes insipidus. The relationship between the hypothalamus and the pituitary gland is exceedingly complex; however, for our present purposes, a relatively simplified account should suffice.

The pituitary gland lies in a recess at the base of the skull called the **sella turcica**. The pituitary gland is connected to the hypothalamus by the infundibulum or pituitary stalk (Figure 8–6). The pituitary gland can be divided into a smaller posterior portion, the **neurohypophysis**, and a larger anterior section, the **adenohypophysis** or the pituitary gland proper. The neurohypophysis is really a continuation of the diencephalon. The axons of the supraoptic and paraventricular hypothalamic nuclei continue into this posterior region of the pituitary via the **infundibulum** or the hypothalmohypophyseal tract and directly release hormones produced by the hypothalamic neurons into the bloodstream (see Table 8–4 for hypothalamic–pituitary hormones).. The two hormones thus released are **vasopressin** (discussed above) and **oxytocin**. The latter is important for uterine contractions and the production of milk. By contrast, the hypothalamus does not have direct, neural connections with the anterior lobe of the pituitary. It releases hormones that are transported by the vascular system to the adenohypophysis (anterior lobe) and where they stimulate the glandular cells in the anterior pituitary. The adenohypophysis in turn releases its own hormones. Among the hormones released by the anterior pituitary are:

- **ACTH** (adrenocorticotrophic hormone) which stimulates the adrenal gland)
- **TSH** (thyrotrophic hormone) which stimulates the thyroid gland
- **FSH** (follicle-stimulating hormone) which is important in the production of sex hormones
- **LH** (luetinizing hormone) which is important in reproduction
- **STH** (somatotrophic hormone) which stimulates growth
- **Prolactin** which stimulates lactation

Hypothalamus

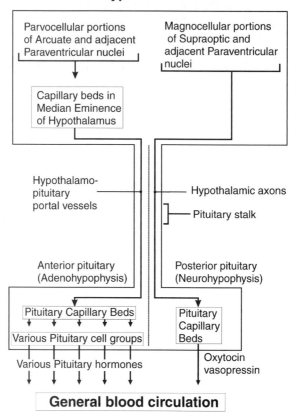

Figure 8–6. Hypothalamic–pituitary connections. The **posterior** portion of the pituitary (neurohypophysis) is innervated by hypothalamic neurons that transport the hypothalamic hormones (oxytocin and vasopressin) down their axons to be released into capillary beds of the posterior pituitary from where they enter the general circulation. By contrast, the capillary beds of the **anterior** pituitary (adenohypophysis) are supplied with hypothalamic hormones (either "releasing" or "inhibitory factors") via a blood portal system from capillary beds in the hypothalamus itself. Once released into the adenohypophysis, these hypothalamic hormones then stimulate pituitary cells to synthesize and secrete their own (pituitary) hormones, which then are released into the bloodstream. **Note:** Some hypothalamic hormones inhibit the production/secretion of pituitary hormones.

Thus, lesions to the hypothalamus can have profound effects on widespread organ systems via disruption of hormonal systems.

Functional Correlates

Trying to delineate the precise function(s) of any limbic or any other brain structure is difficult at best. This is especially true of the hypothalamus. Located deep within the brain, it is not readily accessible to study. More importantly perhaps, as has been pointed out, it is not a homogeneous structure but rather a collection of many individual nuclei with extensive, often overlapping direct and indirect connections to most brain systems and with indirect effects on distant organ systems via neurochemical influences. Most of our information about the hypothalamus has come from experimentation with animals (lesion or stimulation studies) or from analyzing the behavioral effects of naturally occurring lesions

Table 8–4. Hypothalamic–Pituitary Hormones

Anterior Pituitary: Adenohypophysis

Pituitary Hormone	Hypothalamic Regulating Hormone[a]	Main Functions
Growth hormone (GH)	Growth hormone-releasing hormone Somatostatin (inhibits release of GH)	Causes liver and other cells to secrete IGF-I (a growth promoting hormone); Stimulates protein synthesis
Thyroid stimulating H. (TSH)	Thyrotropin releasing H.	Prompts thyroid to secrete thyroxine and triidothyronine, resulting in increasing the metabolic rate
Adenocorticotropic H. (ACTH)	Corticotropin-releasing H.	Causes adrenal cortex to release cortisol, which mobilizes glucose, promotes protein catabolism, prepares body to cope with stress
Prolactin	Prolactin releasing H. Dopamine (inhibits release of prolactin)	Stimulates development of mammary glands, milk production
Follicle stimulating H. (FSH)	Gonadotropin releasing H.	Promotes secretion of estrogen, progesterone in females and testosterone in males; ovulation and spermatogenesis
Luteinizing H. (LH)	Same as for FSH	Similar to FSH; ovarian, sperm maturation

Posterior Pituitary: Neurohypophysis

Hypothalamic Hormone	Hypothalamic Source[b]	Main Functions
Oxytocin	Magnocellular hypothalamic nuclei (primarily supraoptic and anterior paraventricular nuclei)	Uterine contractions during childbirth; stimulates milk production
Vasopressin (ADH)	Same as for oxytocin	In response to increased salt concentrations in blood, increased vasopressin production causes kidneys to increase H_2O retention; in response to low blood volume/pressure, causes arterioles to constrict, increasing blood pressure.

[a] Most hypothalamic releasing hormones come from the arcuate nuclei and adjacent paraventricular nuclei.

[b] Unlike the hormones released to the adenohypophysis, these travel in vesicles down hypothalamic axons where they are released directly into the capillary beds of the neurohypophysis and from there directly into general circulation.

in humans. Among the problems with the former method are that data derived from animal subjects not always may be directly translatable to humans, and even under experimental paradigms one is never certain that the effects of the surgical lesion is confined to the particular nucleus and/or connecting fibers. Adjacent fiber systems and/or neighboring neurons also may be affected.[13] Similarly, stimulation of an area may not represent its normal physiologic operation and/or may involve other structures that are either physically adjacent or connected by neural networks. Additionally, the behaviors that we generally observe are highly complex, and therefore the same brain region may contribute to various behavioral patterns in different ways. In humans, naturally occurring lesions seldom respect anatomical boundaries, and even when they appear to be localized to a discrete region we are still faced with the same limitations described above. Actually, these problems are not peculiar to the study of the hypothalamus and generally apply to lesion studies throughout the CNS. With these caveats in mind, it may be useful to list a few of the syndromes associated with hypothalamic lesions in humans and to review a few of the experimental findings that offer some tentative hypotheses regarding the functional significance of the hypothalamus.

Lesions in the anterior portion of the hypothalamus are likely to result in hyperthermia (failure of the body's cooling mechanisms), while more posterior lesions generally are responsible for hypothermia (a failure of heat conservation). As has been mentioned, lesions involving the supraoptic and/or paraventricular nuclei can result in diabetes insipidus or other electrolyte disturbances. Anterolateral lesions are more likely to result in disturbances of the parasympathetic ANS, while the sympathetic division more commonly is affected by posteromedial lesions. Chronic overeating and obesity, along with inadequate sexual development, has been associated with lesions in the ventromedial aspect of the hypothalamus (see below). Other disturbances in sexual or reproductive functions, as well as sleep disorders, have been associated with various hypothalamic lesions. Acromegaly (*gigantism*), Cushing's syndrome (obesity of abdomen, shoulders, neck and face, along with other metabolic and endocrine disturbances), and hypopituitarism (dwarfism) all have been associated with tumors of the hypothalamus and pituitary gland. Among the more common tumors in this area are acidophil cell adenomas, chromophobe adenomas, and craniopharyngiomas.

Because these and other tumors that affect the hypothalamus and/or the pituitary gland grow in the area of the optic chiasm and optic tracts, visual disturbances, with or without accompanying headaches, are not uncommon. Emotional changes, from rage to increased sexual activity to apathy, also have been attributed to hypothalamic lesions. Less commonly, amnestic disturbances have been linked with discrete lesions that appeared to be centered in the hypothalamic regions (e.g., mammillary bodies, columns of the fornix, or the mammillothalamic tract); however, these regions frequently have been implicated in amnestic disorders with more diffuse lesions, such as in Korsakoff's syndrome.

Numerous, classic experiments in animals have revealed that lesions restricted to lateral hypothalamic nuclei will result in an anorexic syndrome [loss of eating (*aphagia*) and drinking (*adipsia*) responses] which may be so severe that the animal may die if not force-fed and hydrated. Conversely, lesions placed in the ventromedial nuclei of the hypothalamus typically will result in an animal that tends to overeat and becomes obese. The results of stimulation studies show the opposite effects of ablation studies. In these studies, stimulation of the ventromedial nuclei typically produces cessation of eating, while stimulation of the lateral nuclei leads to eating behavior even when the animal appears satiated.

Disturbances of eating and drinking are not the only behaviors associated with these sites. In addition to becoming obese and hypoactive, animals with lesions in the ventromedial nuclei typically will demonstrate aggressive and attack behaviors with minimal provocation (e.g., simply the presence of the experimenter).[14] On the other hand, stimulation in the region of the ventromedial nuclei can have a quieting or calming effect. By contrast, lesions of the lateral hypothalamic (as well as the posterior nuclei) nuclei often have a quieting influence, whereas stimulation is more likely to elicit attack behavior. Perhaps contrary to what might be anticipated, stimulation in the region of the lateral hypothalamic nuclei (actually in many parts of the limbic system, including the medial forebrain bundle) seems to be "pleasurable" to the animal, since it will work tirelessly to deliver stimulation to this area. In contrast, stimulation of the ventromedial nuclei appears to have aversive properties, as the animal will actively attempt to avoid such stimulation. For reviews of these phenomena, see Olds and Forbes (1981), Isaacson (1982), and Joseph (1990, pp. 92–98).

Behavioral disturbances following hypothalamic lesions are not necessarily confined to those described above or to the sites listed. Neither are the behavioral manifestations as simple and straightforward as might be suggested by the above descriptions. The expression of such behaviors often changes over time and may vary depending on environmental or other physiological circumstances (Isaacson, 1982, Chapter 2). Comparable behavioral disturbances following hypothalamic lesions also occasionally can be found in human subjects as a result of injury or disease. The point of the above descriptions is to demonstrate the role of the hypothalamus in the expression of very primitive and basic responses of the organism to its environment. In fact, rage-like behavior can be exhibited by some animals (most commonly demonstrated in cats) deprived of all brain tissue except for the hypothalamus and brainstem; this phenomenon is referred to as "sham rage," since it is thought that the animal could not experience a normal rage reaction in the absence of these higher brain centers.

In summary, the hypothalamus would appear to be directly involved in a wide range of behaviors that are critical for both the survival of the individual and for the survival of the species. Although they are not mutually exclusive, for the sake of simplicity, some of the activities in which the hypothalamic nuclei appear to play a significant role include:

1. Maintaining a homeostatic, internal environment (e.g., oxygen, temperature, and water regulation; circadian rhythms; food intake and utilization).
2. The monitoring and control of the endocrine system (e.g., growth, protein synthesis, sexual responsiveness, and reproduction).
3. Control of the autonomic nervous system, balancing the actions of the sympathetic and parasympathetic branches to meet the demands of both the internal and external environment (e.g., stress or threat).
4. Emotional expression: While the subjective experience of emotions and the selection of an emotional response that is appropriate to external stimuli requires the cooperation of "higher neural networks" (e.g., cortex and other limbic structures), the hypothalamus appears capable of generating basic affective or drive states.
5. Arousal: It appears likely that arousal involves both ascending (external sensory stimulation → reticular system → hypothalamus) and descending (cortical activity → hypothalamus → reticular formation) capabilities. In addition, cortical input is crucial for the mediation of conscious arousal and attention.
6. Memory: This is probably the least well-documented aspect of possible hypothalamic function. While lesions of certain nuclei (particularly the mammillary bodies) and their nuclear connections (e.g., mammillothalamic tracts) have been associated with disturbances of learning and memory (e.g., Korsakoff's syndrome), the critical lesion(s) producing disturbances of memory are still a subject of some debate (Verfaellie & Cermak, 1997; Victor, Adams, & Collins, 1989).

Amygdala

Location and General Anatomy

When viewing the ventral surface of the brain, a small eminence approximately 12 to 15 mm in length is noted on the anteriomedial surface of the parahippocampal gyrus. This protrusion is known as the uncus and marks the general location of the underlying amygdala (see Figures 5–4b; 6–1b–e; 6–2a–c). The amygdala, like the hypothalamus, is not a homogeneous structure but rather consists of a number of smaller nuclear groups. For most practical purposes, it can be divided into three major components: the *corticomedial, central,* and the *basolateral* groups. As we saw in Chapter 5, the corticomedial group has strong connections with the olfactory system. The smaller central group has been closely linked to the hypothalamus and brainstem and is thought to be important for generating autonomic responses. As we shall see, the large basolateral group's extensive connections with the cerebral cortex and the hippocampus appears to reflect its role in attaching emotional significance to external stimuli and learning.

Afferent and Efferent Connections

The amygdaloid complex is subject to extensive cortical and subcortical influences. These include afferent and/or efferent connections with:

1. Olfactory system, including basal–orbital frontal and anterior–medial temporal allocortical areas
2. Other basal–medial nuclei
3. Ventral striatum
4. Hypothalamus
5. Brainstem
6. Thalamus
7. Allocortex and juxta-allocortex
8. Neocortex

In contrast to the corticomedial group's large input from the lateral olfactory stria, the basolateral group does not receive direct input from the olfactory system (it may receive indirect olfactory information from the surrounding primary olfactory cortices).

The amygdala has fairly extensive interconnections (afferent and efferent) with a number of basomedial frontal areas. These include the region in the vicinity of the anterior commissure known as the **substantia innominata** (which, in turn, includes the **nucleus basalis of Meynert**), the **septal nuclei**, the **bed nuclei of the stria terminalis**, and the **nuclei of the diagonal band of Broca**. The fiber pathways that connect the amygdala to these areas include a shorter, more direct route: the **ventral amygdalofugal pathway** and the **stria terminalis**, a long, looping fiber tract that follows the curvature of the lateral ventricles, dorsally over the thalamus (see Figures 6–1a,b; 8–4).

Connections between the amygdala and the basal ganglia are thought to be primarily efferent. These projections generally are to the ventral or "limbic" portions of the striatum, including the nucleus accumbens (see Figure 6–1a). These latter connections constitute a major part of the "limbic feedback loop" discussed in Chapter 6, and would appear to have significant implications for a variety of psychiatric syndromes (Heimer, 2003).

There also are extensive reciprocal connections between the nuclei of the amygdala (particularly the central and corticomedial groups) and the hypothalamus and brainstem. The majority of these connections are via the ventral amygdalofugal pathway. The amygdalofugal inputs extend from the hypothalamus to nuclei of the midbrain, pons, and medulla, with some fibers continuing on to the spinal cord. These efferent pathways are thought to be important in generating appropriate autonomic or visceral responses to emotionally charged situations.

Thalamic input to the amygdala (primarily to the central group) comes largely from the midline nuclei, whereas amygdalothalamic connections are largely from the basolateral group and are projected mainly to the dorsomedial nucleus. It should be recalled that the dorsomedial nucleus, in turn, projects back to the prefrontal cortices.

As previously noted, the corticomedial amygdala receives input from the more "primitive" primary olfactory cortices along the more rostral, medial portions of the temporal lobe. In addition, the juxta-allocortex (from the parahippocampal and cingulate gyrus) appear to project primarily to the more medial or basal portions of the basolateral group. On the output side, there are substantial connections from the amygdaloid complex to the entorhinal cortex and from there to most parts of the hippocampal formation (except perhaps for the dentate gyrus). While the amygdala receives some afferent input from the hippocampal formation, it would appear to be considerably less than the efferent fibers it sends back to this region.

Finally, the more lateral portions of this group appear to receive input from frontal and posterior association areas of the neocortex, as well as the insula. These connections provide the amygdala access to both frontal control and executive functions, as well as to "processed" sensory information. There also are reciprocal connections from the amygdala back to the cortex, although perhaps not always to exactly the same areas from which they came. Most, if not all, areas of the frontal lobes appear to receive extensive efferent input from the amygdala. These connections would seem to ensure that the amygdala plays a significant role in both frontal and limbic feedback loops. As discussed in Chapter 6, these cortical loops have been implicated in mediating basic drives and goal-oriented behaviors, both of which often depend on emotional arousal.

In addition to these various connections with other outside structures, the various nuclear groups within the amygdala also have their own internal connections or communication networks. Thus, while the basolateral group does not have direct access to the olfactory system, it receives input from the corticomedial group. There also are reciprocal connections between the two amygdaloid nuclear complexes by way of the anterior commissure.[15]

Functional Correlates

The short answer to the question of the functional significance of the amygdala is that it appears to play an important role in the **attachment and/or recognition of emotional valences** associated with our sensory experiences. At least this is believed to be one of its major roles (Everitt & Robbins, 1992; Gaffan, 1992; LeDoux, 1989; Rolls, 1995). Before proceeding, however, the reader's attention is directed to a book, *The Amygdala: Neurobiological Aspects of Emotion, Memory, and Mental Dysfunction* (Aggleton, 1992). This book is nearly 600 pages in length. Works of comparable size are available on the hypothalamus, cingulate gyrus, and the hippocampus (Cohen, 1995; Haymaker, Anderson, & Nauta, 1969; Isaacson & Pribram, 1986; Kato, 1995; O'Keefe & Nadel, 1978; Reichlin, Baldessarini, & Martin, 1978; Vogt & Gabriel, 1993). These references are pointed out merely to emphasize the complexity involved in trying to comprehend the behavioral correlates of any of these structures, much less the interactions of all the discrete behavioral systems from the cortex to the brainstem. It also is offered as a forewarning to the reader that the brief summary presented here in no way portends to offer anything near a complete explanation of the role of the amygdala in behavior. The same caveat obviously applies to the other limbic structures to be discussed.

Before attempting to review a few of the more common findings or theories regarding the possible functional significance of the amygdala, it also is helpful to remind ourselves of the inherent limitations in such a task. As was the case with the hypothalamus, the amygdala is not a homologous structure, but rather has multiple nuclei, each of which has its own unique connections and most likely its own behavioral correlates. While there is some commonality of findings in the literature, significant differences may be found depending on various factors. Such factors include the size and exact location of the lesion (or parameters of stimulation), the environmental context and/or time period in which the behavior is observed, and even the species that are the subject of study.

At about the same time that Papez was describing a subcortical circuit to help explain the neuroanatomical substrates of emotion, Kluver and Bucy (1939) were investigating the effects of bitemporal lesions in monkeys. They described the following behavioral syndrome that still bears their names. One of the more striking things they noted was that although these animals appeared to have no significant problems with visual acuity or visual discrimination per se, they nonetheless appeared to have a type of visual agnosia for objects, which they characterized as a *psychic blindness*. This behavior was manifested in two ways. First, whereas normal, intact primates tend to rely heavily on visual feedback to explore and identify things in their environment (including small objects), these lesioned animals tended to repeatedly explore objects by smelling or by placing them in their mouths (termed, hyperorality). Second, previously "feared" objects, when introduced into view, no longer elicited avoidant or aversive responses in these animals. This seemed to be true whether the "object" in question was one that appeared to be genetically programmed to elicit such a response (e.g., a snake) or one that may have been "learned" through experience (e.g., the net or glove used to catch or handle the animal). In these bitemporal-lesioned animals, the normal "fear response" was absent. These animals tended to "explore" (touch, handle, orally incorporate) any and all objects within their environment. This behavior (termed *hypermetamorphosis*) occurred regardless of the stimulus's previous capacity to elicit a withdrawal or fear response.

In addition to the loss of a normal fear response to visual stimuli (e.g., the sight of an unfamiliar human), Kluver and Bucy found that their bilaterally lesioned monkeys not only would tolerate physical contact with humans, but also actively facilitate such contacts (an extremely atypical response for rhesus monkeys). In addition to a loss of fear, these animals evidenced a marked reduction in normal aggressiveness, even when another animal attacked them. Finally, another behavioral abnormality demonstrated by these animals was an increase in sexual interest or sexual responsiveness, which was inferred by their increase in autosexual, homosexual, as well as heterosexual activities (*hypersexuality*).

While the original lesions produced by Kluver and Bucy (1939) were extensive, incorporating much of the lateral and medial temporal cortex as well as the hippocampus, subsequent studies have demonstrated that many of the observed changes in emotional and social behavior could be produced by lesions limited to the amygdala. For example, years later, Downer (1961) sectioned both the interhemispheric commissures and the optic chiasm in monkeys (see also Chapter 9 under "Disconnection Syndromes"), thus ensuring that the input to each eye remained isolated to the same hemisphere. Subsequent lesions then were confined to the amygdala in these animals, but only on one side. The amygdala on the other side was left intact. Such animals would show a normal fear response to appropriate visual stimuli when viewed through the eye on the side of the intact amygdala, but no such response when only using the eye ipsilateral to the amygdalectomy. In addition to amygdalectomies resulting in an abolished or an attenuated fear response to a previously conditioned stimulus, such lesions also have been shown to interfere with the subsequent postsurgical development of such conditioned fear responses (Blanchard & Blanchard, 1972;

Davis, 1992; LeDoux, Cicchetti, Xagoraris, & Romanski, 1990). Thus, the amygdala appears to be important not only for establishing associations between stimuli and associated reinforcements (i.e., affective contingencies), but also for prompting behavioral responses to those stimuli based on previously established affective associations (Gray, 1995: Rolls, 1995). In a related finding, it was demonstrated that following lesions of the amygdala in humans, the ability to recognize emotional expressions in faces was compromised, while facial recognition per se remained intact (Adolphs, Tranel, Damasio, & Damasio, 1994, 1995).

In addition to reducing the potential for conditioned fear responses, many of the other behavioral disturbances observed in the Kluver–Bucy syndrome are not limited to monkeys (Kling & Brothers, 1992). For example, reduced aggressiveness or an increase in docility also can be found following lesions confined to the amygdaloid nuclear complex in rats, cats, monkeys, and humans. In fact, based on the findings of Kluver and Bucy and others, surgical lesions of the amygdala have been used as a means of reducing aggressive or assaultive behavior in humans (Terzian & Ore, 1955; Narabayashi et al., 1963; Aggleton, 1992). Although infrequently employed now, such procedures, known as *psychosurgery* (surgery whose primary purpose is to effect changes in behavior or affect), typically involve(d) lesioning of limbic structures, pathways that interconnect limbic structures, or disconnection of the frontal lobes from the limbic system (frontal leucotomies) (Diering & Bell, 1991; Valenstein, 1977).

While lesions of the amygdala tend to have a "calming" effect for most animals, possibly by raising the threshold for affective responding ("emotional blunting"), stimulation generally has the opposite effect. Depending on a number of variables, including the specific site of the stimulation, the species involved, and the environmental circumstances, stimulation may lead to (1) a cessation of ongoing activities and increased alertness or watchfulness, (2) fearlessness or escape behaviors, and/or (3) aggressiveness or attack behaviors (Ursin & Kaada, 1960; Egger & Flynn, 1963; Zbrozyna, 1972). Often such stimulation is accompanied by increased sympathetic activity. It is interesting to note that violent behaviors associated with seizures are extremely rare,[16] although occasionally reported (Ferguson, Rayport, & Corrie, 1986; Mark & Ervin, 1970). On the other hand, subjective feelings of fear, anxiety, or impending doom are not uncommon auras in patients with seizures, particularly those associated with temporal lobe foci (Gloor, 1972; Spiers, Schomer, Blume, & Mesulam, 1985).[17]

Finally, there has been an ongoing debate within psychiatry for at least the past 30 years as to the possible relationship between seizure disorders, specifically those whose foci are in the temporal lobe (thus whose impact likely would include the amygdala), and certain behavioral or psychiatric disturbances. Most commonly, it has been argued that when compared with baseline populations or to patients with seizure disorders where the seizure focus is *non-temporal*, patients with right "temporal lobe epilepsy" tend to have a higher incidence of manic–depressive-type disorders, whereas left temporal focal lesions increase the probability of schizophreniform-type psychoses. It also has been argued that the presence of chronic seizure disorders with temporal lobe foci increase the likelihood of various behavioral or "personality" features such as hyperreligiosity or increased preoccupation with moral issues, decreased sexual interest, increased irritability, excessive rumination, obsessiveness, and/or cognitive–emotional rigidity. While on the whole there does appear to be some relationship between certain types of seizure disorders, increased incidence of psychopathology, and interictal behavior,[18] this entire area of research remains controversial (Bear & Fedio, 1977; Reynolds & Trimble, 1981; Sherwin, 1981; Bear et al., 1982; Hermann and Whitman, 1984; Spiers et al., 1985; Post, 1986; Adamec & Stark-Adamec, 1986; Stark-Adamec & Adamec, 1986).

Septal Area

Location and General Anatomy

The septal region is located in the posterior, inferior, and medial portion of the frontal lobes, immediately rostral to the anterior commissure and just below the rostrum of the corpus callosum (see Figure 6–1a; 6–2c–e). This area represents a somewhat poorly defined set of nuclei. Hence, in humans and other higher-order primates, the terms septal *area* or septal *region* often is substituted and refers to this general anatomical location without further subdivision into specific individual nuclei [although a more medial and a lateral division are often identified (Andy & Stephen, 1968)]. **Note:** Care should be taken to avoid confusing the septal area or nuclei with the septum pellucidum. The septum pellucidum consists of two thin sheaths of tissue that provide the medial separation of the anterior horns of the lateral ventricles and do not contain any nerve cells.

Afferent and Efferent Connections

At an earlier point in the evolution of the central nervous system, the septal nuclei probably were contiguous, or nearly so, with the posterior portion of the hippocampal gyrus. As the telencephalon (cerebral cortex) gradually expanded, these two structures anatomically were separated and came to occupy their present relative positions in the primate brain. However, several long C-shaped pathways carrying fibers to and from the septal area still reflect the evolutionary and functional relationships of these two structures. As previously noted, one of these C-shaped pathways is the **stria terminalis**, a fiber tract curving around and over the dorsolateral aspect of the thalamus carrying both afferent and efferent fibers between the amygdala and septal areas. The other, the **ventral amygdalofugal pathway,** establishes a more direct route between the amygdala and the septal area. Another major C-shaped pathway that also already has been discussed is the **fornix**. While it is primarily considered an efferent pathway from the hippocampal formation to the mammillary bodies, it also gives off collaterals (known as *precommissural fibers*) to the septal nuclei as it begins its descent to the hypothalamus. The fornix also carries afferent fibers from the septal area back to the thalamus and hippocampus.

The hippocampal gyrus also appears to have another dorsal connection with the septal region. There is a small gray band of tissue, the **indusium griseum**, which runs from the hippocampal region anteriorly along the dorsal surface of the corpus callosum (between the corpus callosum and the cingulate gyrus). Embedded within this band of neural tissue are two fiber tracts, the **medial** and lateral **longitudinal striae** (of Lancisi), which also make connections with the septal region. There is one final C-shaped or curved dorsal pathway that also was mentioned in conjunction with the hypothalamus, the **stria medullaris thalami**. This pathway, which is found on the dorsomedial surface of the thalamus, appears to be primarily an efferent fiber tract from the septal nuclei to the habenular (epithalamus) and other midline thalamic nuclei. From the habenular nuclei (which, in turn, also receive input from the hypothalamus and the basal ganglia) projections are carried to the interpeduncular nuclei of the brainstem (via the fasciculus retroflexus) and on to other brainstem nuclei that are important for mediating autonomic arousal.

In addition to the long, curved dorsal pathways discussed above, other, more ventral pathways interconnect the septal nuclei with the basal frontal cortex, amygdala, hypothalamus, and brainstem. In addition to the amygdalofugal pathway, two other previously noted fiber tracts provide ventral connections between the septal nuclei and the surrounding basilar frontal cortex, hypothalamus, and brainstem. These are the **diagonal band of Broca** and the **medial forebrain bundle** (MFB). As a result of connections to these cortical, subcortical, and brainstem sites, the septal area, like the amygdala, would appear to have access to executive, emotional, sensory, and visceral information. Thus, as was true of the hypothalamus and the amygdala, the septal nuclei have substantial connections not only with "limbic" structures but also with the frontal lobes, the thalamus, and the brainstem.

Functional Correlates

If one reviews the literature on lesion and stimulation studies of the septal nuclei in mammals, it would appear that in many respects the results are directly opposite those found in the amygdala (DeFrance, 1976). One of the more dramatic phenomena to be witnessed in a physiological psychology laboratory is the sight of a 180-pound graduate student being chased up a chair by a white rat that may barely go over one pound. For anyone who may not be familiar with them, white laboratory rats, as opposed to their free-roaming ancestors, are typically rather docile creatures (scientists who routinely work with them generally prefer it that way). However, if the septal nuclei of one of these animals are destroyed by stereotaxic lesions, a major behavioral change may be seen. Left undisturbed, the rat may just sit there, appearing perfectly normal. However, given the proper provocation, sometimes as minimal as blowing on its fur or gently tapping on its back with a pencil, it may launch a full-fledged attack on the source of this "assault."

While this response to septal lesions is not limited to rats, neither is it invariably seen in all species. Even when initially present, the full effect is typically temporary and generally accommodates over time or with increased handling (Brady & Nauta, 1953; Ahmad & Harvey, 1968; Paxinos, 1975). In fact, most animals eventually will show a decreased tendency to respond aggressively, but this may be in part dependent on the specific nature of the stimulation or provocation. However, even this initial increased aggressiveness may be blocked by lesions of the amygdala, which as we have seen tend to reduce aggressiveness. Conversely, stimulation of the septal area has been shown to inhibit or abort aggressive behavior (Rubenstein & Delgado, 1963; Siegel & Skog, 1970).

In contrast to the social withdrawal that often can be seen with amygdala lesions, septal lesions can lead to what appears to be an increased need for social contact. At times, this will take the form of seeking physical contact or closeness with either inanimate objects or with representatives of other species that are normally actively avoided (Meyer, Ruth, & Lavond, 1978). In humans, such behavior occasionally has been described as *social stickiness* where, following lesions that encroach on the septal area, an individual may have difficulty responding to normal cues to maintain appropriate boundaries in social interactions (Joseph, 1990). Such behavior also might appear to be inconsistent with the aggressive tendency noted above, except if both are viewed as a failure to inhibit or modulate emotional/social behavioral responses. Such an interpretation also might be consistent with the observation that while passive–avoidance learning is typically impaired, active–avoidance learning is either unaffected or may be slightly enhanced. However, given the range and variability of behavioral responses both within and between species that can be found following lesions in the septal area, caution is advised against attributing any unimodal function, including response inhibition, to the septal nuclei.

In addition to inhibiting aggression, stimulation of the septal region in humans has been reported as "pleasurable" (Olds, 1958; Heath, 1959). Given a choice, the septal region is one area where animals may work to provide continuing stimulation, although again this effect may vary depending on the exact placement of the stimulating electrode. It is possible that such stimulation results in the release of endogenous opioid peptides from amygdaloid–hypothalamic areas. Stimulation of the septal region at times also will produce certain autonomic changes, a finding consistent with its anatomical connections to the hypothalamus and brainstem.

Hippocampal Formation

Location and General Anatomy

When viewing the ventral surface of the brain, one can identify the gyrus that lies on the most medial aspect of the temporal lobe. This area, separated from the more lateral occipitotemporal or fusiform gyrus by the collateral sulcus, is the **parahippocampal gyrus** (see Figure 5–4b). It even appears visually distinct from the adjacent cortical gyri in the fixed brain, having a somewhat broader and flatter appearance. The hippocampal formation is located within the parahippocampal gyrus. While not visible on the ventral surface, the hippocampal formation is very distinct on both coronal and horizontal sections (see Figures 6–1e–g; 6–2b–g). It can be seen as an enfolding of tissue on the medial surface of the parahippocampal gyrus along the hippocampal sulcus lying on the floor of the temporal horns of the lateral ventricles. Viewed in cross section, it has a spiral-type appearance, not unlike what is seen in a cinnamon roll. The *hippocampal formation* is divided into several anatomical subdivisions: the **subiculum**, the **dentate** gyrus, and the **hippocampus proper** (Figure 8–7).[19] The hippocampus proper derives its name from its resemblance to the shape of a seahorse [it is also referred to as *Ammon's* horn(from the Egyptian god with ram's horns) for the same reason], although as can be seen the dentate gyrus and the subiculum also take on a similar curved appearance. The hippocampus proper traditionally is divided into four contiguous sections based on slight differences in cellular structures. These sections, as seen in Figure 8–7, are referred to as CA1 through CA4 (with "CA" being an abbreviation for "cornu ammonis" or Ammon's horn). In what appears to be a remnant of the expansion

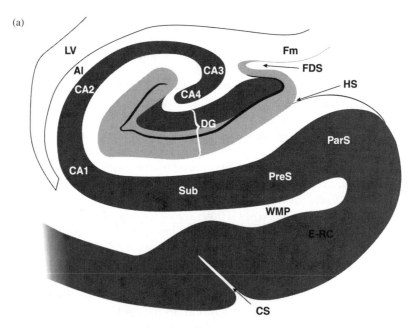

Figure 8–7. (a) Cross section of the hippocampal formation and parahippocampal gyrus. (b) The phylogenetic and ontogenetic development of the hippocampal formation as a result of the enfolding of the medial temporal cortex (adapted from P. Gloor, 1997). (c) through (e) The hippocampal formation in progressively deeper sagittal MRI images, and (f) on an axial cut at the level of the midbrain. Abbreviations: Al, alveus; CA1 → CA4, contiguous sections of the hippocampus proper; CS, collateral sulcus; DG, dentate gyrus; E-RC, entorhinal cortex; FDS, fimbriodentate sulcus; Fm, fimbria; HS, hippocampal sulcus; LV$_{IH}$, lateral ventricle, inferior horn PreS, presubiculum; ParS, parasubiculum; Sub, subiculum; WMP, white matter pathway.

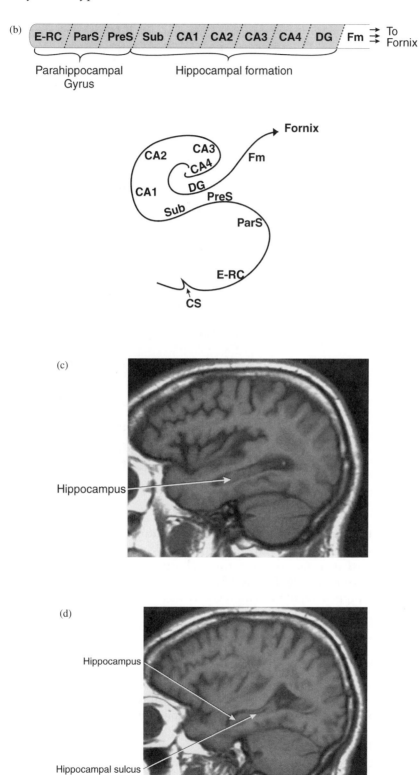

Figure 8–7. *(Continued)*

(e)

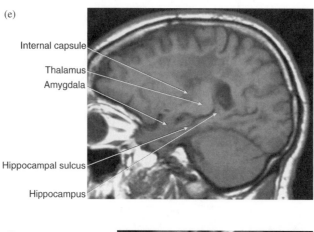

Internal capsule

Thalamus

Amygdala

Hippocampal sulcus

Hippocampus

(f)

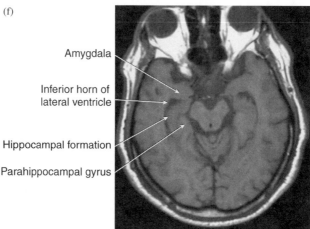

Amygdala

Inferior horn of lateral ventricle

Hippocampal formation

Parahippocampal gyrus

Figure 8–7. *(Continued)*

of the cortex around the diencephalic core, the most caudal portion of the hippocampal formation is continuous with the indusium griseum, which is a thin band of gray matter that lies over the medial surface of the corpus callosum. It extends from the hippocampus to the region of the septal nuclei. Finally, unlike the neocortex, which consists of six layers (see Chapter 9), the hippocampal formation is classified as **allocortex**, which consists of only three layers. The presubiculum and parasubiculum constitute a transitional zone between the hippocampal formation (allocortex) and the entorhinal (neocortical) tissue.

Afferent and Efferent Connections

Although some of the following will be a repetition of pathways and connections discussed elsewhere in this chapter, the major afferent and efferent fiber tracts interconnecting the hippocampal formation with other brain centers will be reviewed from the focal point of the hippocampus. Generally speaking, the connections of the hippocampal formation can be divided into three broad categories: cortical, subcortical (including the brainstem), and the contralateral hippocampus. While the internal connections of the hippocampal formation tend to be unidirectional, most of its external connections are reciprocal, conveying both efferent and afferent information from cortical and subcortical structures.

Cortical Connections

Most cortical projections eventually reach the hippocampal formation, primarily via the perforant pathway (see Figure 8–8) after having first synapsed in the entorhinal cortex (Brodmann's area 28). However, some cortical inputs may first synapse in the perirhinal areas prior to entering the entorhinal cortex, while others may proceed directly to the subiculum or presubiculum (Nieuwenhuys, Voogd, & van Huijzen, 1988). The hippocampal formation is believed to receive input from all secondary and tertiary cortices. This includes the juxta-allocortices of the cingulate, insular, and orbitofrontal areas, as well as the unimodal and multimodal association area of the frontal, parietal, temporal, and occipital lobes (see Figure 8–9). One prominent corticohippocampal connection is the cingulum, a long fiber pathway that travels with the cingulate gyrus (e.g., Figures 6–1f, 6–2f). Synapsing in the entorhinal cortex, the cingulum carries information not only from the cingulate gyrus, but probably ultimately from many other parts of the cortex that project to the cingulate gyrus (see below). Whereas the hippocampus receives information visual, auditory, somatosensory, and visceral information largely through unimodal association areas, it might be recalled that the olfactory system also has relatively direct connections with both the amygdala and the entorhinal cortex. In turn, both the amygdala (see below) and entorhinal cortex have direct connections with the hippocampal formation. These apparently relatively direct connections (i.e., bypassing the neocortex) between the hippocampus and the olfactory system might help account for what often seems to be the relative strength of olfactory memory. Additional discussion of the olfactory system and behavior can be found below.

The subiculum (what might be considered the outermost layer of the hippocampal formation) serves as the primary output channel for the hippocampus. It sends projections back out to the entorhinal cortex. As illustrated in Figure 8–9b, from the entorhinal cortex, efferent connections are established with more or less those same cortical areas that provide input into the entorhinal cortex/hippocampal formation. Thus, the hippocampal formation either directly or indirectly through the entorhinal cortex has access and provides feedback

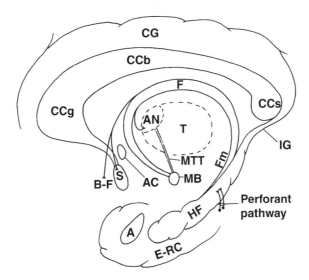

Figure 8–8. Hippocampal formation in relation to other limbic structures. Abbreviations: A, amygdala; AC, anterior commissure; AN, anterior nucleus of the thalamus; B-F, basofrontal region; CC, corpus callosum (b, body; g, genu; s, splenium); CG, cingulate gyrus; E-RC, entorhinal cortex; F, fornix; Fm, fimbria; HF, hippocampal formation; IG, indusium griseum; MB, mammillary bodies; MTT, mammillothalamic tract; S, septal area; T, thalamus.

to secondary sensorimotor, multimodal sensory, frontal executive, and limbic systems. These reciprocal cortical connections, primarily via the entorhinal cortex, likely serve as the basis for cortical long-term storage of information (i.e., memory).

Subcortical Connections

In addition to the extensive cortical inputs just described, the hippocampus receives afferent fibers from a number of subcortical structures. As was the case with the cortical inputs, these subcortical afferents appear to enter the hippocampus by way of the entorhinal cortex or the subiculum. Among the more prominent connections are those from the amygdaloid complex, septal region, hypothalamus, and thalamus. While the connections between the amygdala and the nearby hippocampus are apparently rather direct, the others are more circuitous. The septal region (septal nuclei, nuclei of the diagonal band) appears to primarily utilize the fornix (described below) to send fibers back to the hippocampus. With regard to the hypothalamus and thalamus (primarily the anterior and midline nuclei), recall the earlier description of Papez's circuit. The mammillary bodies (having received input from the hippocampus via the fornix) send projections to the anterior nucleus of the thalamus via the mammillothalamic tract. The anterior nucleus, in turn, sends projections to the cingulate gyrus and cingulum, which project back to the hippocampus. Other hypothalamic input may also return via the fornix.

The main source of subcortical efferents emerging from the hippocampus is the **fornix.** Containing over a million fibers, the fornix represents one of the more prominent pathways within the CNS (Figures 6–1b–f; 6–2c–h). The efferent fibers that make up the fornix appear to be derived from all three components of the hippocampal formation: the subiculum, hippocampus proper, and the dentate gyrus. Along the medial surface of the hippocampal formation there is a thin layer of white matter or nerve fibers (known as the **alveus**) that eventually coalesce to form the fimbria. The fimbria is a broad band of fibrous tissue lying along the dorsal surface of the hippocampal formation that resembles a foot and eventually evolves to become the *crus* of the fornix (Figures 6–1g; 6–2h; 8–8). As each of these fiber tracts proceeds rostrally, they come together just above the dorsomedial surface of the thalamus (and just below the corpus callosum) to form the body of the fornix (Figure 6–1f). At its anterior end, the fornix again splits, forming the *columns* of the fornix (Figures 6–1b; 6–2c–e). The fibers making up the columns themselves can be split into the pre- and postcommissural fibers (those passing in front of and behind the anterior commissure). The postcommissural fibers, which derive primarily from the subiculum, form the main projections to the mammillary bodies and the anterior nuclear groups and the lateral dorsal nuclei of the thalamus. The precommissural fibers, by contrast, which primarily have their origin in the hippocampus proper, tend to project mostly to the septal nuclei, the area of the diagonal band of Broca, nucleus accumbens, and other basal frontal areas. Some of the postcommissural fibers continue on to various midbrain nuclei. In addition to the fornix, the **longitudinal stria of Lancisi** represents another smaller pathway between the septal area and the hippocampal formation that travels along with the indusium griseum. Figure 8–9 offers a highly schematic view of the relationship of the hippocampal formation to some of the structures discussed above.

As was noted earlier, many of the fibers within the fornix also contain afferents to the hippocampal formation from the same nuclei to which it projects. Finally, the fornix also carries commissural fibers that transfer information from the hippocampal formation to the contralateral hemisphere (hippocampal formation and entorhinal cortex) via the hippocampal commissure, which is located just below the body of the fornix. For a more detailed review of the anatomical structure and connections of the hippocampus, the reader again is referred to Gloor's excellent volume on limbic structures (Gloor, 1997, Chapter 6).

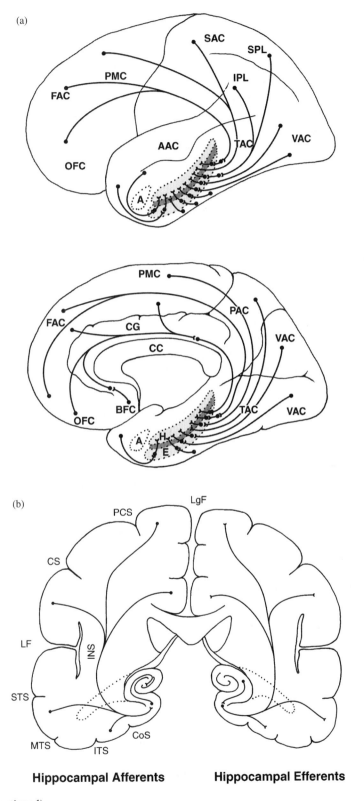

Hippocampal Afferents   **Hippocampal Efferents**

Figure 8–9. *(Continued)*

Functional Correlates

Similar to the other limbic structures considered thus far, the functions of the hippocampal formation are very complex. Given its extensive connections and the fact that it is made up of several distinct structures, it is unlikely that the hippocampal formation mediates a single or circumscribed "function." It likely contributes to various "functional systems." Lesion studies in animals have suggested a number of behavioral effects following lesions of the hippocampal formation depending in part on such variables as the species involved, the animal's previous experience, and environmental or testing conditions. Among the behavioral changes found in animals following lesions of the hippocampal formation have been increases in general activity, problems modulating attentions and arousal, decreased distractibility, increased reactivity to external cues, decreased behavioral flexibility, decreased ability to use spatial or cognitive maps, and impairments in learning and conditioning (Isaacson, 1982; O'Keefe & Nadel, 1978; Routtenberg, 1968). However, the most common and consistent behavioral deficit associated with hippocampal lesions in humans has been problems with memory, particularly the encoding of new memories (Signoret, 1985; Gloor, 1997; Zola, 1997).

Perhaps the most famous patient who has contributed to our modern understanding of the memory function of the hippocampal formation in humans is "HM." This individual was subjected to bilateral surgical ablation of the medial temporal cortices in an attempt to alleviate intractable seizures (Scoville, 1954; Scoville & Milner, 1957). Following the surgery, HM was noted to have a profound anterograde memory deficit or an inability to learn new information (declarative memory), despite being able to benefit from practice on certain sensorimotor tasks (procedural memory). This inability to retain new information was profound and included both verbal and visual–spatial type materials. This syndrome became synonymous with bilateral hippocampal lesions, although the lesions apparently extended well beyond the boundaries of the hippocampal formation and likely included the

◄───

Figure 8–9.   Afferent and efferent cortical connections of the hippocampus. (a) Inputs to hippocampal formation from widespread cortical areas. As shown, while most inputs synapse in the entorhinal cortex prior to entry into the hippocampal formation via the perforant pathways, some proceed directly into the subicular area. Although not illustrated, it is thought that hippocampal–cortical efferents project back to the same general cortical areas from where inputs were derived (as shown in Figure 8–9b). (b) Coronal section illustrating both efferent and afferent hippocampal–cortical connections. As shown, most of these afferent connections likely originate in the subiculum and synapse in the entorhinal areas before going back out to the cortex (hippocampal formation is relatively enlarged). The darker area between the hippocampus proper and the entorhinal cortex represents the subiculum.

A, amygdala	LF, lateral fissure
AAC, auditory association cortex	LgF, longitudinal fissure
BFC, basal frontal cortex	MTS, middle temporal sulcus
CC, corpus callosum	OFC, orbitofrontal cortex
CG, cingulate gyrus	PAC, parietal association cortex
CoS, collateral sulcus	PCS, precentral sulcus
CS, central sulcus	PMC, premotor cortex
E, entorhinal cortex	S, subiculum
FAC, frontal association cortex	SAC, somatosensory cortex
H, hippocampus proper	SPL, superior parietal lobule
INS, insular cortex	STS, superior temporal sulcus
IPL, inferior parietal lobule	TAC, temporal association cortex
ITS, inferior temporal sulcus	VAC, visual association

amygdala. Horel (1978) came to the conclusion that the "temporal stem" (the white matter that connects the amygdala with the temporal neocortex) was the critical lesion producing such symptoms in humans, rather than damage to the hippocampus itself. However, in 1986, the case of another patient ("RB") was reported in which there was significant and permanent anterograde amnesia. In this instance the lesion, which was the result of an ischemic episode, appeared to be confined to CA1 of the hippocampus (Zola-Morgan, Squire & Amaral, 1986). Since that time (Squire, 1986; Squire & Zola-Morgan, 1993; Zola-Morgan & Squire, 1993), using primate models, reached the following conclusions:

1. Lesions confined to the fields of the hippocampus proper probably are sufficient to produce lasting memory loss.
2. As the lesions become more extensive, for example, beginning with the dentate gyrus and subiculum (hippocampal formation) and then extending to the entorhinal and perirhinal tissue, the more dense or severe the memory impairment.
3. Lesions confined to the perirhinal and parahippocampal cortices produced amnestic deficits that were more severe than lesions limited to the hippocampal formation (presumably because much of the cortical input to the hippocampus is funneled through these areas).
4. The inclusion of the amygdala in these lesions added little, if any, to the memory impairment in the experimental situations.

A variety of neuropathological conditions have been associated with bilateral lesions of the medial temporal areas, although the exact locus and extent of the lesion, and hence the nature and severity of the memory disturbances, will vary in individual cases. Included among these conditions are viral, particularly herpes, encephalitis (Drachman & Adams, 1962; Rose & Symonds, 1960), generalized ischemia, or anoxia (Zola-Morgan, Squire, & Amaral, 1986; Volpe & Hirst, 1983), occlusion of the posterior cerebral artery system (Victor, Angevine, Mancall, & Fisher, 1961; DeJong, Itabashi, & Olson, 1969; Benson, Marsden, & Meadows, 1974), primary degenerative disease processes (Hyman, Van Hoesen, & Damasio, 1990; Ball, 1979), and epilepsy (Gilbert, 1994; Adamec & Stark-Adamec, 1986; Babb et al., 1984). While definitive evidence appears to be lacking, transient global amnesia is suspected to result from temporary hypoperfusion of the medial temporal regions bilaterally.

The feature that is common to all these syndromes is difficulty encoding new memories (*new learning* or *anterograde amnesia*). It primarily is a disturbance *of episodic or declarative memory*, that is, the ability to recollect an experience in real-time and real-space parameters. Thus, for example, individuals might have difficulty recalling something they saw, heard, or experienced at a particular time, in a particular place, or under particular circumstances.[20] If reminded of the stimulus, event, or content of the to-be-remembered experience, patients may express some partial recollection or sense of familiarity but not be able to put it into his or her personal space–time continuum. That is, they may not recall where or when, or even if they saw, heard, or experienced the event in question. Nor will they be able to associate any personal meaning to it. With bilateral medial temporal lesions, all modalities and types of sensory input are likely to be affected.

Other types of memory are less likely to be affected. Thus immediate *memory*, or what might be termed "sensory" memory, generally is undisturbed. Thus individuals with medial bitemporal lesions may do perfectly well on digit span or even recite back most of the elements of a story or word list to which they were just exposed. *Procedural* memory, that is, the memory for how to do things, usually is preserved so that a patient with bilateral hippocampal lesions may "learn" from practice how to perform a new sensorimotor task, although he or she may never remember having seen or done it before.[21] *Remote memory* or retrograde *amnesia*, that is, memory for events prior to the onset of the memory disorder, is more variable. Some individuals, even those with profound anterograde amnesia, may have essentially normal retrograde memory (or only a very restricted retrograde deficit which is isolated to a brief period just before the onset of the amnestic episode). Conversely, other patients with bilateral medial temporal lesions may have much more significant retrograde losses, sometimes extending back over decades. Zola (1997) reports several cases that suggest the difference may be in the degree to which the lesion extends beyond the hippocampal boundaries. Finally, semantic memory, that is, the memory for words and certain other types of overlearned information, also generally is fairly well preserved in the medial temporal amnesias.[22]

At this point it is important to note that amnestic disorders (i.e., more or less global impairment of new and possibly remote memory) also can result from lesions that lie outside the medial temporal region. As noted in the previous chapter, memory disorders have been associated with lesions affecting the dorsomedial and other midline thalamic nuclei, which can be global, if bilateral (see also Markowitsch, 1988). Basal forebrain lesions, especially those resulting from rupture of anterior communicating artery (ACA) aneurysms, typically present with severe, global memory deficits (Alexander & Freeman, 1984; DeLuca & Cicerone, 1989; Vilkki, 1985; Volpe & Hirst, 1983). However, perhaps the classic amnestic disorder is **Korsakoff's syndrome** (Butters & Cermak, 1980; Victor, Adams, & Collins, 1989). Frequently, but not invariably, associated with chronic alcohol abuse and thiamine deficiency, this syndrome, which may follow a more acute Wernicke's encephalopathy,[23] is characterized by varying degrees of anterograde memory impairment, a temporal gradient of retrograde memory loss (the more recent the time period, the denser the memory loss), and confabulations ("filling in the blanks"). While deficits in other cognitive areas (e.g., problem solving, executive abilities, visual–spatial skills) may be demonstrated, they generally pale by comparison with the memory deficits (Verfaellie & Cermak, 1997).

The critical lesion(s) producing the amnestic deficit in Korsakoff's syndrome is a matter of continuing debate (Bauer, Tobias, & Valenstein, 1993; Zola-Morgan & Squire, 1993). Pathological changes in Korsakoff's commonly are seen in the thalamus (usually in the dorsomedial nuclei, the mammillary bodies, and in the periventricular gray in the diencephalon and midbrain). While there is some evidence that the thalamic lesions may be more critical, the role of the mammillary bodies, mammillothalamic tracts, the anterior nucleus of the thalamus, or other related lesions cannot be ruled out at this time. As noted earlier, all of these structures have primary and/or secondary connections with the hippocampal/medial temporal area. Thus one might readily hypothesize that lesions that intrude upon this system adversely might affect memory processes. Finally, since confabulations are more likely to be encountered particularly in the more acute stages of Korsakoff's and ACA lesions and are relatively rare in pure medial temporal syndromes, it is likely that involvement of the basal forebrain or other frontal systems are responsible for this behavior (Damasio et al., 1985; Phillips, Sangalang, & Sterns, 1987).

Unilateral lesions affecting the hippocampal formation or related medial temporal structures (or unilateral thalamic lesions) do not produce the type of global amnesia seen with bilateral lesions but rather are likely to result in either verbal memory deficits (with left-sided lesions) or "nonverbal" (e.g., visual spatial) memory deficits following right-hemisphere involvement (e.g., Speedie & Heilman, 1982, 1983; Milner, 1968, 1972; Andrews, Puce, & Bladin, 1990; Christianson, Saisa, & Silfvenius, 1990; Frisk & Milner, 1990; Loring et al., 1991; Cohen, 1992; Saling et al., 1993; Gainotti, Cappa, Perri, & Silveri, 1994; Kaplan et al., 1994).[24] While the (sensory) modality of input generally is immaterial in such cases, lesions that disconnect the hippocampus from modality-specific cortical areas may result in material-specific, as well as modality-specific, memory problems. For example, a patient may have difficulty remembering verbal information presented orally but not in written form.

Finally, for the sake of completeness, it also should be noted that patients might experience difficulties on specific memory tests as a result of a variety of focal cortical lesions. Posterior cortical lesions (i.e., lesions of the parietal, temporal, or occipital lobes) can compromise the processing or integration and analysis (perception and understanding) of the to-be-remembered material, hence making learning more difficult. Here the patient is more likely to experience difficulty in initial acquisition of the material, but unlike hippocampal or diencephalic lesions, show little additional decline following delayed recall. More frontally based lesions, on the other hand, are more likely to impede complex learning where guided attention and the planning or organizational strategies are required (e.g., long word lists as opposed to memory for a paragraph). Recall may be adversely affected by problems with selective retrieval and/or perseveration.

Cingulate Gyrus

Location and General Anatomy

The **cingulate gyrus** is most prominent on a midsagittal section of the brain (Fig. 8–10), but also is seen on both coronal and axial sections (see Figures 6–1 and 6–2). While the majority of the gyrus lies directly above the corpus callosum, anteriorly portions of it continue around the genu of the corpus callosum where it blends into the subcallosal gyrus of the medial orbitofrontal cortex. Posteriorly, the gyrus wraps around the splenium of the corpus callosum where it connects with the parahippocampal gyrus (via the isthmus of the cingulate gyrus). It is bounded dorsally by the cingulate sulcus and ventrally by the callosal sulcus. Within the fold of the gyrus itself, as seen in Figures 8–10e, lies a band of white matter that is known as the **cingulum**. This pathway contains both afferent and efferent, as well as association and projection fibers, representing both corticocortical and corticosubcortical connections (see below). The cingulate gyrus is classified as mesocortex and represents a transition between the allocortex of the hippocampal formation and the neocortex. While a number of different types of cytoarchitecture structure have been identified within the cingulate gyrus, for most clinical and experimental purposes frequently it is simply divided into an anterior (Brodmann's areas 24, 25, 33, and 32)[25] and a posterior region (areas 23, 29, 30, and 31). Given its inclusion in "Papez's circuit" and its extensive connections with the hippocampal gyrus, the cingulate gyrus generally is considered to be a part of "the limbic system."

Afferent and Efferent Connections

Perhaps the best known of the cingulate connections are those included in Papez's circuit. To review, this circuit went from the hippocampal formation → fornix → mammillary body → mammillothalamic tract → anterior nucleus (of the thalamus) → cingulate gyrus and back to the hippocampal formation.[26] We now know that the cingulate gyrus has connections with virtually all regions of the cerebral cortex (most of which are likely reciprocal), as well as extensive subcortical, brainstem, and even spinal connections. The anterior cingulate, as will be seen, primarily appears to be involved in modulating affect, motivation, and the selection and/or modulation of behavioral (voluntary) and autonomic responses and has extensive connection with the amygdala, orbitofrontal area, prefrontal ("executive"), and primary and secondary motor cortices. The posterior cingulate, which is thought to play a more substantial role in learning or memory, has extensive connections with parietal and posterior temporal cortices, the insula, and indirectly with the hippocampus via the parahippocampal

(a)
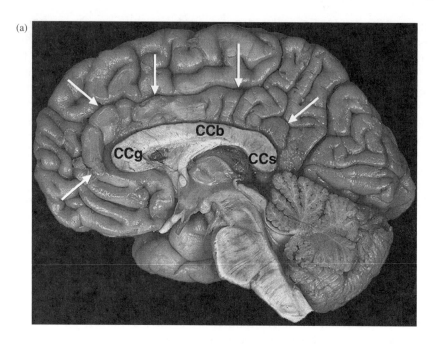

(b)
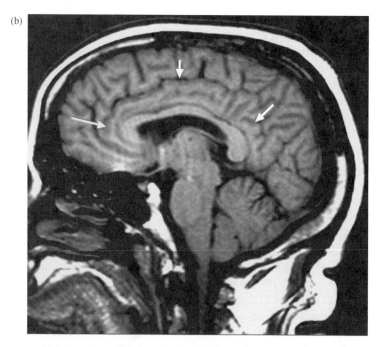

Figure 8–10. Cingulate gyrus as seen on (a) midsagittal brain section, on (b) midsagittal and (c) supracallosal axial MRI images. Cingulum can be seen on slightly more (d) parasagittal and (e) axial MRI images. Abbreviations: CCb, CCg, & CCs, the body, genu, and splenium of corpus callosum: CG cingulate g., Cg cingulum. Brain images were adapted from the *Interactive Brain Atlas* (1994), courtesy of the University of Washington.

(c)

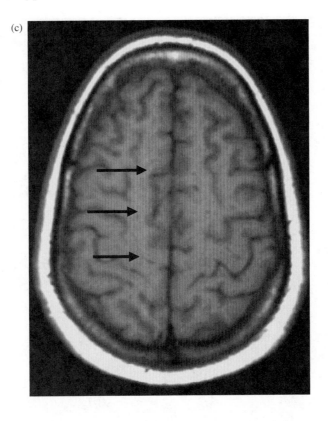

(d)

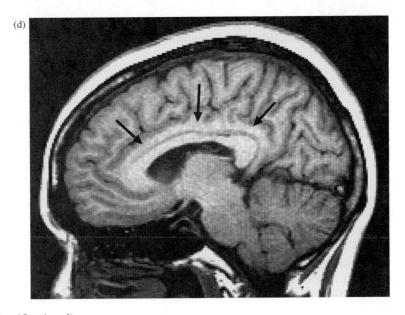

Figure 8–10. *(Continued)*

(e)

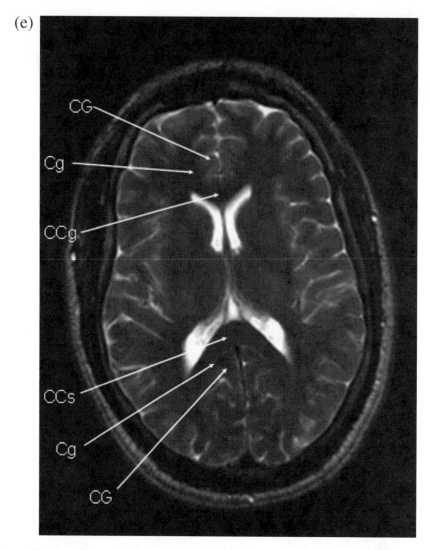

Figure 8–10. *(Continued)*

gyrus and entorhinal cortex of the hippocampal formation. Additional sensory input is suspected to reach the cingulate via the claustrum. As noted in Chapter 6, the cingulate gyrus, particularly the anterior cingulate, also has connections with the ventral striatum as part of the limbic circuit involving cortical–basal ganglia–thalamic feedback loops. It also since has been discovered that in addition to the anterior nucleus, various other thalamic nuclei, including the lateral, ventral, and dorsomedial groups, and the intralaminar and midline nuclei all appear to have some cingulate projections. In addition to these thalamic connections, the cingulate gyrus also has efferent projections to the midbrain (red nucleus), pons, and spinal cord. Finally, there appears to be two-way connections between the anterior and posterior portions of the cingulate gyrus itself which allows for reciprocal feedback between its effector and sensory/memory mechanisms. (Buchanan & Powel, 1993; Dum & Strick, 1991; Mega, Cummings, Salloway & Malloy, 1997; Mufson & Pandya, 1984; Musil and Olson, 1988; Olson and Musil, 1992; Van Hoesen, Morecraft & Vogt, 1993).

As noted above, many of these connections travel through the cingulum, the large fiber pathway that runs the length of the cingulate gyrus. It contains not only internal cingulate connections but serves as a connection between the frontal cortices and cingulate gyrus with the parahippocampal gyrus (and ultimately with the hippocampal formation). Thalamocingulate and other cingulofugal fibers also make use of this pathway. Thus, the cingulum represents a two-way communication system made up of both long and short association and projection fibers, containing both afferent and efferent fibers thattransmit information to and from the cingulate to most of the sites mentioned above. Finally, the cingulate gyri are interconnected via callosal fibers (Table 8–5 provides a brief review of some of the major limbic pathways).

Functional Correlates

Despite its substantial size, the role of the cingulate cortex in behavior remains relatively elusive. While this statement easily could be made with regard to any number of CNS structures, it seems particularly apropos of the cingulate gyrus. As will be discussed shortly, there certainly is no shortage of behavioral correlates that have been associated with either

Table 8–5. Summary of Major Pathways Affecting Limbic Structures

Cingulum: Interconnects subcallosal frontal areas with anterior and posterior cingulate gyrus with hippocampal formation.

Diagonal band of Broca: Connects subcallosal frontal and septal areas with the amygdala.

Dorsal longitudinal fasciculus: Interconnects the hypothalamus with the brainstem.

Fasciculus mammillary princeps: Pathway exiting from the mammillary bodies that bifurcates into the mammillothalamic and mammillotegmental tracts.

Fasciculus retroflexus (habenulointerpeduncular tract): Provides a connection between the habenular nuclei of the thalamus and the interpeduncular and other brainstem nuclei, especially those concerned with autonomic activity.

Fornix: A very prominent pathway between the hippocampal formation and the septal region, thalamus, basal forebrain, and mammillary bodies.

Hypothalamohypophyseal tract: Neuronal pathway connecting the hypothalamus with the neurohypophysis or posterior lobe of the pituitary gland.

Indusium griseum: This pathway includes the medial and lateral longitudinal stria of Lancisi and connects the hippocampal formation with the septal area.

Lateral olfactory stria: Connects the olfactory bulbs with the corticomedial aspect of the amygdaloid complex.

Mamillotegmental tract: Provides a connection between the mammillary bodies and the brainstem reticular formation.

Mamillothalamic tract: Connects the mammillary bodies to anterior nucleus of the thalamus

Medial forebrain bundle: Interconnects the orbital frontal and septal area with hypothalamus and with brainstem nuclei.

Perforant pathway: Interconnects the entorhinal cortex with the adjacent hippocampal formation.

Stria medullaris thalami: Discrete, midline pathway between the septal and habenular nuclei.

Stria terminalis: A very long, curved pathway connecting the amygdala to the septal nuclei, preoptic areas, and hypothalamus.

Ventral amygdalofugal pathways: Rather than a discrete pathway, this designation represents a diffuse fiber system that connects the amygdala (probably primarily the basolateral portion) and the surrounding pyriform cortex to the basal forebrain and subcallosal regions (including septal nuclei), the hypothalamus and dorsal thalamus, and the anterior hippocampal regions.

(**Note**: Most of these pathways probably reflect reciprocal connections. The use of the word "to" suggests the probable major direction of the pathway, otherwise strong reciprocal connections are likely.)

lesions or suspected dysfunction of this structure. Devinsky, Morrell, and Vogt (1995) have suggested that it is precisely this association with such a multiplicity of functions that has helped cloud the question of its behavioral significance. They suggest that the cingulate gyrus may serve as a link and/or modulator among different functional networks. At a more basic level, although a single appellation, *cingulate gyrus*, is applied to this entire expanse of tissue, it should be noted that this gyrus is composed of both agranular (anterior region) and more granular (posterior portion) type cortex. As we have just seen, these different regions have different patterns of connections, further suggesting a probable lack of uniformity of function. Finally, at least in humans, establishing structural–behavioral correlates is complicated further by the fact that, as in other cortical areas, naturally occurring lesions generally do not respect its anatomical boundaries. With these caveats in mind, we can at least explore some of the suspected behavioral phenomena associated with this intriguing region of the brain.

Paul MacLean (1986), it might be recalled, was the originator of the concept of the "limbic system" and took an interesting approach to trying to discover the role of the cingulate gyrus in behavior. He noted that the cingulate gyrus is a mammalian development, not present in reptiles. Thus, he asked, "What major classes of behavior seem to differentiate reptiles from even primitive mammals?" He concluded that there were three: (1) *play*, (2) *maternal care of the young*, and (3) *vocal communication* (in particular, of the young in response to maternal separation).[27] However, most current theories about cingulate function in humans appear to derive largely from either on pathological data or imaging studies.

Anterior Cingulate–Behavioral Associations

One of the more well-documented behavioral syndromes associated with naturally occurring midline frontal lesions generally involving the anterior cingulate gyrus is apathy or diminished motivation. This phenomenon sometimes also is referred to as **abulia**, or lack of will, drive, or initiative. Such individuals often manifest varying degrees of psychomotor retardation and might appear depressed (anhedonic), except that they do not manifest dysphoric affect or acknowledge depressive-type cognitions (Faris, 1969; Freemon, 1971; Laplane, Degos, Baulac, & Gray, 1981; Degos, Fonseca, Gray, & Cesaro, 1993). While they may formulate plans or intentions, these are often never carried out, especially if sustained effort is required. The ultimate extreme of this state is akinetic mutism. Here the patient, while conscious and apparently alert, essentially is mute with little if any spontaneous goal-directed movement, except perhaps for visual tracking behavior. While this latter condition has been reported in brainstem/diencephalic lesions, it tends to more commonly be associated with bilateral, medial frontal lesions. (Stuss & Benson, 1986; Devinsky & Luciano, 1993; Devinsky, Morrell, & Vogt, 1995).

While the above pathological conditions are related to lesions thought to affect the cingulate gyri, surgical lesions that actually improve other conditions also provide some clues as to their potential behavioral contributions. Most notable in this regard are obsessive–compulsive disorders (OCD), anxiety, and chronic pain syndromes. Since at least the middle of the last century, anterior cingulate lesions have been shown to be effective in treating what otherwise had proved to be intractable and highly disruptive obsessive–compulsive symptoms. In addition to OCD, other chronic anxiety disorders also have been shown to have a positive response to these procedures (Whitty, Duffield, Tow, & Cairns, 1952; Tow & Whitty, 1953; Ballantine et al., 1975, 1987; Valenstein, 1977, 1980; Jenike et al., 1991; Insel, 1992; Devinsky & Luciano, 1993; Hay et al., 1993; McGuire et al., 1994). Similarly, anterior cingulotomies and cingulumotomies have proved to benefit patients suffering from chronic pain syndromes (Foltz & White, 1962; Devinsky & Luciano, 1993). Such patients typically report that while they still perceive or experience pain, it no longer is as disturbing to them. In other words, it is as if the somatosensory perceptions are unchanged, but the affective response to those sensations has diminished.

As noted above, apathy without other common manifestations of depression can be associated with lesions involving the anterior cingulate. However, classic depressive syndromes also have been linked, at least in part, with the cingulate gyrus. Numerous functional imaging studies consistently have shown metabolic changes in the frontal and paralimbic regions associated both with the onset and recovery from depression (Ebert & Ebmeier, 1996; Ketter et al., 1996). While multiple, interconnected neural networks have been identified as participating in these behavioral changes, including cortical and limbic feedback loops involving striatal and diencephalic structures, the anterior cingulate cortex appears to play a prominent role. Specifically, it has been demonstrated that depression tends to be associated with *hypometabolism* of the more dorsal regions of the cingulate gyrus and portions of the frontal and parietal cortices, while subgenual portions of the cingulate gyrus (BA 25) and ventral frontal and hypothalamic areas tended to be *hypermetabolic*. Successful treatment of depression, on the other hand, was found to be characterized by just the opposite, namely increased metabolism in these dorsal regions and decreased metabolism in the ventral or subgenual cingulate and adjacent regions (Mayberg, 1997; Mayberg et al., 2000). Following up on these earlier studies, Mayberg and her colleagues have shown in preliminary studies that deep brain stimulation of the subgenual area of the cingulate gyrus (BA 25) in patients with otherwise intractable depression may prove beneficial where drugs and other treatments have failed (Mayberg et al., 2005). While the exact mechanisms behind this antidepressant effect are still unclear, such stimulation has the effect of reducing cerebral blood flow to the region, an effect comparable to that seen in standard treatments for depression when they are effective.

Although not as consistent or well documented as the above (especially inhumans), a number of other behavioral changes have been associated with lesions, stimulation, or other pathological changes in the anterior cingulate and/or adjacent frontal cortices. Among these have been schizophrenia, increased emotionality, neglect or attentional deficits, complex partial seizures, visceral (autonomic reactions), and Gilles de la Tourette syndrome (Benes, 1993; Devinsky & D'Esposito, 2004; Devinsky & Luciano, 1993; Devinsky, Morrell, & Vogt, 1995; Diering & Bell, 1991; Haznedar et al., 2004; Kunishio & Haber, 1994; Levin & Duchowny, 1991; Mazars, 1970; MacLean, 1993; Malamud, 1967; Spence, Silverman, & Corbett, 1985; Vogt, Finch, & Olson, 1992).

In summary, the anterior cingulate gyrus (BA 24, 25, 33, and 32) appears to be involved in the wedding of affect, cognition, and behavioral expression. Devinsky and his colleagues (1993, 1995, 2004) emphasize the need to integrate emotion and cognition as a means of providing motivation or drive. The nature and strength of those emotions might help dictate the direction, intensity, and tenacity of the behavioral response. Affectively driven behavior generally entails not only cognitive choices, but also behavioral (motor) displays or responses. Such responses might include skeletal–motor responses such as smiling, frowning, laughing, crying, or cursing. Equally important are visceral–motor responses. This perhaps most clearly is seen in situations where fear or anger predominate. In these situations, autonomic (sympathetic) activation is essential in preparing the organism for a "flight or fight" response. Vogt, Finch, and Olson (1992) characterize this as the "executive" role of the anterior cingulate cortex and point out that its interconnections with amygdala, frontal cortices, and skeletal and visceral–motor systems place it in an ideal position to mediate (modulate) such functions.

In this context, one also can offer potential explanations for some of the behavioral correlates of hypo- and hypermetabolic states of the anterior cingulate regions. For example, if the anterior cingulate is important in the interface between emotion and cognition, then it is easy to see how disruption of certain components of this network might result in depression, loss of drive or motivation, neglect or inattention to potentially important environmental stimuli, hypokinesia, and in more extreme cases akinetic mutism. Failure to perceive pain as "distressful" (e.g., following cingulectomies) similarly can be viewed as a failure to properly wed the "emotional" aspect of pain to its cognitive representation. On the other hand, increased or disinhibited anterior cingulate activity might account for excess or overflow emotional and/or visceral arousal. This has been hypothesized to possibly play a role in the development of such disorders as OCD, other anxiety disorders, and tics and conversely their amelioration following surgery to these regions.

Posterior Cingulate

As noted earlier, the posterior cingulate is connected primarily with the parahippocampal gyrus and the parietal and temporal cortices. Unlike the anterior cingulate, it has no direct connections with the amygdala. Thus, as might be expected, it is thought to have little if any direct role in emotionally guided behavior. Vogt, Finch, and Olson (1992) suggest that the posterior cingulate is more involved in visual–spatial and somatosensory attention or awareness and memory. While these authors review evidence suggesting that the posterior cingulate likely is involved in memory for visual–spatial information, given the interconnections between the anterior and posterior cingulate cortices, it might be suspected that it also might play a role in the encoding of emotionally driven behaviors and/or the environmental context in which they occur. However, it should be noted that no studies addressing this issue could be found in a review of the literature.

OLFACTORY SYSTEM

As noted at the beginning of this chapter, many structures generally included when discussing the "limbic system" at one point collectively were referred to as the *rhinencephalon* or *smell brain* because of their close relationship with the olfactory system. Although less emphasis is placed on the olfactory connections when discussing these structures today (at least in primates), the olfactory connections are still present and play an important role in many aspects of emotionally charged behaviors or drive states. Evolutionarily speaking, olfaction, perhaps more so than hearing, vision, or somesthesia, has a history of being closely associated with approach–avoidance situations (i.e., with strong affective valences). Olfaction was critical in alerting the organism to the presence of food as well as threats (e.g., fire, predators). Whether prey or predators, olfaction also was important in mating, territoriality, and social identification and communication. While perhaps less critical to basic survival when compared with the animals that roam the plains and savannas, olfaction and affective drive states still retain their strong links in humans. The smell of a freshly baked apple pie or bacon sizzling in the pan usually has the capacity to elicit a strong (if not always prudent) approach response, while putrefaction normally elicits revulsion. Olfaction still plays an important role in sexual/social functions as witnessed in part by the multibillion-dollar perfume and cologne industry. While the teleological significance of other odors may be less clear, they nonetheless can be associated with positive responses (e.g., the fragrance honeysuckles or a pine forest). Finally, as with other types of stimuli, more neutral olfactory sensations (e.g., the smell of the salt sea air or the scent of a burning candle) may take on strong positive or negative associations, depending on the previous experiences of the individual. As was noted in Chapter 5, the reexperiencing of certain odors readily can be imbued with strong emotionally laden recollections.[28] The reverse also may be true, a strong emotional memory of an event may evoke associated olfactory images. Rather than repeating the behavioral and anatomical bases that may underlie these phenomena, the reader is referred back to Chapter 5 where this topic was first introduced.

SUMMARY: CONTRIBUTIONS OF THE LIMBIC STRUCTURES TO BEHAVIOR

Having reviewed the basic anatomy, afferent and efferent connections, and suspected functional correlates of individual components of the "limbic system," it might be useful to speculate as to how these structures interact with other parts of the brain to regulate day-to-day behavior. While this is an important clinical and heuristic consideration, attempting to address such issues again is tantamount to trying to decipher the mystery of the brain. At best, one hopes to develop a few broad theoretical models or generate a few tentative hypotheses that might prove useful in forwarding our conceptualization of these functional relationships. Certainly, whatever models are generated, whether they attempt to define the possible interactions between two discrete areas of the brain or attempt a broader overview of brain systems as a whole, most assuredly they inevitably will represent a gross oversimplification of the actual reality. While two or more areas of the brain may work in concert to help effect a particular behavior, most behaviors including those that at first glance might appear to be rather "simple" typically are highly complex and likely involve an activation of all parts of the brain. When discussing behaviors as complex as drives, emotions, and memory, the more convoluted the problems become and the more they defy description or categorization. Thus, the following is intended as only a very elementary starting point, but one that might assist in providing a conceptual framework that may facilitate our attempt to understand some of the mysteries of brain–behavior relationships.

As a starting point, it might be suggested that one of the teleological goals behind the evolving brain was to permit organisms to gain increased understanding and mastery of the environment. As will be explored in Chapter 9, the expansion of the posterior association cortices in humans has resulted in an enhanced ability to integrate multimodal sensory input. This, in turn, has facilitated the development of abstract thought and ideas, the ability to manipulate spatial and temporal concepts, and to develop language by which these thoughts, concepts, and ideas can be communicated. At the same time, the evolution of the frontal association area has enabled us to plan, to anticipate future consequences, and to choose or prioritize among competing goals or ideas. Once a goal or plan of action has been selected, these association cortices enable us to determine how best to accomplish or execute that goal, monitor the progress of our actions, and where necessary alter the goal or the plan of action.

However, despite these advances, certain fundamental processes of the central nervous system remain relatively unaltered. We still must satisfy our basic biological and social needs. The capacities to attach emotional significance to and learn from our experiences still are essential adaptive mechanisms. Finally, *there needs to be some emotion or drive that initiates (motivates) us to pursue those goals (i.e., the will to act) in the first place and then sustains (energizes) those actions (i.e., keeps us on task). It would appear that it is the limbic structures that are primarily responsible for these latter tasks.*

As we have seen, the hypothalamus, one of the earliest parts of the brain to develop, appears to be responsible for monitoring and initiating basic, biological drive states. One example is the maintenance of internal homeostasis, such as ensuring proper oxygen, glucose, and water levels in the blood and cells. These functions are accomplished by creating internal drive states that "arouse" and "energize" the organism to engage in behaviors designed to meet these needs or reduce the drive state. Simple examples would include the urge to surface after swimming underwater for an extended period or to search for food or drink when one is hungry or thirsty. Other drive states that are less directly related to maintaining internal homeostasis would appear "more complex" (i.e., likely involve other limbic structures to a greater extent). One example of the latter is the organism's need to defend itself against external threats or predators, which frequently results in the need to develop flight or fight responses. Another such drive is the need to ensure the survival of the species, hence the need to respond to sexual cues and stimuli that lead to the reproductive process. These drives not only are primitive, but can be very strong [all are near the top of Maslow's (1954) hierarchy of needs]. Witness, for example, states of panic and the strength of the drive to escape if trapped in an imminent, life-threatening situation. Under such circumstances, people often are reported to act "irrationally," that is, their actions would appear to be driven more by instinct under the influence of primitive brain mechanisms than by logical judgment as mediated by the tertiary cortical zones (Gorman, Liebowitz, Fyer, & Stein, 1989).

While under most normal (i.e., non-panic) circumstances, most drive states likely are heavily influenced by previous learning and experience, but can be modulated or even temporarily suppressed as a result of goals or drive deriving from higher, more recently developed brain centers. However, it is likely that the hypothalamus (and pituitary gland) still plays a crucial role in responding to drive states. For example, in order to prepare the organism to meet the physical demands involved in securing needs or drives, it often is necessary to increase the organism's heart rate and flow of oxygen to make more blood accessible to the peripheral musculature. These latter reactions help prepare the organism for fight or flight or, in the case of the predator, the chase. These functions are accomplished through the activation of the sympathetic nervous system, with parasympathetic arousal being necessary after the emergency (i.e., to conserve energy and/or digest the meal). Thus, hypothalamic arousal is closely linked, both behaviorally and neuroanatomically, with brainstem systems, including general arousal (ARAS) and activation of the autonomic nervous system. This hypothalamic–pituitary system might be conceived of as reflecting an "id" type of response, with the organism responding in a reflexive manner to both internal and external stimulus cues.

As organisms evolved, they developed increasingly sophisticated telereceptors to warn them of danger or signal the presence of stimuli that would satisfy their appetitive drives, such as thirst, hunger or sex. One of the earliest of these senses likely was the sense of smell. However, in order to benefit from these external cues, it was essential that the organism learn and remember from one instance to the next the meaning of these olfactory cues and the appropriate response to them. This, in turn, required close and fairly direct connections between these stimulus cues and those neural centers that were responsible for initiating and driving such behavior, that is, the hypothalamus. Second, it obviously was beneficial to ensure that a more or less permanent memory of this association would be established for future reference. The early (phylogenetically speaking) connections between the olfactory system and the hypothalamus would appear to reflect this intimate link between the sense of smell and these basic drive states. Likewise, the eventual development of the hippocampal formation and its close connections with both the hypothalamus (currently represented by the fornix) and the olfactory system (by way of the more primitive corticomedial portion of the amygdala and the surrounding pyriform cortex) would ensure that the organism could benefit from its prior experiences with similar stimuli. Such interactions might have laid the foundation for true "emotional" development.

Along with the evolution of the other senses, particularly vision and hearing, came the development of the supporting neocortex, which would allow for increasingly finer and more complex perceptual discriminations. Given the capacity to make such discriminations, the organism not only could make more effective use of its environment, but also was less at its mercy. The organism could begin to choose between stimuli or make individual judgments as to the more preferable response. Along with the ability to make finer and finer discriminations likely came the ability to make decisions as to whether the organism needed or wanted to respond at all. The organism could learn to differentially associate specific stimuli with different drive states under different circumstances based on its prior experience. For example, did this stimulus represent a threat the last time it appeared under these circumstances? Along with this increase in encephalization came an increase in social organization and complexity that also required choices or decisions based not only on internal or environmental stimuli but also on social context.[29]

For this system to work effectively, however, several things were important. First, it would have helped if these stimuli were imbued with some type of affective valence to signify their relative importance. If specific stimuli were associated with the reduction or temporary satisfaction of a primary drive state, such as hunger, these stimuli might acquire a "positive" (emotional) valence. Other stimuli may have become associated with self-preservation (e.g., an attack from a predator) and henceforth evoke fear, apprehension, or defensiveness or other "negative" affective valences. It equally was important for other stimuli to remain affectively neutral, as the organism cannot always be running either toward or away from every stimulus that reaches consciousness. It has been suggested that parts of the limbic system, particularly the amygdala, with its corticosensory connections on the one hand and its hypothalamic connections on the other, largely may be responsible for adding these emotional valences to specific stimuli.

However, it would do the organism little good to associate an experience with a particular drive reduction or emotional state if it could not retain that information from one time to the next. Hence, the organism needed to be able to learn, remember, and retrieve the information gained from such encounters and to be able to modify previous associations as circumstances changed. This latter purpose appears to be served by the extensive connections between the hippocampal regions and the posterior or sensory cortices. Of course, the organism did not "need to remember" all sensory experiences, just as it did not need to attach emotional significance to all sensory experiences. The connections between the amygdala and the hippocampus assisted in identifying those stimuli that were "important to remember," that is, the stronger the emotional impact of the stimulus, the greater the need (or likelihood) of its being remembered and the greater the strength of its connection. McGaugh (1992) has shown, for example, that the presence of adrenaline (i.e., emotional arousal) is important in laying down new memories.[30] The cingulate gyrus also is connected to the hippocampal formation and, as mentioned above, may be responsible for higher-order, social behaviors or interactions that need to be integrated into memory.

Once these connections between affective valences and particular stimuli were established (probably through classic conditioning) and stored, the stage was set for external (and/or "remembered") stimuli to initiate drive states. Certainly some fairly rigid and "preprogrammed" (instinctual) response patterns were "built into" many organisms (e.g., light-dimming detectors in the frog), but it was unlikely that these necessarily represented true emotional responses.[31] However, such associations that have been derived from personal experiences may have clearly manifested motivational (emotional) properties. Thus, such stimuli would have had the capacity to trigger emotional responses that then in turn evoked or "drove" an approach, avoidance, or other idiosyncratic response from the organism. This arrangement would have allowed an "integrated" or "meaningful" multi-sensory images to elicit learned emotional, autonomic responses or "externally" triggered drive states in addition to "internally" driven (homeostatic) mechanisms. The potential benefit to the organism should be obvious. Having previously encountered (and survived) a potentially dangerous situation, an immediate "alarm" should have been activated, alerting the entire nervous system to adopt either an avoidance or defense response posture should the same situation again present itself. We now know that the same stimuli often can elicit highly individualized (learned) responses, differing not only in intensity from one person to another, but also in direction or valence. For example, the lyrics of a particular song or a visit to a particular place may elicit either a "pleasant" or "painful memory" depending on one's past experiences.[32]

The neuroanatomical substrates for such associations include connections between the heteromodal or tertiary cortices of the second functional unit (see Chapter 9) and the amygdala, cingulate gyrus, and parahippocampal gyrus, including the hippocampus proper. In turn, these paralimbic areas are connected to the hypothalamus with its own autonomic connections. It also is possible that certain stimuli may elicit an acquired emotional response, such as fear, without any conscious "memory" of its connection to previous life experiences. In this case, while the connections between the second functional unit and the paralimbic structures would appear to remain intact, the connections between these areas and episodic memory stores or the memory stores themselves may be dysfunctional. While one theory would suggest that functional "repressive" mechanisms block these memories from reaching conscious levels (Joseph, 1990), this also could reflect classic conditioning without the benefit of episodic memory for the event(s).

Commonly, the organism is faced not by one but by numerous converging external and internal stimuli. Many of these stimuli may have opposing valences or convey conflicting information. The organism must weigh and balance all this information to determine whether to respond or not. If the organism "decides" to respond, the precise nature and intensity of the response must be determined.[33] Thus, there is a need for a system of counter-balancing these various stimuli or conditions, that is, the capacity to override or inhibit responses. We have seen such a system not only within the hypothalamus itself (e.g., the ventromedial and lateral hypothalamic nuclei), but also with various other structures such as the interplay between the amygdala and the septal nuclei and the effects of anterior cingulate lesions on drive states. It has been suggested that perhaps one of the main roles of many of the limbic structures is to inhibit or modulate the influence of hypothalamic drives. Thus, in keeping with the earlier analogy, if the hypothalamic system can be seen as representing the "id," then this second (limbic–sensory) system might be seen as reflecting "ego" functioning, that is, integrating present reality to best serve the immediate needs of the organism.

At times, conflicting impulses (valences) need to be resolved by weighing the sum total of internal and external stimuli (i.e., the immediate circumstances) and "making an informed judgment." At times, this also may require considering not only immediate but future consequences, "higher-order," or longer-term goals. This subject will be discussed in greater detail in The Third Functional Unit in Chapter 9, Part III.. For now the following provides a simplified example. Suppose a monkey is glucose deficient (hungry) and engages in food-seeking behavior. It spots some fruit in a tree. The fruit triggers a positive emotional response, leading to increased salivation (a descending visceral response). However, as the monkey approaches, it sees not only the fruit, but also a snake. The visual image (percept) of the snake also is conveyed to the paralimbic structures which elicits a quite different emotional response (fear), along with a competing set of visceral responses (increased heart rate, pupillary dilation). One could speculate as to what might happen next. If the monkey is only slightly hungry, the fruit is not ripe, and the snake is not only very large but also is curled up next to the fruit, chances are that the "negative" drive state (fear) will overwhelm the positive (sight of food). In such cases there might be no contest as to which behavioral response takes precedence. However, the hungrier the monkey, the riper the fruit, and the smaller and more distant from the fruit is the snake, neither emotional state (drive) may clearly dominate. At that point, the third (executive) functional unit is left with a decision to make or a plan of action to contemplate. Such a decision likely would take into account not only the sensory information available, but also the respective behavioral or affective relevance of those stimuli, memories of previous encounters, and/or "predicted outcome" and the general state of the organism (e.g., degree of hunger).

Finally, if the organism is motivated to act, then there needs to be a mechanism to translate that impulse into action. It may be here that the association between the limbic structures (particularly the anterior cingulate) and the prefrontal, frontal, and basal ganglia become increasingly important. For example, whereas it may be postulated that the function of the posterior cortices and paralimbic structures is to integrate sensory input and "evaluate" their relative valences in terms of previous associations (information that may incline the organism to respond in a particular manner), the function of the prefrontal association cortices may be geared to subordinate these instinctual or learned behavioral patterns in response to specific stimulus conditions to supraordinate, goals, plans, strategies, or anticipated long-range consequences.[34] In this manner, the input of the prefrontal association cortex to the limbic system may be conceptualized as that of a general overseer (or "superego") for drive states. In humans, this may mean submitting more primitive drive states or contemplated actions to tests of social judgment and/or inhibiting them long enough to consider potentially competing, long-range goals or consequences.

However, these hypothesized prefrontal–limbic interactions are probably only part of a larger, two-way neuronal highway. While the prefrontal–limbic systems may serve to inhibit or facilitate certain drive states depending on selected goals, the goals themselves (or more properly the actions necessary to effect these goals) must be initiated and maintained often in the face of distractions, obstacles, competing drives, or simple inertia. Certain stimuli, whether internal or external, may be imbued with strong affective drive states (either by "nature" or by previous associations); yet unless these drive states can affect motor and executive systems, they will not be converted into action. The same can be said for the "higher-order" goals, plans, or ambitions we normally associate with prefrontal, executive cortex. Without "affect" to drive them to fruition, they remain as just that: unfulfilled "plans," "goals," or "ambitions." It is suspected that this "drive" may be initiated at the limbic and reticular levels and interact (e.g., via the cingulate loop[35]) with the frontal and basal ganglia systems. Consistent with this hypothesis, lesions affecting this loop might be expected to result in apathy and/or akinesis (Devinsky & Esposito, 2004; Devinsky, Morrell, & Vogt, 1995; Mega & Cummings, 1994; Cummings, 1993).

In addition, the frontal lobes may impose specific strategies on learning and memory via connections with the hippocampus (e.g., "chunking" or associative strategies during the learning process and selective "search" strategies during retrieval). These and other possible interactions between the neocortex and the "limbic system" and the impact of lesions affecting these connections and hemispheric differences in emotional responses will be explored in greater detail in Chapter 9. In that chapter we will explore the role of arousal

Table 8–6. Summary of Major Functional Correlates of the Limbic System

Olfaction: Once known as the *rhinencephalon*, olfaction represents only a small part of the overall functional significance of the limbic system, particularly in primates. In humans, anteromedial temporal lobe seizure foci may produce olfactory auras (likely from stimulation of the corticomedial portions of the amygdaloid complex). In humans, smells often trigger affect-laden memories and are still often linked to primary approach/avoidant response patterns or drive states.

Homeostasis: The hypothalamus, in conjunction with the pituitary gland, plays a central role in maintaining internal homeostasis and modulating autonomic and endocrine activity. Lesions involving these structures can result in diabetes, eating disturbances, problems with temperature regulation, and growth, metabolic, and sexual (reproductive) abnormalities.

Preservation of the individual and of the species: In addition to maintaining internal homeostasis, interactions among the various limbic structures are critical in protecting the individual from external threats and in establishing social bonds and relationships that are conducive to the survival of the species. This entails recognizing and responding to affectively meaningful external stimuli, including social cues and communications. Emotions and drive states are integral aspects of these processes.

Emotions and drive: Central to most modern notions of a "limbic system," drive and emotion are closely related. **Emotion** is the affective–physiological response to a stimulus (external or internal). **Drive** reflects impetus to initiate an operant response (approach or avoidant). Although commonly linked, the two appear dissociable.

  While multiply determined, the amygdala is the limbic structure most closely associated with emotions. It appears the amygdala is necessary in establishing affective–stimulus bonds, whereas both the amygdala and hypothalamus are crucial to the affective–physiological response. Stimulation of the amygdala often leads to exaggerated affective responses, while lesions frequently result in attenuated emotional responses. The septal region, although its functions are even less clear, apparently has a modulating influence on the amygdala, and stimulation of this region may result in the release of endorphins ("pleasure center"). The hypothalamus also may play an important role in the expression of emotion as lesions of the medial nucleus or stimulation of the lateral or posterior nuclei can result in aggressive behaviors.

  The anatomical substrates for drive are less well understood. The hypothalamus probably is critical in ensuring proper "drive" to maintain internal equilibrium, as well as in certain aspects of sexual behavior. While emotions can be powerful motivators, sustained drive can be seen in the absence of any clear state of emotional "arousal." In humans, cognitive elements, including abstract ideas and long-range plans, often are at work, although likely not independently of limbic structures. Clinically, pathological inertia is most commonly seen in cases of bilateral mesial frontal lesions, especially where the anterior cingulate gyri are involved.

Memory: Although not exclusively related to "limbic functions," memory is closely associated to the limbic system, both anatomically and functionally. The presence of emotions often strengthens memory traces. Conversely, such memories often can rekindle emotions. The survival value of being able to store and recall emotionally salient events is readily apparent. Anatomically, the hippocampal formation is central to this process (bilateral destruction of this structure leads to severe amnestic deficits), although other structures have also been implicated.

Kluver–Bucy syndrome: Bilateral lesions of the medial temporal lobe were found to produce a characteristic pattern of behavior among lesioned monkeys. Among the behaviors noted were:

1. Decreased aggressiveness, along with increased "tameness"
2. Decreased "fear" or aversion responses to stimuli that would normally evoke such behavior, for example, snakes
3. Increased "sexual" behavior, indiscriminate mounting behaviors
4. Increased submissiveness, decreased social dominance
5. Increased activity
6. Increased oral tendencies, all objects explored by the mouth

At least part of this syndrome can be seen in humans with bilateral lesions to this area, as is the case in Pick's disease.

and motivation in the context of integrating the actions of the three functional units of the brain: arousal, gnostic, and executive. (Table 8–6 summarizes major limbic functions).

Endnotes

1. A more detailed review of the history of thought regarding the central nervous system is presented in Chapter 9.
2. The spelling of this nucleus varies among different authors, some choosing "*mammillary*," others opting for "*mamillary*."
3. In 1937, it may have been more fashionable (or, perhaps, more "politically correct") to view females as being less encumbered with the same degree of sexual passion or drive as males?
4. Owing to the confluence of all these various basic visceral mechanisms in this area, MacLean thought this could help explain the

   "overlapping of oral and sexual behavior [as] more than a fortuitous circumstance. In this part of the brain . . . it is possible to conceive how sexual incitations could stimulate a crude, diffuse feeling of visceral yearning that would make the individual seek to mouth and incorporate the object of its desire. According to intensity, the sex–hunger pattern might lead anywhere from gentle kissing to the deviate forms of oral–sexual behavior. . . . Likewise, the hunger–rage pattern susceptible to sexual firing might express itself in all gradations from aggressive, sadistic behaviour to sex-murder and mutilation."

   He subsequently described this area as representing the "id, the beast or sin in man and noted that it was probably the role of the frontal lobes [the superego?] to "stand guard over this region"; a concept that still seems, at least partially, viable: witness the case of Phineas Gage, the 19th-century gentleman who had a metal tamping bar driven up through his face and skull wiping out a sizeable chuck of his frontal lobes, was described as being a very hard working, sober, and generally pleasant fellow before the accident; he reportedly became slovenly, indolent, capricious, impulsive, and given to fits of profanity afterward (see: Blumer & Benson, 1975).

5. This theme of the role of the limbic system and the unconscious most recently was revisited by Rhawn Joseph in his book, *Neuropsychology, Neuropsychiatry and Behavioral Neurology* (see Suggested Readings)
6. These structures represent what might be characterized as the "short list" of limbic structures. Because many of these structures also have strong associations with yet other structures (e.g., both the habenular hypothalamic nuclei have direct connections with brainstem nuclei), it is easy to expand the definition of "limbic structures" to include many other nuclei (Nieuwenhuys, Voogd, & van Huijzen, 1988; Nieuwenhuys, 1996).
7. Clearly, certain types of fairly complex social structure and social behavior, including communication, is still possible with extremely primitive nervous systems, as witnessed by bees and ants.
8. One hardly needs to rely on man, or even lower-order primates, to provide an example of "emotional" development. Dogs have served as close companions to man for thousands of years. While a certain symbiosis can help account for this long-term cohabitation—the dog provided an early warning system and protection against intruders in return for scraps from the human's kill—this alone probably cannot account for the closeness of this bond over the millennium. A more likely explanation lies in the emotional similarities shared by these two creatures. Anyone who has owned a house-dwelling dog has seen the richness of his or her emotional repertoire. Under the

proper circumstances, they can show what, in humans, might be characterized as joy, playfulness, excitement, affection, jealousy, irritability, aggression, fear, anxiety, guilt, and depression.

9. It would not be in the interest of the organism to become generally energized, or to engage in the strenuous activity of food-capturing behavior every time it saw an antelope on the hoof. If there were sufficient antelope around, it would soon die of exhaustion. Thus the behavioral relevance (drive state) associated with the stimulus will depend on other limbic mechanisms (e.g., state of hunger). On the other hand, there may be certain stimuli (e.g., the silhouette of a hawk or the outline of a snake) that are "programmed" to trigger an emotional (fear) response in the chicken or monkey, regardless of the state of the organism or immediate environmental circumstances.

10. Mesulam differentiates a zone of transitional cortical tissue that includes the caudal orbitofrontal cortex, the insula, the temporal pole, the parahippocampal gyrus, and the cingulate gyrus, from more cytoarchitecturally primitive tissue which includes the septal nuclei, the substantia innominata, the pyriform cortex, the amygdala, and the hippocampus proper (hippocampal formation). He designates the former as the *paralimbic areas*, while the latter is termed the *limbic zone*.

11. Holstege (1992) describes an emotional–motor system, related to the classically described limbic system, which is thought to mediate certain affective responses, such as emotionally driven vocalizations.

12. See Chapter 11 for a more detailed review of the neurotransmitters, including their source, major pathways, and probable functional significance.

13. An example is the lateral hypothalamic nuclei, which is traversed by the medial forebrain bundle and is likely affected by lesions of this area.

14. Lesions in the septal area also may result in increased rage reactions in some species. However, whereas lesions to the amygdala will counteract the effect of septal (lesion) aggressiveness, such lesions will have no impact on the aggression resulting from ventromedial hypothalamic lesions.

15. For additional reviews of the various subdivisions of the amygdala and its connections, see Pierre Gloor (1997) and Nieuwenhuys, Voogd, & van Huijzen (1988).

16. A notable exception may be the **episodic dyscontrol syndrome** or "pathological intoxication." While this syndrome would appear to resemble a type of ictal phenomenon, its specific etiology and/or neuroanatomical substrate remain unclear (Monroe, 1986).

17. The most common effects of either lesions or stimulation of the amygdala reported in the literature tend to be changes in aggression or fear type responses. This might lead to the speculation that the amygdala primarily may be responsible for processing "negative" emotions and that other structures process "positive" reinforcements. Actually, the amygdala appears to be involved in the formation of both positive and negative reinforcements (Gaffan, 1992; Rolls, 1995). As pointed out by LeDoux (1995), one possible explanation for the plethora of "negative" emotions reported is that such responses generally are easier to measure and are more amenable to experimental manipulation.

18. **Interictal behavior** refers to those behavioral patterns or traits that are manifested by the seizure-prone patient in between the actual seizure (ictal) episodes. One explanation that has been advanced for these longer-term changes has involved the concept of **kindling** (Goddard, McIntyre, & Leech, 1969). Broadly defined, kindling refers to the fact that repeated, low level (i.e., subclinical) electrical stimulation of the brain eventually will result in permanent changes to neural tissue. Hence, one assumption is that the repeated, subclinical, interictal abnormal electrical discharges that routinely occur in seizure patients with temporal lobe foci eventually produce changes in limbic struc-

tures which result in behavioral changes (Spiers, Schomer, Blume, & Mesulam, 1985; Post, 1986).

19. The hippocampus proper has been divided into subfields, commonly referred to as CA1 through CA4. CA1 represents the outermost segment of the hippocampus, merging with the subiculum. CA3 and CA4 (sometimes simply referred to as CA3) is the innermost portion of the hippocampus proper, lying adjacent to the dentate gyrus. The subiculum itself also can be subdivided into the subiculum proper (contiguous with CA1 of the hippocampus), the presubiculum, and the parasubiculum which lies adjacent to the entorhinal cortex. Some authors (e.g., Amaral & Insausti, 1990) include the entorhinal cortex as part of the hippocampal formation, although unlike the rest of the hippocampal formation, it has a six-layered cytoarchitectural structure that is associated with the neocortex, rather than the three layers characteristic of allocortex.

20. Depending on the nature and extent of the lesion, the observed deficit may be quite profound, so that the patient essentially is unable to lay down any new memories for persons or events, regardless of the length or frequency of exposure to the stimuli. In other instances, the deficit may be demonstrable, not absolute, and/or may show some gradual improvement over time.

21. There was a classic scene from a TV comedy series years ago in which one of the characters (Jim), whose brain was ostensibly fried from years of drug abuse, found himself at a small, but elite party of art patrons. The scheduled piano soloist for the evening having failed to arrive, Jim "volunteers" to provide entertainment. After going through a series of apparently randomly chosen notes (to the horror of his friend and escort who was utterly embarrassed), Jim proceeds to play, unerringly, a complex concerto. Surprised, his friend remarks, *"I didn't know you knew how to play the piano."* To which Jim responds, *"neither did I!"*

22. Again, depending on the particular etiology and the individual case, this pattern may vary. Anoxic lesions, for example, could affect other cortical or subcortical systems. Alzheimer's disease has extensive cortical involvement. In such cases, not only remote memories, but also procedural, semantic, or even immediate memories may be affected depending on the sites involved and the severity of the disease process.

23. Wernicke's encephalopathy typically has an acute or subacute onset manifested by ataxia, ophthalmoplegia, and mental confusion.

24. In the author's experience, while right temporal lesions likely will produce deficits in learning many visual spatial tasks while leaving verbal memory fairly intact, left temporal lesions are more likely to have an adverse effect on both verbal and visual spatial memory. This may result, at least in part, from the fact that humans typically tend to use verbal encoding strategies to remember, regardless of the nature of the material.

25. While most authors seem to identify only Brodmann's areas 24 and 25 as comprising the anterior cingulate region, others (e.g., Devinsky, Morrell, & Vogt, 1995; Vogt, Nimchinsky, Vogt, & Hof, 1995) include areas 33 and 32, with the latter being considered a transitional region between cingulate and frontal cortices.

26. Subsequent studies have demonstrated that the cingulate cortex likely does not project directly to the hippocampus, but rather to other areas of the parahippocampal gyrus (e.g., to the presubiculum and perirhinal cortices), which in turn project to the hippocampus via the entorhinal cortex and perforant pathway (Shipley & Sørensen, 1975). However, the cingulate cortex does receive afferent input directly from the hippocampus.

27. Newly hatched crocodiles make vocalizations to which female crocodiles respond by digging them out of the nest and carrying them to the water. Hungry baby birds, which

are believed to have evolved from a reptilian tree, are highly vocal when left alone in their nests, but,at least in part this may be their way of stimulating feeding behaviors from their parents. But in either case, it is not clear that these sounds are in response to maternal separation per se.

28. Perhaps reflecting the importance of this connection, at least phylogenetically, it is noted that olfaction alone, among the major external sensory receptors, has direct input into the hippocampus. Information derived from visual, auditory, or somatosensory input reach the hippocampus only after being processed by multimodal association cortex (Gloor, 1997).

29. **Note:** Complex social organization or behavioral patterns do not necessarily require complex cortical systems as witnessed by such "social" insects such as ants or bees. What the brain does allow is for more flexibility or "freedom of choice," that is, the capacity to learn from personal experience.

30. Such may be the explanation as to why those who were adults at the time vividly may emember what they were doing when they heard of the bombing of Pearl Harbor or the assassination of President Kennedy, or other events that were associated with strong personal feelings, even though some of the remembered details had little to do with the event itself.

31. This does not mean that all instinctual stimulus–response associations are devoid of emotional content. For example, there is something about the perceptual gestalt of the newborn or the sight of a decapitated body that tends to elicit a common behavioral response among most primates that would appear to have strong affective component and appears to be relatively independent of prior experiences.

32. The affective valence ("emotional coloring") associated with a particular set of stimulus parameters may not be absolute but dependent on circumstances. For example, the sight of a stack of 100-dollar bills being placed in one's palm might elicit very different reactions depending on (1) whether it is construed to be a gift or a bribe, and (2) the moral character of the recipient.

33. For example, if a "tasty morsel" appears sauntering lazily along on the horizon but the predator's belly is full, the arousal value of that stimulus will be affected by visceral and then hypothalamic feedback. Conversely, the sight or smell of a lion normally might trigger panic in the gazelle, but if certain conditions are met (e.g., the lion makes no moves in its direction) flight is unwarranted and wasteful.

34. It has been suggested that the orbitofrontal cortex is involved in the correction of previously learned reinforcement contingencies in situations where such responses are no longer reinforcing. This might help explain the failure to withhold responses to nonrewarded stimuli following certain frontal lesions in animals (Rolls, 1995).

35. See Chapter 6.

REFERENCES AND SUGGESTED READINGS

Adamec, R.E. & Stark-Adamec, C. (1986) Limbic hyperfunction, limbic epilepsy, and interictal behavior: Models and methods of detection. In: Doane, B.K. & Livingston, K.E. (Eds.), *The Limbic System: Functional Organization and Clinical Disorders.* New York: Raven Press, pp. 129–145.

Adolphs, R, Tranel, D, Damasio, H, & Damasio, A. (1994) Impaired recognition of emotion in facial expressions following bilateral damage to the human amygdala. *Nature, 372,* 669–672.

Adolphs, R, Tranel, D, Damasio, H, & Damasio, A. (1995) Fear and the human amygdala. *Journal of Neuroscience, 15,* 5879–5891.

Aggleton, J. (Ed.) *The Amygdala: Neurobiological Aspects of Emotion, Memory and Mental Dysfunction.* New York: Wiley-Liss Inc, 1992.

Aggleton, J.P. (1992) The functional effects of amygdala lesions in humans: A comparison with findings from monkeys. In: Aggleton, J. (Ed.) *The Amygdala: Neurobiological Aspects of Emotion, Memory and Mental Dysfunction.* New York: Wiley-Liss Inc., pp. 485–503.

Ahmad, S.S. & Harvey, J.A. (1968) Long-term effect of septal lesions and social experience on shock-elicited fighting in rats. *Journal of Comparative and Physiological Psychology, 66,* 596–602.

Alexander, M.P., & Freeman, M. (1984) Amnesia after anterior communicating artery aneurysm rupture. *Neurology, 34,* 752–757.

Andrews, D.G., Puce, A., & Bladin, P.F. (1990) Post-ictal recognition memory predicts laterality of temporal lobe seizure focus: Comparison with post-operative data. *Neuropsychologia, 28,* 957–967.

Andy, O.J. & Stephen, H. (1968) The septum in the human brain. *Journal of Comparative Neurology, 133,* 383–410.

Babb, T.L., Lieb, J.P., Brown, W.J., Pretorius, J., & Crandall, P.H. (1984) Distribution of pyramidal cell density and hyperexcitibility in the epileptic human hippocampal formation. *Epilepsia, 25,* 721–728.

Ball, M.J. (1979) Topography of Pick inclusion bodies in hippocampi of demented patients. *Journal of Neuropathology and Experimental Neurology, 38,* 614–620.

Ballantine, H.R., Levy, B.S., Dagi, T.F., & Giriunas, I.B. (1975) Cingulotomy for psychiatric illness: Report of 13 years experience. In: Sweet, W.H., Obrader, S., & Martin-Rodriguez, J.G. (Eds.) *Neurosurgical Treatment in Psychiatry, Pain, and Epilepsy.* Baltimore: University Park, pp. 333–353.

Ballantine, H.R., Bocukoms, A.J., Thomas, E.K., & Giriunas, I.B. (1987) Treatment of psychiatric illness by stereotactic cingulotomy.*Biological Psychiatry, 22,* 807–817.

Bard, P. (1928) A diencephalic mechanism for the expression of rage, with special reference to the sympathetic nervous system. American *Journal of Physiology, 84,* 490–515.

Bear, D.M. & Fedio, P. (1977) Quantitative analysis of interictal behavior in temporal lobe epilepsy. *Archives of Neurology, 34,* 454–467.

Bear, D.M., Levin, K., Blumer, D., Chatman, D., & Reider, J. (1982) Interictal behavior in hospitalized temporal lobe epileptics: Relationship to other psychiatric syndromes. *Journal of Neurology, Neurosurgery and Psychiatry, 45,* 481–488.

Benes, F.M. (1993) Neurobiological investigations in cingulate cortex of schizophrenic brain. *Schizophrenia Bulletin, 19,* 537–549.

Benson, D.F., Marsden, C.D., & Meadows, J.C. (1974) The amnestic syndrome of posterior cerebral artery occlusion. *Acta Neurologica Scandinavia, 50,* 133–145.

Blanchard, D.C., & Blanchard, R.J. (1972) Innate and conditioned reaction to threat in rats with amygdaloid lesions. *Journal of Comparative and Physiological Psychology, 81,* 281–290.

Braak, H., Braak, E., Yilmazer, D., & Bohl, J. (1996) Functional anatomy of the human hippocampal formation and related structures. *Journal of Child Neurology, 11,* 265–275.

Brady, J.V. & Nauta, W.J.H. (1953) Subcortical mechanisms in emotional behavior: Affective changes following septal lesions in the rat. *Journal of Comparative and Physiological Psychology, 46,* 339–346.

Brodal, A. (1981) *Neurological Anatomy.* New York: Oxford University Press (Chapter 10, The olfactory pathways, the amygdala, the hippocampus, the "limbic system").

Buchanan, S.L., & Powel, D.A. (1993). In: Vogt, B.A. & Gabriel, M. *Neurobiology of Cingulate Cortex and Limbic Thalamus.* Boston: Birkhäuser, pp. 381–414.

Butters, N. & Cermak, L. (1980) *Alcoholic Korsakoff's Syndrome: An Information-Processing Approach to Amnesia.* New York: Academic Press.

Carpenter, M.B. and Sutin, J. (1983) *Human Neuroanatomy.* Baltimore: Williams and Wilkins (Chapter 18, "Olfactory Pathways, hippocampal formation and amygdala").

Christianson, S-A., Saisa, J, & Silfvenius, H. (1990) Hemispheric memory differences in sodium amytal testing of epileptic patients. *Journal of Clinical and Experimental Neuropsychology, 12,* 681–694.

Cohen, N.J. (1995) *Memory, Amnesia, and the Hippocampal System.* Cambridge, MA:MIT Press.

Cohen, M. (1992) Auditory/verbal and visual/spatial memory in children with complex partial epilepsy of temporal lobe origin. *Brain and Cognition, 20,* 315–326.

Cummings, J.L. (1993) Frontal-subcoortical circuits and human behavior. *Archives of Neurology, 50,* 873–880.

Damasio, A.R., Graff-Radford, N.R., Eslinger, P.J., Damasio, H., & Kassell, N. (1985) Amnesia following brain forebrain lesions. *Archives of Neurology, 42,* 263–271.

Davis, M. (1992) The role of the amygdala in conditioned fear. In: Aggleton, J. (Ed.) *The Amygdala: Neurobiological Aspects of Emotion, Memory and Mental Dysfunction*. New York: Wiley-Liss Inc., pp. 255–305.

DeFrance, J.F. (Ed.) (1976) *The Septal Nuclei*. New York: Plenum Press.

Degos, J.D., da Fonseca, N., Gray, F., & Cesaro, P. (1993) Severe frontal syndrome associated with infarcts of the left anterior cingulate gyrus and the head of the right caudate nucleus. *Brain, 116*, 1541–1548.

DeJong, R.N., Itabashi, H.H., & Olson, J.R. (1969) Memory loss due to hippocampal lesions. *Neurology, 20*, 339–348.

DeLuca, J. & Cicerone, K. (1989) Cognitive impairments following anterior communicating artery aneurysm. *Journal of Clinical and Experimental Neuropsychology, 11*, 47.

Devinsky, O. & D'Esposito, M. (2004) Emotion and the limbic system. In: Devinsky, O. & D'Esposito, M. (Eds.) *Neurology of Cognitive and Behavioral Disorders*. New York: Oxford, pp. 330–371.

Devinsky, O., & Luciano, D. (1993) The contributions of cingulate cortex to human behavior. In: Vogt, B.A. & Gabriel, M. (Eds.) *Neurobiology of Cingulate Cortex and Limbic Thalamus. A comprehensive handbook*. Boston: Birkhäuser, pp. 527–556.

Devinsky, O., Morrell, M.J., & Vogt, B.A. (1995) Contributions of anterior cingulate cortex to behavior. *Brain, 118*, 279–306.

Diering, S.L. & Bell, W.O. (1991) Functional neurosurgery for psychiatric disorders: A historical perspective. *Stereotactic and Functional Neurosurgery, 57*, 175–194.

Doane, B.K and Livingston, K.E. (1986), (Eds.) *The Limbic System: Functional organization andclinical disorders*. New York: Raven Press.

Doane, B.K. (1986) Clinical psychiatry and the physiodynamics of the limbic system. In, Doane, B.K. & Livingston, K.E. (Eds.) *The Limbic System: Functional organization and clinical disorders*. New York: Raven Press, pp. 285–315.

Downer, J.L. (1961) Changes in visual gnostic functions and emotional behavior following unilateral temporal pole damage in the split-brain monkey. *Nature, 191*, 50–51.

Drachman, D.A., & Adams, R.D. (1962) Acute herpes simplex and inclusion body encephalitis. *Archives of Neurology, 7*, 45–63.

Dum, R.P., & Strick, P.L. (1991) The origin of corticospinal projections from the premotor areas in the frontal lobe. *Journal of Neuroscience, 11*, 667–689.

Ebert, D. & Ebmeier, K.P. (1996) The role of the cingulate gyrus in depression: From functional anatomy to neurochemistry. *Society of Biological Psychiatry, 39*, 1044–1050.

Egger, M.D. & Flynn, J.P. (1963) Effect of electrical stimulation of the amygdala on hypothalamically elicited attack behavior. *Journal of Neurophysiology, 26*, 705–720.

Everitt, B.J. & Robbins, T. (1992) Amygdala-ventral striatal interactions and reward-related processes. In: Aggleton, J. (Ed.) *The Amygdala: Neurobiological aspects of emotion, memory andmental dysfunction*. New York: Wiley-Liss Inc., pp. 401–429.

Faris, A.A. (1969) Limbic system infarction. A report of two cases. *Neurology, 19*, 91–96.

Ferguson, S.M., Rayport, M., & Corrie, W.S. (1986) Brain correlates of aggressive behavior in temporal lobe epilepsy. In: Doane, B.K. & Livingston, K.E. (Eds.) *The Limbic System: Functional organization and clinical disorders*. New York: Raven Press, pp. 183–193.

Foltz, E.L & White, L.E. (1962) Pain 'relief' by frontal cingulumotomy. *Journal of Neurosurgery, 19*, 89–100.

Freemon, F.R. (1971) Akinetic mutism and bilateral anterior cerebral artery occlusion. *Journal of Neurology, Neurosurgery, and Psychiatry, 34*, 693–698.

Frisk, V & Milner, B. (1990) The relationship of working memory to the immediate recall of stories following unilateral temporal or frontal lobectomy. *Neuropsychologia, 28*, 121–135.

Gaffan, D. (1992). Amygdala and the memory of reward. In, Aggleton, J. (Ed.). *The Amygdala: Neurobiological aspects of emotion, memory and mental dysfunction*. New York: Wiley-Liss Inc., pp. 471–483.

Gainotti, G. (Ed.) (1989) Emotional behavior and its disorders. In:Boller, F., & Grafman, J. (Eds.), *Handbook of Neuropsychology*. New York: Elsevier, Vol 3, Section 6.

Gainotti, G., Cappa, A., Perri, R., & Silveri, M.C. (1994) Disorders of verbal and pictorial memory in right and left brain-damaged patients. *International Journal of Neuroscience, 78*, 9–20.

Gilbert, M.E. (1994) The phenomenology of limbic kindling. *Toxicology and Industrial Health, 10*, 343–358.

Gloor, P. (1972) Temporal lobe epilepsy: Its possible contribution to the understanding of the functional significance of the amygdala and of its interactions with the neocortical-temporal mechanisms. In: Eleftheriou, B.E. (Ed.) *The Neurobiology of the Amygdala*. New York: Plenum Press, pp. 423–457.

Gloor, P. (1990) Experiential phenomena of temporal lobe epilepsy: facts and hypothses. *Brain, 113,* 1673–1694.

Gloor, P. (1991) Neurobiological substrates of ictal behavioral changes. In: Smith, D., Treiman, D., Trimble, M. (Eds.), *Advances in Neurology, 55,* 1–34.

Gloor, P. (1997) *The Temporal Lobe and Limbic System.* New York: Oxford University Press. .

Gloor, P., Olivier, A. Quesney, L.F. Andermann, F. & Horowitz, S. (1982) The role of the limbic system in experiential phenomena of temporal lobe epilepsy. *Annals of Neurology, 12,* 129–144.

Goddard, G.V., McIntyre, D.C., & Leech, C.K. (1969) A permanent change in brain function resulting from daily electrical stimulation. *Experimental Neurology, 25,* 295–330.

Goldberg, E. (1984) Papez's circuit revisited: Two systems instead of one? In: Squire, L. & Butters, N. (Eds.), *Neuropsychology of Memory.,* N.Y.: New York: Guilford Press, pp. 183–193.

Gorman, J.M., Liebowitz, M.R., Fyer, A.J., & Stein, J. (1989) A neuroanatomical hypothesis for panic disorder. *American Journal of Psychiatry, 146,* 148–161.

Gray, J.A. (1995) A model of the limbic system and the basal ganglia: Applications to anxiety and schizophrenia. In: Gazzaniga, M.S. (Ed.), *The Cognitive Neurosciences.* Cambridge, MA: MIT Press, pp. 1065–1176.

Hay, P., Sachdev, P., Cumming, S., Sidney-Smith, J.S., Lee, T., Kitchener, P. & Matheson, J. (1993) Treatment of obsessive-compulsive disorder by psychosurgery. *Acta Psychiatrica Scandinavica, 87,* 197–207.

Haymaker, W. and Anderson, E. (1993) Disorders of the hypothalamus and pituitary gland. In: Joynt, R. (Ed.), *Clinical Neurology.* Philadelphia: J.B. Lippincott.

Haymaker, W., Anderson, E., & Nauta, W.J.H. (1969) *Hypothalamus.* Springfield, IL: Thomas Publishers.

Haznedar, M.M., Buchsbaum, M.S., Hazlett, E.A., Shihabuddin, L., & Siever, L.J. (2004) Cingulate gyrus volume and metabolism in the schizophrenia spectrum. *Schizophrenia Research, 71,* 249–262.

Heath, R.G. (1959) *Studies in Schizophrenia.* Cambridge, MA: Harvard University Press.

Heilman, K. & Satz, P. (1983) *Neuropsychology of Human Emotion.* New York: Guilford Press.

Heimer, L. (2003) A new anatomical framework for neuropsychiatric disorders and drug abuse. *American Journal of Psychiatry, 160,* 1726–1739.

Hermann, B. & Chambria, S. (1980) Interictal psychopathology in patients with ictal fear. *Archives of Neurology, 37,* 667–668.

Hermann, B. and Whitman, S. (1984) Behavioral and personality correlates of epilepsy: A review, methodological critique and conceptual model. *Psychological Bulletin. 95,* 451–497.

Holstege, G. (1992) The emotional motor system. *European Journal of Morphology, 30,* 67–79.

Horel, J.A. (1978) The neuroanatomy of amnesia: A critique of the hippocampal memory hypothesis. *Brain, 101,* 403–445.

Horel, J.A., Keating, E.G., & Misantone, L.J. (1975) Partial Kluver-Bucy syndrome produced by destroying temporal neocortex or amygdala. *Brain Research, 94,* 347–359.

Hyman, B.T., Van Hoesen, G.W., & Damasio, A.R. (1990) Memory-related neural systems in Alzheimer's disease: An anatomic study. *Neurology, 40,* 1721–1730.

Insel, T.R. (1992) Toward a neuroanatomy of obsessive-compulsive disorder. *Archives of General Psychiatry, 49,* 739–744.

Isaacson, R.L. (1982) *The Limbic System.* New York: Plenum Press.

Isaacson, R.L., & Pribram, K.H. (1986) *Hippocampus,* Vol. 3 & 4. New York: Plenum Press.

Jenike, M., Baer, L., Ballantine, T., Martuza, R., Tynes, S., Giriunas, I., Buttolph, L. & Cassem, N. (1991) Cingulotomy for refractory obsessive-compulsive disorder. *Archives of General Psychiatry, 48,* 548–555.

Jones, B. & Mishkin, M. (1972) Limbic lesions and the problem of stimulus-reinforcement associations. *Experimental Neurology, 36,* 362–377.

Joseph, R. (1990) The limbic system: Emotion, laterality, and unconscious mind. In: Joseph, R. (Ed.), *Neuropsychology, Neuropsychiatry and Behavioral Neurology.* New York: Plenum Press, pp. 87–137.

Joseph, R. (1992) The limbic system: Emotion, laterality, and unconscious mind. *Psychoanalytic Review, 79,* 405–456.

Kaplan, R.F., Meadows, M.E., Verfaelie, M., Kwan, E., Ehrenberg, B.L., Bromfield, E.B. & Cohen, R.A. (1994) Lateralization of memory for the visual attributes of objects: Evidence from the the the posterior cerebral artery amobarbital test. *Neurology, 44*, 1069–1073.

Kato, N. (Ed.) (1995) *Hippocampus: Function and clinical relevance: Proceedings of the Satellite Symposium of the 4th IBRO World Congress of Neuroscience.* New York: Elsevier Science.

Ketter, T.A., George, M.S., Kimbrell, T.A., Benson, B.E., & Post, R.M. (1996) Functional brain imaging, limbic function, and affective disorders. *The Neuroscientist, 2*, 55–65.

Kling, A.S. & Brothers, L.A. (1992) The amygdala and social behavior. In: Aggleton, J. (Ed.), *The Amygdala: Neurobiological aspects of emotion, memory and mental dysfunction.* New York: Wiley-Liss Inc., pp. 353–377.

Kluver, H. & Bucy, P.C. (1939) Preliminary analysis of functions of the temporal lobes in monkeys. *Archives of Neurology and Psychiatry, 42*, 979–1000.

Kotter, R. & Stephan, K.E. (1997) Useless or helpful? The "limbic system" concept. *Reviews in the Neurosciences, 8*, 139–145.

Kotter, R. & Meyer, N. (1992) The limbic system: A review of its empirical foundation. *Behavioral Brain Research, 52*, 105–127.

Kunishio, K. & Haber, S.N. (1994) Primate cingulostriatal projection: Limbic striatal versus sensori-motor striatal input. *The Journal of Comparative Neurology, 350*, 337–356.

Laplane, D., Degos, J.D., Baulac, M., & Gray, F. (1981) Bilateral infarction of the anterior cingulate gyri and of the fornices. Report of a case. *Journal of the Neurological Sciences, 51*, 289–300.

LeDoux, J.E. (1989) Cognitive-emotional interactions in the brain. *Cognition and Emotion, 3*, 267–289.

LeDoux, J.E. (1991) Emotion and the limbic system concept. *Concepts in Neurosciences, 2*, 169–199.

LeDoux, J.E., Cicchetti, P., Xagoraris, A., & Romanski, L.M. (1990) The lateral amygdaloid nucleus: Sensory interface of the amygdala in fear conditioning. *Journal of Neuroscience, 10*, 1062–1069.

Levin, B. & Duchowny, M. (1991) Childhood obsessive-compulsive disorder and cingulate epilepsy. *Biological Psychiatry, 30*, 1049–1055.

Loring, D.W., Lee, G.P., Meador, K.J., Smith, R., Martin, R.C., Ackell, A.B., & Flanigin, H.F. (1991) Hippocampal contribution to verbal recent memory following dominant-hemisphere temporal lobectomy. *Journal of Clinical and Experimental Neuropsychology, 13*, 575–586.

Luria, A.R. (1966) *Higher Cortical Function in Man.* New York: Basic Books.

MacLean, P. (1964) Psychosomatic disease and the "visceral brain": Recent developments bearing on the Papez theory of emotion. In: Isaacson, R. (Ed.), *Basic Readings in Neuropsychology.* New York: Harper and Row, pp. 181–211.

MacLean, P. (1986) Culminating developments in the evolution of the limbic system: The thalamocin-gulate division. In: Doane, B.K. & Livingston, K.E. (Eds.), *The Limbic System: Functional organization and clinical disorders.* New York: Raven Press, pp. 1–28.

MacLean, P. (1993) Perspectives on cingulate cortex in the limbic system. In: Vogt, B.A., & Gabriel, M. (Eds.), *Neurobiology of Cingulate Cortex and Limbic Thalamus.* Boston: Birkhauser. pp. 1–15.

Malamud, N. (1967) Psychiatric disorder with intracranial tumors of the limbic system. *Archives of Neurology, 17*, 113–123.

Mark, V.H. & Ervin, F.R. (1970) *Violence and the Brain.* New York: Harper & Row.

Mark, L.P., Daniels, D.L., & Naidich, T.P. (1993). The fornix. American *Journal of Neuroradiology, 14*, 1355–1358.

Markowitsch, H.J. (1988) Diencephalic amnesia: A reorientation toward tracts? *Brain Research Review, 13*, 351–370.

Maslow, A. (1954) *Motivation and Personality.* New York: Harper.

Mayberg, H.S. (1997) Limbic-cortical dysregulation. A proposed model of depression. In: Salloway, S, Malloy, P, and Cummings, J.L. (Eds.), *The Neuropsychiatry of Limbic and Subcortical Disorders.* Washington, DC:, American Psychiatric Press, Chapter 12, pp. 167–177.

Mayberg, H.S., Brannan, S.K., Tekell, J.L., Silva, A, Mahurin, R.K., McGinnis, S, & Jerabek, P.A. (2000) Regional metabolic effects of fluoxetine in major depression: Serial changes and relationship to clinical response. *Society of Biological Psychiatry, 48*, 830–843.

Mayberg, H.S., Lozano, A.M., Voon, V., McNeely, H.E., Seminowicz, D., Hamani, C., Schwalb, J.M., & Kennedy, S.H. (2005) Deep brain stimulation for treatment-resistant depression. *Neuron, 45*, 651–660.

Mazars, G (1970) Criteria for identifying cingulate epilepsies. *Epilepsia, 11*, 41–47.

McGaugh, J.L. (1992) Affect, neuromodulatory systems, and memory storage. In: Christianson, A-S (Ed.), *The Handbook of Emotion and Memory: Research and theory.* Hillsdale, NJ: Lawrence Erlbaum Associates, Inc., pp. 245–268.

McGuire, P.K., Bench, C.J., Frith, C.D., Marks, I/M., Frackowiak, R.S.J., & Dolan, R.J. (1994) Functional anatomy of obsessive-compulsive phenomena. *British Journal of Psychiatry, 164,* 459–468.

Mega, M.S. & Cummings, J.L. (1994) Frontal-subcortical circuits and neuropsychiatric disorders. *Journal of Neuropsychiatry, 6,* 358–370.

Mega, M.S., Cummings, J.L., Salloway, S., & Malloy, P (1997) The limbic system: An anatomic, phylogenetic, and clinical perspective. In: Salloway, S, Malloy, P, and Cummings, J.L. (Eds.), *The Neuropsychiatry of Limbic and Subcortical Disorders.* Washington, DC: American Psychiatric Press, Chapter 1, pp. 3–18.

Mesulam, M. (1985) Patterns in behavioral neuroanatomy: Association areas, the limbic system, and hemispheric specialization. In: Mesulam, M. (Ed.), *Principles of Behavioral Neurology.* Philadelphia: F.A. Davis, Chapter 1, pp. 1–70.

Meyer, D.R., Ruth, R.A., & Lavond, D.G. (1978) The septal cohesiveness effect. *Physiology and Behavior, 21,* 1027–1029.

Milner, B. (1968) Visual recognition and recall after right temporal-lobe excision in man. *Neuropsychologia, 6,* 191–209.

Milner, B. (1972) Disorders of learning and memory after temporal lobe lesions in man. *Clinical Neurosurgery, 19,* 421–446.

Monroe, R.R. (1986) Episodic behavioral disorders and limbic ictus. In: Doane, B.K. & Livingston, K.E.(Eds.), *The Limbic System: Functional organization and clinical disorders*New York: Raven Press, pp. 251–266.

Mufson, E.J. & Pandya, D.N. (1984) Some observations on the course and composition of the cingulum bundle in the Rhesus monkey. *Journal of Comparative Neurology, 225,* 31–43.

Musil, S.Y., & Olson, C.R. (1988) Organization of cortical and subcortical projections to anterior cingulate cortex in the cat. *Journal of Comparative Neurology, 272,* 203–218.

Narabayashi, H,, Nagao, T., Saito, Y., Yoshida, M., & Nagahata, M. (1963) Sterotaxic amygdalotomy for behavioral disorders. *Archives of Neurology, 9,* 1–16.

Nieuwenhuys, R. (1996) The greater limbic system, the emotional motor system and the brain. In: Hoslstege, G., Bandler, R., & Saper, C.B. (Eds.), *Progress in Brain Research,* Vol 107. Elsevier Science B.V.

Nieuwenhuys, R., Voogd, J., & van Huijzen, Christiaan (1988) *The Human Central Nervous System: A synopsis and atlas—Third Revised Edition.* New York: Springer-Verlag.

Nolte, J. (1993) *The Human Brain: An introduction to its functional neuroanatomy.* St. Louis: Mosby Year Book (Chapter 16, "Olfactory and limbic systems").

O'Keefe, J. & Nadel, L. (1978) *The Hippocampus as a Cognitive Map.* New York: Oxford University Press.

Olds, J. (1958) Self-stimulation of the brain. *Science, 127,* 315–324.

Olds, M.E. & Forbes, J.L. (1981) The central basis of motivation: Intracranial self-stimulation studies. *Annual Review of Psychology, 32,* 523–574.

Olson, C.R., & Musil, S.Y. (1992) Topographic organization of cortical and subcortical projections to posterior cingulate cortex in the cat: evidence for somatic, ocular, and complex subregions. *Journal of Comparative Neurology. 237,* 237–260.

Papez, J. (1937) A proposed mechanism of emotion. Archives of Neurology and Psychiatry, 38, 725–744. Reprinted in Isaacson, R. (Ed.) (1984) *Basic Readings in Neuropsychology,* New York: Harper and Row, pp. 87–109.

Paxinos, G. (1975) The septum: Neural system involved in eating, drinking, irritability, muricide, copulation, and activity in rats. *Journal of Comparative and Physiological Psychology, 89,* 1154–1168.

Paxinos, G. (1990) Hippocampal formation. In: Paxinos, G. (Ed.), *The Human Nervous System.* New York: Academic Press, pp. 711–755.

Phillips, S, Sangalang, V. & Sterns, G., (1987) Basal forebrain infarction: A clinicopathologic correlation. *Archives of Neurology, 44,* 1134–1138.

Post, R.M. (1986) Does limbic system dysfunction play a role in affective illness? In: Doane, B.K. & Livingston, K.E. (Eds.), *The Limbic System: Functional organization and clinical disorders.* New York: Raven Press, pp. 229–249.

Reichlin, S., Baldessarini, R.J. & Martin, J.B. (1978) *The Hypothalamus*. New York: Raven Press.

Reynolds, E.H. & Trimble, M.R. (1981) *Epilepsy and Psychiatry*. New York: Churchill Livingston.

Rolls, E.T. (1986) Neural systems involved in emotion in primates. In: Plutchik, R. & Kellerman, H. (Eds.), *Emotion: Theory, research and experience*. Vol. 3. Biological foundations of emotion. New York: Academic Press, pp. 125–143.

Rolls, E.T. (1990) A theory of emotion and its application to understanding the neural basis of emotion. *Cognition and Emotion, 4*, 161–190.

Rolls, E.T. (1995) A theory of emotion and consciousness, and its application to understanding the neural basis of emotion. In:, Gazzaniga, M.S. (Ed.), *The Cognitive Neurosciences*. Cambridge, MA: MIT Press, pp. 1091–1106.

Rose, F.C. & Symonds, C.P. (1960) Persistent memory defect following encephalitis. *Brain, 83*, 195–212.

Routtenberg, A. (1968) The two arousal hypothesis: Reticular formation and limbic system. *Psychological Review, 75*, 51–80.

Rubenstein, E.H. & Delgado, J.M.R. (1963) Inhibition induced by forebrain stimulation in monkey. *American Journal of Physiology, 205*, 941–948.

Saling, M.M., Berkovic, S.F., O'Shea, M.F., Kalnins, R.M., Darby, D.G., & Bladin, P.F. (1993) Lateralization of verbal memory and unilateral hippocampal sclerosis: Evidence of task-specific effects. *Journal of Clinical and Experimental Neuropsychology, 15*, 608–618.

Sawle, G.V., Lees, A.J., Hymas, N.F., Brooks, D.J., & Frackowiak, R.S.J. (1993) The metabolic effects of limbic leucotomy in Gilles de la Tourette syndrome. *Journal of Neurology, Neurosurgery, and Psychiatry, 56*, 1016–1019.

Scoville, W.B. (1954) The limbic lobe and memory in man. *Journal of Neurosurgery, 11*, 64–66.

Scoville, W.B. & Milner, B. (1957). Loss of recent memory after bilateral hippocampal lesions. *Journal of Neurology, Neurosurgery, and Psychiatry, 20*, 11–21.

Sherwin, I. (1981) Psychosis associated with epilepsy. *Journal of Neurology, Neurosurgery and Psychiatry, 44*, 83–85.

Siegel, A. & Skog, E. (1970) Effects of electrical stimulation of the septum upon attack behavior elicited from the hypothalamus in the cat. *Brain Research, 23*, 371–380.

Signoret, Jean-Louis, (1985) Memory and amnesias. In: Mesulam, M. (Ed.), *Principles of Behavioral Neurology*. Philadelphia: F.A. Davis, pp. 169–191.

Speedie, L., & Heilman, K.M. (1982) Amnesic disturbance following infarction of the left dorsomedial nucleus of the thalamus. *Neuropsychologia, 2*, 597–604.

Speedie, L., & Heilman, K.M. (1983) Anterograde memory deficits for visuospatial material after infarction of the right thalamus. *Archives of Neurology, 40*, 183–186.

Spence, S. Silverman, J.A. & Corbett, D. (1985) Cortical and ventral tegmental systems exert opposing influences on self-stimulation from the prefrontal cortex. *Behavioural Brain Research, 17*, 117–124.

Spiers, P.A., Schomer, D.L., Blume, H.W., & Mesulam, M-M. (1985) Temporolimbic epilepsy and behavior. In: Mesulam, M-M (Ed.), *Principles of Behavioral Neurology*. Philadelphia: F.A. Davis, pp. 289–326.

Squire, L.R (1986) Mechanisms of memory. *Science, 232*, 1612–1619.

Squire, L.R., & Zola-Morgan, S. (1993) The medial temporal lobe memory system. *Science, 253*, 1380–1386.

Stark-Adamec, C. & Adamec, R.E. (1986) Psychological methodology versus clinical impressions: Different perspectives on psychopathology and seizures. In: Doane, B.K. & Livingston, K.E. (Eds.), *The Limbic System: Functional organization and clinical disorders*. New York: Raven Press, pp. 217–227.

Strauss, E., Risser, A., & Jones, M.W. (1982) Fear responses in patients with epilepsy. *Archives of Neurology, 39*, 626–630.

Stuss, D. T. & Benson, D.F. (1986) *The Frontal Lobes*. New York: Raven Press.

Sweet, W.H., Ervin, F., Mark, V.H. (1969) The relationship of violent behavior in focal cerebral disease. In: Garattini, S. & Sigg, E. (Eds.), *Aggressive Behavior*. New York: John Wiley & Sons.

Terzian, H, & Ore, G.D. (1955) Syndrome of Kluver and Bucy in man by bilateral removal of temporal lobes. *Neurology, 5*, 373–380.

Tow, P.M. & Whitty, C.W.M. (1953) Personality changes after operations of the cingulate gyrus in man. *Journal of Neurology, Neurosurgery, and Psychiatry, 16*, 186–193.

Trimble, M.R. (1984) Disorders of the limbic system. *Integrative Psychiatry, 2*, 96–102.

Ursin, H., & Kaada, B.R. (1960) Functional localization within the amygdaloid complex in the cat. *Electroencephalography and Clinical Neurophysiology, 12,* 1–20.

Valenstein, E.S. (1977) The practice of psychosurgery: A survey of the literature (1971-1976). In: *Psychosurgery.* The National Commission for the Protection of Human Subjects of Biomedical and Behavioral Research.

Valenstein, E.S. (ed) (1980) *The Psychosurgery Debate: Scientific, legal and ethical perspectives.* San Francisco: W.H. Freeman and Company.

Van Hoesen, G.W., Morecraft, R.J., & Vogt, B.A. (1993) In Vogt, B.A. & Gabriel, M. (1993*) Neurobiology of Cingulate Cortex and Limbic Thalamus.* Boston: Birkhäuser, pp. 249–284.

Verfaellie, M. & Cermak, L.S. (1997) Wernicke-Korsakoff and related nutritional disorders of the nervous system. In: Feinberg, T.E. & Farah, M.J (Eds.), *Behavioral Neurology and Neuropsychology,* New York: McGraw-Hill, pp. 609–619.

Victor, M, Adams, R.D., & Collins, G.H. (1989) *The Wernicke-Korsakoff Syndrome.* Philadelphia: F.A. Davis.

Victor, M., Angevine, J.B., Mancall, E.L., & Fisher, C.M. (1961) Memory loss with lesions of the hippocampal formation. *Archives of Neurology, 5,* 244–263.

Vilkki, J. (1985) Amnestic syndromes after surgery of anterior communicating artery aneurysms. *Cortex, 21,* 431–444.

Vochteloo, J.D. & Koolhaas, J.M. (1987) Medial amygdala lesions in male rates reduce aggressive behavior. *Physiology and Behavior, 41,* 99–102.

Vogt, B.A., Finch, D.M., Olson, C.R. (1992) Functional heterogeneity in cingulate cortex: The anterior executive and posterior evaluative regions. *Cerebral Cortex, 2,* 435–443.

Vogt, B.A. & Gabriel, M. (1993) *Neurobiology of Cingulate Cortex and Limbic Thalamus.* Boston: Birkhäuser.

Vogt, B.A., Nimchinsky, E.A., Vogt, L.J., & Hof, P.R. (1995) Human cingulate cortex: Surface features, flat maps, and cytoarchitecture. The *Journal of Comparative Neurology, 359,* 490–506.

Volpe, B.T., & Hirst, W. (1983) The characterization of an amnestic syndrome following hypoxic ischemic injury. *Archives of Neurology, 40,* 436–440.

Volpe, B.T. & Hirst, W. (1983) Amnesia following the rupture and repair of an anterior communicating artery aneurysm. *Journal of Neurology, Neurosurgery, and Psychiatry, 46,* 704–709.

Whitty, C.W.M, Duffield, J.E., Tow, P.M. & Cairns, H. (1952) Anterior cingulotomy in the treatment of mental disease. *Lancet, 1,* 475–481.

Yakovlev, P.I. (1948) Motility, behavior and the brain: stereodynamic organization and neural correlates of behavior. *Journal of Nervous and Mental Disease, 107,* 313–335.

Zbrozyna, A.W. (1972) The organization of the defense reaction elicited from the amygdala and its connections. In: Eleftheriou, B.E. (Ed.), *The Neurobiology of the Amygdala.* New York: Plenum Press, pp. 597–606.

Zola, S. (1997) Amnesia: Neuroanatomic and clinical aspects. In: Feinberg, T.E. & Farah, M.J. (Eds.), *Behavioral Neurology and Neuropsychology,* New York: McGraw-Hill, pp. 447–461.

Zola-Morgan, S. & Squire, L.R. (1993) Neuroanatomy of memory. *Annual Review of Neuroscience, 16,* 547–563.

Zola-Morgan, S., Squire, L.R., & Amaral, D.G. (1986) Human amnesia and the medial temporal region: Enduring memory impairment following a bilateral lesion limited to field CA1 of the hippocampus. *Journal of Neuroscience, 6,* 2950–2967.

9 THE CEREBRAL CORTEX

CHAPTER OVERVIEW

As noted in the preface, the intended audience for this text primarily was neuropsychologists, behavioral neurologists, and other behavioral scientists with comparable interests. Consequently, our treatment of the cerebral cortex has been greatly expanded relative to most other textbooks on functional neuroanatomy. For convenience, this chapter has been divided into three sections. The first (Part I) provides a brief review of how the cerebral cortex came to be recognized as integral to the expression of complex behavior. Part II will focus on the gross neuroanatomical features of the cortex and related supportive structures, except for the vascular system. The latter will be covered separately in Chapter 10. In addition, Part II will review common syndromes that traditionally have been associated with focal lesions within each cerebral hemisphere. Part III is more theoretical in nature. It attempts to provide a model of brain organization and how the integration of sensory, motor, and emotional or motivational factors result in goal-directed behaviors.

PART I. A BRIEF HISTORY OF LOCALIZATION OF FUNCTION IN THE BRAIN

While the mechanisms by which the brain accomplishes its marvelous feats are still largely a mystery (and probably will remain so for many generations to come), at least it is readily apparent, even to the average person, that the brain itself is central to the thoughts, feelings, and actions that characterize daily experiences. However, this was not always the case. Despite often highly sophisticated advances in other fields in the last 2,000 to 3,000 years (e.g., mathematics), knowledge not only of the operation but even the relative importance of the brain to behavior lagged far behind. As will be pointed out in this section, it was not until the 18th century that the relevance of the brain to purposeful, goal-directed behavior commonly began to be appreciated as a subject of legitimate scientific investigation. Perhaps surprisingly, despite eons of traumatic head injuries, apparently it was not until the following century that specific cognitive functions became clearly identified with focal lesions in the brain. Because it is always interesting to understand the roots of any scientific field of study, Part I presents a brief review of how the brain likely was perceived over the millennia. This begins with the early Egyptians and the Greeks and brings us to the modern era (early 20th century) when questions of *how* the brain operates first began to be asked.

Broadly speaking, the brain came to be widely recognized as important in the mediation of conscious, goal-directed behavior in humans at least since the time of the ancient Greeks. However, it might be noted that the brain's **parenchyma** took second place to the cerebrospinal fluid in order of importance, the former not gaining ascendancy until about the 18th century. Although Galen laid the groundwork for experimental studies on the central nervous system (primarily at the level of the medulla and below) with animals in the second century, the next true systematic, scientific experimental studies involving the cortex did not take place until the early 19th century with Flourens. Most historians seem to trace the real beginning of the application of modern scientific methods to the study of such correlations in humans to the publication of Paul Broca in 1861. Shortly afterward, there was a number of documented associations between various "higher cognitive functions" in humans and specific neuroanatomical sites on the cortex. In reaction to these "localizationists," there was another group of scientists (for whom Lashley became one of the primary spokesperson) who felt it was important to view the functional organization of the brain as being more wholistic. This latter group believed that the brain functioned as a whole, especially when carrying out complex behaviors. As we shall see, elements of both of these approaches are incorporated into most current theories.

ANCIENT CONCEPTS OF THE BRAIN

One of the first recorded mentions of the brain was contained in a papyrus discussing medical observations in ancient Egypt. Although the papyrus itself dates back to at least the 17th century BC, some of the accounts it presents may date as far back as 3500 BC. It described what was apparently an open head trauma that may have produced aphasia (failure to speak, although conscious) and may have led to seizures (extreme "shuddering" when the surface of the brain was touched). Despite this account, the early Egyptians, as well as the Hebrews, most likely attributed the control of most cognitive functions to the heart. As might be suspected, the written records from this period were quite scarce and the first recorded systematic observations of the brain and of the debate as to whether the brain or the heart was central to man's awareness and knowledge came with the Greeks.

Although little is preserved from his writings, Alcmaeon (ca. 500 BC) clearly saw the brain as central to the capacity to appreciate perceptions from the outside world (recognizing the connections between the sense organs and the brain), as well as to thought.

Hippocrates (460 to 370 BC), following the medical traditions of his day, believed that various combinations of the four basic humors were responsible for behavior. He also believed that of the four (air, water, earth and fire), air seemed to be the most critical as our source of knowledge of the world around us. However, it was in the brain that air, in its purest form, was first introduced, and hence, it was the brain that was our ultimate source of knowledge and understanding. It was perhaps during this period that for the first time the brain gained acceptance as the organ of perception and intelligence. Of equal importance was an attempt to lay to rest the misconception that epilepsy (as well as other psychological maladies) was a punishment from the gods. Hippocrates concluded that epilepsy instead was a disease of the brain, even noting the contralateral relationship in focal seizures. This focus on the physical, rather than the spiritual, as a cause of illness in humans obviously was a major impetus for the study of modern medicine. The following is an excerpt from this period:.

One ought to know that on the one hand pleasure, joy, laughter, and games, and on the other grief, sorrow, discontent, and dissatisfaction arise only from the brain. It is especially by it that we think, comprehend, see, and hear, that we distinguish the ugly from the beautiful.... Furthermore, it is by the brain that we are mad, that we rave, that fears and terrors assail us—be it by night or by day—dreams, untimely errors, groundless anxiety, blunders, awkwardness, want of experience. We are affected by these things when the brain is not healthy, that is when it is too hot or too cold, too moist or too dry, or when it has experienced some other unnatural injury to which it is not accustomed. (Hippocrates or the Hippocratic writers)

However, not all the Greek intelligentsia of the time adhered to this cephalocentric theory of humans. Aristotle (384–322 BC), certainly one of the greatest thinkers of his time, provided rather detailed anatomical descriptions of the brain. One major difference between his observations and current knowledge was that he thought the substance of the brain itself was devoid of a supply of blood. But this fit well with his theory regarding the function of the brain. He stated, "for where there is no blood, there in consequence is but little heat. The brain then tempers the heat and seething of the heart." Thus for Aristotle, the heart, not the brain, was the center for perceptions, knowledge, and awareness. The senses were thought to be in more direct connection with the heart than with the brain. The main purpose of the brain was to cool the blood from the heart. He explained the fact that the brain in men is larger than in women because the "region of the heart *and* lung is hotter and richer in blood" in the former.

Erasistratus had observed in the 3rd century BC that the cortex of man was more convoluted than it was in animals and he surmised that this probably was related to man's superior intelligence. He reasoned that it was within the brain, specifically within the ventricles, that the *vital spirits* (humors) of the blood were changed into the *animal spirits*, which traveled through the (hollow) nerves and then could effect muscular contractions through distention of the body of the muscle.

Galen (131–201 AD) disputed Erasistratus's claim of the functional significance of the cerebral convolutions, in part because he saw the pituitary gland and the ventricles as perhaps the most critical parts of the brain. However, following Erasistratus, he also thought it was within these areas that the "vital spirits" of the blood, after being influenced by the humors entering the body through the sense organs, were transformed into the "animal spirits," which in turn were responsible for all nervous system activity. As an anatomist, Galen was quite familiar with brain structure. Among other things, he identified and named the meninges, the corpus callosum, the fornix, the superior and inferior colliculi, the pituitary gland, and 11 of the 12 cranial nerves. He also performed experimental lesions of the spinal cord in animals. However, it was probably in his capacity as physician to the gladiators that Galen became quite familiar with the effects of head trauma and was aware that injuries to the brain (and spinal cord) were related to both physical and behavioral changes. Because of his eminence as a physician and the force of his writings, his views on the roles of humors or spirits as the major force behind the function of the nerves persisted until well into the 18th century.

THE MIDDLE AGES

Following Galen, apparently there was little advancement in the knowledge of brain mechanisms until the Middle Ages. What little was written on the subject suggested the ventricles, rather than the brain substance, continued to be considered the most important determinant of behavior. Apparently, at least in part, advancement in the knowledge of the nervous system was hampered by widespread religious and cultural proscriptions against dissection of the human body. This began to change around the time of Andreas Vesalius, who provided the first detailed book on the anatomy of the nervous system in 1543. Vesalius thought the purpose of the convolutions and sulci was to allow the blood supply of the pia mater to extend into the cortical surface more efficiently. Other than that, he felt they "cannot be compared to anything more happily than to clouds as they are usually delineated by either untrained art students or by schoolboys."

Archangelo Piccolomini first discussed the distinction between the gray and white matter of the cortex in detail in 1586, referring to the former as the cerebrum and to the latter as the medulla. He also was the first to use the term "medulla oblongata." The term "cortex" apparently is traced to C. Bauhin in 1616. Sylvius de le Boe first described the lateral fissure in his writings in 1663, although its discovery was attributed to him some years earlier.

Rene Descartes (1596–1650) primarily is thought of as a philosopher and mathematician, but he also was a student of physiology and was the first to attempt to define a reflex pathway. His book, *De Homine*, published posthumously in 1662, was one of the first European textbooks in physiology. While he adhered to many of the Galenic principles of the functioning of the nervous system (e.g., the presence of the animal spirits flowing through the nerves and muscles that mediated perceptions and motion), Descartres differed with the prevailing notions regarding the site of the *sensorium commune* ["judgment" or "common sense"], which generally had been placed within the ventricles. In his dualistic philosophy on the nature of man, he envisioned the body of humans and animals as operating as a complex machine that was fueled by the "fire of the heart." This machine or body of man and the animal spirits, which served as the basis for its perceptions and guided its muscles, was similar in its general makeup and functioning to that of the animals. The difference between humans and animals was the presence of a spiritual soul that provided man with will, knowledge, and the power of introspection. He recognized that the soul was dependent on the body to gain knowledge (via the senses) and as a means of expressing its will (e.g., through movement and speech). He perceived of the soul as being a unitary entity (unlike the tripartite notions of the Greeks) that resided in the body as a whole, but not in any one part. However, there needed to be a single place where the soul and the body interacted because there is a unity to our perceptions, and therefore, the soul must be a unity. Following Galenic thought, Descartes decided that the brain must be the site of this interaction (since it was here that all the senses convened) and that the pineal body was the center of the soul. Descartes contended that, in addition to being the only unitary structure in the brain (a unity demanded by the nature of the soul), the pineal body also was critically located (at the juncture of the anterior and posterior ventricular system). From here, the pineal body could be influenced by the spirits coming in from the senses, through which knowledge was obtained, and in turn could influence the spirits that went out from the brain via the nerves to the muscles, whereby the will could express itself. Descartes did not believe the soul resided in the pineal body, but rather this was the "seat" or the place where the soul interacted with the body. Thus, Descartes was perhaps the first to attempt to precisely localize function within the brain itself.

Thomas Willis (1621–1675), after whom the **circle of Willis** is named (although several others before him had described it), helped to focus increased attention on the cortex as an important anatomical entity. He apparently was the first to use the terms "hemispheres" and "lobes" as they are currently applied to the cortex. As did Erasistratus almost 2,000 years before, he pointed to the increased complexity of the convolutions of the brain in humans as a sign of increased intelligence. He reinforced this thesis through the study of comparative anatomy, contrasting, for example, the relatively smooth cortex of the bird, with the somewhat more convoluted cortex of quadrupeds (which seemed to be capable of more complex behaviors than birds), and finally with the highly convoluted cortex of man. However, the role he attributed to the cortex was the production of the animal spirits.

Despite this focus on the relevance of the cortex with regard to intelligence, little further attention seemed to be afforded the study of the sulci and gyri until around the beginning of the 19th century when Gall and Spurzheim began their studies in phrenology. Perhaps one notable exception was Emanuel Swedenborg who in the early to mid 1700s noted the correct location of the motor cortex. However, his findings were still based on a glandular theory of neurophysiology and there was no further confirmation of his theory of motor localization in the cortex until the works of Fritsch and Hitzig, more than a century later. Although not involving the cortex, one early contribution to the localization of function that occurred during this period involved Julian Jean Le Gallois (1770–1814), one of the first French experimental physiologists of the 19thcentury. In 1811, he reported that the anatomical substrate for respiratory control resided in the brainstem (in rabbits).

The cerebral convolutions and many of the subcortical structures were depicted fairly accurately in an anatomy atlas by Felix Vicq d'Azyr (whose name is associated with the mammillothalamic tract) in 1786. But it was not until 1829 that Luigi Rolando, for whom the central sulcus is named, suggested that the gyri and sulci in humans were not randomly arranged, but rather followed regular, predictable patterns. It might be noted that he was not including the gyri of the frontal lobes, which he thought were characterized by a great degree of individual variation. While Rolando recognized that certain general functions, such as consciousness and voluntary action, were dependent on the integrity of the cortex, he did not attempt to localize any other behaviors to specific cortical areas. In fact, he espoused the traditional belief that the "sensorium commune," which might be considered the highest cognitive ability in humans, was probably "localized" in the brainstem.

In 1839, Francois Leuret, a French anatomist, began publishing his studies on the comparative anatomy of the cortex of mammals. In his comparative study from mouse to elephant (including sea mammals such as the whale), he noted the relationship between the complexity of gyral patterns and the native "intelligence" of the organism. In addition to noting the relationship between gyral complexities and their relative phylogenetic development, Leuret concluded that the "number, form, arrangement and connections of the cerebral convolutions are not formed by accident; each family of animals has the brain shaped in a determined fashion." Dying before he could complete his task, his work was continued by Louis Pierre Gratiolet (1854) who provided us with most of the terminology used today in delineating the cortical surface anatomy. He provided the current names for and clarified the boundaries of the four cortical lobes[1] and supplied the current names of most of the gyri. It might be noted that the names of the lobes were derived from the names of the individual bones of the calvarium (skull) under which they lay. While adhering to the doctrine of his predecessor regarding the general relationship between complexity and intelligence, no specific functions or roles were ascribed to the lobes or the gyri that he named.

Thus into the early 19th century, the issue of the interaction or relationship between the "mind" and the "body" still was being strongly contested. Certainly, there was no consensus regarding localization of cortical functions. The brain or more specifically the cerebrospinal fluid was thought to be the source of cognitive functions or of the "sensorium commune." These functions generally were thought to result from the humors or animal spirits that pervaded or resided in this structure (the brain and/or ventricles). Furthermore, it was not until the end of the 18th century (1786–1792) that Galvani's work establishing the electrical properties of nervous tissue was published.

Despite, the admonitions of Hippocrates 2,000 years earlier, many "disorders of the mind" were still considered to be the result of *supernatural anomalies* (e.g., demonic possession). However, the notion that "intelligence" and other "cognitive abilities" reside in the brain was beginning to emerge, creating a scientific community that was receptive to the modern approaches to the study of the brain. Benjamin Rush in America, William Tuke in England, and Philippe Pinel in France had begun a movement for reform in the perception and treatment of the insane in their respective countries. They argued that the psychic afflictions suffered by these "insane" individuals were akin to a "disease" of the mind, rather than a result of spiritual corruption or supernatural disfavor. Although not relating psychiatric disturbances to a specific disease or dysfunction of the brain, they laid the groundwork for the discovery that abnormalities of the brain (which in one way or another was generally associated with cognition) might be related to anomalies of behavior.

BRAIN LOCALIZATION: THE EARLY YEARS

It was about this time that Franz Joseph Gall (1758–1828) and his student, Johann Spurzheim (1776–1832), entered into the picture. Gall was a highly respected anatomist who emphasized the relative importance of the gray matter of the cortex and discussed its relationship to the underlying white matter. As the story goes, supposedly as a schoolboy, Gall was struck by what he perceived to be a correlation between certain physiognomical characteristics of the face and head and specific mental abilities among his schoolmates. He reportedly noticed, for example, that individuals with protruding eyes seemed to have excellent memories. Later he tested these hypotheses by studying criminals and psychiatric patients to determine whether particular aspects of the shapes of their heads or faces were related to obvious or "atypical" personality traits. Partially with Spurzheim's urging (who saw a potential market for these ideas), Gall began to study normal individuals as well. He began with the theories of Thomas Reid and Dugald Stewart of the Scottish school of psychology. Reid and Stewart detailed a variety of "faculties," "traits," or "powers" that alone or in combination were thought to explain most of human behavior or personality. Gall reasoned that it was impossible for a "homogeneous" brain to be responsible for such diverse behavioral traits or faculties. Instead, Gall suggested that it was much more reasonable to assume that the brain was composed of multiple, though interconnected organs, each of which was responsible for a separate observed behavioral characteristic. These "organs," in effect, were delineated groups of cells that made up all or portions of the cortical gyri. His theory was based on the following two assumptions. First, if a faculty, power, or trait was prominent or well developed in an individual, the area of the brain responsible for that trait also must be "well developed" or "prominent" (the converse would hold true if the trait was relatively lacking in the individual). Second, if an area of the brain responsible for such a trait was prominent (or lacking in development), that portion of the skull that covered that region likewise would reflect this difference by being more "prominent" (bulging), or in the case of underdevelopment, show a relative depression. That is, the configuration of the skull would reflect the configuration of the underlying brain. Thus, the study of the "bumps" of the brain, or **phrenology**, was born.

Although the theories of Gall and Spurzheim were quickly discredited among most scientific circles, largely a result of the efforts and persuasiveness of Spurzheim, they did produce a large popular following and managed to garner some respect in the scientific community. Numerous phrenological societies and journals flourished in the United States and abroad. The *Journal of Phrenology* lasted until 1912, approximately a century following Gall's original publications. Despite their extremist positions, Gall and Spurzheim's contributions are noteworthy. Gall was the first to ascribe importance to the gray matter of the cortex. He also was the first to suggest that different parts of the cortex made differential contributions to human behavior, particularly higher cognitive functions. This was quite a dramatic and perhaps somewhat courageous departure from the notion of a "sensorium commune" that had persisted since ancient Greece. Because of their insights regarding these new possibilities of brain organization, Gall and Spurzheim are credited with having marked the beginning of the study of brain localization.

In 1824, Flourens did manage to quiet the "phrenology" movement for a while through his experimental work on animals. First, he noted that he was unable to elicit movement in the dog by manual stimulation of the anterior cortex; it was only by stimulating the quadrigeminal plate that a motor response was obtained. What he felt were his most telling experiments, however, were his studies involving the extirpation studies of the cortex of the pigeon. Basically, after removing the cerebral hemispheres of these birds, he noted that the birds remained in what he called a *state of perpetual sleep*. They would open their eyes if disturbed, but then quickly return to their quiescent state. While no spontaneous movements were observed, if food were placed in their beaks they would swallow it and if they were thrown into the air they would fly (although they would bump into walls). Flourens concluded that when deprived of the entire cerebral cortex, these animals lost all higher cognitive abilities, including sensation, judgment, memory, and volition, while retaining basic motor skills. Additional experimentation revealed that if the cerebellum was removed, the animal was no longer capable of coordinated movement. Thus, Flourens concluded that the cerebellum was primarily responsible for all motor activity.

Using careful scientific methods and precise surgical extirpations of the cortex, Flourens sought to determine whether specific functions were localized to discrete areas in the cerebral hemispheres or whether the cortex as a whole subserved specific functions. He concluded the following:

1. If certain sized lesions were made, regardless of its localization on the cortex, no behavioral effects were noted, suggesting that the remaining, intact cortex could adequately carry out all functions.
2. As larger amounts of cortical tissue were removed, behavioral deficits (sensation, perception, judgment, memory, and volition) began to become evident, and with sufficient loss of brain mass might be lost completely. However, he noted that all of the various functions seemed to decline (or be lost) at the same rate, at the same time, again regardless of the particular site of the lesion. He deduced that although all of these functions reside in the cerebral hemispheres, they must be represented equally throughout the cortex. He did seem to recognize that *each of the various sense organs* might have had a distinct localization within the brain.
3. He also noted that when a function was lost as a result of a circumscribed cortical lesion, it frequently would be restored over time, suggesting a sort of *equipotentiality* principle.

In conclusion, Flourens noted that "the cerebral lobes are the exclusive site of sensations, perceptions and volitions [but these functions] concurrently occupy the same areas in these organs. Therefore the ability to feel, to perceive, and to desire constitute only one essentially single faculty."

Next came Jean-Baptiste Bouillaud, a French physician who was a student of phrenology. In opposition to Flourens' conclusions, he noted that disturbances of motor functions were common and well-known sequelae of cerebral lesions in clinical populations. Furthermore, consistent with the claims of Gall and Spurzheim, he noted, in a paper presented before the Royal Academy of Medicine in 1825, that there was a relationship between disturbances of motor speech and the anterior frontal lobes. He did not relate these impressions, however, specifically to the dominant left hemisphere. His assertions moreover were not well received at the time due to the distrust and disdain associated with any theory supporting phrenology. However, 36 years later, Ernest Auburtin, the son-in-law of Bouillaud and himself a physician, argued for his father-in-law's position about the anterior frontal lobes being responsible for speech. He reported a patient with a traumatic cranial defect over the frontal lobes. When gentle pressure was applied to the frontal cortex of this patient, speech was halted, but returned when the pressure was removed. He felt this case supported Bouillaud's original conclusions.

He also cited other previous cases of disturbed language that eventually revealed frontal pathology. He had a second, critically ill patient with progressive loss of speech, despite preserved comprehension and general intelligence. He predicted, with both his theory and reputation on the line, that on autopsy a focal lesion would be found in the frontal cortex. Among those who heard Auburtin's presentation was an anthropologist and physician, Pierre Paul Broca. As it happened, Broca was familiar with a patient who in addition to suffering from epilepsy also had lost his ability to speak many years earlier, repetitively uttering only the word "tan." He was being seen by Broca following progressive right-sided weakness and cellulitis in his right leg. Reportedly, knowing his interest in this area, Broca had consulted Auburtin on the case. Shortly after Auburtin's formal presentation and predictions, Broca's patient died. Broca performed an autopsy and found a lesion in the left frontal region as has been predicted. Broca's findings were presented at the next meeting of the Anthropological Society. Several months later, another of Broca's patients died who also had a history of expressive language deficits and in whom the lesion was more restricted to the third and portions of the second frontal convolution. In his presentation to the Anatomical Society of Paris later that same year, Broca stated more definitively that the "faculty of expressive speech" is a result of lesions in the "anterior lobes" of the brain. While he thought the second or third frontal convolutions were perhaps the more critical sites, he noted that more confirmatory data were necessary. He did manage to point out that his findings nevertheless were inconsistent with (and thus distancing himself from) phrenological theory that placed this faculty much more anteriorly. From this point on, theories of cortical localization were reestablished. Interestingly, during these early presentations, Broca apparently said nothing about hemispheric localization, but by 1863, after collecting eight additional cases, Broca began to suspect the leading role of the left hemisphere in language. However, it was not until 1865 that he reported, after also noting the frequent absence of speech problems with right frontal lesions, the dominance of the left hemisphere for language. At the time he gave credit to Dax who had suggested the same thing in 1836.

Soon, other clinically based findings supporting localization of function were being reported. Hughlings Jackson (1835–1911), who might well be considered the father of modern neurology, added further credence to the localization of speech from his own clinical observations, despite disagreeing with Broca on the nature of aphasia. From his studies of focal epilepsy, he concluded that these focal seizures likely resulted from focal disturbances in the gray matter of the cortex and varied according to the extremity initially involved. While he recognized that certain regions of the brain seemed to be responsible for certain behavioral effects (e.g., speech, movement), he also cautioned against a strict localization approach, that is, the identification of "centers" for various cognitive functions. He suggested that the nervous system was organized on different levels of vertical organization. Therefore, functions could be represented in different ways at different levels, each contributing to the whole in its own unique manner. Needless to say he was a bit ahead of his time and these theories of brain organization or localization were not widely accepted.

Karl Wernicke (1848–1904) published his doctoral dissertation in 1874 in which he suggested two speech "centers": an anterior or **motor center** that was responsible for the expression of speech (Broca's center) and a posterior or **sensory center** that was responsible for the storage or recognition of auditory images. He placed the latter in the superior temporal gyrus. He later also discussed a third type of aphasia (conduction aphasia) that resulted from a lesion that interfered with the connection between these two primary centers. In describing the latter, he appeared to be the first to recognize the behavioral relevance of different regions of the brain being able to work together in a cooperative fashion, rather than there simply being an independent center for each activity. In doing so, he began laying the groundwork for Geschwind's (1965) discussions of disconnection syndromes in the second half of the next century.

Taking another tack, Eduard Hitzig (1838–1907), Gustav Fritsch (1838–1927), and David Ferrier (1843–1928) used experimental methodology (electrical stimulation and ablation) to map out the motor cortex in animals. Hitzig and Fritsch, working together in a bedroom of Hitzig's home, mapped out the motor cortex of dogs using the electric current from a galvanic pile. They discovered experimentally what Hughlings Jackson had arrived at clinically, namely that the more anterior portions of the cortex were associated with motor functions, while the posterior cortices evoked no motor response when stimulated. They also reaffirmed the principle of contralateral representation. To further substantiate their findings, ablation studies later were performed on those cortical areas that elicited motor responses when stimulated. They found that, following such lesions, contralateral motor limb weakness could be demonstrated.

Ferrier basically replicated the work of Hitzig and Fritsch, but carried it a step further by being much more precise and detailed in his technique and observations. Whereas the former generally produced gross movements of the limbs, Ferrier was able to demonstrate the sensitivity and selectivity of very small muscle groups to cortical stimulation, typically stimulating cortical sites only millimeters apart. Working largely with monkeys, he was able to map out in great detail not only what we now refer to as the frontal motor cortex, but was able to demonstrate regions in the parietal and superior temporal regions that also produced discrete motor responses.

Despite continuing opposition, the trend to establish "maps" for the cortical localization for both simple sensorimotor as well as "higher cognitive abilities" continued well into the 20th century. However, caution was being urged against overly simplifying localization, not only as a reaction to phrenology, but by those who thought that brain organization and function might be much more complex. Among those making the latter argument was the prominent neurologist, Hughlings Jackson, and some of the newer antilocalizationists. These included Friedrich Goltz who had arrived at a mass action interpretation after studying the effects of cortical lesions in dogs; Constantin von Monakow who argued for the effects of diaschisis; and Karl Lashley who studied the effects of progressively larger lesions in rats. Illustrative of the push to identify "centers" of cognitive and behavioral traits or capacities was the functional maps of Kleist (1934) (Figure 9–1). Later, others such as Kurt Goldstein recognized what appeared to be the inescapable conclusion that certain functions, particularly sensory and motor functions, were clearly localizable and tried to reconcile the two approaches. For example, he argued that certain capacities (e.g., basic sensory and motor functions) likely were localized in the peripheral portions of the cortex, whereas the more central cortical areas were responsible for the more abstract and categorical aspects of behavior and operated more on an equipotentiality principle.

Thus, by the beginning of the 20th century, the principle of localization was well established, even if not universally accepted. The next step was to determine in greater detail the pattern of this localization, or perhaps more precisely the functional organization of the brain. Were there indeed "centers" for certain behaviors such as writing or reading? If so, what was the nature of the interaction among these various "centers"? Were certain behaviors more dependent on the functioning of the brain as a whole? Was there some unitary principle(s) that reflected the roles or functions of the various lobes? What was the difference between the hemispheres with regard to the control or expression of behavior? Was this difference simply a matter of differences in the type of behaviors being mediated

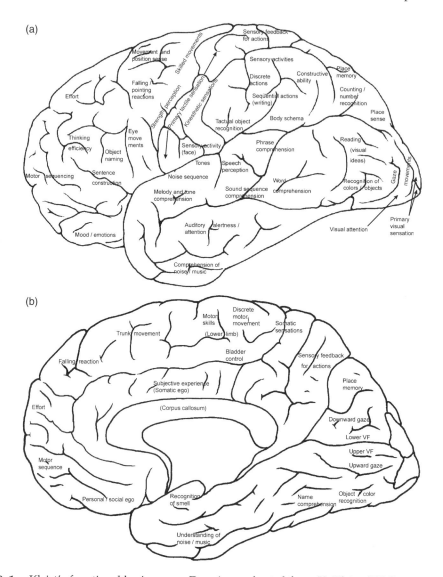

Figure 9–1. Kleist's functional brain maps. Drawings adapted from K. Kleist (1934).

or were the hemispheres organized and did they function in fundamentally different ways? These are the types of questions that characterized much of 20th century neuropsychology and behavioral neurology and which much of Part II will attempt to address.

Endnote

1. Here there seems to be some disagreement in the literature. Some authors attribute the naming of the lobes to Gratiolet. Others suggest that the names of the four major lobes initially were attributable to *Chaussier* (1807) who substituted the names "frontal," "temporal," and "occipital" for the terms "anterior," "inferior," or "medial" and "posterior" divisions first employed by Varolio in 1573, with the delineation of a "superior" (parietal) division being a later development.

CHAPTER 9 ◆ PART II

PART II. NEUROANATOMY AND FUNCTIONAL CORRELATES OF THE CEREBRAL CORTEX

CHAPTER 9, PART II OVERVIEW

As we have progressed from the peripheral and spinal nerves to the diencephalon and limbic structures, the extraordinary complexity of the human nervous system repeatedly has been emphasized. But, as the saying goes, "you ain't seen nothing yet." The cerebral cortex alone probably contains in excess of 10 billion neurons. If we consider that each of these neurons may have as many as 1,000 or more connections, the total number of possible interactions is staggering. When considering the task of trying to understand anything of such cosmic proportions, one British neurologist reportedly observed that if our brain were simple enough for us to understand, then we would be too simple to understand it. In Part I of this chapter, a brief history of the evolution of the appreciation of the brain as the organ responsible for behavior and cognition was presented. With the recognition that we have barely scratched the surface in our understanding of this marvelous organ, the remainder of this chapter hopefully will provide the student with a frame of reference with which they might approach the study of brain pathology and its effect on behavior. Toward this purpose, Part II will focus on some of the major internal and external structural features of the cerebral cortex, as well as a general overview of its functional organization and known behavioral correlates (Part III will continue this investigation by exploring how discrete functional areas might interact to produce integrated behavioral responses).

Knowledge of neuroanatomical nomenclature and structural relationships of the brain are essential when discussing pathology or functional correlates. Part II begins by reviewing the support structures of the cerebral cortex, specifically the skull, meninges, and ventricular system. All three are involved in various pathological processes and knowledge of the latter is crucial in the interpretation or understanding of neuroimaging. The other key supporting elements of cortical functions are the vascular and neurochemical systems. Because of the detail required to do justice to these two systems, separate chapters (Chapter 10 and Chapter 11) will be devoted exclusively to them.

The neocortex itself has both natural and somewhat arbitrary divisions. The most obvious of these are the two cerebral hemispheres and the gyri and sulci. The more arbitrary divisions are the four lobes of the brain and the insular cortex. Although only observable under microscopic techniques, for approximately the past 100 years it has been recognized that the various gyri and lobes also are characterized by variations in their cellular structures, variations that appear to reflect functional differences. Also not obvious on gross inspection but clearly visible on either dissection or on neuroimaging are the white matter (axonal) pathways that interconnect the hemispheres, lobes, gyri, as well as provide a means for the neocortex to communicate with subcortical structures and eventually with the peripheral nervous system.

In addition to a review of basic neuroanatomy, Part II begins to explore the functional organization of these various cortical structures. One of the first things we will discuss is what happens when communications are disrupted between one part of the brain and another. While probably relatively common in cases of brain injury, in certain instances such phenomena create classic (*disconnection*) syndromes that not only are useful for purposes of clinical localization but also provide important insights into the functional organization of the brain. We also will be reviewing hemispheric differences, both in terms of anatomy and function. For the most part, it is believed that the brain functions as a whole. That is, both hemispheres and all four lobes of the brain are likely engaged in carrying out most behavioral activities. However, it also seems likely that each hemisphere and each subregion of the brain is making its own unique contribution to the behavior as a whole. In this section, we will discuss what appear to be the relative contributions of each hemisphere to commonly identified behaviors or syndromes. What is going to be of particular relevance here is not simply trying to identify the locus of a lesion based on a particular behavioral syndrome (current imaging techniques generally do that fairly well), but rather knowing what type of behavioral disturbances to explore given an identified lesion.

By the end of this section, the reader should be able to identify the meningeal layers and discuss their functional/pathological significance. Similarly, one should be able to discuss the ventricular system and discuss ways in which it may be of clinical relevance. One also should be able to localize and name the major gyri and sulci in the brain, identify the boundaries (where possible) of each of the lobes, and discuss the major types of white matter pathways within the brain. Finally, the reader should be able to identify and discuss the major behavioral symptoms and syndromes commonly associated with functional disturbances of the right and left hemispheres (assuming typical patterns of dominance).

CEREBRAL ORGANIZATION: AN INTRODUCTION

As was noted in Part I of this chapter, Broca's presentations on aphasia in the 1860s provided the long-needed impetus (and scientific credibility) for the systematic study of the cerebral cortex as the repository for complex mental and emotional behaviors. Over the next 100 years, many neurobehavioral scientists attempted to develop increasingly refined functional maps of the cortex. These functional cortical maps were derived primarily through the use of ablation and stimulation techniques in animals, the study of naturally occurring and surgical lesions in man, electrical recordings, and the tracing of neuroanatomical pathways. By the early part of the 20th century, to many of the "localizationists," the situation must have looked fairly promising. The general parameters of the sensorimotor cortices were fairly well outlined. Many aspects of speech and language had been generally localized to the perisylvian areas of the left hemisphere. In 1924, Gerstmann first identified a quartet of symptoms consisting of finger agnosia, agraphia, right–left confusion, and acalculia which were present in a single patient. By 1930, after having collected additional cases, he was convinced that these four symptoms (later authors often included visual–spatial constructional deficits as a fifth) represented a specific neurobehavioral syndrome associated with lesions of the dominant angular gyrus, a syndrome that still bears his name (**Gerstmann's syndrome**).

Despite these early discoveries, other investigators felt that the promise of the localizationists was not to be easily fulfilled. While recognizing the capacity to identify certain basic motor and sensory "centers," these individuals contended that the more complex the behavior, the more difficult it was to "localize." If the cortical representations of elementary sensory and motor activities were relatively easily identified, attempting to localize one's capacity for abstraction, concept formation, judgment, learning and memory, and other "intellectual" and problem-solving behaviors proved more elusive. Even with regard to speech and language, naming ability appeared to be fairly widely distributed in the left hemisphere.[2] It has been only relatively recently that the "nondominant" hemisphere has been identified as critical in mediating certain affective valences of speech and language (Ross, 1981).

Constructional disability, which will be discussed in greater detail below, represents a good example of the problems in attempting to localize a specific neurobehavioral symptom. Broadly defined as a difficulty in the visual–motor reproduction of two- or three-dimensional spatial patterns, constructional ability initially was thought to represent a relatively circumscribed behavioral skill mediated by the "dominant" or left hemisphere. Later it came to be viewed as a major indicator of "nondominant" or right hemisphere pathology, specifically of the parietal lobe. However, by the 1950s it was well established that constructional disability could result from lesions of either hemisphere and the search was on to qualitatively differentiate hemispheric involvement by types of errors made. Now we seem to recognize that constructional disability is a highly complex behavior that is not readily "localized" to any single hemisphere or lobe of the brain. The more complicated the task, such as copying the Rey–Osterrieth figure or reproducing a difficult pattern of Kohs blocks (see below), the more likely that planning, organization, problem solving, and self-monitoring abilities, in addition to visual–spatial skills will be involved.

Perhaps it was a realization of the inherent complexity of the brain and behavior, as well as the limitations of a strict localization theory, that prompted a number of prominent neurobehaviorists to seek alternate solutions to the secrets of the functional organization of the brain. One such solution was to view the brain from a more wholistic standpoint. While acknowledging that certain more elementary or even intermediate types of behavior might be capable of being "localized" within discrete cortical regions, it was believed that many higher-level cognitive abilities reflected the operation of the brain as a whole and could not be separated into individual components. As we have seen, Karl Lashley, with his theories of **mass action** and **equipotentiality**, probably reflected the epitome of this position. However, others, notably Kurt Goldstein, also were convinced that certain functions, such as abstractive abilities, relied on the integrity of the brain as a whole. This philosophy, in part, seemed to underlie attempts by a number of investigators to develop single or one-dimensional tests of brain injury, particularly from the 1940s into the 1960s.

While both the strict localizationist and more wholistic theories have made important contributions to our understanding of the brain, neither adequately explains all the facts. Nevertheless, by nature we tend to be *chunkers* and *labelers*. Perhaps, in part, because of this proclivity, a localization bias has remained fairly pervasive in our thinking and in our research. This trend is reflected, for example, in many current texts of neuropsychology and behavioral neurology. Under the headings of frontal, parietal, temporal, or occipital lobes, they often list functions, symptoms, or psychometric test results thought to be associated with those particular lobes or specific areas within the lobes of the brain. Despite its limitations when applied to functional neuroanatomy, even a relatively simple localizationist approach has its merits. In addition to providing convenient and more or less easily manageable labels for exceptionally complex phenomena, these synopses or generalizations may serve as guideposts in generating working hypotheses or making certain types of clinical predictions in diagnostic investigations. Assume, for example, that a patient presents with a particular functional deficit that is thought to be associated with a particular region of the brain or that the site of a lesion is already known from imaging studies. In such instances, the clinician normally will want to carefully assess for other functional deficits typically associated with that particular brain region.

The advent of static and functional neuroimaging techniques has provided a renewed impetus in the ongoing search for improved functional maps of the cortex. In contrast to the static neuroimaging techniques, such as computerized tomography (CT) and magnetic resonance imaging (MRI) scans, which simply show the structural status of the nervous system, functional neuroimaging techniques, such as regional cerebral blood flow (rCBF), positron emission tomography (PET), single photon emission tomography (SPET), and functional MRI (fMRI), provide a glimpse into the actual functioning of the brain. One limitation is that functional neuroimaging technology still looks at behavior rather macroscopically. For example, subjects are asked to engage in certain types of mental activities while localized cortical areas with the highest level of glucose utilization at that time are noted. It is then inferred that the cortical area with the greatest glucose utilization is "critical" for the experimental task elicited. However, even such seemingly simple mental activities as naming pictured objects or performing calculations are the result of the integration of a number of more elementary processes.

To use what is obviously a very limited analogy, suppose we look at the brain as a collection of stores. We can allow the different lobes to represent different types of stores. Thus, the frontal lobe could represent a hardware store and the parietal a grocery store. Different areas within each of the lobes may represent sections within that store. Hence, for the frontal lobe, the dorsolateral, orbital, medial, premotor, supplementary motor, and primary motor areas could represent the plumbing, electrical, lumber, gardening, paint, and tool sections of the store. While these distinctions among the various areas may be useful for general descriptive purposes, this methodology fails to identify the individual items (i.e., elementary functions) contained therein. In this one respect, this analogy might approximate our current status with regard to the functional neuroanatomy of the brain. Perhaps with increasing advances in functional scanning technology and through subtraction techniques using overlapping zones, we may continue to refine our functional maps.

However, progress in the quest to understand brain–behavior relationships ultimately may depend not so much on technological advances as in how we define and measure function. Take the example of constructional disability used earlier.[3] As noted, constructional deficits may become manifest as a result of lesions in diverse cortical areas, including the anterior or posterior portions of either hemisphere. In this sense, constructional disability could be described more as a behavioral *symptom*, such as a fever, a common sign reflecting various possible etiologies. In this respect, the manifest behavior (constructional disability) may be the result of the confluence of multiple, more elementary behavioral phenomena such as visual perception, spatial judgment, motor integrity, somatosensory (kinesthetic) feedback, planning ability, self-monitoring capacity, and the simultaneous integration of any combination of these. Additionally, each of these behaviors in turn may represent a constellation of even more elementary "functions." If we begin with the premise that there are highly circumscribed (micro) functions that indeed may reside in or are mediated by very circumscribed areas of the cortex, we then can conceptualize most observable behavior (such as constructional ability) as the final result of the successful integration of all these more elementary "functions" distributed in various parts of the brain. Such an approach would then readily account for the fact that constructional deficits may be observed following lesions to various cortical sites. At the same time, knowing that separate regions of the brain make their own unique contributions to the overall behavior allows for the possibility of qualitative differences in the manifestation of that behavior. In turn, this could help identify the source of the disturbance by an analysis of the particular elementary functional disturbances or nature of the behavioral deficit.

This basically is the approach espoused and initially elaborated by Aleksandr Luria (1966) in his work, *Higher Cortical Functions in Man*, and more recently discussed by Heilman and Valenstein (2003) and Mesulam (2000). On the one hand, Luria's approach can be viewed as a strict localizationist approach, merely attempting to break down behaviors into much smaller chucks (a more microscopic perspective). At the same time however, it incorporates some key wholistic tenets, focusing on the integration of the brain as a whole, or at least multiple parts thereof, in the production of behavior. Although it has been more than 40 years since Luria's original manuscript was published, no subsequent investigations have offered a substantially different model for conceptualizing brain–behavior relationships.

This chapter will discuss the functional anatomy of the cerebral cortex from this general perspective, borrowing heavily from Luria, Mesulam, and other writers. Let the reader be warned from the outset that there are limitations to this approach. Perhaps the main one is its inherent complexity. There are no simple lists of functional–anatomical correlates to memorize. Rather, as was done in previous chapters, the hope is to present a way of looking at the brain that emphasizes both its horizontal (anterior–posterior; right–left) and vertical (cortical–subcortical–subtentorial) organization.

As we go through this chapter, it may seem that there is a somewhat greater ease in discussing left as opposed to right hemispheric functions. Although we experience and respond to the world around us in a multidimensional fashion, we tend to conceptualize and explain those experiences in verbal terms. Furthermore, even though the processing of verbal information by the brain is incompletely understood, at least language itself is perceived as being a very logical, sequential, and orderly phenomena or process, readily lending itself to rational analysis (i.e., it is easy to use words to explain words). In contrast, right or "nondominant" hemisphere functions appear predominately (if not exclusively) nonverbal, intuitive, and gestaltist and as such do not readily lend themselves to verbal explanations or encoding, even if the "insights" were available. Another factor may be that the functions subserved by the dominant left hemisphere are more discretely localized, and therefore behavioral dissociations are easier to isolate and identify than those of the nondominant right hemisphere, where inherent functions, such as affective processing and attention seem to be more widely distributed.

Because of our limited understanding of the brain and its cognitive operations, we recognize that the following discussions may raise more questions than answers. However, it is our hope that reviewing a theoretical model of the function and organization of the brain will promote a deeper analysis of behavior and appreciation of the need to constantly test and refine such models in our clinical practice. However, before beginning to explore this particular model, it may be useful to review some of the more salient structural aspects of the cerebral cortex and their possible behavioral implications.

SUPPORT STRUCTURES OF THE CEREBRAL CORTEX

Although not technically part of the neuroanatomy of the cerebral cortex proper, we shall begin our discussion by reviewing the skull and the roles of the meninges and ventricular system, all of which play a vital role in the normal functioning of the brain. The vascular system and the biochemical aspects of brain function will be covered in separate chapters.

The Skull

Structurally, the brain is really a fairly delicate organ. In its natural state, it has a consistency not too dissimilar from a jellyfish. Collectively, the cranial vault, meninges, and ventricles with their cerebrospinal fluid help preserve the structural integrity of the brain. However, as we shall see, under certain conditions, even these protective structures can have an adverse effect on the brain. The skull provides the first line of protection. Once the individual bony plates that make up the skull are completely fused in early childhood, they offer a fair amount of resistance to outside physical forces that might otherwise traumatize the brain. However, being rather tightly enclosed in this solid structure at times can create its own problems. Two of the more common examples include acceleration–deceleration injuries and increased intracranial pressure. If one is traveling in a linear direction and the skull comes to an abrupt stop (as when the forehead hits the windshield in a motor vehicle accident) the brain tends to keep moving forward, slamming into the interior surfaces of the cranium, producing either coup (on the site of the initial impact) and/or contracoup (produced by the brain rebounding off the inside of the skull opposite the side of the impact) injuries or contusions. Often such injuries are most acute in the frontal and temporal poles where the physical forces are more concentrated given the curvature of the brain and the bony cavity in which it sits.

In the adult brain, there is very little extra space inside the skull and the only "opening" is the foramen magnum at the base of the skull through which the brain stem exits. In situations where there might be swelling of the brain, such as in the case of closed head trauma, large infarcts (which are often accompanied by edema), major hemorrhages, infections, or hydro-cephalus, the brain has virtually no room to expand. The resulting shifting of intracranial content strains axonal connections and puts increased focal pressure on certain parts of the cortical tissue which produces further disruptions of function. If this swelling is of suffi-cient magnitude, especially if it occurs rather rapidly, the medial portions of the temporal lobes can be literally squeezed down through the tentorial notch (see next section). This produces what is referred to as "uncal herniation" through the tentorium which can affect the structures in and around the midbrain, resulting in third nerve deficits, long tract findings or paresis (due to compression of the cerebral peduncles), or symptoms related to involvement of the posterior cerebral arteries. Increased pressure can affect subtentorial structures either directly, as a result of mass lesions in the posterior fossa (i.e., cerebellum, and brain stem), or indirectly by too rapid a reduction in spinal fluid pressure, such as conducting a spinal tap in an individual with increased intracranial pressure. If the medullary centers in the brain stem that control vital functions such as respiration and heart rate are compromised by herniation of the cerebellar tonsils through the foramen magnum, death may result.

The Dura Mater

Within the skull, the brain is further protected by the **meninges**, particularly by the outermost, the **dura mater** (or simply the *dura*), and the middle layer, the **arachnoid**. The innermost layer or the **pia mater** (or *pia*) lies immediately in contact with the brain tissue itself. It is a very delicate, thin membrane that attaches some of the extremely fine blood vessels to the surface of the cortex. By contrast, the dura is a relatively dense, tough fibrous tissue that adheres to the inner surface of the skull. While the dura can easily be cut, it would be extremely difficult to pull apart by hand. It lines the entire cranial vault, except for the area around the midbrain where it forms the tentorial notch. In addition to covering the outer surfaces of the brain, the dura also dips into the longitudinal fissure between the hemispheres, extending to the vicinity of the corpus callosum. This portion is known as the **falx cerebri** (Figure 9–2). The dura also takes another deviation, creating a separation between the cerebellum and the occipital lobes. This latter portion of the dura is called the **tentorium** (tent) **cerebelli** (or simply, the tentorium). Thus the term *supratentorial* refers to the brain above the level of the midbrain or above the tentorium cerebelli and the *posterior fossa*

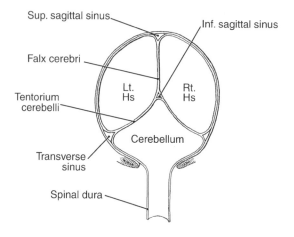

Posterior View

Figure 9–2. Schematic representation of the dura mater providing an external covering for the brain and spinal cord. The figure illustrates the falx cerebri, which forms a partition between the dorsal portions (above the corpus callosum) of the cerebral hemispheres, and tentorium cerebelli, which separates the occipital lobe and the cerebellum in the posterior fossa. Also illustrated is the continuation of the dura into the vertebral foramen of the spinal cord where it surrounds the spinal cord and the spinal nerve roots. Not shown is the tentorial notch, which is an opening around the base of the brain forming a passage for the midbrain. Also shown are several of the venous sinuses that are embedded within the dura.

encompasses those structures that lie below (*subtentorial*) the tentorium cerebelli (i.e., the brain stem and cerebellum). These dural structures help support the brain within the skull.

In addition to this structural support, the dura also serves a critical circulatory function. At several convergence zones that follow the bony contours of the inner table of the skull, the dura separates into two layers creating channels or **sinuses** through which venous blood and cerebrospinal fluid (CSF) flow (Figure 9–3). The major sinuses include the

1. **Superior sagittal sinus:** formed along the superior longitudinal fissure at the top of the brain, it serves to collect blood from surface veins, as well as a point of reabsorption for the cerebrospinal fluid.
2. **Inferior sagittal sinus:** found along the free edge of the falx cerebri, it collects blood from deep cerebral veins.
3. **Straight sinus:** connects the inferior sagittal sinus with the superior sagittal sinus.

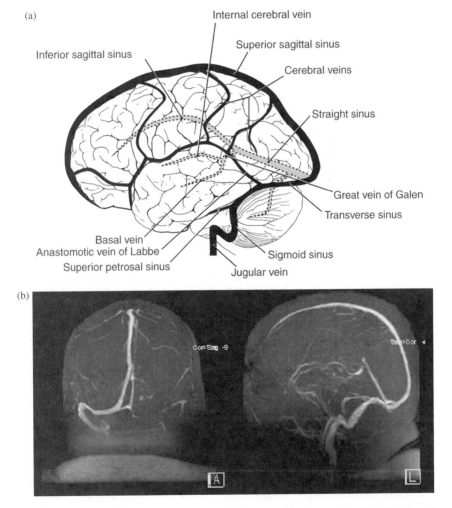

Figure 9–3. (a) Major external (solid, dark) and internal (dotted, lighter) veins draining the cerebral hemispheres. (b) A-P and lateral views of venous drainage on MRA. (a) Adapted from DeArmond, Fusco, and Dewey (1989). (b) Courtesy of Dr. Jose Suros,

4. **Transverse sinuses:** begins at the confluence of the superior sagittal and transverse sinuses and runs laterally along the outer edges of the tentorium cerebelli.
5. **Sigmoid sinuses:** a continuation of the transverse sinuses, they soon empty into the jugular veins, which return the deoxygenated blood to the heart.

While the major types of cerebrovascular events will be reviewed once again in Chapter 10, those that specifically involve the meninges deserve mention here. Two such events related to the dura are **epidural** and **subdural hematomas.** The dura itself is supplied by blood vessels that lie between the dura and the skull. Since the dura attaches to the inside of the skull, there is no space between the two, but there is a *potential epidural space.* If one of the meningeal arteries that supply the dura is ruptured, typically due to skull fractures (less commonly due to laceration of the dural sinus), bleeding can occur between the dura and the skull within this potential space, which results in an **epidural hematoma**. If arterial blood, which is under a fair amount of pressure, is involved, a sizeable hematoma can develop fairly rapidly. Although the blood does not come into direct contact with brain tissue, it can create a rapid and often fatal increase in intracranial pressure (recalling the skull does not allow room for expansion) unless surgically drained. In addition, the hematoma is confined to the epidural space along the skull sutures, as the dura is fixed to the cranial sutures, thus creating a "lens" type appearance on imaging studies. Because the development of this type of hematoma can develop hours (occasionally days) following a head injury, it is important to carefully observe victims of significant head trauma for changes mental status.

A more common, although typically less disastrous, condition can occur when there is a tearing (usually as a result of closed head trauma) of the bridging veins between the arachnoid and the dural sinuses in the subdural space between the dura and the arachnoid. This can result in a **subdural hematoma**, creating a pool of blood between the dura and the arachnoid layers. Since it typically involves venous blood under lower pressure than arterial blood, the bleeding usually is much slower and more self-limited, although the blood can track along the contours of the brain as this space is not restricted by the dural attachments at the sutures. Chronic alcoholics and the elderly are especially prone to subdurals for two reasons. First, with the chronic, heavy use of alcohol or with aging, there tends to be atrophy of the brain that is thought to cause stretching of these bridging veins, making them more vulnerable to the shearing forces of traumatic injury. Second, both alcoholics and the elderly are more prone to head traumas as a result of falls or violence perpetrated against them.

While a subdural hematoma can produce neurological or neuropsychological symptoms depending on its size, it is not uncommon to find evidence of old subdurals on autopsy that never were reported by the patient. This may be due in part to the fact that the cortical atrophy that typically accompanies these conditions also results in more potential space around the brain that can accommodate larger volumes of blood without compression and resultant brain injury.

Other relatively common pathologies involving the dura include meningiomas and infection. **Meningiomas** are very slow growing encapsulated tumors that arise from meningeal tissue, most commonly along the base of the brain or along the falx (see Figure 9–4). As these tumors do not involve the brain parenchyma itself ("extrinsic" tumor), their symptoms often can be rather subtle, especially in the early stages. Their presence commonly is noted because of the pressure they exert on nervous tissue. However, because they are so slow in developing, allowing the brain to compensate for their presence, unless they impinge directly on a cranial nerve, they may grow to considerable size without producing the mass effect (decompensation) created by the more rapidly developing hematomas. **Meningitis**, whether bacterial, viral, or fungal, represents an inflammation or infection of the meninges. These infectious or inflammatory processes can lead to any number of problems, including:

1. Focal swelling of the brain tissue itself (encephalitis).
2. Diffuse, nonobstructive hydrocephalus as a result of inflammation of the subarachnoid layer, which blocks the outflow of CSF (see below).
3. Infarctions secondary to involvement of the blood vessels.
4. Cranial nerve deficits.

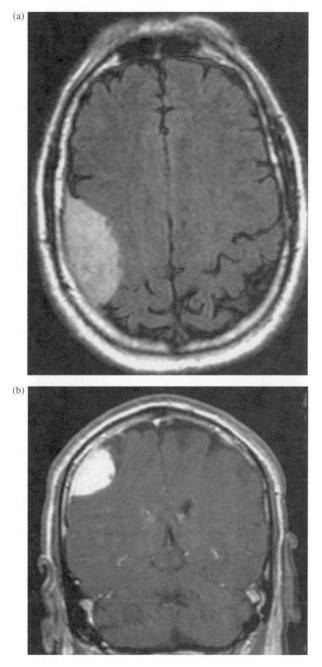

Figure 9–4. (*Continued*)

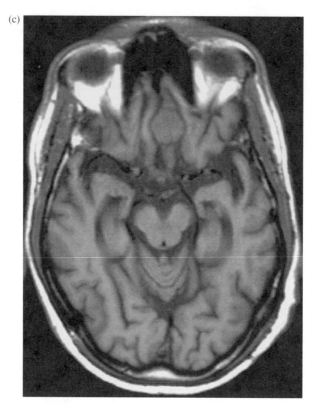

Figure 9–4. (a,b) Meningiomas arising from the convexity and from (c) the olfactory groove.

It should be noted that the dura is pain-sensitive, with the anterior and middle fossa being supplied by the trigeminal nerve and the posterior fossa by the second and third cervical nerves. Brain tissue itself is not pain-sensitive. Thus certain types of headaches that involve intracranial irritants (e.g., bleeds, tumors, infections) generally are thought to result from irritation of the meninges, not the brain itself.

The Arachnoid Layer

The arachnoid represents the middle meningeal layer. It adheres to the inner surface of the dura but creates a space (*subarachnoid space*) between itself and the pia beneath it (Figure 9–5). Since the dura and the arachnoid do not follow the pia into the sulci, the subarachnoid space tends to be greater in these areas. There also is excessive space around certain other concavities of the brain, such as in the area of the peduncles, the corpora quadrigemina, the cerebellar–medullary junction, and the optic chiasm, which are referred to as **cisterns**. The subarachnoid space, in addition to containing most of the surface arteries and veins, contains CSF. The larger cisterns, which therefore contain greater amounts of CSF, show up nicely on both MRI and CT scans. One of the net effects of this fluid-filled subarachnoid layer is that it appears to provide an additional cushion and/or supportive structure for the brain and spinal cord.

  A number of pathological conditions are associated with the subarachnoid layer. Subarachnoid or porencephalic **cysts** or collections of CSF that do not communicate with the ventricular system can develop following trauma, hemorrhage, tumors, or surgery. Clinically, they may present with headaches, seizures, or other focal findings secondary

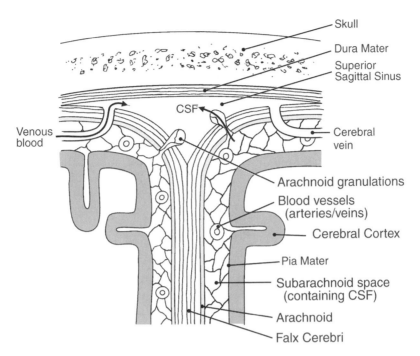

Figure 9–5. Arachnoid space, illustrating relationship to dura and pia and entry for venous blood and CSF into the superior sagittal sinus.

to the pressure they exert on surrounding brain tissue. On occasion such cysts may arise developmentally, usually in the middle fossa (temporal region). Despite possibly growing to great size and associated with agenesis of the temporal lobe, such cysts may be clinically silent and only serendipitously be discovered. Figure 9–6 shows such a lesion in a young man with totally normal physical and mental development and unremarkable findings on neuropsychological examination. Leakage or rupture of surface vessels, such as from

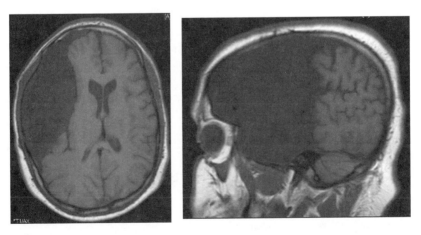

Figure 9–6. Subarachnoid cyst. Despite its huge size, this cyst was virtually asymptomatic clinically, both in terms of cognition and sensorimotor functions. Note the lack of effacement of cortical sulci adjacent to the cyst and the minimal midline shift.

arteriovenous malformations, aneurysms, or other causes, can result in a **subarachnoid hemorrhage**. Since in this case the presence of the blood in the subarachnoid space can produce localized compression of adjacent parenchyma, vasospasms, edema, and absorption delays for the CSF, behavioral effects may be more pronounced and more focal than with subdural hematomas.

The Pial Layer

The pia is the very thin, delicate membrane encapsulating fine blood vessels that closely adheres to the surface of the brain, following all the gyral and sulcal patterns. At times differentiating the pia from the innermost arachnoid layer is difficult, and in fact the two sometimes simply are referred to collectively as the *leptomeninges*. Finally, it should be noted that these same meningeal layers also continue into the vertebral column surrounding the spinal cord.

The Ventricular System

In the course of embryonic development, the central nervous system starts out as a long, hollow neural tube. The cerebral cortex develops at one end of this tube, with the remainder becoming the brain stem, cerebellum, and the spinal cord. However, vestiges of the central, hollow portion of this neural tube remain in the fully developed CNS as the spinal canal and the ventricular system of the brain. The latter consists of the two lateral ventricles, the third and fourth ventricles, and their associated passageways. As we shall see, the fourth

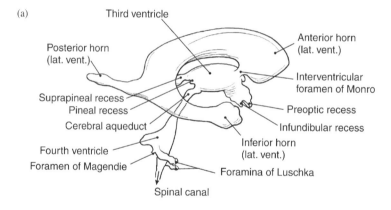

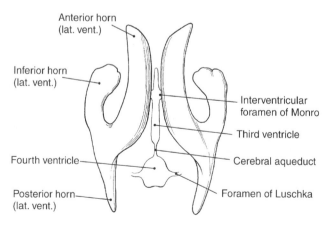

(b) A

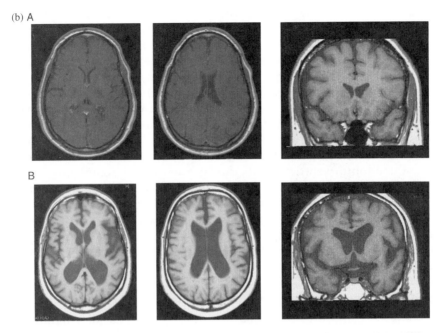

B

Figure 9–7. (a) Representation of ventricles as seen on lateral (top) and superior views. Flow of CSF is from lateral ventricles to the third via the interventricular foramina, then to the fourth via the cerebral aqueduct. From the fourth ventricle, some CSF may enter the spinal canal, while most is extruded into the subarachnoid space in the posterior fossa where it surrounds the exterior surfaces of the brain and spinal cord prior to being picked up by the arachnoid villi and deposited in the venous sinuses. Adapted from Carpenter and Sutin (1983). (b) Normal (top row) and moderately enlarged ventricles (bottom row) as seen on neuroimaging.

ventricle is continuous with the spinal canal. The ventricles, along with the subarachnoid spaces (cisterns) are filled with CSF. As noted earlier, one functional contribution of this system may be to help support and cushion the CNS; however, there also is evidence that the CSF helps maintain proper electrolytic and/or metabolic balance in the extracellular fluids of the brain. For our current purposes, an understanding of the basic anatomy and mechanics of the ventricular system helps to provide a basis for understanding various types of neuropathology.

The largest of the ventricles are the two lateral ventricles located bilaterally in the cerebral hemispheres. As can be seen from Figure 9–7, these may be conceived of as beginning in the frontal lobes on either side of the midline, just beneath the genu of the corpus callosum and are bounded laterally by the head of the caudate nucleus. This area represents the **anterior** or **frontal horns** of the lateral ventricles. As the ventricles proceed posteriorly, they enlarge somewhat (the **body** of the lateral ventricles) and then curve down, out, and forward again into the medial portions of the temporal lobes (the **temporal horns**). That portion of the body of the ventricle that curves to begin forming the temporal horns is called the **trigone** or **atrium**. From these latter regions, two smaller cavities, known as the **posterior or occipital horns**, extend into the parietal–occipital cortices.

On the ventromedial aspect of each of the lateral ventricles, just anterior to the thalamus are two openings, known as the **interventricular foramen** or the **foramen of Monro**, which allow drainage of CSF from the lateral ventricles of the left and right cerebral hemispheres into the single, midline third ventricle (see Figure 6–1c). The third ventricle is surrounded by the thalamus and basically separates the paired nuclei of the thalamus (or thalami) and the hypothalamus, except for a small region (which is present in most, but not all people) where the two thalami are connected: the **massa intermedia**. At the ventral and posterior extent of the third ventricle, there is a small canal that passes down through the midbrain and opens up into the fourth ventricle. This passage, known as the **cerebral aqueduct**, allows for CSF communication between the third and fourth ventricles. (see Figure 4–4a)

The fourth ventricle lies in the brain stem, extending from the level of the pons, slightly into the medulla. For the most part, the floor of the fourth ventricle is made up of the pons and upper medulla, while its roof is the cerebellum (although there are actually two thin sheets of tissue known as the superior and inferior medullary veli which cover much of the dorsal surface of the ventricle). There are several openings of significance in the fourth ventricle. First, the fourth ventricle, at its caudal extension, is continuous with the spinal canal that reflects the original neural tube. Second, there are three openings that allow the CSF to get outside the ventricular system and into the subarachnoid space, the importance of which will be discussed shortly. At the two points where the fourth ventricle widens out in somewhat of a diamond shape, there are two openings: the **foramina of Luschka**, located laterally, and the **foramen of Magendie,** medially located in the caudal portion of the roof of the ventricle.

The Choroid Plexus and Cerebrospinal Fluid Circulation

Within all the ventricles are glandularlike structures known as the **choroid plexus**. The choroid plexus produces most of the CSF. The cerebrospinal fluid, which is the clear fluid that is found both within and around the outside surface of the brain and spinal cord, can play a crucial role in the diagnosis of many types of neuropathology. Analysis of its composition, typically obtained via a spinal tap or lumbar puncture, can assist in the diagnosis of such CNS problems as infections, hemorrhages, tumors, and multiple sclerosis. However, there also can be problems associated with CSF, the most common being problems of blockage and/or reabsorption. Cerebrospinal fluid is constantly being produced, fully replenishing itself about once every 5 to 6 hours. Obviously, the CSF must be excreted in order to accommodate the demand for additional fluid production. To accomplish this process, the fluid produced in the lateral ventricles flows into the third ventricle through the interventricular foramina, and from the third into the fourth via the cerebral aqueduct. Once in the fourth ventricle, it flows into the subarachnoid spaces of the brain stem through the foramina of Luschka and Magendie. From there it can flow into the subarachnoid spaces of the spinal canal or through the dural opening around the brain stem (tentorial notch) into the subarachnoid space over the convexities of the hemispheres, where most of the absorption takes place via the arachnoid villi (or granulations). These villi, which are specialized groups of cells, protrude into the dural sinuses and are particularly dense along the superior sagittal sinus. It is through these villi that the CSF is passed into the venous system and out of the brain (see Figure 9–5).

Problems occur when some of these passages or apertures fail to properly develop at birth or when they later become compressed or occluded (such as might occur with an intracranial mass). This situation can produce a **noncommunicating** (*obstructive*) **hydrocephalus** when the CSF within the ventricles cannot communicate or flow from one ventricle to another, and thus cannot flow out into the subarachnoid space to be reabsorbed. Noncommunicating hydrocephalus results from an obstruction of the flow of CSF from the lateral ventricles, through the aqueduct, or at the outlets of the fourth ventricle. Another type of problem can occur if the problem occurs at the point of production (choroid plexus) or reabsorption by the arachnoid villi. This latter condition is referred to as a **communicating hydrocephalus** (the CSF retaining its capacity to flow or *communicate* with the subarachnoid space). Meningitis and subarachnoid hemorrhages are two of the more common causes of communicating hydrocephalus. Bleeding into the subarachnoid space also can interfere with the ability of the villi to transport CSF. Meningitis can have a similar effect as a result of residual scarring or adhesions. In certain cases, these processes also can interfere with the ability of the CSF to flow from the brain stem cisterns through the tentorial notch (because of adhesions in this area) into the subarachnoid spaces over the hemispheres.

Additional Diagnostic Considerations

Analysis of the shape, size, displacement, or in some cases, absence of the ventricles on CT or MRI scans can have major diagnostic utility. In the days before the CT scan, the pneumoencephalogram was a frequently used diagnostic tool. It involved displacing the CSF in the ventricles with air. The air, being considerably less dense that the surrounding tissue, provided a nice profile of the ventricular system with standard skull X-rays. One of the major drawbacks to this procedure is that it also frequently left the patient with a severe headache that could last for days. Current neuroimaging techniques (CT or MRI scans) provide noninvasive means of visualizing the ventricular system. Although a normal leftward asymmetry of the lateral ventricles has been well documented, significant compression and distortions of the lateral ventricles or one of its horns can suggest either the presence of a mass or "mass effect" lesion (if compressed) or focal atrophy (e.g., from an old infarct) if the ventricle is dilated. This same principle also may help differentiate between a recent versus an old infarct (new infarcts either may compress the ventricle due to edema or leave them relatively unaffected). Lateral displacement of the ventricles from midline also is suggestive of mass effect from a space-occupying lesion. Distortions of the normal shape of the ventricles also can be of diagnostic significance. For example, the anterior horns typically have a boomerang type appearance due to the lateral encroachment by the head of the caudate nucleus. In **Huntington's** disease, in which there is early atrophy of the caudate, the lateral surfaces of the frontal horns will take on an uncharacteristic convex shape.

Enlargement of the ventricles, along with prominent cortical sulci, suggests generalized cortical atrophy. While generalized atrophy often accompanies specific degenerative dementias, of themselves enlarged ventricles are not necessarily diagnostic of a clinically or behaviorally identifiable neuropathological process. It is not uncommon to find large ventricles in the elderly (and occasionally in the young) with no indications of compromised mental status. On the other hand, the presence of enlarged, symmetric ventricles with effacement (diminution) of the cortical sulci may suggest **normal pressure hydrocephalus** in an adult with appropriate clinical symptoms. Hyperdensities ("bright spots" on imaging studies) within the body or temporal horns of the lateral ventricles and in the third ventricle simply may reflect normal calcification of the choroid plexus that occurs with age.

THE CEREBRAL HEMISPHERES

We believe that anatomists first began to examine the human brain at least some 2,500 to 3,000 years ago. One of their first observations must have been to note that the surface of the brain (the cerebral cortex) was divided into two, seemingly identical, hemispheres, each being characterized by extensive, apparently random, convolutions (sulci and gyri). As was seen in Part I of this chapter, it was not until the 19th century that either of these concepts was seriously challenged. First came the realization that some gyral patterns are generally consistent among individuals. By the second half of the 19th century, the functional asymmetry of the hemispheres was becoming accepted in the scientific community. Although anatomical asymmetries between the cortical hemispheres were discussed in the late 1800s, it was not until *Geschwind and Levitsky's* study in 1968 (see below) that the notions of functional and anatomical asymmetries were truly integrated.

The Cerebral Gyri and Sulci

Before discussing hemispheric asymmetries, it may be useful to briefly review some of the structural features of the cerebral hemispheres. Probably the most obvious are the gyri (ridges) and sulci (grooves) on the surface of the brain. Although study of the surface of the human cerebral cortex demonstrates that there is considerable individual variability in patterns of gyrification (morphology) within and between the cerebral hemispheres, the major gyri and sulci generally are similar across individuals, making it possible to identify specific gyri and sulci within the brain. Knowledge of these patterns provides the clinician and scientist with a topographical or structural road map of the lobes of the brain, and it is in these terms that the findings of clinical neuropsychology, behavioral neurology, and functional neuroanatomy frequently are framed. Hence, a familiarity with the names and locations of the major gyri and sulci is crucial to the study of brain–behavior relationships (see Figure 9–8a–d). However, again a word of caution: although one may diligently memorize the diagrams presented here, when looking at representations in other texts or even an actual human brain, one may still find himself or herself struggling to identify certain specific gyri or sulci due to some minor individual variability.

It is this morphology or pattern of sulci and gyri that gives the human brain its characteristic convoluted appearance. Normal aging or certain disease processes, most notably primary degenerative disorders, can cause a loss of neuronal tissue resulting in shrinking of gyri (and consequent enlargement of the sulci), making them appear more prominent and easier to study. Such changes can be easily observed on MRI or CT scan.[4] Before exploring some of the specific sulci and gyri, it is interesting to consider how this particular morphology developed in man.[5] Clearly, the convoluted brain is not unique to humans. Most, if not all, mammalian brains also are characterized by some degree of convolution, albeit very rudimentary in some instances, such as in rodents, while those of the porpoise and the great apes rival that of humans in terms of the number or complexity of gyral patterning. Brain size seems to be largely a function of two rather independent factors: body size and behavioral complexity (*animal intelligence*, if you wish). Both seem to play an important role. In the case of the elephant, for example, although generally considered to be a highly "intelligent" animal, it is not thought to be more intelligent than the human. Its body size, then, would seem to largely account for the fact that its brain is at least three times the size of man's. Yet the brain cavities of some of the dinosaurs, despite their immense bulk, are thought to have been quite small. Thus, once body size is taken into account, behavioral complexity also would seem to be related to the size of the brain and its gyral complexity. The underlying principle would appear to be that "more brain" is necessary in order to carry out more sophisticated and adaptive behaviors. As we shall see later in this chapter, it is those areas of the cerebral cortex that mediate the highest levels of integrative functions (the tertiary association areas) that show the greatest amount of relative development in man. The convoluted cortex thus would allow for increasingly larger cortical surface area and more complex brain morphology that still fits into a manageable-sized head. It might be interesting also to note that as the human fetus develops there is a corresponding increase in gyral development, another instance where ontogeny recapitulates phylogeny.

Both gyri and sulci may vary in size (e.g., length, depth, or breadth). However, regardless of size, a gyrus is still referred to as a gyrus, not so with sulci. A very large (deep) sulcus is referred to as a fissure. There are only two such sulci associated with the cerebral cortex. The first is the very deep fissure that runs longitudinally between the two cerebral hemispheres known as the **superior longitudinal fissure**, or the **superior sagittal fissure**. The other is the **lateral** or **Sylvian fissure** (after Sylvius de le Boe who described it in the 17th century). Present bilaterally, the lateral fissure represents the enfolding of the frontal and parietal cortices along its dorsal margin and the temporal cortex ventrally. Its anterior end is readily identified as the division between the temporal and frontal lobes. The posterior extension of the lateral fissure is surrounded by the **supramarginal gyrus**, which is how the latter

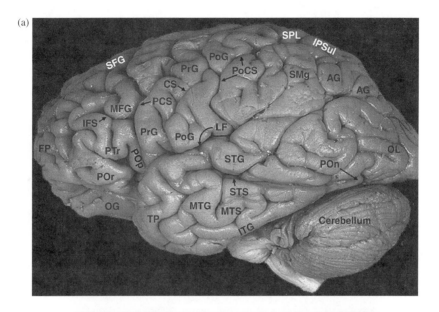

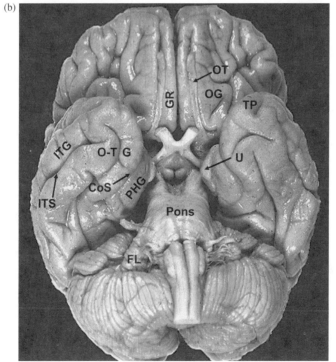

Figure 9–8. *(Continued)*

(c)

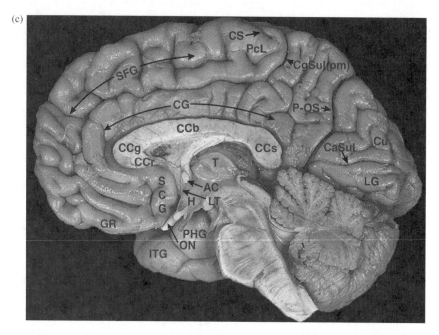

Figure 9–8. Lateral, ventral, and medial surfaces of the brain with major gyri and sulci. Brain images were adapted from the *Interactive Brain Atlas* (1994), courtesy of the University of Washington.

Lateral, Ventral & Medial Surfaces of the Brain

Gyri	Sulci	Other Features
AG, angular gyrus	CS, central sulcus	AC, anterior commissure
C, cuneus	CaSul, calcarine sulcus	CCb, body of corpus callosum
CG, cingulate gyrus	CgSul, cingulate sulcus	CCg, genu of the corpus
Fl, flocculus	CgSul$_{(pm)}$, cingulate sulcus, pars	callosum
GR, gyrus rectus	marginalis	CCr, rostrum of corpus callosum
ITG, inferior temporal gyrus	CoS, collateral sulcus	CCs, splenium of corpus
LG, lingual gyrus	IFS, inferior frontal sulcus	callosum
MFG, middle frontal gyrus	Ipsul, intraparietal sulcus	F, fornix
MTG, middle temporal gyrus	ITS, inferior temporal sulcus	FP, frontal pole
OG, orbital gyrus	LF, lateral fissure	H, hypothalamus
OTG, occipitotemporal (fusiform) gyrus	MTS, middle temporal sulcus	LT, lamina terminalis
PC, precuneus	PoCS, postcentral sulcus	OC, optic chiasm
PcL, paracentral lobule	P-OS, parietooccipital sulcus	OL, occipital lobe
PHG, parahippocampal gyrus	PrCS, precentral sulcus	ON, optic nerve
PoG, postcentral gyrus	STS, superior temporal sulcus	OT, olfactory tract
POp, inferior frontal gyrus, pars opercularis	SulCC, sulcus of the corpus	POn, preoccipital notch
POr, inferior frontal gyrus, pars orbitalis	callosum	SPL, superior parietal lobule
PrG, precentral gyrus		T, thalamus
PTr, inferior frontal gyrus, pars triangularis		TP, temporal pole
SCG, subcallosal (parolfactory) gyrus		U, uncus
SFG, superior frontal gyrus		
SMg, supramarginal gyrus		
STG, superior temporal gyrus		

may be identified (see Figure 9–8a). The cortical areas around the rim of the fissure and penetrating into its depths are known as the **opercular cortex** (e.g., *frontal operculum*, *temporal operculum* or *parietal operculum* depending on the particular region). It is within the temporal opercular region that the **Heschl's gyrus** lies, the primary auditory projection cortex and the **planum temporale**, both of which become important in the discussion of language. As seen in Figure 9–9, the lateral (Sylvian) fissure in a typical adult human brain is composed of five segments:

1. Anterior horizontal ramus extending from A to B (AHR)
2. Anterior ascending ramus from B to C (AAR),
3. Posterior horizontal ramus from B to D (PHR)
4. Posterior ascending ramus from D to F (PAR)
5. Posterior descending ramus from D to E (PDR)

The central sulcus (see below) also can be used to divide the PHR of the Sylvian fissure into a pre- and postcentral segment. These divisions of the lateral fissure are important in relation to functional cortical regions and asymmetries, which will be discussed later. If the lateral fissure is pried apart (or if you look at either an axial or coronal section), deep within the fissure is found a section of "hidden" cortex called the **insula** (see Figure 9–11).

Of other major sulci worthy of special comment (see Figure 9–8), one is the **central sulcus** (the older eponym, **sulcus of Rolando**, is seldom used today). The central sulcus (CS) demarcates the **precentral** (primary motor) from the **postcentral** (primary somatosensory) **gyrus** and represents the boundary between the frontal lobes and parietal lobes. On a postmortem brain, the central sulcus can be recognized by identifying the two adjacent gyri in the central region of the lateral cortex, which are primarily oriented in a vertical direction. These are the pre- and postcentral gyri with the central sulcus between them.

Within the parietal lobe, approximately two thirds of the way up is a horizontally oriented sulcus (not always easy to find). This is the **intraparietal sulcus** which separates the superior parietal lobule from the inferior parietal lobule. The inferior parietal lobule includes both the supramarginal gyrus and angular gyrus, both of which are prominently mentioned in neuropsychological literature.

In pursuing the surface anatomy of the cortex from the lateral to the inferior surface of the temporal lobe, several horizontal (although not necessarily continuous) sulci are

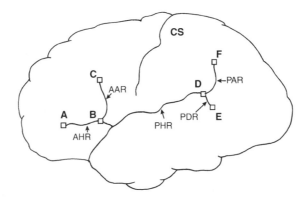

Figure 9–9. Schematic view of various branches of the lateral (Sylvian) fissure. Abbreviations: AAR, anterior ascending ramus; AHR, anterior horizontal ramus; PAR, posterior ascending ramus; PDR, posterior descending ramus; PHR, posterior horizontal ramus. As will be seen later, the angle of the posterior ramus typically differs in the right versus left cerebral hemispheres.

encountered. Respectively, these are the **superior** (STS), **medial** (MTS), and **inferior temporal sulci** (ITS) and the **collateral sulcus**. If one follows the superior temporal sulcus back as it curves up into the inferior parietal lobule, it becomes surrounded by the **angular gyrus.** The **superior, middle,** and **inferior temporal gyri** each lie immediately above their respective sulci. Thus, the superior temporal gyrus lies between the superior temporal sulcus and the lateral fissure (sulcus), the middle temporal gyrus between the superior and middle temporal sulci, and the inferior temporal gyrus between the middle and inferior sulci. Between the inferior temporal and collateral sulci the **occipitotemporal** or **fusiform gyrus** lies. Finally, just medial to the collateral sulcus is the **parahippocampal gyrus**, located on the ventromedial surface of the temporal lobe.

On the lateral surface of the frontal lobes are two horizontal sulci, the **superior** and **inferior frontal sulci**, which separate the cortex of the anterior frontal convexity into the superior, middle, and inferior frontal gyri. The inferior frontal gyrus is composed of the **pars orbitalis, pars triangularis,** and **pars opercularis** (the latter two comprising **Broca's area**). Generally, these two frontal sulci and their three divisions are much easier to see on coronal sections rather than on the lateral surface.

On the medial surface, the fairly deep **parietal–occipital sulcus** marks the boundary between the parietal and occipital lobes. The **callosal** and **cingulate sulci** also are quite prominent above the corpus callosum and **cingulate gyrus**, respectively. Finally, the **calcarine sulcus** also rather easily can be identified. It divides the primary visual cortex of the occipital lobe into its more dorsal portion, the **cuneus** (the term "gyrus" is not used here) and **lingual gyrus** located on its ventral bank.[6]

Lobes of the Cerebral Cortex

The cerebral cortex (neocortex) of the telencephalon (forebrain) consists of four distinct lobes within each hemisphere: the frontal, parietal, temporal, and occipital lobes. Developmentally, the skull in which the brain is housed is formed from a number of separate bones that eventually fuse to form the solid, bony covering of the brain. Even casual inspection of an adult skull reveals the suture lines formed by the union of these bones. As noted in Part I of this chapter, it is commonly accepted that when Gratiolet provided us with the modern names for these lobes in the mid-19th century, they were so named because of their location under the respective bones of the skull. Perhaps, in part, because of this rather arbitrary designation, initially no attempt was made to attach any specific functional significance to the various lobes. Over time, however, there has been a natural tendency to assign unique behavioral phenomena to each of the lobes, a process that has met with some limited success. The general anatomical boundaries of the four lobes are depicted in Figure 9–10.

The Frontal Lobe

The frontal lobes are the largest of the human brain, comprising approximately one third of its lateral surface. The frontal lobe is the only lobe of the brain for which fairly clear anatomical boundaries exist. Its anterior, superior, and anterior–inferior boundaries are defined by the respective limits of the cranial vault. Its posterior limit is the central sulcus. The inferior portion of the frontal lobe clearly is separated from the temporal lobe by the lateral (Sylvian) fissure. Like the parietal and occipital lobes, its medial surface extends down into the superior longitudinal fissure. Both structurally and functionally, the frontal lobe has been divided into a number of distinct but often overlapping areas. As noted above, the frontal lobe can be divided horizontally by the superior and inferior frontal sulci into the superior, middle, and inferior frontal gyri. Vertically, the frontal lobe can be divided into a posterior portion, often termed simply the *frontal* or *frontal motor* cortex and a larger, anterior portion that is called the *prefrontal* cortex. This division is based on similar cytoarchitectural structure in these respective regions. The "frontal" region consists primarily of *agranular* cortex (enhancement of layers III and V), while relative enlargements of layers II and IV (frontal granular cortex) characterize the "prefrontal" region (see below). It should be noted that even within these two general cytoarchitectural categories there are multiple subdivisions that are thought to have unique functional properties. It is common,

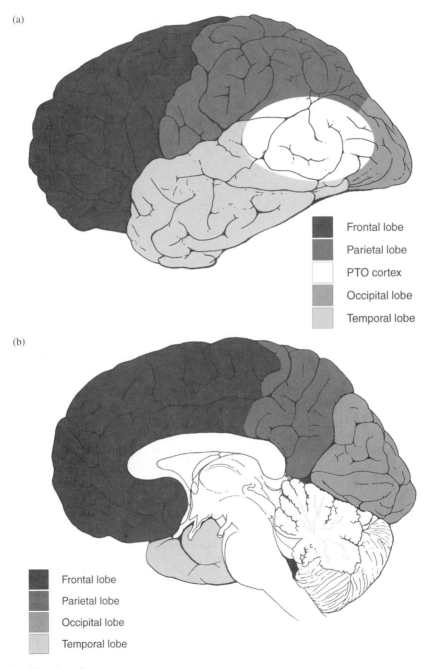

(a)

Frontal lobe
Parietal lobe
PTO cortex
Occipital lobe
Temporal lobe

(b)

Frontal lobe
Parietal lobe
Occipital lobe
Temporal lobe

Figure 9–10. (*Continued*)

(c)

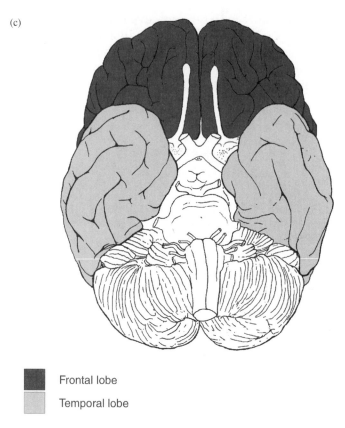

■ Frontal lobe

□ Temporal lobe

Figure 9–10. Lobes of the cortex. Figures illustrate the relative positions of the four lobes on the (a) lateral, (b) medial, and (c) ventral surfaces of the brain. Note the white area on Figure 9–10a. Because the boundaries of the parietal, temporal, and occipital cortices are indistinct as they converge on the lateral surface, this area is sometimes referred to as "PTO" cortex (see text below).

for example, to distinguish between dorsolateral, orbital and inferiomedial, and dorsomedial areas of the frontal granular cortices.

Albeit overly simplistic, for now the frontal lobes usually are described as playing a critical role in carrying out the *executive* functions of the brain. The more anterior portions of the frontal lobes, particularly the dorsolateral cortices, are thought to be primarily responsible for higher-order behaviors, for example, deciding *when, where, why, how,* and *if* one should respond in certain situations. The orbital and inferiomedial frontal cortices appear to be important in processing and/or modulating internal drive states, whereas the dorsomedial granular cortex appears crucial for maintaining optimal arousal and motivation. The primary motor and motor association cortices are thought to be responsible for the final organization, control or modulation, and implementation of the actual motoric response.

The Parietal Lobes

The anterior extent of the parietal lobe is demarcated by the central sulcus, which also denotes the posterior boundary of the frontal lobe. On its medial surface, the parietal lobe is separated from the occipital lobe by the parietooccipital sulcus. The separation of the parietal from the occipital lobe on the lateral convexity can be approximated by imagining a line extending from the parietooccipital sulcus on the medial surface to the preoccipital notch. The latter is a small sulcal indentation on the ventrolateral surface of the brain that also approximates the posterior extent of the temporal lobe. Except for at its more anterior extent along the posterior horizontal ramus (PHR) of the lateral fissure, it is difficult to differentiate the boundary between the parietal and temporal lobes (see discussion of PTO cortex below). The posterior–inferior extent of the parietal lobe on the lateral surface of the brain can be approximated by extending an imaginary line from the PHR of the lateral (Sylvian) fissure to a perpendicular line drawn upward from the preoccipital notch. The main surface features of the parietal lobe are the postcentral gyrus (primary somatosensory cortex) and the division of its more posterior portions into the **superior** and **inferior parietal lobules** by the intraparietal sulcus. As noted above, the inferior parietal lobule is composed of the **supramarginal gyrus** [Brodmann's area (BA) 40] and the **angular gyrus** (BA 39).

If the frontal lobes are seen as primarily responsible for controlling if, when, why, and how we respond to situational demands, the posterior cortices, including the parietal lobes, may be viewed as being responsible for **collecting, encoding, integrating,** and **storing** sensory information upon which the frontal lobes rely to carry out their activities. In part, the parietal lobes appear to take the lead in processing and interpreting somatosensory input and developing an appreciation of internal (body schema) and extended or external space.

The Temporal Lobes

The dorsal limits of the anterior temporal lobe are easily identified by the PHR of the lateral fissure. The cranial vault essentially defines the anterior and inferior extents of the temporal lobes. The posterior extent of the temporal lobe is not defined by a precise sulcus, but rather by the "imaginary lines" described above that separate the parietal and temporal lobes on the lateral surface of the brain. Important regions of the temporal lobe include **Heschl's gyrus** (primary auditory cortex) and auditory association cortex, which includes the **planum temporale** in the temporal operculum; the **superior, middle,** and **inferior temporal gyri**; and the **occipitotemporal** (*fusiform*) **gyrus**. On the inferiomedial surface of the temporal lobe lies the **parahippocampal gyrus,** which contains the **hippocampal formation**. On the medial aspect of the anterior portion of the parahippocampal gyrus is the **uncus,** a small bulge on the surface of the brain that marks the general location of the **amygdala** lying beneath this surface feature (see Figures 5–4b; 6–1b).

The temporal lobes perhaps most commonly are associated with the processing of auditory input and with the encoding of memory. However, among other possibilities, the temporal lobes also are believed to play a substantial role in the processing of affective information, language, and in certain aspects of visual perception.

The Occipital Lobes

The main portion of the occipital lobe lies on the medial surface of the hemisphere. As previously noted, its main surface feature is the calcarine sulcus, which separates the cuneus (above) from the lingual gyrus (below). The area around the calcarine sulcus represents the primary projection area for the optic radiations. The functional–anatomical significance of the calcarine sulcus is that the visual cortex immediately above it (**cuneus**) processes information that comes from the (contralateral) inferior visual field, while the **lingual gyrus** mediates input from the superior visual field (see Chapter 5). One way to keep this straight is to think of the visual cortex as not only reversing the visual fields horizontally (contralateral representation), but also vertically (in the contralateral fields: see Figure 5–6).

Behaviorally, the occipital lobes are linked primarily to visual perception, including color, form, and motion. However, when considering the interactions among the occipital, parietal, and temporal lobes, occipital lobes also are seen as critical for a host of spatial, linguistic, and object recognition functions.

"PTO" Cortex

Because the area where the parietal, temporal, and occipital lobes converge on the lateral surface of the hemispheres is difficult to delineate, this area often is referred to as "PTO" (parieto-temporo-occipital) cortex. This area is primarily composed of higher-order hetero-modal association cortex (see below) that serves as a point of convergence for unimodal inputs from adjacent auditory, visual, and somatosensory secondary association cortices. The linking of concurrent sensory inputs is thought to provide a basis for creating highly complex, multimodal percepts. These, in turn, provide a foundation for language, abstraction, problem-solving and other complex cognitive activities. Consider, for example, the following. While the visual cortex enables us to accurately perceive the written word "fleur," it is through the interaction of the temporal and occipital lobes that one learns to associate this visual pattern with a particular sound. However, if one is unfamiliar with French, the word "fleur" still will be devoid of meaning. However, once we learn that "fleur" means "flower" (through subsequent visual, auditory, or other associations), then it immediately takes on a whole new, rich dimension, complete with visual images, olfactory, tactile, and perhaps even symbolic associations as a result of other previously established, multimodal connections.

The Insular Cortex

Although not typically considered to be a part of the frontal, temporal, or parietal lobes, the **insula** or the *Island of Reil* is a small area of cortex buried deeply within the lateral (Sylvian) fissure. On an axial or coronal section (Figure 9–11), the insular cortex can be identified lying lateral to the putamen, separated from this basal ganglia structure by two thin fiber tracts,

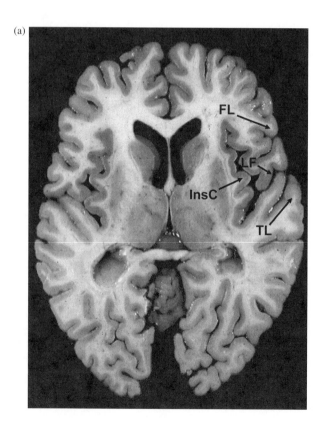
(a)

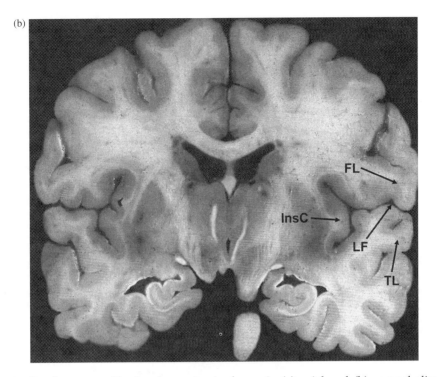

Figure 9–11. Insular cortex. The insular cortex is shown in (a) axial and (b) coronal slices in the depths of the lateral fissure. It is bounded medially by the claustrum and the extreme and external capsules. Brain images were adapted from the *Interactive Brain Atlas* (1994), courtesy of the University of Washington.

the external and extreme capsules and the claustrum (the thin band of gray matter that is sandwiched in between these two capsules). Compared to the other cortical areas, relatively little is known about the specific function of the insula in humans. It certainly does not seem to be a vestigial structure as it shows increased development in apes and humans, and like the other parts of the cortex it is highly convoluted. Several shorter gyri characterize the anterior portions, with one or two longer gyri making up the posterior aspect. The insula has extensive connections to neocortical, limbic, and paralimbic structures. The insula is linked to somatosensory, visual, and auditory association areas, as well as to the prefrontal areas and primary motor cortex. While specific functions are difficult to ascribe to the insula, its anterior portions appear to be more closely related to olfactory and gustatory sensations and to autonomic (especially, gastrointestinal) activities, whereas more posterior portions may be more responsive to visual, tactile, and auditory input. Because of its association with limbic and paralimbic centers, it has been postulated (see Mesulam & Mufson, 1985) that the insula may be an important link in the process of imbuing sensory experiences with emotional valances and/or responding affectively to sensory stimuli. The insula's possible role in certain epileptic and psychiatric conditions also has been suggested (Bauman, 1992). More recently, lesions in the insula were found to reduce nicotine cravings (Naqvi, Rudrauf, et. al., 2007).

Cytoarchitectonic Organization of the Cerebral Cortex

As was noted in the section on the history of cerebral localization, one potentially promising approach to studying the functional organization of the brain was the discovery of the cytoarchitectural organization of the cerebral cortex. Variations in the pattern and organization of the cellular processes in the different cortical areas define the cytoarchitectural organization of the brain. Although the microscope was available to Gall in the early part of the 19th century, he cautioned that there was "little to be learned" from an analysis of the cells of the brain. However, at the beginning of the 20th century, it also was noted that while all the "neocortex" was similar in that six distinct "layers" of cell organization generally could be identified, the pattern, prominence, or distribution of certain types of cells varied from one area of the cortex to another. On the assumption that changes in this pattern of cellular structure (cytoarchitecture) reflect differences in function, Brodmann (1909) analyzed these variations and mapped out the entire cortex, assigning numbers to each of approximately 50 contiguous regions where such a change in structure was detected (Figure 9–12). However, Brodmann himself never actually used his system of cytoarchitectural classification to define areas of functional specialization in the brain; this was to come much later. In publishing his own version of a cytoarchitectural map of the cortex in 1929, von Economo, although identifying approximately twice the number of regions defined by Brodmann, was able to reduce the categories to five basic types of cortex. Vogt and Vogt (1919) found what they believed were over 200 distinctive cytoarchitectural areas in the cortex. However, it is Brodmann's classification system that has been most

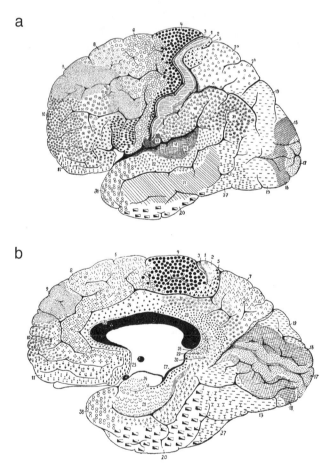

Figure 9–12.   Brodmann's classification of cortical areas based on cytoarchitectural variations. From P. Bailey and G. Von Bonin (1951). With permission of the University of Illinois Press.

extensively used as a means of labeling cortical areas. Brodmann's classification system will be discussed more thoroughly in Part III of this chapter.

The Six-Layered Neocortex

The exterior layers of the neocortex (gray matter) contain neurons, neuroglia (supporting cells), and blood vessels. Underlying the gray matter is the white matter, which consists of axonal processes connecting the cortical mantle with other parts of the central nervous system (the various white matter pathways will be discussed in greater detail below). The neurons making up the gray matter of the cortical mantle can be divided into two major types: **pyramidal** and **nonpyramidal neurons**. Pyramidal neurons primarily are located in layers III and V. These layers, especially layer V, are the major source of cortical efferents (output fibers). In contrast, layers II and IV (granule cell layers) are composed primarily of nonpyramidal neurons (stellate, polymorphic, granule, and Golgi type II cells) and are the major sites of cortical afferents (inputs). As noted above and illustrated in Figures 9–13 and 9–14, the relative distribution of these cells form the six distinctive layers upon which cytoarchitectonics are based. However, the relative prominence of these various layers, as well as the relative prominence of certain types of cells within each layer, will vary depending on the area of cortex involved and its general function. Areas of the cortex that are characterized by a relative absence of the larger pyramidal cells and a greater abundance of stellate or granule cells and small pyramidal cells (as in the primary sensory cortices) are known as **granular cortex**. Areas in which larger pyramidal cells predominate are referred to as **agranular cortex** (the latter are typical of motor output areas). Collectively, both of these more extreme types of tissue are referred to as **idiotypic** or **heterotypical** cortex, whereas the more "balanced" cortical areas, in which the six layers are more clearly differentiated, are known as **homotypical** cortex.

From the pial surface inward, the six cortical layers are:

I. The **molecular** or **plexiform layer,** which is the outermost layer (closest to the surface of the cortex). It contains relatively few cells but a rather dense collection of dendritic processes. This layer is the site of termination of cortical-projecting fibers from the thalamic association nuclei.

II. The **external granular layer,** which consists of a large number of closely packed, small granule and pyramidal neurons whose dendrites terminate in the molecular layer and the majority of whose axons terminate in deeper cortical layers, while additional axons enter the white matter where they project to other cortical areas. Along with layer III, this layer is the termination site of most corticocortical projections.

III. The **external pyramidal layer,** consisting of medium pyramidal neurons whose dendrites largely ascend to the first layer, while the majority of the axons enter the white matter as either association or commissural fibers. In addition to being a major source of efferent fibers to other cortical areas, the external pyramidal layer is a major termination site for afferent fibers coming from other areas of the cortex.

IV. The **internal granular layer,** consisting primarily of closely packed stellate and granule cells. Most of these intrinsic neurons have short axonal processes that stay within the same layer. Other longer axons may go to other layers or enter the white matter. This layer is the site of termination of most of the thalamocortical fibers originating in the specific thalamic nuclei.

V. The **internal pyramidal layer,** consists primarily of medium- to large-sized pyramidal neurons, along with some intrinsic granular cells. Again the dendritic processes of the pyramidal neurons tend to ascend to the molecular layer, although

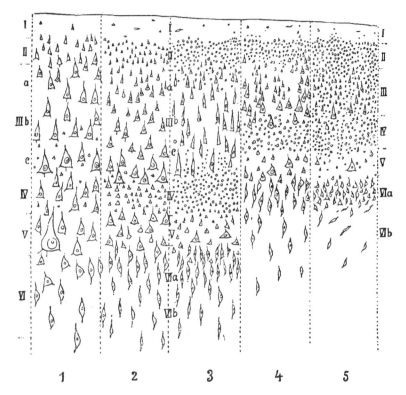

Figure 9–13. Schematic representation of the five fundamental types of isocortex according to von Economo (1929). Roman numerals represent the cortical layers and Arabic numerals indicate the five basic type of neocortex:

1. "Motor cortex": prominence of pyramidal cells ("agranular"), especially prominent in layers III and V. Some very large pyramidal cells (Betz cells) present. Found primarily in Brodmann's areas 4 and 6.
2. "Frontal association cortex": pyramidal cells still quite obvious in layers III and V, but more granular cells appear, especially in layers II and IV. Typical of the prefrontal and some posterior association areas.
3. "Posterior association cortex": even greater development of the granular layers (II and IV). Seen in the cortices of the inferior parietal lobule and superior temporal gyrus.
4. "Polar cortex": has well-differentiated layers representative of homotypical cortices like 2 and 3, but thinner. Found in orbital frontal and surrounding the primary visual cortices.
5. "Primary sensory cortex": represents idiotypic (heterotypical) cortex that, like "motor cortex," has poor differentiation among the layers, but most consists of granule type cells (koniocortex). Present in primary visual (17), auditory (41,42) and somatosensory (3,1,2) cortices.

From P. Bailey and G. Von Bonin (1951). With permission of the University of Illinois Press.

many will terminate in layer IV. The axonal processes of the pyramidal cells enter the white matter as either association or primarily as projection fibers that proceed to areas outside the cortical mantle, for example, the basal ganglia, the thalamus and other subcortical nuclei, the limbic lobe, the cerebellum, and the spinal cord.[7]

VI. The **multiform or fusiform layer,** is made up chiefly of spindle-shaped cells, as well as stellate and granular cells. Some dendrites will ascend to the molecular layer, but most will either terminate in layer IV or remain in layer VI. Thus both the cells of

a   b

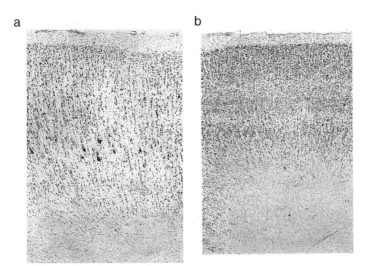

Figure 9–14. Examples of (a) agranular (motor) and (b) granular (sensory) cortex taken, respectively, from the precentral and visual cortices. Note the presence of several large Betz cells in the motor cortex. From P. Bailey and G. Von Bonin (1951). With permission of the University of Illinois Press.

layers V and VI (whose axons project both within and outside the cortex) come into direct, extensive contact with thalamic projections terminating in layer IV. Many of the axons of the spindle cells end up as either projection or association fibers. The stellate cells (intrinsic neurons) are thought to be a main source of the shorter association fibers that connect the cells of one gyrus with those of an adjacent one. Fibers originating in the intralaminar nuclei of the thalamus tend to terminate in this layer. (Table 9–1 provides a summary of the major features of these six cortical layers).

The more highly differentiated **neocortex**, which contains these six layers (not always clearly distinct), comprises about 90% of the cerebral hemispheres. In contrast, more primitive **allocortex** contains three distinct layers of cells and includes the olfactory cortex (**paleocortex**), the hippocampus, and dentate gyrus (**archicortex**). There is a third cortical region, which includes the cingulate cortex and portions of the parahippocampal gyrus and is transitional from allocortex to neocortex, that contains three to six layers depending on

Table 9–1. Summary of Features of Cortical Layers

I	**Molecular:** Consists of dendrites of pyramidal cells and a few intrinsic neurons, receives input from thalamic nuclei
II	**External granular:** Consists of small pyramidal cells and intrinsic neurons, receives from and projects to other cortical areas
III	**External pyramidal:** Consists of medium pyramidal cells, source of association and commissural fibers
IV	**Internal granular:** Consists of intrinsic neurons (stellate and granular cells), receives input from specific thalamic nuclei
V	**Internal pyramidal:** Contains large pyramidal cells, source of association and projection fibers
VI	**Multiform:** Contains intrinsic and pyramidal neurons, primary source of U-fibers to adjacent gyri.

the location. This transitional cortex sometimes is referred to as mesocortex or paralimbic cortex. Using these general cytoarchitectural divisions as a starting point, Mesulam (2000b) categorized cortical tissue into five basic subtypes that also may possess more or less distinct functional properties. As outlined in Figure 9–15, these are:

1. **Idiotypic (heterotypical) cortex**, consisting of "granular" (sensory) and "agranular" (motor) areas
2. **Homotypical (unimodal) isocortex**
3. **Homotypical (heteromodal) isocortex**
4. **Mesocortex or paralimbic cortex**
5. **Corticoid** and **allocortical** or *limbic* **regions**

Each of these five basic cortical subtypes will be discussed below.

Corticoid Regions

Corticoid (cortexlike) structures involve areas in or around the basal forebrain that do not manifest clear cellular layering, but nevertheless appeared to be part of the neocortex. Included in this group were the **septal nuclei**, **substantia innominata** (which includes

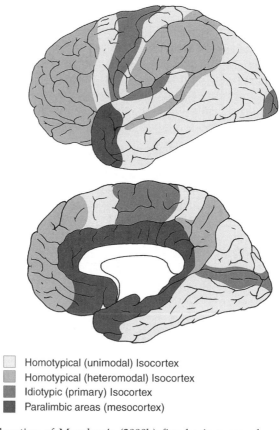

☐ Homotypical (unimodal) Isocortex
▨ Homotypical (heteromodal) Isocortex
▨ Idiotypic (primary) Isocortex
■ Paralimbic areas (mesocortex)

Figure 9–15. General location of Mesulam's (2000b) five basic types of cortex. Not shown are the deeper, more primitive cortical tissues defined by Mesulam as "corticoid" (e.g., septal nuclei, parts of the amygdala, and adjacent basal frontal areas) and "allocortical" (e.g., hippocampal formation). Figure adapted from Mesulam (2000b).

the **nucleus basalis of Meynert**, the source of acetylcholine), and the **amygdala**. *Allocortex* refers to a structurally and phylogenetically primitive type of cortex. While allocortex shows a clear layering effect, it consists of only one or two layers of cells. Two subtypes of allocortex are described (1) those areas in the anterior temporal lobe that serve as the primary cortical projections site for the **lateral olfactory stria** (*pyriform or paleocortex*), and (2) the **hippocampus proper** (*archicortex*). All of these areas tend to have extensive connections with the hypothalamus and are considered to be some of the more primitive parts of the "limbic system" (or the brain as a whole). As was seen in Chapter 8, these areas are involved in very basic drive states and survival mechanisms.

Mesocortical Regions

The "meso" or paralimbic cortex consists of the **cingulate gyrus**, portions of the **parahippocampal gyrus** and temporal pole, the **insula**, and the **caudal orbitofrontal cortex**. As noted above, these tissues are characterized by three to six layers of cells and represent a transition from the more "primitive" allocortex and the more highly developed neocortex. Functionally, these areas appear to be related to "higher-order drive states," or motivation (e.g., maternal instincts, socialization), learning and memory, and autonomic activity (Mesulam, 2000b).

Homotypical Isocortex

Homotypic isocortex and idiotypic cortex constitute the neocortex, which in humans represents by far the largest majority of cortical tissue. As in all neocortical tissue, these regions are characterized by a six-layered cellular structure. The major structural difference between the homotypical and idiotypic cortex is that in the **homotypical cortex** the six layers tend to be more distinct, whereas in the **idiotypic cortex** there is greater blurring or greater degrees of uniformity in the cellular layers (particularly II though V). The homotypic cortex is further divided into the unimodal and heteromodal association cortices. Although there apparently are some subtle differences between the unimodal and heteromodal cortices cytoarchitecturally, the major differences appear to be in their pattern of connections.

Homotypical Unimodal Isocortex

Unimodal or modality-specific cortices are comparable to Luria's "secondary" association areas. They receive input only from a single sensory modality (e.g., auditory, visual, or somatosensory) and only from the idiotypic cortex with which they are associated (or from other areas within the same modality-specific unimodal region). The output of the unimodal cortices in large part (though not exclusively) is to the heteromodal areas. These unimodal cortices seem to be responsible for further processing of specific sensory input, and hence, lesions to these areas generally will result in modality-specific deficits, depending on the specific area involved. Three unimodal sensory association areas are typically identified: auditory (along the superior temporal gyrus), visual (peristriate, midtemporal, and inferior temporal areas), and somatosensory (parts of the postcentral gyrus and superior parietal lobule). The premotor areas anterior to the precentral gyrus also generally are considered to be part of a motor equivalent of the sensory unimodal cortices. These cortices always lie interposed between the "primary" (idiotypic) and the "tertiary" (heteromodal) cortices.

Homotypical Heteromodal Isocortex

Heteromodal (*tertiary association*) cortices, in contrast to the unimodal areas just discussed, receive input from multiple modalities via the unimodal or "secondary" association cortices (apparently they receive no direct input from the "primary" or idiotypic areas). Because of the fact that they are privy to and can integrate this diversity of information, these areas seem to be associated with cross-modal cueing and associations, abstractive capacity, and what are commonly termed "higher-level executive functions." The inferior parietal lobule (angular and supramarginal gyrus) and the prefrontal cortex (that portion of the frontal lobe that lies anterior to the premotor area) constitute the primary heteromodal cortices in humans. As can be seen in Figure 9–15, portions of the superior parietal lobule and the temporal cortex (BA 37) also may have multimodal connections and are considered homotypic heteromodal isocortex by some anatomists.

Idiotypic Isocortex

Idiotypic (also called *primary*) cortex differs from homotypical isocortex both with regard to its cytoarchitectonic structure, connections, and hence, functional role. In homotypic isocortex, there is a clearer differentiation of the cortical layers (especially layers II through V) based on the predominance of cell types in the respective layers. For example, stellate or granule cells tend to be more manifest in layers II and IV, while pyramidal cells are relatively more common in layers III and V. In primary sensory (idiotypic) cortex, stellate cells tend to predominate in layers II through V, whereas in primary motor (idiotypic) areas pyramidal cells tend to be more prominent.

Idiotypic Motor Cortex.   Since pyramidal cells tend to be associated with longer fiber tracts (as are found in the motor pathways), it makes sense that the primary motor area (precentral gyrus or BA 4) is characterized by this relative proliferation of pyramidal cells. Because of the relative lack of stellate or granule cells, the primary motor cortex often is referred to as *agranular* cortex. The latter term also occasionally is applied to the premotor (unimodal homotypical isocortex), which also tends to have a greater ratio of pyramidal to stellate cells. As will be discussed in Part III, the primary motor cortex appears to represent the final common upper motor pathway for executing voluntary motor responses.

Idiotypic Sensory Cortex.   By contrast, the cortical regions that are the major cortical projection sites for the medial and lateral geniculates and the VPL and VPM nuclei of the thalamus have very few pyramidal cells. They represent those cortical areas concerned with the initial processing of incoming auditory, visual, and somatosensory information, respectively; hence, the term, *primary sensory* cortices. As noted above, due to the preponderance of stellate cells, the cortical layers in these regions have a finer "granular" appearance, and thus, also are referred to as *granular* [8] or *koniocortex* (meaning *"dust-like"* – see Figure 9–14). These areas include:

1. **Primary visual cortex** (BA 17): located in the occipital lobe surrounding the calcarine fissure on both its superior (cuneus) and inferior (lingual gyrus) bank.
2. **Primary auditory cortex** (BA 41 and probably part of 42): located primarily within the temporal operculum (Heschl's gyrus) on the superior surface of the superior temporal gyrus.
3. **Primary somatosensory cortex** (BA 3,1,2): located on the postcentral gyrus, between the central and postcentral sulci.
4. **Primary olfactory cortex**: less well defined, but as was seen in Figure 5–4b, generally is thought to include the basal forebrain in the region of the anterior perforated substance or basal nuclei, portions of the amygdala (cortical amygdaloid nucleus), and adjacent pyriform and entorhinal cortices. This olfactory cortex differs from the other primary sensory cortical regions in a couple of important ways. First, these areas involve limbic and more primitive allocortex. Second, the sensory input is more direct, rather than being relayed through the thalamus.

In addition to the horizontal lamination of the cortex into cellular layers, a vertical or columnar organization of cells has been identified [9] (Szentagothai, 1979; Szentagothai & Arbib, 1974). Initially identified in sensory areas of the cortex, all cells (within these vertical units, only fractions of a millimeter in diameter) were found to respond to a particular type (and locus) of stimulation. These vertical columns or units are thought to represent one of the more elementary functional units in the cortex. Short axons within these vertical columns make connections to neurons in adjacent columns to form distinct closed chains or *modules*. In turn, these modules are interconnected with other modules that may reside in divergent areas of the cortex, providing a neural network for more complex (cortical) "functions." Thus, distinct cortical "functions" each may be represented by particular patterns of interconnected modules, making up these larger (macro) functional units. While the same modules or subsets of modules might be shared by other macrofunctional units, each of these larger behavioral (functional) units would have their own unique set of modules. The preceding forms the basic premise for **distributed systems** as a model or theory of cortical organization (see Mountcastle, 1979, 1997; Cytowic, 1996, pp 80–84; Cummings & Coffey, 1994; Duffy, 1984). This approach, which will be addressed again later in this chapter, would help explain how a single lesion can affect a variety of behavioral functions, and conversely how lesions at various cortical sites may disrupt a given behavior.

FUNCTIONAL ORGANIZATION OF CEREBRAL CORTEX: OVERVIEW

Clearly, there is some correlation between certain cytoarchitectural features and function. Brodmann's divisions have survived in part because of their fairly good correspondence with areas of the cortex with known functional distinctions. This is particularly true with regard to primary and secondary cortical zones. Thus we have come to associate the primary motor cortex with (Brodmann's) area 4; the motor association cortex with area 6; frontal eye fields with area 8; the primary visual cortex with area 17; the visual association cortex with areas 18 and 19; the primary somatosensory cortex with areas 3, 1, and 2; the primary auditory cortices with areas 41 and 42; and the auditory association cortex with areas 21 and 22. Within the "tertiary" or heteromodal association cortices, it becomes more difficult to associate specific "functions" with specific Brodmann areas, although reference often is made to areas 39 and 40 in conjunction with the angular and supramarginal gyri, respectively.

However, despite what would appear to be some degree of anatomical validity for Brodmann's classification systems, this means of classification failed to offer a comprehensive explanation of cortical function. A unifying theory still was needed into which these anatomical findings could be integrated. Some type of framework incorporating structural data with known clinical correlates needed to be developed to begin to understand how the brain integrates elementary sensory input into meaningful, complex, multimodal perceptions and concepts, thus allowing the organism to respond in a complex, goal-directed manner. The discussion of such a working model will be the primary focus of Part III of this chapter. For now, a few additional observations on possible unifying principles of cortical organization may be helpful.

Sensory Input

One approach to studying the organization of the neocortex is to consider the levels of processing required from sensory input to behavioral output. We might begin by considering the sensory input. Externally driven behaviors typically are elicited by sensory input. This input can be visual (an object or a written word), auditory (environmental sounds, words, or music), somesthetic (texture, temperature, or shape), or chemical (the smell of a flower or taste of a strawberry). As noted earlier, each of these sensory inputs is processed at a basic level in the corresponding primary cortical sensory area represented by idiotypic, granular cortex. In turn, these primary cortical projection areas have important common anatomical relationships that are important to subsequent levels of processing. For example, each of these primary sensory cortices is adjacent to and has extensive connections with its corresponding modality-specific (unimodal) association area. These connections allow for increasingly complex levels of information processing. Specifically, primary visual cortex (BA 17) is adjacent to visual association cortices (BA 18 and 19) in the occipital lobe, primary auditory cortex (area 41) is adjacent to auditory association areas (BA 22 and probably part of 42)[10] in the temporal lobe, and primary somatosensory cortex (BA 3, 1, 2) is adjacent to somatosensory association cortex (BA 5 and 7) in the superior parietal lobule. Additional processing of modality-specific input occurs within each of these unimodal (secondary) sensory association areas such that recognizable and reproducible percepts are formed, although specific meanings or significance may not be attached at this stage. Next, each of these modality-specific areas converges onto adjacent heteromodal association cortices (e.g., the supramarginal (BA 40) and angular (BA 39) gyrus). It is within these heteromodal areas, through the richness of cross-modal sensory associations, that these unimodal percepts are imbued with complex meaning and symbolism, forming the basis for higher-order functions such as concept formation and abstractive capacities.

Thus far, this model of neocortical organization has been discussed with reference to externally driven sensory inputs. In addition to these externally driven sensory inputs (auditory, visual, and somatosensory), internally driven inputs are processed and can be experienced at the sensory level. For example, recalling the memory of a negative experience might arouse subjective feelings of fear or shame and involve activation of more primitive limbic structures.

Motor Output

The motor output system of the cerebral cortex appears to be organized in a manner similar to the sensory input system but, in some respects, in reverse. In sensory systems, information can be conceived as being processed from the bottom (from primary or idiotypic sensory cortex) up (to heteromodal cortex). In the motor or executive system, it is perhaps more logical to think about the flow of information from the heteromodal, higher-order frontal granular cortex to the primary or idiotypic motor cortex (i.e., from more complex to less complex levels of processing). If the sensory systems (posterior cortices) provide the organism with the information it needs to survive and prosper, it is the frontal (executive–motor) systems that decide where and how that information needs to be translated into actions. Thus, the information that was processed by the homotypic isocortex of the parietal, temporal, and occipital lobes must be made available via association pathways (discussed below) to the heteromodal frontal association cortices. Here the information is weighed in light of the current needs and circumstances of the organism, immediate or long-term goals, and other available information to decide if and how the organism should respond. Once a particular response is decided on, this information then is conveyed from the heteromodal to the unimodal frontal cortex or premotor areas (e.g., BA 6, 8, 44, 45) where the response is programmed and orchestrated. The final step in this process is for the idiotypic motor (agranular) cortex or primary motor area (BA 4) to actually execute the planned motor response.[11]

Clinical Implications

How does this model of neocortical organization affect the analysis of deficits associated with cortical lesions? Again, this topic will be addressed in greater detail later, but a brief overview in the context of the present discussion may be useful.

Two distinct cortical visual pathways have been described: a ventral (object vision) pathway and a dorsal (spatial vision) pathway (Ungerleider & Mishkin, 1982). Each of these visual pathways consists of multisynaptic circuits that flow from primary visual to visual association to heteromodal association cortex, as described above. The ventral stream flows from occipital to temporal heteromodal cortex following the course of the inferior longitudinal fasciculus and/or the inferior occipitofrontal fasciculus (see immediately below), while the dorsal stream flows from occipital to parietal heteromodal cortex following the superior longitudinal fasciculus. The corticocortical connections of these two visual streams in part determine the functional specificity of each pathway. Within each of these visual pathways, primitively formed visual information flows from primary visual cortex (BA 17) to adjacent visual association areas (BA 18 and 19) for further processing into visually integrated percepts. The visual information then is thought to flow into one of the two visual streams: the ventral object vision pathway or the dorsal spatial visual pathway. Whereas the ventral occipitotemporal pathway is thought to be crucial for object identification, the dorsal occipitoparietal pathway is considered crucial for the visual location of objects. Each pathway has connections to more remote cortical regions that determine associated visual functions. For example, the ventral visual pathway has connections to limbic structures in the temporal lobe that may enable the individual to associate visual objects with emotion and memories. Both pathways likely connect to frontal cortices to mediate motor responses.

This organization and pattern of connections of these pathways determine the nature of the specific deficits produced by discrete cortical lesions. Such lesions will reflect the organization and levels of processing that occur within the lesioned region. In humans, ablative or destructive lesions to the primary visual cortex (BA 17) produce the characteristic visual field defects as described in Chapter 5. Bilateral lesions to primary visual cortex may result in "cortical blindness." Excitatory lesions (e.g., seizures) to primary visual cortex (although relatively rare) may result in poorly formed visual hallucinations consisting of amorphous colors or flashes of light. Destructive lesions at the next level of processing, that is, involving the visual association cortex, produce more complex and heterogeneous deficits that differ behaviorally depending on which pathway (dorsal or ventral) is lesioned and depending on the specific location within visual association cortex. For example, lesions affecting the more ventral pathways may result in deficits in visual object recognition or color agnosia, whereas a more dorsally situated lesion might result in problems with visual spatial perception or reading. All deficits, however, share specificity for processing of visual information.

Lesions that are generally limited to heteromodal association cortex tend to produce more complex behavioral, cognitive, and emotional–affective disturbances, while preserving more basic perceptual and motor capacities. For example, destructive lesions affecting PTO areas or the frontal granular cortex may result in disturbances of such skills as propositional speech, language comprehension, arithmetical and abstractive abilities, visual–spatial construction, problem solving, "executive" abilities, learning and memory, or even breakdowns in emotional/social behaviors. Excitatory lesions (seizures) lesions to heteromodal association cortices (more commonly in temporal regions) can produce well-formed visual images or visual hallucinations.

The cortical visual pathways presented above have been studied quite extensively and offer a model for discussion. However, many of the structural and functional organizational principles can be applied to the other sensory systems.

Limitations

Models of cortical organization and the distribution of neural networks that mediate precise cognitive functions contribute to an understanding of brain–behavior relationships. Although the models of cortical organization and interconnections presented here and in other sections of this chapter provide a schema for understanding general patterns of behavior, the application of these anatomic and neuropsychological principles to lesion localization in humans has its limitations. Ablation or destructive lesion paradigms, which study the correlation of lesion sites with specific behavioral deficits, have been the basis for the study of many higher cortical functions. However, the drawbacks of these paradigms are important to consider. First, damage to the human brain rarely respects cytoarchitectonic boundaries. Consequently, multiple cytoarchitectonic regions usually are involved in a single lesion, which may account for the diversity of behavioral deficits observed on clinical examination. In addition, deficits normally associated with a particular cortical region can be produced by lesions remote to the site as a result of disturbance of other structures or sites (both cortical and subcortical) that are part of the distributed system for that particular function. In addition, lesions to *inter*hemispheric (commissural) or *intra*hemispheric (association) pathways may produce **disconnection syndromes,** which can result in a wide variety of cortical behavioral deficits. Such syndromes, which have been well articulated by Geschwind (1965), will be discussed at the end of the next section. Finally, most neuropsychological tests used to assess neurobehavioral syndromes are themselves multidimensional and may show impairments for different reasons in different subjects. Despite these limitations, the study of clinical cases with discrete lesions represents an important methodology to study brain–behavior relationships and is an important clinical tool.

CORTICAL PATHWAYS AND THEIR CLINICAL SIGNIFICANCE

Most observed behavior results from the integrated functioning of the brain and the rest of the nervous system, with each part making its own unique contribution. In order for such integration to occur, there must be a constant sharing of information (intercommunication) within the system and an infrastructure that supports such communication. With respect to the cerebral cortex, two general communication systems are readily identified. One might be considered a more diffuse, analogue-type system that establishes the background state or cortical tone within which the second system operates. The neuronal mechanisms for this first system are largely represented by various neurochemical pathways that will be reviewed in Chapter 11. The second system may be characterized as a more discrete, digital communication network that conveys highly specific information. It is this latter system that will be the focus of this section. One should be mindful that while this analogy makes a useful dichotomy for the following discussion of cortical pathways and disconnection syndromes, it does so at the risk of vastly oversimplifying the process.

Types of Pathways

The interchange of information involving the cerebral cortex is mediated largely by collections of vertical and horizontal axonal fibers. Vertical connections consist of ascending and descending pathways between cortical and subcortical, brain stem, or spinal nuclei, while horizontal communications are represented by either intrahemispheric or interhemispheric connections. Respectively, these constitute the three major types of information highways within the brain: **projection, association,** and **commissural pathways.** Whether looking at coronal, axial, or parasagittal images of the brain, the extent of these white matter pathways beneath the cortical mantle readily can be appreciated. A basic awareness of these pathways is essential in attempting to understand brain–behavior relationships and clinical neuropathology.

A few basic cortical connections already have been presented in the discussion of cytoarchitectonics, including the fact that:

1. Sensory information is conveyed to the idiotypic sensory cortex from peripheral receptors after synapsing in the thalamus (via ascending projection pathways).
2. After being processed in the primary sensory cortex, this information is passed on to the homotypic sensory cortices (via short association fibers).
3. Reciprocal feedback with the frontal granular and agranular cortices (via longer association pathways) then allows for utilizing previous and current sensory input to decide if a response is indicated, and if so how that response might best be planned, monitored, and executed.
4. Finally, the primary motor cortex actually executes the behavioral response (via descending projection fibers).

What was not mentioned earlier was the fact that at both the sensory or perceptual level and at the executive or motor level there is a sharing of information across the midline (via commissural pathways). Both cerebral hemispheres are normally recruited in the above processes, along with other subcortical structures (e.g., basal ganglia), via ascending and descending projection pathways.

In previous sections and chapters, we already have mentioned the critical role each of these various cortical zones or subcortical nuclei play in behavior and the potential impact of lesions to these various areas. In this and later sections (particularly the one on disconnection syndromes) the focus will be on the effect of lesions which disrupt communications between or among these various cortical and subcortical sites as a result of impingement on one or more of these pathways. For now, consider the following analogy. You have a television set that is in perfectly good condition, the nuclear power plant 20 miles away is up and running, and the geosynchronous communications satellite is sending perfectly good signals to your local cable company. However, if the underground cables (conducting electrical power or TV signal) interconnecting any of these three locations are disturbed, your TV viewing will be adversely affected. This is basically what happens with many brain lesions. While some lesions may be more or less limited to the cortical mantle (gray matter) or to subcortical nuclei, most also encroach on fibers of passage as well. Other lesions or disease processes (e.g., lacunar infarcts, leukoencephalopathy, and multiple sclerosis) primarily may involve white matter or fiber tracts. The behavioral effects of lesions that interfere with normal communications within or between the hemispheres (i.e., association or commissural fibers) result in what commonly are referred to as **disconnection syndromes**. Some of the more common disconnection syndromes will be reviewed shortly. For now, let us consider the three major types of fiber tracts within the brain. These are (1) **commissures**, (2) **association pathways**, and (3) **projection pathways**.

Cerebral Commissures

By definition, *commissural fibers connect a region or area on one side of the brain either with its homologous area, or a closely related area, on the opposite side of the brain.* This arrangement allows for rapid and effective interchange of information between the two hemispheres and has a number of important functions. First, one side of the brain often is privy to unique sensory input that may be important to share with the other cerebral hemisphere. A simple example is when we put our left hand in our pocket to find something. The sensory feedback via the dorsal columns, medial lemniscus, and ventral posterior nucleus of the thalamus initially is directed to the right hemisphere only. Literally, if the right hand is to know what the left one is doing, this must be accomplished through the commissures.[12] As will be discussed shortly, one hemisphere may be better adapted to perform certain behavioral or cognitive functions (the leading or "dominant" hemisphere for that function). Thus, if the hand ipsilateral to the hemisphere that is "dominant" for a particular function wishes to assist in performing that function, the information or directions about how best to carry out such tasks needs to be conveyed to the contralateral hemisphere (the one controlling the hand ipsilateral to the "dominant" hemisphere). As we shall see, ideomotor apraxia provides a good example of such specialization and interhemispheric communication.

In addition to taking a leading role in directing different aspects of either our internal or external experiences, each hemisphere also seems to process information in very different ways. This duality serves a complementary function and provides for a greater richness of experience. Examples might include "semantic" versus "prosodic" (affective) components of speech and language, or focusing on "gestalt" versus "local" internal details in constructional abilities. In these cases, each hemisphere makes unique contributions to the total behavioral response. Since efficient behavior typically requires an integration of both types of information, communication between the hemispheres is essential. One such purpose may be to maintain some type of internal or experiential balance. It is not uncommon to hear people talk about "right-brain" versus "left-brain" individuals. Theoretically, "right-brained" individuals tend to act on a "gut-level, instinctual, emotional, immediate, overall impression" basis, whereas the "left-brained" individual takes a more logical, measured, or considered approach to problems or situations. Think of it as the difference between Captain Kirk and Spock in the original *Star Trek* series. This is obviously a gross oversimplification of a very complex phenomenon. Whatever element of truth there may be to such a "right-brain/left-brain" dichotomy, in the vast majority of situations it is likely that the two hemispheres constantly interact, each contributing its own unique influence, in an effort to maintain some balance or equilibrium in terms of how we respond to our environment. Again, it would appear that the commissures are important in facilitating such communications or interactions. In some respects, the role of the commissures and the corpus callosum in particular can be dramatically demonstrated by clinical cases where there has been a substantial disruption of these crossing fibers. However, even in cases of apparently complete commissurectomies, there still is a remarkable unity of experience, suggesting other subcortical mechanisms also play an integral role in this process (e.g., Liederman, 1995). A review of some of these specific findings will be presented below under Disconnection Syndromes.

In the human brain there are a number of such commissures. The **posterior** commissure, located at the upper end of the brain stem beneath the anterior portion of the pineal body, is a rather small commissure[13] that appears to be largely involved in carrying information regarding visual reflexes. Even smaller is the **habenular** commissure, interconnecting the habenular nuclear groups on the posterior–dorsal aspect of the thalamus. The **fornical** or **hippocampal** commissure allows for interhemispheric communications among various limbic structures, particularly the hippocampus. The **anterior** commissure, another relatively small commissure that lies in the area above the optic chiasm in the third ventricle, transfers information from one anterior temporal region to comparable areas in the opposite hemisphere (see: Figure 9–16). This pathway also appears to carry olfactory information.

However, all the above commissures combined pale by comparison in terms of sheer magnitude to the **corpus callosum**, the major commissure that interconnects the two cerebral hemispheres. Lying between the lateral ventricles below and the cingulate gyrus above, in an anterior–posterior dimension, the corpus callosum stretches for approximately half the length of the cerebral hemispheres. The corpus callosum is clearly visible on coronal sections throughout its length as the large band of white matter crossing the midline immediately below the cingulate gyri (Figure 9–17). On a midsagittal section, its five divisions are clearly visualized. Beginning anteriorly, these divisions include the **genu** (knee), which represents the anterior curvature or bend of the commissure, and the **rostrum,** which is the ventral or inferior continuation of the corpus callosum. The main posterior portion that continues over the superior surface of the lateral ventricles is called the **body**, with the slight narrowing

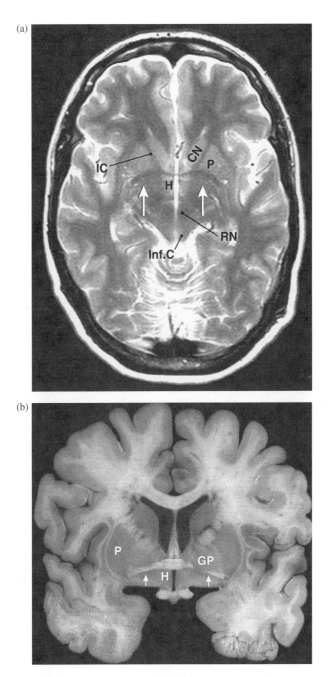

Figure 9–16. Anterior commissure as seen on (a) axial MRI and (b) coronal brain section (arrows). Note: Axial cut is at the level of the midbrain. Abbreviations: CN, caudate nucleus; GP, globus pallidus; H, hypothalamus; IC, internal capsule (anterior limb); Inf C, inferior colliculus; P, putamen; RN, red nucleus. Brain image (b) was adapted from the *Interactive Brain Atlas* (1994), courtesy of the University of Washington.

at its posterior extent being referred to as the **isthmus**. Finally, the more bulbous posterior end of the corpus callosum is referred to as the **splenium**.

For the most part, like the other commissures, the corpus callosum connects either homologous or closely related cortical areas in each cerebral hemisphere (Zaidel et al., 1990;

Zaidel, 1995). There are two notable exceptions. The primary visual cortex (BA 17) is connected with areas 18 and 19 of the opposite hemisphere. There apparently are few if any direct callosal connections between the primary visual cortices. Also, there appears to be minimal if any connection between the primary motor cortex that mediates hand movements with the comparable area in the opposite hemisphere. Actually, in retrospect, this seems quite logical. By necessity, the hands often need to function quite independently,

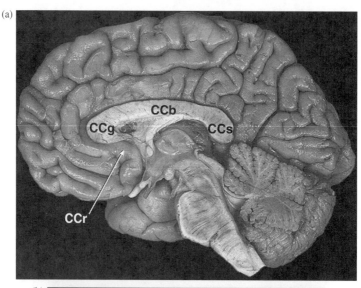

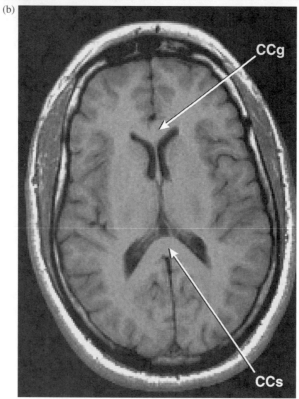

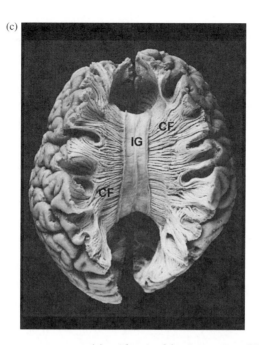

Figure 9–17. Corpus callosum as seen on (a) midsagittal brain section, (b) axial views MRI image, and (c) blunt dissection. Abbreviations: CCb, CCg, CCr, and CCs represent the body, genu, rostrum and splenium of the corpus callosum, respectively; CF, commissural fibers; IG, indusium griseum. Brain image was adapted from the *Interactive Brain Atlas* (1994), courtesy of the University of Washington. Blunt dissections (c) of the brain from Gluhbegovic, N. and Williams, T.H. (1980). Used with permission.

as witnessed in playing any number of musical instruments or other work or engaging in bimanual activities (typing on a keyboard). Too much direct feedback from one hand to the other might interfere with this independent operation. In the case of the eyes, since the two hemispheres are processing information from two visual fields, it would seem important to keep such information free from contamination.

Association Pathways

Association pathways, by contrast, interconnect different parts of the same hemisphere. Such association pathways may be very long, in which case they are typically termed *fasciculi*. In other instances, they may be relatively short, perhaps simply consisting of U-shaped (*arcuate*) fibers connecting one gyrus with an adjacent gyrus, or lateral or horizontal connections within the gyrus itself (e.g., *bands of Baillarger*). Just as it is important for one hemisphere to share information with the other, it also is important that different areas within the same hemisphere be in communication with one another. For example, it may be simply a matter of the cortical motor neurons that control the movements of the fingers in using a screwdriver needing sensory feedback (e.g., stereognosis, pressure or resistance, proprioception, kinesthesia, and visual alignment with the screw). On perhaps a more complex level, the prefrontal cortex requires access to ongoing, current (perception) and previous (memory) information in the planning and execution of a response. In either case, there has to be some mechanism for getting this information from one part of the hemisphere to another. This is probably accomplished, for the most part, by association fibers.

Any attempt to provide a detailed description of all such association pathways, given their complexity and the fact that some are still poorly defined, would be impractical if not impossible. For our present purposes, it is sufficient to outline a few of the major identified fasciculi. What is critical is for the reader to develop some appreciation of the presence of these intrahemispheric connections (both short and long), the general functional purposes they serve, and the potential effects that lesions of these pathways can exert on behavior. And their general locations can be seen in Figure 9–18a. These pathways generally are conceived as interconnecting the more anterior portions of the brain with more posterior parts. While this is indeed correct, it also should be noted that many fibers likely enter

(a)

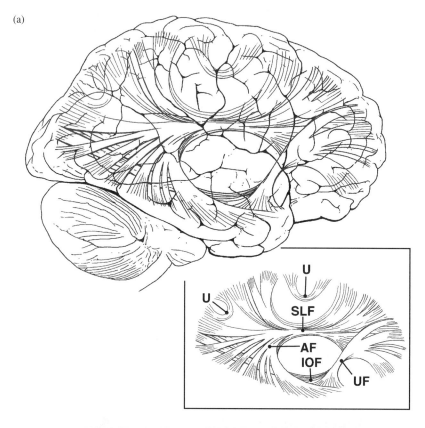

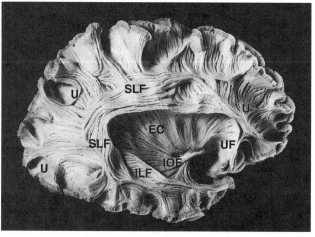

(b)

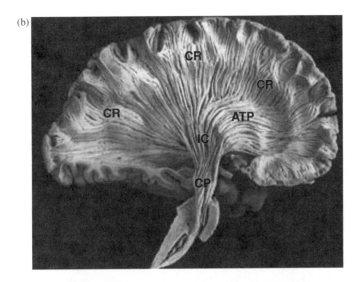

(c)

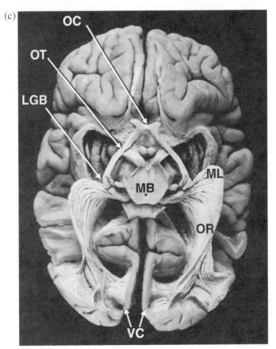

Figure 9–18. Association and projection pathways. Schematic drawing and photographs showing (a) association pathways within the brain, (b) coronal radiata forming the internal capsule, and (c) optic radiations (a major projection pathway). Abbreviations: AF, arcuate fasciculus; ATP, anterior thalamic peduncle; CP, cerebral peduncle; CR, coronal radiata; IC, internal capsule; ILF, inferior longitudinal fasciculus; IOF, inferior occipitofrontal fasciculus; LGB, lateral geniculate body; MB, midbrain; ML, Meyer's loop; OC, optic chiasm; ON, optic nerve; OR, optic radiations; OT, optic tract; SLF, superior longitudinal fasciculus; U, U-fibers; UF, uncinate fasciculus; VC, visual cortex. Photographs are adaptations of blunt dissections of the brain from Gluhbegovic, N. and Williams, T.H. (1980). Used with permission.

and exit throughout the course of these fasciculi. These fibers connect the frontal lobe with the parietal, posterior temporal, and occipital lobes, or the frontal with the temporal and occipital lobes. One such fasciculus is the rather prominent **superior longitudinal fasciculus**, portions of which are referred to as the **arcuate fasciculus**, and as we shall see, may be involved in the disconnection syndrome of conduction aphasia, as well as in some aspects of ideomotor apraxia. The **uncinate fasciculus**, which travels with the more anterior portions of the **inferior occipitofrontal fasciculus**, interconnects the more ventral and orbital frontal cortices with the anterior temporal regions. There is also a **superior occipitofrontal fasciculus,** which on a coronal section can be seen dorsolateral to the head of the caudate nucleus. Another large association pathway that previously was discussed is the **cingulum**. This pathway connects the frontal lobes and septal regions with the cingulate gyrus and more distally with the parahippocampal areas. Finally, an **inferior longitudinal fasciculus** also frequently is described in neuroanatomical texts. However, the existence of an inferior longitudinal fasciculus, which previously was described as a direct connection between the occipital and temporal lobes, has been challenged (Tusa & Ungerleider, 1985). Instead of a long, continuous fasciculus, it is suggested that these areas are indirectly linked by a series of short, "U" fibers that interconnect various portions of the occipital and inferior temporal cortex.

Projection Pathways

The third general types of white matter pathways within the cerebral cortex are the projection fibers. Recall that commissures and association fibers basically connect one cortical area with another. Projection fibers by contrast represent connections (projections) either from the cortex to noncortical structures or to the cortex from noncortical structures. The former are descending projection pathways, while the latter represent ascending pathways. Directly or indirectly, most of the major projection pathways have been reviewed in previous discussions of the motor and sensory systems, basal ganglia, cerebellum, and thalamus. Nonetheless, a brief review may be useful. Among the projection pathways, the most prominent are the **internal, external,** and **extreme capsules, cerebral peduncles,** and **optic radiations**. The **internal capsule** (see Chapter 7 for more detail on its internal structure) consists of both ascending and descending fibers. The descending fibers include projections to the neostriatum (caudate nucleus and putamen). The **cerebral peduncles** represent a caudal continuation of the internal capsule (see Figures 9–18b and 1–6b). They consist of corticobulbar, corticopontine, and corticospinal fibers that mostly end up on various brain stem nuclei, in the cerebellum (after synapsing with the pontine nuclei), or on the anterior horn cells of the spinal cord. Ascending fibers within the internal capsule represent the various thalamocortical projections. Recall that, with the possible exception of olfaction, all information we obtain from the outside world is funneled to the cortex via the thalamus. Even the cerebellum, basal ganglia, and other subcortical structures involved in the control of movement project back to the cortex via the thalamic nuclei.[14] A major thalamocortical projection pathway are the **optic radiations**. As was seen in Chapter 05, the optic radiations represent the visual pathways from the lateral geniculates to the primary visual cortices on the banks of the calcarine sulci (see Figures 9–18c and 5–5).

These various projection pathways, similar to the commissures and the long association fibers, tend to consist of both tight, compact bundles as well as flared arrays of fibers. The descending fibers start out from diffuse cortical areas and narrow as they enter the internal capsule, whereas the opposite is true for the ascending fibers. They start out, mostly from the thalamus, in a more compact arrangement and then branch out as they head to their cortical destinations. This flaring out is what constitutes the **coronal radiations**. By contrast, the commissures and the projection fibers tend to form more compact bundles in the middle of their course and diffuse out at each end. The reason for emphasizing this general feature is that if damage occurs where these fibers of passage are more densely packed, the resulting behavioral deficit can be quite substantial, even from a small lesion. One common example is hemiplegia following a lacunar infarct in the posterior limb of the internal capsule. Figure 9–19 illustrates all three types of pathways in an axial brain section.

While we tend to think of lesions of projection fibers as producing motor or sensory deficits, they in fact can produce a whole host of behavioral deficits. For example, white matter lesions that undercut thalamofrontal connections may result in a *frontal lobe syndrome*, whereas disconnections between the temporal, parietal, or occipital lobes and the thalamus may lead to symptoms of aphasia, neglect, or other sensory–perceptual disturbances. As was discussed in Chapter 5, homonymous quadrantic visual field cuts commonly are associated with lesions that encroach on the optic radiations. Although the term *disconnection syndrome* most commonly is applied to disruptions of the commissures or association pathways, in effect lesions of the projection pathways also may be viewed as a type of disconnection as well. In this case, they disrupt the normal communications between the cortex and

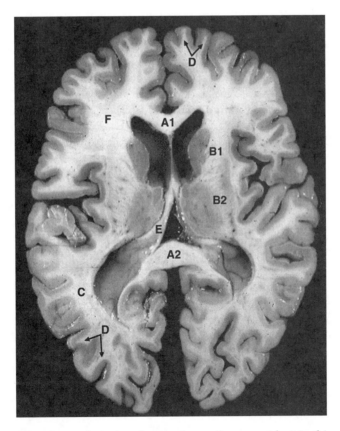

Figure 9–19.   Commissural, association, and projection pathways evident in this axial view through the thalamus. (A1) genu and (A2) splenium of the corpus callosum. (B1) anterior and (B2) posterior limbs of the internal capsule, (C) optic radiations, and (E) fornix all represent projection pathways. (D) Marks short arcuate (association) fibers connecting adjacent gyri. The (F) ventral extension of the centrum semiovale is essentially a mixture of commissural, association and projection fibers. Brain image adapted from the *Interactive Brain Atlas* (1994), courtesy of the University of Washington.

the subcortical, brain stem, or spinal nuclei, and thus produce functional deficits without necessarily directly impinging on gray matter.

DISCONNECTION SYNDROMES

If one of the basic assumptions in neuroanatomy is that different areas (parts) of the brain subserve different functions, for complex behavior to occur it is essential that these areas communicate with one other. The exact nature of this communication still is unclear. One possibility is that some aspect of the information that is processed in one area(s) is conveyed or transferred to another area(s) of the brain, which in turn further processes this integrated information in the context of other information it has received from yet other parts of the brain or nervous system. As this process is repeated, multiple brain areas are recruited (either simultaneously or successively), leading to an integrated response (Goldman-Rakic, 1988).[15]

Another and seemingly more likely hypothesis is that rather than an actual "transfer" of information from one part of the brain to another, there is a unique linkage (temporal–spatial integration) of various areas of the brain, which is specific to the current condition of the organism, stimulus, or stimulus–response pairing. Such temporal–spatial patterns then may be evoked and/or combined with other patterns in subsequent situations as stimulus conditions warrant (see, for example, Liederman, 1995). Regardless of the actual mechanisms involved in the recruiting of diverse brain areas to produce an integrated response, central to both these hypotheses are that different areas of the brain must be in communication with one another. The integrity of the long axonal pathways is critical to this process. Earlier in this chapter three general types of long axonal pathways within the brain were discussed: commissures, association, and projection pathways. Disconnection syndromes result from the disruption of these pathways, effectively shutting off communication between or among areas that they normally interconnect. However, traditionally it is only when corticocortical connections are involved (i.e., commissures or association pathways) that the term *disconnection syndrome* has normally been applied.

Lesions producing such syndromes may result from any of a number of neurological conditions, although disconnection syndromes most frequently are associated with strokes and tumors. Some of the more dramatic cases have been the result of surgical lesions, especially the so-called *split brain* studies in which the corpus callosum is cut, typically in cases of severe and intractable seizure disorders to prevent the spread of excitation from one hemisphere to the other. Occasionally, there are individuals in whom there is an agenesis of the corpus callosum. Interestingly enough, such individuals rarely show evidence of a disconnection syndrome, perhaps because of the enhancement of other pathways (e.g., anterior commissure) or greater bilateral representation of function (Bogen, 1993).

Although the basics of disconnection syndromes were established very early in the history of neurology, the modern appreciation of this phenomenon largely was advanced by Geschwind's paper entitled, "Disconnexion Syndromes in Animals and Man" (Geschwind, 1965). While theoretically at least there may be a large number of disconnection syndromes, for illustrative purposes only those that are most commonly recognized will be discussed here. These include ideomotor apraxia, alexia without agraphia, conduction aphasia, and callosal transections (split-brain preparations).

Ideomotor Apraxia

First described around the turn of the 20th century, ideomotor apraxia may result from lesions of either association or callosal fibers, although its expression will be somewhat different depending on the actual site of the lesion. Ideomotor apraxia is defined as an *inability to carry out a discrete, previously learned, skilled movement, most commonly to verbal command, in the absence of any primary language or sensorimotor deficits.* Such commands typically involve asking the patient to carry out either a transitive (e.g., demonstrating the use of a tool) or intransitive (e.g., a symbolic gesture) action using either the upper extremities or buccofacial musculature. Examples might include demonstrating the use of a hammer or screwdriver, sucking through a straw (transitive gestures), or saluting or sneezing (intransitive). The disturbance at times may be somewhat subtle, as evidenced by a distortion of fine, distal movements or improper orientation of the hand when demonstrating the use of a screwdriver. Typically tested in the absence of the actual object, performance may improve if the object is provided.

To provide a brief and very schematic review of the neuroanatomy of ideomotor apraxia, consider the following command: "Show me how you would use a hammer with your left hand." Assuming normal patterns of dominance, according to Heilman and Rothi (1993) the following systems not only must be in place and working, but must be in communication with one another:

1.   Auditory input and comprehension (Heschl's gyrus and left superior temporal gyrus).
2.   Memory engrams for the particular motor skill (left inferior parietal lobule).
3.   Left motor association area (premotor cortex).
4.   Right motor association area.
5.   Right motor cortex.
6.   Corticospinal projection fibers.[16]

Thus, lesions affecting the arcuate fasciculus, an association pathway connecting parietotemporal and frontal cortices, potentially could result in a bilateral apraxia, as the cortical areas thought to be responsible for the sensorimotor engrams governing these skills are "disconnected" from the frontal motor cortices. The latter are hypothesized to be essential in their final organization, initiation, and/or execution.

In contrast, a lesion affecting callosal fibers connecting the left motor association area to the homologous area in the right hemisphere may result in an isolated left-sided apraxia. In this latter case, since the information regarding both the command and the sensorimotor template for these actions can be communicated to the left motor association cortex, the right hand would be able to carry out the requested action. Because the right motor association cortex is dependent on the specific information (or cooperation) from the left hemisphere via the callosal fibers from the left motor association cortex, the left hand will be unable to properly execute the task to verbal command alone. However, it may improve with imitation or if the actual object is provided. The left hand may carry out other spontaneous actions with little or no apparent difficulty.

Finally, in the case of lesions encroaching on the left motor association areas, as is frequently seen in Broca's aphasia, it is not uncommon to find right-sided weakness and left-sided ideomotor apraxia. This is consistent with the scenario described above where the right motor association areas are dependent on the integrity of their counterparts on the left. The presumption is that if it were testable, the right hand would show similar deficits. Although they arrive at slightly different conclusions regarding the cortical areas primarily responsible for the storage and/or initiation of the motor engrams responsible for the learned actions tested in ideomotor apraxia, both Geschwind (1975), as well as Heilman and Rothi (1993) and Heilman, Watson, and Rothi (1997) provide excellent, detailed descriptions of this particular disconnection syndrome. Figure 9–20 summarizes this schema as outlined by Heilman and his associates.

APRAXIA

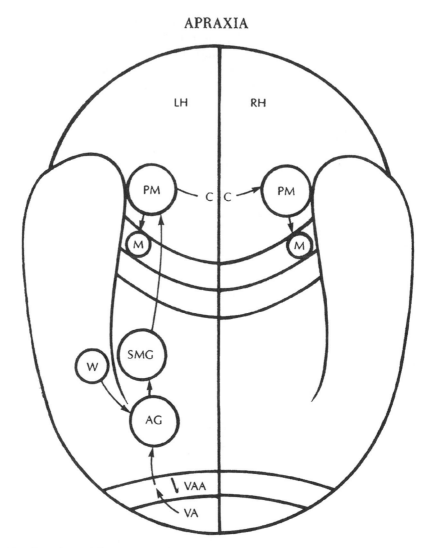

Figure 9–20. Drawing of the major pathways thought to be involved in ideomotor apraxia. From Heilman and Valenstein (2003), with permission of the author. Abbreviations: AG, angular gyrus; CC, corpus callosal fibers; LH, left hemisphere; M, primary motor area; PM, premotor area; RH, right hemisphere; SMG, supramarginal gyrus; VAA, visual association area; VA, primary visual cortex; W, Wernicke's area.

Alexia without Agraphia

Dejerine recognized this disconnection syndrome as early as the late 1800s. Generally the result of an infarction of the left posterior cerebral artery (see Chapter 10), this syndrome typically results from a lesion of the **splenium** of the corpus callosum and the left (or language-dominant) occipital cortex, the latter resulting in a right homonymous hemianopia (Geschwind, 1965; Damasio & Damasio, 1983). The patient retains the ability to write because the left temporal, parietal, and frontal cortices essential to this process remain intact. However, the input of lexical information is limited to the right hemisphere (because of the right visual field deficit) and this information cannot be shared with or communicated to the left temporal speech areas for decoding because of the disconnection of the relevant callosal

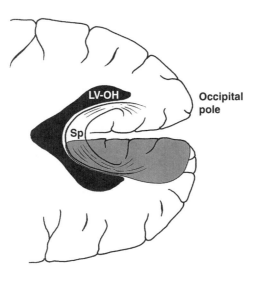

Figure 9–21. General location of lesion typically associated with the syndrome of alexia without agraphia (dominant visual cortex and splenium of corpus callosum). Abbreviations: LV-OH, occipital horns of the lateral ventricles; Sp, splenium of the corpus callosum.

pathways (see Figure 9–21). Hence, the patient is able to write, but if later presented is unable to read what he or she has written. Interestingly enough, the patient may have little or no difficulty naming objects seen by the right occipital cortex. One possible explanation is that other association pathways may be utilized in object recognition that proceed more anteriorly before making trans-hemispheric, callosal connections.

Conduction Aphasia

One classic aphasic disorder that frequently has been associated with a disconnection syndrome is conduction aphasia. This disorder is characterized primarily by marked distur-bances of repetition in the presence of reasonably intact (or at least relatively better) language comprehension, adequately articulated speech with abundant literal paraphasic errors, and disturbances of confrontation naming. The decoding of auditory verbal input into phonetically meaningful units (e.g., phonemes, words) appears to take place largely in the temporal areas surrounding Heschl's gyrus or the primary auditory cortex. In order to repeat what was heard, perceived speech must be converted into phonological codes and conveyed to the area of the frontal operculum. The arcuate fasciculus was thought to be one of the critical pathways over which this information is communicated, and if interrupted, repetition deficits might occur (Geschwind, 1965) (see Figure 9–22). However, identifying the anatomical basis for this disorder has proved not to be that simple. While lesions affecting the arcuate fasciculus were believed to be associated with conduction aphasia, this was not always found to be the case (Albert et al., 1981, Benson, 1979, 1993; Benson & Geschwind, 1985); more recently, Damasio et al. (2000) have suggested that at least one of two cortical areas (the supramarginal gyrus or the dominant primary auditory cortices) are typically compromised.

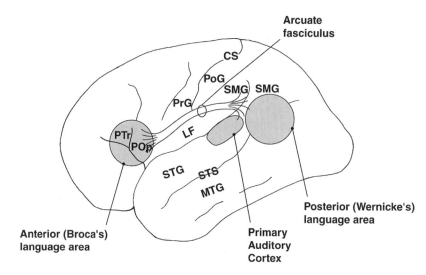

Figure 9–22. General language areas associated with conduction aphasia. Lesion most commonly thought to involve arcuate fasciculus, supramarginal gyrus, and/or primary auditory projection areas. Abbreviations: CS, central sulcus; LF, lateral fissure; MTG, middle temporal gyrus; PoG, postcentral gyrus; Pop, inferior frontal lobe, pars opercularis; PrG, precentral gyrus; PTr, inferior frontal lobe, pars triangularis; SMG, supramarginal gyrus; STG, superior temporal gyrus; STS, superior temporal sulcus.

Conduction aphasia might be contrasted with two other classic aphasic syndromes: Wernicke's and transcortical sensory aphasias. While repetition similarly is compromised in Wernicke's, comprehension is usually no better, if not worse. In the latter, the connections between the secondary auditory association areas and the frontal cortex are presumed to be intact. However, either the connection between the secondary auditory association area and the semantic association areas of the posterior temporal regions (and/or the semantic association areas themselves) are thought to be disrupted. Thus, in transcortical sensory aphasia, repetition is relatively intact, while by contrast auditory comprehension notably is impaired.

Callosal Syndromes

When discussing disconnection syndromes, the most fascinating examples probably are those provided by the split-brain studies of Gazzaniga, Bogen, and Sperry (1962, 1965, 1967). This procedure typically involves the severing of all or part of the corpus callosum and occasionally other commissural fibers as well. For some time this procedure was thought to have no significant behavioral consequence as many patients failed to evidence noticeable change in their normal activities following the early recovery period (Akelaitis, 1944). However, in the early 1960s Bogen, Sperry, and Gazzaniga proved that major behavioral consequences indeed could be demonstrated if only the right questions were asked of the patient.

Initially, Bogen and his colleagues found that if information were restricted to one hemisphere (e.g., through tactual input to one hand only or to one visual field through tachistoscopic presentation), the other hemisphere did not appear to know what had taken place. For example, if an object were placed in the patient's right hand, the left hand could not retrieve its match from assorted objects. If an object or picture of an object were placed in a subject's left hand or projected to the left visual field, the patient (i.e., the left hemisphere) could not name the object, despite the fact that the left hand (right hemisphere) accurately could demonstrate its use. If the anterior commissure is also severed, the patient may be able to pick out with his left hand objects associated with a particular smell presented to his left nostril but fail to be able to name the smell.[17]

As a result of their studies, these investigators provided additional support for some of the theories of hemispheric specialization. They were able to demonstrate, for example, that the left hemisphere normally was dominant for language. Although the right hemisphere evidenced some capacity to comprehend simple words, it could not initiate speech or produce intelligible spontaneous writing.

Following such surgical disconnections, while the left hand is likely to evidence ideomotor apraxia, it also is likely to perform better than the right in tests of visual–spatial constructional ability. Again, this would suggest right hemisphere superiority for these tasks. Most of these findings, along with specific suggestions for examining patients suspected of having callosal syndromes, are described in greater detail by Bogen (1993).

If all this appears fairly straightforward and predictable, consider the studies by Cronin-Golomb (1986) and Sergent (1987). Here "split-brain" patients were presented with independent visual stimuli to each hemifield. If asked whether the two stimuli were identical or not, the patients were unable to respond above chance. However, if asked to arrive at some judgment regarding the two stimuli, such as whether they were categorically related (e.g., a shoe and a sock) or whether the total number of dots in both hemifields combined were greater or less than 10, the patients performed reasonably well. This was despite the fact that they still could not verbalize what stimuli had been presented to their left visual field! Although Liederman (1995) interprets these findings as evidence of subcortically mediated, implicit awareness, others have discounted the reliability or validity of such findings (Corballis, 1994; Kingstone & Gazzaniga, 1995).

Reconsider the studies of Downer (1961) with macaque monkeys discussed in Chapter 8. Using split-brain preparations in which the optic chiasm also was cut (sagittally), thus restricting vision in each eye to the ipsilateral hemisphere, these monkeys were subjected to unilateral temporal lobectomies. Macaques normally do not particularly care for humans and frequently become agitated when they come into view. However, Kluver and Bucy (1939) had demonstrated that following bilateral temporal lesionsthese monkeys tended to be docile in the presence of humans (as well as showing indifference to a variety of other previously frightening stimuli). If the monkeys in Downer's study were fitted with a patch over the eye ipsilateral to the temporal lesion and then exposed to humans, they responded like a normal macaque (e.g., showed signs of agitation). However, if the eye ipsilateral to the lesion were the only one left uncovered, they remained placid at the sight of humans. Do such animals have one or two brains? In his chapter cited above, Bogen (1993) reports how in some instances the two hemispheres (when disconnected) appear to be working at cross-purposes or with "two minds at the same time." For example, one patient reportedly was observed buttoning his shirt with one hand and unbuttoning it with the other. Yet for the most part what is intriguing is that in most situations the behavior of such individuals appears perfectly normal, with the deficits generally becoming evident only under specific test conditions.[18]

Finally, one must attempt to resolve the discrepancies between naturally occurring and surgically induced lesions. While surgical lesions of the corpus callosum routinely produce the symptoms described above, naturally occurring agenesis of the corpus callosum does not. In contrast, naturally acquired lesions of the anterior portion of the corpus callosum (e.g., stroke, tumor) often lead to left-sided apraxia, but surgical lesions restricted to this region fail to produce this effect (Risse et al., 1989).

These latter phenomena once again demonstrate the mysteries of the nervous system. Again, the possible organization and integrated operation of the brain will be discussed from a broader perspective in Part III. The reader is forewarned, the goal will not be to provide definitive answers, but rather to offer a heuristic model that may lead to a better appreciation of the complexities of brain-behavior relationships and hopefully generate a conceptual framework with which clinical problems can be approached. For now, attention will be turned to structural and functional asymmetries.

HEMISPHERIC SPECIALIZATION

Anatomic Asymmetries

The existence of functional asymmetries in the human brain has been accepted as an axiomatic principle in neurology for well over a century. It commonly is accepted that the left hemisphere is specialized for language and other symbol-based functions and is probably the leading hemisphere in carrying out certain skilled motor activities and in mediating certain aspects of body awareness (body schema). In contrast, the right cerebral hemisphere seems to be specialized for *nonverbal*[19] functions, including directed attention, emotional–affective processing, certain aspects of musical abilities and in particular visuospatial or visuoperceptual abilities. However, as will be seen, while this dichotomy works reasonably well for most right-handers, the situation becomes more complicated with left-handers and there also appear to be gender differences (for review, see Hellige, 1993; Molfese & Segalowitz, 1988; Geschwind & Galaburda, 1984). With the advent of more reliable neuroradiographic and neurobehavioral techniques for in vivo investigations of the brain, it has become increasingly apparent that there often is considerable individual variation in the structural and functional organization of the cerebral hemispheres. Before exploring functional hemispheric differences, it may be helpful to explore what currently is known regarding structural hemispheric differences.

Postmortem Studies

Although cerebral dominance has been studied for over 140 years, the neuroanatomical substrates for lateralized behaviors have received only recent and relatively scant attention. At the turn of the 20th century, anatomists observed some measurable gross anatomical asymmetries of the lateral sylvian fissure and language-related cortical areas (von Economo & Horn, 1930; Eberstaller, 1890; Cunningham, 1892; Pfeifer, 1936). In fact, some of these anatomists were reporting these asymmetries at the same time that neurobehaviorists like Paul Broca and Carl Wernicke were advocating that functional brain asymmetries existed. The anatomists, however, generally felt that the gross asymmetries observed were not significant enough to account for the suspected behavioral asymmetries. It was not until 1968 that anatomical asymmetries of language-related cortex were proposed to be related to known functional asymmetries. Geschwind and Levitsky (1968) measured the **planum temporale** (Figure 9–23), which constitutes part of Wernicke's area, in 100 postmortem brains and found that the left planum temporale was larger in 65% of the brains studied compared to only 11% in which the right was larger. Although individual language dominance and hand preference was unknown in their sample, given that language is functionally lateralized to the left hemisphere in most right-handed individuals, Geschwind and Levitsky proposed that the asymmetry of the planum temporale was compatible with the known functional asymmetries, and therefore might represent the neuroanatomical substrate for language. This notion is supported by the fact that the planum temporale encompasses both auditory association areas, which are important in higher-order processing of language input and part of Wernicke's area, which when lesioned produces fluent aphasia with impaired comprehension, repetition, and naming. In addition, subsequent studies (Tezner et al., 1972; Witelson & Pallie, 1973; Wada, Clarke, & Hamm, 1975; Galaburda et al., 1987)

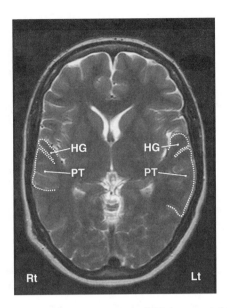

Figure 9–23. Approximation of the planum temporale (PT) in an axial MRI showing the asymmetry between right and left hemispheres. Also shown is Heschl's gyrus (HG).

have replicated Geschwind and Levitsky's findings and demonstrated that the predominant leftward asymmetries of the planum temporale exist as early as the 29th gestational week. The finding that asymmetries of speech-related cortex are present before language acquisition suggests that the human brain has biologically determined, anatomical substrates that are preprogrammed for the asymmetric representation of speech–language functions in most right handers.

The asymmetries of the planum temporale reflect in part asymmetries of the lateral sylvian fissure. As described earlier in the section on cerebral sulci and gyri, the sylvian fissure has five segments (see Figure 9–9). The longest segment is the posterior horizontal ramus (PHR) of the sylvian fissure. The length and distribution of this segment has been measured in studies of asymmetries of the sylvian fissure as far back as the 19th century. Specifically, in a majority of cases studied, the **left posterior horizontal ramus** (PHR) was found to be longer on the left (Eberstaller, 1890) and less angled than the PHR on the right (Cunningham, 1892). More recently, in examining 36 postmortem brains, Rubens, Mahowald, and Hutton (1976) confirmed the asymmetries of the PHR reported by these earlier investigators and demonstrated that these sylvian fissure asymmetries were related to asymmetries found in the planum temporale and in the inferior parietal lobule. It is reasonable to assume that these asymmetries of the sylvian fissure and adjacent cortical areas provide the anatomical substrate not only for language, but perhaps for other hemispheric functional differences as well. For example, there is a relative expansion of the anterior portions of the parietal operculum and a reduction in the posterior portions of the parietal operculum in the left cerebral hemisphere when the sylvian fissure is longer in the left hemisphere. In contrast, the parietal operculum is enlarged posterior to the sylvian fissure in the right hemisphere due to the shorter sylvian fissure length and more angled terminal upswing to the sylvian fissure (Figure 9–24). These hemispheric differences in asymmetries of anterior and posterior portions of the parietal operculum appear to support the notion that the enlarged anterior left parietal regions may be more crucial for left hemisphere dominant functions, such as language and praxis. In contrast, the enlarged posterior portions of the

(a)

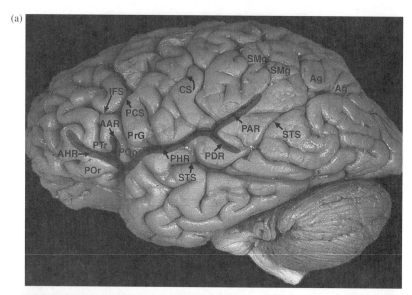

(b)

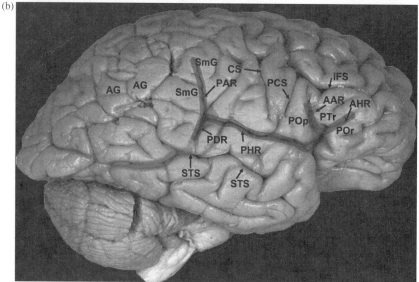

Figure 9–24. As illustrated in the figure, the most noticeable difference in the lateral fissure between the two hemispheres is the sharper rise in the posterior ascending ramus in the (b) right hemisphere. Abbreviations: AAR, anterior ascending ramus; AHR, anterior horizontal ramus; PAR, posterior ascending ramus; PDR, posterior descending ramus; PHR, posterior horizontal ramus (see Figure 9–8 for additional keys). Brain images adapted from the *Interactive Brain Atlas* (1994), courtesy of the University of Washington.

right parietal operculum may be more important in visuospatial and attentional systems, which are more commonly lateralized to the right hemisphere.[20]

Anatomical asymmetries of the frontal operculum have been more difficult to study. Eberstaller (1890) first described in detail the anatomy and complex morphology of the rami that comprise the inferior frontal gyrus. He noted that the frontal operculum is highly convoluted and that the lateral surface anatomy may not reflect the true relationships of the rami that form the frontal, language-related cortex. Before asymmetries of this region are discussed, the anatomical boundaries of Broca's area need to be described. As also can be seen in Figure 9–24, the inferior frontal gyrus is divided into three constituent parts by the anterior rami of the sylvian fissure: **pars opercularis**, **pars triangularis**, and **pars orbitalis**. Broca's area is composed of two adjacent regions: **pars triangularis** and **pars opercularis**. Pars triangularis (Brodmann's area 45) is bounded superiorly by the inferior frontal sulcus, inferiorly by the anterior horizontal ramus (AHR), and posteriorly (caudally) by the anterior ascending ramus (AAR). Thus, the base of the "triangle" is formed by the inferior frontal gyrus and the sides are formed by the AHR and AAR. The size of pars triangularis depends on the angle that these rami form with each other. Within the angle formed by the AHR and AAR, there may be two or more sulci that reach the surface and give this region the appearance of multiple convolutions within the larger triangular gyrus. Pars opercularis (Brodmann's area 44) is located adjacent to the pars triangularis but in a more caudal or posterior location. Pars opercularis is bounded superiorly by the inferior frontal sulcus, anteriorly by the AAR, and posteriorly by the precentral sulcus. In many cases, pars opercularis is divided into halves by the diagonal sulcus, which is a branch of the sylvian fissure. In some cases the diagonal sulcus is absent or branches directly from the AAR. The foregoing discussion illustrates the complex morphology of Broca's area. Given that the planum temporale, which constitutes part of the posterior speech region (Wernicke's area) is anatomically a flat triangular plane on the surface of the superior temporal gyrus, asymmetries in the latter have been easier to measure. In contrast, attempts to measure asymmetries in Broca's area have met with more limited success, although the pattern of gyrification of the third frontal convolution was noted to be more elaborate and perhaps deeper in the left hemisphere on postmortem brains (Wada, Clarke, & Hamm, 1975; Falzi et al., 1982; Nikkuni et al., 1981; Albanese et al., 1989). These findings were suggestive that an anatomical asymmetry existed in the region of Broca's area such that the total surface area of this region was larger in the left hemisphere. This again would support the notion that this area on the left is critical for some aspects of speech–language functions.[21] For additional reviews of this subject, see Witelson and Kigar (1988).

In Vivo Studies

Anatomical asymmetries derived from post-mortem brains have contributed significantly to our understanding of anatomical brain asymmetries of language related cortex that probably reflect some aspects of hemispheric specialization for language functions. However, two major limitations to these studies are the uncertain correlation between structural asymmetry and lateralization of language function in any given case and the need to wait for the individual to die before the brain could be studied. In order to overcome these limitations, a noninvasive way to study structural differences in individuals while they were alive and a way to more definitively document language lateralization (and handedness) were needed. While handedness easily can be established for most individuals, language was another matter. The advent of the Wada procedure in 1949 (see Wada & Rasmussen, 1960) was one solution to part of the problem. By selectively injecting a barbiturate into one of the internal carotid arteries, and hence selectively sedating one cerebral hemisphere, it was possible to derive a reasonably reliable estimate of the leading hemisphere for language functions. The other factor in the equation was to find a way to visualize the brain in the patient while he or she was still alive. The development of pneumoencephalography, cerebral angiography, and particularly advanced radiographic techniques, such as computerized axial tomography in the 1970s and magnetic resonance imaging scans in the 1980s, provided investigators with the necessary tools. By combining these methodologies (brain imaging and Wada testing), it became possible to explore these structure–function relationships more directly.

Pneumoencephalography, a radiographic method that involves injecting air into the ventricular system and then radiographically visualizing the contours of the lateral ventricles on lateral or coronal views was one of the earlier techniques available for visualizing the brain. Although the clinical information provided by the pneumoencephalograms was rather limited, early studies demonstrated asymmetries of the occipital horns of the lateral ventricles. The most common pattern was a longer left occipital horn in right-handed individuals (McRae, 1948). However, McRae, Branch and Milner (1968) failed to demonstrate any relationship between language dominance determined by selective hemispheric anesthesia (Wada testing) and occipital horn length. Two possible explanations were suggested. First, it was noted that the exact length of the occipital horns derived from pneumoencephalography might be unreliable in some cases due to difficulty in visualizing the anterior extent of the lateral ventricles. Second, the patient population was composed of individuals with intractable epilepsy who may have had early brain injury that resulted in anomalous brain organization or reorganization of biologically preprogrammed of language functions.

Cerebral angiography was another investigative tool. A lateral view of the arterial phase of the middle cerebral artery (MCA) provides an estimate of the length, angle, and morphology of the sylvian fissure. Hochberg and LeMay (1975) found that this angle differed in right- and left-handers. Whereas the most common pattern in right- handers was the expected greater angulation of the right sylvian fissure, this pattern was observed in only 21% of the left-handers. These findings were thought to support the notion that "typical" patterns of asymmetry are more common in right-handers but less common in left-handers. From these data, however, it still was unclear whether these anatomical asymmetries correlated with hemispheric specialization of language.

The advent of **computerized tomography** (CT scans) and later **magnetic resonance imagery** (MRI) allowed the brain to be viewed in situ with much greater clarity and accuracy (Pieniadz & Naeser, 1984). In a series of CT scan studies, LeMay (1976, 1977) noted typical patterns of skull-based asymmetries that differed in right- and left-handers. In right-handers the most common pattern was to have a greater frontal protuberance or *petalia* on the right, with a greater occipital protuberance on the left. In contrast, left-handers less commonly exhibited this "typical" pattern and were more likely to have an "atypical" left frontal and/or a right occipital protuberance. Other investigators confirmed these handedness differences (Chui & Damasio, 1980; Koff et al., 1986). As we shall soon see (under Anomalous Dominance), atypical patterns of hemispheric dominance may offer some advantages, for example, greater recovery of acquired language disturbance following strokes.

With the development of the volume acquisition MRI, which involves the rapid acquisition of thin, contiguous images, the brain readily could be visualized in vivo in ways that previously were inaccessible using conventional CT or MRI scans or even postmortem specimens. In particular, regions of the brain, such as the planum temporale, could be accurately measured using these methodologies with techniques pioneered by Steinmetz and colleagues (1989). They were able to demonstrate that asymmetries of the planum temporale differ by hand preference, such that right-handers have an increased incidence and magnitude of leftward asymmetry (greater surface area) in contrast to left-handers who have a reduced leftward asymmetry (Steinmetz et al., 1989, 1991). Foundas and colleagues (1994) subsequently measured the planum temporale on volumetric MRI scans of patients who had selective hemispheric anesthesia (Wada testing) performed for language lateralization. A leftward asymmetry of the planum temporale was found in all 11 subjects who had language lateralized to the left hemisphere. The one subject, who had a marked rightward asymmetry of the planum temporale, had language lateralized to the right hemisphere. Although these data must be interpreted cautiously given the clinical population of patients and small sample size, these data support the notion that the planum temporale plays an important role in language dominance and that anatomic asymmetries measured on volumetric MRI may predict some aspects of language dominance.

The planum temporale was not the only anatomic asymmetry that could be shown to predict language dominance using volumetric MRI measures. The depths of the convolutions that constitute Broca's area (e.g., pars triangularis) also were more accessible to measurement using the MRI than on routine postmortem investigation. Using these techniques with a group of normal right- and left-handers, Foundas, Leonard, and Heilman (1995) found that seven of eight right-handers examined had a leftward asymmetry of the pars triangularis and one had equal measures. In contrast, the pars triangularis was larger on the left in three of eight left-handers, was equal in one, and larger on the right in four of eight left-handers. Furthermore, when the pars triangularis was measured on volumetric MRI scans of patients who had undergone Wada testing for language localization, nine of the ten patients with language lateralized to the left had a leftward asymmetry of the pars triangularis. The one patient with language lateralized to the right hemisphere had a significant rightward asymmetry of the pars triangularis (Foundas et al., 1996). These data offer further evidence that cortical asymmetries (in this case, the pars triangularis) underlie some aspects of language lateralization.

Functional Specialization

Handedness

In discussing the functional aspects of hemispheric specialization, both historically and at present, probably the two concepts that most frequently come to mind are language and handedness. Hand preference is an easily observed behavior and will be discussed in some detail. Before reviewing some of the data on the relationships among handedness, language, and hemispheric specialization, it might be useful to briefly define handedness. When speaking of handedness, one is referring to the fact that one side of the body is "better" (i.e., more facile) in performing certain skilled tasks, such as writing (perhaps the most commonly used index of handedness), throwing (or kicking) a ball, and using a spoon or a fork. The implication is that this preference is a result of genetics or innate hard wiring in the nervous system rather than a practice effect or a concession to a contralateral insult or impairment early in life. However, unilateral preference is not an all-or-none phenomenon. An individual may show preferences for the use of a particular eye (e.g., when sighting a rifle), a particular ear (for listening on the phone), a particular foot (for kicking a ball), or a particular hand (when writing). Although these preferences are often on the same side in right-handers, they are dissociable. Even the preferred hand may not be consistently favored for all skilled manual tasks. This seems to be especially true of left-handers who as a group often are not consistently left-handed. In fact, it is not uncommon for left-handers to preferentially use the right hand for many unimanual tasks. Furthermore, many left-handers may be more skillful or dexterous with their "nondominant" right hand for certain activities (Satz, Achenbach, & Fennel, 1967). Although it remains to be empirically tested, it may be that the left-handers who are more *anomalous* (less consistently left-handed) also are more anomalous with regard to hemispheric specialization of language and visuospatial abilities (also see below under Concepts of Cerebral Dominance, and anomalous Dominance).

As noted above, several structural asymmetries between the cerebral hemispheres have been established. However, it is not always clear to what extent these differences may reflect dominance for language, rather than handedness per se. With regard to the anatomical bases for language, as we have seen it appears that the perisylvian "language" areas, such as portions of the inferior frontal gyrus and the planum temporale, may be enlarged and more complex on the left. Comparable findings have not been as firmly established for other primary motor areas other than those related to oral motor speech. Several investigators have studied anatomical asymmetries of cortical and subcortical structures and motor pathways that are associated with lateralized motor functions, including handedness. For example, White et al. (1994) found a trend for increased neuronal density in the area surrounding the central sulcus; Kooistra and Heilman (1988) reported that the globus pallidus was larger in the left hemisphere[22]; and Kertesz and Geschwind (1971) found an asymmetry in the crossing of the corticospinal fibers at the decussation of the pyramids (fibers from the left hemisphere tended to cross above the level of those from the right), although the reliability of the later phenomenon has been called into question by other authors (Witelson & Kigar, 1988). Asymmetries in the central sulcus (in right-handers) and the size of the pars opercularis (in both right- and left-handers) were found to be correlated with handedness using MRI data (Foundas et al., 1998a,b).While these findings may be suggestive of an anatomical correlate of handedness, the precise relationship of these reported asymmetries to lateralized motor behaviors is speculative and what constitutes the neurological basis for handedness remains a mystery. Is it a matter of increased neuronal or dendritic density, complexity of associations, an interaction of practice effects, or other factors that account for the increased skill demonstrated by the dominant hand? These are questions that remain to be investigated.

The next question to be addressed is the relationship between handedness and language localization. In 1865, Dax postulated the doctrine of cerebral dominance which contends that language is mediated by the cerebral hemisphere opposite to the preferred hand for writing. This hemisphere was considered to be the "dominant" hemisphere, while the hemisphere ipsilateral to the preferred hand was considered to be the "nondominant" hemisphere. According to Dax's doctrine, right-handers would have language lateralized to the left hemisphere, while left-handers would have language lateralized to the right hemisphere. Numerous clinical and neurobehavioral studies have demonstrated that this doctrine is not completely correct. What has been found is that indeed language is lateralized to the left hemisphere in anywhere from about 95 to 98% of right-handers (Milner, 1974; Rossi & Rosadini, 1967). However, left-handers are not the mirror image of right-handers. About 70% of left-handers have language lateralized to the left hemisphere, suggesting dissociation between hemispheric dominance for manual hand preference and language in these individuals. The hemispheric localization of language for the remaining 30% of left-handers is divided between those in whom language appears to be organized primarily in the right hemisphere versus those in whom language seems to be bilaterally represented (Herron, 1980).[23]

Nonetheless, there are more left- than right-handers with language functions lateralized to the right hemisphere or bilaterally represented, a condition referred to as anomalous dominance (Hellige, 1993; Molfese & Segalowitz, 1988; Geschwind & Galaburda, 1984). Thus, while Dax essentially was correct with regard to right-handers, there still is a fair degree of uncertainty with respect to patterns of dominance and handedness, especially in left-handed individuals.

Concepts of Cerebral Dominance

Speculation about the existence and nature of hemispheric or cerebral dominance is at least as old as speculation about functional differences between the cerebral hemispheres (Broca, 1861; Dax, 1865; Wernicke, 1874) and has remained a major focus of scientific study into the present (e.g., see Davidson & Hugdahl, 1995). The term "dominance" can be used in a more restricted or in a very broad sense. In its more classically and narrowly defined form, we refer to the dominant hemisphere, meaning the hemisphere that is primarily responsible for understanding and expressing language and, in right-handers at least, the hemisphere controlling the "dominant" hand. This is the context in which the term thus far has been used in this chapter. Because of the preeminence of language with regard to studies of hemispheric specialization, this generally will be its connotation in the following discussions on anomalous dominance. However, the term "dominance" also can be used in a more general manner. In this context, when speaking of "cerebral dominance," it is fair to ask, *dominance for what?* One might just as readily speak of "dominance" for visuospatial abilities, directed attention, musical abilities, emotional responsiveness, social perceptiveness, or other equally relevant skills or abilities. The implication here is that skills, functions, or abilities, other than language, also may be differentially organized or distributed in the two cerebral hemispheres. This differential organization will be discussed in greater detail later, but first, it may be useful to once again apply the general principle of **distributed systems** (Mountcastle, 1979) in the current context of cerebral dominance.

At the most basic level, the concept of cerebral dominance appears relatively simple: a particular given function, such as language, primarily is organized in, or mediated by, a particular hemisphere. Behaviorally, this would imply that if critical areas of the "dominant" hemisphere were damaged, that particular function should be expected to suffer. Conversely, if the opposite, "nondominant" hemisphere is lesioned, regardless of the site of the injury, no disruption of that function should result. Thus, an infarction of the proximal branches of the left middle cerebral artery in a right-handed individual should produce significant aphasic deficits, whereas a comparable lesion of the right middle cerebral artery should not. While, for the most part, this may be true, even with regard to language (which, at first blush, would appear to be a highly lateralized function in most individuals), things are not quite that simple.

One cerebral hemisphere may be associated with a particular behavioral function (such as language) or preferentially processing certain types of information, and hence, said to be "dominant" for that cognitive process. However, components of most behavioral functions or types of information processing are likely to be distributed between both cerebral hemispheres. There is now ample reason to believe that most cognitive processes require some input from each cerebral hemisphere, although one hemisphere may be the "leading" or dominant hemisphere. Hughlings Jackson originally proposed this idea in the latter part of the 19th century. This conceptualization of hemispheric specialization with distribution of cognitive processes across the hemispheres utilizes principles that were supported by both the *localizationists* (such as Joseph-Jules Dejerine) and the *wholists* (such as Pierre Marie) in the early part of the 20th century.[24]

Currently, the idea that the left cerebral hemisphere is specialized for all speech and language functions has been replaced by the view that specific components are mediated by the left and right hemispheres. Whereas the propositional and linguistic aspects of speech production, phonetic decoding, syntax, and semantic language functions are thought to be mediated by the left hemisphere, certain nonverbal aspects of speech appear to be primarily controlled by the right hemisphere (Meyers, 1994). These nonverbal speech/language functions include (1) the use of inflections or intonations in spoken language to convey specific emotional tones; (2) comprehension of similar inflections or emotional aspects in the speech of others, (3) certain metalinguistic aspects of language such as the appreciation of humor and metaphors and understanding the underlying message or context of complex communication, and (4) the overall organization of written or spoken prose (see Right Hemisphere Specialization).

In addition to language skills, different components of visuospatial processing also are thought to be relegated to both the left and right cerebral hemispheres. Whereas the right hemisphere likely processes global (wholistic) aspects of visuospatial input, it is believed that the left hemisphere processes "local features" (Delis et al., 1988). Edith Kaplan (1988) and colleagues (Milberg, Hebben & Kaplan, 1986; Kaplan et al., 1991) have demonstrated these hemispheric differences in processing visual–spatial information by analyzing the different approaches used by left versus right hemisphere-damaged patients to solve the Block Design and Object Assembly subtests of the Wechsler Adult Intelligence Scale. She found that whereas the left hemisphere-damaged patients attended to the outer contour or "gestalt" on the Kohs blocks and tended to preserve the outer configuration (matrix), the right hemisphere-damaged patients attended to the inner details or "features" of the design and often would break the outer configuration. Thus, the left hemispheric patients were utilizing the preserved right hemispheric "global" or gestalt abilities to solve the block design, while the right hemispheric patients seemed to be relying more on specific details or "local features" (functions better served by the intact left hemisphere). Similarly on Object Assembly, right hemisphere patients had more difficulty in recognizing the overall configuration of the puzzle, while the left hemisphere patients reportedly had difficulty with the details. Earlier, a number of investigators found those types of errors (e.g., left-sided neglect, fragmentation, loss versus preservation of gestalt, amount of elaboration) tended to differentiate graphomotor copies of geometric figures completed by individuals with either right or left hemispheric lesions (Arena & Gainotti, 1978; Gainotti & Tiacci, 1970; Hecaen & Assal, 1970; Warrington et al., 1966; Arrigoni & DeRenzi, 1964; Piercy et al., 1960)

Finally, it also has been suggested that not only do the hemispheres differ in the types of information they process, but in fact may organize and process that information in fundamentally different ways. The left hemisphere is thought to organize information in a more focal or discrete manner, thus enabling it to use a more linear or sequential and logical approach, which may explain its greater facility in handling detail. In contrast, the right hemisphere appears to rely on a more diffusely organized, simultaneous, gestaltist model, which would better prepare it for perceiving more global patterns, configurations, or relationships (DeRenzi & Faglioni, 1967: Semmes, 1968; Delis et al., 1988). Such a functional organization and operational strategy is believed to be compatible with abstractive capacities and rational thought characteristic of the left hemisphere and the emotional, nonlinear, intuitive approach that often is attributed to the right hemisphere. If different hemispheric functional approaches are indeed the case, this model may help explain why the organization of language in both hemispheres, while enhancing the verbal skills of certain groups or individuals, may do so at the expense of developing extraordinary visuospatial abilities and vice versa. Data that would support these ideas, which will be reviewed next, are still controversial and circumstantial.

Anomalous Dominance

Handedness and dominance for language typically are mediated by the left hemisphere in the vast majority of right-handed individuals. Thus, a lesion to the left cerebral hemisphere, particularly one that involves the middle cerebral artery distribution, might be expected to produce an **aphasia** (a disorder of language), **hemiparesis** (a disorder of motor strength, dexterity, and skill) affecting the dominant hand, and somewhat less consistently a bilateral limb **apraxia** (a disorder of learned skilled limb movements). Since this pattern of brain organization is typical of most right-handers, it generally is considered to be the "dominant" or typical pattern of brain organization. In contrast, any pattern of organization that differs from this "typical" pattern commonly is considered to be "anomalous" and in some cases "dysfunctional." Such anomalous patterns are more likely to occur in left-handers. As noted earlier, it is estimated that approximately 30% of left-handers exhibit some form of anomalous dominance for language (i.e., language being organized either primarily in the right hemisphere or represented bilaterally). Although anomalous dominance can occur in right-handers, it probably is quite rare. Other associations that have been reported to be related to anomalous patterns of hemispheric organization of language are female gender, mixed hand preference (ambidexterity), and family history of sinistrality. Less well documented is writing position, with the inverted hand position reportedly being more frequently associated with left-hemispheric dominance for language (see Geschwind & Galaburda, 1985; Herron, 1980).

One intriguing finding with regard to hemispheric specialization has centered on possible gender differences both in the structure and functional organization of the cortex. While there probably is considerable overlap or individual variation, it has been suggested that differences in cognitive functions may exist between the sexes when these two groups are considered as a whole, and that these differences reflect not merely social or cultural factors but basic differences in the way cognitive functions are organized in the cerebral hemispheres (Geschwind & Galaburda, 1984; McGlone, 1977, 1986). Specifically, it has been proposed that language may tend to develop earlier, become more highly developed, and may be more bilaterally represented in women. In contrast, it has been suggested that in men language tends to be more unilaterally organized, resulting in a relatively greater development of visuospatial skills as compared to language (Halpren, 1991). Although there still is considerable controversy surrounding this issue, it always is interesting to speculate how, from a Darwinian standpoint, this might occur. For example, one possible scenario is that if the males of the species were indeed the primary hunters and defenders of their family units or tribes, it is easy to imagine how enhanced visuospatial abilities may have had important survival value (and hence, a valuable trait in natural selection). Whatever factors determine handedness and rate of cerebral maturation of the hemispheres, it is likely that gender differences and hormonal influences are not without some effect. (Annett, 1985, 1995).

Lending support to these speculations of functional and developmental difference between the sexes, certain gender-based, structural differences also have been identified. Gender differences in human brain anatomy were first reported by Crichton-Browne (1880), who suggested that the weights of female cerebral hemispheres were more symmetric than male brains. Wada, Clarke, and Hamm (1975) in a study of 100 adult and infant postmortem brains reported that variations in the typical leftward asymmetry of the **planum temporale** was more likely to occur in males than in females. Using current in vivo technology, males were found to have a more significant leftward asymmetry of the planum temporale, in contrast to females where the expected leftward asymmetry was reduced (Witelson & Kigar, 1992; Kulynych et al., 1994). Similarly, a functional greater asymmetry in the frontal speech areas was demonstrated in male as opposed to female subjects using functional neuroradiology techniques (Shaywitz et al., 1995).

While what has been described as the more "typical" pattern of brain organization (strong left hemispheric lateralization for language and strong right hemispheric lateralization for visuospatial skills) would appear to offer some advantages, there may be a few unique, even if not universally expressed, advantages to anomalous patterns of dominance. For example, it has been suggested that left-handers may suffer less severe aphasic syndromes, on average, and recover more quickly (Subirana, 1964, 1969; Luria, 1970; Benson & Geschwind, 1985; Pieniadz et al., 1983). As has been mentioned, the tendency for greater bilateral representation of speech may give certain individuals a greater verbal advantage. There would appear to be a disproportionate number of highly successful professional females who are left-handed as well as individuals who perform exceptionally well on the Scholastic Aptitude Test (Benbow, 1988). It also has been suggested that while additional delays in the maturation of the left hemisphere in males may lead to an increased incidence of language-related problems (e.g., stuttering, learning disabilities), it also may lead to the development of extraordinary nonverbal talents or skills in such individuals (Geschwind & Galaburda, 1987; Smith et al., 1989). Although such ideas are provocative, additional research is needed before their validity and/or the specific factors that are responsible for these effects can be more definitively established.

Left Hemisphere Specialization

Language

The special association between the left hemisphere and language functions has been repeatedly emphasized throughout this chapter. However, so far, the precise nature of those functions has not been elaborated. While an analysis of the possible functional organization of language and its interaction with other aspects of behavior will be offered in Part III, at this point, it might be useful at least to list in greater detail the linguistic and other contributions of the left hemisphere. A good starting point for this discussion might be to recall the beginnings of language localization. As noted earlier, in 1861, Paul Broca described eight right-handed patients who lost the facility of speech and were found to have left hemispheric lesions. Broca thought that the left third frontal convolution, including the **pars triangularis** (*Brodmann's area 45*) and **pars opercularis** (*Brodmann's area 44*), was critical for speech, since this portion of the frontal operculum was lesioned in his patients. In 1874, Carl Wernicke demonstrated that a lesion in the posterior portion of the left **superior temporal gyrus** (*Brodmann's area 22*) produced a syndrome different from that described earlier by Broca. Whereas Broca's patients lost the ability to speak fluently but retained auditory comprehension, Wernicke's patients retained the ability to speak but were impaired at the level of auditory speech comprehension. These two observations basically set the two anchors for the role of the left hemisphere with regard to localized language functions. Although far from reflecting pure dichotomies of function, Broca's findings have come to represent the external production of speech or language, whereas Wernicke's findings generally symbolize the ability to internally utilize or process language or linguistic symbols. Lesions that directly affect these areas or lesions of association pathways to or from these areas produce deficits of language function thought to be mediated by the left hemisphere. These include:

1. Spontaneous production of oral (speech) or written language.
2. Reproducing patterns of speech or language generated externally (e.g., repetition or transcription).
3. Generating linguistic associations (e.g., word finding and confrontation naming).
4. Making meaningful associations from linguistic stimuli symbols (e.g., auditory or written comprehension).
5. Using internal language to solve problems or reach new insights (e.g., abstract thinking).

While each of these functions will be reviewed briefly below, for a more in-depth coverage of language disorders the reader is referred to the works of Albert et al. (1981), Benson (1979), Goodglass (1993), and Goodglass and Wingfield (1997).

Production of Oral or Written Language.   Normal spontaneous speech begins with the intent to communicate followed by the internal organization of the thought, access to the words to be used in expressing the thought or idea and their phonetic representations (word sounds), the initiation of the intention, and finally the actual production (articulation) of speech. Spontaneous writing makes similar demands, except rather than requiring the external articulation of phonemes, the phonemes are converted into written symbols (graphemes).

In typical dominance patterns, most of these language functions are mediated primarily by the left hemisphere. Whether the left, right, or both hemispheres are "responsible" for the *intent* to communicate is unclear. However, the failure to initiate spontaneous communication typically has been associated with left anterior (frontal) lesions (usually termed *dynamic* or *transcortical motor* aphasia). As demonstrated by split-brain studies (see, for example, Zaidel, 1985), access to words for expression and phonetic or graphic representation appears fairly restricted to the left hemisphere. Depending on location, lesions of the left hemisphere may result in either a paucity of words (most commonly associated with *nonfluent* or *Broca's* type aphasia), an inability to access the desired word (*verbal or semantic paraphasia*), or an inability to access its proper phonetic representation (*literal* or *neologistic paraphasia*). The latter are more characteristic of *fluent* or *Wernicke's* aphasia. While the left hemisphere also appears to play the primary role in converting the phonemic forms of language into intelligible speech sounds or written symbols [including the graphomotor and orthographic (spelling) components], damage to the right hemisphere has a more limited impact on speech and writing. Since the musculature controlling speech output (tongue, larynx, pharynx) is bilaterally innervated, interference with these centers or pathways in the right hemisphere can result in disturbances of articulation (*dysarthria*) without an underlying disturbance of language per se. The effects of right hemisphere lesions on writing typically are limited to problems with spatial organization and occasionally perseveration or repetition of individual letters. More subtle problems with the organization of either written or oral discourse also may be affected by damage to the right hemisphere and will be discussed in greater detail under "Right Hemisphere Functions."

Language Reproduction.   In contrast to language production, language *reproduction*, in its broadest sense, refers to the ability to reproduce language in either the same or alternate form from which it was perceived. Typically, when we think of this aspect of language, we think of the repetition of spoken language or the transcription of spoken or written language. However, reading aloud (as opposed to silent reading for comprehension) also may be considered language reproduction. Disturbances in repetition can result from faulty sensory integration or comprehension, production disturbances, disruptions between sensory input and sensory integration areas (e.g., *pure word deafness, pure alexia,* or *alexia without agraphia*), or disconnections between the anterior and posterior language areas. The latter is commonly referred to as *conduction* aphasia and was discussed under the section Disconnection Syndromes. Written transcription or reproduction may involve either writing to dictation or transcribing print into script. As described, these functions appear to be mediated solely by the left hemisphere. Except for possible problems with articulation (dysarthria) or the spatial or perseverative errors described above, or reading disturbances secondary to hemispatial neglect, right hemisphere lesions typically have little or no impact on these language reproduction abilities.

Word-finding Ability.   The ability to associate a "word" with either an internal (thought or recollection) or external (perception) representation of an object or idea is a fundamental function of language. Creating these associations (i.e., words) and then retrieving them, either spontaneously or on cue (e.g., when a patient is asked to provide the name for a particular object, action, color, attribute), appear to be skills relegated to the left hemisphere in cases where normal patterns of dominance are present. Studies on split-brain preparations in humans reveal virtually no capacity of the right hemisphere to carry out these activities (Bogen, 1993).

Word Recognition.   In addition to being able to generate or retrieve a word when needed (verbal expression), linguistic communication also demands that when a word is perceived, either aurally (auditory comprehension) or visually (reading comprehension), its meaning and/or associations are understood (verbal comprehension). Language comprehension may be broken down further into its semantic and syntactic components. **Semantics** refers to the ability to make concrete associations between an individual word and its referent (e.g., the word "rose" conjures up the image of a particular type of flower; "running," a particular type of activity; or "red," a particular color). **Syntax** refers to the ability to understand the relationship between words, which in turn conveys additional meaning to the communication. For example, the sentence, "Jane hit Bob on the head," has less syntactic complexity than, "Mary's mother's brother [Uncle Bob] was hit by his father's mother" [Grandma Jane]. While the left hemisphere clearly is dominant for comprehending both semantics and syntax, again in split-brain studies the right hemisphere has been shown to have some limited semantic capacity and even more limited ability to process syntax independent of the left hemisphere (Bogen, 1993; Zaidel, 1985). However, somewhat paradoxically, in the presence of an intact left hemisphere, right hemispheric damage may lead to significant difficulties in appreciating subtle or thematic aspects of communication, especially when metaphors or sarcasm are employed (see Right Hemisphere Functions). Table 9–2 provides a summary of commonly identified aphasic syndromes.

Internal Use of Language. Language not only is used for communicating with others, it also is used internally. It serves as an important base for abstract reasoning and problem solving. While both hemispheres contribute to the development of new and creative insights into the world around us, many of the problems presented to us on a day-to-day basis are represented in verbal terms. Even if not, we often try to assign words to our ideas, motivations, imaginings, and conflicts in order to analyze, manipulate, and weigh their various permutations and potential outcomes. Strictly speaking, what we define as rational thought and abstractive capacities appear to be the application of formal linguistic principles to a particular problem. Again, while the split-brain work has suggested that the right hemisphere certainly is capable of problem solving and decision making (in certain circumstances, apparently even more efficiently than the left hemisphere), it appears that it is the left hemisphere that mediates such thought processes in most individuals.[25]

Ideomotor Praxis

As discussed earlier under "Disconnection Syndromes" (p. 328), one functional disturbance often associated with the left hemisphere is **ideomotor apraxia**. The term, *praxis*, refers to the accurate execution of an action. Therefore, *apraxia* refers to the inability to carry out an action. However, this failure should not be the direct result of a loss of primary sensory or motor disturbances, an inability to understand what is being requested, or a general cognitive defect. Because the term apraxia has been applied to a wide variety of other perceptual–motor, visuospatial, and other difficulties (thus creating the potential for confusion), Table 9–3 lists and briefly reviews various ways in which the term apraxia has been used.

According to Geschwind (1975), ideomotor apraxia is the classic and perhaps only legitimate form of apraxia. Typically considered a disconnection-type syndrome, ideomotor apraxia results in the inability to properly carry out a previously learned skilled movement. The movements disrupted in ideomotor apraxia may involve transitive ("Show me how you would use a hammer") and/or intransitive ("Make a salute") gestures, and depending on the site of the lesion may be expressed in one or both hands. Inability to carry out an action requiring the use of the facial muscles ("Pretend to blow out a candle"), is called **buccofacial apraxia**, but this is thought to be simply another form of ideomotor apraxia by some investigators (Kimura, 1982), although these behaviors often are dissociable, and therefore may be mediated by partially independent neural networks (Raade, Rothi, &

Heilman, 1991). Typically, the most sensitive procedure is to have the patient pantomime an action, having provided only the verbal command. Patients with less severe pathology may improve markedly if allowed to use the real object. As noted above, in order to qualify as an apraxic deficit, the inability to properly execute the command should not result from basic disturbances in motor or sensory skills or from the inability to comprehend instructions.

Table 9–2. Aphasic Syndromes

Broca's Aphasia (Non-fluent, Motor, Expressive aphasia)
>   **Speech:** Effortful, dysarthric, agrammatic, telegraphic, non-fluent (sparse output) speech; anomic errors, literal paraphasias, if present (distinguish from dysarthria)
>   **Repetition:** Impaired, but typically better than spontaneous speech
>   **Auditory Comprehension:** Some impairment, but typically much better than expressive abilities, particular difficulties with complex syntax
>   **Reading:** Difficulty reading aloud, paraphasic errors. Reading comprehension similar to auditory
>   **Writing:** Similar to speech in terms of output and errors; misspellings; poorly formed block letters
>   **Lesion:** Frontal opercular area, with subcortical extension in more severe cases

Wernicke's Aphasia (Fluent, Sensory, Receptive aphasia)
>   **Speech:** Effortless, circumlocutory (loss of substantive, content words); frequent anomic errors; paraphasic errors common (literal, verbal, neologistic)
>   **Repetition:** Impaired
>   **Auditory Comprehension:** Significant impairment, may respond better to simple, whole body commands
>   **Reading:** Significant deficit, both aloud and for comprehension
>   **Writing:** Similar to speech in terms of content; graphically fluent, cursive style, but often meaningless
>   **Lesion:** Posterior superior temporal gyrus

Conduction Aphasia (Central, Afferent-motor aphasia)
>   **Speech:** Fluent, though blocking is possible due to recognition of anomic errors. Literal paraphasias most common, Stock phrases easily produced
>   **Repetition:** Marked impairment relative to other aspects of language
>   **Auditory Comprehension:** Some impairment, especially for formal, syntactically complex commands, fair with conversational speech
>   **Reading:** Aloud: marked paraphasic errors common; Comprehension: similar to auditory comprehension
>   **Writing:** Fluent, cursive script, with excessive paraphasic and spelling errors
>   **Lesion:** Arcuate fasciculus underlying supramarginal gyrus

Transcortical Motor Aphasia (Dynamic aphasia, Anterior isolation syndrome)
>   **Speech:** Markedly reduced output without marked dysarthria; output frequently limited to single words or common, repetitive phrases
>   **Repetition:** Relatively intact. Corrects grammatical errors and completes sentences
>   **Auditory Comprehension:** Fair for social, conversational speech, may have difficulties with complex syntax of formal commands
>   **Reading:** Aloud: deficient - doesn't produce; Comprehension: fair, except for complex syntax
>   **Writing:** Non-fluent, agrammatical
>   **Lesion:** Rostral and dorsal to "Broca's area," possibly deep

Transcortical sensory aphasia (Isolated speech area syndrome)
>   **Speech:** Normal articulation and rate, but speech empty and circumlocutory; may be echolalic
>   **Repetition:** Relatively good, echoes instructions or questions; fails to correct grammatical errors or complete sentences
>   **Auditory Comprehension:** Severe impairment, similar to Wernicke's

Chapter 9 ♦ Part II

Reading: Aloud: variable; Comprehension: severe impairment
Writing: Severe impairment
Lesion: Posterior parietotemporal; auditory association area intact

Anomic aphasia (Amnestic aphasia - controversial whether exists as separate aphasic syndrome)
Speech: Normal articulation with frequent word-finding pauses; loss of content words make speech "empty" and circumlocutory; verbal paraphasias common
Repetition: Normal or near normal
Auditory Comprehension: May be mildly deficient, especially for isolated nouns
Reading: Variable
Writing: Variable
Lesion: Variable; angular gyrus common, if initially presents as pure anomia with other Gerstmann's signs.

The above "Syndromes" may vary considerably, both in terms of type and severity of symptoms depending on extent of lesion, individual differences, and recovery.

The underlying premise behind this phenomenon is that the learned sensorimotor codes (*engrams*) for carrying out these actions either have been directly impaired as a result of a lesion or have been disconnected from the primary motor areas (see also Disconnection Syndromes).

Table 9–3. Other Disorders to Which the Term "Apraxia" Is Commonly Applied

Constructional apraxia Refers to the inability to reproduce geometric designs, patterns, or an assembly of three-dimensional objects either by drawing or through the manipulation of actual objects, such as the Block Design subtest from the Wechsler intelligence scales (see Critchley 1969). Although constructional deficits may result from lesions of either the right or left hemisphere, it more commonly is seen and often is more severe following right hemispheric damage. While qualitative differences occasionally can be found between groups of right and left hemisphere-damaged patients, except for unilateral neglect, these differences are not very reliable predictors of hemispheric involvement in individual cases (for review, see Hecaen & Albert, 1978).
Dressing apraxia Describes an inability or difficulty in dressing oneself. Somewhat more common in right hemispheric lesions, dressing apraxia often is seen in association with unilateral neglect and/or severe visuospatial or visuoperceptual deficits (Brown, 1972).
Gait or frontal apraxia Describes the characteristic gait problems in which there is particular difficulty in the initiation of movements of the foot or legs (Meyer & Barron, 1960). Also referred to as a "magnetic gait," this disorder typically results from conditions that involve the frontal lobes bilaterally and/or disrupt descending pathways (e.g., normal pressure hydrocephalus).
Apraxic agraphia Describes difficulties forming letters in writing and may occur independently of other types of apraxia. Assuming normal patterns of cerebral dominance, this type of agraphia typically is associated with left hemisphere lesions (Roeltgen, 1985; Hecaen, 1978).
Ideational apraxia Refers to a inability to carry out a sequence of actions utilizing real objects (e.g., folding a letter, putting it in an envelope, sealing and affixing a stamp to it). Most commonly associated with either extensive bilateral posterior lesions or a generalized dementia, ideational deficits usually are accompanied by significant perceptual disturbances (e.g., apperceptive visual agnosia), visuospatial deficits, aphasia, and/or general mental incapacitation.
Limb-kinetic apraxia Refers to the inability to make fine, rapid, or precise distal movements with or without the use of an object. This difficulty usually is confined to a single limb and affects only the side of the body contralateral to the lesion and is not related to dominance. Although the exact site(s) of the lesion causing this difficulty is still debated, the contralateral premotor and/or somatosensory cortices are the most likely sites (Heilman & Rothi, 1993).

While bilaterally expressed ideomotor apraxia is not uncommon with left hemisphere lesions in right-handed patients, bilateral ideomotor apraxia virtually never occurs following isolated right hemisphere lesions (in right-handers). However, isolated left-sided deficits occasionally may be found with right hemisphere lesions,[26] usually following lesions to transcallosal pathways. Taken together, these findings suggest that in right-handers these motor "engrams" may reside solely in the left hemisphere or be distributed bilaterally, but never (as far as is known) solely in the right hemisphere. With left-handers, the situation is less clear. Based on far fewer case studies, the assumption is that ideomotor praxis tends to be mediated primarily by the left hemisphere (or bilaterally) in left-handers as well; however, unlike right-handers, this function may reside solely within the right hemisphere. Thus, for right-handers, ideomotor praxis is never totally dissociated from language, but occasionally this may be true with left-handers (for review, see Heilman & Rothi, 1993).

Most of the other forms of apraxia, other than ideomotor and possibly apraxic agraphia, typically are associated with and likely in large part result from other elementary disturbances in motor, perceptual, or cognitive abilities. Hence, some authors, following Geschwind's suggestion that the term *apraxia* be restricted to ideomotor type deficits, have begun to use different designations in referring to these other phenomena (e.g., *constructional disability*, *dressing difficulties*).

Calculations

Numbers, like letters, are abstract symbols to which specific meaning has been attached and can be manipulated or rearranged into a variety of sequential patterns according to specific logical rules. Just as words that have become disconnected from their referents or have lost their associative value cannot be understood or used effectively to express a thought or idea, so too numbers may be deprived of their meanings (Benson & Denckla, 1969). As with words, the loss or deficit often is likely to be partial and transient rather than absolute. However, unlike language in which correct interpretations often can be made without complete comprehension, this is not true of mathematics. Disturbances in the interpretation and/or manipulation of mathematical symbols are known as **dyscalculias.** Just as grammar in language follows certain rules, arithmetic or the ability to calculate requires both access to and the ability to apply the rules governing computational operations. A disturbance of the latter, called **anarithmetria**, may occur independently of the ability to appreciate the symbolic aspect of numbers, although elements of both usually are present in the same patient. Disturbances in the reading, writing, or appreciation of the symbolic significance of numbers, inability to compare the relative value of numbers, or a primary disturbance in the ability to carry out basic calculations commonly reflect left hemisphere dysfunction (Levin et al., 1993; McCloskey et al., 1991).

However, in addition to their symbolic aspects, calculations and arithmetical problem solving also require other cognitive abilities, not necessarily restricted to the left hemisphere, that may affect such tasks (Kahn & Whitaker, 1991; Boller, 1985; Hecaen, 1962). For example, carrying out arithmetical operations involving multiple digit numbers require that they be done in a prescribed spatial order. As will be discussed shortly, disturbances of visual–perceptual abilities may result in what has been termed **spatial dyscalculia** (see Visual–Perceptual Abilities under Right Hemisphere Specializations). In a similar vein, unilateral spatial neglect may produce errors in calculation if numbers on the affected side are ignored. Some arithmetical word problems require a higher-level conceptual analysis and reasoning ability that may tap the integrity of bilateral frontal or more diffuse brain systems (see Luria, 1966, pp, 463–467).[27] Depending on their nature, level of difficulty, and how they are presented, problems involving calculations may be affected by disturbances of attention, concentration, or working memory. Finally, not only are education and motivation critical factors, especially when more complex mathematical operations are required, but a developmental deficit involving arithmetical abilities, independent of any recent acquired brain disease, also may contribute to impaired calculations.

Gerstmann's Syndrome and Disturbances of Body Schema

Josef Gerstmann (1940) described a symptom complex of **finger agnosia**, **right–left disorientation**, **acalculia**, and **agraphia** that he associated with lesions of the left angular gyrus. This quartet of symptoms has come to be known as *Gerstmann's syndrome* (see Critchley, 1966; Benson & Geschwind, 1985). It also has been noted that when these four symptoms are present, **constructional deficits** also likely are to be seen (Stengel, 1944; McFie & Zangwill, 1960). Although it is controversial as to whether or not this original group of four symptoms constitutes a true syndrome (Benton, 1961; 1977b; Critchley, 1969; Geschwind & Strub, 1975; Poeck & Orgass, 1966; Roeltgen et al., 1983), it generally has been agreed that if all four (or five, if constructional deficits are included) symptoms are present, there is a strong probability that the lesion involves the left angular gyrus (Geschwind & Strub, 1975). Writing and calculations already have been established as primarily left hemispheric phenomena. On the other hand, finger agnosia and right–left disorientation constitute part of what has been termed a *disturbance of body schema* (Frederiks, 1969; Benton & Sivan, 1993).

In childhood, we develop a sense of personal right–left orientation. Not only do we become aware that the right and left sides of our bodies are different (e.g., that one hand is stronger or better at doing certain things than the other), but we also learn to readily distinguish between them (i.e., personal right–left orientation) and to identify them by name (i.e., "right" versus "left"). Only later do we become facile in identifying right versus left in others (extrapersonal right–left orientation). Benton's (1959) review of both the developmental and neuropathological correlates of right–left orientation and finger agnosia is still one of the more comprehensive treatments of this subject.

Finger agnosia is the ability to distinguish and identify individual fingers. If truly present as part of Gerstmann's syndrome, it should present bilaterally (i.e., the patient should have equal difficulty identifying fingers on either hand). Finger agnosia actually is considered to be one facet of **autotopagnosia**, or the inability to identify one's own body parts. However, the inability to name or identify parts of the body (other than the fingers) in the absence of aphasia or other generalized deficits (such as severe unilateral neglect) is exceedingly rare. Complete finger agnosia is not particularly common and generally some fingers (e.g., thumb, little, and index) are easier to identify than others (middle and ring). Testing for finger agnosia involves asking the patient to name or point to a named finger or to otherwise identify (e.g., by matching) fingers by touch or sight.

In terms of anatomical correlates, while there does appear to be some relationship between disturbances of body schema (such as finger agnosia and right–left disorientation) and the left hemisphere, one should be cautious in drawing conclusions from these observations. Both right–left disorientation and finger agnosia, if demonstrated in the presence of aphasic disturbances, simply may be a function of the language disturbance (Poeck & Orgass, 1966). Both right–left disorientation and finger agnosia also may be present in right (or bilateral) hemispheric lesions, especially if general confusion is present. If finger agnosia is tested using tactual stimulus procedures and unilateral deficits are observed, disturbances in the contralateral somatosensory fields are a more likely explanation. Finally, while personal right–left disorientation in the absence of language or other cognitive deficits indeed may reflect a disturbance of body schema (and hence, possible left hemisphere involvement), isolated extrapersonal right–left confusion probably is more indicative of right hemisphere dysfunction (Ratcliff & Newcombe, 1973). For further review of disorders of body schema, see Benton & Sivan (1993); Denes (1989); Ogden (1985); Frederiks (1969); Benton (1959); Critchley 1969.

Verbal Learning and Memory

While both hemispheres contribute to most learning and memory tasks, there is substantial evidence that the left hemisphere assumes greater importance when learning verbal information (Christianson, Saisa, & Silfvenius, 1990; Cohen, 1992; Frisk & Milner, 1990; Gainotti, Cappa, Perri, & Silveri, 1994; Loring et al., 1991; Saling et al., 1993). This left hemisphere superiority for verbal learning undoubtedly is related to the fact that language tends to be organized predominately in the left hemisphere. However, performance on memory tests is not strictly dichotomized. Often left hemispheric lesions will be accompanied by difficulties on visuospatial, as well as verbal memory tasks (the converse is true with right hemispheric lesions, although perhaps to a lesser degree). At least two potential explanations come to mind. The first is that as highly verbal creatures we tend to verbally encode information, including nonmeaningful, geometric designs, such as the Rey-Osterrieth figure or those from the Wechsler Memory Scales. If one is unable to fully utilize such verbal cues as a result of aphasic deficits, their performance might be expected to suffer when compared to controls. Conversely, precisely because of the availability of verbal cues, some right hemispheric patients may be able to perform better than they might otherwise if such cues were less accessible (e.g., using visuospatial stimuli less amenable to verbal labeling). A second explanation for impaired memory for certain visuospatial memory tasks in some left hemispheric-lesioned patients may be a function of the presence of visuospatial constructional deficits that may occur in association with lesions of the left posterior cortices, particularly the left angular gyrus (Loring et al., 1988). Certainly, other more general factors associated with either frontal lobe impairments or loss of abstract attitude also may negatively impact on learning, regardless of the content.

Color Naming and Association

As might be expected, the left hemisphere, particularly the inferior occipitotemporal region, is thought to mediate color naming in the absence of more elementary color imperception which could interfere with this ability. However, in a more abstract sense, the ability to associate colors with particular objects also appears to be mediated primarily by the left hemisphere. For example, the color red is associated with the inside of a watermelon, while green is associated with the outside. Although disturbances in the ability to make these latter associations (even when a nonverbal testing format is used) often are associated with color-naming problems, the two abilities typically are considered to be separate functions (DeRenzi & Spinnler, 1967; Faglioni et al., 1970).

Right Hemisphere Specialization

Recall that the left hemisphere is thought to process information in a logical, analytical, sequential manner and to attend to details necessary for the mediation of propositional speech and related phenomena. By contrast, it has been suggested that the right hemisphere not only is responsible for very different kinds of mental activities (e.g., nonverbal, spatial–perceptual, affective, creative, divergent), but that the right hemisphere may operate in a fundamentally different manner than the left (DeRenzi & Faglioni, 1967; Semmes, 1968; Harris, 1978; Mesulam, 2000b; Delis et al., 1988; Joseph, 1988, 1990). The right cerebral hemisphere is thought to function in a more wholistic or gestalt type mode, processing multiple types of information in a more simultaneous or global manner. Such processing would allow for an immediate, "gut-level," or "instinctual" overall impression or analysis of a situation without having to "stop and think about it." Thus, it may be that the right hemisphere facilitates "seeing the big picture" or grasping the meaning of a situation without being bogged down or possibly even distracted by details. It also has been suggested that the right hemisphere may be more adept in handling novel situations that call for greater flexibility and innovative responding (Goldberg, Podell, & Lowell, 1994; Kittler, Turkewitz, & Goldberg, 1989). Finally, as the right hemisphere more often has been associated with emotional processing, one might speculate that it also is the right hemisphere that tells us that there is something about a situation that "doesn't feel right," although we cannot exactly identify the problem (i.e., "we can't put it into words").

At the beginning of this section on Hemispheric Specialization, it was noted that the right cerebral hemisphere is specialized for nonverbal functions such as emotional–affective processing, visuospatial abilities, spatial memory, musical abilities, directed attention, and unilateral neglect. Each one of these cognitive processes briefly will be discussed in turn, with specific attention to possible anatomic bases of behavioral disturbances in each of these domains.

Emotional–Affective Processes

Mood and affect are multiply determined or influenced. Social or psychological events, including physical illness or disability; changes in hormonal or neurochemical transmitter concentrations; or direct damage to those neurological structures that mediate emotions may all impact on an individual's mood (internal feeling) or affect (external expression), or the relative lack thereof. For purposes of the present discussion, we will restrict our focus to the neuroanatomical substrates of mood and affect. However, even with this limitation, the complexity of deciphering the neurological correlates of affect or emotion readily becomes manifest. For example, as was noted in the previous chapter, subcortical (limbic and hypothalamic), as well as cortical structures play a major role in emotions (e.g., Liotti & Tucker, 1995). This discussion will be limited to those aspects of emotional–affective processing that are thought to be mediated by cortical systems directly or by components of the limbic system that are modulated by cortical inputs.

Emotional Expression in Communication. Emotional "messages" as displayed by postures and facial expressions are powerful communication tools throughout much of the animal kingdom. While the development of language adds a whole new dimension to human communication, we have not divorced ourselves from our phylogenetic roots. Not only do we continue to employ some of the same bodily and facial expressions to communicate our feelings as our four-legged ancestors, but even our language itself has become imbued with and enhanced by affective coloring. There is a large volume of clinical research that suggests that the right cerebral hemisphere plays a predominant role in emotional expression in general and in its association with verbal communication in particular.

In comparison to listening to an actor on the stage or a politician, much of our day-to-day conversations at first glance may appear rather flat and uninspired. What little emotional affect we perceive in our own speech might seem to impart little additional significance to the message conveyed by the words we use. However, this is generally far from the case. The tone (*emotional coloring*) that normally accompanies speech adds a richness (even though at times subtle) to our verbal communications and enhances social interactions. Recall your own reaction to someone who speaks in a total monotone, without facial expression (which itself could have been a deliberate attempt to convey a particular message).

In addition to simply enhancing the aesthetics of communication, the presence of emotional tone provides added clarity and emphasis to our words. The manner (tone) in which something is said often conveys as much if not more information about the attitude, urgency, or meaning behind the communication as the words themselves. In fact, in some cases, the words and the affect or tone are discordant and the true meaning is derived not from the words but in the manner in which they are spoken. It is this discrepancy that typically forms the basis for sarcasm, ridicule, or at times humor. In these situations, the ability to interpret the tone is more critical than the literal translation of the words themselves.

As implied above, this "affective coloration" of speech has two fundamental components: expressive and receptive. We add emotional accents to our own speech appropriate to the mood or meaning we wish to convey (expressive), and at the same time it is imperative that we accurately interpret the mood and meaning behind the communications of others (receptive). Take, for example, the question, "You bought us two season tickets?" It is semantically neutral. However, depending on whether the speaker is (1) a football fan, (2) lives in a city with a losing franchise, or (3) deeply concerned about the family's financial situation, the affective tone with which this sentence is uttered may be quite different, and hence, may convey vastly different messages to the listener. Disturbances in the ability to either add the intended emotional tone to one's speech or to interpret the tone added by others has been termed **aprosodia**. It has been proposed that these affective–emotional components of language in the right hemisphere mirror similar expressive and receptive verbal language deficits in the left hemisphere (Ross, 1981, 1985; Gorelick & Ross, 1987). Thus, depending on the site of the lesion in the right hemisphere, one might expect to find a predominately **motor** (in the area of the frontal operculum), **sensory** (posterior temporal operculum), **conduction** (arcuate fasciculus), **transcortical** (anterior or posterior watershed), or **global** type deficits. Thus, a patient with motor aprosodia might manifest difficulty imparting proper affective or emotional inflections to his spontaneous speech. In a sensory type aprosodia, the individual might have difficulty recognizing the affective intonation used by the speaker (e.g., anger, sadness, or delight), and hence, may misinterpret the underlying meaning of the communication. These have been termed **disturbances of affective prosody.**

In addition to affective intonations, the meaning of a spoken statement may be changed by variations in the stress or pitch applied to a given word or the use of critical pauses between certain words. In the example used above ("You bought us two season tickets?"), note how the meaning is altered as stress is applied to different words in the sentence. This latter phenomenon is referred to as **linguistic prosody**. While the right hemisphere generally is thought to be dominant for both affective and linguistic prosody, it generally is agreed that syndromes of aprosodia are more widely distributed and clinically variable than aphasic disturbances (Heilman et al., 1975, 1984; Tucker et al., 1977; Schlanger et al., 1976; Weintraub et al., 1981; Ross, 1997).

Facial expression and "body language" (i.e., nonsymbolic gesturing)[28] also facilitate and contribute to communication of emotional states. It has been suggested that the right hemisphere also is dominant both for the recognition of facial expressions (Adolphs et al., 1996; Bowers et al., 1985) and for emitting facial expressions appropriate to the situation (Borod et al., 1988). These deficits appear to be independent of disturbances of visuoperceptual abilities or facial recognition in general, both of which are often associated with right hemisphere injuries.

Experience of Emotion Outside of Language.   While there is a general consensus that prosody and the recognition of facial expression is most likely mediated primarily by the right hemisphere, the potential leading role of the right hemisphere in controlling emotions in general is more controversial. It has been noted, for example, that patients with left hemispheric lesions (particularly anterior strokes) are more likely to be depressed, while those with right hemispheric damage are more likely to be seen as euphoric or indifferent (Gainotti, 1972; Starkstein et al., 1987). It is a common clinical observation that despite the presence of dense aphasia, it often is easy to ascertain the mood of the left hemispheric patient. Also, once any initial depression has lifted, patients with left hemispheric lesions are perceived as being more likely to maintain their premorbid affective bonds with their families. In contrast, right hemispheric patients often become emotionally aloof, cold, or indifferent to previously close family members. Although neither of these types of reactions by any means is universal, this perceived tendency has led to the assumption that the right hemisphere may be the more dominant of the two in controlling or engaging negative emotions (withdrawal responses), while the left hemisphere may be responsible for more positive emotions (approach responses) (Sacheim et al., 1982; Davidson, 1995).

However, alternate explanations have been offered to help account for these apparent discrepancies in emotional tone or mood following unilateral brain injuries. It has been suggested, for example, that the loss of the ability to communicate (most notably with left anterior lesions) readily could account for a patient's depressed mood (Ross & Rush, 1981), while diminished awareness of deficits and/or reduced capacity to express emotions (motor or expressive aprosodia) might help explain a right hemisphere-lesioned patient's apparent indifference or limited emotional responsiveness. Nonetheless, while recognizing the probable contribution of both hemispheres, there appears to be a consensus that the right hemisphere plays a special if not leading role in the processing, expression, and/or experiencing of emotions (Gainotti, 1997; Heilman & Satz, 1983: Heilman, Bowers, & Valenstein, 1993; Kolb & Whishaw, 1990, pp. 607–642; Liotti and Tucker, 1995; Morrow et al., 1981).

Additional Contributions to Language Processes. It has been noted that, while propositional language is thought to be mediated by the left hemisphere, both affective and linguistic prosody appear to be primarily under the direction of the right hemisphere. However, there are other elements of speech or language that may be related to or affected by changes in the functional integrity of the right cerebral hemisphere. One of these is the appreciation of context. Most sentences, whether declarative, imperative, or interrogatory (prosodic variations aside), are probably fairly straightforward and easily comprehended. However, this is not always the case. Longer, more complicated narratives are difficult to process word-by-word, sentence-by-sentence, or perhaps even paragraph-by-paragraph. If narratives are to be followed and fully understood or appreciated, they have to be broken down in terms of their major and possibly secondary themes. Discovering such themes provides a framework or structure into which the individual elements of the discourse can then be woven (i.e., placed into context). However, the need to discover and adhere to this structure applies not just to comprehension on the part of the listener, but also to the organization of the speaker. In listening to a story or discourse (especially if complex), individuals suffering damage to the right hemisphere have been described as having difficulty:

1. Identifying its point or moral.
2. Discerning major or central themes from minor details.
3. Deciphering incongruities.
4. Appreciating humor (both verbal and visual).
5. Drawing inferences or conclusions.
6. Interpreting metaphors, idioms, or sarcasm.[29]

Conversely, when engaging in conversation, individuals with right hemispheric lesions may use a lot of words but convey relatively little information, stray off on tangents rather than getting to the point, and have difficulty taking turns or allowing others equal time (Garner et al., 1975, 1983; Wapner et al., 1981; Diggs & Basili, 1987; Hough, 1990; Meyers, 1993, 1994).

The "indirect command" or "indirect request" provides a good model for appreciating the occasional failure of right hemispheric patients to take into account social context and how this may impact on verbal communications. An indirect command is a communication in which the initiator, in opting for a question or simple declarative rather than an imperative statement, nonetheless expects the listener to respond in a particular fashion given the particular situational or environmental cues. Thus, in visiting a friend on a warm summer day and seeing a beer commercial on television, one might say, "Boy, that sure looks good!" Given the circumstances, obviously the statement is meant to be interpreted as, "I sure would like a beer right now, do you happen to have a cold one handy?" This would be the "pragmatic" interpretation. However, the same statement also might be interpreted quite literally, in which case the response might be, "Yes, I guess it does, but personally I prefer Cokes" (without an offer of either forthcoming). Such a response, which is more typical of right then left hemispheric-lesioned patients (Foldi, 1987), appears indicative of their failure to integrate nonverbal, contextual cues in social communication.

Given the notion of "distributed functions" as the most likely explanation of brain organization, one must be cautious in attributing perceived deficits in the highly complex behaviors outlined above as being exclusively attributable to a right hemisphere dysfunction.[30] However, right hemisphere involvement in or possibly control over these aspects of language are understandable. Rather than strictly adhering to the formal rules of propositional speech in formulating or comprehending language, these behaviors require an awareness or simultaneous processing of a multiplicity of social (environmental) cues, divergent thinking, and in general an appreciation of the gestalt, features that are more commonly associated with the right hemisphere. Table 9–4 summarizes some the right hemisphere's impact on communication.

Table 9–4. Right Hemisphere Contributions to Communication

Expressive affective prosody: verbal
Imparting the desired emotional valence to one's speech through appropriate affective intonations (e.g., "sounding angry" when you are indeed upset).

Expressive linguistic prosody: verbal
Ability to vary the meaning of spoken expression through the use of appropriate tonal inflection to key words or phrases (e.g., "YOU (the perennial cheapskate) bought us two tickets to the football game?" versus "You BOUGHT (when I could have gotten them free at the office) us two tickets to the football game?" versus "You bought us two tickets to the FOOTBALL game (when you know I don't like football)?"

Expressive prosody: non-verbal
Expression of mood or emotional state through non-verbal means (e.g., facial expression, body posture).

Receptive affective prosody: verbal
Deciphering the meaning of oral verbal communication or mood of the speaker by accurately interpreting tonal inflections.

Receptive linguistic prosody: verbal
Appreciation of humor, sarcasm, or emphasis in the speech of others

Receptive prosody: non-verbal
Ability to accurately decipher the mood of others, or the "message" they are sending by accurately interpreting facial expressions or body postures.

Meta-aspects of oral or written communication
- Organization of oral or written discourse - "staying on course, or keeping to the point."
- Responsive to social cues (e.g., turn-taking, awareness of the impact one's discourse is having on others-via verbal and non-verbal receptive prosody).
- Discerning the "point" or "theme" in the discourse of others.
- Interpreting the "message" in light of the overall context or circumstances in which it was delivered.

Finally, other more general cognitive–perceptual deficits also generally associated with right hemispheric damage may affect other language-related abilities. In reading, for example, individuals with right hemispheric-based severe visual–spatial deficits may have difficulty making smooth transitions from one line to the next, although this difficulty also might result from disturbances in visual tracking abilities. If unilateral neglect is present (see below), the individual may ignore words on the left side of the page or the initial letters or prefixes to words (e.g., seeing the word "key chain" as "chain"). Writing may begin toward the center of the page, well away from the left margin, and may tend to slant upward at about a 15° angle (especially on unlined paper). There may be a tendency to perseverate certain letters, especially double le*tt*ers (e.g., le*tt*ers) when writing (Ardila & Rosselli, 1993; Roeltgen, 1993). Obviously, either failure to grasp the central theme of what is being read, or rambling, nonfocused written compositions may result in problems similar to those discussed earlier in relation to spoken discourse.

Spatial-Perceptual Processing

Beyond left-sided deficits in sensorimotor abilities, the behavior most commonly associated with the right hemisphere is the integration and processing of spatial–perceptual information. In turn, spatial–perceptual abilities most frequently are associated with tasks that tap visual information, such as measures of "visual spatial," "visual–constructive," or "visual–perceptual" abilities. However, as with communication skills, the situation is far more complex. First, while it appears that the right hemisphere plays a dominant role in certain aspects of spatial judgment or analysis, this superiority probably transcends any single modality. Congenitally blind individuals, for example, develop good topographical maps of their environment without the benefit of visual input. Spatial disturbances may also affect tactile performances (DeRenzi & Scotti, 1969; Corkin, 1978). More importantly, even those tasks that appear to be predominately visual–spatial in nature, such as the Kohs blocks from the Wechsler intelligence scales, frequently are disrupted by left hemispheric lesions (Kaplan, 1988; Kaplan et al., 1991). A brief review of various types of spatial–perceptual tasks that traditionally have been associated with right hemisphere processes may help to illustrate the right hemisphere's role in spatial–perceptual processing, while at the same time demonstrating the hazards in making broad generalizations with regard to hemispheric specialization.

Visual–Spatial Ability. While there are numerous formal and informal tests that make visual–spatial demands on an individual (Paterson & Zangwill, 1944; Walsh, 1994, pp. 251–279; Benton & Tranel, 1993; Lezak, 1995, pp. 385–417), most are either functionally very complex or readily lend themselves to verbally mediated solutions. Use of such tests in clinical settings without sufficient controls or systematic manipulation of the constituent variables (factors) often can confound attempts to arrive at some understanding of hemispheric specialization. For example, **topographical orientation** or **topographical memory** in its most basic form represents a spatial map or mental blueprint that allows an individual to appreciate from memory where a particular location or object exists relative to those around it. However, tests of these abilities (e.g., localization of cities on a schematic map, route description, or drawing floor plans or maps of familiar places) may be confounded by verbal associations, constructional disabilities, or hemispatial neglect that make it difficult to tease out the pure spatial element. While a right hemisphere superiority often is suggested for the spatial aspect of many of these topographical memory tasks, deficits often are observed following unilateral lesions of either hemisphere, probably because of the multiplicity of factors involved (Benton & Tranel, 1993; DeRenzi, 1997a).

One task that appears to be a somewhat purer (i.e., less contaminated) measure of visual–spatial ability is judgment of angularity or **line orientation**. This approach was originally developed by DeRenzi et al., (1971) and later formalized by Benton et al., (1975) and was found to be strongly associated with the right hemisphere, that is, deficits on this task occurred with a much greater frequency following unilateral right-sided lesions (Benton et al., 1983). Although typically used as a measure of visual–spatial ability, judgment of line orientation also can be adapted for tactile presentation (Meerwaldt, 1982), where it retains its right hemispheric superiority.

Visual–Constructive Ability.   Although visuoconstructive tasks can be classified as a measure of visual–spatial ability, they merit independent consideration because of their common constructive component and because they constitute some of the more well known and frequently used measures in neuropsychology and behavioral neurology. Unfortunately, these tasks also are among the most misunderstood. Visuoconstructive measures represent a highly diversified group of tasks. These might include:

1. Drawing familiar patterns, such as a clock or a daisy.
2. Copying (drawing) two- or "three-dimensional" designs, (e.g., a house, or a cube).
3. Drawing the floor plan of one's house or hospital unit.
4. Recreating geometric patterns using the Kohs blocks (e.g., Block Design from the Wechsler intelligence batteries).
5. Reconstructing two- or three-dimensional arrays using matchsticks or assorted wooden blocks (Critchley, 1969; Benton et al., 1983).
6. Reassembling fragmented pictures (e.g., Object Assembly from the Wechsler intelligence batteries).

As we will see later, increasing the demands on the spatial analysis, planning, and organization can increase the level of difficulty on some of these tasks. One way of increasing the visual–spatial complexity is by attempting to incorporate a three-dimensional perspective in two-dimensional space, as is the case in trying to draw a solid cube,[31] a house from an oblique angle, or other quasi-three-dimensional design (see Figure 9–25). The Rey-Osterreith figure illustrates another way of increasing complexity by simply increasing the number of individual elements in the design.

All of the above tasks likely tap slightly different aspects of visuospatial ability. Depending on level of complexity, some are less demanding than others, while others might be more

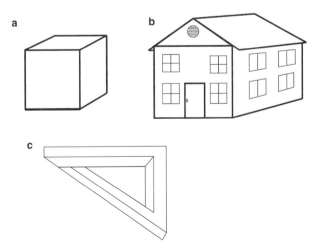

Figure 9–25.   Examples of graphomotor copying tasks tapping visual spatial construction ability.

conducive to verbal mediation (e.g., Object Assembly).[32] The multidimensional nature of most of these tasks readily is apparent, and hence, potentially subject to factors other than pure visual–spatial ability. In general, drawing tasks probably are more sensitive to disturbances of motor praxis than those requiring object manipulations. Basic visual disturbances, especially syndromes involving unilateral neglect (see below), may adversely affect performance on all these measures. While the successful completion of most of these tasks will benefit from good organizational and planning ability, some will be less affected by a piecemeal, trial-and-error approach. Virtually all of these tasks require some degree of self-monitoring ability. Some visuoconstructive tasks obviously require a fair degree of problem-solving skills, and as a result have been incorporated into formal intelligence tests (e.g., Block Design, Object Assembly). However, even relatively simple drawing tasks may be affected by sociocultural and intellectual factors (Strub et al., 1979). Because of the multiplicity of factors involved, it is not surprising that (1) the intercorrelations among these various tests are far from perfect (Benton, 1967; Dee, 1970), (2) they are extremely sensitive to most types of brain injury, and (3) are often impaired following either right or left hemispheric lesions.

With regard to hemispheric localization, originally constructional deficits were thought to reflect damage to the left parietal lobe. Later they came to be recognized as more of a right hemisphere syndrome when it was believed that right-sided damage (again, in the parietal area) led to more frequent and more severe constructional deficits. Eventually, there was a realization that constructional deficits could occur with lesions to either hemisphere (Warrington, 1969). Although parietal lobe involvement still was thought to be critical, attempts were begun to differentiate laterality by qualitative analyses of drawings. Deficits in construction resulting from right hemispheric lesions were thought to be characterized by greater fragmentation in the designs, a loss of gestalt and three-dimensional perspective (if present), and a tendency to ignore features on the left side of the design (left unilateral spatial neglect). The basic problem in right hemisphere deficits was thought to result from a breakdown in spatial integration or perception. Constructions by patients with left hemisphere lesions frequently were described as manifesting a breakdown in planning and executive abilities (apraxia), resulting in overly simplified but spatially more intact (greater perseveration of the gestalt) designs that were more likely to show difficulties in replicating internal details (see DeAjuriaguerra & Tissot, 1969; Hecaen & Assal, 1970; Kaplan, 1979). While these observations may have considerable merit, with the exception of noting the presence of consistent unilateral neglect, these signs have not proven sufficiently robust to provide definitive markers in individual cases (Black & Strub, 1976; Arena & Gainotti, 1978; Benton & Tranel, 1993). Finally, while parietal lesions, particularly bilateral parietal pathology (as is frequently present in Alzheimer's disease), may result in more classic, severe instances of constructional deficits, frontal lesions also may produce deficits on these tasks [presumably, in part, due to deficits in organization, planning, and self-monitoring (Benson & Barton, 1970; Luria & Tsvetkova, 1964)].

Visual–Perceptual Abilities. Most visual–perceptual activities involve various degrees of spatial analysis and integration. Printed words and (multidigit) numbers depend on an accurate perception of the spatial order of their constituent units (letters or arithmetical symbols) for their proper interpretation. Pure color discrimination may be one notable exception to this rule. Historically, there has been a frequent tendency to associate impairments on a wide variety of "nonverbal, visuoperceptual" tasks with right hemisphere dysfunction. In addition to discrimination of line orientation noted above, other examples of such tasks have included:

1. **Discrimination among unfamiliar faces** (as opposed to recognition of familiar faces or prosopagnosia) (Benton, 1990; Benton & Van Allen, 1968; DeRenzi, 1997b).
2. **Incomplete, fragmented, overlapping figures**, or **familiar objects viewed from an unusual perspective** (Boyd, 1981; DeRenzi & Spinnler, 1966; Warrington & Taylor, 1973).
3. **Mental rotations of patterns or figures in space** (Butters & Barton, 1970; Ratcliff, 1979; Fischer & Pellegrino, 1988; Benton & Tranel, 1993**).
4. **Localization of objects in space** (Warrington & Rabin, 1970; Hannay, Varney, & Benton, 1976).
5. **Discrimination of complex visual patterns** (Benton et al., 1983; Corkin, 1979).
6. **Visual mazes**, **map making**, and **route finding ability** (Benton et al., 1974; see also Lezak, 2004).
7. **Depth perception** (Carmon & Bechtoldt, 1969).

While such tasks often have been identified as being particularly sensitive to right hemispheric pathology (i.e., deficits on these various tasks appear to occur with somewhat greater frequency following right focal lesions), as noted by Lezak (2004) in her review of these tasks, deficits often can be found following lesions to either hemisphere. In fact, deficits in most if not all of these tasks tend to be most severe in cases of bilateral, posterior disease. This implies that both hemispheres contribute to the perceptual–cognitive operations demanded by these various tasks.

However, in keeping with the commonly accepted theories of hemispheric asymmetry, the presumption is that the contributions made by each hemisphere are different or unique. A good example may be the **Hooper Visual Organization Test** (HVOT). This test consists of fragmented schematic drawings of common objects that are to be identified (Figure 9–26). Some items, such as the candle (item No. 18) and the cat (No. 20) readily can be identified by simply analyzing a single fragment, a task easily accomplished by the left hemisphere. On the other hand, the correct identification of items such as the flower (No. 21) and the broom (No. 30) are greatly facilitated by the individual's ability to mentally integrate the fragments, arguably a process better suited to the right hemisphere. In fact, while potentially disrupted by various brain injuries, the HVOT has not proven to be selectively sensitive to right hemisphere pathology, probably at least in part because so many of the items can be recognized using left hemispheric strategies. However, as noted by Sergent (1995), another difficulty in using any such measure to unlock the secrets of hemispheric function is that we often are attempting to reduce exceeding complex, multidimensional phenomena to a single, bipolar principle. While such endeavors may provide useful insights, we are still a long way from piecing together the puzzle.

Sergent's admonitions notwithstanding, one rather clever strategy that has been used in an attempt to tease out these different hemispheric mechanisms is the global versus local perceptual paradigm devised by Navon (1977) that consists of large letters (or other figures) composed of strings of contrasting smaller letters (or figures). Recognition of the larger letter (**global processing**) was thought to better reflect right hemispheric, gestaltist mechanisms, while recognition of the smaller, constituent letters (**local processing**) seemed to evince the analytic capacities of the left hemisphere (Figure 9–27). While some studies using such stimuli in focal brain-injured populations have provided support for this hypothesis (Delis, Robertson, & Efron, 1986), again the results have not always been consistent, suggesting that other intervening factors also need to be considered (Brown & Kosslyn, 1995; Hellige, 1995). One of the recurring lessons to be derived from these various studies of visual perception (also true of most other mental abilities) is that both cerebral hemispheres normally work in concert when processing (and responding to) everyday stimulus input, regardless of its

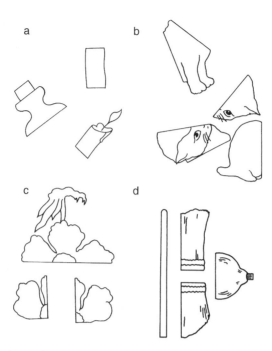

Figure 9–26. Examples from the Hooper Visual Organization Test that might be identified from a single fragment (a,b) versus those in which integration of all elements is commonly necessary (c,d). Material from the VOT copyright © 1957 by H. Elston Hooper. Reprinted by permission of the publisher, Western Psychological Services, 12031 Wilshire Boulevard, Los Angeles, California 90025, *www.wpspublish.com.* Not to be reprinted in whole or in part for any additional purpose without the expressed, written permission of the publisher. All rights reserved.

nature. The dichotomies that we establish in describing certain tasks as being mediated by or representative of the right or left hemispheres are largely our own inventions, not nature's. Still, this does not prevent us from asking the question, "what does each hemisphere (lobe or gyrus) contribute to this behavior and how is the behavior altered when that particular contribution is missing?"

There are certain visual–perceptual syndromes that typically only become manifest following bilateral insults to the brain. **Balint's syndrome** is one example (ideational apraxia, which was discussed earlier, being another). Balint's syndrome is primarily characterized by (dorsal) **simultanagnosia** (the inability to perceive more than one object at a time) and

Figure 9–27. "Global" versus "local" processing. Figures constructed for illustration purposes.

marked disturbances of spatial disorientation (e.g., difficulty in localizing objects in space, impaired depth perception, and loss of topographical memory). It is a relatively rare disorder that generally results from bilateral parietooccipital lesions. Despite what may be normal visual acuity, such individuals virtually are unable to negotiate their visual environment (Holmes & Horax, 1919; Rafal, 1997a).[33]

As noted earlier, even certain aspects of symbolic, linguistic abilities such as reading, writing, and calculations may be affected by "spatial" deficits. For example, certain aspects of syntax may require a spatial type analysis (e.g., "over," "above," "after," and "before"). As noted earlier, disturbances in reading, writing, and arithmetic also may be associated with right hemispheric lesions. Mathematical operations are carried out in two-dimensional space and follow certain spatial rules. Whether performed mentally or in writing, complex calculations require precise spatial operations, such as (1) keeping the columns and rows of numbers in correct alignment, and (2) proceeding in an orderly spatial fashion, both horizontally and vertically. While simple calculations that rely on previously memorized associations may not be readily affected, the likelihood of error increases with the use of multidigit numbers. An even greater demand may be placed on spatial integrity when such calculations are presented and must be performed mentally, rather than in writing. While these types of deficits (**spatial dyscalculia**) have been more commonly associated with right hemisphere lesions (Cohn, 1961; Hecaen & Angelergues, 1961, also cited in Hecaen & Albert, 1978; Rosselli & Ardila, 1989), "spatial" errors have been identified following lesions of either hemisphere (Dahmen et al., 1982; Collignon, Leclerq & Mahy, 1977, also cited in Levin et al., 1993; Grafman et al., 1982). Finally, errors in estimation of magnitude also have been associated with right hemisphere lesions (Dehaene & Cohen, 1991).

Spatial (Nonverbal) Memory

As the right hemisphere long has been perceived as being the "dominant" hemisphere for nonverbal, spatial, particularly visual–spatial abilities, the superiority of the right hemisphere in mediating nonverbal, visual–spatial memory has been well established in clinical lore. Certainly, some of the earlier studies by Corkin (1965), Kimura (1963), and Milner (1965, 1968, 1972) helped to support this notion. More recently, the notion of separate verbal and visual–spatial working memories, with the latter being identified primarily with the right hemisphere in PET studies, has provided further credence to this hypothesis (Baddeley, 1994; Jonides et al., 1993). However, while some more recent studies have been able to demonstrate a hemispheric effect on tests of visual/spatial memory in clinical populations (Christianson, Saisa, & Silfvenius, 1990; Cohen, 1992; Gainotti, Cappa, Perri, & Silveri, 1994), the effect appears to be neither consistent nor robust. The problem is that when the performance of patients with lesions lateralized to one hemisphere are compared on most commonly used tests of visual–spatial memory, the difference in scores often is either insignificant or there is such overlap between the two groups that the predictive validity of these measures in localizing lesions is not very good. This includes the visual–spatial memory tests from the Wechsler Memory Scale, one of the more widely used psychometric measures of memory (Chelune & Bornstein, 1988; Naugle et al., 1993). For a relatively recent review of the capacity of other specific tests or measures of visual–spatial memory relative to differentiate focal hemispheric lesions, the reader is referred to Lezak (2004). However, the bottom line is that most studies utilizing these measures fail to clearly discriminate right versus left hemispheric-lesioned patients on the basis of overall level of performance.

This is not to suggest that certain patterns of performance or qualitative features are without clinical merit. For example, marked and consistent discrepancies between verbal and visual–spatial abilities, including "non–verbal" tests of learning or memory indeed may be indicative of right hemisphere pathology. This is particularly likely where performance on predominately "verbal" tasks are markedly and consistently superior to "visual–spatial" measures, especially if there are clear signs of left spatial neglect. Problems are encountered when isolated or borderline findings are used to make such predictions without considering other factors or possibilities.[34]

Thus, the failure of memory tests to reliably differentiate right from left hemispheric-lesioned patients may not lie so much with the theory as with the practice. It may well be that the right hemisphere, particularly the hippocampus and surrounding regions (e.g., dentate gyrus, subicular region, and entorhinal and perirhinal cortices) are critical for encoding and retrieving certain types of visual–spatial or related information. However, as noted, most of the commonly used visual–spatial, "nonverbal" memory tests actually are multidimensional and capable of substantial verbal encoding. Impairment of this process in left hemispheric patients may reduce their performance, while utilization of these verbal cues may help right hemispheric patients compensate for their "nonverbal" deficits. Similarly, problems with visual perception or perceptual–motor (constructional) difficulties associated with left hemisphere damage (see above) may result in impaired performance on these tests. Finally, it is possible that some of the studies that have failed to find significant differences between right and left hemispheric-damaged patients on these measures themselves may be flawed. It has been demonstrated that the exact locus and extent of damage to the hippocampal region is critical in the amount of memory impairment that might be expected (Zola, 1997). In clinical studies, such precision in establishing right versus left hemispheric groups is virtually nonexistent.

Musical Abilities

Musical abilities provide another example of hemispheric specialization and hemispheric cooperation. The appreciation as well as the expression of musical melodies frequently has been associated with the right hemisphere (Basso, 1993; Benton, 1977a; Gates & Bradshaw, 1977a; Henson, 1985). The preservation of one's ability to reproduce familiar melodies, despite extensive disturbances of expressive language following dominant hemispheric lesions, repeatedly has been documented (Smith, 1966; Yamadori et al., 1977). On the other hand, music, like language, is composed of individual, temporally sequenced notes, each capable of being analyzed with regard to specific individual characteristics such as pitch and timbre, functions that would appear to be more in keeping with the suspected operations of the left hemisphere. Yet, it would seem that well-trained musicians and composers would be considerably more adept in processing music in this fashion than the majority of us who lack such training or abilities. In fact, there is evidence that trained musicians indeed may rely more heavily on the left hemisphere for processing certain aspects of music when compared to nonmusicians (Bever & Chiarello, 1974; Shannon, 1984). However, reliably ascribing specific aspects of musical abilities to either the left or right hemisphere in this group has proven difficult (Sergent et al., 1992). It also has been noted that familiar pieces of music may be more readily recognized by the left hemisphere (Gates & Bradshaw, 1977b). Thus, not only may different aspects of music more readily lend themselves to processing by one hemisphere versus the other, but also the strategies by which these various musical elements are approached also may be important in determining which may be the leading hemisphere in a given situation. In summary, while both the right and left hemisphere apparently are involved in the expression and perception or appreciation of music regardless of level of training, the specific contributions of each are still somewhat of a mystery.

Directed Attention and Unilateral Neglect

Patients with certain unilateral lesions, particularly if the lesion was of acute onset (e.g., strokes) may display problems that commonly are interpreted as difficulties with attention to or awareness of the contralateral side of the body or the contralateral hemispace. Also referred to as problems of directed attention or unilateral neglect, these symptoms are considerably more likely to occur following damage to the right side of the brain than the left.[35] The problem may be manifested in terms of external hemispace, hemibody awareness or hemibody stimulation, and/or hemibody movement.

Hemispatial Neglect. In its most classic presentation, **hemispatial neglect** is characterized by the patient's failure to report or attend to visual stimuli that are present in the hemispace contralateral to the lesion (in the absence of primary sensory defects). For example, following a right hemispheric lesion the patient may fail to report objects or movement presented in the affected left visual field or left hemispace by the examiner. He or she also may fail to complete the left side of drawings or puzzles and ignore food on the left side of the tray and words or letters on the left side of a page. At times, it may be difficult to distinguish between a left homonymous hemianopia and a severe left visual neglect as in either condition the patient may fail to spontaneously respond or attend to visual stimuli confined to the affected hemifield. However, severe neglect, in the absence of a visual field defect, occasionally may be identified by directing the patient's attention to the affected field, for example, informing the patient that the stimulus will be presented on that side and that the patient needs merely to report *when* (as opposed to *if*) the stimulus occurs. Also, patients with field cuts without neglect normally search the affected field if given an opportunity, so they may perform normally on cancellation tasks and typically will fail to show the other signs of neglect listed above.

More common than the failure to report single stimuli presented to the affected field, hemispatial neglect often is manifested by the presence of extinction (suppression) on **double simultaneous stimulation** (DSS). In double (bilateral) simultaneous stimulation, the patient may respond normally (or near normally) when stimuli are presented singly and independently to one or the other hemifield or hemispace but fail to respond to stimuli on the side contralateral to the lesion when comparable stimuli are presented simultaneously to both fields. Depending on the severity of the neglect, such a failure to respond may be absolute

(a)
```
H  S    A   N   U   F      M      D   S   C    S   E   Q     S   J    F   X
   G   F  M   E   I   C   S   K   A   E     C   M       B   S    T
     O    P  S  R  N  X  Y  A  L  S  Z    A   R    L    U
     O  S    C   M       B   P  R  N  F    C  W  S      R   S
       K   A  I   C   P   F  S  B  E  W     S    L   P        E
     X  J    S   G   M   E   O    R  P  N    R   F     D   S      I
       E   V      S   T   S   E  F     T     K    C      Z   A    F
   S   N   J   F  D  M     D   B  E  Q   S     J  C  V     T    X
     E  G     V   M   E   F     S   K  D  E    M      F  D  E      S
   G   B    T    S    V R     X    E  F     A      X   S      W  I
   F  S    C   M   S    U    O   P  B    S  D    C  S      O   B  R  P
L   K   A    C     F  T    B     R     N  D     C      F    T    R
     S    L    P   R    E    P  S      B      W      Z      O   S
   E  G   M    S   O   S  X   N    R  Z     T   S   I      V   B
H   S   C    E    T   K     C      Z   F    S   J    R  D   F
   K    E     O   S     A   N    F   Z    D   S    K  V      E    H   S
S    F      Y     W    T  C    S        E      N     M  F   S      M  T
   D  B  Q    B   F   J     M    S  B    X   H    Y  V    S   H
E  G    S   M      J  I    S   K  F     E  C    M    K    E  U
F   C   B     X     O      S  R    N  X    E   L    A   S     W  N
   X  S     C    M   T    B      H   T     S   F    D    T      S
   C    Z      S   U   F   V  D  M     D  C     S  E    Q     B  C
S   J     E    F   M     E       C   S   K  M    U  C    A  M       F
   G   I    O     P      R    N  X    E    L   A  Z      P   G  S
O   S    C    M   T   B   P  S  F    D        C  W    Z     R   L
   F   A   P   X    S   T    Q   I  G      N S      P  H    S     F  U
```

Figure 9–28. (*Continued*)

(b)

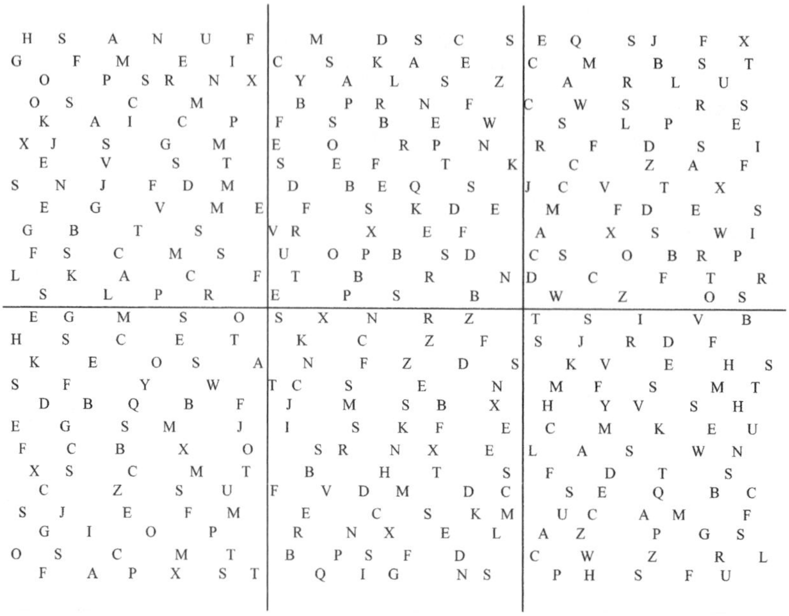

```
  H   S       A       N       U   F         M       D   S   C       S   E   Q       S J       F   X
G     F   M       E       I     C   S   K   A       E     C   M       B   S       T
        O       P   S R       N   X   Y       A   L       S   Z       A       R   L       U
    O   S       C       M           B   P   R   N       F   C   W   S       R   S
        K   A   I       C       P   F   S       B   E       W       S       L   P       E
    X J       S       G       M   E   O       R   P       N       R   F       D   S       I
        E       V       S       T   S   E   F       T       K       C       Z   A       F
  S   N   J       F   D   M       D   B   E   Q       S     J   C   V       T   X
        E   G       V       M   E   F       S   K   D   E   M       F   D   E       S
    G   B       T       S     V R       X       E   F       A       X   S       W   I
    F   S       C   M   S       U   O   P   B       S   D   C   S       O   B   R   P
  L     K   A       C       F   T       B       R       N D   C       F   T   R
      S       L   P       R   E       P       S       B   W       Z       O   S
    E   G       M   S       O S   X   N       R   Z       T   S   I       V   B
    H   S   C   E       T       K   C       Z   F       S   J       R   D   F
    K       E       O   S       A   N   F   Z       D       S   K   V       E       H   S
  S   F       Y       W   T C   S       E       N   M   F       S       M   T
        D   B   Q       B   F   J       M   S   B       X   H       Y   V       S   H
  E   G       S   M       J   I       S   K   F       E   C   M       K       E   U
  F   C       B       X       O       S R       N   X       E   L   A   S       W   N
    X   S       C       M   T   B       H   T       S   F       D   T       S
      C       Z       S   U   F   V   D   M       D   C   S   E       Q       B   C
  S   J       E       F   M   E       C       S   K   M   U   C       A   M       F
    G   I       O       P   R       N   X       E   L   A   Z       P   G   S
O   S       C       M   T   B   P   S   F       D   C   W       Z       R   L
  F       A   P   X       S T   Q       I   G       N   S   P   H       S       F   U
```

(c)
_____   _____   _____

_____   _____   _____

_____   _____   _____

_____   _____

_____   _____

_____

_____

_____   _____   _____

Figure 9–28. Cancellation and line bisection tasks. The cancellation task contains a total of 60 "S"s and 30 "F"s, 20 and 10, respectively, in each lateral third of the figure and 10 and 5, respectively, in each sixth. The reader is free to reproduce and enlarge figures for their personal use.

(i.e., occurring on 100% of the trials) or relative (occurring on a significant percentage of the trials).

Two common bedside methods of directly testing for hemispatial neglect involve (1) visual and auditory confrontation exams (e.g., the presentation of unilateral or bilateral (DSS) stimuli, in a random order, to each side of the body or hemispace), and (2) the use of visual cancellation and/or line bisection tasks (Figure 9–28). However, as noted above, signs of hemispatial neglect may be observed in other situations. For example, the patient may fail to attend to someone standing on his/her affected side, even though both ears are monitoring the sound of their voice, which is inputted into both hemispheres. In drawing symmetrical figures or objects, there may be an omission or degradation of details in the affected hemispace. Again, in the case of a right hemispheric lesion, the patient routinely may begin working on the right side in a task like Block Designs and may make significantly more errors on the left side of the design. The patient may have trouble negotiating external space (getting lost on the ward) because of their inability to use landmarks in their affected field. At times, such patients even may evidence difficulty recalling from memory details of familiar scenes when the information to be recalled from their visual imagery falls in the affected field.

As the patient begins to recover from the acute brain insult that resulted in the unilateral neglect syndrome, there normally is a noticeable improvement in the symptoms. This is commonly first noted in "predictable" situations, that is, situations in which the patient knows to "expect" stimuli in both spatial hemispheres. For example, the patient "learns" to search the left side of his tray for his cup of coffee, or learns to scan to the left until he reaches the end of the page before reading a line (thus, he no longer omits words on the left side of the page). There also may be improvements on cancellation and line bisection tasks or with confrontation exams where DSS is employed. However, more subtle deficits may persist, especially in situations where the patient is confronted by a multiplicity of stimuli, where he/she is expected to respond quickly, or where there is less predictability regarding the presence or absence of a particular stimulus. This is why, despite the presence of more or less normalized responses on clinical examination, one might be cautious about advising a patient with a recent history of unilateral visual neglect that it is safe to resume driving once the more obvious symptoms of neglect dissipate on formal examination.

Hemibody Awareness. In addition to impaired awareness of external space, lateralized lesions also may result in impaired perception of stimuli impacting on or diminished awareness of one half of the body itself. This impairment can take one of several forms: (1) lack of awareness or suppression of tactile stimuli, (2) lack of or diminished awareness of impaired capacity (denial of illness), or (3) failure to recognize one's own limb(s). As in the case of hemispatial neglect, these phenomena are more likely to be seen following acute lesions to the right hemisphere.

Neglect of contralateral tactile stimulation in the absence of primary sensory disturbances, as with vision, may occur with either single stimulation (less sensitive) or DSS (more sensitive). Occasionally, single tactile stimulation applied contralateral to the lesion may result in reports of the stimulus being perceived (subjectively displaced) to the homologous region on the ipsilateral side (of the lesion). This phenomenon is known as **allesthesia**. In using DSS, the patient may be touched in homologous areas or in nonhomologous areas (e.g., right face–left hand). While this latter procedure is very sensitive to subtle signs of neglect, it is not uncommon for nonlesioned controls to make errors on this task; hence, repeated measures with adequate comparison of the right and left sides of the body are particularly important when using nonhomologous stimulation.

Tactile neglect, as described above, probably occurs with approximately the same frequency as visual neglect and may or may not occur simultaneously with neglect in other modalities (vision or hearing). On rare occasions one may encounter a more unusual phenomenon where the patient may deny ownership of the affected limb(s). While not actively denying that the affected limb belongs to them, other patients may effectively fail to attend or ignore one side of their body (or parts thereof) when engaging in such activities as dressing or shaving. One patient with a right hemisphere stroke, for example, managed to put on his robe; however, when asked to tie the belt together, he immediately secured the end on his right side but never searched past midline for the other end hanging from the left side. An alternate explanation is that this failure could represent an example of directional hypokinesia or a hesitancy to move into the hemispace contralateral to the lesion (Coslett et al., 1990).

In the initial period following a lesion, a patient, while not disowning a paretic limb, may deny that there is anything wrong with it. This is known as **anosognosia**. These more dramatic symptoms of denial generally occur immediately after an acute event, such as a cerebral vascular accident. However, even as these symptoms begin to subside over time, the patient still may underestimate the significance or potential impact of their disability. For example, although acknowledging a persisting weakness on one side, if asked the patient may indicate he anticipates being able to return to work (which requires strenuous manual labor) after "a few days of rest." This latter phenomenon is referred to as **anosodiaphoria**.

Diminished or Distorted Motor Responses. In addition to loss of sensory awareness (either internal or external), the patient with a lateralized lesion may show a diminished tendency to (1) spontaneously utilize the hand contralateral to the lesion, (2) move either limb in the affected hemispace (**directional hypokinesia**), or (3) deviate the eyes toward the affected hemispace. For example, another right hemispheric stroke patient was searching for his soft drink can that was on the left side of his tray. The author picked up the can and held it in his right hand while facing the patient (thus, in the patient's left hemispace). When the patient was informed that the examiner was holding the soft drink in his hand, the patient immediately focused on the examiner's left hand (patient's right hemispace). After he expressed some bewilderment, the patient was informed that the can was in the examiner's *other* hand and was asked to find it. Despite multiple verbal cues as to the likely location of the "other" hand, the patient's eyes never deviated to the left of midline in his search, despite evidence of his ability to move his eyes in that direction when following a moving target.

At times it may be difficult to determine whether the patient's failure to respond may result from a lack of awareness of one hemispace (e.g., not searching to the left for the soft drink can since that "space" no longer existed for the patient) or whether there is simply an inertia to move into that hemispace. Either or both behaviors may be present in different patients. Heilman, Watson, and Valenstein (1995) review several techniques by which these features might be dissociated. Understanding these phenomena are critical in neuropsychological or neurobehavioral examinations since many of the tests and measures used require either a visual (e.g., Picture Arrangement) or tactual–motor (e.g., Sequin-Goddard formboard) exploration of (and response to) right and left hemispace.

As noted, while all of the above behaviors can occur with lesions in either hemisphere, these various manifestations of unilateral neglect are most likely to be found following right hemispheric lesions. In a way, this is consistent with what has been said previously about the role of the right hemisphere in behavior. If the right hemisphere is charged with the responsibility of arriving at an immediate global, gestaltist impression of a situation, not by initially attending to or analyzing its specific details (the hypothesized role of the left hemisphere), but by attending to the whole situation at once, then it also has to be able to immediately attend to or pick out those features of the situation that are immediately most salient. To effectively accomplish this task, it must attend in a general way to stimuli in both fields or on both sides of space, that is, it must be capable of directing attention both in the contralateral as well as in the ipsilateral half of space. Conversely, the left hemisphere, not so compelled, may tend to "limit its attention" to only the contralateral space. Hence, the right hemisphere may be better capable of compensating for the contralateral attentional loss following left hemisphere lesions than is the left hemisphere following right hemisphere lesions (Mesulam, 2000c; Heilman, 1995; Heilman, Watson, and Valenstein, 1995). For readers interested in a more detailed explanation of these neglect syndromes, in addition to these sources, one may wish to review Critchley (1969) and Rafal (1997b),

Visual Imagery

Mental imagery is a complex construct that requires explicit definition prior to a discussion of the theories regarding hemispheric specialization. Mental imagery involves the perception of an image without sensory input. The mental image of an object, such as a coffee cup, is conjured up in the "mind's eye" from memories of the visual representation of either a prototypic or a specific coffee cup (depending on the "sharpness" of the image). Mental images also can be experienced in other sensory modes, like auditory or gustatory.

It once was thought that the right hemisphere primarily was responsible for generating visual imagery. Presently, the more predominant opinion seems to be that specific components of imagery are preferentially processed in the right and the left cerebral hemisphere. Kosslyn (1988) has suggested that two broad types of processes are required to generate a visual image. These processes include (1) the activation of stored memories of parts of the visual image, and (2) the arrangement of these parts into the proper configuration using a coordinate system. It is argued that the process of accessing the parts of the object to be imaged are predominantly in the domain of the left cerebral hemisphere, while the process of spatially arranging these parts into the whole object would be in the domain of the right cerebral hemisphere. Although still controversial, it seems that the integrity of visual association cortex is crucial for the preservation of visual imagery. Patients with cortical blindness and lesions limited to primary visual cortex have been demonstrated to have intact visual imagery (Chatterjee & Southwood, 1995), suggesting that primary visual systems are not critical for the generation of visual images. For a review of this topic, the reader is referred to Sergent (1990).

FUNCTIONAL HEMISPHERIC SPECIALIZTION: FINAL THOUGHTS

Just as different parts of the brain in its anterior–posterior axis (e.g., the frontal versus the parietal lobes) have different structural features and subserve different functions, so too do the two hemispheres. Actually, it is likely that such functional divisions (beyond contralateral sensorimotor representation) also characterize subcortical structures such as the thalamus, basal ganglia, and limbic structures (e.g., Liotti & Tucker, 1995). However, our knowledge of these functional divisions or distinctions is still very rudimentary. For example, given a typical right-handed individual, it is known that the left hemisphere is primarily responsible for processing prepositional speech and language. However, is there a fundamental difference in the function, structure, organization, or interconnectivity of the cells of the left hemisphere (distinct from those on the right) that serve as the basis for this linguistic advantage? That is, do the left and right hemispheres process information in fundamentally different ways? Besides propositional language, what are the other major functions of the left hemisphere and is there a more general way of describing or encompassing these functions? If the left hemisphere primarily is the "language" (at least in its semantic or propositional aspects), how is the function(s) of the right hemisphere best characterized? To what extent are right and left subcortical structures specialized? While it is known, for example, that lesions affecting the left thalamus or basal ganglia are more likely to produce aphasic-like syndromes than comparable lesions on the right, is this the result of primary functional differences in these structures themselves or are these effects simply a result of their cortical connections? While some tentative hypotheses have been offered for some of these questions, others have been barely explored.

Sergent notes that it is difficult to define the different roles played by each hemisphere in a relatively circumscribed task like facial recognition. It certainly is unlikely that we could ever identify "a single, bipolar principle that would encompass the functional properties of the two hemispheres..." (Sergent, 1995, p. 178). Recognizing these limitations, the most prudent approach may be to emphasize the complexities involved in describing hemispheric specialization rather than to attempt definitive statements about the functions of a specific hemisphere. It needs to be emphasized that the descriptions provided in the preceding sections are the result of attempts to define and categorize extremely complex and often incompletely understood phenomena. The problem may be especially compounded when trying to discuss right hemisphere functions. For example, if we are substantially correct in asserting that the right hemisphere primarily is responsible for certain "nonverbal" and "gestaltist" aspects of behavior, we find ourselves trying to analyze, describe, categorize, and verbally label behaviors that by their very nature are not amenable to these types of verbal analysis or description. Second, and perhaps most importantly, it is important to keep in mind that behavior itself is not dichotomized, but rather the result of an integrated brain, functioning in an integrated manner, to produce an integrated response. Whether one is referring to a predominately "left hemispheric behavior" such as language or to what might appear to be a predominately "right hemisphere behavior" such as visual–spatial memory, it is clear that both the left and right hemispheres, acting in a complementary manner, contribute to the final output.

As noted earlier, Mountcastle's (1979) concept of a "distributed system" or distributed functions provides an excellent model for describing this behavioral interaction on a hemispheric level. To briefly review, the basic premise of this concept is that there probably are small groups of cells that perform very specific and circumscribed (micro) functions. Even the simplest, elementary behavioral response (or "macro" function) is a result of the interconnection and coordinated response of a specific subset of these groups of cells (or microfunctions) representing varied and diffuse cortical and subcortical areas. Another behavioral response may share some of these same microfunctions, while encompassing others not used by the first. At the hemispheric level, a comparable phenomenon likely occurs. Different neural networks both at a micro- and macrolevel are likely incorporated from both hemispheres for each and every behavioral response. The best we can hope for is to attempt to tease out the unique contributions of each hemisphere (or portion of each hemisphere) by careful analysis of and attention to the behavior in question, the circumstances under which it occurs, and the stimulus and response demands placed on the organism at a given place and time. Through the study of the differential effects of unilateral lesions, we attempt to discern not only the different behavioral effects of hemispheric lesions, but also the often subtle differential effects on a given behavior (e.g., differences in constructional tasks between right and left hemispheric damaged patients). Conversely, in observing or analyzing behavior, there needs to be an awareness of the complexity of the behavior and the multiplicity of factors that contribute to it, including factors that represent potential contributions of both hemispheres. Attention to the "spatial" aspects of mathematical calculations or to the benefits of internal "verbal" cueing in copying a complex geometric design are two, very simple examples.

In the next and final section of this chapter we will review the functions of the primary, secondary, and tertiary cortical zones and explore possible ways in which the cortex processes information and effects behavioral responses.

Endnotes

2. A recent study suggests that the anatomical basis for different classes of words may be much more specifically localized than previously thought (Damasio et al., 1996).

3. The concept of "function" is a construct that humans impose on behavior. What we define as "functions" are not necessarily unitary, invariant, or homogeneous operations of the brain.

4. As noted earlier, there is a fairly weak association between the appearance of this type of cortical atrophy in the elderly and mental status changes. An elderly individual may show signs of significant atrophy on a CT without any significant loss of cognitive ability. In the earlier stages of Alzheimer's disease, there may be significant mental impairment with little or no evidence or cortical atrophy.

5. For a more detailed treatise on this subject, see Evolution and Development of the Cerebral Cortex in *Trends in Neuroscience* (Special Issue) Vol. 18 (9), 1995.

6. The calcarine sulcus passes right through the primary visual projection area. The visual cortex above it (cuneus) processes information that comes from the inferior visual field (contralaterally), while the lingual gyrus is responsible for the superior field.

7. In addition to fibers that enter and exit the cortex vertically, horizontal bands of fibers interconnect neurons in adjacent cortical fields. While present in all cortical layers, they tend to be most prominent in layers IV and V, especially in the primary sensory areas. Generally known as the **bands of Baillarger**, in primary occipital cortex (where they are readily seen by the naked eye), they are called the **line of Gennari**.

8. Occasionally, one might see the term "frontal granular cortex" used with reference to the "prefrontal" or "heteromodal" frontal cortices. Although these frontal cortical areas are not characterized by the same degree of granularity as is found in the primary sensory areas, relatively speaking they are more granular than premotor and primary motor areas (in the posterior portion of the frontal lobes).

9. A distinction needs to be made between these vertical columns (which are fully within the cortex itself), and the broader concept of vertical organization within the brain and CNS as a whole. Again, the latter refers to the fact that integrated feedback loops involving cortical, subcortical, brain stem, cerebellar, or spinal cord structures are essential in effecting specific behaviors.

10. There seems to be some uncertainty whether area 42 should be considered a primary or secondary association area. While parts of 42 receive direct projections from the medial geniculate nuclei, it also receives input from area 41. In addition, area 42 does not show the typical granular pattern characteristic of the other primary sensory cortices (Nolte, 1993, p. 376; Carpenter & Sutin, 1983, p. 678). Regardless, the secondary (unimodal) auditory association cortex in the left hemisphere incorporates Wernicke's area.

11. The reader should recognize this as not only a speculative and highly schematized account of what might transpire at the cortical level, but also, at least for the moment, ignores the contribution of subcortical structures.

12. In theory, there may be other ways for one hemisphere to cue the other than through commissural connections, but this certainly is the most common way this occurs.

13. The posterior commissure is one example where, despite being called a commissure, many of the fibers connect nonhomologous areas of the brain (e.g., oculomotor nuclei with pretectal nuclei). Strictly speaking such fiber systems, which cross the midline to nonhomologous areas on the opposite side, are termed "decussations."

14. Some projection fibers from the olfactory and limbic structures have direct connections with the cortex, rather than going through the thalamus.

15. These concepts will be discussed in greater detail in Part III. For now, they briefly are introduced as to illustrate several classic syndromes to be discussed bellow.

16. According to both Geschwind (1975) and Heilman and Rothi (1993) it appears that the motor engrams for overlearned, skilled movements normally reside in the dominant hemisphere. Thus, for the nondominant left hand to demonstrate such movement, the information must cross from the left to the right hemisphere. The assumption is that this sharing or transfer likely takes place via the corpus callosum between the left and right motor association cortices.

17. In a related example cited by one of the early investigators, slightly embarrassing pictures were flashed to the right hemisphere of a split-brain patient. Although being aware of an emotional reaction to the stimulus (blushing), the patient's left hemisphere was at a loss to explain their reaction, even denying that anything had been seen.

18. While "normalized" behavior in such patients is often explained by cross-cueing between the hemispheres, findings such as this help to raise the question of whether at least in split-brain patients there is one mind or two, a question raised early in the course of these investigations (Sperry, 1968), but as of yet still not satisfactorily resolved.

19. Although commonly used in this context, the designation of function by means of a negative antonym ("nonverbal") likely reflects our limited understanding and/or capacity to succinctly describe the role played by the right hemisphere. This topic will be revisited later.

20. In addition to these gross structural differences, asymmetries of cytoarchitectonic areas within auditory association cortex (Galaburda & Sanides, 1980) and inferior parietal lobule (Eidelberg & Galaburda, 1984) have been reported.

21. As was the case with the temporoparietal areas, certain cytoarchitectonic differences were noted in the frontal regions that might be suggestive of increased complexity in the left frontal speech areas (Shiebel, 1984; Hayes & Lewis, 1993).

22. It should be recalled that the basal ganglia in general (including the globus pallidus) are involved in a broad range of cognitive as well as sensorimotor activities.

23. It is generally presumed that included among left-handers is a small percentage of "pathological left-handers." This would represent those individuals who despite being genetically destined to have language and handedness organized in the left-hemisphere became left-handed (right or mixed hemispheric dominant for language) as the result of early pathology affecting the normal development of the left hemisphere. While there also may be "pathological right-handers," statistically these should be even rarer.

24. This functional dichotomy of the cerebral hemispheres also has been eagerly embraced by the popular press with notions of "right brain" or "left brain" potential (e.g., Edwards, 1979).

25. Before proceeding, we should be reminded that these dichotomies exist more in our mind than in reality and that the brain functions as a whole. Simple introspection provides convincing evidence that even one's performance on tasks that at first glance might appear largely visuoperceptual in nature (e.g., Raven's Progressive Matrices or memory for geometric designs) are heavily influenced by verbal mediation. Although perhaps not as obvious, the converse likely holds equally true for tasks that appear primarily verbal. Thus, when looking at the effects of unilateral or focal brain injury, rather than witnessing how a damaged area of the brain distorts performance on a given task what we may actually be observing is how the rest of the brain carries out a that task without the normal input of the lesioned area. This point will be a major focus of discussion in Part III of this chapter.

26. Alf Brodal, MD, a well-known neuroanatomist, in reporting his own personal experience with a right hemispheric stroke that resulted in a left hemiparesis, describes his subsequent attempts to tie his own tie, a previously well-practiced skill. He writes,

The appropriate finger movements were difficult to perform with sufficient strength, speed and co-ordination, but it was quite obvious...that the main reason for the failure was something else. Under normal conditions the necessary numerous small delicate movements had followed each other in the proper sequence almost automatically, and the act of tying when first started had proceeded without much conscious attention. Subjectively the patient [Brodal] felt as if he had to stop because "his fingers did not know the next move" ...It was felt as if the delay in the succession of movements (due to paresis and spasticity) interrupted a chain of more or less automatic movements. Consciously directing attention to the finger movements did not improve the performance; on the contrary it made it quite impossible" (Brodal, 1973, p. 679).

27. For illustrative purposes, the following are examples of arithmetical word problems generated by one of the current authors (JEM) in mental status examinations: (1) "If a man has $4 more than his son and, together, they have $13, how much does each have?" (2) If you have four quarters and half that many dimes, how much money would you have?" While the first was intentionally designed to be mathematically simple but conceptually difficult, the latter was intended to be both arithmetically and conceptually fairly simple. The former turned out to be exceedingly difficult for most normal controls, even many college graduates, while the latter proved conceptually difficult for a high percentage of an adult population from limited socioeconomic backgrounds upon which it was tested.

28. "Symbolic gesturing" such as raising a clenched fist (anger or defiance) or a "thumbs up" (praise or approval) is an ideomotor-type response and as such is more likely to be mediated by the left hemisphere (Goodglass & Kaplan, 1963).

29. McDonald and Pearce (1996) provide a good discussion of various elements involved in the detection of sarcasm and the role of the frontal lobes in this process.

30. Assuming the presence of a normal as opposed to an anomalous pattern of dominance, which of course one can never be fully certain.

31. A solid cube generally provides a more stringent test of constructional praxis than asking the patient to reproduce a "transparent" cube. This is because most people have learned how to accomplish the latter by simply drawing two overlapping squares and connecting the respective corners. At times, a patient will use this same strategy to produce a solid cube, but if it happens to face in the wrong direction from the model (or if the model is changed), they often are unable to proceed. A cautionary note: the ability to perform this task, like many others used in neuropsychological investigations, shows considerable variation in a "normal" population and care should be taken to avoid interpreting all performance deficits as "pathological."

32. Individual Object Assembly subtest items from the Wechsler IQ tests may offer varying degrees of cues that are more accessible to the left hemisphere. For example, the "local" cues or internal details present on the "elephant" may be readily recognized by the left hemisphere, thus facilitating its solution. Whereas, the "butterfly" (WAIS-III) provides no such cues.

33. **Dorsal simultanagnosia**, which is associated with Balint's syndrome and results from bilateral occipitoparietal lesions, is differentiated from **ventral simultanagnosia**. The latter, which typically is associated with unilateral, left occipitotemporal lesions, is characterized by greater flexibility in shifting from one visual percept to another, although the ability to integrate the individual visual elements into a meaningful perceptual whole is still impaired (see Endtnote 50).

34. Here we are focusing on cognitive/perceptual processes. However, one should recall that generally the most reliable signs of hemispheric lesions are lateralized changes

in sensorimotor functions, followed by aphasic deficits. Even here, one needs to be aware of two obvious caveats. The first is to rule out more caudal lesions along the neuroaxis or peripheral pathology (in the case of sensorimotor deficits), and second, left-handedness (in the case of aphasic deficits). In the latter situation, recall that while the odds still favor left-hemisphere involvement, the probability is reduced compared to natural right-handers.

35. While neglect seems to be most frequently associated with lesions involving the posterior, heteromodal association cortices, it also can be seen following lesions to the frontal cortex, subcortical structures, and even upper brain stem.

PART III. THEORIES OF FUNCTIONAL ORGANIZATION OF THE BRAIN

CHAPTER 9, PART III OVERVIEW

Hopefully, some of the things that separate the scientist–practitioner from the average technician are an appreciation of the principles underlying the observed phenomena and the ability to integrate new phenomena into useful and meaningful constructs based on an understanding of unifying theoretical models. While theories have abounded in physics and mathematics for centuries, complete understanding of true unifying principles often remains elusive. In physics, for example, it has been the search for a model that would explain and integrate the strong nuclear force that binds the subatomic particles in the nucleus of the atom with those of the weak nuclear force accounting for radioactivity and the electromagnetic force that explains the orbits of electrons. As suggested earlier, the neurobehavioral sciences are still in their relative infancy and we are likely much further from discovering such unifying principles. This, however, should not dissuade us from the search, for it is from the search itself that knowledge and understanding progresses. In this final section on the cerebral cortex, there is an attempt to provide some very tentative ideas into what may prove to be one of nature's greatest mysteries: the functional organization of the brain. No great insights or revelations are promised. To the contrary, most of the ideas presented here are not new. Rather they are derived, for the most part, from models introduced by earlier theorists, notably Luria, Mesulam, Geschwind, Heilman, Mountcastle, among others. What is hoped is simply to provide the reader with a framework by which he or she may think of brain function (or dysfunction) in a dynamic fashion, to hone and further define their own conceptualization of how the brain responds to internal and external cues and demands, and of how thoughts become translated into actions. Even as the final solutions continue to evade us, we trust that the joy will be in the search as much as in the discovery.

In this section, following Luria's model, we divide the brain into three primary functional units: those basically responsible for attention and arousal (motivation), processing and storage of information, and expressive behavior. With regard to the first functional unit, the role of subcortical structures and the limbic system in initiating or maintaining cortical arousal briefly will be explored. Next, we will discuss how sensory information (primarily vision, audition, and somesthesia) might be transformed from elementary sensations to meaningful, unimodal percepts, and ultimately to complex, multimodal concepts. We shall discuss how the latter may allow for abstractive abilities and other higher-level cognition. Along the way, we also shall review how lesions affecting the various stages in this process may result in specific behavioral syndromes. Finally, considerable attention will be devoted to the frontal lobes or what has been described as the "executive system." Here we will explore the suspected neuroanatomical substrates of decision-making, including the individual's response to such questions as "if, when, where, why, and how should I respond?" In the course of trying to answer such questions, we will address the suspected role of the prefrontal cortex in arousal, intention, attention, inhibition, planning and organization, comparison of performance to goals, and modulation or control of basic drives, and again how disturbances of these functions might be expressed behaviorally. Finally, we will examine how plans and intentions likely are translated into overt motor behavior via the primary and secondary motor cortices.

Thus, at the conclusion of this chapter, one should have a better understanding of the neuroanatomy of the cerebral cortex and its general functional correlates. In addition, hopefully one will begin to consider the brain's potential organization and how even the simplest of actions are likely the result of complex interactions involving wide expanses of brain tissue. Conversely, one might derive a fuller appreciation of how diverse lesions may differentially affect similar behaviors. Again, we are merely scratching the threshold in our attempt to understand this magnificent organ, but the deeper the puzzle, the more intriguing the search.

INTRODUCTION

In Part II of this chapter a number of the better known associations between specific symptoms or behaviors and specific regions of the cortex were reviewed . As we have seen, these correlations initially were established through the study of the effects of focal brain injuries (and more recently supported by technological advances in brain mapping, such as the PET scan and functional MRI). While these approaches provide valuable clinical and scientific information, a fundamental question still remains: "Is there any type of unifying theory to help account for these observations?" Beginning with Broca and continuing into the early part of the 20th century, most of the aphasic, apraxic, and agnosic syndromes that we know today were identified. In some respects, however, these findings might be viewed as the new and improved (albeit more clinically relevant and scientifically sound) "phrenology." The names of the functions were changed and the cranial bumps were gone, but in many respects the principle remained pretty much the same, namely to identify "centers" for various behavioral functions that could be associated with specific areas of the cortex. In this endeavor, behaviors were matched or associated with specific gyri or cytoarchitectural areas simply as a matter of clinical correlation. In these earlier stages of discovery, there seemed to be relatively little effort to understand or explain *why* that particular gyrus or area was necessarily involved with a particular function.

Using only this correlation approach, it was difficult to explain why lesions in different cortical areas, sometimes even in different hemispheres, appeared to produce comparable deficits on similar tasks. Also, while specific cortical areas might be fairly reliably associated with certain specific signs, symptoms, or behaviors, as the behaviors or tasks became more complex, such predictions often became less reliable. Beyond the strict "localizationists" and those espousing the theory of "mass action and equipotentiality," there were few other attempts to develop a unifying theory to account for the observed relationships between structure and function.[36] It might be noted that even during the late 1800s when enthusiasm for localization of function was perhaps at its peak, Hughlings Jackson began to question what localization actually meant. From the very outset, Jackson argued for a vertical organization of behavior. He suggested that behavioral functions were likely organized in different ways, at different levels of the nervous system, and as a result the localization of a *symptom* did not necessarily mean the localization of a *function*. Despite Jackson's contributions, overall there was little progress in the early part of this century to integrate the emerging clinical data into a more unified theory of brain organization. Perhaps the first truly comprehensive attempt to do this is reflected in the efforts of Aleksandr Romanovich Luria (1966) in his book, *Higher Cortical Functions in Man*. First published in Russia in 1962, this work still serves as the most comprehensive, macroscopic model of brain organization available to date.

The third part of this chapter will attempt to present a heuristic model, based in part on the theories outlined by Luria, for conceptualizing behavioral functions and their organization in the brain, particularly with regard to the cortex. This model hopefully will provide the reader with a tool to systematically analyze a broad range of behaviors, including performance on neuropsychological measures.

Defining "Function"

Luria gave credit to Hughlings Jackson and to Pavlov for redefining the concept *function*, thus setting the stage for the concept of **dynamic localization.** Luria suggested that the first step in developing a theory of brain organization and function was to understand how behavior itself ("function") was defined or conceptualized. The early localizationists tended to look at "functions" as circumscribed, self-contained behavioral units that were controlled by equally circumscribed, self-contained groups of cells in a particular area of the brain. Luria pointed out that while such a description might be adequate when speaking of the "function" of the pancreas (i.e., to produce insulin), as behavior became more complex, this approach to defining function became less meaningful. He suggested that the behaviors that we normally attribute to the brain, behaviors such as speech, perception, arithmetical calculations, and learning, are infinitely more complex and should be referred to as **functional systems**. Luria's notion was that such behaviors are composed of many elements, involving multiple cortical (and subcortical) units, acting in a coordinated fashion to effect the desired goal. Depending on time and circumstance, some of the individual elements (brain units) called upon might vary, although the behavioral goal should remain constant.

To illustrate, if someone is asked to point to a picture of a flower, what "function" is being tapped? One possibility is that there is a group of "flower recognition" neurons located 237 millimicrons due east of the end of the superior temporal sulcus that carries out this task all by itself, which we "strongly suspect" is not the case. Although oversimplified, the following scenario may be closer to what actually happens. First, there must be a general level of awareness or cortical activation so that the nervous system is capable of responding. That awareness or "attention" must be focused on the examiner and the test materials. Assuming that the peripheral auditory (and visual) mechanisms are intact, the examiner's instructions must be processed cortically. This is not a single-step process. In the case of the command "point to the flower," the incoming auditory stimuli need to be decoded into individual phonemes and then encoded into phonemic groupings that can be recognized as a specific word, "flower." The "word" must be associated with a variety of past experiences (e.g., shape, color, smell, and texture) to give the word meaning. In this latter process, a certain amount of abstraction or synthesis of all individual experiences is required so that such diverse shapes and colors as represented by the rose and a dandelion will be considered as potential choices. Simultaneously, visual input must be decoded in the striate cortex and then encoded into an integrated pattern that separates figure from ground. This "percept" also is compared to previous experiences that may elicit multiple associations, including tactile, olfactory, as well as the auditory association with the word "flower." Probably somewhere between the idiotypic cortices of the occipital pole and the middle of the superior temporal gyrus (the primary visual and auditory reception areas), these two processes come together so that the auditory percept and the visual percept are recognized as relating to the same entity. This is still not the end of the process. This "information" is then conveyed to the prefrontal cortices where, after "consulting" with parts of the limbic system to check on the prevailing mood state and current "feelings" with regard to the immediate situation and/or this particular examiner, a "decision" is made as to whether a response is to be made and what the nature of that response might be. Only if and when a response is chosen are signals then conveyed through the premotor to the motor cortex to effect the desired movement. The smooth, accurate execution of this movement also requires the integration of subcortical (e.g., basal ganglia, cerebellar) and cortical (occipitoparietal) feedback.

A central idea in much of Luria's writing is that behavior can be disrupted at any point ("point" meaning either behavioral or structural) in the stimulus–response process. Hence, while different areas of the brain may have uniquely contributed to each stage in this process, for the behavior to be smoothly and efficiently executed multiple areas of the brain acting in concert are required. This basically is the same principle behind the notion of **distributed systems** that was discussed earlier in this chapter. The corollary is that depending at which stage or locus in the brain problems might have been encountered, the nature of the behavioral disturbance varies. When such variations are taken into account, clues are provided as to which areas of the brain or what pathways might be involved. Similar analyses also are useful in identifying what elementary behavioral process(es) are likely disturbed. Such information not only allows for a better understanding of immediately observed deficits, but also allows for better prediction of potential difficulties across multiple tasks or situations.

To summarize, Luria (1973) suggested that the study of brain–behavior relationships should begin with two critical steps. The first requires an examination of the particular "mental activity" (symptom, behavior) in question to determine what specific factors (i.e., more elementary functions) are essential for its successful execution. The second is to analyze the expression of that behavior in the brain-injured individual, paying particular attention to the types of errors made, in order to develop hypotheses as to the specific nature of the underlying deficit(s) in a given case. These insights then are combined with information regarding the specific focus of the lesion across cases, comparing the common elements or factors that appear to result from lesions in a given cortical area, as well as analyzing how similar behaviors are differentially affected by lesions in different cortical areas. Over time a picture should begin to emerge both as to how "functions" are mapped within the brain and how these individual functions are integrated to produce specific, observed behaviors.

Before proceeding, however, two additional points are worthy of mention. The first is that while at first glance this seems to be simply a resurrection of 19th century localization theories, there are some significant differences. The earlier theories focused on *molar* (more complex) behaviors that were thought to be more or less self-contained within a limited cortical area. The latter approach focuses on discrete, *molecular* (more elementary) behaviors distributed throughout the brain that are combined and recombined in a multitude of varying organizational patterns and that we observe as complex behavior. Second, even these patterns or interconnected networks of microbehavioral elements are not necessarily fixed and immutable within or between individuals. Different individuals may have learned different strategies or approaches to accomplish a given task. For the same individual, dynamic changes may occur over time with learning or experience (e.g., when first learning to read, one often attends to the learned acoustical patterns associated with particular letters; however, once well learned this process is discarded or reduced in importance).[37]

FUNCTIONAL ORGANIZATION OF THE BRAIN: LURIA'S MODEL REVISITED

In searching for a theoretical model to conceptualize, and hence, approach a fuller under-standing of the higher level mental activities of the brain, Luria began by dividing the brain into three functional (quasivertical) units and the cortex into three horizontal zones. Volitional mental activity was thought to require the integrity of these three functional units with the various cortical zones working in concert with one another. The cortical zones, to be covered in a later section, provide a framework for understanding the progressive analysis and integration of behavior. More immediately, the focus will be on the three major functional units of the brain, which are (1) attention and arousal, (2) information processing, and (3) executive functions. The first unit, which Luria described as the unit responsible for **regulating cortical tone**, will be divided into three separate components: (1) the reticular activating system, which is responsible for general, nonspecific arousal; (2) the thalamus, which is largely responsible for selective attention and cortical modulation; and (3) the "limbic" structures, which is involved in emotion, motivation, or "goal-directed" arousal, as well as learning and memory. The second or **gnostic** functional unit encompasses the temporal, parietal, and occipital lobes and is thought to be responsible for obtaining, processing, and storage of information from the outside world. Finally, the third or **executive** unit, which primarily involves the frontal systems, can be characterized as being concerned with programming, regulating, and verifying mental activity.

The following discussion largely represents a synopsis of the theories as originally set forth in Luria's *Higher Cortical Functions in Man* (1966), and in *The Working Brain* (1973). Since that time, his approach to understanding the functional operation of the brain has continued to evolve through the work of such individuals as Norman Geschwind, Marsel Mesulam, Kenneth Heilman, and Edith Kaplan, among others. The remainder of this chapter will attempt to revisit and elaborate on those ideas.

THE FIRST FUNCTIONAL UNIT: ATTENTION AND AROUSAL

Directed and Non-directed Attention: The Reticular Activating System and the Thalamus

Goal-directed behavior generally is predicated on the capacity for focused and sustained attention. In order to achieve and maintain this state, two more basic elements are required: (1) a certain level of optimal arousal of the cerebral cortex, and (2) adequate drive or motivation. Diminished arousal (e.g., in states of lethargy, stupor, or obtundation) or hyper-excitability (e.g., excessive anxiety or panic) both can interfere with efficient mental opera-tions. Lack of sufficient drive states also can result in the failure to initiate behavior (despite adequate alertness or arousal) or result in the inability to sustain and direct behavior long enough to accomplish its intended goal. Because of the importance of optimal levels of arousal underlying all successful mental activities, the reticular activating system (RAS) and the thalamus were thought to be the cornerstones of this first functional unit. Since the structure and function of the RAS and the thalamus were discussed in some detail in Chapters 4 and 7, only a few general comments will be made here.

The RAS and the thalamus are thought to be responsible for the tonic or "state-dependent" arousal of the two other functional units of the brain. In turn, these regions are responsive to external stimuli, as well as internal influences (e.g., hypothalamic or cortical feedback). We have all experienced the *arousal* value of an "unexpected" tap on our shoulder or a sudden loud noise. By contrast, an "expected," "nonthreatening" stimulus might evoke relatively little response, unless of course it takes on added signal value as a function of past experiences, expectations, or enhanced vigilance. The latter might be present, for example, in anxiety or paranoid states. This type of signal enhancement generally is the result of input from higher-level functional units, or is produced at a higher level within the first functional unit (e.g., limbic system). Such descending influences, which can be either excitatory or inhibitory, are often state-dependent. For example, the smell of food will have a different arousal value on the RAS in a sated versus a hungry organism. Similarly, the recollection or thought of a test the next day might have different "arousal" value in the well-prepared versus the not-so-well-prepared student.

As noted in Chapter 7, the role of the thalamic nuclei in the activation of the cortex is not completely understood. Whereas different nuclei may subserve somewhat different functions, collectively these nuclei appear to be responsible for altering cortical tone or arousal by serving as gating or filtering mechanisms for incoming stimuli. Recall that the thalamus apparently selectively screens out (or selectively attends to) those sensory inputs or cortical processes most critical at the moment. In addition, some thalamic nuclei may serve to selectively modulate or "shift attention" to different cortical zones depending on circumstances. If this type of selective attention and activation did not occur, the organism might either (1) attempt to attend to everything at once, or (2) selectively attend to stimuli primarily based on intrinsic properties of the stimulus (e.g., stimulus intensity or novelty) rather than on its teleological relevance. While the latter does occur to some extent, maximal efficiency of cognitive/behavioral functioning demands that under normal circumstances attention is focused on stimuli related to a narrow range of problems or goals at any given time. Because the function of this first unit is so essential to the optimal operation of the second and third functional units, the patient's level of arousal must be taken into account when considering any apparent breakdown in "higher cortical behaviors" such as learning and memory, perception, or problem solving. The inability to selectively attend to relevant or to screen out irrelevant stimuli and the subsequent compromise of cognitive efficiency appears to be part of the problem in diverse pathological conditions. These may include conditions where there is diminished levels of arousal (e.g., severe depression), excessive levels of arousal (e.g., manic or anxiety states, amphetamine abuse), and/or disruptions of the filtering system such as would appear to be the case in such diverse conditions as agitated delirium, head trauma, or attention deficit disorders.

Drive: The Role of the Limbic System in Arousal (Motivation)

The third subdivision of this first functional unit[38] primarily consists of structures generally associated with the "limbic system." While these structures and their functions were discussed in detail in Chapter 8, it may be useful to briefly revisit the limbic contributions of this first functional unit and preview a few key relationships between the first functional unit and the second and third units.

Although Luria primarily focused on the second (gnostic) and third (executive) functional units, he recognized that the "activating" and/or "motivating" influences of the first unit were an essential component of behavior. He postulated that changes in homeostasis as well as external stimuli were important in cortical "arousal" and in initiating (i.e., prompting) behavioral responses. These responses were thought to be mediated primarily via their influence on the ascending RAS. In addition, he recognized the potential for the brain itself, specifically the third functional unit (the "executive system"), to evoke behavioral responses, as in the establishment of goals and the formulation of the plans by which they might be achieved. However, he also appreciated the fact that the "energy" (drive, motivation) to implement these activities probably was derived not from the third functional units itself but rather from phylogenetically older structures or more intermediate cortex comprising the medial zones of the cerebral hemispheres. These areas, which now frequently are referred to as the "limbic system," were thought to serve an important link between the cortex and the RAS. He stated:

the principle function of these brain zones [limbic structures] is not communication with the outside world (the reception and analysis of information, the programming of actions), but the regulation of the general state, modification of tone, and control over the inclinations and emotions (Luria, 1973, p. 60).

Today, these limbic structures still are associated with such constructs as "drive," "energy," "will," or "motivation." As Luria suggested, the influence of these limbic structures, including the hypothalamus, in prompting and sustaining goal-directed behavior is extensive. As noted, the stimuli to which the structures of this first functional unit respond may be either internal or external and may involve (1) the regulation of basic biological homeostatic mechanisms, (2) instinctually guided behaviors, (3) acquired or learned response patterns, or at the highest level (4) abstract or long-range goals or ideas. Their impact is both ascending (affecting the telencephalon)[39] as well as descending (having an influence on the brainstem and the RAS). This influence may elicit positive (approach) or negative (avoidance) or even relatively neutral response patterns, depending on the overall state of the organism or the environmental context in which the stimulus occurs.

By way of brief review, hunger or thirst provides a good example of basic biological needs or homeostatic drives that may be "initiated" by the first functional unit at the level of the hypothalamus. The "drive" state initiated at this level subsequently recruits the second and third systems to satisfy this need by implementing a search for and/or coordinated response to potential sources of food or water. "Nest building" and the feeding and protection of one's offspring are examples of "instinctual behaviors" that are thought to be mediated by more recently developed limbic structures, such as the cingulate gyrus in vertebrates, and whose analogues can be seen in human parental behaviors. Highly species-specific avoidance-type responses also seem to be built into most individuals in the animal kingdom. Whether it is a sudden "dimming" stimulus (for the frog), the silhouette of a flying hawk (for the chicken), or perhaps the sight of a decapitated body (for primates, including humans), there normally is a strong drive to take some type of evasive action. The capacity of certain stimuli to elicit approach or avoidance responses based on the previous experiences also is well known. The sight of a loved one following a prolonged absence versus that of a neighborhood bully by a child may elicit equally strong, but highly divergent emotional and behavioral responses.

In the examples presented above, the stimuli, which lead to a state of arousal or a response drive, relate primarily to the first functional unit itself (homeostasis) or to stimuli associated with the second functional unit (i.e., external stimuli that have acquired a specific emotional coloring, either as a result of the preprogramming or past experiences of the individual). However, there is one other important source of drive or emotional valance listed above, namely that which is derived from more abstract or long-range goals or plans. It is the latter that usually is specifically linked to the third (executive) functional unit (the frontal lobe) and perhaps is most readily appreciated in humans. Immediate stimuli, initially perceived as affectively neutral or even imagined or anticipated stimuli (including abstract ideas), can take on motivational significance (i.e., have arousal value) because of their relationship to long-range goals or plans. For example, one may have the plan to attend college. The goal eventually may be to land a well-paying job in order to have financial security and all the other things this provides. To secure this goal, you need not only the initial motivation to get started (e.g., going to college), but sustained drive and energy throughout the process to meet the more intermediate goals (e.g., passing the next exam). The drive or emotion necessary to sustain these efforts on a day-to-day basis likely requires the mediation of limbic structures.

While experimental evidence for this type of frontal–limbic interaction is difficult to obtain in humans, there is strong clinical evidence of its existence. It is not uncommon to find patients, particularly those with bifrontal lesions, who repeatedly make elaborate plans to return to work or to engage in other meaningful, productive endeavors yet never demonstrate the motivation or initiative to carry out these plans. If they manage to actually get started, they often falter, especially if difficulties are encountered. Thus, it is not the ability the make plans that is affected by such lesions, but the motivation, energy, or emotional drive to execute and sustain them. It is easy to speculate that the critical lesion in such cases is the severing of the connections between the dorsolateral frontal lobes and the limbic or paralimbic structures. As we have seen, further evidence is derived from the observation that lesions that are more or less restricted to the anterior cingulate gyrus or dorsomedial frontal cortices also have been shown to produce marked apathy or inertia in patients.

In addition to providing the drive for future goals or plans, these frontal–limbic connections play a role in many other aspects of day-to-day behavior. One of these roles is the motivational or emotional influence of learned socialization on behavior. Social approval (or disapproval) normally is a very powerfully motivating force, whether it is encoded as the social mores or religious beliefs of the larger community or in the idiosyncratic customs or expectations of one's peer group. In these cases, often the reinforcement does not have to be immediately present. Simply the memory (which also may be mediated by certain limbic structures) of past, more direct social reinforcements (e.g., a hug or a smile, or perhaps a scolding) or the capacity to predict the possibility of such future reinforcements, whether positive or negative, usually is sufficient to guide the behavior. As we will see, frontal lesions can lead to a breakdown in these patterns of "emotional control" (reinforcement) over behavior. Patients with certain types of frontal pathology, especially if the orbitofrontal areas are implicated bilaterally, may behave as if they are "indifferent" to the social consequences of their behavior. They may become crude, impulsive and, at times, grossly socially inappropriate. An example of the latter would be the frontally impaired gentleman, previously considered to be very refined, who stopped to urinate in public while accompanying his wife out to dinner. While the "indifference" to personal hygiene and grooming sometimes seen in "frontal" patients also might be interpreted as a failure in socially derived motivation, it may reflect a disturbance of even more primitive, instinctual phenomena, as similar behavior can be seen in lower-order primates with frontal lesions.

These observations emphasize a critical aspect of frontal–limbic connections, namely the importance of reciprocal control or modulation between these cerebral systems. Phylogenetically, it is likely that most basic drive states are preprogrammed for more or less immediate gratification. If the organism is hungry, it seeks food; if it is frightened, it flees; if its young are threatened, it attacks; if it is sexually stimulated, it seeks fulfillment. However, as we have seen, it may not always be in the best interest of the individual to immediately respond to each and every stimulus or drive state the moment it is experienced. One may decide it is better to delay immediate gratification in deference to a future, acquired (abstract or social) goal or objective. Conversely, one might choose to confront rather than avoid a frightening or unpleasant situation now in order to achieve a long-term goal or to avoid the possibility of an even greater unpleasant future consequence (e.g., "Do I want to satisfy my need for sleep or try to stay up and study so I can pass the exam?").

Not infrequently, the immediate gratification of two or more current drive states may be incompatible. In these cases, the individual must determine which goal has the greater priority or perhaps derive a strategy whereby one drive might be satisfied without sacrificing the other. This prioritization or assigning of values to different drive states is thought to be a primary function of "higher cortical centers," specifically the third functional unit. The role of the third functional unit will be discussed in greater detail later in this chapter. For now the reader should simply be aware that lesions can disrupt either the source of this capacity for decision making or judgments (i.e., the frontal lobes) or its reciprocal connections with the limbic structures. Either might produce a variety of clinical pictures that may result in behavioral disinhibition, poor judgment, inability to delay gratification, or impulsivity. In addition to the brief examples presented above, the following section offers a glimpse into a few other situations where a disturbance of these frontal–limbic connections might impact behavior.

Other Behaviors Likely Associated with Frontal–Limbic Disconnections

Pseudobulbar Affect

Patients with bilateral frontal damage (either cortical or subcortical) may evidence increased emotional lability under certain stimulus conditions. For example, they may be moved to crying in situations that elicit feelings of tenderness or sympathy for others or by discussions that may remind them of their own losses. While such a reaction is not necessarily inappropriate, many individuals have learned that certain emotional displays, such as crying, are not "expected" in certain situations. As a result, the expression of certain types of emotions (e.g., crying) are commonly suppressed, particularly in public, and often even in situations where it may be considered appropriate, such as funerals. Patients with bilateral frontal or subfrontal lesions commonly are not so inhibited. Such patientsoften are said to "show emotions," despite the absence of any apparent underlying affect (the so-called *pseudobulbar affect*).[40] While the cause of the displayed affect (usually crying) may not be immediately obvious, on careful questioning a specific stimulus or underlying mood consistent with the affect being expressed usually can be uncovered. A disconnection between the frontal and limbic systems, resulting in diminished frontal control or modulation of these emotional drive states would appear to be a likely contributor to this condition. The fact that the mood may be very transient and the provocation minimal not only might help explain the significance of such disconnections in eliciting this effect but also might explain the patient's reticence to admit to breaking down given such a minimal provocation. As a corollary to this hypothesis, it might be noted that children also are less likely to suppress their emotions. On the one hand, it might be argued that this is because they have not yet learned the social stigma attached to such emotional expression. However, their relative lack of constraint with regard to emotional expression also might be related to developmental delays in the myelination of the frontal lobes.

Sociopathy

The capacity to link previously experienced feelings to the anticipated outcome or consequences of our own behavior in part allows one to develop a sense of empathy for others. Empathy is not simply a cognitive awareness of the connection between behavior and affect (an awareness that the sociopath or "con man" clearly manipulates to his [or her] advantage), but the capacity to vicariously "experience" the emotion. For true sociopaths, one might suspect a "hard wiring" defect (e.g., a type of learning disability) that interferes with either the capacity to fully experience certain emotions (limbic deficit) and/or an inability to establish these connections as a result of frontal or frontal–limbic aberrations. This hypothesis, however, is not meant to discount the role of environment and learning in shaping these behaviors.

Real Life versus "Abstract" Judgment

The role of the third functional unit in modulating limbic emotional or drive states becomes an important consideration in the mental status examination of brain-injured patients. It is common practice to estimate a patient's judgment by asking how he or she would respond in given situations, for example, "What should you do if you noticed a fire in a movie theater?" In this instance, we might witness a breakdown in "judgment" simply due to cognitive impulsivity. However, the essential element that typically is missing in this situation is the presence of intense emotional stimulation. It is one thing to ask a patient with frontal–limbic lesions how they *should* respond in a hypothetical situation that at the moment is relatively free of any strong affective pull versus observing how they would respond in the actual situation with its normally attendant emotional arousal (e.g., fear). In the context of the mental status examination, the patient merely has to fall back on previously learned cognitive associations or at best reasoning capacity. However, in real-life situations, there has to be an active subordination of the initial behavioral impulse elicited by the emotional or affective pull of the situation (e.g., to flee in the case of a fire) in service of another overriding goal or consideration, namely the superordinate goal of avoiding panic and ensuring the safety of others.

THE SECOND FUNCTIONAL UNIT: INFORMATION PROCESSING

The Three Horizontal Zones of the Second Functional Unit

The second functional unit encompasses the parietal, occipital, and temporal lobes, excluding the medial limbic and paralimbic structures. Luria refers to this functional unit as the "gnostic" portion of the brain. While the first unit was responsible for arousal and drive, the second functional unit is believed to be primarily responsible for the reception, analysis, and storage of information, largely but not exclusively obtained from the outside world. This input, which primarily comes from the sensory modalities (e.g., visual, auditory-vestibular, somatosensory, gustatory, and olfactory), first needs to be "decoded" from the bioelectrical impulses carried from the peripheral receptors to the brain. Next, this information then must be encoded into meaningful units, analyzed, and integrated with other previous or concurrent information to enable the organism to arrive at some understanding of its immediate environment. These data then are cataloged and stored to provide a "library of information" through which we gradually increase our capacity to appreciate, anticipate, and manipulate our environment. Finally, it is this data bank that facilitates the development of superordinate concepts based on common and overlapping elements. In humans, at least, these processes likely also provide the substrate for abstract thought and the communication of ideas.

A key to understanding the second functional unit involves an appreciation of its division into primary, secondary, and tertiary (horizontal) zones.[41] Mesulam (2000b) describes these functional regions as "modality-specific" or "unimodal" association cortex when referring to Luria's secondary zone and "higher-order" or "heteromodal" association cortex when referring to the tertiary zone. Both the secondary (modality-specific) and the tertiary (heteromodal) cortices also are referred to as "homotypical" cortex, while the first or primary zones have been designated as "idiotypic" cortex (Mesulam, 2000b).

Before proceeding to a more in-depth look at these various zones within the second functional unit, a brief review of several concepts touched on earlier might be useful. Each of the three major senses represented in the second functional unit (vision, hearing, and somesthesia) has its own primary and secondary zones (the third zone, by definition, being multimodal).[42] Each primary zone of the second functional unit consists primarily of granular or koniocortex cortex. These zones represent the primary cortical projection areas for each of the respective senses via the specific sensory relay nuclei of the thalamus. The cells in these areas respond only to the sensory input from that modality and are responsible for the initial decoding of that information. The comparable zone in each hemisphere selectively may respond to the spatial source of the incoming stimuli (i.e., right versus left fields, particularly vision and touch). However, at this level there would appear to be much less (if any) functional differentiation as to the exact nature of the stimuli, that is, it is not certain that there is clear hemispheric specialization in the decoding of information at the level of the primary zones. Hemispheric differentiation in the processing of sensory input becomes increasingly evident as information gets channeled through the secondary and tertiary zones. Thus, lesions involving the primary zones theoretically will result in primary sensory disturbances (e.g., blindness, deafness, and loss of tactile sensation). However, because of the redundancy that is built into the system, with the exception of lesions involving the primary visual cortices and the resulting visual field cuts, unilateral lesions generally have limited effects. The cortical projections of these primary zones are to their respective secondary zones that surround or are adjacent to each of the primary zones. It is likely that these projections to the secondary zones are contralateral as well as ipsilateral.

The secondary zones, in turn, would appear to be responsible for the further organization of the output of their respective primary cortical projection areas into more highly integrated, meaningful percepts such as recognizable phonemes, words, or visual or tactile images or percepts. Information processing in each of the secondary zones still is thought to be largely, if not totally, restricted to each of the respective sensory modalities. That is, the secondary zones are modality-specific (unimodal association cortex) and damage restricted to these areas lead to modality-specific, perceptual deficits, while elementary sensations may remain essentially intact. As opposed to the primary zones, hemispheric specialization likely is to be observed at this level. Again, there likely is to be interhemispheric transfer of information between these homologous zones, as well as to the third functional units in each hemisphere.

Finally, each of these secondary, or unimodal, association areas provides convergent input to the tertiary zone(s) of the second functional unit. In these regions, evoked potential responses can be elicited from multimodal stimuli, perhaps even in individual cells. Very little of the knowledge we have about the outside world is strictly unimodal. As we step into a hot shower, we appreciate the temperature of the water on our skin (*somesthesia*), as well as observing the steam or the fog forming on the windows (*vision*). Similarly, we associate the volume or amount of water being used by the force with which it strikes our skin (*mechanical, pressure receptors*), as well as by the intensity of the sound made by the water as it falls to the floor of the tub (*audition*). Even if our experience is not immediately multisensory, we can call on past experiences to build on such associations. For example, as a result of past experiences, by simply seeing a chair, one can make an estimate of what it will feel like to sit on it or how heavy it will be if one tries to lift it. It is in the tertiary zones of the second functional unit within each hemisphere that the processed, integrated percepts from each of the senses can be combined, compared, and associated with one another. In turn, it is this higher-order, multimodal integration of information that, as noted above, allows for the full richness of associative knowledge of the outside world and that in humans forms the basis for deriving abstract concepts. As a result, damage to these areas are not restricted to any single modality, and it is in these tertiary regions of the second functional unit in which hemispheric specialization is most evident. Additional examples of how multimodal sensory integration may take place in this tertiary zone will be presented below as the various modalities are individually discussed.

Auditory Processing in the Second Functional Unit

The Primary and Secondary Auditory Zones

The primary cortical projection area for the auditory system and the medial geniculate nuclei is Heschl's area, located along the Sylvian fissure on the transverse gyrus of Heschl

(Brodmann's area 41), with the adjacent area 42 apparently also receiving some direct projections from the medial geniculates. Area 22, which constitutes a portion of the superior temporal gyrus, immediately surrounds areas 41 and 42. Area 22 (the posterior portion of which is commonly referred to as Wernicke's area) likely comprises the major part of the secondary auditory cortex or auditory unimodal association cortex. Parts of area 21 also at times are included as part of this secondary zone.

Unlike the primary visual and somatosensory cortices, there is an admixture of contralateral and unilateral pathways throughout much of the auditory system beginning with fibers leaving the cochlear nuclei. Hence, by the time the auditory projections (which originate in the organ of Corti) reach the cortex, approximately 60% of the fibers represent contralateral pathways, with the remaining 40% consisting of unilateral fibers. The primary auditory cortex is characterized by an organized pattern of "topographical" representation (not unlike the visual and somatosensory systems). In the case of audition, however, this representation becomes a matter of *tonotopic* organization, with lower frequencies being located at different cortical sites than higher frequencies. The function of the primary auditory cortex likely involves the preliminary decoding and transformation of incoming auditory information and storage of this information (probably short term) prior to higher-level analysis. The secondary or unimodal association auditory cortices then are thought to be responsible for decoding (analyzing) and encoding (synthesizing) the auditory input into meaningful or recognizable "chunks" or percepts. This latter process appears to be hemispherically differentiated, for example, phonemes and words in the left hemisphere and nonverbal patterns in the right hemisphere.

Effects of Lesions to the Primary Auditory Cortex.   As a result of the bilateral projections to the auditory cortex, unilateral hearing deficiencies following unilateral cortical damage to Heschl's area tend to be very subtle. The only difficulties that may be observed in such lesions are increased difficulty with sound localization and a mildly elevated threshold for sound perception in the contralateral ear. Even the results of dichotic listening tasks, which have been used to study such lesions, often are seen as inconclusive. Although rare, central (cortical) deafness may result from bilateral lesions of Heschl's gyrus or ascending subcortical white matter pathways. Left-sided (dominant hemisphere) lesions, if limited to the primary auditory cortex, typically do not result in any observable language-related deficits. In addition to reflecting the general lack of functional lateralization in the primary cortical zones, this also suggests that auditory input to Heschl's area in the right hemisphere is not funneled exclusively to the primary auditory cortex of the left hemisphere via commissural pathways. Rather, there probably are interhemispheric connections between the primary zones and the contralateral secondary auditory zones. There would appear to be evidence of a comparable arrangement in the visual system. In the visual system, lesions isolated to the left primary visual cortex producing a right visual field defect do not impair reading ability for stimuli in the left visual field unless the interhemispheric connections are disrupted by extension of the lesion into the splenium of the corpus callosum (see: *alexia without agraphia,* p. 329).

Effects of Lesions to the Secondary Auditory Cortex.   A major cortical lesion, which is restricted to the secondary auditory association area of the left hemisphere, would account for the somewhat rare phenomenon of pure word deafness or verbal auditory agnosia. In this syndrome, the patient has intact hearing; can identify nonverbal, environmental sounds; can speak, read, and write normally; but cannot understand or repeat spoken language (a comparable deficit with nonverbal sounds may occur with nondominant hemispheric lesions). Less complete lesions of this area would be expected to produce increased difficulty in processing phonemic material without a complete loss of capacity. Thus, while simple, highly familiar, single words might be understood, anything that puts additional stress on the system might lead to impairment. Examples might include someone speaking in a thick, foreign accent, listening in the presence of background noises or in a group where multiple people may be speaking at once, or having to process a string of words or multistep commands, especially if they are not well articulated or complex. These types of deficits reflect problems that still are at a relatively basic perceptual level. If such a lesion were to take place prior to or during the early stages of language or speech development, especially if bilateral, one would expect that a sound-based language (as opposed to a visual-based language) either would, never develop or would be extremely compromised. We might expect such results because the basic building blocks for aural language, that is, the appreciation of phonemes and the capacity to integrate or group phonemes into word units, have been disrupted.

Audition and the Posterior Heteromodal (Tertiary) Cortices

As noted above, the secondary or unimodal auditory association cortex serves an intermediary role, transforming the relatively elementary sensory input to primary auditory cortical zone (Heschl's gyrus) into more refined, organized, wholistic (yet still unimodal) percepts (i.e., words or word-sounds).[43] At this point, however, these word-sounds, while perhaps recognizable (i.e., familiar sounding), probably are devoid of meaning or reference. The next step in the auditory process is for this information in the unimodal, homotypic cortex (secondary zone) to be transmitted to or linked with the posterior, heteromodal association cortex of the tertiary zone. Here the information can be integrated with information coming from the unimodal, secondary zones of other sensory systems. It is probably only then as a result of this convergence of concurrent and/or previous multisensory experiences at this tertiary level that the word-sound takes on its full semantic, associative value (meaning).

Consider, for example, how a child learns a language. It is not simply from hearing the words, but rather from associating the word-sound with another sensory experience, frequently but not invariably involving a second or a third sensory modality. When we first heard the word "hat," it likely was associated with being shown either an actual hat or a picture of one (visual input) or having one being placed on one's head (as in, "Here, let's put on your hat": visual and tactile input). Similarly, we came to associate the word "sun" with an exceedingly bright, round body that is well beyond our reach and that gives off light and warmth (visual, tactile, and visual–tactile or "spatial" percepts). Later, on learning that the sun is composed of burning gases, we added the visual and tactile images of the words "fire" and "gas" to our sensory images of the word "sun," thereby enlarging our original semantic/sensory associations. Without such multimodal associations, certain concepts or word-meanings would be difficult if not impossible to grasp.[44]

With experience, the depth, variety, and complexity of these associations increased geometrically. Thus, the sentence, "The man tipped his hat," might elicit a very specific scene from our past experience or more likely a whole host of more general associations. These might include "man": human, male, adult; "hat": Stetson, fedora, tophat; "tipped": person likely wearing it on his head at the time; hand rising with thumb and forefinger touching its brim, deflecting it slightly downward, perhaps accompanied by a slight nod of the head. Again, based on previous sensory experiences (perhaps watching old movies on television), one may make additional inferences beyond the immediate verbal–visual associations to the words themselves. For example, one might infer the presence of a woman and a certain type of breeding or character in the man himself.

It is such diversity of learned associations, which are mediated by the tertiary, heteromodal cortices, that not only enrich our language but also establish the basis for convergent and divergent thinking and the development of higher-order abstractive capacities. For example, the word "hat," having been associated with such a wide variety of specific stimuli, allows us to conceptualize certain common features that all such associations share. Thus, we come to a general "abstract concept" of what constitutes a hat, such that we recognize both a paper bag and cooking pot turned upside down on the head of a child as a "hat."

So far we have tried to trace the verbal stimulus from the primary auditory reception area to the secondary cortex where it develops its phonemic morphology to the tertiary association cortex where it derives its full semantic and logical–grammatical meaning. This process also works in reverse. One may generate the idea of "something that cuts" that triggers the association with the phonemic pattern "knife". Similarly, if shown a picture of a knife (assuming the relative intactness of the visual association cortices to ensure an accurate visual percept), because of its previous pattern of associations in the heteromodal cortex we may immediately know "what" it is. Simultaneously, perhaps, the visual image also elicits the verbal word form "knife." Similar associations are built up between the lexical or written form of the word "knife" and its phonemic or spoken form.

These various intermodal associations, via the heteromodal association area(s), would seem to form the basis for the various disconnection syndromes that provide important cues in our attempt to develop a theory or understanding of the structural/functional organization of the brain. Thus, we witness the dissociation between the phonemic words for and the perception of color that may be distinct from the phonemic words for and the perceptions of visual forms (objects) or body parts. We also witness individual differences between one's inability to spontaneously recall the phonemic form of the word for an object versus one's ability to recognize (or inability to recognize) a word when either phonetic cueing or the entire word itself is provided. These differences would appear to relate to which associative feedback loops are disrupted by the lesion. Developmental factors also may be important to our understanding of these relationships. For example, children for whom the lexical–phonetic associations are particularly vulnerable during the learning stage will show marked decrements in their comprehension if they are required to hold their tongue between their teeth (thus hindering their ability to "sound out" the word) while reading. Even as adults, we often resort to sounding out a difficult or less familiar word prior to attempting to spell it.

Effects of Lesions to the Tertiary Cortex.   Lesions that encroach upon the tertiary cortex of the dominant hemisphere would be expected to interfere not only with the semantic aspects of language (word-meaning), but abstractive capacities as well. While some variation in symptoms can be seen depending on the exact locus of the lesions, this in fact is what tends to be seen in aphasic syndromes that involve more extensive portions of PTO cortex, such as **Wernicke's** and **transcortical sensory aphasia.** Luria (1973) also describes the tertiary or heteromodal cortex, especially portions of the left parietal lobe, as critical for the under-standing of what he terms "quasispatial," logical–grammatical relationships. He describes these relationships at times as being dissociable from the more purely semantic associations described above. As a result, lesions that involve more parietal or occipitoparietal portions of this tertiary cortex might be expected to produce a syndrome in which the meanings (semantic associations) of individual words may be relatively well preserved, while the meaning of the phrase or sentence, which is derived from logical and/or grammatical relationships expressed by the verbal construction as a whole, may be impaired. An example of the latter might be the sentence: "Bill's brother's father saw his father's brother," or the command: "Touch your ear before your nose, but after your chin."[45]

Audition and Other Functional Units.   While much of the output of the unimodal, auditory association cortex, like the other unimodal cortical association areas, is to the heteromodal (tertiary) parietal–temporal cortex, some fairly direct connections between these unimodal, auditory association areas and frontal association cortices likely exist as well. This is suggested, in part, by the syndrome, **transcortical sensory aphasia**, in which despite severe auditory comprehension deficits repetition of auditory verbal input is relatively preserved. **Conduction aphasia** also may be considered to provide evidence of the same or similar pathways, although from a different perspective. In its classic form, while the patient with conduction aphasia evidences relatively well-preserved auditory (and reading) comprehension, both repetition of heard speech as well as spontaneous articulations (e.g., confrontation naming) are disturbed. This impairment in repetition (or spontaneous speech), which is characterized by literal paraphasias, would appear to reflect a breakdown between those cortical areas responsible for phonological analysis of speech and the frontal motor speech areas. Lesions responsible for this syndrome traditionally have been thought to involve association pathways underlying the supramarginal gyrus (the **arcuate fasciculus**), suggesting that the auditory, unimodal association areas must be connected to the frontal speech areas. These connections may be fairly direct, or what appears more likely also may involve the inferior parietal lobule (postcentral and/or supramarginal gyrus), with the latter providing the kinesthetic link between acoustical patterns and motor programming for the actual articulation of phonemes.[46]

In addition to being necessary for repetition (and articulation), feedback loops between the second and third functional units also likely play a role in language comprehension, especially the frontal heteromodal association cortex. This would appear evident, for example, in situations that call for critical analysis of complex commands or instructions, especially those involving the type of convoluted, logical–grammatical relationships cited above. We have all had the experience of reading a paragraph (or listening to an instruction) only to realize that we probably failed to appreciate its full significance or realize that it "didn't seem to make sense" in light of other information available to us. In such cases, we make a conscious attempt to review all or part of what we have heard or read in an attempt to decipher its true meaning (e.g., "Given what I know, that [communication] didn't make sense. I better clarify it before I start"). As will be seen later, this type of active, goal-directed analysis, self-monitoring, comparator operation is the type of activity traditionally associated with the prefrontal cortex.

In speaking of the processing of auditory information (and language), thus far the emphasis has been simply on the decoding or semantic interpretation of speech as a means of communicating ideas or information. However, language often is intended not simply to communicate information but to stir passions or emotion. Language has the power to make us cry, laugh, arouse sympathy, or ignite anger or resentment. Obviously, this is a complex process that likely takes place at a variety of levels, involving multiple areas of the brain. But, it would appear that at least some of these responses are mediated by connections between the heteromodal cortex and limbic/paralimbic structures. For example, upon hearing the word "knife," we may conjure up a variety of visual and tactile images, such as how it looks and how it feels when held. But if we had the misfortune to be mugged at knifepoint recently, this word also may elicit emotional reactions such as fear, anger, and helplessness. The capacity of music and the cry of a baby to stir certain emotional responses also are familiar phenomena that link components of the first functional unit to auditory processes.

Processing "Nonverbal" Auditory Input

This section on audition thus far has focused on the auditory perception of verbal stimuli or language. In large part, this was due to the fact that we probably have a lot more information about verbal auditory perception (language) than we do about nonverbal auditory perception. However, many of the same type of secondary and tertiary associative processes that take place in the language-dominated, left hemisphere probably also take place in the right hemisphere, involving either "non-language," environmental sounds, certain aspects of music, or "non-semantic" aspects of speech. For example, just as the combination of certain phonetic sounds (words) eventually become matched to or associated with certain visual stimuli (e.g., the word "dog" with the visual image of a dog), non-linguistic sounds independently become associated with other transmodal stimuli. Thus, the sound of barking also becomes associated with visual images of dogs. As in language, both generalizations and specificities of associations are possible through the building up of experiences. We learn to associate a variety of different types of barks as possibly emanating from a dog(s). With practice, we also can learn to differentiate the bark of a dog from the bark of a seal, to distinguish the bark of one dog as opposed to another, and even the particular "meaning" behind a certain bark. While we frequently are aware of "verbalizing" (at least internally) such interpretations, the initial processing and understanding of such information (associations) may be mediated by the right or "nondominant" hemisphere. In addition to such "environmental sounds," many aspects of music also are thought to be mediated by the right hemisphere, especially for one not well trained in music for whom certain notes or musical phrases are not readily encoded in symbolic fashion.

Finally, even within the realm of language, the auditory processing capacity of the right hemisphere would appear extremely crucial for understanding language in its proper social context. This of course has to do with the accurate perception (secondary or unimodal association cortex) and interpretation (heteromodal or tertiary cortex) of the affective or emotional tone of the discourse. Again, just as we learn to associate phonemic patterns with objects, qualities of objects, or action states through simultaneous intermodal stimulation, we also become adept at discerning certain "nonverbal" messages depending on tone of voice, inflections, and timing of the discourse. If the affective tone is consistent with the verbal message, we still can obtain valuable information about the level of intensity or seriousness of the speaker. If the two are incompatible, depending on the emotional tone, we may choose to ignore the message (as in a joke) or to interpret it very seriously, but in a manner opposite to that dictated by the semantic connotations (sarcasm).[47] These nonverbal, auditory cues then may be reinforced in light of other multimodal information available at the time, including visual cues (posture, gestures, facial expression) and the overall situational context. As in the case of semantic information, communication between the anterior and posterior heteromodal cortices, as well as with limbic/paralimbic structures, are integrated as part of the total behavioral experience.

Visual Processing in the Second Functional Unit

The Primary and Secondary Visual Zones

The primary cortical projection area for the lateral geniculates by way of the optic radiations is Brodmann's area 17. Located on the medial surface of the occipital poles along the banks of the calcarine fissure, this area commonly is referred to as the striate cortex because of the horizontal stripe of myelinated fibers transversing this region (**line of Gennari**). The secondary association cortex surrounding area 17 variously is referred to as the **parastriate** cortex (area 18) and the **peristriate** cortex (area 18 and 19). As will be recalled from Chapter 5, each occipital pole receives input from the lateral half of the ipsilateral eye and the nasal half of the contralateral eye. Effectively this means that the primary visual zone in the left hemisphere processes visual information coming from the contralateral right visual field and vice versa for the right hemisphere. Additionally, the projections to the superior banks of the calcarine fissure (**cuneus**) are from the superior portions of the retina (inferior visual fields), whereas the inferior bank (**lingual gyrus**) receives input from cells in the inferior portion of the retina (superior visual fields).

As is the case with the other sensory systems, the role of the primary visual cortical projection zone is not clear. Again, it is suspected that area 17 generally is responsible for preliminary transformation and/or integration of incoming data that facilitates the eventual processing and combining of the data into more highly integrated and organized percepts by the secondary zones. Lesions of this primary zone in humans typically lead to homonymous field cuts or hemianopia in the contralateral visual fields. On the other hand, stimulation of these areas typically produces the subjective experience of fairly primitive or elementary visual phenomena such as poorly or unformed flashes of lights or colors. These usually are restricted to the contralateral visual field.

In discussing the visual unimodal association cortices, Mesulam (2000b) departs somewhat from Luria in extending this region from areas 18 and 19 to also include areas 21, 22, and 37 in the middle and inferior temporal gyri. Within this larger region, smaller areas can be defined that seem to respond to different aspects of visual processing, for example, object form, color, spatial localization, and movement. Another general distinction is made between the more posterior peristriate cortices and the more anterior ventral temporal regions. Both project to their respective contralateral areas in the opposite hemisphere and to the heteromodal association cortices. While both also project anteriorly to the frontal cortex, the main projection of the peristriate cortex would appear to be to the frontal eye fields (area 8), while the temporal region has more extensive projections to the frontal hetero-modal cortex. Another major difference in terms of projections is that, while the majority of the input to the peristriate cortex is from the primary visual projection area (17), the more anterior temporal region receives most of its projections from the peristriate area. From a functional standpoint, it would appear that, while remaining unimodal in character, these more anterior association areas seem to be adapted to handling more complex visual perceptions and the posterior areas more elementary visual percepts.[48] Finally, the more anterior (temporal) visual association areas appear to have more extensive projections to the limbic and paralimbic structures than do the more posterior peristriate areas.

A key to understanding the role of the unimodal visual association areas and the effects of cortical lesions in these regions is to appreciate the diversity that is present within this system. While primates in general, including humans, have a well-developed visual system, they do not have the sharpest vision in the animal kingdom. That distinction likely is reserved for the birds of prey. However, whatever primates may be lacking in acuity seems to be more than made up for in the complexity and richness of the visual experience. While the full extent of this functional diversity probably is beyond our grasp, for the sake of discussion it may be useful to attempt to identify a few of its more basic dimensions, at least in very broad terms. First of all, like other predators, primates have binocular vision, which allows for more accurate judgments of distance and spatial relationships. As is true even of certain lower-order vertebrates, such as frogs, there also would appear to be specific motion detectors within the visual cortex. Another feature that we share with many other animals is our capacity to distinguish forms, shapes, and patterns. Within this latter domain, there still is considerable diversity: from distinguishing a circle from a cross, to appreciating the subtle nuances necessary for facial recognition, to instantly recognizing the complex patterns that characterize the individual words on this page, even when they are "mesplled." Apart from this, and unlike some other mammals, we have the capacity to recognize and appreciate subtle nuances of color.

The fact that certain sites within the visual cortex differentially respond to different types of visual stimuli and that lesions may variously affect these different aspects of vision suggest that even within the secondary association areas different pathways as well as levels of visual organization are likely involved (Van Essen & Maunsell, 1983). As noted earlier, Ungerleider and Mishkin (1982) identified at least two of these apparently independent systems. A more dorsal, occipitoparietal system that appears to play a more predominant role in the spatial localization of an object and a more ventral, occipitotemporal pathway that is more concerned with object identification. In keeping with the general principle outlined earlier, there may be different types and levels (degree of complexity) of perceptual organization (i.e., percepts) that take place or different aspects of the stimulus field to which select groups of visual cells may respond. However, as long as these processes take place within the secondary association areas themselves, they remain unimodal in nature. This notion is important in trying to comprehend lesion effects.

Effects of Lesions to the Secondary Cortex.   As suggested above, depending on its particular location, lesions restricted to the unimodal visual association areas may result in a variety of deficits. However, there are common features that generally tend to characterize such lesions. First, the person is not blind. While there may be a visual field cut (as a result of a lesion encroaching on the optic radiations or portions of the primary visual cortex), the deficit should be able to be demonstrated in the intact field. Second, the deficit should be one of perception or the synthesis of primary visual input, not simply an inability to name, identify, or interpret what is being seen.[49] Thus, the patient may have difficulty on matching to sample or simple discrimination type tasks. This may entail matching or discriminating objects or pictures of objects, symbols (e.g., letters), spatial arrays, faces, or colors. Because it is difficult to clinically assess quadrantic or even hemifield deficits of this type, they would most likely be identified only following bilateral lesions. Even though the lesions would not necessarily have to be symmetrical (one lesion might involve the primary visual cortex producing a hemianopia; the other lesion limited to the unimodal association cortex), in practice such isolated deficits tend to be identified only on rare occasions. Again, depending on the specific locus of the lesion within the secondary visual cortices, different types of perceptual problems may become manifest. For example, difficulties with pure color perception are more likely to involve ventral (occipitotemporal regions), whereas pure symbol (number, letter) discriminations are more likely to be affected by occipitoparietal lesions. If the lesion is in the more posterior portions of the secondary visual cortex, the patient may evidence greater selective impairment of more elementary visual attributes such as the perception of movement or shape. By contrast, more anterior lesions may impair the synthesis of more complex visual features such as the perception of a face or an object.

These visual–perceptual disturbances may be quite severe or relatively mild. Extensive, bilateral lesions, as might be expected, tend to produce greater deficits than more restricted or circumscribed, unilateral lesions. In severe cases, the patient may show virtually no capacity to integrate or process visual input and demonstrate virtually no recognition or capacity to copy or match visual stimuli of either a particular class (e.g., color, objects, or faces) or multiple classes. In less severe or more restricted disturbances, the patient may be able to process individual parts of the picture but not perceive or integrate as a whole, a phenomena referred to as **simultanagnosia**.[50] Patients with mild apperceptive disturbances may be able to accurately perceive actual objects better than pictures or sketches (simple line drawings) of objects. In turn, pictures or line drawings that are presented in a perceptually unencumbered fashion likely will be perceived more readily and more accurately than pictures of objects presented in such as way as to produce visual "interference." Examples of the latter might include "degraded" images, pictures overdrawn with crosshatched or parallel lines, drawings that have overlapping outlines or "cluttered" backgrounds, or familiar objects presented from an unfamiliar perspective. Even for "unencumbered" stimuli, the speed and efficiency of visual processing may be compromised, especially as the stimuli become more complex. Thus, it is possible that modality-specific memory problems (e.g., memory for visual stimuli) could result not from a disconnection with limbic structures (see below), but rather from the inefficient processing of the information at the perceptual level.

However, "pure" lesions that are strictly confined to the secondary visual cortex tend to be either (1) somewhat rare or (2) if present, not clinically obvious, especially if unilateral. The effects of lesions involving these unimodal association areas nonetheless may manifest themselves in other ways. Since higher-level, integrative activities (e.g., those involving the tertiary zones) are predicated on the integrity of elementary perceptual processes, disturbances of the latter may affect the former. Thus, cognitive and perceptual–motor activities involving learning and memory, problem solving, linguistic, or constructional abilities that depend on efficient and/or accurate visual perception may be slowed, limited, or impaired. The degree and nature of the impairment will depend in part upon the severity, extent, and nature of the perceptual difficulties. Some of these effects are reviewed under Hemispheric Specialization in Part II and in the following section.

Vision and the Posterior Heteromodal (Tertiary) Cortices

Primates have evolved into highly "visual" creatures. One can easily witness the difference in the reactions of a cat or a dog versus a monkey that is placed in front of a television set. Despite quite adequate visual abilities, only on rare occasions will most dogs attend to a television screen, and if they do it will be only momentarily and typically only if the picture is accompanied by the barking of another dog. Many primates, on the other hand, will attend for long periods to almost anything that is showing.[51] One result of this proclivity is that visual input plays a major role in our multimodal associations. For example, from a very early age, there is a selective affective response in infants to face-like stimuli. Early exploration and manipulation of the environment largely is through visual–motor associations. These associations will be reviewed below. For now, the immediate focus is on the multisensory associations that take place in the tertiary zone of the second functional unit. At least in humans, some of the more obvious examples center on language development. While it is possible for language to develop in the absence of such multisensory input, as witnessed by the story of Helen Keller, such limitations make it a much more arduous undertaking. As noted above, in the normal course of language development, the first associations are largely sounds paired to visual images. Later, at a more symbolic level, we learn to associate sounds to visual letters and patterns of letters (written words).

Visual–tactile associations also are routinely made. Having touched, held, and used a knife, we can make some preliminary judgments regarding the weight, feel, texture of a particular knife simply by seeing a picture of it. We also have seen how, through multiple associations, the brain is able to "abstract" common features from a class of visual objects to form higher-order concepts. Similarly, through visual-tactile associations we come to "know" an object. Thus, although perhaps never having actually seen it before, one is able to "see" a long piece of obsidian, which is being held in a certain manner, as a "knife."

In addition to assisting in the multimodal "identification and classification" of objects, somatosensory or sensorimotor–visual associations also important in a way that we might not typically consider: the development of *spatial concepts*. As we move our body through space, either as a whole (as in walking), or in part (as in moving our arms), we generate a concept of both personal and extrapersonal space. By pairing these concepts with visual input, especially with binocular vision and concepts of size and the apparent converging of parallel lines, this concept of "space" is enhanced to include not only to objects that are within versus just outside our reach but to "traveling" distances and relational (i.e., three-dimensional) space. Audition also plays its part as we also learn to judge distance by the relative volume (or echoes) of certain sounds.

Obviously, some visual–visual, auditory–auditory, or tactile–tactile associations also are possible. Assume that a child's first exposure to knives was watching his mother use one in preparing to cook. Without being told anything or actually using it herself, by seeing the knife and witnessing the use to which it is being put the child would learn something about knives and what they can do. However, a more complete appreciation of its qualities only comes from having used a knife one's self, just as the mother's admonition that "knives are dangerous" (or "the stove is hot") unfortunately never may be fully appreciated until one actually cuts or burns one's self.

Colors are another common example of visual–visual associations. Except in exceedingly rare circumstances (e.g., certain hallucinogenic drugs) or congenital conditions, the experience of color does not directly trigger other, non-visual sensory associations. However, we do make direct associations between color and other visual images. For example, it is difficult to think about a slice of watermelon without picturing a multitude of colors (the green rind, the black seeds, and the red pulp). At the same time, given a color, certain visual images readily come to mind (e.g., the color red may elicit pictures of cherries, strawberries, apples, and maybe even convertibles).[52] Regardless, of whether such associations are intramodal or intermodal, the assumption is that they all likely take place in the tertiary association cortices.

In summary, it is through these pairings of visual, sensorimotor, and other primary sensory feedback that we are able to predict, even from a distance, that the grass in the shade of a big oak tree is likely to be cool and soft, or upon hearing the song of a mockingbird or the roar of a motorcycle to visualize the source of these sounds even though they may not be immediately before us. These associations also lay down a basis for the verbal identification (naming), classification, and categorization of objects. Finally, the multilevel associations that take place in the posterior tertiary zones allow for the development of higher-order concepts (such as, time and space) and provide the foundations for abstract ideas.

Effects of Lesions to the Tertiary Cortex on Visual Processes.   The effects of lesions to the primary visual cortex (visual field cuts) and to those involving the unimodal association cortex (perceptual deficits) already have been discussed. What then are the effects of lesions to the tertiary, heteromodal areas to visually mediated functions? Before attempting to answer this question, we need to recall that there would appear to be at least three fundamental ways that behaviors associated with these areas could be affected: (1) by disruption of the input to these areas (e.g., lesions to the secondary cortical zones or their connections), (2) lesions that affect output or feedback loops (e.g., connections with frontal systems), or (3) lesions affecting the tertiary zones themselves. The visual agnosias, which will be discussed next, in part may reflect this first type of deficit. The effects of loss of feedback from the third functional unit (frontal systems) also will be addressed below. For now, consider the effect of lesions to the posterior tertiary zone on visual functions.

Although the posterior heteromodal association area extends beyond the angular gyrus (i.e., likely includes at least the supramarginal gyrus and portions of the temporal lobe), lesions centering in and on the angular gyrus are most likely to result in symptoms that either have a direct or indirect visual component. This is true in lesions of either the right or left hemisphere. Perhaps one of the clearest indications of this phenomenon is the presence of difficulties on various tasks requiring visual–spatial and visual–perceptual analysis and synthesis. Constructional disabilities, along with difficulties on visual puzzle type tasks [e.g., Object Assembly from the Wechsler IQ tests, reversible operations in space (Butters & Barton, 1970), or judgment of line orientation (Benton, Hannay, & Varney, 1975)] all have been associated with parietal lobe lesions.[53] **Gerstmann's syndrome** (Gerstmann, 1940), which includes dysgraphia (both lexical and apraxic dysgraphia), dyscalculia, right–left disorientation, and finger agnosia, classically is associated with lesions of the "dominant" angular gyrus, as are disturbances in reading. By contrast, dressing difficulties, "spatial dysgraphia," "spatial dyscalculia," and unilateral spatial neglect or visual inattention more frequently are associated with lesions affecting the "nondominant" parietal lobe (Benton & Tranel, 1993; Critchley, 1969, 1966; Walsh, 1994, Chapter 6). As already has been discussed, Balint's syndrome and problems of visual–spatial orientation commonly are associated with bilateral parietal lesions. Even subtle visual–spatial or visual–perceptual problems likely will produce interference on learning and memory tasks that employ such stimuli. Subjective analysis of each of the above symptoms clearly suggests the visual–perceptual or visual–spatial contributions of each.

Associative Visual Agnosias. At this point, it may be important to discuss associative visual agnosias. As previously noted, lesions that disrupt the unimodal visual cortex tend to produce **apperceptive visual agnosias**, which are characterized by the patient's retained ability to see but difficulty perceiving (e.g., matching to sample) visual objects, drawings, or colors. This type of deficit is to be differentiated from **associative visual agnosias** that appear to represent either a more subtle form of unimodal deficit or a type of disconnection between unimodal and heteromodal and/or limbic (memory) areas of the cortex. Several types of visual associative agnosias are commonly identified: (1) visual object agnosia, (2) color agnosia, (3) prosopagnosia (agnosia for familiar faces), and (4) agnosic alexia (inability to recognize letters or symbols).

Visual Object Agnosia. In visual object agnosia, if an object is presented visually to a patient, he or she is unable to name the object, demonstrate its use, or sort it into its proper category. Similarly, the patient is unable to retrieve a named object from among a group of objects. This is despite apparently adequate visual acuity and elementary perception, as might be demonstrated, for example, by the patient's ability to visually match an object to a sample or to draw it. However, if the object is presented in another modality (e.g., tactually or by verbal description) naming and identification are possible. Visual object agnosia is relatively rare in its pure form (i.e., without associated perceptual difficulties or more widespread cognitive deficits) and little is known about its underlying pathology other than it is likely to involve bilateral lesions of the posterior cerebral artery distribution (medial and ventral occipitotemporal areas).[54] Somewhat more common are disturbances in the recognition of faces (prosopagnosia), letters (agnosic alexia), and colors.

Color Agnosia. Deficits involving the perception, naming, and/or recognition of color are fairly convoluted and judging from the literature still are poorly understood. Three separate syndromes relating to disturbances of color perception generally are described: (1) achromatopsia, (2) color anomia, and (3) color agnosia. In order to appreciate color agnosia, it must be differentiated from these other two conditions. **Achromatopsia** basically entails a loss of color vision. In milder conditions (*dyschromatopsia*), colors are described as "dull," "washed out," or "faded." This syndrome typically results from inferior, posterior occipitotemporal lesions involving the lingual and/or fusiform gyri and reflects disturbances of the secondary visual association areas. Depending on whether bilateral or unilateral lesions are involved, the deficit in color perception may be manifested as a full or partial field deficit, with visual depth and form or shape perception being intact. If the lesion is restricted to the left hemisphere, a right, superior quadrantanopia and associated alexia is not uncommon. With bilateral lesions, prosopagnosia is also likely to be present.

As its description implies, **color anomia** reflects an inability to name colors (or to point to colors named by the examiner) in the absence of a more general naming or aphasic disorder.

It may occur independently of problems with color perception or recognition (i.e., the patient still may be able to match or sort colors or even be able to respond to questions about the colors of objects, such as: "What is the color of lettuce?"). This deficit usually is associated with a lesion involving the left mesial, occipitotemporal area, a right-field defect, and the syndrome of alexia without agraphia. Thus, color anomia reflects a *disconnection syndrome* in which color information from the intact left visual field cannot get to the language area in the left hemisphere, and vice versa.

By definition, **color agnosia** is a loss of color knowledge. Again, although relatively rare, a patient manifesting this syndrome would be expected to have difficulty naming or pointing to a named color (in the absence of a more generalized aphasic disturbance). He also would have difficulty matching colors to familiar colored objects (e.g., cherries, lettuce, watermelon), either verbally or visually, despite relatively preserved color perception (i.e., ability to match or to identify the numbers on the Ishihara plates) (Damasio, Yamada, et al., 1980; Damasio, 1985; Meadows, 1974; Tranel, 1997). The problem is that this particular syndrome, at least in its pure form, infrequently is encountered clinically and its anatomical correlates, like that of object agnosia, are ill-defined.

Agnosia for Familiar Faces (Prosopagnosia). As was true of disturbances of color perception, a distinction must be made regarding one's ability to perceive faces, namely the difficulty in recognizing familiar faces versus difficulty in distinguishing unfamiliar faces. These two syndromes are both clinically dissociable and typically have different anatomical substrates. The inability to distinguish between or match unfamiliar faces, for example, as measured by Benton's *Test of Facial Recognition* (Benton, Hamsher, Varney, & Spreen, 1983), is not a true agnosic deficit, but rather seems to reflect a problem of perceptual discrimination. Patients who are unable to recognize familiar faces may perform normally on facial discrimination tasks, and those who perform poorly on tests of facial discrimination may still be able to recognize familiar faces (Benton, 1980, 1990; Benton & Tranel, 1993).

In **prosopagnosia** (agnosia for faces), there is an inability to recognize familiar faces simply by an analysis of general facial features. As in other true agnosias, there is no clear evidence of a major problem with visual perception.[55] The patient suffering from this disorder still may be able to recognize gender and accurately interpret emotional facial expressions. As noted above, such patients are often may be able to perform normally on tests of facial discrimination or matching-to-sample. Prosopagnosia is not the result of a general amnestic deficit, as the patient may readily recognize the person before them by the sound of their voice or sometimes by idiosyncratic visual cues (e.g., clothes, hairstyle, eyeglasses, or perhaps by a scar or mole on the individual's face) (Bauer, 1993; Damasio, Tranel, et al., 2000; DeRenzi, 1997b). In addition to having difficulty recalling information that previously had been associated with a particular face (retrograde deficit), these patients also have difficulty attaching new information to faces (anterograde deficits). Finally, while recognition of familiar human faces typically represent the most obvious functional disruption, many such patients will show comparable deficits in making specific individual identifications of other objects, both animate and inanimate, that is, while correctly identifying a picture of a cat, they may have difficulty identifying their own cat.

The lesion(s) responsible for prosopagnosia and disorders of facial discrimination are still a matter of some debate. Benton and Tranel (1993) suggest that the only lesions responsible for problems with facial discrimination are likely to be posterior and are more likely to be in the right than in the left hemisphere (if in the left, the individual is likely to show signs of receptive aphasia). As previously noted, prosopagnosia is likely to occur in association with achromatopsia when bilateral lesions are present. Most authors agree that bilateral lesions involving the mesial and inferior aspects of the occipitotemporal association cortex (with extension to the underlying white matter association pathways) can produce difficulties in identifying familiar faces. The question is whether bilateral lesions are necessary or can this syndrome result from unilateral lesions? Most seem to agree that the right hemisphere is crucial. DeRenzi et al. (1994) and Hecaen and Angelergues (1962) present evidence for right-sided lesions being sufficient to account for this syndrome, whereas Damasio (1985) presents arguments to the contrary. However, again there would appear to be a general consensus that both a disruption of unimodal association areas and/or some type of disconnection (e.g., white matter involvement) between these association cortices (whether unilateral or bilateral) and other "multimodal" brain areas (most likely limbic structures) are crucial elements in this syndrome.

Agnosia for Written Words (Alexia without Agraphia).   In the syndrome of alexia without agraphia (also known as *pure alexia,* or *pure word blindness*), the patient retains the capacity to recognize individual letters [i.e., he or she can copy letters or even "read" by sounding out (spelling aloud) the letters]. However, the visual representation of the word itself has lost its associative value (meaning). As writing, spelling, and other language skills remain essentially intact, aphasic disturbances cannot account for the observed deficits, although achromatopsia and other color disturbances may be present. While a variety of lesions may produce this syndrome (Damasio & Damasio, 1983; Henderson, 1986), all basically involve a disconnection between the visual cortex and the left angular and superior temporal gyri that mediate the orthographic and semantic aspects of language (see also under Disconnection Syndromes – p. 329).

Summary.   If identification of a given stimulus is restricted to input through the visual modality without evidence of more primary (apperceptive) visual disturbances, then the likelihood of a visual agnosia, which disconnects the unimodal visual areas from cortical or limbic multimodal areas, should be considered. However, in practice, these distinctions may not always be easy. Subtle apperceptive deficits may be present (although not sufficiently contributory), or present and contributory but not clearly identified. If, on the other hand, the lesion is in a heteromodal area, deficits are likely to be observed with multimodal inputs. Another sign of tertiary involvement, especially in the left hemisphere, is loss of abstract attitude. As an example, a number of years ago a patient with a dense aphasia resulting from a large infarct in the perisylvian area was in speech therapy. Using AMERIND, a type of concrete, representational sign language, the patient was being taught to associate hand gestures with pictures of common objects (e.g., a cup, comb, spoon, etc). Although he was highly cooperative and easily replicated the appropriate signs in practice sessions with the speech therapist, when presented with the actual objects there was no evidence of transfer of training (he also failed to spontaneously use these gestures to communicate his needs). Such patients also often fail to appropriately sort objects (e.g., accordingly to use, such as a hammer and a nail), despite showing recognition of individual objects.

Vision and Other Functional Units

Visual–Frontal Connections.   Visual perception is not a passive process. It is not like a camera that simply captures and records whatever image is placed before it. In most instances, perception is a dynamic process, becoming increasingly active as the stimulus complexity increases and/or the response demands become more critical or more refined. The eye actively scans the stimulus or environment, and depending on the situation focuses on (1) inherently salient features in stimulus itself (e.g., the eyes, nose, and mouth of a face), (2) those features that are aberrant, unpredicted, or unusual, or (3) those features that are essential to critiquing the situation or effects of one's action prior to initiating or altering one's response. This selective attention/visual guidance system in large part thought to be mediated by the frontal association cortex and executed through area 8 (frontal eye fields).[56] Luria describes how frontal lesions can impair interpretations of thematic pictures as a result of loss of executive control to guide the visual search (Luria, Karpov & Yarbuss, 1966). While not specific to lesions of the frontal systems, visual search tasks certainly are sensitive to disruptions of this frontal–visual connection (Teuber, 1964).

Visual–Limbic Connections.   By way of introduction, the multimodal images resulting from the confluence of sensory inputs into the posterior tertiary or heteromodal cortex form the ultimate building blocks for our knowledge of the world around us. From these more elementary "blocks," higher- order concepts, generalizations, and abstract thought are developed. In fact, one of the hallmarks of the evolutionary process is the increasing capacity of organisms to better adapt to and manipulate their environment as a function of experience (i.e., an increased capacity to learn). However, from an evolutionary standpoint, perhaps there is even a more basic prerequisite for survival, that is, the capacity to immediately recognize those stimuli that should be approached (e.g., potential food or shelter) and those that need to be avoided. If the fox loses a toe the first time he encounters the hunter's trap, it serves him well to recall and avoid the area less he potentially lose much more than a toe the second time. Thus, at a very basic level, it is not only important that the organism recall or remember a particular stimulus (whether visual, tactual, auditory, or olfactory), but that he be able to attach a particular affective response to that stimulus or experience. For example, did it involve pleasure or satisfaction of a need, or fear and pain? Working in concert, the limbic and paralimbic structures would appear to play a crucial role both in laying down the memory traces for the stimulus (event) and in attaching an affective or emotional significance to it. We already have noted the probable connections between the olfactory and auditory systems and these limbic structures; clearly the same holds true for the visual system.

Evidence of these connections comes from multiple sources. From phylogenetic studies of perception and behavior and simple observation, we know that organisms respond in a highly specific manner to a host of visual stimuli. In many instances these stimulus–response connections appear innate, while others appear to be the result of either classical or operant conditioning. Depending on circumstances, the stimulus may be rather general or highly specific. Frogs, for example, will respond to a sudden dimming of light (shadow) by jumping off the lily pad into the water (a highly adaptive response), while chickens, which will also run for cover if a dark silhouette is passed overhead, will do so if the shape is consistent with that of a soaring hawk but not if it is made to resemble a goose. Certain colors or other visual stimuli will sexually excite members of various species, while other visual stimuli seem to innately elicit fear or revulsion in numerous members of the primate family, including humans (e.g., snakes, decapitated bodies, precipices). People who own dogs frequently witness the classic conditioning that occurs with respect to a leash (a positive response if it generally means going for a walk, a negative response if the leash is only used for visits to the vet or to get a bath). Under certain conditions, humans seem to respond affectively simply to the absolute level of illumination. If one is walking alone down a dark street in a questionable neighborhood, any sudden or unexpected movement likely will trigger an affectively charged, orienting response. Such emotional responses probably are not mediated through the heteromodal cortices; they likely represent a straight shot from the visual association cortex to the limbic system.

Disconnections between the unimodal visual association areas and the limbic/paralimbic structures, or damage to the limbic structures themselves, depending on the nature of the lesion, might be expected to produce at least two rather specific types of problems. The first might be a failure to benefit normally from purely visual experiences. The second could be a failure to recognize or show previous normal affective response to the stimulus, while demonstrating intact perception or discrimination capacity. An example of the former might include diminished capacity to learn certain geometric designs, despite normal copying ability.[57] Prosopagnosia would appear to provide a good example of how previously acquired information may be dissociated from visual input as a result of lesions apparently involving this system. It might be recalled that despite an inability to recognize familiar faces, patients suffering from prosopagnosia generally are able to identify and respond to emotional facial expressions. However, the discrimination of emotional facial expression is not merely a dispassionate, objective exercise. One of the teleological purposes served by facial expressions is to elicit an appropriate emotional response in the observer; hence, the limbic system must be involved for something other than episodic or declarative memory. In fact, lesions to the limbic system, specifically the amygdala, interfere with the appreciation of facial expression (Adolphs et al. 1994). Similar examples have been discussed in reference to the Kluver–Bucy syndrome and Downer's (1961) experiments with monkeys in Chapter 8. As you will recall, in both of these cases, previously aversive visual stimuli no longer elicit an emotional response following lesions to the amygdala or lesions that disrupt visual input to this structure.

Somatosensory Processing in the Second Functional Unit

The Primary and Secondary Somatosensory Zones

The primary somatosensory projection area (SI) is located along the postcentral gyrus, represented by Brodmann's areas 3a, 3b, 1, and 2. Each of these areas reflect their own, well-defined somatotopic organization with the face area being the most ventral (closest to the lateral sulcus), followed dorsally by the hand, arm, trunk, and upper leg. The lower leg and foot are represented along the medial surface of the hemispheres.[58] The major input to this region is from the VPL and VPM nuclei of the thalamus. Recall that, respectively, these nuclei receive input from the medial lemniscus and the spinothalamic tracts (carrying sensations of pain, touch, temperature, stereognosis, and position sense from the trunk and extremities) and the trigeminothalamic tracts (which transmit comparable information from the face, tongue, and throat).[59] Sensations of taste also appear to project to the VPM nuclei, and hence, are likely mediated by the same general cortical area as these other somesthetic sensations.

The primary efferent projections from SI are to:

1. Homologous areas for the "submodalities" of somatosensory perception within the postcentral gyrus of the same hemisphere (e.g., area 1 for the right hand to area 2 for the right hand).
2. Homologous areas for somatosensory perception in the contralateral hemisphere (except for the hands and feet, whose input remains unilateral, at least for SI).
3. Precentral gyrus (primary motor cortex).
4. Unimodal somatosensory association cortices (areas 5 and the more anterior or dorsal portions of 7 in the superior parietal lobule).
5. SII (see below) in the same hemisphere.
6. SII in the contralateral hemisphere.

A second somatosensory area (SII) is also topographically organized; but unlike SI, which appears to primarily receive contralateral somatosensory input, SII has a substantial amount of ipsilateral as well as contralateral innervation. Much of the input to SII, especially to its more caudal regions, seems to represent spinothalamic fibers mediating pain and temperature fibers via the posterior thalamic nuclei (Jones & Powell, 1969a,b). The SII has reciprocal connections with SI and the motor cortices, but unlike SI, it does not appear to have direct connections with the superior parietal lobule.

Brodmann's areas 5 and 7, which constitute the major portion of the superior parietal lobule (see Figure 9–13) appear to represent the main portion of the secondary (unimodal) somatosensory association cortices. However, Kaas (1983) argues that areas 1 and 2 may represent unimodal association areas, while Mesulam (2000b) suggests that the posterior portion of area 7 may be heteromodal in nature. It is generally accepted, however, that the area(s) closest to the central sulcus (areas 3, 1, and 2) likely represent more elementary somatosensory functions, while the more posterior regions, which include areas 5 and 7 (and some believe, even areas 1 and 2), represent higher levels of somatosensory integration. A third, independent, somatosensory association area had been postulated to exist on the medial aspect of the hemisphere involving the mesial portion of area 5 (Penfield & Jasper, 1954).

In many respects, the somatosensory system appears more complex and less well understood than the auditory and visual systems. This should not be surprising given the wide range of stimuli and types of receptors from which somatosensory information is derived.[60] These sensations often are broken down into several broad categories that include mechanical cutaneous (pressure, touch), nonmechanical cutaneous (pain, temperature), and sensations from muscles and joints. For the purposes of this chapter, it may be useful to try to categorize somatosensory processes as a function of the level of integration that would appear to be required. Using this approach, we can divide somatosensory information as (1) elementary, (2) the product of an intermediate level of integration, or (3) the product of a high level of sensory integration.

Those sensations that at least at first glance would appear to be rather primary or elementary include simple touch, gross localization, pressure, vibration (rapid, alternating pressure?), pain, temperature, and kinesthesia (awareness that movement has occurred in a joint or limb). Even though in contrast to the auditory and visual systems, some if not most of these sensations may be mediated either at the subcortical (thalamic) level or by multiple cortical sites (including motor cortices), they are generally considered representative of the sensations processed by the primary somatosensory cortex (i.e., area 3, and probably areas 1 and 2).

At an intermediate level, an integration of these elementary sensations occurs to produce a higher-order perception, although still purely tactile in nature. Examples might include determinations of size, shape, texture, weight, and position sense. Analysis of these perceptions would suggest that they are not immediate, elementary sensations. Whether we are holding a grain of sand between our thumb and forefinger or holding an apple, the determination of size requires a simultaneous comparison of which cutaneous receptors are being stimulated and which are not (i.e., the extent of the stimulation field). In the case of the apple, the degrees to which our fingers are extended (position sense) also provide a clue as to size. Given even larger objects, determination of size by tactile stimulation requires a summation of cutaneous feedback as the hand explores the object (cutaneous and kinesthetic feedback). The determination of shape and texture also is generally determined, not only by pressure and simple touch, but also by a manipulation of the object or movement across the surface of the skin. Perception of weight often is the result of both pressure on the surface of the hand and kinesthetic feedback (the subjective judgment of the weight of an object often is accomplished by slight vertical movements of the hand, testing how much effort is required to overcome gravity). Even position sense is not an instantaneous perception, but often requires some reflection, as is generally greatly assisted by cutaneous and/or kinesthetic feedback.

Proprioception, or the awareness of our body moving in space, appears to represent an even more sophisticated integration of position sense and kinesthesia. Cortical areas 5 and 7 (anterior portion) may be particularly involved in monitoring and integrating proprioceptive information that is directly associated with goal-directed, exploratory behaviors (Mountcastle et al. 1975). Certainly, any type of somatosensory perception that includes an element of discrimination or comparison of any two stimuli would appear to fall within this intermediate category. Examples might be determining which of two stimuli is heavier, smoother, warmer, or larger or whether in fact there are two stimuli instead of one (e.g., two-point discrimination). All these various functions are likely under the control of the unimodal, somatosensory areas (i.e., area 5, the anterior portions of area 7, and perhaps with significant contributions from areas 3, 1, and 2; see below).

At the highest level of integration are the recognition of objects through tactual means (*tactile asymbolia*) and the development of an integrated **body schema**. The latter would include concepts of front and back, up and down, right and left, awareness of the body's orientation in space, its ability to function as a coordinated unit, personal physical integrity (i.e., possession of one's limbs), and an awareness of defect or impairment. These latter functions would seem to represent the contributions of multimodal input, especially vestibular and visual input, and depend heavily on the integrity of the posterior portions of area 7, as well as the supramarginal and angular gyri.

Somesthesia and Motor Functions. In addition to its importance for tactile identification and discrimination, somatosensory feedback serves another important role. As witnessed in the spinal, cerebellar, and basal ganglia systems, there is a close interaction and interdependence between motor and somatosensory functions. In fact, in addition to receiving extensive input from the parietal cortex, the motor cortex itself contains cells that are directly responsive to somatosensory stimuli. While most motor activity, especially that involving the striate muscles, relies on sensory feedback, this is most evident in discrete, voluntary movements. Such feedback is important not only for maintaining muscle tone (largely accomplished at an unconscious level), but also for consciously gauging and controlling the speed, intensity, direction, and duration of movement. It is not unusual for one of the initial symptoms of cerebral disease to be that patients notice that they have a tendency to drop things they are holding in their hand, especially when their attention is distracted. The problem is not motor, but sensory.

As noted in Chapter 3, with practice many of our skilled motor activities seem to become almost automatic and, as a result there is diminished awareness of the role of proprioceptive and cutaneous feedback in the execution of these actions. However, certain circumstances readily can call these processes to one's attention. For example, try writing a sentence with your eyes closed, using your nondominant hand. You become much more aware of the importance of proprioceptive feedback than when writing with your dominant hand. A more dramatic demonstration would involve the elimination of afferent feedback, while preserving motor function. There are at least two circumstances, which most of us have experienced, in which this happens. The first is associated with a trip to the dentist, where the mandibular portion of the trigeminal nerve is anesthetized. The second involves laying or sitting in such a position that the hand or foot "falls asleep." In the former condition, one is aware of the increased difficulty with articulation. While the latter condition also may directly involve muscle groups, it often seems that the impaired sensory feedback is the main limiting factor in carrying out fine motor tasks during the acute phase of this temporary paresthesia. Finally, in trying to explore an unknown or unidentified object through tactile perception alone, not only can one readily appreciate how the manipulation of the objects is guided and directed by cutaneous feedback, but how proprioceptive feedback assists in determining the size and shape of the object to be identified.

Effects of Lesions to the Primary and Secondary Cortices on Somatosensory Processes. In contrast to the auditory and visual systems, lesions affecting the primary somatosensory cortex, whether defined as area 3 or areas 3, 1, and 2 do not typically result in a permanent loss of elementary somatosensory information. Thus, even with substantial lesions of the postcentral gyrus, the sensation of touch, pressure (including vibration), pain, and temperature are not abolished. Although there may be some substantial change in thresholds (i.e., tactual sensitivity), especially for the perception of light touch and temperature, these too may diminish over time. The reason for this relative preservation of sensation again is unclear, but probably reflects either the perception of certain stimuli at a thalamic level, at multiple cortical sites, or perhaps bilaterally (as in SII).[61]

In addition to changes in threshold sensitivity, lesions of SI or the postcentral gyrus result in contralateral sensory impairments of tactile or somatosensory discrimination, again most notable in the more acute stages. Thus, the individual may have increased and perhaps permanent difficulty making judgments regarding the relative size, shape, sharpness, weight, thickness, temperature, texture, or other surface features between two stimuli. Similarly, two-point discriminations, localization (of the stimulus), and position sense may be adversely affected. **Astereognosis**, which probably reflects a higher-level apperceptive deficit (which will be discussed in the next section), also may be present. Extinction or relative neglect also may be present with such lesions, although the latter are more likely to result from inferior or posterior parietal or even frontal–subcortical lesions. However, to reiterate, not all aspects of somatosensory function typically will be suppressed at the same time and the pattern of deficits may vary greatly from one individual to another. Due to the topographical organization of this area, more dorsal lesions selectively may affect the lower extremities, while more centrally placed lesions will affect the upper extremities. As will be noted below, lesions at the base of the postcentral gyrus (near the lateral fissure) will affect the face and articulatory mechanisms.[62]

Lesions along the postcentral gyrus also can impact directly on certain motor activities, including fine motor manual skills and speech. As noted, voluntary motor activity, especially discrete, skilled-motor activities, are constantly being modulated by somatosensory feedback. In the absence or disruption of such feedback, motor skills will be adversely affected. Luria (1973) used the previously coined term, *afferent paresis*, to describe this phenomenon. If such somatosensory feedback to the motor cortex is compromised by lesions affecting the postcentral gyrus, the movements become coarse, clumsy, imprecise, or unrefined. Symptoms may be more pronounced when attempting to handle or manipulate objects without the aid of vision. If the patient is allowed to visually observe the intended action, some improvement may be observed but some difficulty would be expected to persist. The more complex the action, the more likely deficits will become manifest. Thus, writing (penmanship) may be observed to be slower, more laborious, and less precise than before the injury. Another sign, which may reflect a disturbance of afferent feedback, sometimes can be observed when a coin or other small object is placed in the palm of a patient's hand and he or she is asked to identify it by touch alone. The normal response is for the individual to transfer the coin to the fingertips where it more easily and thoroughly can be manipulated in order to discover its identity. Patients with parietal lesions may simply close the hand, allowing the fingertips, moving "en masse," to simply palpate the object as it lies in the palm. While identification may be possible using this technique, it is clearly much less effective.[63]

If the lesion to the postcentral gyrus is more caudal, encroaching upon the face area, speech articulation deficits may be observed. Luria (1966) describes this loss of afferent feedback as *afferent motor aphasia*. He suggested that it primarily results in the confusion or substitution of letters or sounds with similar articulation characteristics (e.g., "pat" for "bat" or "dog" for "log"). The problem is thought to result from impaired feedback regarding the position and/or movements of the organs of articulation (i.e., the tongue and lips).

Lesions affecting the secondary association areas may not always be easily differentiated from those involving the primary somatosensory cortices. Both are likely to present with difficulties in making subtle tactile discriminations and other judgments based on proprioceptive feedback.[64] There may be two explanations for this. From a theoretical perspective, the capacity for making subtle tactual discriminations is predicated on the integrity of elementary perceptions. Despite the lack of total loss of tactile sensation with lesions of the postcentral gyrus, if such lesions disturb or disrupt sensory analysis, then the higher-order functions on which they depend (in this case discriminatory judgment) also will be disrupted. The practical explanation is that the literature does not appear to provide unambiguous data where lesions restricted to specific cytoarchitectural divisions of the somatosensory cortex are compared and contrasted in a systematic fashion in humans. As stated earlier, there still is controversy as to which somatosensory areas represent primary versus secondary cortex. Rather than being a source of discouragement to the practitioner, the fact that the functional organization of this region of the brain is still largely an enigma should stimulate and encourage its clinical exploration. For the interested reader, although originally published in 1953, Critchley's (1969) book on the parietal lobes still offers one of the more comprehensive and clinically useful treatments of somatosensory disorders and their assessment.

Somesthesia and the Posterior Heteromodal (Tertiary) Cortices

The integration of visual, auditory, and tactual information in the posterior heteromodal association cortex as the probable foundation for the development of language, abstract concepts, and the ability to perceive, imagine, or identify objects from a multidimensional perspective already has been discussed. Similarly, the development of our concept of space and spatial relationships in general in large part was attributed to the integration of somatosensory and visual processes. While these observations can help account for many of the higher-level associative deficits seen in parietal lesions, it would be useful to briefly revisit and expand on several of these notions, particularly as they relate to what has been referred to as **disorders of body schema**.

The perception we have of our bodies is central, not only to the perception of the self, but ultimately to the rest of the world around us. From a very early age we not only establish the physical parameters and integrity of our body, but also set up a basic dichotomy: that which is "me" versus that which is "not me" (i.e., the rest of the world around us). We also learn from an early age that through the physical command that we have of our bodies, we can impact and control (at least to some extent) our environment to either satisfy our needs or to avoid being impacted in a adverse manner.[65] Thus, more than the later-developing telereceptors (e.g., vision, hearing), cutaneous feedback and proprioceptors provide the most primitive means of knowing in what manner our bodies are being directly affected by our environment and, along with the capacity for muscular contraction, how we respond to it.

It is from this process of differentiation and manipulation that the concept of body schema would appear to develop. Assisted in part by visual feedback, we learn to identify various parts of our body and their relationship one to another. We also develop awareness that vertically the two halves of our bodies for the most part are symmetrical but nonetheless distinct. Not only do we learn to verbally label these two similar but distinct halves but also to recognize that this spatial directionality (i.e., right versus left) may be applied to external bodies (often by means of a visual reversal). Also, as noted earlier, in addition to the notion of "right and left," concepts of "up and down," "top and bottom," "near and far," "front and back," "above and below" are readily conceived, first in terms of our own bodies and then in relationship to the environment. Thus, the whole notion of body image or body sense would appear to be rooted in our awareness of our bodies, a supramodal (integrated) concept on which even higher-order concepts (such as "space") are developed.

So far, however, we have largely explored only relatively static concepts of body image based on somatosensory feedback. There is yet another whole dynamic dimension that needs to be addressed. In the process of attempting to manipulate our environment, whether this involves moving an object at rest (picking up a fork, or a bowling ball), moving ourselves within our environment (walking or running), or interacting with a moving object (returning a tennis ball or catching a pass), we gain a sense of mastery over our bodies. When combined with vestibular input, we begin to appreciate notions of balance, equilibrium, and stability. When combined with visual input, in particular, these sensorimotor "experiments" also provide for or augment our understanding of concepts such as speed or velocity; mass, density, or inertia; gravity; vectors; effort, energy, or force; and related physical laws. Again these higher-level concepts would appear to be rooted in our awareness of our own bodies and how it responds (motorically) in a three-dimensional environment. It is not difficult to imagine how the parietal lobe, in its juxtaposition between the primary somatosensory– motor cortices and the visual cortex, would be an ideal site for this multimodal integration.

Finally, as was mentioned earlier, in addition to body schema, the parietal lobes also may be important for tactual object recognition and identification. Phylogenetically, this is probably a relatively late development. For most species, object identification likely occurs through other sensory modalities, such as vision, olfaction, or even taste or hearing. While differentiation of the physical characteristics of the terrain may take place through the pads of the feet in some species, it is in primates, with the evolution of the opposable thumb, that this capacity truly developed. Although the tactual modality ordinarily is not the primary means for identifying objects or other stimuli, even in humans, the level of skill that can be developed using this method can be exceptional. This is most evident in cases of the visually impaired who are able to learn to read using Braille.

Effects of Lesions to the Tertiary Cortex on Tactual Processes.   Lesions that impact on the posterior portions of the superior parietal lobule, as well as the inferior parietal lobule (supramarginal and angular gyri), all have been associated with symptoms or syndromes in which there seems to be a somesthetic component. In general these include (1) disturbances in tactual object recognition, (2) disorders of body schema, including neglect, and (3) ideomotor apraxia. While each of these will be reviewed briefly, our purpose is not to provide a comprehensive review of these syndromes and how they might be assessed, but rather to illustrate the multimodal nature of these cortical areas. While we will address some of the confusion that may surround these topics, for a more detailed discussion of these syndromes, the reader again is referred to Critchley (1969) and Hecaen & Albert (1978).

Tactile Agnosia.   Tactile agnosia implies the failure to recognize an object by touch in the absence of more basic disturbances of sensation or perception (or, if perceptual disturbances were found, they would not appear sufficient to account for the failure to identify the object). Thus, for example, a patient, when given a hammer to identify by touch alone should be able to describe its general shape, size, and contours, as well as its weight and texture, perhaps even venturing that it seems to be made of metal, but unable to identify it as a hammer or even as a tool. If given a pencil and paper, he should be able to make a reasonable drawing of it (at which point, he may be able to identify it, given the visual image of the drawing). Obviously, the failure to identify the tactually presented object should not be the result of a more general language problem (he also should be unable to demonstrate its use).[66] In practice, however, such syndromes have proved to be exceedingly rare (Bauer, 1993; Reed & Caselli, 1994).

However, there is potential for confusion here. In the past, the term, **astereognosis**, has been used interchangeably for the terms (and the concept) **tactile agnosia** or **tactile asymbolia**. At times, "astereognosis" still is used to refer to an inability to name or identify an object by touch (i.e., as an associative-type agnosia), which, as was just pointed out, is rare. More recently, there has been a tendency to restrict the use of this latter term to refer to an apperceptive-type deficit, which in fact is much more common and is associated with lesions affecting the primary or secondary somatosensory areas. Yet, even when used in this manner, the term, astereognosis, is subject to some confusion. Strictly speaking, astereognosis refers to the inability to appreciate the physical qualities of an object as a whole by touch alone. The term reflects the broader integration of more elementary percepts, a process typically associated with unimodal association cortex. In this case, it represents an inability to appreciate both the *form* of the object (e.g., its size, shape, or contours), as well as the nature of the material from which it is made [e.g., its density (weight) and texture], in the absence of more gross tactual sensory losses.[67] Because the pathways mediating this ability originate in the dorsal columns, lesions in the spinal cord, brainstem (medial lemniscus), thalamus, or cortex may all result in astereognosis, which most likely is to be expressed unilaterally, often in combination with graphesthesia.

Disorders of Body Schema.   Disorders of body schema, or *asomatognosias*, are a fairly common finding following parietal lesions. Although such disturbances of body schema may be expressed in a variety of ways, they usually involve either a lack of, awareness of, or difficulty identifying specific parts of the body. Probably the most common and certainly the most dramatic of these phenomena involve various types of **unilateral neglect**, either of the body itself or of the affliction from which it may suffer (see also under **Hemispheric Specialization: Right Hemisphere Functions**, p. 362). In both instances, there can be gradations in the severity of the disorder. If present, these disturbances tend to be most severe in the early stages following an acute event (e.g., a stroke), often becoming less noticeable with the passage of time. In the case of neglect, the individual may behave as if one side of the body did not exist. The person may fail to wash, shave, or dress the affected side. If asked to point to a part of the body on the affected side, they may fail to cross midline. If there is no associated hemiparesis, the individual may show no difficulty walking or using both hands in a cooperative fashion to carry out a bimanual task, such as putting on a pair of socks. However, there may be a noticeable inhibition to use the affected extremity (usually the hand and arm) in situations where it might be appropriate.[68] Such patients may fail to respond to tactile stimuli on the left side, especially with bilateral simultaneous stimulation. Occasionally, tactile stimulation of the affected side will be reported to have occurred in the homologous area of the unaffected side (**allesthesia**). Typically, such personal neglect also is associated with neglect of extrapersonal space (i.e., visual neglect on the affected side), although this may not always be the case (Guariglia & Antonucci, 1992).

A related, but less frequently observed symptom of unilateral neglect is denial of unilateral deficits (**anosognosia**) or failure to demonstrate sufficient concern or appreciation for such afflictions (**anosodiaphoria**). Again, most common in the early stages of acute illness, anosognosia typically is manifested by the patient's failure to acknowledge illness, usually a hemiplegia, on the affected side. If asked to move the affected limb, the patient may respond by moving the unaffected limb (**allokinesia**). If the lack of ability to move the hemiparetic limb is pointed out by the examiner, the patient may make up an explanation ("It fell asleep") or even may deny ownership of the affected limb. As the time from the onset of the illness passes, the patient will usually come to acknowledge that the limb is impaired but fail to show appropriate concern or appreciate the consequences of the weakness or paralysis (anosodiaphoria). For example, while acknowledging the existence of the impaired limb, the patient may report that he or she plans to return to work after leaving the hospital, despite the fact that their injury would preclude a return to their former occupation. While both types of unilateral neglect can occur with lesions of either hemisphere, they are more frequently noted following lesions of the nondominant hemisphere (Bisiach & Geminiani, 1991; Hecaen & Albert, 1978, pp. 303–304; Heilman, Watson, & Valenstein, 2003; Weinstein, 1991).

The other disturbance of body schema that appears to reflect a higher-order, integrative deficit, suggestive of tertiary cortical involvement, is **autotopagnosia**. This syndrome refers to a failure or difficulty in identifying body parts and appreciating their relative relationships to one another (Frederiks, 1969). Although not common, when present autotopagnosia involves both sides of the body, thereby not only distinguishing it from unilateral neglect or other more elementary somatosensory deficits, but also adding credence to its designation as a integrative deficit. It typically is manifested by a marked difficulty in naming or identifying specific body parts on command. The difficulty is not simply in reference to one's own body, but to the bodies of others, whether real or pictured. The fascinating thing about this syndrome is its apparent specificity with regard to the body itself, particularly the human body. In their reviews of this disorder, Hecaen and Albert (1978) and Benton and Sivan (1993) point out that such patients may be able to name parts of clothes (e.g., a "sleeve"), but be unable to identify the arm that it covers. While apparently somewhat more variable, these patients may show a greater facility in identifying body parts of animals than those belonging exclusively to humans (e.g., a paw or tail is more likely to be identified than a hand or arm) (Ogden, 1985).

Finger agnosia, the inability to identify or differentiate individual fingers, would appear to be a special case of autotopagnosia, but generally only when expressed bilaterally.[69] As noted clinically by Critchley (1966) and experimentally by Benton (1959), the three "center" fingers (especially, the "middle" and "ring" fingers) are the most likely to be confused. Occurring more frequently than autotopagnosia for the body as a whole, the hand likely represents a particular sensitivity to disruption. This may, in part, reflect the hand's special role in tactile exploration, as well as the fact that the fingers, particularly the central ones, are neither visually nor tactually as distinct as the thumb and little finger. However, as with other clinical measures, identifying body parts, especially finger recognition, is not a unidimensional task. As a result, disruptions may result from factors that must be considered in assessing the potential clinical implications of these findings. As indicated above (see footnote), the presence of unilateral findings or primary somatosensory deficits may suggest very different mechanisms at work.

Right–left disorientation is yet another symptom that commonly is associated with disturbances of body schema. Recall, if you can, when you first learned to discriminate your right side from your left. For most people this is not automatic; you must develop some strategy. It likely took even longer to be able to reliably differentiate right versus left on others, as this involved another, distinct cognitive process. On average, about 10% of the normal population continues to have some difficulty making this distinction (Wolf, 1973). For most others, the earlier strategies (e.g., which arm felt stronger when you "made a muscle") have long been abandoned, replaced by some sort of "intuitive insight" as to which is the right side and which is the left. However, while the concept of right versus left appears rooted first and foremost in one's own body, introspection suggests that this is not simply a somatosensory process, but a "visual–spatial" one as well, at least for most individuals. Finally, from a clinical perspective, a much greater than chance association between finger recognition and right–left orientation frequently has been observed (Benson, 1979; Hecaen & Albert, 1978; Strub & Geschwind, 1974), even by Benton (1959) who casts doubt on the existence of "Gerstmann's" syndrome (Benton, 1961).

Since autotopagnosia, finger agnosia, and right–left orientation often are demonstrated by asking the patient to either name or point to the body part or finger named by the examiner, aphasic syndromes in some instances could help account for the observed deficits (Sauguet, Benton, & Hecaen, 1971). If the patient's difficulty on tests of finger agnosia or right–left orientation primarily is tested through identification or matching using extrapersonal stimuli, the difficulty may stem from problems involving reversible operations in space. If tested only in the tactual modality, finger recognition in particular requires sustained concentration and attention. Hence, any general cognitive deterioration may adversely affect performance (Gainotti, Cianchetti, & Tiacci, 1972). However, as body and finger recognition deficits frequently can be demonstrated using nonverbal means (e.g., drawing, matching to sample) and can be found in patients who appear to suffer neither from aphasia nor general cognitive deterioration (Sauguet, Benton, & Hecaen, 1971; Ogden, 1985), these conditions cannot account for all autotopagnosic-type deficits.[70] In those cases where neither aphasic nor general cognitive deterioration would appear sufficient to account for these deficits, the most common lesion site is in the region of the inferior parietal lobule, particularly in the left hemisphere (Kinsbourne & Warrington, 1962b; Sauguet, Benton, & Hecaen, 1971; Varney, 1984). Although the data are still equivocal, they suggest that indeed there may be an area in the tertiary cortex that is critical for integrating visual and somatosensory information. Such integration allows for the development of a supraordinate concept of the body and its parts, commonly referred to as a *body schema* (Hecaen & Albert, 1978, Pp. 319–330 offer additional theoretical explanations for these phenomena).

While *ideomotor apraxia* appears to represent a very different phenomenon from those just discussed, it sometimes is included under the rubric of disorders of body schema (Goldenberg, 1997). For a more complete discussion of this syndrome, the reader is referred to the section on Disconnection Syndromes (p. 328). For now, what is important to note is that in order to carry out tasks that measure ideomotor praxis (e.g., brushing one's teeth, hammering a nail), a mental template delineating how that action is supposed to be carried out has to have been established as a result of repeated practice or associations derived from use of the objects in question. Clearly there are somesthetic associations (cutaneous and proprioceptive). At the same time, most of us learned these actions by observing others as well as utilizing visual feedback to monitor the effectiveness and efficiency of these actions. Thus, these tactile–visual associations constitute a template ("engram") by which parts of the body are programmed to respond in a four-dimensional environment (the fourth dimension being "time"). According to Heilman and Rothi (1993) it is the parietal (tertiary) association cortex of the dominant hemisphere that serves as the integration point for these engrams, or what they term *praxicons*, to become established. Table 9–5 summarizes some of the various classes of disturbances associated with lesions of the somatosensory areas.

Somesthesia and Other Functional Units

As had been noted, there are strong associations between the motor and somatosensory zones. The third functional unit, particularly the primary and premotor zones (see below), are intricately linked to somesthetic feedback systems as a prerequisite for efficient, coordinated muscular activity, especially goal-directed activity. Cells in the parietal, somatosensory cortex are observed to fire, even if one is just thinking about performing an action. This relationship will be discussed in greater detail later in this chapter in reviewing the motor zones.

Table 9–5. Summary of Somatosensory Disturbances

Perception Difficulty appreciating elementary sensations, such as heat, pain, or simple touch. Generally not permanently disrupted by lesions of the somatosensory cortices.

Sensitivity Changes in intensity of the stimulus required for perception to take place.

Discrimination Difficulty differentiating two stimuli along a single dimension (e.g., which stimulus is more painful, heavier, or hotter, or "two-point" discrimination).

Recognition Difficulty "knowing" what an object is. May reflect an apperceptive deficit (e.g., astereognosis), or, more rarely, an associative deficit (tactual asymbolia).

Body Schema Lack of awareness of one's body or the forces acting upon it. May include unilateral neglect, anosognosia, bilateral finger agnosia, and right–left disorientation.

Sensorimotor "Afferent paresis" or "ideomotor apraxia" resulting from the loss of somatosensory feedback or disruption of visual–tactile engrams for skilled movements, respectively.

As is true of other sensory areas, the somatosensory cortices would appear to have fairly direct connections with the first functional system. While the somatosensory areas are important in providing general information regarding one's external or internal environment in order to plan and execute goal-directed behaviors, cutaneous stimulations also are associated with more immediate alerting or arousal reactions, as well as directly mediating many approach–avoidance responses. Unexpected tactual stimulation can have a dramatic arousal or withdrawal response. The almost violent (as in "excessive") nature of the response in some situations may have something to do with the possibility of imminent danger to the organism.[71] One might envision slipping one's feet under the bed covers at night and feeling something small, furry, and *moving!* Unless you happen to own a lot of hamsters and this is a frequent occurrence, chances are you are not going to continue to palpate the object with your toes, trying to discern what it is before you decide how to respond. Certainly noxious or painful stimuli would be expected to trigger negative affective responses.[72] On the other hand, tactile stimulation also often is associated with positive affective responses. Grooming in animals, in addition to controlling parasites (and perhaps adding a bit of protein to the diet), also has a calming influence and helps establish social bonds. In humans, the fact that many individuals derive pleasure from having their hair stroked or backs rubbed in part may represent residuals of these grooming behaviors and their impact on limbic structures. Although it may be counterconditioned as a result of negative experiences, the strong positive reinforcement normally associated with tactual stimulation during sexual arousal is but another example.

THE THIRD FUNCTIONAL UNIT: EXECUTIVE CONTROL

Overview

The cortical substrates of the third functional unit basically encompasses the frontal lobes.[73] However, the frontal lobes are neither functionally nor anatomically homogeneous. As was seen from Figures 6–5 and 9–13, different parts of the frontal lobes are characterized both by cytoarchitectural variations and differences in their subcortical connections. From a cytoarchitectural perspective, the frontal lobe has been divided into two general types: (1) **agranular**, and (2) **granular cortex**. The agranular cortex, characterized by a well-defined layer III and prominent pyramidal cells in layer V, consists of the **primary motor** (area 4) and **premotor zones** (areas 6, 44, and, to a lesser extent, the more caudal portions of 8).[74] As we shall later see, these two zones roughly correspond to the primary and secondary areas in the posterior sensory cortices in that while they are both unimodal in nature, only the premotor areas have direct connections with the tertiary cortices. It is the area rostral to these two agranular zones that constitutes the **frontal granular cortex**. Also known as the **prefrontal cortex**, these areas lack the pyramidal cell concentrations of the motor and premotor zones, hence taking on the more uniform, granular appearance that is characteristic of heteromodal zones of the posterior lobes. These prefrontal, heteromodal cortices of the frontal lobes receive substantial input from polysensory areas of the posterior cortical areas, and along with the heteromodal zones of the (PTO) cortex appear to represent the areas of the greatest cortical development in the human brain.

The structural aspects of the frontal lobes, along with their interconnections to other brain areas, will be addressed in greater detail below. For now, it may be helpful to briefly preview several functional considerations as they apply to this behavioral unit. If the role of the second functional (gnostic) unit is to gather, process, and store information (primarily about the outside world), the role of the third functional unit may be viewed as putting that information to use to formulate and then carry out the needs, desires, or intentions of the organism with respect to that world. As will be seen, the functions of the frontal lobes are multiple and complex. However, for these preliminary purposes, they can be broken down into two broad categories: (1) deciding *if, where, why, how,* and *when* to respond, and (2) actually going through the motions of carrying out that response. This latter function is mediated by the frontal agranular cortices (motor and premotor areas), while the former is thought to be the providence of the frontal granular (prefrontal or heteromodal) cortices.[75]

In one respect, the third function unit (or frontal lobes) is similar to the second functional unit in that both consist of primary, secondary, and tertiary cortical units arranged to function in a hierarchical fashion with decreasing specificity and increasing higher-order and multimodal capacity. However, as the frontal lobes (or third functional unit) are considered to be the executive unit of the brain, with the motor outcome representing its logical and final common mode of expression, we might think of its organization as being just the opposite of the second functional unit. In gnostic or sensory systems, information initially arrives at the primary sensory zones and is processed there before proceeding to higher levels of perceptual integration in the unimodal association cortices. These unimodal or secondary association areas, in turn, funnel their information to the PTO (heteromodal) cortex, where well-integrated, multimodal images, generalizations, or abstract concepts are established. Conversely, in the frontal executive system, the process would appear to progress from the tertiary to the primary areas. Thus, based on the information provided by the first and second functional units, the tertiary cortex of the frontal lobes selectively attends to and evaluates the relative significance of the available data (including internal and external input, as well as past memories) and balances affective vectors associated with those inputs. This tertiary frontal zone also is thought to be important in devising plans of action that (hopefully) best meet both the immediate and the long-range goals or interests of the organism. Once these plans of action are formulated, the information is forwarded to the premotor zones to coordinate an integrated motor response. The primary motor area, representing the final common cortical pathway for expressing the will of the organism, then executes this motor response.[76]

The remainder of this chapter will focus primarily on three general topics:

1. Reviewing the suspected role of the prefrontal cortex in man in general, including the behavioral effects produced by disturbances in this system.
2. Attempting to differentiate the specific roles of different portions of the frontal heteromodal cortex.
3. Delineating the probable functions of the premotor and motor zones.

However, before proceeding with this task, several general caveats or comments may be in order:

1. As difficult a task as it was to attempt a description of the organization and operation of the first and second functional units, trying to describe the operation of the frontal or executive unit is even more precarious. The best we can hope for is to come away with broad working hypotheses of its role in behavior. The following hypotheses are derived, in large part, from an analysis of the functional deficits seen following "frontal lobe" lesions and to some extent from knowledge of its structural connections with other parts of the brain.

2. While it is common to speak of the "prefrontal cortex" as a homogeneous structure, this is not the case. The frontal lobes are composed of areas with diverse cytoarchitectural structure and with different patterns of afferent and efferent connections. While, the prefrontal cortex often is subdivided into dorsolateral, mesial, and orbital components, attempts to define the functional specificity of these smaller areas have met with only limited success.

3. Although behavioral differences occasionally are identified as resulting from right versus left frontal (hemispheric) lesions, such dichotomies generally have not been as pronounced with regard to the frontal lobes as is the case with the posterior cortices. The effects of bilateral frontal lesions often seem qualitatively different than lesions of the right or left hemisphere alone.

4. The nature of frontal lobe deficits themselves often is difficult to define. Unlike the posterior cortices in which deficits typically are defined in terms of specific patterns of sensory-based, behavioral decrements, most of the deficits observed following anterior insults are truly supramodal in character. Rather than simply responding to a set of stimulus parameters, the prefrontal cortex responds to the entire "context" in which those parameters are set, including current drive states, the affective valence of the stimulus, current goals or behavioral intentions, immediately preceding response sets, as well as anticipated consequences of current response possibilities.[77]

5. As is the case with studies of the posterior heteromodal association areas, the use of animal models to gain insight into the effects of prefrontal lesions probably has limited benefits. In some ways, trying to fully comprehend frontal lobe functions in humans from animal studies may be comparable to trying to understand the mechanics of aphasia from the study of temporal lesions in cats.

In contrast to the approach taken with the second functional unit, where the primary and secondary zones were discussed prior to discussing the tertiary zones, here we will start with the tertiary zone or the prefrontal cortex. This approach was adopted since, as noted above, it seems logical that the flow of information in the third functional unit primarily proceeds from the tertiary to the secondary and primary motor cortices. While the main focus of this section, like the rest of the third portion of this chapter, is devoted to an exploration of the possible role(s) of the prefrontal cortex in behavior, it may be helpful to first briefly review its anatomical parameters.

Anatomy of the Prefrontal Zone

The frontal or granular cortex generally is considered to encompass everything that lies rostral to the premotor cortex and the frontal eye fields. In humans, this represents the majority of the frontal cortices. However, as noted earlier, the prefrontal cortex is not a homogeneous area, neither anatomically nor functionally. While additional subdivisions likely will be identified in the future, for most clinical purposes three to four subdivisions are now commonly identified. As seen in Figure 9–29, these include the **dorsolateral, orbitofrontal, mesial,** and **paralimbic** cortices. In addition to the paralimbic area, the orbitofrontal and ventral mesial cortices all appear to have strong limbic connections.[78]

The major connections to the prefrontal areas would appear to derive from the (1) hetero-modal zones of the posterior cortices, (2) temporal lobes, (3) limbic/paralimbic structures,

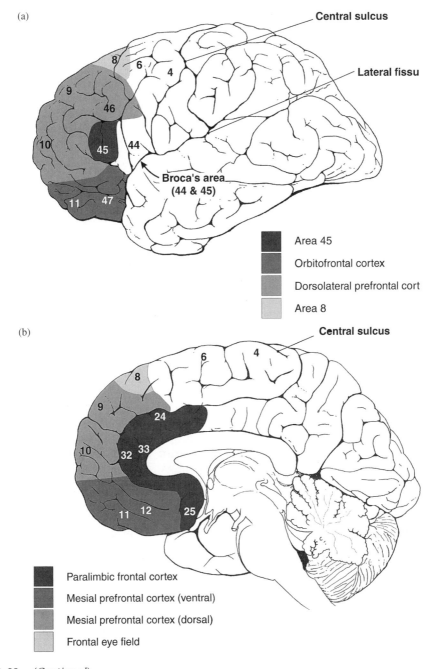

Figure 9–29.   (*Continued*)

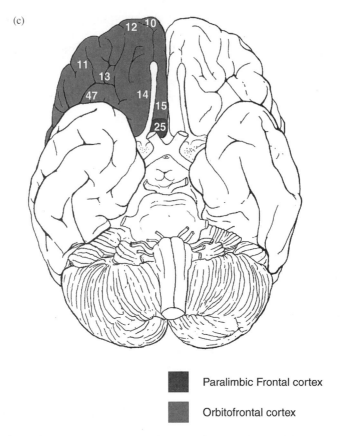

Figure 9–29. "Prefrontal" cortex, (a) lateral, (b) mesial, and (c) ventral views. Demarcations are approximate. Area 45 and the more rostral portions of area 8 (frontal eye fields) are most commonly classified as being part of the prefrontal, heteromodal cortex. Transitional areas include "paralimbic" cortices.

(4) hypothalamus and brainstem, (5) olfactory system, (6) thalamus, and (7) the contralateral prefrontal cortex via the corpus callosum (Stuss & Benson, 1986; Barbas & Pandya, 1991). The posterior tertiary cortex, which carries multimodal and highly integrated information, reaches the prefrontal areas primarily via the **superior longitudinal fasciculus**, whereas information from the more medial or anterior portions of the temporal lobe is carried by the **uncinate fasciculus**. The parahippocampal gyrus in turn also is connected with the prefrontal cortex via both the **inferior orbitofrontal fasciculus** and the **cingulum** (Goldman-Rakic et al., 1984). The orbitofrontal and basomedial portions of the frontal lobes appear to have substantial connections with the amygdala and septal nuclei, as well as with hypothalamic/midbrain nuclei (including the reticular nuclei). The cingulate cortex in turn appears to project to both the dorsolateral and the more orbitomesial portions of the prefrontal cortex (primarily via the cingulum). The olfactory system appears to have both direct and indirect (via the amygdaloid nuclear complex) connections with the basomedial frontal areas. As appears to be true of most cortical areas, the prefrontal region in one hemisphere is in communication with the prefrontal cortex in the opposite hemisphere via commissural fibers. Finally, the prefrontal association cortex in part also is defined by the fact that it is the primary, if not exclusive, cortical recipient of afferent fibers from the dorsomedial nuclei of the thalamus.

While most of the above afferent projections to the prefrontal cortex likely have reciprocal connections with these various structures or sites, there also are prefrontal efferent pathways that likely have no direct afferent counterparts. The best known of these is the connections between the prefrontal cortices and the corpus striatum. While the dorsolateral prefrontal cortex likely sends some fibers to the putamen, the majority of its output is to the head of the caudate. In contrast, the orbitomesial portions of the prefrontal cortex appear to project to the region of the nucleus accumbens or the ventral striatum, along with projections from the amygdala. The dorsolateral prefrontal cortices also have substantial connections with the premotor zones of the frontal cortex, through which motor action can be initiated. Finally, the prefrontal cortex makes an extensive contribution to the corticopontine pathways that after synapsing in the pons proceed to the cerebellum.

Functional Correlates of the Prefrontal Zone

As is true of much of neuroanatomy, hypotheses concerning the role of the prefrontal cortices in behavior are generated through observations of behavioral change following damage or injury to these areas. Perhaps the earliest and most well-known report of frontal lobe damage in the modern era is Harlow's (1868) account of the case of Phineas Gage. Following an injury that certainly compromised his frontal lobes, marked personality and behavioral changes were noted (particularly disinhibition and loss of work ethic), despite the relative preservation of his "intellectual faculties."[79] Other case studies of personality change associated with focal lesions of the frontal lobes in the late 19th and early 20th century reported problems of disinhibition or impaired judgment, puerility, *Witzelsucht* ("cruel or tasteless joking"), lack of concern, and general apathy in such patients (for review, see Benton, 1991). However, despite these early reports, two additional findings were seen as being somewhat "positive." First, patients with lesions confined to the anterior portions of the frontal lobes appeared to suffer neither physical impairments nor loss of basic intellectual capacities, at least as measured by formal IQ tests at the time. Second, studies with primates suggested that surgical lesions of the frontal lobes might have a "calming" effect on behavior. Subsequently, frontal lobotomies or leucotomies became a relatively common treatment option in certain centers for a wide range of behavioral and emotional problems by the 1940s. The justification and continued use of this procedure was facilitated in part by the observation that certain patients who suffered loss or damage of frontal lobe tissue continued to function with apparently minimal behavioral disruptions (e.g., Hebb, 1945). These practices continued into the 1950s and beyond despite reports, often from the surgeons themselves, of the potential deleterious effects of such procedures (Freeman & Watts, 1942; Rylander, 1948; Strum-Olsen, 1946; Partridge, 1950; Tow, 1955). About this same period, evidence continued to mount from independent studies of war-related frontal lobe injuries, documenting and expanding on earlier findings that suggested that the frontal lobes indeed were critical not only for normal behavior but also for certain higher-level cognitive functions (Goldstein, 1939, 1944). It only was after the advent of more effective psychotropic drugs and increased awareness of patients' rights issues that the role of psychosurgery began to come under closer scrutiny (Valenstein, 1980).

In retrospect, these early studies of the frontal lobes from the first half of the 20th century managed to identify what we still perceive to be important behavioral correlates of the frontal lobe, such as:

1. Initiating and persevering in goal-directed behavior, especially with regard to long-range goals.
2. Modulating emotions, basic drives, and instincts to conform to social standards, and developing a sense of self-awareness.

3. Achieving optimal levels of arousal (drive) and sustaining and directing attention.
4. Considering the circumstances and potential consequences of behavior when planning an action.
5. Maintaining cognitive flexibility in problem solving.

In attempting to summarize these ideas, we might say that the role of the prefrontal cortex in humans is to balance, coordinate, and integrate physiological, social, and "existential" needs with external (environmental) demands or conditions to effect desired goals. Although behaviorally there probably is considerable overlap, for purposes of discussion this more general notion of prefrontal cortical functioning in turn can be broken down into two broad categories: (1) planning and execution of cognitive–behavioral programs, and (2) modulating or controlling internal drives and emotions. As will be seen, the former tends to be more associated with the dorsolateral prefrontal cortices, while the latter is usually linked to the orbitofrontal or mesial frontal zones.

In what remains of this section on the prefrontal zones we will attempt to review each of these two broad functional areas in greater detail and to suggest how, in executing these behaviors, the prefrontal areas rely on or influence other portions of the CNS. Although separated for the purposes of discussion, one should keep in mind that cognitive programming and behavioral control do not necessarily function independently of one another in day-to-day activities, but in fact likely complement one another. Finally, examples of specific behavioral deficits that might result from breakdowns or disruptions of these frontal systems will be reviewed.

The Planning and Execution of Cognitive–Behavioral Programs

A good place to start is to ask, "In what situations is the prefrontal cortex most critical?" Although there is no clear, simple answer to this question, we know that many patients with massive frontal lesions, despite some changes and limitations, continue to function, often in an apparently fairly sophisticated, complex, and goal-directed manner. They may be able to take care of their basic biological needs. They carry out many activities of daily living, including dressing, shopping, conversing, engaging in recreational activities, and finding their way around. They are capable of learning, may recall prior events, and may be capable of solving certain types of cognitive tasks. In fact, as previously noted, it was not uncommon for patients with damage to the frontal lobes, including frontal leukotomies, to perform adequately on formal tests of intelligence or other neuropsychological measures (Ackerly, 1937; Freeman & Watts, 1944; Benton et al., 1981; Stuss & Benson, 1986; Damasio & Anderson, 1993). In contrast to the preceding observations, Duncan et al. (2000) report a study in which subjects were administered cognitive tasks while undergoing PET scans. They conclude that the frontal lobes, particularly the dorsolateral areas, possibly reflect the primary neural substrate for Spearman's "g" factor (i.e., general intelligence).

The entire brain, including the prefrontal cortex, likely participates in most if not all cognitive behavior. However, it also is believed that certain areas of the brain probably are more critical for successfully executing certain behaviors or responding effectively under certain circumstances than others. This realization largely has come about by studying the effects of focal brain lesions on various classes of behavior. At the risk of oversimplification, one might consider the following as general examples of types of situations where the integrity of the prefrontal cortices may be most critical.

1. Where it becomes necessary to control or inhibit conflicting or situationally inappropriate drive states.
2. Where it becomes necessary to harness drive states (via inhibition or facilitation) to achieve future or abstract goals.

3. To make appropriate judgments in problem-solving situations where there are conflicting, ambiguous, unfamiliar, vacillating, distracting stimuli or task demands, and/or equally compelling response alternatives.

4. Where persistence is required in maintaining goal-directed behaviors in the face of distracters and/or changes in response contingencies or demands.

While the first two situations would appear to involve critical components of the orbitofrontal and mesial cortices (to be discussed below), the latter two more clearly relate to the planning and execution of cognitive–behavioral programs. In trying to devise an "entertaining," yet meaningful analogy of the type of situation where the dorsolateral cortices might be critical, again the crew of the "Starship Enterprise" comes to mind. Consider the following possible scenario:[80]

While returning from a routine mission to its starbase, the planet Zercot in the Beta Centauri system, and despite traveling at warp speed, any of the first officers could easily command the ship. Even minor problems or malfunctions such as the loss of communications with Starfleet Command when passing through a gamma-charged radiation belt or the overloading of the main engines due to premature hydrogen ionization could be handled by Lt. Uhuru and Scotty, the chief engineer. Captain Kirk remains available, if necessary, but he and the chief science officer, Spock, are spending some leisure time with the ship's computer library classifying and cataloging the new life forms encountered during their last voyage through Sector 12 of the Andromeda galaxy. The presence of Kirk and Spock is not needed for these fairly routine activities. However, as the Enterprise approaches Zercot, a strange phenomenon is encountered. A series of three, extremely bright objects, approximately 50 miles in diameter, come into view from their stationary orbit on the far side of the planet. Preliminary data from the sensors suggest that these "objects" are pure light energy, with no mass and apparently emitting no significant heat. Before further analysis can be made, the light energy suddenly increases by a factor of 10, all but blinding the crew. Simultaneously, Scotty reports that the matter–antimatter reactor is dangerously overheating, causing not only a loss of power, but unless corrected will result in an explosion that will vaporize the vessel. The first impulse of the crew is immediately to activate the shields, but they are not responding, and with the loss of power evasive action is impossible. A message is being received from the planet. It is from the Klingons, who contrary to Federation treaties have invaded the sector and seem to be transmitting from deep within the interior of the planet. They warn the Enterprise that unless they immediately and unconditionally surrender, the starship will be annihilated. A quick check of the ship's phasers reveals they also have been rendered inoperative.

Now would be the time for Captain Kirk and Spock to be summoned to the bridge (i.e., for the frontal lobes to really kick into action). Routine, previously programmed response patterns are no longer adequate. Panic must be avoided and all energies must be focused and channeled into creative strategies to deal with the emergency. A review of all current and historical data, a thorough systems check, and analysis of all conceivable response options, both conventional and unconventional, is essential. However, since this "situation" is unprecedented, despite efforts to anticipate the consequences of contemplated actions, accurate predictions are impossible. Even with such planning, a trial-and-error approach is likely, with each initiative being carefully monitored, as the crew of the Enterprise prepares to quickly adopt new strategies as dictated by Klingon response.

The above example was intentionally extreme. Obviously, the frontal lobes routinely respond under far less dramatic and complex circumstances. However, the potential value of this example is twofold. First, it illustrates that even in the absence of the frontal lobes (in this case represented by Captain Kirk and Science Officer Spock) many complex activities still may be accomplished. Second, it suggests that certain complex cognitive activities depend heavily on the integrity of the prefrontal cortex. Having provided this more general example, it now may be helpful to review in greater detail several aspects of prefrontal (particularly, dorsolateral) behavior that are deemed critical for successfully carrying out complex cognitive-behavioral programs. Although the list certainly is not exhaustive, we shall consider the following:

- **Arousal**
- **Intention** and **initiation**
- **Selective attention**
- **Inhibition**
- **Planning** (strategy) and **execution**
- **Self-monitoring**

Arousal

As is true of the entire cortex, a certain optimal level of arousal is necessary for efficient functioning to occur. Cortical arousal is likely mediated by at least two systems: one specific and one nonspecific (Heilman, Watson, & Valenstein, 1993). Nonspecific arousal would appear to be a function of the diffuse norepinephrine pathways that emanate from the mesencephalic reticular formation (see Chapter 11). More specific arousal in large part probably results from thalamocortical projections. In the case of the prefrontal cortices, these projections originate primarily in the dorsomedial nuclei of the thalamus. The hypothalamus and "limbic system" likewise play an integral part in cortical, particularly prefrontal, arousal. It also appears likely that the prefrontal cortex itself plays a key role in initiating and maintaining arousal under certain circumstances. The respective roles of these various systems and how they can specifically impact one's general behavior will be discussed in greater detail under the heading: **The Modulation of Internal Drives and Emotions** (p. 443). What is important to note here is that any action that requires sustained planning, prolonged execution, and constant monitoring also require sustained arousal or drive to ensure proper levels of cortical activation throughout the process. Any disturbances that result in fluctuating or inadequate levels of arousal significantly may impede these operations.

Intention and Initiation

Closely related to the notion of arousal discussed in the previous section is the concept of creating a stable intention to perform some action or carry out some activity. Both arousal and intention would appear to involve the interplay of limbic mechanisms. With arousal, the emphasis is more on gearing up or energizing the organism to respond to some perceived need or demand, more or less independent of whatever specific course of action that is subsequently to be taken. **Intention** would appear to take this process one step further by combining the notions of will and motivation to create a stable "intention" that the goal be reached. To do so means to be motivated (i.e., have the intention) to develop, initiate, and/or adhere to whatever plan of action is necessary to reach the goal. It means the willingness and effort to bring in whatever resources are essential to this process, including attention, active perception, memory retrieval, planning and organization, and where necessary prolonged mental and/or physical exertion. Obviously, that intention also must be stable over time, at least over the time period necessary to carry out the plan or program until the goal is reached.

An example might be to consider the following situations. You are walking down the street in a brand new suit and your obnoxious neighbor's big, friendly St. Bernard runs up and puts his muddy paws on your shoulders, knocking you down into a large puddle of water. You probably would not need your prefrontal lobes to later recall what happened as you relate the incident to your lawyer. However, if while walking down the street you were to witness a burglar running out your front door with a jewelry box in hand, the proper operation of the frontal lobes would be much more important. In order to remember all the essential details to provide to the police (e.g., his approximate height, weight, and age, hair color, clothes, shoes, whether or not he wore glasses, the make, model, color of his car) you might have to create a stable intention to remember, including rehearsing the facts while waiting for the police to arrive. Providing a patient with a long list of words to recall over several trials is more like the second than the first scenario. In order to be successful, it requires the stable intention to pay particular attention to and/or covertly rehearse those items not recalled on the previous learning trial.

Similar and probably related to disorders of arousal and intention in prefrontal patients are problems of **initiation**. As noted, with problems of arousal and intention, the patient may demonstrate either insufficient motivation to respond or failure to act in such a manner that is consistent with a stable intention to prepare and plan for the successful completion of a complex or sustained activity. Failure of initiation is most often reflected in the failure of the patient to engage spontaneously in goal-directed behaviors. Although the patient may engage in elaborate discussions and make detailed plans regarding future goals (e.g., getting a job, fixing the washing machine, or preparing a Thanksgiving dinner), he or she may never take the necessary steps to carry them out. Even if they do begin, they may abandon these efforts within a short period of time.[81] Other patients with severe, bilateral damage to the prefrontal areas may revert from a very active, productive existence to that of a couch potato. **Transcortical motor aphasia** (a "frontal" type aphasia) appears to reflect a similar dynamic. These patients engage in little if any spontaneous speech, although they are perfectly capable of speech as witnessed by their generally intact repetition, even to the point of becoming echolalic at times.

In summary, the three functions just discussed—,arousal, intention, and initiation of behavior—seem to share a common feature, that is, they all are concerned with ensuring that the thought or idea is converted into appropriate action.[82] The frontal lobes would appear to play a critical role not only in harnessing limbic drives (see below), but also in facilitating the cognitive (conscious) generation of certain drive states and the channeling of drive states into overt, goal-directed activity. Obviously, failure to act or failure to act in a highly efficient, dedicated manner at times maybe attributable to factors other than frontal lobe pathology.[83] Depression, anxiety, passive aggressiveness, and malingering are but a few of the "psychological" factors that could inhibit or interfere with completing long-range goals.

Depending on the severity and/or locus of the pathology, deficits in arousal, intention, or initiation may be relatively subtle or quite profound. In certain cases, catatonia, or what Luria (1973) referred to as **apathetico-akinetico-abulic** syndromes can result from frontal lobe damage. Again, these conditions will be discussed in greater detail below under **The Modulation of Internal Drives and Emotions**, however, a few preliminary anatomical observations can be offered here. While the more severe abulic-like conditions are most likely to be seen following extensive, bilateral frontal lobe damage, variants of these conditions may be associated with lesions involving the dorsolateral convexities, the orbitomesial regions, as well as more dorsomedial lesions. Two general mechanisms can be postulated as possibly contributing to these deficits. One is that frontal lobe lesions (particularly orbital and mesial lesions) disconnect the prefrontal cortices from the energizing or activating influences of the mesencephalic reticular, norepinephrine pathways, the dorsomedial nuclei of the thalamus, and/or the dorsal (cingulate) or ventral (amygdala, hypothalamus) limbic structures. The second possibility is that damage to the dorsolateral convexity itself may disrupt either its capability to harness and channel these limbic influences to carry out behavioral programs or to generate and execute these programs themselves.

Less dramatic, more subtle alterations in drive states easily may go unnoticed or be attributed to premorbid personality or psychological reactions in more restricted or less obvious frontal system damage. Thus, the patient simply may be characterized as "unreliable," "lazy," or "lacking in ambition." In contrast to but occasionally accompanying deficits in sustained intention and initiation are problems of what might be described as "hyperactivity" or extreme restlessness in patients suffering frontal lobe damage. First noted in early research with animals (Benton, 1991), this latter syndrome is most commonly found following orbital or ventromesial lesions (Benson & Stuss, 1982). On rare occasions, more overt manic disorders may occur from lesions that include prefrontal cortices (Clark & Davidson, 1987; Starkstein et al. 1987). The more benign manifestations of restlessness and milder hypomanic states in part may be related to disturbances of attention or inhibition, topics that will be addressed next.

Selective Attention

Unless one is engaged in a sensory deprivation experiment, in most normal waking states, at any point in time we are constantly exposed to a myriad of external stimuli that vie for our attention. Simultaneously, multiple thoughts, feelings, or memories may well up from within that also potentially command our attention. If not impossible, it is certainly inefficient to attempt to attend to all such stimuli as they might arise. We need to be selective in terms of those stimuli that merit our attention at any given moment. The prefrontal cortex would appear to be critical to this selection process.

Before discussing the possible role of the frontal lobes, it is important to recognize that all stimuli by their very nature are not equally compelling. There are various features that may affect the relative salience of a stimulus. **Intensity** of the stimulus is clearly a major factor. Thus, for example, the sudden burst of bright light and booming sound associated with a nearby flash of lightning readily will command our attention simply because of the intensity of the stimulus. In contrast, to use a well-worn example, you were probably not aware of the watch on your wrist or the shoe on your foot until your attention was called to them. **Contrast** also may heighten stimulus value. This may be a discordant note on the piano, the red wine stain on the white carpet, or even the teacher beginning to speak in a soft voice amid the din of the classroom. Uniqueness, novelty, or the "unexpected" (which may be simply other variants of contrast) also frequently commands attention. This may be represented by an atypical use of form or color in the paintings of Picasso or Pollock, an experimental "concept" car at an auto show, or the presence of a cat in church. Any stimulus that has some type of *emotional tag*, whether positive or negative, also is likely to capture our attention. This could range from the smell of a freshly baked apple pie, to a scantily clad model,[84] to the sight of a snake or unconfined blood.

Directed/Voluntary Attention. The above examples, to a large extent, represent stimuli that elicit what might be termed *involuntary attention*. While we may later chose to divert our attention from them, when initially presented they literally "command" our attention. Conversely, we can think of another type of attention, namely *voluntary* or *directed* attention. Voluntary attention often but not invariably involves stimuli that are not necessarily particularly compelling in and of themselves, but rather derive their importance from their relationship or relevance to something else, such as a behavioral goal or plan. The role of the prefrontal cortex may be somewhat different depending on which general class of stimuli is involved, that is, those commanding involuntary attention versus those that are more likely to be the object of voluntary attention.

Luria (1973, p. 216) described how a normal person, when shown a thematic picture, will alter his or her patterns of eye movements (active searching or scanning) in looking at the picture depending on what questions are asked. The idea behind this exercise was that depending on the question the subject would have to scan different parts of the picture that might be expected to provide the answer. For example, if asked about the age of the subjects in the picture, one might scan their facial features, whereas if asked about their social status, attention might be better directed to their clothes or furnishings. Conversely, a patient described as having a "massive" frontal lesion was observed to exhibit a constant pattern of more or less random scanning regardless of the questions posed.

The principle illustrated in the above example easily can be extrapolated to any number of situations. When faced with any task demand or problem, the number of stimulus options (again, both internal and external) to which one possibly might attend usually are considerable. In many cases this does not present a difficulty. As a result of previous experiences with the same or similar demands, the individual simply relies on well-entrenched stimulus–response patterns. However, the situation is different when the solutions are not so clear-cut or when, as noted earlier, one is confronted by ambiguous, conflicting, unfamiliar, vacillating, or equally compelling stimuli or response alternatives for which one has no prior "blueprint" to fall back on. In such situations, one must explore (i.e., selectively attend to) those stimuli that may hold the key to successively completing the task, just as the subjects had to visually explore the picture to look for information to answer the examiner's questions. But to be maximally efficient two things must happen: (1) the search must not be random but rather guided by certain logical strategies or principles, and (2) it must be flexible (i.e., it must be possible to rapidly shift attention from one set of stimuli, thoughts, or memory engrams to another, either as the problem evolves or potential solutions are explored and discarded).

It would appear that the prefrontal cortex is ideally positioned to handle directed attention. Through feedback loops, predominately between the dorsolateral cortices and the second functional unit (i.e., the parietal, temporal, occipital lobes), the frontal lobes could facilitate focusing attention on those perceptual processes, concepts, or memory stores that would appear immediately salient based on current strategies, plans, or feedback.[85] In trying to come up with an example, one is faced with the difficulty of separating out problems of distractibility (which will be addressed shortly), faulty strategy, or inadequate self-monitoring capacity versus problems of selective attention per se. Perhaps, for most practical purposes, these various processes are largely inseparable. However, a clue might be found in analyzing the process that optimally takes place when attempting to learn a list of unrelated words that exceed one's immediate retention span. When a patient is presented with such a list for the first time, a certain percentage of the words are recalled (usually based on primacy or recency effects). When the list is read again, what strategy(i.e.s) might be employed in order to enhance recall? As the word list is being repeated by the examiner, one strategy might be to identify and selectively attend to (i.e., concentrate on or covertly rehearse) those words not recalled on the previous trial. Frontal patients may have difficulty utilizing this strategy. Luria (1966) reported that when presented with such lists, frontal patients tend not to significantly improve over trials.[86]

Distractibility.   In addition to its apparent role in guiding voluntary or selective attention, the prefrontal cortex also is important for controlling involuntary attention. As noted in the introduction of this topic, one constantly is being bombarded with multiple stimuli from a variety of sources, again both internal and external. The overwhelming majority of these stimuli simply represent "noise" in relation to whatever task is at hand. The student who is trying to complete a difficult exam is best served if he or she can refrain from attending to "extraneous" stimuli, such as the sounds of conversations in the hallway, the bird perched on the window ledge, or thoughts of either last night's or tomorrow night's date. This process represents the flip side of selective or directed attention. Attentional capacity is limited; most people effectively can attend to only one or two things at a time. If one must direct their attention to a particular set or class of stimuli, especially if prolonged or intensive concentration is required, this is most effectively accomplished if other potentially distracting stimuli can be ignored (i.e., prevented from intruding on one's consciousness). This capacity to ignore irrelevant stimuli then would appear to be an integral part of the role of the frontal lobes in planning and executing behavioral programs. In fact, problems of distractibility as a consequence of frontal lobe lesions have been repeatedly reported in both humans and animals (Brutkowski, 1965; Chao & Knight, 1995; Fuster, 1997; Hecaen & Albert, 1978; Stuss & Benson, 1986; Woods & Knight, 1986). If present, distractibility can lead to significant impairments on a wide variety of clinical tests, especially those requiring more intense levels of concentration and attention, such as learning and memory tasks. One of the more common signs suggestive of problems with distractibility on the clinical exam is loss of mental set. This might be seen, for example, on the Wisconsin Card Sorting test, on serial reversal tasks (such as serial 7s), or reciting the months in reverse order. On the former, the subject might forget the current sorting principle if pulled by the saliency of another stimulus. On the above serial reversal tasks, the individual may lose track of the task as demonstrated either by subtracting by another number or reverting to the more accustomed task of reciting the months in the forward order. Figure 9–30 appears to illustrate this same phenomenon on clock drawing. Here the patient became immediately distracted by the insertion of the number "12" and continued the number series until running out of room.

Motor impersistence (Heilman & Watson, 1991), which is the inability to sustain an action (such as keeping one's eyes closed or tongue protruded) over a specified period of time, and **utilization behavior,** which is the tendency to pick up any nearby object and begin to use it either without being asked or when circumstances would not dictate such behavior (Lhermitte, 1983: Shallice & Burgess, 1991), both may be seen as expressions of distractibility.

Perseveration. We routinely selectively attend to those internal or external events that are relevant to current problems or task demands and filter out irrelevant stimuli. In addition, we also must be able to abandon or shift our fixation on a stimulus once it is no longer relevant or when circumstances normally would dictate that attention should be focused elsewhere. This finding is less frequently identified as an outstanding finding in the clinical literature (Ackerly, 1937, as reported by Benton, 1991, being one notable exception). However, **perseveration**, which is not an uncommon finding following frontal injuries, may reflect

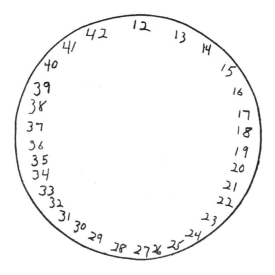

Figure 9–30.   Loss of set on clock drawing task.

difficulties in shifting attention (Sandson & Albert, 1987). As will be discussed below, perse-veration also may be associated with difficulty inhibiting prepotent response sets and can be demonstrated on a wide range of tasks. While it may be associated with various types of pathology, it is perhaps most frequently associated with lesions of the prefrontal zone or underlying structures (Iverson & Mishkin, 1970; Luria, 1966; Sandson & Albert, 1987).

Neglect. Finally, it should be noted that symptoms of unilateral neglect (see earlier discus-sions), while more frequently associated with the posterior cortices (Heilman, Valenstein, and Watson, 1983) also may accompany lateralized frontal (or anterior cingulate) lesions (Crowne, 1983; Damasio, Damasio, & Chui, 1980; Heilman & Valenstein, 1972).[87] Frontal neglect, which likely reflects a special form of an attentional disorder, can be demonstrated in all sensory spheres. However, it most readily may be evidenced in visual neglect of contralateral space (see Figure 9–31) and certainly may be exacerbated by involvement of the frontal eye fields. The neglect need not be complete or consistent. In fact, neglect secondary to frontal lobe lesions often is subtler or less dramatic than neglect syndromes seen following parietal–temporal lesions. As is true of neglect in general, left-sided neglect following right hemisphere lesions appears to be more common than with lesions of the left hemisphere.

Summary. The frontal lobes would appear to instrumental in (1) facilitating our ability to focus or concentrate only on those stimuli that are currently relevant to the task at hand, (2) maintaining that focus as long as the situation dictates (i.e., not allowing ourselves to become distracted by "irrelevant" stimuli), and (3) being able to appropriately shift attention as task demands or contingencies change. However, one also must keep in mind that in any given situation it may be difficult not only to differentiate attentional problems of "organic" versus "functional" origin, but also difficult to differentiate or separate out other functional disturbances, such as planning errors, disinhibition, impaired self-monitoring capacity, or more general arousal difficulties. All may be associated with frontal pathology, and all may present behaviorally as disturbances of attention.

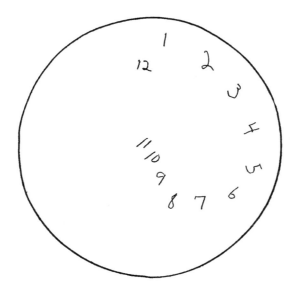

Figure 9–31.   *(Continued)*

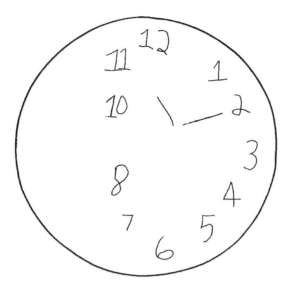

Figure 9–31. Examples of left visual neglect on clock drawing.

Inhibition

Damage to the prefrontal areas often is associated with problems of disinhibition or failure to inhibit inappropriate responses. Both behaviorally and anatomically, there would appear to be justification for dividing problems of inhibition into three broad categories. The first generally involves problems in inhibiting biological or affectively driven responses, particularly in situations where such expressions might be considered socially inappropriate. Clinically, such behavior is usually termed **behavioral disinhibition**. The second might be characterized as difficulty inhibiting cognitive response sets that may have been appropriate at one point but which are no longer appropriate or effective. A related phenomenon might be the inability to distance oneself from a particular stimulus or stimulus cue in order to produce an appropriate response. The third type of disinhibition might be conceived of as difficulties inhibiting kinetic programs once they are initiated, again despite the fact that they are no longer useful. Problems of "behavioral disinhibition" will be addressed below under the heading **Modulation of Internal Drives and Emotions**. Here the focus will be on problems of inhibiting cognitive and motoric response sets.

Cognitive Disinhibition.   The problem in inhibiting inappropriate cognitive response sets is in itself a multidimensional phenomenon. While there may be some benefit in subsuming this class of behaviors under a single rubric, many factors likely contribute to the observed deficits. Hopefully, this will become clear as we proceed through the next few paragraphs. On a very basic level, many if not most of our behavioral responses are stimulus-bound. Our response in a given situation is dictated in large part by the stimulus parameters that constitute the particular situation at any given point in time. In some instances our response choice will be predicated on similar stimulus–response patterns that proved effective in the past. This has a certain practical value. Instead of having to sort through or individually analyze the myriad of potential responses available, we can respond quickly and efficiently in a more or less preprogrammed manner based on previous successful experiences. However, there also are times when such approaches may not be the best choice.

Normally, if the selected response proves ineffective or if the stimulus parameters (the situation) are sufficiently unique, a new trial-and error-approach will be adopted. If this new approach is successful (i.e., produces the desired result or otherwise positively reinforced), it will continue; otherwise it likely will be discarded in favor of yet another approach. Choosing what response to make in a given situation and deciding when and under what circumstances certain responses need to be abandoned in favor of new, more effective responses are what might be termed **executive functions,** and are thought to be largely mediated by the prefrontal, dorsolateral cortices.

With damage to the prefrontal (especially, dorsolateral) cortices this "executive" process can be disrupted. As suggested above, this can happen for a number of different reasons. First, a particular stimulus or aspect of a situation may trigger an association with a previously learned response. Two examples of the types of errors that might be made in this situation are failure to take the entire situation into account or simply difficulty in inhibiting a "prepotent" (overlearned) response. The patient who described the two pictures in Figure 9–32 as a "pipe" and a "plumber's helper" failed to inhibit his "first impression" based on partial information, ignoring the remaining salient features of the stimuli.

In the Stroop Test, where the names of colors (e.g., "red" or "blue") are printed in different colored inks, the patient may persist in giving the color designated by the words rather than the color of the ink in which the word is printed as instructed. In this latter situation, the patient accurately may retain the verbal instructions, but have difficulty inhibiting the prepotent response in face of the pull of the stimulus. This latter phenomenon also might be termed becoming *stimulus-bound*.

Additional examples might include the patient who either places the hands of a clock pointing at the "10" and the "11," or places an extra "10" between the 11 and the 12 when asked to set the time to "ten after eleven" (see Figure 9–33). It should be noted that the kinds of errors shown in these clocks also might result from failures of self-monitoring capacities (see below). Finally, patients with frontal lobe pathology will occasionally experience difficulty with the following task. If asked to "Close your eyes and raise the hand opposite

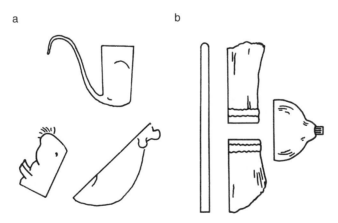

a   b

Figure 9–32. Examples of figures from the Hooper Visual Organization Test that might result in incorrect response by failing to take into account all available information prior to responding, identifying (a) as a "pipe" and (b) as a "plunger." Material from the VOT copyright © 1957 by H. Elston Hooper. Reprinted by permission of the publisher, Western Psychological Services, 12031 Wilshire Boulevard, Los Angeles, California, 90025, www.wpspublish.com. Not to be reprinted in whole or in part for any additional purpose without the expressed, written permission of the publisher. All rights reserved.

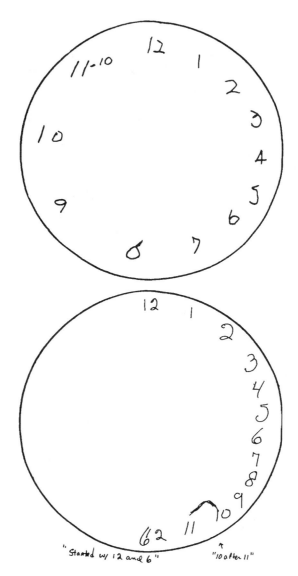

Figure 9–33. Examples of what might be interpreted as stimulus-bound responses on a clock drawing task.

the one I touch," they may persist in raising the hand touched by the examiner (i.e., have difficulty in overcoming the tactile stimulus). This response can occur despite the capacity to accurately repeat the instructions.

Closely associated with the type of deficit described above, a second type of "executive" error may be linked to **perseveration**. As noted earlier, perseveration on occasion may eflect difficulty shifting attention, but in some situations, it also simply might be the case of getting stuck in a particular response set. In this latter situation, the prepotency does not necessarily represent a well-ingrained stimulus–response pattern (as was saying *blue* regardless of the color of the ink when the letters b-l-u-e were presented). Rather, the patient might adopt the first readily accessible response that comes to mind based on its recent use in another context. Thus, if in response to the question, "How are an apple and a banana alike?" the patient responds, "They are both good to eat," he may provide a similar answer (e.g., "both are edible") to the next question, "How are a fish and a bird alike?"

Perseveration, as a sign of failure of inhibition, also frequently occurs in prefrontal patients when "previously reinforced" response patterns no longer are appropriate because of changes in the task demands or changes in the tasks or goals themselves. As suggested above, very few tasks or situations are truly static; most are fluid and dynamic. Stimuli and/or response contingencies constantly change, either as a result of our (the organism's) actions or as a result of forces independent of our actions. Alternately, as one goal or stage within the process of achieving a goal is accomplished, there is a need to shift to the next stage or goal. Response strategies that are initially adopted in the attempt to accomplish a particular goal ultimately may prove ineffective. In all these instances, behavioral (mental) flexibility is required. While self-monitoring, which will be discussed shortly, certainly is critical in ascertaining response effectiveness on an ongoing basis, behavioral responses must remain fluid and flexible. Perseveration would appear to reflect a loss of this behavioral or mental flexibility.

The Wisconsin Card Sorting Test (WCST) (Grant & Berg, 1948; Milner, 1963) and the Category Test (Boll, 1981) are examples of formal psychometric measures that specifically tap this type of mental or behavioral flexibility. Although differing in their stimuli and presentation format, both tasks essentially selectively reinforce a particular response pattern. However, once that response pattern is well entrenched, the patient is required to adopt a different response to what are essentially comparable if not (in the case of the WCST) identical stimuli. On such tasks the patient with significant frontal lobe pathology may continue to rely on previously reinforced response patterns, even though they are obviously no longer effective (Robinson et al., 1980: Osmon & Suchy, 1996).[88] Because of their sensitivity to "loss of mental flexibility" and/or perseverative response tendencies, these two measures frequently are designated as "frontal lobe tests." While they indeed are frequently sensitive to lesions of the prefrontal cortex, it is important to bear in mind that like most tests that are sensitive to brain injury, these are multifaceted tests and as such are subject to disruption by lesions in various parts of the brain (Anderson et al. 1991; Grafman et al., 1990; Wang, 1987). Conversely, absence of deficits on such tests does not preclude the possibility of lesions in a given area of the brain, including the frontal lobes.

Sorting tests, such as those found in the Delis-Kaplan Executive Functions System (published by Harcourt Assessment) and the stimuli in Figures 9–34a and 9–34b offer additional examples of tasks that also would appear to tap what frequently has been characterized as mental flexibility. This capacity might be defined as the ability to look at the same set of data from different perspectives or to explore alternate solutions to an old problem. This requires one to abandon, at least temporarily, previous notions and to be creative in searching for new ones. In the two figures below, each item has something in common with each other item in the array. One's task simply would be to identify ten ways in which the five items within each array share a common feature.

A final example of cognitive response–inhibition difficulty might be the situation where there is no immediate or clear-cut solution to a given problem. Suppose, for example, the problem in question would appear to offer several potentially equally viable responses, at least initially. In order to ensure the maximal probability of success, the most appropriate response might be to delay (inhibit) responding until such time as either the problem can be analyzed in greater detail or the various response possibilities can be given more careful consideration with regard to their potential appropriateness and/or anticipated consequences. Patients with marked frontal pathology often demonstrate a persistent disinclination to delay their response under such circumstances, resulting in perceived poor judgment and impulsive behaviors. While such behaviors can occur under a variety of

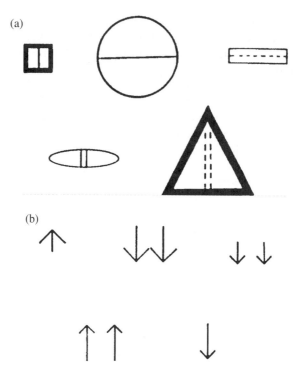

Figure 9–34. "Mental flexibility" type task for bedside exam.

circumstances, even among "normals," the problem likely is to be exacerbated in the presence of certain frontal lobe lesions, especially in basal or orbital lesions and when strong emotions or biological urges are present, topics that will be addressed independently later.

In the situations just described, the ability to keep in perspective both one's immediate and overall goals are critical. Anticipation of possible consequences or predicted outcomes prior to actually carrying out a response should constitute an essential preliminary filtering process. Often the "favored" (prepotent) response might have to be abandoned in lieu of one that, although perhaps less routine, may have a higher probability of success given the circumstances. Thus, response inhibition is not an isolated process. To be maximally effective, it also must involve other aspects of frontal lobe functioning including selective attention, maintaining a stable intention, planning, and self-monitoring ability.

Motor Disinhibition.   Before discussing planning and self-monitoring ability, we need to briefly review what might be considered a third type of behavioral disinhibition, that is, **difficulty inhibiting motor action programs once they have begun**. Luria (1965, 1966, 1973) described at least two general types of motor perseveration or failures of inhibition in writing or drawing tasks. The first, as illustrated in Figure 9–30, seems to represent more of a cognitive-type deficit. In this type of deficit the "action program" is contaminated by intrusion of "stereotypic" components from either preceding programs or preceding aspects of the current program. For example, when asked to copy an array of three designs as part of a memory-for-designs task (Figure 9–35), patients with lesions affecting the prefrontal lobes might carry over elements of the first design (e.g., the circles) into the second or third figures. While occasionally seen during the copying part of the task, it is even more likely to

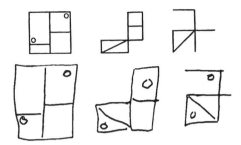

Figure 9–35. Example of perseveration on a memory-for-designs task.

occur during the immediate and delayed memory phases. Similar phenomena might occur if the patient is asked to carry out some type of alternating, sequential motor task. As seen in the second example in Figure 9–36, the more complex pattern dissolves into a tendency to repeat one element of the task.

In contrast, for other patients the problem may be even more basic. There may be no disintegration of the action program itself; rather the problem is that once the motor action is begun, there is a breakdown in the mechanisms that normally would terminate the motor action once it is completed. As a result, the action continues in a perseverative, meaningless fashion well beyond that which is requested. Figure 9–37 provides an example of the latter. In another instance, a patient was asked to write three sentences about how parents should raise or treat their children. The patient filled up the half page allotted for this task and then, turning the page over, proceeded to fill up most of the next page with (on target) sentences until finally stopped by the examiner. When subsequently asked, the patient had retained the original instructions ("Write three sentences...") but apparently had become "stuck in set." Whereas the disintegration of the motor program itself is suggestive of dorsolateral involvement, the inert repetition of the elementary motor responses, although relatively rare, when seen are more likely to result from deep frontal lesions that affect frontal–striatal feedback loops.[89]

Planning (Strategy) and Execution

Because of their relatively recent phylogenetic expansion, the prefrontal cortices had come to be thought of as more or less the seat of man's higher cognitive or intellectual ability. Hence, as noted earlier, the discovery that many frontal lobectomy patients did not appear substantially impaired on IQ tests (see Stuss & Benson, pp. 197–198) must have come as a bit of a surprise to many investigators. In trying to break down the individual factors that are relevant to the role played by the prefrontal zones in higher cognitive functioning and problem-solving behavior (which would appear to underlie "intelligence"), Luria suggested that they become particularly critical in situations where the problem itself fails to outline the response patterns necessary to arrive at its solution or where one simply cannot fall back on previous experience. Thus, the prefrontal zones were thought to be critical in situations where one must (1) carefully analyze the problem, searching for critical clues (while avoiding

Figure 9–36. Perseveration and "simplification" in a Lurian type alternating pattern task.

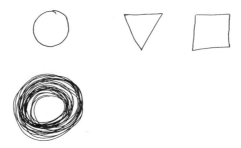

Figure 9–37. Inability to terminate motor action once initiated on a simple copying task.

being distracted by noncritical ones), (2) anticipate the potential consequences of various response options, and (3) develop a general plan of action or strategy that then must be systematically executed in a orderly, sequential manner. After all of the above have been accomplished, one still has to assess the outcome of the action to ensure that it has had its desired effect, or that either the previous goal is still viable, or "rules of the game" have not changed.

This type of analysis and planning ability obviously transcends a large range and variety of situations, both in terms of what one might experience in his or her normal environment and those that might be presented psychometrically (i.e., in neuropsychological test batteries). Yet, at the same time, we often are struck by how well individuals with demonstrable frontal lobe lesions adapt and respond, not only to environmental demands, but also on formal neuropsychological test batteries. This can be seen with both smaller focal lesions and following more extensive lesions involved in psychosurgical procedures. In fact, unless marked by significant personality changes such as increased disinhibition or apathy (which are more likely to result from orbital and mesial lesions) or the individual is engaged in occupations that require extensive and elaborate planning and mental flexibility on a daily basis (i.e., where one simply cannot fall back on more established routines or past experiences), many lesions that are restricted to the dorsolateral prefrontal cortices often may go relatively unnoticed.

In part because of the difficulty in demonstrating planning and other related cognitive deficits on clinical examination, there has been a long history of attempting to identify or develop specific psychometric measures that would enhance such predictions. These efforts have met with limited success. Among the tests that appear to be most commonly listed as probably measuring planning ability are the Porteus Mazes (or similar mazes) and "tower" tests (Damasio & Anderson, 1993; Fuster, 1997; Lezak, 2004; Mapou, 1995; Stuss & Benson, 1986).[90] While traditional visual maze-type tests have proved sensitive to frontal lobe impairment (Milner, 1965; Smith, 1960), the "planning strategy" required for successful performance is relatively straightforward and remains more or less constant, not only throughout the task but also from one problem to the next. Basically this strategy involves delaying the response at each choice point in order to mentally trace the alternate available pathways, thus ensuring that the one selected does not lead to a blind alley. However, impulsivity, failure to anticipate the consequences of a particular response, or "forgetting" to employ this strategy is not the only source of potential error on mazes. Impaired performance also also may result from an inability to profit from past experience (errors) or failure to adhere to the "rules," such as cutting through walls. These latter difficulties at least in part are likely to reflect difficulties with self-monitoring (see below).

"Tower" tests such as the Tower of London (Shallice, 1982) and the Tower of Toronto (Saint Cyr & Taylor, 1992) tasks would appear to benefit heavily from planning ability, especially for the more difficult problems. As can be seen in the accompanying illustration, the subject is required to anticipate the potential effect of each of several possible moves on his or her ability to accomplish the final goal.[91] Despite some correlation between maze or tower tests and prefrontal pathology, the ability of these tests to identify frontal lobe lesions are not exceptionally robust (Shallice & Burgess, 1991; Levin, Goldstein, et al., 1991). This also is true of many if not most measures that commonly are referred to as "frontal lobe tests". Figure 9–38 shows a variation on these tower-type tasks. While it certainly is possible to reach the desired goal by trial and error alone, developing some type of plan or strategy clearly facilitates finding its solution. Before exploring some possible reasons for the frequently marginal association between these and similar-type tasks and frontal pathology, it might be useful to review planning ability as it can influence other types of tasks.

Visual–Spatial Constructions and Planning Ability.   Although visual–spatial construction difficulties generally are associated with disturbances of the second functional unit, specifically the parietal (parietal–occipital) cortices, lesions of the dorsolateral prefrontal cortex also may result in impaired performance on such tasks (Benson & Barton, 1970; Black & Strub, 1976; Luria & Tsvetkova, 1964). The reasons are not difficult to appreciate if one considers the processes involved, especially in the more complex constructional tasks. Think about building a house or planning a garden. One does not start by nailing 2 x 6s together for the rafters before the walls are framed, nor does one simply start to put seeds in the ground in a random fashion. Before either project is begun, there usually is a plan and strategy in place, not only in terms of laying out the basic parameters, but also at least a general idea of the best sequence in which to proceed.

The execution of many visual–spatial construction tasks can be greatly facilitated by following a similar approach. For example, in attempting to copy the Rey-Osterreith figure, one is greatly benefited by starting out with a systematic approach, such as first drawing the

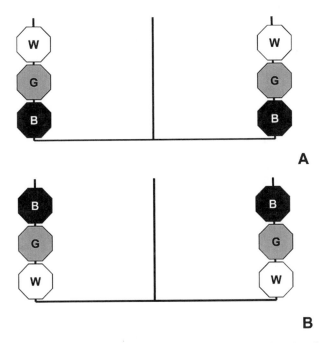

Figure 9–38.   Starting with arrangement A, the goal is to end up with B by shifting the pieces one at a time to either a center or end stack. No stack can ever exceed three objects. (*This test can be tried safely at home using any three identifiable pairs of stackable, crash-resistant objects*).

central rectangle, the diagonals, and the vertical and horizontal divisions before attempting the internal details. Patients with significant frontal pathology, as well as those with right hemisphere lesions in general, are more likely to be observed adopting a more "piecemeal" approach to this task (Figure 9–39). Similarly, asking patients to produce drawings showing the floor plans of their homes generally will be greatly facilitated if they begin with an outline of the perimeter of the building and subsequently putting in the various rooms and hallways rather than beginning with a single room and adding the others as they go.

"Planning errors" also may be seen in other constructional tasks. Figure 9–40 illustrates examples of such errors in clock drawing. As might be expected, it was found that planning errors on this task are more likely to be elicited if the patient is provided with a large clock face with which to work and, with the exception of the "12," is forced to write in the numbers in sequence (Mendoza et al., 1989, 1993). Although not always as obvious as in the case of the Rey figure or clock drawing, planning errors also may be present in block reconstruction-type tasks. While patterns, such as shown in Figure 9–41a & b, are less likely to be affected by planning deficits, clinical experience has suggested that those in which the boundaries of the constituent blocks are not immediately discernable (e.g., Figure 9–41c & d) are more vulnerable to such errors. When presented with the latter patterns, one is more likely to benefit from a preliminary analysis of the design to establish those boundaries. It is this type of preliminary analysis and planning ability that often is compromised in patients with prefrontal lesions. Illustrative of this deficiency is the fact that frontal lesion patients are much more likely than posterior lesion patients to profit from the external cues provided by placing a grid placed over the design as compared with posterior patients (Luria & Tsvetkova, 1964).

Learning and Memory and Planning Ability.   Lesions that are restricted to the dorsolateral prefrontal cortices do not appear to impact on the ability to establish new memory associations per se or the ability to retrieve old memories. This does not mean that frontal lobe lesions have no impact on learning and memory. Deficits in delayed recall following lesions to the dorsolateral cortex in primates (Jacobson, 1936; Oscar-Berman, McNamara, & Freedman, 1991) was one of the earliest and most persistent experimental findings of frontal pathology, and more recently certain experimental procedures using "conditional memory" paradigms were found to be sensitive to such lesions (Petrides, 1985, 1990).[92] Luria (1973, pp. 211–212) also notes that prefrontal lesions can interfere with the learning process in part as a result of an inability to "create stable motives...and to maintain the active effort" to learn.

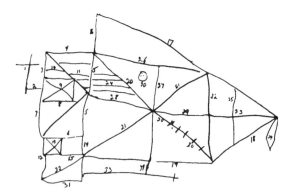

Figure 9–39.   Adopting a piecemeal approach in copying complex designs such as the Rey-Osterreith figure increases the probability of significant distortions of the original design.

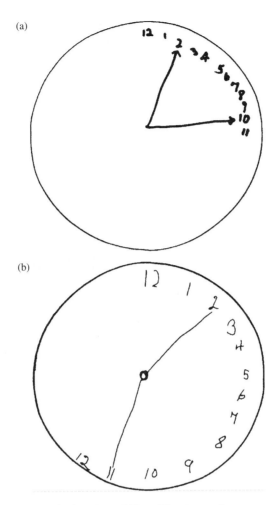

Figure 9–40. Clock drawing and planning ability. The type of error seen in (b) might easily be mistaken for left visual field neglect rather than a planning error. If the patient has adequate self-monitoring capacity, such "planning" errors may be recognized and eventually corrected. Requiring the patient to write the numbers "in order" on a large clock face places a greater demand on spatial planning.

Again clinical observation would suggest that patients with dorsolateral frontal lesions generally perform better when presented with tests that measure logical verbal memory (e.g., paragraphs or stories) than free verbal learning (e.g., word lists), especially those that exceed one's immediate memory span. It is easy to speculate why this might be the case. Paragraphs have a built in structure of their own and do not need to be imposed from without. By contrast, when presented with a long (10 to 15) list of words to remember, such is not the case. If given a list of totally unrelated words to learn, such as the Rey Auditory Verbal Learning Test (RAVLT) (Lezak, 2004), it is impossible to recall all the words after a single trial. However, since these tests typically are administered over several trials, it is possible to improve, but some planning or strategy is essential. As noted previously, one such strategy is to concentrate more intensely on those words not recalled on the previous trial as the words are repeated over subsequent trials. Another strategy would be to actively create meaningful associations among various words. But this takes planning. A variation

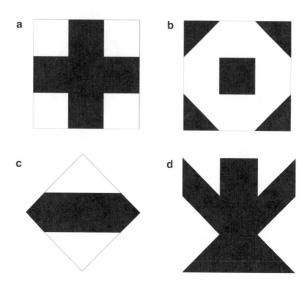

Figure 9–41. Planning ability and solving Kohs' blocks type tests (Kohs, 1919). In figures (a) and (b), one more easily can discern the placement and orientation of the nine individual blocks used in creating the patterns shown, especially (a). In (c) and (d) this is much less clear, requiring more preliminary analysis and planning to be maximally efficient.

of this type of test is the California Verbal Learning Test (CVLT) (Delis et al., 1987). Like the RAVLT, the subject is presented with a long (16) list of words. However, the words in the CVLT can be grouped into four general categories (e.g., tools, spices). Although the patient is not advised of this fact beforehand, if the patient "discovers" this fact and attempts to group the items into these categories prior to recall, performance should be enhanced. Again, this involves the patient having to discover and utilize this strategy (plan) in order to benefit.

The same principle (e.g., focusing more intensely on elements not previously recalled) applies to more complex tests of nonverbal memory in which successive learning trials might be given. An additional consideration would apply when using a highly complex figure like the Rey-Osterreith as a test of memory. As noted above, right hemisphere as well as certain frontal patients are somewhat more likely to take a piecemeal approach in simply copying the design prior to being asked to copy it from memory. Even so, despite using a disorganized approach in copying this design, the patient still may produce a reasonably accurate facsimile, especially if the posterior cortices are intact. However, even if they manage a decent reproduction of the figure, a less-than-systematic approach will make it more difficult to reproduce the design from memory, since "chunking" becomes more difficult. Thus despite producing a "reasonably" representative figure as seen in Figure 9–42a, the number of "elements" to be remembered is greater than the 18 to 20 elements (depending on how one designates individual elements) should a more "organized" approach have been adopted (e.g., starting with the central rectangle and the central vertical, horizontal, and diagonal lines). "Memory" for the design is even more likely to show impairment if the piecemeal approach results in a significant distortion of the original, as seen in 9–42b.

Borrowing and expanding on a term used by Dobbs and Rule (1987), Shimamura, Janowsky, and Squire (1991) define the use of planning, organizational strategies, self-monitoring, and initiation (remembering to remember) in order to facilitate the learning of new information or the recall of previously learned information as *prospective memory*. The use of this term emphasizes the notion that efficient memory is an active and not merely a

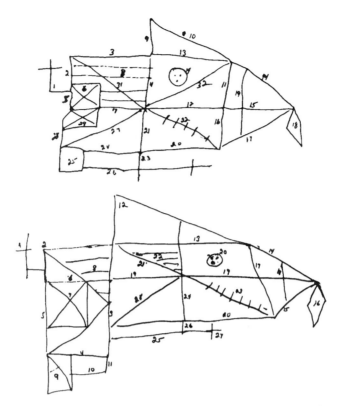

Figure 9–42. Poor planning in copying of the Rey-Osterreith figure can result in weaker "memory scores" either as function of (a) having more individual elements to remember or due to (b) the distortion of the original design.

passive process. While the frontal lobes may not be involved in the encoding of the memory engrams themselves (most likely this is a function of the hippocampal system), the frontal lobes play an important role in setting the stage for this process (Baddeley, 1986).[93]

Arithmetical Word Problems and Planning Ability. Patients with lesions to the prefrontal cortices generally will have little difficulty with arithmetical word problems when the nature of the mathematical operations to be performed is clearly delineated in the way the problem is presented. Consider, for example, the following problem: "How much would 12 tickets cost if each ticket sold for $3.25?" It does not require any sophisticated reasoning to recognize that one simply must multiply $3.25 times 12. While it is possible that a patient with limited arithmetical skills or attention–concentration difficulties might encounter significant difficulties with the above problem, it is unlikely that the difficulty would stem primarily from an analysis and planning error.

However, consider the following problem: "If a man has $4.00 more than his son and together they have $13.00, how much does each have?" Here the problem necessitates preliminary analysis in order to determine ("plan") what operations need to be carried out. Although the calculations themselves in this latter problem are relatively simple, this represents the type of problem for which patients with lesions of the dorsolateral cortex might be expected to show particular difficulty.[94] Most patients, including those with nonfrontal pathology, will simply try to find two numbers that add up to 13 that are separated by "4." They typically will try "5" and "8," or "9" and "4," obviously, neither of which will be correct. In order to efficiently solve this problem, one must first truly analyze it and then devise a plan or strategy for its solution.

Despite its modest difficulty, the above problem still may not be suitable for many patients. However, it illustrates the type of planning that often goes into arithmetical word problems and how frontal lesions might compromise performance on such tasks. Luria (1973, p. 339) offers the following example: "A candle is 15 cm long; the shadow from the candle is 45 cm longer. How many times is the shadow longer than the candle?" Not only does this problem require some forethought ("analysis and planning"), but also one must inhibit the tendency to impulsively respond, "Oh, that's easy, the shadow is three times longer than the candle."

Additional Measures Involving Planning Ability.   In addition to the tests mentioned above, many complex cognitive measures utilize varying degrees of analysis and strategic planning. Disruptions of these processes, while not necessarily precluding the possibility of successfully completing these tasks, may impair the efficiency with which they are executed. Picture Arrangement, one of the subtests from the Wechsler IQ tests, might be a good example of one such task. Success on this particular test requires not only systematically scanning each of the individual pictures to search for relevant cues, but the ability to systematically envision alternate scenarios while simultaneously considering the antecedents and consequences of human behavior. The "tinker toy" test devised by Lezak (1982) also can entail a fair amount of planning and organization. Those who stop and think (plan) what they want to make are expected to do better than those who simply begin to more or less randomly put pieces together. Another nonverbal test that also benefits from developing an organized plan of attack is the Sequin-Goddard formboard (also called the Tactual Performance Test from the Halstead–Reitan battery). One eventually may succeed by more or less randomly trying to fit each block on the board. However, identifying a particular block or shape on the board and then systematically searching for its match (cutout with the same shape) is a much more effective strategy.[95]

Another simple test designed to tap planning ability is shown in Figure 9–43. The task is to generate an itinerary that would involve traveling to all the cities listed visiting one, but only one, city twice. The goal would be to create a trip that would log as *many* miles as possible. Thus the would be to inhibit the natural tendency to draw routes that connect more or less adjacent cities, while traveling back and forth across the country as often as possible.

Figure 9–43.   Map test. A task requiring simple planning ability (see text for explanation.

Verbal Planning Tasks.   Although often more subtle, a variety of verbal tasks also relies on developing and adhering to an organized strategy. Perhaps some of the clearer examples of this are the semantic word generation (fluency) tasks (e.g., animals, items in a grocery store). In animal naming (Randolph et al., 1993), an effective strategy might be to sequentially picture animals in various setting or subcategories (e.g., farm, zoo, African, animal, pets, or birds). For a grocery store, one could take an imaginary trip down the produce, meat and dairy, frozen, canned goods, or condiments isles. Such approaches likely would be more effective than thinking of items at random.

Success with verbal abstractions tends to be correlated with education and intelligence and many might be answered relying on well-learned prior associations. However, a previously unfamiliar pairing (e.g., windmill and river) might require a systematic review of multiple independent associations (a form of planning behavior) prior to discovering a common potential for generating energy. Finally, even certain complex verbal commands, such as "Touch your nose before your chin but after your ear") require forethought, before responding.

Problems of Measurement and Diagnosis.   If "planning ability" indeed is a primary function of the prefrontal lobes, why then is it so difficult to devise a test of planning ability that is selectively sensitive to frontal lobe lesions? The answers are probably multiple, but briefly exploring two of them at this point may be useful in thinking in broader terms about the relationship between "mental functions" and brain localization and their assessment. Perhaps the most challenging test of "planning" and "strategy" that is commonly available is the game of chess. As it is played by those who are truly proficient (the authors not included), it is primarily a game of strategy. Perhaps except for some classic opening gambits, moves must be planned well in advance,based not only on your current position, but also predicated on the moves you believe your opponent will make in response to your own (if these predictions fail to hold, new strategies must then be developed). Unfortunately chess is not practical to use as a clinical measure as it is too complex and difficult. As a result, while it might be a very sensitive test of the dorsolateral prefrontal cortex, it likely would not be very specific to such damage. For example, an individual may have difficulty succeeding because (1) by nature he is reckless and impulsive rather than thoughtful and deliberate, (2) external distractions may cause him to lose sight of the original plan, or (3) he simply may be not too bright and easily can be outmaneuvered by a more astute opponent.

The reason for going into such detail is that many of the tests that are construed as measures of frontal lobe functioning (including planning ability), like chess, tend to be fairly complex or multidimensional in nature. As noted previously, such tasks tend to be sensitive to lesions in multiple areas of the brain (Grafman et al., 1990; King & Snow, 1981) or other psychological processes (Watts et al., 1988).[96] Not all tests of frontal lobe function necessarily are highly complex. Luria, in fact, was noted for devising relatively simple techniques to assess brain functions (Christensen, 1975). However, in the latter case we have the opposite problem: tests that might be more specific, but less sensitive.

In trying to identify tests that are sensitive to deficits in planning (or other symptoms of prefrontal dysfunction), not infrequently one encounters patients with "frontal lesions" who perform normally on "frontal lobe tests" (Shallice & Burgess, 1991; Damasio, Tranel, & Damasio, 1991; Eslinger & Damasio, 1985). This point leads us to other potential explanations as to why certain tests may not appear especially sensitive to frontal pathology. Certainly one hypothesis is that either the test itself is not as demanding as one might hope or the patient is able to figure out an alternative way of solving the problem (perhaps using previous experiences). However, there probably is an even simpler reason. Just as we traditionally may have thought of the parietal or temporal lobes or the basal ganglia in monolithic terms, often so too is the case with the frontal lobes. Although the current tendency is to divide the prefrontal areas into a few anatomical regions (as were done in this chapter), in fact, almost certainly there are many more areas that make specific functional contributions. As is the case with other systems, if lesions spare certain critical areas of the prefrontal cortex, much more so an entire hemisphere, it is possible that many tests may remain relatively unaffected.[97]

While at present we may lack formal precise, practical, and reliable tests that tap planning ability, this does not mean that such problems necessarily are difficult to isolate and identify. What one should keep in mind is that planning ability is not a test construct but rather a process. As we have just seen, evidence of planning and organizational deficits may appear in a variety of tasks, many of which on the surface would seem to be designed to measure quite disparate functions. Understanding how such tasks may be impacted by prefrontal pathology may help illustrate this process at work.

Before proceeding to the next section, once again two caveats are in order. First, it is important to remember that all the tasks described above are multidimensional and as such may present difficulties for individual patients for many different reasons. This may include "psychological" factors, limited intelligence or education, as well as any number of either elementary or complex cognitive and perceptual difficulties resulting from lesions affecting other functional units of the brain. Thus, not every person who has difficulties on these or comparable tasks suffers from lesions of the prefrontal cortex. Such a diagnosis, if made, should come only after a careful review and integration of all available psychometric, historical, and medical data. The above examples are provided only as models for thinking about planning ability and the role of the frontal lobes in this process. Second, it is equally important to keep in mind that not all patients with lesions encroaching on the prefrontal lobes will always show impairment on any or all of these tasks. If critical areas of the frontal lobes are not involved or if the patient can rely on previous experiences or other compensatory mechanisms, he or she may achieve relative success on some of these tasks. Obviously, the more extensive the lesion(s), particularly if the dorsolateral cortices are involved bilaterally, the more likely deficits on these tests will become manifest. However, in such instances, the behavioral history itself should provide strong indicators of frontal pathology, independent of formal assessments.

Self-Monitoring Behavior

One final aspect of the frontal "executive" functions to be discussed is the capacity to spontaneously monitor one's own behavior. Self-monitoring, in this context, refers to one's capacity and propensity to monitor his or her response(s) to a problem or task based largely on the original goal, intervening circumstances, and approximate success in meeting that goal. The purpose of such monitoring is to ensure that:

1. Any preliminary steps or responses are in keeping with or serve to forward one's original plan or goal.
2. The original plan or strategy, as well as the goal itself, remains credible, given either the inadequacy of the response, changes in the nature of the problem, or the current circumstances with which one is faced.
3. The final response is internally consistent with other sources of previous or current information.
4. The action or response, in fact, accomplishes the intended purpose.

This process has been described as a *comparator* function. Thus, one of the roles of the frontal lobes might be to "compare" the results of one's analyses (of the problem), actions, plans, or strategies with behavioral outcomes as they relate to our ultimate goals.[98]

The benefits of such an activity should be obvious. Problem solving or coping with one's environment is necessarily a very dynamic process. Just as we cannot expect that our analysis of a given problem or situation always will be perfect, neither can we expect that our initial plans or strategies will be the best nor that the responses will always be on target or perfectly executed. Independent of whatever action taken, circumstances or environmental demands change, often necessitate a change in plans, strategies, response patterns, or even in the goals (or subgoals) themselves. Constant feedback is required to ensure that our responses and goals stay on course and remain appropriate. Additionally, there always is some degree of uncertainty in most situations. Response B in large part will depend on the result of having executed response A. Whether consciously or subconsciously, we often employ some form of "hypothesis testing" approach to our environment. If our response proves to be inappropriate or lacking in its anticipated consequences, reevaluations are required. On a simpler level, consider what movement would be like in three-dimensional space if we did not have access to visual and proprioceptive feedback from our muscles and joints. In the motor system, this feedback is mediated largely by the dorsal columns (although the anterolateral or spinothalamic and visual pathways also contribute). In the dorsolateral prefrontal system, feedback is obtained via reciprocal connections with the second (and probably parts of the first) functional systems. Of course, all self-monitoring activity also ultimately is predicated on the proper concern or motivation to produce an adequate response.[99]

Observing the concrete results of our actions or responses once they have been executed is certainly one means of self-monitoring. However, it is not the only or even necessarily the best way of carrying out self-monitoring activities. In many situations, once an overt response has been made, the possibilities for correction or compensation are greatly reduced; it is like writing with ink on a piece of fine parchment. A much better procedure might be to compare an anticipated response (plans or strategies) either with the possible long-range consequences of that action, not only vis-a-vis the immediate intended goal, but also its potential effect on other ancillary, unstated, or less immediate goals. In either case, this ability to mentally consider the potential consequences of an action prior to its execution requires another aspect of frontal lobe functioning that was previously introduced, namely, inhibition. Thus, in summary, one might ask

1. "Has my response been properly executed" (or, if intended, what are the odds that it can be?).
2. "Has my response accomplished its intended purpose (i.e., was it correct?) or can I reasonably anticipate that it will be?"
3. "Did my response have any unintended and/or undesirable effects that need to be corrected or might I reasonably expect such unintended effects?"
4. "Does my response (goal, plan, or strategy) need to be changed, either due to its ineffectiveness, inappropriateness, or changes in circumstances?"

Failures in self-monitoring represent frequent, although not necessarily invariable, sequelae of prefrontal lobe damage. When present, they usually cut across a wide variety of tasks or situations (see below for examples). Regardless of their level of severity, their behavioral significance should not be underestimated. The frontal patient who evidences little or no insight into his performance errors (poor self-monitoring of completed responses) might make no spontaneous attempt to correct or compensate for his or her errors or to solicit assistance from others. Inadequate responses not only will be allowed to stand unchallenged or uncorrected, but subsequent behaviors, which might be predicated on fallacious assumptions resulting from the original erroneous response, might only compound the error. Failure to monitor or consider the consequences of responses before they are executed frequently will result in impulsive, inappropriate behaviors.

Although, when present, obvious deficits in self-monitoring behaviors are frequently considered pathognomonic of frontal lobe disease, as with the other signs suggestive of frontal pathology some caution in making this interpretation is advised. Other factors could be contributing to the observed behavior. Going back to the analogy of the relationship between muscular contractions and proprioception, movement disturbances might result either from lesions that affect the motor system directly or the sensory feedback loops. Similarly, in order to effectively carry out self-monitoring activities, the frontal lobe is very much dependent on the posterior cortices to provide accurate sensory feedback, and the limbic system and the first functional unit to provide access to an ongoing memory of events and experiences, optimal arousal, and sustained drive or motivation. Patients who suffer some compromise of these latter systems, their connections with the frontal lobes, or a generalized deterioration of the brain (as in Alzheimer's disease) might be expected to show either limited or more generalized compromise of these essential feedback loops. For this reason, sometimes it is useful to speak of disturbances of **frontal systems** rather than lesions of the frontal lobes without clear, indisputable evidence of such pathology.

The following illustrate some instances in which deficits in self-monitoring ability might represent contributing factors in common neuropsychological measures of mental status:

Self-Monitoring Errors and Memory.   Some of the "memory difficulties" exhibited by frontally impaired patients might reflect self-monitoring errors. For example, if when asked to provide the name of the current president or the current year, the patient responds with an answer that clearly is out of date, or when asked what he did the previous night (which was spent in the hospital) the patient responds with a story about attending a retirement party for a friend, deficits in self-monitoring should be suspected. Similarly, problems with self-monitoring ability might be reflected in situations where, when asked to recall a long list of words or a paragraph, the patient might repeat items from previous tasks (perseverations) or simply include numerous non-list words or ideas or images that never were present in the original story. These latter failures, which may be classified as *contaminations* or *confabulations*, might result from a failure to inhibit prepotent (or otherwise inappropriate) responses, combined with a subsequent failure to adequately self-monitor (Mercer et al., 1977; Shapiro et al., 1981). Similar perseverations or intrusion errors also might be seen in nonverbal memory tasks (Vilkki, 1989).

As noted above, when attempting to learn a large amount of material over repeated trials, a good strategy is to selectively attend or covertly rehearse those items or elements not recalled on the previous trial. Such a strategy, however, is predicated on awareness of and attention to the adequacy of the previous response (self-monitoring).

When testing memory for complex geometric designs, one occasionally may want to "test the limits" of a patient who shows significant immediate retention deficiencies by affording multiple learning trials. Such a patient might be given three or four additional trials with timed exposures (e.g., 10 to 15 seconds) to the stimulus, resulting in very limited if any increased retention. Under such circumstances, the author then might inform the patient that he or she will be allowed to "look at the design for as long as you want." Recognizing their difficulty in learning the task, most patients will continue to look at the design as long as permitted (it is usually withdrawn after 60 to 75 seconds). However, others will indicate they have "got it" after only 10 seconds or less (clearly they did not). Where lack of cooperation does not seem to have been an issue, the latter reasonably might be said to have had some type of self-monitoring deficiency.

In addition to the potential for failing to detect or correct performance errors on tests of memory, patients with frontal lobe pathology may fail to appreciate the presence of significant memory deficits per se. Frontal damage itself is not usually associated with severe memory disorders (i.e., general amnestic syndromes). However, in the presence of additional frontal findings, such as might be seen following anterior communicating artery aneurysms or Korsakoff's syndrome patients are more likely to either fail to appreciate or substantially minimize the severity of their memory disturbances (for review, see Schacter, 1991).

Self-Monitoring Errors and Visual–Spatial Constructions. Although major difficulties on visual–spatial constructional tasks generally are associated with lesions of the posterior cortices, some constructional or other visual–perceptual tasks may be compromised following damage to the prefrontal regions. As noted earlier, such errors may result, for example, from impaired planning ability, becoming stimulus-bound, or perseveration. However, regardless of whether the basic deficit is parietal or frontal, it is easy to see how deficits in self-monitoring ability might contribute to either the initial production of such errors or the failure to recognize and/or spontaneously self-correct errors once they have been made. For example, as seen in Figures 9–44 and 9–45, errors may result from the failure to compare one's performance with either external (Kohs' block construction) or internal (clock drawing) models. In attempting block construction, part of the planning and executive process involves comparing the end product of one's response to the original goal (in this case, the external model) to determine whether the two are identical and if not making an attempt to correct the response. In the case of clock drawing there typically is no external model; here the patient must rely on an internal model (perceptual memory of a clock face), but the self-monitoring process otherwise is the same.

Clearly, not all failures to recognize or correct errors necessarily reflect a deficiency in self-monitoring ability. Patients with severe perceptual difficulties (especially, unilateral visual neglect) may be hampered in their ability to identify performance errors. This often will be true also of patients with generalized dementia, severe attention deficits, or as noted earlier even patients with certain emotional or psychiatric disorders. Differential diagnosis largely is based on clinical judgment and observation after having taken into account the patient's performance across multiple tasks and situations.

Self-Monitoring Errors and Other Higher Cognitive Tasks. Obviously, breakdowns in self-monitoring capacities can influence performance and become manifest in any of a variety of cognitive tasks or situations. Take, for example, the arithmetical word problem presented

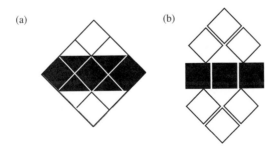

(a)   (b)

Figure 9–44. Illustration of the type of self-monitoring error that can occur in a Kohs' block design type task. While more or less reproducing the general internal detail of the design, there can be a failure to appreciate the external (3 × 3) configuration of the design and to recognize the discrepancy between the model (a) and the reproduction (b).

(a)

(b)

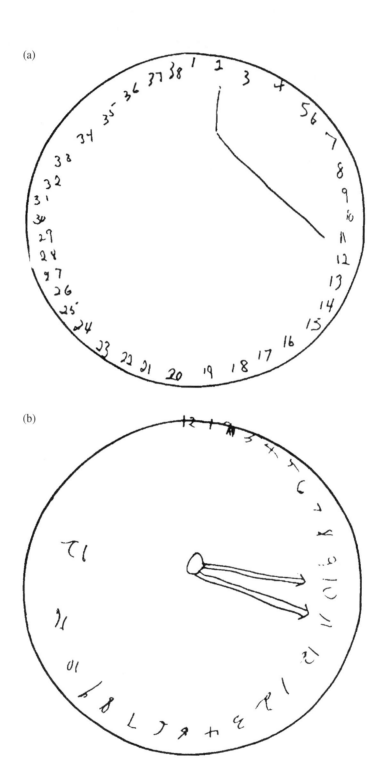

(c)

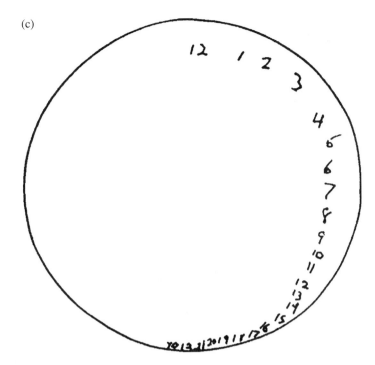

Figure 9–45. Errors in self-monitoring in clock drawing. Superfluous numbers are seen in all three clocks, although none were trying to produce a 24-hour military clock (as is occasionally seen). In (a) hand lengths are reversed and lack a central pivot point. In (b), the hands are not set at "10 after 11" as requested. In (c) there was no attempt to set the time. None of the patients appreciated their errors.

earlier: "If a man has $4.00 more than his son, and together they have $13.00, how much does each have?" While it is not unusual for patients to make a preliminary guess, perhaps coming up with either "eight dollars and five dollars," or "nine dollars and four dollars" (each of which add up to $13.00), most patients usually discover their error when they check their response against the original problem. They then discover that neither response fulfills the original condition, namely that there needs to be a $4 difference between the two amounts. While they still may be unable to arrive at the correct solution, at least they recognize that their initial responses are incorrect. Frontal patients with deficits in self-monitoring ability may never take this second step to discover their error.

The following represent what would appear to be additional examples of this phenomenon. In administering the Information subtest from the WAIS-R to a patient in Chicago who was suspected of experiencing frontal degeneration, he was asked to make a guess as to the current population of the United States. His response was "two million." He then was asked to guess the current population of the Chicago metropolitan area at the time. He correctly estimated that it was "approximately five million." After being told that his latter response was correct, he was then asked to make another estimate of the population of the United States. He again said "two million." Going through the above exercise a second time produced the same results. While his second response (regarding the US population) might be interpreted as simply perseveration, in effect he failed to compare the correct information he gave for Chicago to his response for the country as a whole.

Another Chicago patient with a frontal tumor was being seen in the examiner's office. It was winter and a steam heater that was directly under the window was making a slight "ticking" noise. The patient remarked that "it must be raining," mistaking the sound being made by the steam heater for raindrops. However, this window, which was only a few feet away and directly in front of the patient, had its curtains fully extended, clearly exposing a perfectly beautiful, blue, cloudless sky. Here the patient was drawing a conclusion based on a single source of sensory input (auditory), totally ignoring a second source of information (visual) that was readily available to him.

A third example was a patient who had suffered significant frontal lobe injuries as a result of a head trauma. This was a patient whom I had known for some time and with whom I often conversed. One day, on noticing a picture of my child on the wall, he asked how long I had been married. Jokingly, I replied "59 years." Without missing a beat, he inquired as to what we had done to celebrate our 50th wedding anniversary. At this point I was still in my early forties and despite my graying hair I dare say I would not normally have been taken to be 30 years older than my actual age. (Even when this obvious discrepancy was brought to his attention, he still was unable to reconcile the facts with which he was presented with his own observations. It never occurred to him that I might have been joking about being married for 59 years).[100]

In all three cases, all of which involved patients with evidence of frontal lobe damage, the patients failed to "compare" their responses or perceptions with other, obviously contradictory information immediately available to them. This was true whether such information could be gleaned from their senses (second and third cases) or from their own knowledge base (first case).

Deficits in Self-Monitoring and Executive Abilities: Concluding Thoughts. While this section has focused on the limited ability of certain frontally impaired patients to monitor the appropriateness or adequacy of specific responses, particularly in regard to whether they satisfy immediate goals or needs, self-monitoring also can be viewed in the broader context of self-awareness or insight. In this context, we might consider both awareness of deficit and awareness of the impact of one's behavior on others. Frontal lobe patients, especially those with more significant lesions, frequently are deficient in both these areas.[101] With regard to awareness of deficit, the problem typically is not so much one of failing to recognize or appreciate physical handicaps or limitations (anosognosia) as much a failure to appreciate cognitive changes or limitations. In addition, although frontal lobe patients frequently manifest marked personality and behavioral changes (Blumer & Benson, 1975; Lishman, 1973), they commonly fail to acknowledge the pathological nature of these changes or to recognize the negative impact these changes may have on others. As a result of this lack of insight, combined with their frequent lack of drive or motivation (points that will be addressed in greater detail in the following section), patients with significant frontal lobe impairment often have marked difficulty in social and occupational rehabilitation (Prigatano, 1991).[102]

Stuss & Benson (1986, pp. 244–249) emphasize the preeminence of self-awareness and executive ability in the hierarchy of human behavior in general and of the function of the frontal lobes in particular. Although for didactic purposes we may separate out individual aspects of these more global behavioral concepts, such as attention, intention, planning and goal-selection, inhibition, and self-monitoring ability, we should be aware that in practice they often are difficult to dissociate. It is as if these behavioral elements are like the individual threads woven into a complex tapestry. Even though at times it may be possible to isolate the separate components, their full richness and meaning can be appreciated only when observing the behavior as a whole.

Thus far we have focused mainly on self-awareness in relation to the executive aspects of human behavior. We have just seen how a failure to compare the appropriateness of a response to determine whether it served its intended purpose and/or was compatible with previously selected internal or external goals or criteria indeed is central to disorders of planning and execution. It was further noted that such disturbances typically were associated with lesions of the dorsolateral convexities of the frontal lobes. Next, we will explore a very different type of problem with self-awareness, one that also may lead to deficits in self-monitoring capacity, although for very different reasons. As alluded to previously, the deficits to be discussed in the following section are generally derived from a lack of genuine concern about specific future or long-range consequences and/or generalized apathy. Typically associated with orbitofrontal or mesial frontal pathology, as will be seen these latter deficits often appear to result from a disconnection between limbic structures and the dorsolateral frontal zones.

The Modulation of Internal Drives and Emotions

Differentiation from "Executive" Functions

In reviewing the role of the prefrontal cortex, the focus thus far has been on the more "executive" aspects of behavior. The preceding discussions have included such concepts as directed attention; freedom from distractibility; mental flexibility; planning ability; and the propensity to monitor and self-correct responses, particularly as they apply to cognitive-type tasks. However, well before these executive behavioral functions were correlated with frontal pathology, clinical case studies revealed that lesions of the frontal lobes were associated with more general disturbances of "personality," emotion, and drive states. Whereas "executive" functions came to be linked to lesions of the dorsolateral cortices, destruction of the orbital and mesial aspects of the frontal lobe more often were related to disruptions of "social" behavior. Such behavioral disturbances might range from apathy or depression to mania, restlessness, and/or (social) disinhibition; from increased agitation or irritability to a childlike euphoria; or from social withdrawal to sociopathy (Blumer & Benson, 1975; Hecaen, 1964; Holmes, 1931; Freeman & Watts, 1942; Greenblatt, Arnot, & Solomon, 1950; Levin, Benton & Grossman, 1982; Lishman, 1968; Malloy & Duffy, 1994; Petrie, 1952; Rylander, 1948; Tow, 1955). As we will see, most of these disturbances seem to be related to failures in properly accessing or modulating basic drive states or emotional impulses. As such, the lesions that result in these disturbances generally are thought to result from either (1) a disconnection syndrome between limbic/hypothalamic structures and the dorsolateral frontal zones, (2) disturbances of limbic structures directly or those deep mesial or ventral cortical zones with which they interface, or (3) a disruption of those neurochemical pathways that regulate the activities of the frontal zones.

There is an extensive literature on the subject of personality and behavioral change associated with damage to frontal systems. For the purposes of this chapter, however, the majority of these findings can be subsumed under two general rubrics: the effective utilization of drive states (arousal) and the inhibition of competing drive states or emotions (social inhibition). In the remainder of this section we will explore the probable role of the prefrontal cortex in the modulation of these functions, some of their probable anatomical substrates, and the behavioral effects produced by lesions to these areas.

Utilization of Drive States (Arousal)

The Role of the Reticular Activating System. Typically we think of the first functional unit, specifically the ascending reticular activating system (ARAS) through its thalamic connections, as being initially responsible for diffuse cortical activation (arousal) (Moruzzi & Magoun, 1949). However, for our present purposes, it is important to differentiate among (1) general cortical activation or arousal, (2) specific or selective cortical arousal, and (3) drive states. The brainstem ARAS would appear to be concerned with alerting the cortex in a general manner, either informing the cortex that "there is something happening out there that may be important, pay attention," or "there are some needs that have to be met, let's get on the stick and do something about it." While the level or intensity of this alerting response may vary depending on the situation, qualitatively (in terms of telling the cortex what to attend to) its influence is likely quite limited. This alerting or arousal of the cortex by the ARAS probably is mediated both via its influence on the intralaminar nuclei of the thalamus, as well as through diffuse neurochemical pathways (e.g., norepinephrine).

How does the ARAS know when and to what degree the cortex needs to be alerted or aroused? As previously mentioned, there probably are three basic mechanisms at work. The first, is external stimulation. Most, if not all, sensory receptors have fairly direct connections with or provide input to the brainstem. Thus, pain, simple touch, sound, light, or movement (vision), or even the smell of coffee and bacon in the morning depending on the circumstances may have strong arousal value.[103] Internal biological needs also may trigger a response. Thus, the cortex can be aroused during states of hunger, thirst, oxygen depletion, or other homeostatic deviations that are essential to the basic biological integrity of the organism. While the initiation of such responses may come directly from somatic receptors of the autonomic nervous system, they are likely supported and reinforced by hypothalamic connections (e.g., the medial forebrain bundle, the dorsal longitudinal fasciculus, or the mammillotegmental tract).

However, it also is likely that the cortex itself, particularly the prefrontal lobes, through feedback loops itself can have an activating influence on the ARAS, helping to maintain optimal levels of arousal under given conditions. Consider, for example, the following situations. One is walking or traveling through a questionable part of a large city late at night. He might be thinking, "While I don't see anything to be afraid of, I know that the circumstances are potentially dangerous and I need to be on guard." Or, take the case of a college student after a night on the town with perhaps a few too many beers, who suddenly realizes that she is ill-prepared for a test the next day. Despite the sedating influence of the alcohol, she manages to arouse herself sufficiently to hit the books for an hour (the influence of more abstract or long-range goals in generating arousal). One would strongly suspect that in these two latter instances the frontal lobes probably play a significant role in generating cortically initiated activation of the ARAS, which in turn may activate the cortex in general. Among the pathways potentially involved in such a situation again are the medial forebrain bundle, connecting the basal forebrain with the brain stem, as well as the stria medullaris of the thalamus and the habenulointerpeduncular tract. Combined, these provide alternate routes to the midbrain from the basal and mesial frontal areas.

The Role of Frontal–Thalamic Connections in Selective Attention or Arousal.   At any one point in time, our external senses constantly are being bombarded by a multiplicity of stimuli. Numerous internal thoughts or recollections also may be vying for our attention. There has to be some central gating mechanism that serves to control or direct attention, basically telling us what is important to attend to under the circumstances. As noted earlier, the prefrontal lobes likely play a key role in this process, especially with regard to more purely "cognitive" type tasks, and disruption of this system is likely to result in increased distractibility, neglect, or difficulty shifting response set (mental flexibility), or selective or sustained attention (concentration). However, rational considerations are not the only factors that determine the focus of our attention. Certainly affective drive states (motivation), which are more closely linked with the orbital and mesial portions of the prefrontal lobes, also help shape the direction of attentional processes. Hence, either directly or indirectly, the orbital and mesial frontal areas likely participate in selective attention.

While the anatomical mechanisms that might mediate selective attention are not clear, at least three possibilities readily come to mind. As suggested earlier, one is the possibility that the frontal ("executive") cortex might influence directly the posterior ("gnostic") cortices via corticocortical association pathways. The second and third possibilities both involve frontal–cortical influences on thalamocortical projections. Theoretically, this might occur in one of two ways. First, recall that the frontal association cortex projects heavily to the head of the caudate, which in turn projects to the thalamus. While there is a tendency to see these corticostriatal–thalamocortical loops as being "closed" (i.e., projecting back to the same general areas from which they originated), there also are thought to be "open" circuits that allow for much broader influences. Mega and Cummings (1994) postulate that these circuits are crucial in all aspects of behavior, including selective attention. Finally, the prefrontal lobes have direct connections with the thalamus. Although these connections primarily are to the dorsomedial nucleus, which in turn projects back to the frontal cortices, all corticothalamic fibers pass through the reticular nuclei of the thalamus, which basically surrounds the lateral surface of the dorsal thalamus. Since these reticular nuclei subsequently project back to the other thalamic nuclei (the only thalamic nucleus whose efferent connections remain within the thalamus), the input from the frontal cortex has the potential for influencing vast cortical regions via the various thalamic relay and association nuclei (see Mesulam, 2000b, Chapter 3).

The Role of Frontal—Limbic Connections in Drive and Motivation.   In Chapter 8 the role of various limbic structures (including the anterior cingulate gyrus, the amygdala, the hypothalamus, and possibly the septal area) in biological drives and emotions were extensively discussed. Also, as previously suggested, such drives or emotions provide the impetus, fuel, or energy to initiate and sustain all types of goal-directed activities. From a philosophical perspective, we can view all behavior as a combination of sensation, intellect, and will. For present purposes, *will* can be defined simply as the intention, desire, or motivation to initiate and/or sustain an action to obtain a desired goal (i.e., the result of operant, as opposed to classical, conditioning). The goals in question may be abstract (to be a "good person") or concrete (to satisfy one's thirst), immediate (to flee from imminent danger) or remote (to study hard in order to graduate and secure a good paying job several years from now). Again, as noted in Chapter 8, there is good reason to believe that "will" or "drive" derives from limbic structures. So, what is the role of the frontal cortex in this process?

In simpler organisms there is little need for sophisticated frontal systems. Many, if not most, behaviors are the product of either innate preprogramming ("fixed action patterns") or simple operant conditioning. As organisms became more complex, however, so too did the behavioral responses and/or the goals from which they were able to choose. It seems reasonable then to assume that frontal systems may have developed for two basic purposes: (1) to better meet the challenge of choosing, planning, and executing increasingly complex or varied response options in order to secure particular behavioral goals, and (2) to make more efficient choices when confronted with competing drives or goals (see below). Thus, while the prefrontal systems probably are not themselves the source of the basic physiological or "psychological" drives or "will," they harness and utilize these drive states to determine if, when, and how to best respond to a given situation and to persist in those efforts despite potential hardships or difficulties.

Although the exact nature of this interface between the prefrontal cortices and the energizing influence of the limbic structures is still a matter of some speculation, there is evidence primarily from lesion data to suggest that mesial–frontal–limbic connections are crucial to this process. Again, from the discussions in Chapter 8, you may recall that basic homeostatic drives (such as hunger), innate or learned fear responses, as well as generalized emotional responsiveness can be altered by lesions that directly affect the hypothalamus, amygdala, and anterior cingulate regions, respectively. This finding should not be surprising if, as suggested, limbic structures primarily are involved in generating these affective drive states.

Depressive-type States. It also has long been observed in both animal and human subjects that lesions of the prefrontal regions, some of which also likely involve the anterior cingulate gyrus, can result in apathy and hypokinesis, depending on the severity of the lesions (Blumer & Benson, 1975; Freeman & Watts, 1942; Fuster, 1997; Greenblatt et al., 1950; Hecaen, 1964; Luria, 1966; Rylander, 1948; Tow, 1955). While such responses have been reported in lesions (usually bilateral) involving the dorsolateral frontal cortices, mesial lesions appear more likely to result in diminished emotional responsiveness. Thus, it may well be that we are dealing with a diffusely projecting, interactive motivational system that encompasses limbic, mesial, and dorsolateral frontal areas and is essential for sustained, goal-directed executive activity.

Because of the nature and pattern of the affective changes that often are found following major insults to the frontal lobes, particularly the mesial and dorsolateral regions, the term *pseudodepression* has been applied to these behavioral states (Blumer & Benson, 1975). In certain frontal lobe patients, the appellation certainly seems apropos. Spontaneous emotional expression and individual initiative may be markedly diminished. While such patients may render lip service to carrying out work-related tasks or accomplishing long-term goals, in fact, they may do very little on a day-to-day basis, even to the point of appearing hypokinetic on occasion. They may show little interest or pleasure in former pursuits (*anhedonia*) and demonstrate reduced inclination to attend to their own personal hygiene. Hence, such patients often look and act "depressed," although as a rule they do not display dysphoric mood or affect or express feelings of worthlessness, negative preoccupations, or suicidal ideations that more frequently characterize typical "clinical" (psychiatric) depressions.

However, in considering this general behavioral syndrome, several qualifications are in order. First, depending on the extent and nature of the deficits, the clinical picture may show considerable variation. Often the individual may not evidence the level of inertia described above, but rather demonstrate subtler dysfunction. Thus, the patient may attend school or return to work, but with less intensity, drive, or enthusiasm than before. Projects may be initiated but either not carried through to completion or completed in a less conscientious manner. At times, evidence of mild to moderate apathy may be intermixed with emotional lability and irritability. Finally, while certain drive states (especially those associated with more abstract or long-term goals) may appear reduced, other more basic biological drives (e.g., sexual) may appear to be enhanced.[104]

Certainly, one also must consider the possibility that a depressive-type syndrome may result from the "psychological" reaction of having sustained a brain injury, perhaps with some related loss of function and/or change in psychosocial status. However, this explanation cannot account for many of the behavioral changes found after frontal lesions. In addition to the fact that behaviorally many of these individuals do not resemble "psychiatrically depressed" patients, hemispheric differences in the expression of depressive-type syndromes have been noted, specifically a greater incidence in "depressive" symptomatology following left frontal lesions (Gainotti, 1972; Eastwood, Rifat, Nobbs, & Ruderman, 1989; Robinson & Price, 1982). While one possible explanation for depression following left frontal lesions is the co-occurrence of Broca's aphasia (Robinson & Benson, 1981), depressive conditions have not only must been found in the absence of Broca's aphasia, but have been positively associated with lesions of the more anterior portions of the left frontal lobe that are less likely to produce major expressive disturbances (Robinson et al., 1984; Robinson & Szetela, 1981; Sinyour et al., 1986). Finally, left subcortical lesions that affect structures directly associated with frontal systems (e.g., anterior limb of the internal capsule and the head of the caudate nucleus) also have been associated with increased incidence of depression in stroke patients (Starkstein, Robinson, Price, 1987; Starkstein et al., 1988). In the final analysis, there would appear to be at least two possible neuroanatomical explanations for these findings of increased apathy and/or depression following lesions to the frontal lobes. One is, as suggested above, a disconnection between the frontal executive system and critical limbic structures that are critical in initiating drive and emotion. The other possibility (and the two are not necessarily mutually exclusive) is that certain frontal lesions may disrupt ascending noradrenergic and serotonergic brainstem pathways (Starkstein & Robinson, 1989).[105]

Inhibition of Competing Drive States or Emotions (Social Inhibition)

In the preceding discussions, mention was made of the difficulties patients with frontal lobe lesions might encounter when faced with distracting but irrelevant stimuli or prepotent response sets. Because of failures of inhibition in these situations, patients may experience difficulty in staying on task or fall into perseverative response patterns. Much more problematic for certain frontal lobe patients, or at least for those around them, are difficulties in inhibiting affective or emotional responses or emotional drive states. It is not clear whether these two problems represent an overall problem with inhibition in general following frontal lobe lesions or whether they reflect fairly independent processes. The fact that phenomenologically they are frequently also dissociated might argue for at least partially independent mechanisms. While we do not intend to belabor this point, what is important to note is that behaviorally and possibly anatomically affective or emotional disinhibition presents as a very different type of problem from the inhibition of cognitive response sets previously discussed.

The failure to inhibit affectively or emotionally charged behavioral responses in situations where it normally would be appropriate to do so represents one of the classic syndromes associated with frontal lobe pathology. In reviewing this behavior, it first might be noted that there is obvious teleological value to affective drive states and the behaviors with which they tend to be associated. Such drive states often are essential either for the propagation of the species or important for the well-being or survival of the individual. Examples of affectively driven behavior might include sex, hunger and thirst, fear, anger, or aggression. Perhaps even humor and criticism have evolved as means of social communication and jockeying for social position. However, more often than not, the organism is faced not with a single goal or drive but with multiple and often competing or mutually exclusive drives. In humans at least many of these drives reflect more abstract or long-range needs rather than more immediate, biological urges. Thus, as humans and society have evolved, many behavioral expressions, especially those based on what might be described as more "primitive" emotional drive states are not given indiscriminate free rein, but rather are often subjugated to commonly accepted systems of internal checks and balances (the process we know as *acculturation*). As previously noted, this latter process would appear to be mediated primarily if not exclusively by the frontal lobe as part of its planning and self-monitoring capacities. Among the operations carried out by the frontal lobes prior to acting on various drives or impulses are the following considerations:

1. Is this the *time* and *place* to respond (to this impulse/emotion)?
2. What might be the *long-range consequences* of such a response?
3. If negative consequences occur, should I consider an alternate response or delay responding for a better time, place, or circumstance?
4. Is responding to this drive or impulse consistent with my other immediate or long-range goals?
5. What impact might this action have on others (empathy, compassion)?

Pseudopsychopathy. Patients with frontal lobe damage, particularly when the damage affects the orbitofrontal regions, often act as if they fail to weigh these considerations prior to acting. As a result, they are seen as disinhibited, indiscreet, impulsive, or lacking social judgment. Blumer and Benson (1975) characterized this behavioral syndrome as *pseudopsy-chopathic*, indicating the apparent disregard for social convention or future consequences. Also, as noted earlier, such patients often exhibit a tendency to behave in a childish, petulant manner and may be given to blunt, inconsiderate remarks or inappropriate humor (*Witzel-sucht*). A classic example of such behavior was the case of Phineas Gage, a case that is still cited in many modern psychiatric and neurobehavioral texts and still elicits scientific curiosity (Damasio et al., 1994). Briefly, Phineas Gage reportedly was a somewhat reserved, pleasant, hard-working man prior to having a tamping iron driven through his face and skull, destroying a substantial portion of his frontal lobes. The following is a brief quote from the description of his case at the time (Harlow, 1868):

The equilibrium or balance, so to speak, between his intellectual faculty and animal propensities, seemed to have been destroyed. He is fitful, irreverent, indulging at times in the greatest profanity (which was not previously his custom), manifesting but little deference for his fellows, impatient of restraint or advice when it conflicts with his desires, at times pertinaciously obstinate, yet capricious and vacillating, devising many plans of operation, which are no sooner arranged than they are abandoned in turn for others appearing more feasible. A child in his intellectual capacity and manifestations, he has the animal passions of a strong man (cited from Blumer & Benson, 1975, p. 153).[106]

The problem in such individuals may not necessarily lie in their cognitive ability to make accurate judgments about the potential consequences of their behavior. Rather, the difficulty may be in the patient's tendency either (1) not to be concerned about making such judgments in the first place, or even if it were possible to make these rational deliberations on a cognitive level, (2) to act as if these deliberations or the consequences of the actions were immaterial. In practice, the cognition of patients who exhibit serious problems with impulse control frequently can be demonstrated to be relatively intact (see Eslinger & Damasio, 1985), as are their emotional drives or affective responsiveness. The problem in the case of certain orbitofrontal-lesioned patients is that these two processes seem be disconnected, most likely as a result of a disruption of frontolimbic pathways. Davidson, Putnam, and Larson (2000) review evidence suggesting that disturbances in the orbitofrontal cortex and other prefrontal–limbic connections may be linked to aggressive behaviors, especially more impulsive (as opposed to premeditated, planned) aggressive outbursts.

From a psychological perspective, these fronto–limbic connections might be viewed as bridging those processes that, in psychoanalytic jargon, are connoted by the *id* and the *superego*. Under this model, the role of the frontal lobes ("superego") is to temper the limbic impulses ("id"). When lesions interfere with this process, either by severing or disrupting these connections or by damaging areas of the frontal cortex directly, there is a loss or disruption of these mediating influences. Consequently, what might be seen in such patients is poor judgment, inability to delay gratification or impulsivity, social inappropriateness, affective lability or disinhibition, restlessness, and/or "unbridled euphoria." For a review of various psychiatric conditions associated with frontal lobe pathology, see Malloy and Duffy (1994).

From a neurobehavioral standpoint, we can couch these phenomena in slightly different terms. For example, we might say poor judgment typically results from a failure (absolute or relative) to anticipate or consider possible negative or long-term consequences or learned social prohibitions or to compare probable outcomes with personal goals or interests. For this process to be effective, there must be some way for the various behavioral alternatives faced by the patient to be directly linked to the anticipated emotional consequences of each potential behavioral response. Thus, in considering potentially available responses, it would appear to be quite advantageous (as a part of planning and self-monitoring functions) to vicariously experience the possible emotional ramifications of those choices. For this to happen, the various "mental images" and/or past memories associated with these responses need to be "linked to" relevant "emotional tags." Damasio and his colleagues (Damasio, Tranel, & Damasio, 1990, 1991; Bechara, Tranel, Damasio, & Damasio, 1996) refer to this process as the activation of "somatic markers." Anatomically, it is thought that this process likely minimally involves connections among the orbitofrontal cortices, the ventroanterior portions of the basal ganglia (e.g., nucleus accumbens, ventral pallidum), the anterior and medial temporal regions (including the amygdala), possibly the hypothalamic areas, and the dorsomedial nuclei of the thalamus. For a review, see Chapter 6, section Lateral Orbitofrontal Circuits and Limbic Circuits (this volume) (Mega & Cummings, 1994; Davidson, Putnam, & Larson, 2000).

Additional Clinical Considerations. As noted, impairment of judgment or impulsivity in patients with frontal lobe lesions most often involves the inability to inhibit inappropriate impulses or perhaps more properly impulses that given the time, place, or circumstance in which they occur would be inappropriate to express. Again, we should keep in mind that this is not merely a cognitive exercise or simply an impairment of "cognitive judgment." For example, the patient with frontal injuries may respond perfectly appropriately to the question "What one should do if, while in a theater, they are the first to see smoke or fire?" When posed during the course of a mental status examination, there is no strong affective pull to respond otherwise. On the other hand, if that person were actually in that situation (e.g., where there was a real fire), his or her response might be quite different.

This raises another important question. If a person placed in the situation described above were to yell, "Fire," or simply run out of the theater without taking the time to warn or notify anyone, does that mean they are likely suffering from frontal lobe damage? It is important to remember that the world is full of people who routinely tell inappropriate jokes, make "thoughtless" but hurtful or embarrassing comments and observations about others in their presence; who constantly act in ways that would suggest they fail to consider the long-range consequences of their behavior; who frequently act in an impulsive, often self-destructive manner; or who otherwise disregard social convention or fail to use "good judgment" when it comes to modulating or curbing their emotional propensities. For example, are the individuals who habitually blow most of their paychecks on drugs, liquor, gambling, sex, jewelry, shoes (or whatever) suffering from frontal lobe damage? Certainly the possibility of a frontal type of learning disability readily comes to mind in trying to explain the behavior of many of these individuals.[107] Obviously, however, many other learning and environmental factors also may help explain such behavior. What is important from a clinical or diagnostic standpoint is to note whether these behaviors reflect a change from the patient's normal personality or response pattern.

It should be noted that poor judgment or behavioral improprieties do not necessarily reflect an all-or-none situation. The extent or likelihood that someone with frontal lobe damage will act out in an inappropriate manner will depend on a variety of factors. First, and perhaps most importantly, what areas of the frontal lobes or pathways are affected? As a rule, such behavioral problems are more likely to occur following basal–medial or orbitofrontal lesions, particularly if they are bilateral. However, lesions in this area do not always produce such deficits and lesions in other areas also may be associated with problems of judgment and impulsivity. The age of the patient also may be important. A frontal lesion in an adolescent or young adult is more likely to result in behavioral disturbances than a comparable lesion in an older adult. It is suspected that the normal impetuosity and higher energy levels of youth may have an additive effect in potentiating such acting out. Circumstances may also dictate to some extent the expression of impulsive behavior. For example, the patient who needs to relieve his bladder or bowels generally will do so in an appropriate place and manner, especially when it is convenient to do so. However, if such a place is not convenient, then the patient may not go to any great pains to delay gratification he will not be deterred by the anticipation of social consequences, choosing instead to relieve himself in whatever place is handy (*disinhibition*).

Another consequence of the reduction of cortical control mechanisms is that the patient is more at the whim of his or her immediate drives or affective states, rendering the individual more emotionally labile, restless, and perhaps appearing no longer "in control." Finally, because the frontal–limbic connections discussed above represent a two-way street, disruptions of these connections also may help explain why certain cognitive considerations are less able to take on appropriate affective coloration. We already have discussed one possible consequence of this, namely apathy or reduced drive or initiative. Another possible consequence is that without the normal inhibitory influences of the frontal lobes (which can, in turn, give rise to more somber considerations), there may be an imbalance of affective or drive states. Such a formulation might help explain the childlike euphoria that often is seen in frontal lobe-damaged patients.

Other Specific Prefrontal Pathology and Frontal Dementias. The prefrontal cortex and related structures, like all parts of the brain, potentially are subject to a wide variety of pathological processes. These can range anywhere from small, focal lesions to relatively large, diffuse, or multifocal processes. While the potential etiology of such lesions essentially are similar to that which might be found in other parts of the brain (e.g., vascular accidents, tumors, penetrating and closed head injuries, infections, metabolic, demyelinating and degenerative disorders), certain pathologies have an increased predilection for the frontal lobes. Deceleration head trauma and certain type of degenerative diseases are common examples. As is true elsewhere in the brain, the exact nature and extent of the resulting behavioral disturbances following such lesions typically depends on a host of factors. This would include, for example, the size, extent, and exact location of the lesion, its specific etiology or pathological process, and the nature of onset and/or progression. At times, even the age and sex of the individual could be a major factor in the clinical manifestation of the disorder. While it is beyond the scope of this chapter to review all the potential types of lesions or disease processes, it might be useful to mention a few specific dementia syndromes that commonly affect the frontal lobes.

When we think of dementia, typically the first disease process that comes to mind is Alzheimer's. This is understandable since it appears to be the most common form of the degenerative processes and of dementias as a whole. While it can and does affect the frontal lobes, in the earlier stages it tends to have its most profound effects on the parietal and temporal lobes and in particular the hippocampus (Cummings & Benson, 1992). However, there are other degenerative disorders that tend to primarily affect the prefrontal cortices from the very beginning. While generally referred to as **frontal lobe degeneration of the non-Alzheimer type (FLD)** (Brun, 1987; Gustafson, 1987), actually a number of somewhat related degenerative disorders have been identified as having an early impact on the frontal lobes. These include **frontotemporal dementia (FTD)**, **Pick's disease**, and **progressive aphasia (PA)** (Brun & Gustafson, 1999; Cummings & Benson, 1992; Filley & Cullum, 1993; Gustafson, Brun & Risberg, 1990; Kertesz, Davidson, & Munoz, 1999; Miller et al., 1991; Neary, Snowden, Northen & Goulding, 1988; Neary et al., 1998). What most of these frontal dementias tend to have in common are early

Table 9–6. Early Clinical Characteristics of "FLD"

General Behavior	Social disinhibition, poor judgment, impulsivity, diminished insight, decline in personal hygiene, hyperorality,[a] hypermetamorphosis[a] (utilization behavior).
Affective Behavior	Emotional blunting (apathy, "depression," loss of empathy, interpersonal warmth), combined with emotional lability and inappropriateness (e.g., aggressiveness, irritability, puerility); diminished concern and/or drive with regard to long-range goals.
Speech and Language	Variable, from reduction in spontaneous output to excessive talkativeness (inappropriately so); output may be stereotypic, perseverative, echolalic, confabulatory; verbal paraphasia and/or anomia are common. Receptive typically better than expressive language.
Cognition/Perception	Distractibility, mental rigidity, diminished "executive functions" and problem-solving abilities; complex learning and selective retrieval may be compromised, but routine memory grossly intact, along with basic visual–spatial abilities.
Motor	Restlessness, "hypomania," stereotypic, meaningless repetitions, frontal release signs, incontinence may be present.

[a] Components of the Kluver–Bucy syndrome.
Adapted from Neary et al., (1998), Brun & Gustafson, 1999.

changes in social judgment, personality, insight, and/or affect. While disturbances of speech or language also tend to become manifest early on, unlike Alzheimer's, memory and visual–spatial skills typically are *relatively* well preserved in the early stages of these disorders.[108] Table 9–6 presents the main clinical features of FLD. In addition, several other degenerative diseases are identified as significantly affecting frontal–subcortical systems. These include **Huntington's disease (HD)**, **progressive supranuclear palsy (PSP)**, and **dementia with Lewy bodies (DLB)** (Brun & Gustafson, 1999; Cummings, 1990; Duke & Kaszniak, 2000; McKeith, 1999; McKeith et al., 1996; Wechsler et al., 1982). Table 9–7 compares and contrasts some of the main clinical features among the more common degenerative dementias.

LOCALIZATION OF PREFRONTAL LOBE FUNCTIONS

As noted in the introduction of this section, the prefrontal regions of the frontal lobes are composed of cytoarchitecturally diverse cortex. Each of these areas is characterized by varying patterns of cortical and subcortical connections (Fuster, 1997; Grafmann, Holyoak & Boller, 1995; Passingham, 1993). The prefrontal cortex also is divided into the right and left cerebral hemispheres. As a result of these cytoarchitectural, connective and hemispheric differences it is likely that functional differences follow. Some of this functional differentiation was alluded to in the foregoing discussion in reference to the rather broad divisions within the prefrontal cortex, namely the dorsolateral, orbital, and mesial zones (see Table 9–8).

Caution is advised, however, when attempting to draw diagnostic conclusions regarding either the specific locus of a lesion or even the presence or absence of a prefrontal lesion based on behavioral findings alone. First, whether looking at "personality" (i.e., gross behavioral traits) or performance on specific cognitive tasks, rarely are we witnessing behavior that is the result of a circumscribed cortical area. As previously discussed, most

Table 9–7. Comparing Early Dementia Syndromes

	SDAT	PD	DLB	FTD	Pick's
Memory	Impaired	Slow; improved recognition	Intact early	Poor, if effortful	Relatively intact
Language	Dysnomic, empty speech, weak comprehen	Hypophonia, micrographia	Intact early	Variable output,[a] dysnomia	Early anomia
Visual–Spatial	Typically impaired	? Mildly impaired	Impaired	Relatively intact	Relatively intact
Executive Functions	No selective impairment	Slow in shifting	Impaired	Diminished judgment, initiation	Impaired
Behavior/ Affect	Relatively preserved[b]	Frequent depression	Freq. visual hallucinations	Disinhibited, apathetic, labile	Disinhibited, labile, Kluver-Bucy
Motor	Intact	Bradykinesia rigidity, tremors	Like PD, but w/o tremors	Intact	Intact

[a] Frequently diminished output, stereotypic utterances
[b] Anxiety, depression may be seen early while insight is relatively preserved
Legend: SDAT, Senile dementia, Alzheimer's type; PD, Parkinson's disease; DLB, dementia with Lewy bodies; FTD, frontotemporal dementia; Pick's (dementia)

behaviors are the result of distributed processes that not only involve multiple regions of the prefrontal cortices, but also nonfrontal cortices and subcortical areas of the brain. For example, we have just discussed the importance of frontolimbic connections in guiding or modulating certain behaviors. Other cortical–subcortical circuits or connections involving the basal ganglia, thalamus, and other subcortical structures were reviewed extensively in Chapter 6. Included among these were specific and apparently independent circuits incorporating the dorsolateral, orbitofrontal, and mesial portions of the prefrontal cortices. Lesions or disease processes affecting any portions of these neural circuits could result in behavioral disturbances similar to those found with lesions directly affecting those cortical areas (Cummings, 1990, 1995; Litvan, 1999). While advancements in neuroimaging continue to offer new insights into structural/behavioral correlates, the variable and erratic nature of naturally occurring lesions and the still less-than-precise ability of such techniques to identify the exact anatomical boundaries of dysfunctional tissue limit our insights into those correlates, and hence, clinical predictability.

Another problem in trying to pinpoint frontal lobe functions relates to the nature of frontal lobe functioning itself. Unlike the first and second systems that have relatively specific areas of function that can be identified and more or less directly assessed (such as perception, language, memory, visual–spatial skills, or even basic arousal), the frontal lobes seem to impact or interact with all these functional systems. Furthermore, many of the neuropsychological tests or measures that are commonly used to assess the integrity of the prefrontal cortex are themselves multidimensional tasks and can be adversely affected by lesions outside the frontal areas (Anderson, Damasio, Jones, & Tranel, 1991; Benson, 1979; Grafman, Jones, & Salazar, 1990).[109] As was emphasized by Kaplan (1990), the key to identifying

Table 9–8. Frontal Lobe Syndromes

Syndrome Associated with Lesions of the Dorsolateral Convexities
General Behavioral
Apathetic, "depressed"
Self-neglect (e.g., personal hygiene)
Distractibility/Impersistence
Failures of initiation
Instability in pursuing abstract, long-range goals
Cognitive–Behavioral
Poor planning
Limited use of innovative strategies
Reduced "mental flexibility"
Decreased ability to "shift mental sets"
Difficulty inhibiting prepotent response sets
Perseveration
Frequent loss of mental set
Diminished "abstractive ability"
Reduced "self-monitoring" capacity
Diminished performance on fluency, "motor programming" tasks
Poor judgment (inadequate cognitive-perceptual analysis)

Syndrome Associated with Lesions of the Orbitofrontal Areas
General Behavioral
"Personality" changes
Disinhibition or diminished impulse control
Poor "social" or moral judgment
Irritability/explosiveness
Emotional lability
Limited insight into behavioral deficits
Restless/hypomanic/euphoric
Facetiousness, puerility, "Witzelsucht"
Anosmia[a]
Cognitive–Behavioral
Occasional learning/memory difficulties with confabulations

Syndrome Associated with Lesions of the Medial Frontal Areas
General Behavioral
Apathy, diminished drive[b]
Hypokinesis, diminished spontaneity
"Akinetic mutism" (in more severe states)
Abulia
Incontinence
Gait disturbances

[a] The olfactory bulbs and tract lie on ventral surface of orbitofrontal cortex,
  hence bilateral damage to this area may result in anosmia.
[b] May be more profound than with lesions of the dorsolateral zones.
Adapted in part from Duffy & Campbell (1994); Cummings (1985); Chapter 6.

frontal pathology is not so much a matter of determining which tests or tasks are impaired but how the tasks are approached and in analyzing the nature of the deficits or errors.

Finally, not all behavioral sequelae normally associated with frontal lobe pathology necessarily are the result of structural lesions such as tumors, strokes, infections, or traumas that might be identified using neuroimaging procedures. The prefrontal cortices are extensively supplied by a variety of neurochemical transmitters, including norepinephrine, serotonin, dopamine, and acetylcholine. Selective depletion of these neurochemical transmitters (as are found in some of the disease states mentioned above) can result in specific neuropsychological or neurobehavioral deficits as different regions of the prefrontal cortex may be selectively supplied by the different neurotransmitters, hence subject to differential compromise (Campbell, Duffy, & Salloway, 1994; Glick et al., 1982; Fuster, 1997; Javoy-Agid et al., 1989; Litvan, 1999; Mega & Cummings, 1994; Mesulam, 1988; Oscar-Berman, McNamara, & Freedman, 1991; Weinberger, Berman, & Zec, 1986; also see Chapter 11, this volume).

Hemispheric Differences

Unlike smaller, unilateral lesions in the posterior heteromodal (tertiary) association cortices that often can result in obvious and occasionally almost pathognomonic deficits, unilateral lesions of the prefrontal cortices often are far more subtle in their clinical presentation. As a general rule, many of the more dramatic "frontal lobe syndromes" are likely the result of larger, bilateral lesions.[110] One notable exception to this rule, of course, is when the posterior, agranular areas (i.e., Brodmann's areas 4, 6, and 44) become involved and appendicular motor or speech difficulties become evident. These latter symptoms will be reviewed separately below. The presence of "dynamic" or transcortical motor aphasia, which may be associated with lesions involving the "prefrontal," usually is indicative of left frontal damage (Freedman, Alexander, & Naeser, 1984; Luria & Tsvetkova, 1968).

As was mentioned above, while left frontal (especially polar) lesions have been associated with depressive syndromes more frequently than the right, the presence of depression per se is not a reliable index of either brain injury or frontal or hemispheric impairment. Similarly, while disturbances of certain types of verbal, semantic memories have been associated with left frontal lesions and nonverbal, episodic memories with the right frontal lobe (Buckner et al., 1995; Kapur et al., 1995; Milner, 1982; Petrides, 1985), such findings are not particularly robust indices of either frontal or hemispheric damage. Among cognitive measures, tests of verbal versus nonverbal fluency (Benton, 1968; Jones-Gotman & Milner, 1977; Ruff et al., 1994) have proven to be one of the more reliable means of differentiating left from right frontal patients when the two groups are compared in controlled studies. However, these tasks, while sensitive, are not specific to frontal pathology (Stuss & Benson, 1986).[111] Finally, as noted in Part II of this chapter, Goldberg, Podell, and Lowell (1994) suggest that the specialization of the prefrontal lobes may be a function of whether the information being processed is "novel" or dependent on "external" contingencies (the right hemisphere) versus based on "internal" programs or previously learned schemas (left hemisphere). Thus, while a variety of neuropsychological tests have been associated with frontal pathology (see Kimberg, D'Esposito, & Farah, 1997; Lezak, 2004; Malloy & Richardson, 1994; Stuss & Benson, 1986; Walsh, 1994), again, except for those that focus on various motor skills (e.g., Christensen, 1975; Luria, 1966), none would appear to reliably differentiate right versus left frontal lesions. While the general distinctions between right and left hemispheric function discussed under Hemispheric Specialization in Part II continue to have relevance for the frontal lobes (i.e., there may be some material-specific distinctions between operations carried out by the right versus left prefrontal cortex), judgments based solely on such criteria are unlikely to be highly reliable. A recent battery of tests designed to measure mental flexibility shows promise as a means of assessing executive functions mediated in large part by the dorsolateral frontal zones lobe deficits (Delis, Kaplan, & Kramer, 2001).

Summary of Prefrontal Lobe Functions

As we have seen, both in regard to structure and function, the frontal granular cortex is far from being homogeneous. In comparison to the posterior association cortices, the functions of the prefrontal zones generally are more difficult to define in concrete terms and generally more difficult to assess. Most formal tests that currently are employed because of their purported sensitivity to frontal lobe lesions are not highly specific, being sensitive to lesions elsewhere in the brain. For the most part, it is only through informed and considered analysis of both mental status examination data and the spontaneous social behavior that specific features of frontal lobe syndromes begin to emerge. While by no means all-inclusive, the following is an attempt to summarize some of the key features regarding the probable role of the prefrontal cortices in human behavior. The prefrontal zones, which include the dorsolateral, orbitofrontal, and mesial frontal areas, would appear to share primary responsibility for:

1. Harnessing limbic drive to create and maintain stable intentions to achieve long-range, abstract, social goals.
2. Developing and executing higher-order, adaptive behavioral plans, programs, or strategies based on appropriate analyses, especially when faced with unique, challenging situations.
3. Considering response alternatives and weighing their probable effectiveness in view of previous experiences, current contingencies, and ultimate goals.
4. Monitoring one's progress, not only in view of the original goal, but also in terms of the results of the preliminary responses, changing conditions, or environmental demands.
5. Adapting or modifying either one's goal or plan of action as a result of feedback or circumstances.
6. Appreciating and balancing immediate, concrete needs or demands with more abstract or long-range needs and interests.
7. Inhibiting tendencies to respond in a perseverative, capricious, impulsive, or reflexive fashion.

Far from unraveling the mystery of the frontal granular cortex (or, for that matter, any other part of the brain), the preceding discussions at best provide a narrow window that hopefully provides a limited glimpse into the types of operations for which the prefrontal lobes are responsible. For more extensive coverage of this intriguing topic, the reader is referred to the following works: Luria's *Higher Cortical Functions in Man* (1966); Stuss and Benson's *The Frontal Lobes* (1986); Fuster's *The Prefrontal Cortex* (1997); Miller and Cummings' *The Human Frontal Lobes* (1999), Grafman, Holyoak, and Boller's *Structure and Function of the Human Prefrontal Cortex* (1995); Levin's *Frontal Lobe Function and Dysfunction* (1991), and Frackowiak et al., *Human Brain Function* (1997). In addition, the following represent a few of the many book chapters and articles that one might find informative: Damasio and Anderson (1993); Duke and Kaszniak (2000); Kimberg D'Esposito, and Farah (1997); Joseph (1990); Walsh (1994); and Stuss and Benson (1984).

EXECUTION OF MOTOR PROGRAMS

We commonly envision the ultimate goal of the executive unit (i.e., the frontal lobes) of the brain to be the carrying out ("execution") of a voluntary behavioral response. Most commonly, this involves some overt motor response. Such responses might range from relatively "simple" motor acts such as visually scanning an array of stimuli or pressing a button to more complex, highly integrated skilled responses such as speaking or playing a game of tennis. Discussions in the preceding section focused on the role of the frontal lobes in the overall planning, selection, initiation, or in some cases, the *inhibition* of behavioral responses.[112] Here the focus will be on the process of translating those goals, intentions, or directives into concrete motor responses. While as we have seen the former is mediated primarily by the prefrontal, granular cortex, the actual execution of the behavioral (motor) response is more properly the function of the **agranular** divisions of the frontal lobes. Located posterior to the prefrontal areas, the agranular cortex of the frontal lobe generally is considered to be composed of the **premotor (PM)**, **supplementary motor (SMA)**, and **primary motor (MI)** areas and the frontal eye fields.[113]

Before reviewing the general anatomy and suspected roles of the various divisions of the agranular motor cortex in greater detail, it might be useful to have at least in very broad terms an overall picture of the relationships that seem to characterize the various components of the frontal or executive system. The best analogy that comes to mind is that of the military (or, for that matter, any large corporate organization). Within the military, we normally think of the formulation of a battle plan as ultimately being the responsibility of the commanding general. In formulating these plans, he (she) needs access to and should review information regarding not only the available resources, equipment, and supplies, but also the mood and/or morale of the troops as well as any "gut feelings" about the situation (i.e., a review of internal milieu). In addition, information is needed regarding the status and position of the enemy forces, as well as a good understanding of the strategic situation in which both armies are engaged (external milieu). Finally, there needs to be an awareness of how the enemy might be expected to proceed based on historical precedents (memory). In this analogy, the "general" represents the prefrontal or granular cortex, which gathers essential information from the posterior cortices (concerning the "external milieu"), the limbic structures ("internal milieu"), as well as from personal memory.

When all the necessary information has been gathered and evaluated, the "general" (prefrontal cortex) formulates a plan of action. This includes not only determining the plan of attack, but also deciding when, where, and under what conditions the plan will be executed. Once executed, an ongoing assessment is necessary to determine how the battle plan is unfolding and what if any changes in strategy or objectives are indicated. Lastly, it also is typically the responsibility of the general to determine when or under what conditions the battle plan is to be terminated or aborted. Again, the above functions could be compared to the planning, monitoring, and related actions of the prefrontal zones of the executive unit discussed above.[114]

However, once the plan of action is formulated and the order for its initiation[115] has been given, it still must be executed or carried out. This is more properly the function of the agranular frontal cortex. We might think of this formal execution of action as a two-stage process. One stage is the responsibility of the medial and lateral premotor areas, with the other being the responsibility of the primary motor cortex.

Continuing with our military analogy, while the general (the prefrontal cortex) generates the overall plans, strategies, and timetables, for the campaign to be successfully carried out, most of the details need to be worked out at the battalion and company levels. This first stage might include making preliminary preparations to ensure that all units are ready to respond; coordinating support operations among the various units; and perhaps even making minor adjustments in the overall plan, depending on unexpected fluctuations in internal or external conditions or the results of immediately preceding operations. Such decisions typically are the responsibility of intermediate level officers. Parallels can be seen in the biological organism. While the overall plan may be to initiate a search for food, this usually involves extensive timing and coordination of individual muscles and appendages, constant sensory feedback, and continuous postural and limb adjustments in response to changes in gravity and in the target of the response (e.g., eye–limb coordination). As we shall see, these are functions that are largely, although probably not exclusively, controlled by the "premotor" zones.

In the military, while the planning and strategizing takes place at the division, battalion, or company levels, all of this eventually funnels down to the individual platoons or squads. Here it falls to the enlisted personnel to carry out the battle plan in a concrete fashion. This would be analogous to the second stage of the behavioral response where the enlisted personnel who actually execute the planned actions represent the primary motor cortices. For the most part, primary motor cortex is responsible for triggering the upper motor neurons that compose the corticobulbar and corticospinal tracts. It is this firing that results in the discrete, skilled action that represents the culmination of the directive that originated in the prefrontal cortex. Just as in the military, no platoon or squad is charged with carrying out the entire battle plan, but rather each is assigned a very specific objective, so too the primary motor cortex is divided into discrete units, each of which is devoted to specific muscle groups. Finally, just as the foot soldier must have access to direct, although usually very limited and circumscribed feedback about the effect of his or her initiatives, motor responses also require such feedback. In the case of the premotor area, such feedback comes not from all the senses, but is more or less limited to somatosensory information from the postcentral gyrus.

It is important to make one final point with regard to our military analogy before proceeding. While the chain of command generally may be viewed as that described above, in actuality to ensure maximal efficiency the lines of communication are necessarily quite diverse and the nature of the cooperative interactions should be quite flexible. Information needs to be able to flow not only down the chain of command but both laterally and back up the chain. Similarly, individual components of the frontal executive system (e.g., prefrontal, premotor, supplementary motor, and primary motor) are not only extensively interconnected, thus allowing for constant feedback and coordination of activities within the executive system, but most also are interconnected with both the posterior (gnostic) cortices and subcortical structures (e.g., thalamus, basal ganglia, cerebellum, limbic structures). These latter connections ensure that these frontal systems have access to sensory information, internal (e.g., motivational) states, and critical inhibitory influences not only at the planning stage, but also as required during the preparatory and execution phases.

With this broad framework in mind, we can examine several of the areas commonly identified as making up the frontal agranular cortex. Again, the focus will be on defining basic anatomical and functional correlates and where possible their interrelationship to the executive system as a whole. However, the reader should be advised that despite the renewed and often intense attention that some of these areas have enjoyed in recent years, particularly the supplementary motor areas, there still is considerable controversy, not only with regard to their basic functions, but even concerning their anatomical features (Marsden et al. 1996; Wise et al. 1996).

THE AGRANULAR FRONTAL CORTEX

The agranular regions of the frontal cortex generally are thought to encompass Brodmann's areas 4, 6, and at least the posterior portions of area 8 on the lateral and medial surfaces of the frontal lobes. Area 44 (*pars opercularis*) also typically is associated with the agranular cortex (Damasio, 1991; Kaufer & Lewis, 1999; Mesulam, 2000b; Nieuwenhuys, Voogd, & van Huijzen, 1988); see Figure 9–46. These cortical areas are referred to as "agranular" because of the relative absence of the granular layers (II and IV) and the prominence of the pyramidal cell layers (III and V).[116]

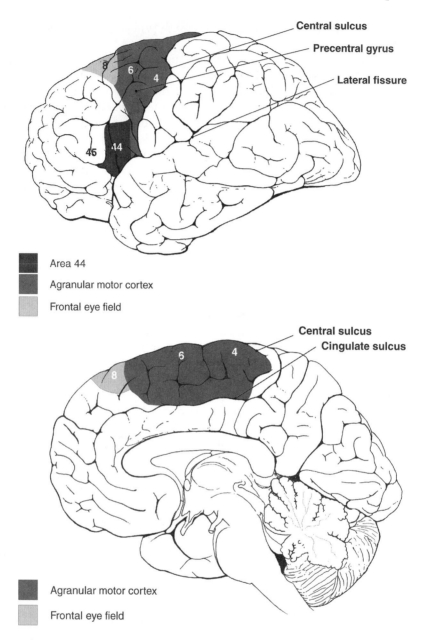

Figure 9–46. Approximate location of the frontal "agranular" cortices on the (a) lateral and (b) medial surfaces of the hemispheres. Area 44, which includes pars opercularis of the inferior frontal gyrus, and the more posterior or caudal portions of area 8 commonly are characterized as being more representative of the unimodal (agranular) motor cortex.

Despite this common feature, as with the prefrontal granular cortex, the frontal agranular cortex is neither structurally (e.g., with regard to its cytoarchitectonics and connections) nor functionally homogeneous. In the monkey, no fewer than ten distinct "motor areas" have been identified (Freund, 1996b). In humans, for most practical purposes the agranular cortex can be divided into three to five separate areas. The most common divisions include the (1) **primary motor cortex** (MI), (2) **premotor area** (PM), (3) **supplementary motor area**

(SMA), and (4) **frontal eye fields** (area 8). Area 8 has been noted to be heterogeneous in its cell structure, with its more rostral portions being more characteristic of the prefrontal, heteromodal cortices, while more caudally it is more similar to premotor cortex (Akert, 1964; Mesulam, 2000b). These caudal areas seem to have more extensive connections with the supplementary motor area (see below). The SMA itself has been subdivided into an *anterior zone* and a *posterior zone*. The latter is referred to as the **supplementary motor area proper** (SMA proper), while the more anterior division is known as the **presupplementary motor area** (pre-SMA).[117]

By way of a brief overview, although several agranular areas (SMA proper, PM, MI) all send projections to the brainstem and spinal cord, only MI is generally associated with discrete, highly differentiated voluntary actions. On the other hand, the SMA and PM areas generally are thought to play a critical role in the planning, sequencing, and integration of motor output leading to the overall coordination of gross motor behavior. Area 8 is thought to be primarily responsible for voluntary gaze, such as in active, selective, or directed scanning of one's external environment. Finally, none of these motor systems work in isolation. These regions are characterized by extensive interconnections and receive extensive input from and provide output to multiple cortical and subcortical structures. For example, in addition to receiving input from the prefrontal zones [which, as previously indicated, probably are responsible for the initiation (or inhibition) of all voluntary action], these motor areas also receive extensive input from the posterior cortices. The latter provide ongoing feedback regarding both tactile and proprioceptive information necessary for coordinated movement and object manipulation. These premotor areas also receive visual–spatial information that is essential in manipulating the external environment or the targeting of external stimuli. Finally, via the thalamus, these cortical motor systems are intimately linked to the basal ganglia and the cerebellum to form elaborate feedback mechanisms that are essential in motor learning and coordination. We will now examine the potential roles of each of these cortical motor areas in greater detail.

Supplementary Motor Area

Anatomy

The supplementary motor area (SMA) is represented by that portion of the premotor cortex (Brodmann's area 6) that lies on the medial surface of the hemispheres. During early electrical mapping of the motor cortex by Penfield and Welch (1951), this area was noted to have a somatotopic organization that was independent of those found on the lateral surface of the frontal lobes and that with sufficient stimulation would elicit complex motor responses. As noted above, recent investigators have further subdivided this region into the more posterior SMA proper and the more anterior pre-SMA zones (Picard & Strick, 1996; Rizzolatti, Luppino, & Matelli, 1996; Tanji, 1994).

Connections

The SMA cortex as a whole receives considerable input from the posterior somatosensory cortices, as well as from the prefrontal, premotor (PM), and primary motor (MI) areas.[118] The supplementary motor area also is an integral part of the basal ganglia and cerebellar feedback loops as witnessed by inputs from VL and VA nuclei of the thalamus. The output of SMA primarily is directed toward the lateral premotor (PM) and primary motor (MI) areas (including some bilateral connections). A relatively small contingent of fibers proceeds directly to the brainstem and spinal cord (mostly from SMA proper). Like most cortical areas, there also are projections to the neostriatum (caudate and putamen) (Bates & Goldman-Rakic, 1993; Luppino, Matelli, Camarda, & Rizzolatti, 1993; Chauvel, Rey, Buser, & Bancaud, 1996; Wiesendanger et al., 1987; Wise, 1996).

According to Rizzolatti, Luppino, and Matelli (1996) research interest in SMA sharply increased following the observation that electrical activity (a phenomena known as the *Bereitschaftspotential*) in the medial frontal zones routinely could be detected prior to the onset of voluntary motor activity (Deecke, 1987). This electrical activity in SMA can be detected even if one is only "thinking about" or "planning" a movement (regardless of whether the movement is ever carried out) (Freund, 1991; Roland, 1987). Despite the recent increased attention given to the SMA (e.g., Bates & Goldman-Rakic, 1993; Halsband et al., 1994; Luders, 1996; Mushiake et al., 1991; Picard & Strick, 1996; Tanji, 1994), there still is considerable controversy regarding the SMAs role in behavior and how its functions are distinguished from the lateral premotor (PM) cortex (Marsden et al., 1996; Tanji & Shima, 1996), and the primary motor (MI) area (Freund, 1996a).

Effects of Lesions/Stimulation

The SMA have been implicated in preparatory responses, both for simple and complex response patterns (Freund, 1996a). Animals with SMA lesions generally have greater difficulty in learning tasks that involve more complex, sequential motor patterns or where there is a paucity of external cues to guide behavior. Thus, it has been suggested that SMA may be critical in carrying out complex, sequential tasks that rely heavily on previously learned patterns of motor responses (sensorimotor engrams) (Mushiake, Inase, & Tanji, 1991; Passingham, 1993; Sergent, Zuck, Terriah, & McDonald, 1992; Tanji & Mushiake, 1996; Tanji & Shima, 1996). Lesion and stimulation studies in humans have offered some additional clues as to the functional significance of SMA. Naturally occurring lesions in humans (which unfortunately lack anatomical precision) have been reported to result in varying degrees of decreased voluntary or spontaneous motor activity in the contralateral limb (hemiakinesia), decreased speech output or facial expression, and occasionally unilateral neglect (Bleasel et al., 1996; Brust, 1996; Damasio & Van Hoesen, 1980; Freund, 1991, 1996b). Extensive cortical lesions that involve the medial frontal granular cortex may be associated with the **"alien hand syndrome"** (Freund, 1996a; Stuss & Benson, 1986, p. 87). Bilateral SMA lesions may result in akinetic mutism (Freund, 1991, 1996a). SMA lesions in humans also may be associated with difficulties performing sequential or rhythmic tasks in the contralateral hand or with difficulty coordinating bilateral reciprocal or simultaneous action patterns (Dick et al., 1986; Freund, 1987; Halsband et al., 1993). By contrast, stimulation of SMA generally results in more proximal, tonic movements, especially of the eyes and contralateral upper extremity. The most common description involves an elevation of the contralateral arm, followed by a turning of the head and eyes "as if following the movement of the hand" (i.e., a fencing posture) (Brust, 1996; Chauvel et al., 1996; Freund, 1996a). If stimulation occurs during speech or other motor activity, these activities will usually cease.

Summary. In broad terms, the medial frontal agranular cortex appears to play a critical role in the preparation and initiation [119] of a motor response. The SMA also appears to be important in the preparation and execution of complex, well-learned motor response patterns. Additional clarification and differentiation may be afforded by reviewing the apparent distinctions between pre-SMA and SMA proper, to which we now shall turn our attention.

SMA Proper versus Pre-SMA

On the basis of differences in cytoarchitecture, responses to electrical stimulation, and their afferent and efferent connections, SMA can be divided into "SMA proper," which lies rostral to area 4 on the medial surface of the frontal lobe, and "pre-SMA," which lies anterior to SMA proper (Matsuzaka, Aizawa, & Tanji, 1992; Rizzolatti, Luppino, & Matelli, 1996; Zilles et al., 1996). Although both are characterized by independent somatotopic organizations and overlapping patterns of anatomical connections, there are notable differences between the two. For example, while both receive input from the VA, VL, and DM nuclei of the thalamus, the pre-SMA area receives a greater proportion of its input from VA and DM, whereas VL is the major contributor to SMA proper. In terms of cortical connections, pre-SMA has greater input from the prefrontal granular cortex and anterior cingulate gyrus, while SMA proper has a greater somatosensory component. Only SMA proper appears to have direct connections with both MI and the spinal cord (Luders, 1996; Luppino et al. 1993; Picard & Strick, 1996; Rizzolatti, Luppino, & Matelli, 1996; Wise, 1996). The boundary between pre-SMA and SMA proper in humans is roughly demarcated by a vertical line drawn from the anterior commissure, perpendicular to a line connecting the anterior and posterior commissures (Luders, 1996; Zilles et al., 1996) (see Figure 9–47).

The functional distinction between pre-SMA and SMA proper is still a matter of some conjecture. However, as a result of recent studies, both in humans and in other primates, some tentative hypotheses have been offered. The pre-SMA areas may serve as an important link or transition between the prefrontal and cingulate cortices and the executive motor areas (Rizzolatti, Luppino, & Matelli, 1996). More specifically, the pre-SMA may be critical in the final planning, preparation, or final decision phase immediately prior to actual movements (when increased activity in this area is most prominent), whereas the SMA proper may be more highly correlated with the actual execution phase (i.e., increased activity during the movement itself) (Halsband, Matsuzaka, & Tanji, 1994; Matsuzaka, Aizawa, & Tanji, 1992; Passingham, 1996). It has been hypothesized that pre-SMA may be more critical for motor response patterns that are less automatic, characterized by greater cognitive demands or that require greater flexibility in terms of response alternatives (Picard & Strick, 1996; Sergent et al., 1992; Shima et al., 1996). SMA proper, by contrast, is thought to play a special role in

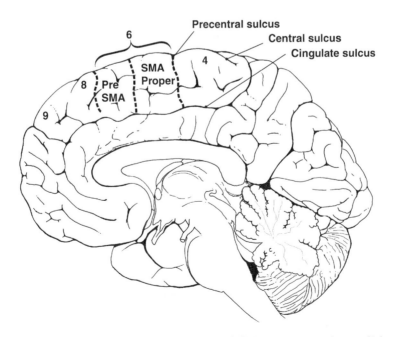

Figure 9–47. Approximate boundaries of pre-SMA and SMA proper on the medial surface of the cerebral hemisphere.

the execution of sequential movements, especially those that are more heavily practiced or overlearned (Marsden et al., 1996; Picard & Strick, 1996; Sergent et al., 1992). It also has been proposed that SMA proper may be important in coordinating axial and/or proximal musculature during overt motor activity (Rizzolatti, Luppino, & Matelli, 1996). As noted above, lesions of SMA often result in decreased spontaneous motor activity, decreased bimanual coordination, limb-kinetic-type apraxias, or "alien hand" syndromes (Bleasel, Comair, & Luders, 1996; Brust, 1996). However, in humans, discrete lesions that are isolated to pre-SMA or SMA proper are sufficiently rare so as not to provide a good model for the systematic study of the functional differences between these two anatomical areas.

Premotor Area

Anatomy

The premotor area most commonly is described as encompassing that portion of Brodmann's area 6, which lies on the dorsolateral surface of the cerebral cortex just anterior to the primary motor cortex (area 4). Other adjacent cortical areas that have similar cytoarchitectural features, particularly area 43, 44, and, less commonly, area 45,[120] are included in this premotor area (Jouandet & Gazzaniga, 1979; Stuss and Benson, 1986; Nieuwenhuys, Voogd, & van Huijzen, 1988; Freund, 1996b). As noted previously, area 8 (frontal eye fields) sometimes is included as part of the premotor cortex. Perhaps more commonly it is seen as an independent motor area, in part due to its distinct functional role. The premotor area, like the other "motor" areas, shows enhancement of layers 3 and 5 but lacks the large Betz cells that are characteristic of the primary motor cortex.

Connections

The premotor area has three general sources of input outside of the other cortical motor areas that are important in appreciating its probable behavioral functions. They are the prefrontal cortex, the posterior association areas (particularly areas 5 and 7), and subcortical input from the basal ganglia and cerebellum via the ventral anterior and the ventral lateral nuclei of the thalamus (Dum & Strick, 1991; Passingham, 1993; Fuster, 1997). These connections appear important for (1) the initiation of behavioral motor programs,[121] (2) sensory feedback in guiding and modulating movements, and (3) the timing and coordination of movements. Significant reciprocal connections also are established with SMA, but probably not the frontal eye fields, The major efferent output of the premotor cortices are to the primary motor cortex (ipsilateral), while transcallosal fibers connect the premotor cortex with its comparable area in the opposite hemisphere. These callosal connections are instrumental in explaining sympathetic dyspraxia following lesions of the premotor area of the dominant hemisphere (see **Ideomotor Apraxia** under **Disconnection Syndromes** p. 328) Geschwind, 1965; Heilman & Rothi, 1993. The premotor cortex also sends fibers caudally to the spinal cord. The latter fibers involve extensive connections in the brainstem that in turn give rise to the various ventral motor tracts in the cord (e.g., reticulospinal tract). These connections likely serve to help maintain postural stability while engaging in other motor activity (e.g., throwing a ball).

Effects of Lesions

The exact role of the premotor area and how it differs from that of the supplementary motor area remains somewhat unclear. Lesions involving the premotor cortex can result in (1) transient weakness, (2) diminished or slowing of spontaneous movement, and (3) limb-kinetic apraxia, or what Luria describes as *loss of kinetic melody* (Luria, 1966, 1973; Freund & Hummelsheim, 1985; Freund, 1991, 1966b; Halsband et al., 1993). These latter phenomena can be defined as the disruption of learned or habitual complex motor tasks as a result of difficulties in making smooth, fluent transitions from one phase of the action or motion to the next. While the goal and general form of the action remains intact, it may be carried out in an awkward or clumsy fashion. Such deficits of manual dexterity are contralateral to the lesion.[122] If the ventral premotor area, particularly area 44 of the dominant hemisphere, is affected, motor speech is likely to show impairment.[123]

A rare, but intriguing finding associated with lesions of the premotor area is the perseveration of elementary motor responses. For example, if asked to draw a circle, the patient may get stuck in set, repetitively drawing overlapping circles (Fig 9–37). When present, this symptom likely represents deep or more extensive lesions, probably including the basal ganglia (Luria, 1973; Stuss & Benson, 1986). While motor deficits following lesions of the premotor area are likely to affect any overlearned motor task requiring dexterity (e.g., playing a musical instrument, typing, drawing or writing), deficits can often be demonstrated by simply requesting the patient to tap out alternating rhythms or perform other complex, sequential movements, whether unilaterally or requiring reciprocal bilateral coordination (Christensen, 1975). In these instances one should be attending to differences in the fluidity of movement beyond that which might be explained by handedness. However, similar deficits also may be found following lesions to SMA (Halsband, Ito, Tanji, & Freund, 1993).

Feedback Mechanisms and the Premotor Cortex

For skilled movements to achieve maximal effectiveness there needs to be the capacity to make rapid, smooth transitions from one discrete movement to another so that each flows easily into the next. This is particularly evident as certain basic or automatic (overlearned) sequences of movements become well established. Yet motor programs must be flexible enough to respond to either changes in goals or plans (e.g., frontal programming) and to conform to the environment in which these actions occur. Both the SMA and the premotor areas rely extensively on both internal and external feedback. While the exact nature and role of such feedback are uncertain, some speculation is possible. As noted earlier, there are at least three primary sources of feedback to which we might attend. These include prefrontal, posterior cortical, and subcortical systems.

Feedback with Prefrontal Systems.   Motor actions are designed to carry out a specific plan, goal, or purpose. Once the action has been initiated, it should follow that plan until either it is accomplished or circumstances dictate that the goal be changed. Thus, the actions must stay "on target" via constant monitoring to ensure that the actions are accomplishing their intended purpose. The connections between the prefrontal cortex and the premotor area are likely critical in transmitting these initiatives and directives to the primary motor cortex for execution. While there probably are multiple cortical regions within the motor system engaged in this process (e.g., SMA, basal ganglia), since the premotor cortex, along with the primary motor area, has the most detailed somatotopic organization, it is likely that the premotor area is critical for the more precise or detailed aspects of the preparatory response.

Feedback with Subcortical Systems.   The execution of skilled movements requires a smooth transition from one component of the movement to each succeeding one, while maintaining an optimal overall muscular tone. Gross motor activities often require bilateral postural adjustments to maintain balance and ensure maximal efficiency (power) of movement. Even more discrete, fine motor activities often require the coordination of both sides of the body (e.g., typing, playing a musical instrument). The premotor area, with its extensive connections with the basal ganglia, the cerebellum, and the contralateral premotor area via the corpus callosum, appears ideally suited to integrate these various influences to effect this fluid coordination of individual movements.

Thus, for example, if the hand area of the premotor zone and/or its connections is affected, writing may lose its automatic character. Letters may be slavishly and clumsily reproduced. Script may give way to printing, since less fluid transitions between letters are required. Problems may be seen if the patient is required to produce rapid, alternating movements or rhythms using either the affected hand or both hands. If the lesion is more ventrally located, especially on the left side, a dysarthric-type speech may result. The patient may have difficulty making transitions from one articulation to another, especially when asked to repeat multisyllabic words or strings of phonemes requiring major, rapid transitions in the use of the lips, tongue and soft palate such as "bah-kuh-lah."

Feedback with Posterior Systems. The effective execution of movement is very much dependent on sensory feedback.[124] First and foremost, as each movement builds upon the last, one needs to maintain awareness of the position and movement of one's body in space, as well as the force and velocity of movement (e.g., proprioceptive and kinesthetic feedback). With the loss of such feedback, movements become coarse and awkward, especially if unaided by vision. Efficient manipulation of objects requires somatosensory feedback concerning the physical and dynamic qualities of the object itself. Witness the difficulties experienced in attempting to button a shirt or retrieve an object when one's hand "falls asleep." As previously noted, the dorsolateral premotor cortex has extensive connections with the superior parietal lobule. Thus, it would appear that the dorsolateral premotor cortex is critical in evaluating highly integrated somatosensory information either prior to or during movements initiated by the primary motor cortex.

Many movements take place in three-dimensional space, whether it is throwing a dart, catching a ball, writing or drawing, or simply reaching for an object lying on the table. The accuracy of such movements depends on feedback from the posterior association cortices. While there appears to be no consistent evidence of major direct connections between the dorsolateral premotor areas and the inferior parietal lobule, area 7 likely relays some visually processed information.[125] What is apparent is that the dorsolateral premotor area likely has access to visual as well as tactile feedback in that one of the more consistent conclusions regarding possible functional differences between the dorsolateral premotor area (PM) and SMA is that PM has been shown to be particularly active when motor responses are guided or directed by external cues, including visual stimuli (Rizzolatti, 1987; Roland, 1987; Tanji, 1987, 1996).

Summary.   While both the SMA and dorsolateral premotor (PM) cortex probably are critical in the preparation and organization of the motor response, given the more precise somatotopic organization present in PM, the latter may assume a more direct role in the final coordination of discrete motor responses. The type of deficits seen following lesions of PM would lend support to this hypothesis. Finally, there has been some evidence to suggest that PM plays a critical role when the motor response is heavily dependent on external cues.

Frontal Eye Fields

Anatomy

The frontal eye fields (FEF) lie on the dorsolateral surface of the frontal lobe in the middle portion of the middle frontal gyrus, just anterior to area 6. Commonly referred to as area 8, the frontal eye fields and area 8 are not coterminal, but the FEF makes up a substantial part of area 8. The FEF sometimes is included as part of the "premotor cortex," but unlike area 6, it has no direct connections with the primary motor area (4). A "supplementary" motor area serving the FEF has been identified lying in or near the midline (Passingham, 1993).

Connections

Like the other frontal motor areas, the FEF has feedback loops through the basal ganglia and thalamus.[126] Of major interest here are its cortical and brainstem connections. The FEF has both ipsilateral and contralateral connections with the prefrontal granular cortex. As is true of the other premotor areas, these connections are important in the initiation (execution) of motor behaviors (in this case, eye movements). In the case of vision, such activity also is critical in establishing and maintaining selective attention to the external environment. The other major cortical inputs to the FEF are from the posterior cortices, especially those parietal and occipitotemporal areas that are important for processing visual information (however, there is no direct connection with the primary visual area) (Barbas, 1988; Fuster, 1997). There is substantial input from the posterior eye fields in area 7 (Cavada & Goldman-Rakic, 1989).

While there probably are reciprocal corticocortical connections, the efferent connections of primary concern are those to the brainstem, particularly those to the superior colliculi, the pretectal area, and either directly or indirectly to the **vertical** and **horizontal** gaze centers. The former is represented by the rostral interstitial nucleus of the medial longitudinal fasciculus and the latter by the paramedian pontine reticular formation. Apparently there are minor if any direct connections with the motor nuclei of the third, fourth, or sixth cranial nerves that innervate the extrinsic muscles of the eye.

Function

The frontal eye field is responsible for *voluntary* eye movements such as might be initiated in active visual searching or in voluntarily directing one's attention to a specific visual stimulus or portion of the visual field. Following the general principle of contralateral representation, the right FEF is responsible for voluntary conjugate gaze directed toward the left visual hemifield and vice versa for the left FEF. In general, this area behaviorally is consistent with area 6, which is responsible for mediating somatomotor responses to carry out the goals established by the prefrontal cortex. The tracking of moving objects (e.g., the opticokinetic reflex) or shifting one's gaze in response to sudden, unexpected noises are mediated by the posterior gaze centers and by reflex mechanisms at the level of the superior colliculi.

Pathology

Just as lesions that affect area 6 result in detectable motor deficits, so too do lesions that affect the FEF. Excessive stimulation of area 8 (as might occur with focal seizures) usually result in the eyes being conjugately driven to the side opposite the seizure focus (i.e., "looking away from the lesion"). Following the termination of the seizure, the eyes may be found deviated toward the lesion. In nonexcitatory, structural lesions (e.g., (stroke) involving the FEF unilaterally, the eyes may show a deviation toward the side of the lesion at rest. Although the eyes frequently may be brought to midline when attention is called to them, even with voluntary effort the patient may have difficulty achieving and/or maintaining conjugate gaze toward the opposite hemifield. Marked contralateral inattention or neglect has been reported in monkeys following unilateral lesions of the FEF. Such neglect or inattention may not necessarily be limited to visual stimuli and may be produced by frontal lesions other than those affecting area 8 (Mesulam, 2000c; Stuss & Benson, 1986).

Finally, despite difficulties in voluntary gaze that may be seen following frontal lesions (e.g., in response to verbal commands to "look" toward the contralateral field), unlike lesions of the PPRF in the brainstem there is no paralysis of conjugate gaze. This readily 'can be demonstrated by having the patient follow a moving object into the affected field or clapping loudly to that side. Under these circumstances the eyes will show the ability to shift, relying on the posterior gaze centers and brainstem reflexes.

Primary Motor Cortex (Area 4)

With the primary motor cortex, we reach what might be considered the final common pathway of the executive system. The axons of area 4 neurons synapse on motor nuclei in the brainstem and on the anterior horn cells in the spinal cord. In terms of our earlier military analogy, this region of the brain can be said to represent the enlisted men and women who ultimately are responsible for carrying out the plans and directives that were conceived, initiated, organized, and coordinated at higher levels in the chain of command.[127]

Anatomy

The primary motor cortex is represented, for the most part, by the precentral gyrus, although in its more ventral aspects much of area 4 lies within the folds of the central sulcus. Area 4 can be distinguished histologically from the other cortical motor areas by the presence of large Betz cells, predominately in layer V. Initially thought to represent the origin of corticospinal fibers, it is now known that these Betz cells reflect only between 1 and 3% of the corticospinal fibers found in the medulla.

Area 4, like area 8 and the postcentral gyrus, is characterized by a well-defined somatotopic organization. The face and oral musculature are located in the most ventral aspects of the precentral gyrus, followed dorsally by the representation of the hands, arms, shoulder, and trunk. The cells that eventually supply input to the lower legs and feet are located along the medial aspects of the hemispheres.[128] The amount of cortex devoted to a given area of the body is proportional to the degree of fine motor control that can be exerted by that body part. Hence, the cortical area reserved for the muscles of the hands and face are quite extensive in humans relative to the other parts of the body. These expanded areas are a function of the increased number of columnar cell units that are allocated to the increased number of discrete motor units, thus permitting increased fine motor control. Such control is further facilitated by equally fine point-to-point feedback from the somatosensory cortex.

Connections

The primary motor cortex has three major sources of input: (1) motor association cortex, (2) somatosensory cortex, and (3) subcortical projections. The major inputs from the motor association cortices are from SMA proper and from the dorsolateral portion of area 6 (i.e., premotor cortex). Somatosensory projections primarily come from those areas that represent proprioception (as opposed to cutaneous) feedback as well as input from area 5, which conveys more highly integrated somatosensory information. Such feedback is instrumental in modulating the force, direction, and accuracy of movements. Finally, as opposed to the motor association cortices that have extensive feedback from the basal ganglia, the subcortical input to area 4 seems to be shifted more to cerebellar input (via the posterior portions of the ventral lateral thalamic nuclei), although some basal ganglia input through the VA nucleus is likely present (Jones, 1987; Passingham, 1993).

As we learned earlier, the motor cortex sends fibers to the striatum (primarily the putamen) and likely has reciprocal connections with the cortical areas from which it receives projections (e.g., the somatosensory cortex, the premotor area and SMA proper, as well as the cerebellum). These various connections provide the anatomical substrate for ongoing feedback to the motor cortex. Those efferent fibers of the primary motor cortex that synapse on the motor neurons of the brainstem and spinal cord probably account for only a little over 30% of the corticospinal fibers (the rest coming from the premotor and somatosensory cortices). The neurons of the primary motor cortex are distinguished by the fact that they exert control over discrete, voluntary motor responses, especially those requiring fine motor skills (Kuypers, 1987).

Effects of Lesions

As noted above, the primary motor cortex represents the final common cortical pathway by which the brain is able to initiate voluntary, goal-directed motor activity, including discrete, fine motor skills. It is through the primary motor cortex that the brain is able to communicate with, exercise its "will," and exert physical control over the external world. Discrete lesions that affect only the primary motor cortex (to the exclusion of either premotor or somatosensory areas or deep fiber pathways) are relatively rare. However, several general observations can be gleaned from the clinical and experimental literature. These are:

1.   In the case of small, focal lesions, the resulting deficits will be contralateral, and will differentially affect those areas of the body represented by the site of the lesion.
2.   Following the onset of the (acute) lesion, there will be a period of flaccid paralysis, followed by an increase in tone and spasticity, although the degree of spasticity may be less than that seen with lesions of the internal capsule.
3.   Pathological reflexes consistent with upper motor neuron lesions (e.g., Babinski, Hoffmann's) may be present.
4.   Although some initial recovery often is seen following reductions in edema or restorations of blood flow, residual deficits are common and usually are more prominent distally than proximally.
5.   The major functional deficits found after lesions to the primary motor cortex are reduced strength and decreased control of movement.

A FINAL WORD

This concludes the chapter on the cortex. What it tried to convey is that the brain probably acts as a whole, with widely distributed neural networks synchronously activated to produce any volitional behavior. While some tasks likely place a greater demand on some parts of the brain than others, most tasks are indeed highly complex, at least from the perspective of the various brain systems involved. Even the most simple tasks likely require the simultaneous cooperation of major portions of the three functional systems discussed. Disturbances in any of these systems or subcomponents thereof may result in behavioral deficits. Our task as cognitive neuroscientists is to analyze patient behaviors (including the results of neurobehavioral tests) and to search for patterns that enable us to better understand the specific or more elementary breakdowns (deficits) in brain function. To do so not only serves to enhance our understanding of the brain itself, but likely puts us in a better position to understand our patients and more accurately predict the type of difficulties they may expect to encounter in their day-to-day lives.

Endnotes

36.   This is not to suggest that, despite whatever advances may have been made in the last 50 years, we are much beyond having barely scratched the surface of this uniquely complex organ!

37. This readily can be demonstrated by observing the difficulties a young child may encounter if he or she attempts to read or write while holding their tongue fixed between their teeth, thus hampering their ability to "sound out" the word.

38. While adhering to the three units as outlined by Luria, this subdivision of the first unit can be construed as having some capacity for arousal via the organism's need to maintain a homeostatic balance, instinctual drives to ensure both personal and species survival, as well as acquired emotional valances. However, the arousal mechanisms mediated by the limbic system would appear quite different from those of the RAS. Consequently, it seems very reasonable to speculate that the limbic structures might be construed as constituting a separate functional unit, one that, among other things, could be characterized as regulating drives and emotions.

39. In addition to influencing "higher cortical centers," it is likely that these structures in turn also may be subject to descending influences from the cerebral cortices, a point that will be addressed later.

40. As originally defined by Wilson (1924), certain areas of the brainstem, if deprived of cortical influence (as in bilateral lesions of the corticobulbar tracts), spontaneously may manifest facial expressions typically associated with certain emotional states, despite the lack of internal emotion, a condition he described as pseudobulbar palsy. While it is possible that certain bilateral, subcortical lesions indeed may produce this effect, emotionality lability ("emotional incontinence") probably is a more common explanation when subcortical (or cortical) frontal lesions are present.

41. The term "horizontal" zone refers to the supposition that once the sensory information is transmitted to the various primary zones, the transformations of the data mentioned above, which proceed from the primary to the tertiary zones, would appear to be largely a cortical phenomena without necessarily involving additional subcortical structures or input.

42. Smell and taste may have similar primary and secondary zones but their sites of cortical representation probably are much smaller and along with their functional analysis are less well understood.

43. For simplicity's sake, we will confine ourselves to auditory language at this point. Non-language, auditory inputs will be considered below.

44. While the words are readily available, the meaning of specific word-names for colors or the concept of color itself cannot be grasped fully by the individual who is congenitally blind.

45. However, these are fairly complex linguistic constructions, and as a result could be affected by other cognitive difficulties as well. Also these are by no means the only symptoms associated with lesions of the posterior tertiary cortex of the dominant hemisphere (see, for example, Gerstmann's syndrome). Many of these symptoms and syndromes will be discussed elsewhere in this chapter.

46. It is beyond the scope of this work to attempt to present a detailed description of the neuroanatomical and behavioral substrates of language and its syndromes. Table 9-4 (p. 347) provides a brief summary of classically defined aphasic syndromes. However, the interested reader is referred to more detailed treatments of this topic that can be found in books by Benson, (1979) and Albert et al., (1981); chapters in Feinberg and Farah, (1997) and Heilman & Valenstein, (1993); as well as an article by Heilman, Tucker, and Valenstein, (1976) that offers an interesting heuristic model for thinking about the organization of various components of language within the brain.

47. These "nonsemantic" aspects of language and the putative role of the nondominant hemisphere were reviewed in greater detail under Emotional-Affective Processes, Chapter 9, Part II, p. 352).

48. For example, it is noted that certain cells in this region are more likely to respond to integrated visual gestalts or percepts such as faces. If stimulated, these more anterior areas are more likely to elicit complex, well-formed visual hallucinations such as images of objects, animals, people, at times including sequences of actions involving such images. On the other hand, stimulation of the more posterior, peristriate areas tends to produce more elementary patterns of light, color, or movement.

49. The perceptual deficits associated with lesions to these secondary association areas often are referred to as **apperceptive agnosias**, in contrast to **associative agnosias** that result from lesions of the heteromodal areas. However, a perceptual (apperceptive) deficit, if sufficiently severe may preclude or interfere with the visual identification of an object or stimulus. Apperceptive agnosias frequently are distinguished from the associative variety by the fact that in apperceptive agnosias the patient not only is unable to name the stimulus (although naming should be preserved if the object or stimulus is presented in another modality, such as tactually or by verbal description), but also unable to draw or copy the figure or object. In contrast, in purely associative agnosias, while naming (or other signs of recognition) is impaired, the patient may retain the ability to draw the object in question. However, some caution is advised in relying solely on these procedures as the act of drawing itself is a higher-level integrative activity and could be present in associative disorders.

50. Simultanagnosia, as noted earlier, is normally associated with Balint's syndrome (although it may occur as a more isolated symptom) and bilateral occipitoparietal lesions. Typically, simultanagnosia (or dorsal simultanagnosia) presents as an inability to attend to or focus on more than one object or aspect of a visual picture or array at a time. Consequently, objects, pictures, or visual scenes may be misidentified or misinterpreted as a result of this failure to attend to all the visual information present (Luria, 1959; Rafal, 1997a). This problem may be exacerbated if there are simultaneous lesions affecting the frontal eye fields that further impair visual searching. Kinsbourne & Warrington (1962a) identified a somewhat more benign version of this disorder (ventral simultanagnosia). Resulting from left occipitotemporal lesions, the individual has difficulty determining how one part of a complex picture or visual percept relates to another, and thus, is unable to accurately interpret it (Cytowic, 1996, pp. 433–436; Bauer, 1993, pp. 224–226). This latter form of simultanagnosia may be associated with particular types of reading disturbances (i.e., letter-by-letter reading, as opposed to normal "whole word" recognition).

51. With the possible exception of most daytime talk shows and prime time sitcoms! Also, this is not meant to imply that members of other mammalian, avian, or even reptilian groups do not possess keen visual skills. Members of the canine family, for example, rely on vision not only for spotting and pursuing prey, but also for "communication" (the attitude of the lips, ears, tail, head, and back all provide extremely important cues regarding intention and social status). The difference in primates may be their greater tendency to visually explore their environment for its own sake.

52. Multisensory images commonly are secondarily (indirectly) associated with color in that, if asked to picture something red, we may visualize an apple. In turn, the image of a juicy red apple may secondarily elicit tactile, gustatory, olfactory, and even auditory associations. Interestingly enough, colors may be associated in a more direct, although less concrete or tangible manner with mood states (limbic structures), a fact not lost on designers and decorators.

53. With the exception of judgment of line orientation that is significantly more highly correlated with lesions of the "nondominant" hemisphere, disruptions of most such visual–spatial tasks can occur following lesions to either hemisphere. However, it also

should be noted that even constructional tasks (such as drawing geometric designs, or reproducing two- or three-dimensional block-type constructions) are complex tasks and, as such, may be disrupted following diverse cortical lesions.

54. One possible explanation for the robustness of visual object recognition is the more diverse associations the brain makes to visual objects (hence, a greater degree of redundancy that is built into the system). Consider that while many objects have direct somatosensory or sensorimotor associations (we sit in a chair, we manipulate a hammer), the same cannot be said for colors, faces, or letters. Such associations may provide additional alternative pathways by which information may be transferred from one part of the brain or one hemisphere to another. This also might help explain why object naming is generally preserved in alexia without agraphia (see Disconnection Syndromes).

55. While there may be some evidence of apperceptive deficits, these generally would not appear sufficient to account for the severity of the deficit manifested. Also, interestingly enough, despite a lack of any conscious awareness of the person to whom a previously familiar face belongs, the patient may show signs of "unconscious recognition," that is, may perform well above chance in a forced-choice paradigm (Sergent & Poncet, 1990; Diamond et al., 1994).

56. As will be noted later, two of the major functions of the frontal tertiary cortices are (1) planning and organization, and (2) self-monitoring, which involves comparing actual with expected outcomes. Both of these activities require constant, directed sensory feedback. Certainly, even simple tasks requiring hand–eye coordination also must rely on visual connections with the sensorimotor cortices.

57. The Visual Reproduction subtest of the Wechsler Memory Scale-Revised, a test that involves the reproduction of geometric designs, was dropped from the primary memory indices in the new WMS-III (Psychological Corporation, 1997) in large part because it failed to adequately differentiate right hemispheric patients (Chelune & Bornstein, 1988; Naugle et al., 1993). One possible explanation for these findings is that the stimuli used in this test (as in many tests of "visual memory") are easily verbally encodable. Hence, despite lesions to the supposedly more "visual" right occipital–temporal–limbic pathways, the left hemisphere (which also has the capacity to lay down visual memories) may have been further "assisted" by verbal encoding. This example is presented to show the inherent difficulty in trying to isolate stimulus–response patterns in the real world.

58. Although each of these cytoarchitecturally diverse cortical projection areas in the postcentral gyrus likely represent different aspects of tactile sensation (submodalities), the exact functional correlates of each are still a matter of speculation. However, area 3a, which lies in the depth of the central sulcus, appears to represent a transitional zone between area 4 and the rest of the somatosensory cortex and likely receives input from muscle (spindles), joints, and Golgi tendon organs. There is some debate as to whether information supplied to 3a reaches a level of conscious awareness. Area 2 seems to respond, at a conscious level, to muscle and joint receptors and would appear to be important in judgments about size, shape, kinesthesia, and position sense. Area 1 appears to respond preferentially to more rapid conducting and surface receptors and may be selectively attuned to texture discrimination. The role of area 3b, which lies along the posterior bank of the central sulcus, appears less clear but it may be responsive to slower conducting, cutaneous receptors (e.g., temperature), as well as capable of mediating mechanical cutaneous input (Kaas, 1983; Warren, Yezierski, & Capra, 1997).

59. While information from both the medial lemniscus and the spinothalamic tracts apparently projects to SI, the lemniscal system, which is better equipped to process discrete somatosensory stimuli, especially information that is the result of active tactile exploration or manipulation, may be preeminent in SI.

60. Among the receptors that provide feedback to the somatosensory cortex are Meissner's corpuscles, Pacinian corpuscles, Ruffini endings, Merkel's disks, Krause bulbs, free nerve endings of various types, including those attached to hair follicles, Golgi tendon organs, muscle spindles, and joint receptors.

61. Lesions of SII tend to produce more subtle, bilateral sensory deficits.

62. Cortical lesions are not necessary to produce these deficits. Thalamic lesions or lesions affecting the thalamocortical pathways may produce similar symptoms. Lesions that affect the spinal–cortical pathways (e.g., the medial lemniscus and the spinothalamic tracts) generally will have a much more profound effect on these elementary somatosensory perceptions than will thalamic lesions.

63. In this, as in many of the signs and symptoms described here, one must be cautious in drawing anatomical conclusions. Not only may similar deficits arise from various vertically placed CNS lesions but they also may arise as a result of lesions to other functional units. In this case, lesions affecting motor programming must also be considered.

64. There is reason to suspect that area 5 and perhaps parts of 7 may be critical for the more elaborate and complex integrative tactile perceptions that rely heavily on proprioceptive, as well as mechanical cutaneous information, particularly as these might relate to guiding motor responses. Also, the more posterior the lesion, the greater the likelihood that vestibular and visual interactions begin to play a critical role (Hecaen & Albert, 1978, Chapter 6).

65. One definition of a narcissist is the person who, while acknowledging that it is theoretically possible for the universe to continue in existence after his demise, does not believe it actually will happen; after all, what would be the point! Even if one does not subscribe to that philosophy, it still is in the interest of most, if not all, sentient organisms to manipulate the environment to meet their needs.

66. Occasionally, patients may be found who are unable to name an object held in the nondominant hand but accurately can demonstrate its use. This does not represent an agnosia, but more likely a disconnection from the dominant hemisphere.

67. **Amorphognosis** refers to the inability to appreciate the form of the object, while **ahylognosia** refers to the inability to determine its substance (material). Theoretically, in astereognosis, the patient should be able to make individual, independent judgments of size, shape, weight, or texture but have difficulty integrating all of them at once. According to Critchley (1969), in practice, however, it is rare to find astereognosic deficits without some evidence of more elementary defects.

68. For example, when asked to right or draw something on a sheet of paper, the right-handed individual usually will rest the left hand on the top of the paper to steady it. Patients with left-sided neglect may fail to do this.

69. Unilateral finger agnosia more likely suggests an underlying disorder other than that related to autotopagnosia (Gainotti & Tiacci, 1973). The manner in which the deficit is elicited may provide some clue. For example, if the deficit is present unilaterally, but only under conditions where visual feedback is unavailable, an elementary somatosensory defect should be ruled out. If present under both visual and tactile conditions, unilateral neglect may account for the findings.

70. Goldenberg (1997), in his review of disorders of body schema, concludes that aphasic disturbances probably account for most of these type deficits in left hemisphere lesions,

while general mental impairments likely account for such findings associated with right hemisphere disease.

71. Unlike vision, hearing, and olfaction, which are all telereceptors, tactile stimulation implies that your body space already has been invaded. One of the authors (JEM) vividly recalls an incident from his childhood that demonstrated this phenomenon. It was a warm spring afternoon and the author's dog, a Doberman pinscher, was quietly sleeping under the shade of a tree. He decided it might be "fun" to see if he could "sneak up" on the sleeping dog. Having successfully accomplished this "mission," he gently but firmly poked the dog on his back. "Instantly" the Doberman was on his feet with his fangs bared and his ears back. Fortunately, just as quickly visual, olfactory, or possibly auditory cues kicked in and the dog resumed his normally friendly posture.

72. Recall from Chapter 8 that anterior cingulate lesions have been known to abolish these negative responses to painful stimuli, at least temporarily. While the patient may continue to report that he or she experiences "pain," it no longer appears to be disturbing. This phenomenon, traditionally referred to as **pain asymbolia**, actually has been reported with various thalamocortical and cortical lesions, most frequently in the area of SII in the dominant hemisphere (Hecaen & Albert, 1978); however, Geschwind (1965) hypothesized that the critical element in all these cases was a disconnection between the cortex and the limbic system

73. While the parietal and temporal lobes of the brain have substantial connections with the basal ganglia and limbic structures, the frontal lobes and the prefrontal cortex in particular have rather special connections with these subcortical structures. Recall that the prefrontal cortex has extensive connections to the basal ganglia, particularly the head of the caudate (see Chapter 6) and are directly associated with the basal and mesial limbic structures (see Chapter 8). As will be seen, these latter connections would appear to exert a major modulating influence on the first functional unit. Since lesions that encroach upon these subcortical connections (including thalamic projections) may adversely affect the functioning of the third unit, at times, reference is made to lesion processes that disrupt *frontal systems*. This terminology is a reminder that the frontal lobes (as well as other parts of the brain) do not operate in isolation, but may be significantly affected by lesions that technically may be outside the cortical boundaries of the frontal lobes themselves.

74. The posterior portions of area 6 sometimes are included as part of the primary motor area, while the more anterior part of area 8 is occasionally classified as heteromodal cortex (see Mesulam, 2000b, pp. 14, 23).

75. The role of this prefrontal, granular cortex was summarized by Luria (1973, pp. 79–80) in the following passage:

> Man not only reacts passively to incoming information, but creates intentions, forms plans and programmes of his actions, inspects their performance, and regulates his behavior so that it conforms to these plans and programmes; finally he verifies his conscious activity, comparing the effects of his actions with the original intentions and correcting any mistakes he has made.

76. This does not imply, however, that motoric responses are always the direct goal or consequence of activity in the prefrontal cortices. The immediate outcome of this activity at times may remain strictly within the mental realm. It also should be noted that the frontal cortices do not act in a vacuum. While the frontal heteromodal cortex likely provides the means for devising plans or strategies, it is the posterior lobes of the brain that supply the data on which the frontal lobes operate. Each would be virtually useless without the other.

77. Because deficits associated with prefrontal regions of the brain often are so context dependent, it sometimes is difficult to devise appropriate laboratory tests that measure content-dependent behaviors. For example, judgment (inhibition) routinely is tested in mental status exams by asking the patient, "What would you do if you saw smoke and fire in a theater"? "Frontally impaired" patients may give perfectly adequate responses in the emotionally neutral setting of the bedside exam but behave quite differently if actually placed in that affective-laden situation.

78. There is no uniform consensus as to these designations or the exact cytoarchitectural areas that constitute them. For example, sometimes it is suggested that areas 24 and 32 could be included as part of the mesial prefrontal cortices (see Benson & Stuss, 1986, p. 16; Benton, 1991, p. 17), while others would classify these two areas along with area 25 as constituting paralimbic cortices (Mesulam, 2000b). On the lateral surface, there also has been debate as to the classification of areas 44 and 45 (otherwise referred to a "Broca's area"). Again, there has been some debate as to how these two areas should be classified (e.g., see: Jouandet & Gazzaniga, 1979). However, the consensus seems to be that, cytoarchitecturally, area 45 is more consistent with the heteromodal, prefrontal cortex, while area 44 is more representative of the unimodal, premotor cortex (Damasio, 1991; Kaufer & Lewis, 1999; Mesulam, 2000b; Nieuwenhuys, Voogd, & van Huijzen, 1988). Finally, as Brodmann's maps do not include the orbital regions of the frontal lobe, the designations for this region were derived from other sources (see H. Damasio, 1991, p. 96; Robin & Macdonald, 1975).

79. For additional detail, see Harlow's 19th-century description of his patient, Phineas Gage (page 448) and Macmillan (1986).

80. Since the original Star Trek characters are most familiar to the author, they were chosen for this example, trusting the younger reader can bridge this obvious "generation gap."

81. Again, see Harlow's description of Phineas Gage (p. 448).

82. Thoughts (plans, schemas, ideas) can be considered a type of "action" and certainly often involve considerable "psychic energy." Consequently, it is not only the translation of ideas into physically executable programs that may suffer, but also the generation of the psychic programs or thoughts themselves.

83. Often the problem is not simply inaction, but the choice of the behavioral goal that is selected to be acted on. For example, it is not uncommon for more long-range, esoteric, or abstract goals to be eschewed in favor of more immediate, primitive, or biological needs. These issues, which relate to another aspect of prefrontal lobe function, will be addressed later.

84. A fact that is well known to most advertising executives.

85. Because of the direct connections between the dorsolateral prefrontal cortex and the frontal eye fields (area 8) and the premotor zones (area 6), visual and tactile exploration of specific aspects of external stimuli easily can be manipulated. Although hearing is not readily capable of such manipulation, certain animals prick up their ears (lacking that capacity, humans may cup their hands behind their ears). Both of us may turn our heads toward a sound to provide a clearer shot for the sound waves to strike the tympanic membrane. Also, see under The Modulation of Internal Drives and Emotions, "The Role of Frontal–Thalamic Connections in Selective Attention or Arousal" for other potential neuroanatomical substrates of selective attention.

86. Again, one could argue that the problem might lie not in deficits of selective attention but in a defective strategy that includes the need for directed or selective attention. Although this easily could develop into a circular argument, the important point is that this attention–perceptual process, which appears to be mediated by frontal systems, is not random but rather is guided and directed by the goals or behavioral programs

operating at the time in order to provide information critical to or consistent with the completion of that goal.

87. Unilateral neglect also may be produced by various subcortical lesions, including lesions of the thalamus, basal ganglia, and midbrain (Heilman, Watson, & Valenstein, 1993).

88. At times a patient may initiate and persevere with a particular response pattern, despite the fact that it is incorrect (i.e., fails to be reinforced).

89. Both of these failures of inhibition also may be seen in speech and writing or other motor programs other than drawing tasks. Also see Sandson and Albert (1987).

90. While the Wisconsin Card Sorting Test and the Category Test also are listed under tests of planning ability by Stuss and Benson (1986), these latter tasks, as noted earlier, are probably better measures of loss of mental flexibility or even decreased ability to profit from feedback (Goldstein & Green, 1995).

91. In some ways these tests are similar to the old word problem we encountered in school that goes something like this: A man had a duck, a sack of corn, and a fox that he needed to transport across a river. His boat was small and could carry only one extra item (in addition to himself) across at a time. However, if he left the duck and the corn together, the duck would eat the corn, and if he left the duck and the fox alone together, the fox would eat the duck. So how could he arrange to get all three safely across?

92. In contrast, several pathological syndromes that either primarily or secondarily may affect the orbital zones of the frontal cortex have been associated with significant amnestic syndromes, often with an accompanying tendency to confabulate. The most notable of these conditions is aneurysms of the anterior communicating artery (ACoA) (Alexander & Freedman, 1984; Damasio et al., 1985; Deluca & Diamond, 1995; Fasanaro et al., 1989; Gade, 1982; Green et al., 1995; Talland et al., 1967; see Chapter 10). Likewise, Korsakoff's syndrome (Butters & Cermak, 1980: Butters et al., 1987), which clinically has much in common with the amnestic syndrome associated with ACoA aneurysms, commonly affects the dorsomedial nucleus of the thalamus whose main cortical projection site is the frontal lobes. Victor, Adams, and Collins (1971) suggest that it is these thalamic lesions that may be primarily responsible for the observed memory deficits in Korsakoff's (see also Cummings, 1990, pp. 63–64).

93. Other more specific constructs or difficulties that have been associated with frontal lobe pathology include increased sensitivity to interference, or disturbances of attention and concentration (Chao & Knight, 1995; Stuss, 1991), defective working memory (Goldman-Rakic, 1987a; Kimberg et al., 1997), difficulty with recency judgments or temporal order (Milner & Teuber, 1968; Milner, 1971), difficulty with source memory (Janowsky et al., 1989), and a failure to inhibit irrelevant responses (Luria, 1966; see section on Self-monitoring).

94. If the solution to this problem does not immediately come to you, do not panic. Chances are your prefrontal cortex is just fine. This problem has temporarily stumped more than a few college graduates.

95. For those readers who may not be familiar with these measures, Picture Arrangement consists of a series of cartoon-style pictures that are presented in a jumbled order. When properly sequenced, they depict a logical, typically humorous, story. In the "tinker toy" test, the patient is presented with predetermined set of pieces from a Tinkertoy set, with the instructions to "make something." Scoring is based on complexity, meaningfulness, and its ability to perform some "function" (e.g., the ability to roll). The Sequin-Goddard formboard consists of ten cutouts of various geometric shapes (e.g., star, diamond, square, cross, etc.) in a board approximately 12 x 18 inches. While blindfolded and never having seen the board the subject is required to locate and place ten wooden blocks into their corresponding shapes (cutouts) on the board.

96. While such tests are indeed highly sensitive to brain injury, poor performance is not necessarily indicative of brain injury. A number of years ago, one of the authors (JEM) administered the Wisconsin Card Sorting Test to an intern, both as a means of teaching the test and to provide a sense of empathy for test-taking by patients. The intern in question was a very bright student yet he achieved zero categories on the test! As it is now many years later and this particular intern is still a very successful and articulate clinician, the possibility of some occult lesion is highly remote. In this case, the problem if anything seemed to be that the student was searching for more esoteric solutions rather than settling on the obvious.

97. With a few notable exceptions, such as the liver, heart, pituitary gland, and the digestive tract, much of our body is fairly symmetrically represented. Whether or not it reflects the evolutionary plan, this bilateral symmetry does allow for some biological redundancy. This is seen to a large extent in the brain and perhaps to a greater extent in the prefrontal cortices than elsewhere. For most individuals the secondary and tertiary zones of the posterior hemispheres show significant behavioral specificity. While bilateral lesions of the occipitoparietal cortices typically will result in visual–perceptual difficulties much worse than that seen with unilateral lesions, unilateral lesions of the angular gyrus, for example, may themselves produce very dramatic syndromes. While we will later review symptoms associated with unilateral lesions of the prefrontal cortex, as a general rule the prefrontal cortex appears more robust in this regard. Unilateral damage is less likely to result in the marked behavioral deficits. When obvious behavioral disturbances resulting from frontal lobe pathology are present, the damage is likely to be bilateral.

98. A related construct that sometimes is applied to patients with frontal impairment is "lack of insight," a concept that we will return to later in this section. However, as with many such metasymptoms, one must be cautious in attributing poor insight solely to frontal lesions. The term "insight" can take on a host of meanings but in a neuropsychological context it most commonly refers to a diminished capacity to appreciate one's deficits or shortcomings. In turn, among other things this can refer to one's failure to acknowledge physical symptoms, such as a hemiparesis (**anosognosia**) or more general behavioral or cognitive changes. However, at times, specific performance deficits may be due to sensory or perceptual limitations. The latter might include, for example, the failure to appreciate major errors in the attempt to draw a clock as a result of unilateral neglect. As previously noted, while anosognosia for hemiplegia most commonly has been associated with posterior lesions of the right hemisphere, a variety of cortical and subcortical lesions including lesions of the frontal lobes have been associated with this syndrome (Bisiach & Geminiani, 1991; Feinberg, 1997; Heilman, 1991). Poor insight into or lack of apparent concern about behavioral or cognitive changes or deficits frequently is associated with generalized dementias (Cummings & Benson, 1992; McGlynn & Kaszniak, 1991).

99. There may be reasons other than frontal lobe pathology for being relatively unconcerned about the adequacy of one's response. Depression, agitation, and acute pain are a few of the more common circumstances in which a patient may respond in an uncritical manner. In cases of malingering, the patient intentionally may provide erroneous responses.

100. In this case one can observe several potential signs of frontal pathology: poor comparator function, disinhibition (inappropriate familiarity), decreased ability to appreciate humor and/or concrete thinking, symptoms that also had been reported by others.

101. For a more thorough review of this topic, the reader is referred to Chapters 5 and 7 to 9 in Prigatano and Schacter's (1991) excellent volume that brings together a varied collection of works on deficits in awareness following brain injury.

102. Successful rehabilitation generally is based on several broad principles, including (1) the recognition that a problem exists, (2) the ability to develop a plan or stable intention to work on the problem, (3) the ability to utilize feedback, and (4) the drive or motivation to persevere in the effort to overcome it. With frontal lobe damage one or all of these basic abilities may be impaired. Thus rehabilitation for many of these individuals is like trying to build a house with limited or defective tools. Consider, for example, the potential difficulties posed by impaired self-monitoring capacity in a work setting.

103. Even at this level, however, descending cortical influences likely play a role, since alerting responses will be influenced by such factors as novelty and habituation (Sharpless & Jasper, 1956).

104. The problem may be not so much one of *increased* sexual drive as much as *reduced* inhibition of sexual impulses, a topic that will be addressed in the following section.

105. There also has been considerable interest in recent years over the possible connection between obsessive–compulsive disorders and disturbances in frontal–subcortical (particularly orbitofrontal–caudate) circuits that might normally regulate limbic arousal (Baxter et al., 1987; Malloy & Duffy, 1994; Mega & Cummings, 1994; Otto, 1992; Tallis, 1997; Zald & Kim, 1996a,b).

106. It should be noted that while "behavioral disinhibition" was part of his clinical picture, he (Phineas Gage) also manifested other aspects of frontal pathology, including what appeared to be a loss of cortically induced drive.

107. As noted above, certain types of impulsive, aggressive behaviors are suspected of being linked to frontal system impairment (Davidson, Putnam, and Larson, 2000). Although the evidence has been equivocal, it also has been suggested that sociopathy in general may reflect some type of learning disability or hard-wiring deficit involving the frontal lobes (Damasio, Tranel, & Damasio, 1991; Gorenstein, 1982; Kandel & Freed, 1989; Lapierre, Braun, & Hodgins, 1995; Lueger & Gill, 1990; Meyers et al., 1992; Price, Daffner, Stowe, & Mesulam, 1990).

108. One exception to the above rule is the syndrome of progressive aphasia, in which disturbances of speech and language (nonfluent, agrammatic speech, with marked paraphasic and anomic errors) typically precedes more gross behavioral changes.

109. Take, for example, the command, "Touch your left ear with your right hand." One should be able to generate at least a half-dozen or more "factors" that potentially might cause a failure to properly execute this apparently simple request. Contrast this with a test like the Category or Wisconsin Card Sorting tests.

110. Because of the nature of frontal lobe pathology, even when behavioral changes are detected early, they often are attributed to psychological or psychiatric factors rather than neurological disease. This is particularly true if the pathology has a slow, insidious onset and primarily affects the orbital–mesial areas of the prefrontal cortex.

111. It has been suggested (Malloy & Richardson, 1994) that tests of verbal fluency that require the subject to generate words beginning with particular letters of the alphabet (Benton, 1968) rather than provide words within a semantic category (e.g., fruits or animals) are reportedly more sensitive to frontal lesions.

112. It is important to keep in mind that initiating a particular response frequently involves the inhibition of potentially competing actions or response alternatives. In addition, the selected "response" may be *not to respond* (i.e., not to "go with" your initial or prepotent impulse to act or respond in a particular manner). Whether such response inhibition involves single muscle groups or more elaborate behavioral patterns, it ultimately

is based on an interaction between the prefrontal and agranular cortices (Goldman-Rakic, 1987b, p. 193).

113. Like the rest of the brain, these areas are extremely complex, both structurally and functionally. Only the highlights will be presented here. For those readers who may be interested in more detail, the following works might be recommended: *Motor Areas of the Cerebral Cortex* (Ciba Foundation, 1987); *Advances in Neurology*, Vol. 70: Supplementary Sensorimotor Area (Luders, 1996); *The Frontal Lobes and Voluntary Action* (Passingham, 1993).

114. Just as some routine "housekeeping" functions may be carried out in the military with little if any direct input from the General, so too certain "executive functions" may require little if any direct input from the prefrontal zones. One example might be reflex visual tracking behavior resulting from direct connections among the frontal eye fields, the occipitoparietal or peristriate areas, and the brainstem (e.g., Wise, Boussaoud, Johnson, & Caminiti, 1997).

115. The meaning of the term "initiation" (of action) may vary depending on context. In reference to the prefrontal areas, it refers to the *intention* or *will* to act. The prefrontal zones can be said to "initiate" behavioral (including, motor) programs in that they initiate the process that culminates in a certain action. However, the prefrontal zones do not directly produce (initiate) the depolarization of the corticobulbar or corticospinal fibers. The latter is the responsibility of the primary motor zones.

116. In addition to cytoarchitectonic differences, the frontal granular and agranular cortices also tend to have different thalamocortical connections. The prefrontal or granular areas are generally is associated with the nucleus medialis dorsalis (or "dorsomedial nucleus), whereas the motor and premotor areas are linked to the ventral lateral nucleus of the thalamus (Akert, 1964).

117. Picard and Strick (1996) also identify two additional motor areas on the medial surface along the banks of the cingulate sulcus ("cingulate motor areas"), but relatively little is known about their behavioral significance.

118. The pre-SMA area receives most of its input from the prefrontal area, while the SMA proper has extensive inputs from the primary motor

119. Again, "initiation" in this context does not refer so much to the organism's decision as to whether, when, or how it will respond to a class of stimuli or a set of circumstances as a whole (that most likely is the province of the frontal granular cortex). Rather, it refers to the SMA's role in the mechanics of initiating, planning (preparing to respond), and executing the more specific or elementary motor response patterns once the "go" signal is given to act at a particular time and/or in a particular manner (Freund, 1991).

120. Areas 44 and 45 generally are considered to constitute Broca's area.

121. Goldman-Rakic (1987b) suggests that since the parietal connections with the premotor area are much more substantial than those from the prefrontal area (at least in the monkey), a substantial portion of the prefrontal lobe's influence on the premotor area may be via the parietal lobe. However, in the ensuing discussion following Goldman-Rakic's proposal, other panel members suggest that the situation may be different in humans.

122. **Sympathetic apraxia** (apraxia of the left hand following left hemispheric lesion) is not uncommon following lesions that produce a Broca's-type aphasia and right-sided hemiparesis. However, in such cases the critical site of the lesion is uncertain. According to Heilman and Rothi (1993) lesions that are restricted to the premotor cortex have not been shown to result in ideomotor apraxia of the limbs, although Heilman and his colleagues reported several cases of bilateral ideomotor apraxia following left-sided SMA lesions (Watson, Fleet, Rothi, & Heilman, 1986).

123. While Broca's aphasia may be present in lesions affecting the premotor area, again a more extensive lesion is usually present often extending more anteriorly and/or deeper involving the basal ganglia (Alexander, Benson, & Stuss, 1989).

124. It has been suggested that SMA is associated with movement that is internally initiated. By contrast, the premotor area is more responsive to behaviors elicited by external stimuli. Tanji (1996) rejects this notion but does acknowledge that the premotor area appears critical when external cues help guide the behavior.

125. Although deficits following lesions confined to the premotor area would appear to result more from disruptions of subcortical and/or somatosensory as opposed to visual, feedback loops, the reverse is not true. Lesions that affect the posterior cortices often can result in motor responses that are poorly guided in space, although the action itself may be intact. One example may be difficulty maintaining well-organized, horizontal lines in writing that can be seen following certain parietal lesions. One possibility is that such lesions disrupt frontal–motor connections.

126. In contrast to area 6, which receives substantial input from the VA and VL nuclei of the thalamus, area 8 receives most of its input from the dorsomedial nucleus of the thalamus, similar to much of the prefrontal granular cortex. There also appears to be some connections with the pulvinar that is associated with processed, multisensory information.

127. The frontal eye fields also come close to meeting this description, the primary differences being that (1) the actions they control are intransitive, and (2) their influence on the extrinsic muscles of the eye are indirect.

128. As will be seen in the next chapter, this makes the legs and feet more susceptible to diseases affecting the anterior cerebral arteries and less susceptible to lesions of the middle cerebral artery.

REFERENCES

Ackerly, S. (1937) Instinctive, emotional, and mental changes following prefrontal lobe extirpation. *American Journal of Psychiatry, 92*, 717–729.

Adolphs, R., Damasio, H., Tranel, D., Damasio, A.R. (1996) Cortical systems for the recognition of emotion in facial expressions. *Journal of Neuroscience, 16*, 7678–7687.

Adolphs, R., Tranel, D., Damasio, H, & Damasio, A. (1994) Impaired recognition of emotions in facial expressions following bilateral damage to the human amygdala. *Nature, 372*, 369–372.

Akelaitis, A.J. (1944) A study of gnosis, praxis and language following section of the corpus callosum and anterior commissure. *Journal of Neurosurgery, 1*, 94–102.

Akert, K. (1964) Comparative anatomy of frontal cortex and thalamofrontal connections. In: Warren, J.M. & Akert, K. (Eds.) *The Frontal Granular Cortex and Behavior.* New York: MCGraw-Hill, pp. 372–396.

Albanese, E., Marlo, A., Albanese, A., & Gomez, E. (1989) Anterior speech region: Asymmetry and weight-surface correlation. *Archives of Neurology, 46*, 307–310.

Albert, M.L., Goodglass, H., Helm, N.A., Rubens, A.B., and Alexander, M.P. (1981) *Clinical a Aspects of Dysphasia.* New York: Springer-Verlag Wein.

Alexander, M.P., Benson, D.F., Stuss, D.T. (1989) Frontal lobes and language. *Brain and Language, 37*, 656–691.

Alexander, M. & Crutcher, M.D. (1990) Functional architecture of basal ganglia circuits: Neural substrates of parallel processing. *Trends in Neurosciences, 13*, 266–271.

Alexander, M. & Freedman, M. (1984) Amnesia after anterior communicating artery rupture. *Neurology, 34*, 752–757.

Anderson, S. W., Damasio, H., Jones, R.D. & Tranel, D. (1991) Wisconsin Card Sorting Test performance as a measure of frontal lobe damage. *Journal of Clinical & Experimental Neuropsychology, 13*, 909–922.

Andrews, D.G., Puce, A., & Bladin, P.F. (1990) Post-ictal recognition memory predicts laterality of temporal lobe seizure focus: Comparison with post-operative data. *Neuropsychologia, 28*, 957–967.

Annett, M. (1985) *Left, Right Hand and Brain: The right shift theory.* London: Erlbaum.

Annett, M. (1995) The right shift theory of a genetic balanced polymorphism for cerebral dominance and cerebral processing. *Cahiers de Psychologie Cognitive, 14*, 427–480.

Ardila, A. & Rosselli, M. (1993) Spatial agraphia. *Brain and Cognition, 22*, 137–147.

Arena, R. & Gainotti, G. (1978) Constructional apraxia and visuospatial disabilities in relation to laterality of lesions. *Cortex, 14*, 463–473.

Arrigoni, G. & DeRenzi, E. (1964) Constructional apraxia and hemispheric locus of lesion. *Cortex, 1*, 170–197.

Baddeley, A. (1986) *Working Memory.* Oxford: Oxford University Press.

Baddeley, A. (1994) Working memory: The interface between memory and cognition. In: Schacter, D.L. & Tulving, E. (Eds.), *Memory Systems 1994.* Cambridge, MA: MIT Press, pp. 351–367.

Barbas, H. (1988) Anatomical organization of basoventral and mediodorsal visual recipient prefrontal region in the rhesus monkey. *Journal of Comparative Neurology, 276*, 313–342.

Barbas, H., & Pandya, D.N. (1991) Patterns of connections of the prefrontal cortex in the Rhesus monkey associated with cortical architecture. In: Levin, H.S., Eisenberg, H.M., & Benton A.L. (Eds.), *Frontal Lobe Function and Dysfunction.* New York: Oxford University Press, pp. 35–58.

Basso, A. (1993) Amusia. In: Boller, F. & Grafman, J. (Eds.), *Handbook of Neuropsychology,* Vol. *8.* Amsterdam: Elsevier, pp. 391–410.

Bates, J.F. & Goldman-Rakic, P.S. (1993) Prefrontal connections of medial motor areas in the rhesus monkey. *Journal of Comparative Neurology, 336*, 211–228.

Bauer, R.M. (1993) Agnosia. In, Heilman, K. & Valenstein, E. *Clinical Neuropsychology.* New York: Oxford University Press, pp. 215–278.

Bauman, M.L. (1992) Neuropathology of autism. In: Joseph, A.B. & Young, R.R. (Eds.), *Movement Disorders in Neurology and Neuropsychology.* Oxford: Blackwell Scientific, pp. 662–666.

Baxter, L.R., Jr., Phelps, M.E., Mazziotta, J.C., Guze, B.H., Schwartz, J.M. Selin, C.E. (1987) Local cerebral glucose metabolic rates in obsessive–compulsive disorder. *Archives of General Psychiatry, 44*, 211–218.

Bechara, A., Tranel, D., Damasio, H. & Damasio, A.R. (1996) Failure to respond autonomically to anticipated future outcomes following damage to the prefrontal cortex. *Cerebral Cortex, 6*, 215–225.

Benbow, C.P. (1988) Sex differences in mathematical reasoning ability in intellectually talented preadolescents: Their nature, effects and possible causes. *Behavioral and Brain Sciences, 11*, 169–232.

Benson, D.F. (1979) *Aphasia, Alexia, and Agraphia.* New York: Churchill Livingstone.

Benson, D.F. (1993) Aphasia. In: Heilman, K. & Valenstein, E. (Eds.), *Clinical Neuropsychology.* New York: Oxford University Press, pp. 17–36.

Benson, D.F. (1994) *The Neurology of Thinking.* New York: Oxford University Press.

Benson, D.F. & Barton, M. (1970) Constructional disability. *Cortex, 6*, 19–46.

Benson, D.F. & Denckla, M.B. (1969) Verbal paraphasia as a source of calculation disturbance. *Archives of Neurology, 21*, 96–102.

Benson, D.F. & Geschwind, N. (1985) Aphasia and related disorders: A clinical approach. In: Mesulam, M. (Ed.), *Principles of Behavioral Neurology.* Philadelphia: F.A. Davis Co., pp. 193–238.

Benson, D.F. & Stuss, D.T. (1982) Motor abilities after frontal leukotomy. *Neurology, 32*, 1353–1357.

Benton, A.L. (1959) *Right-Left Discrimination and Finger Localization.* New York: Hoeber-Harper.

Benton, A. (1961) The fiction of the Gerstmann syndrome. *Journal of Neurology, Neurosurgery and Psychiatry, 24*, 176–181.

Benton, A. (1967) Constructional apraxia and the minor hemisphere. *Confina Neurologica, 29*, 1–16.

Benton, A.L. (1968) Differential behavioral effects in frontal lobe disease. *Neuropsychologia, 6*, 53–60.

Benton, A. (1977a) The amusias. In: Critchley, M. & Henson, R.A. (Eds.), *Music and the Brain.* London: William Heinemann, pp. 378–397.

Benton, A. (1977b) Reflections on the Gerstmann syndrome. *Brain and Language, 4*, 45–62.

Benton, A.L. (1980) The neuropsychology of facial recognition. *American Psychologist, 35*, 176–186.

Benton, A. (1990) Facial recognition 1990. *Cortex, 26*, 491–499.

Benton, A.L. (1991) The prefrontal region: Its early history. In: Levin, H.S., Eisenberg, H.M., & Benton A.L. (Eds.), *Frontal Lobe Function and Dysfunction.* New York: Oxford University Press, pp. 3–32.

Benton, A. Hamsher, K., Varney, N. & Spreen, O. (1983) *Contributions to Neuropsychological Assessment.* New York: Oxford University Press.

Benton, A.L., Hannay, H.J., & Varney, N.R. (1975) Visual perception of line direction in patients with unilateral brain disease. *Neurology, 25,* 907–910.

Benton, A, Levin, H., & Van Allen, M. (1974) Geographic orientation in patients with unilateral cerebral disease. *Neuropsychologia, 12,* 183–191.

Benton, A. & Sivan, A.B. (1993) Disturbances of body schema. In: Heilman, K. & Valenstein, E. *Clinical Neuropsychology.* New York: Oxford University Press, pp. 123–140.

Benton, D.F., Stuss, D.T., Naeser, M.A., Weir, W.S., Kaplan, E.F., Levine, H. (1981) The long-term effects of prefrontal leukotomy. *Archives of Neurology, 38,* 165–169.

Benton, A. & Tranel, D. (1993) Visuoperceptual, visuospatial, and visuoconstructive disorders. In: Heilman, K. & Valenstein, E. *Clinical Neuropsychology.* New York: Oxford University Press, pp. 165–213.

Benton, A. & Van Allen, M. (1968) Impairment in facial recognition in patients with cerebral disease. *Cortex, 4,* 344–358.

Bever, T.G. & Chiarello, R.J. (1974) Cerebral dominance in musicians and nonmusicians. *Science, 185,* 537–539.

Bisiach, E. & Geminiani, G. (1991) Anosognosia related to hemiplegia and hemianopia. In: Prigatano, G.P. & Schacter, D.L. *Awareness of Deficit after Brain Injury: Clinical and theoretical issues.* New York: Oxford, pp. 17–39.

Black, F.W. & Strub, R.L. (1976) Constructional apraxia in patients with discrete missile wounds of the brain. *Cortex, 12,* 212–220.

Bleasel, A., Comair, Y. & Luders, H.O. (1996) Surgical ablations of the mesial frontal lobe in humans. In: Luders, H.O. (Ed.), *Advances in Neurology, Vol. 70: Supplementary Sensorimotor Area.* Philadelphia: Lippincott-Raven, pp. 217–235.

Blumer, D. & Benson, D.F. (1975) Personality changes with frontal and temporal lesions. In: Benson, D.F. & Blumer, D. (Eds.), *Psychiatric Aspects of Neurologic Disease.* New York: Grune & Stratton, pp. 151–170.

Bogen, J.E. (1993) The callosal syndromes. In: Heilman, K. & Valenstein, E. (Eds.), *Clinical Neuropsychology.* New York: Oxford University Press, pp. 337–407.

Boll, T.J. (1981) The Halstead-Reitan neuropsychological battery. In: Filskov, S.B., & Boll, T.J. (Eds.), *Handbook of Clinical Neuropsychology.* New York: Wiley, pp. 577–607.

Boller, F. and Grafman, J. (Eds.) (1985) *Handbook of Neuropsychology.* New York: Elsevier, Vol 1–5.

Boller, F. & Grafman, J. (1988) Acalculia. In: Vinkin, P.J., Bruyn, G.W. & Klawans, H.L. (Eds.), *Handbook of Clinical Neurology,* Vol. 1. Amsterdam: North Holland, pp. 473–481.

Borod, J., Koff, E., Perlman-Lorch, M., Nicholas, M. & Welkowitz, J. (1988) Emotional and nonemotional facial behavior in patients with unilateral brain damage. *Journal of Neurology, Neurosurgery and Psychiatry, 51,* 826–832.

Bowers, D., Bauer, R.M., Coslett, H.B. & Heilman, K.M. (1985) Processing of faces by patients with unilateral hemispheric lesions. I. Dissociation between judgments of facial affect and facial identity. *Brain and Cognition, 4,* 258–272.

Boyd, J.L. (1981) A validity study of the Hooper Visual Organization Test. *Journal of Consulting and Clinical Psychology, 49,* 15–19.

Brazier, M.A.B (1984) *History of Neurophysiology in the 17th and 18th Centuries.* New York: Raven Press.

Brazier, M.A.B (1988) *History of Neurophysiology in the 19th Century.* New York: Raven Press.

Broca, P. (1861) Remarques sur le siege de la faculte du langage articule, suives d'une observation de aphemie. *Bulletin Des Societe Anatomique de Paris, 2,* 330–357.

Brodal, A. (1973) Self-observations and neuro-anatomical considerations after a stroke. *Brain, 96,* 675694.

Brown, J. (1972) *Aphasia, Apraxia and Agnosia: Clinical and theoretical aspects.* Springfield, IL: Charles C. Thomas, pp. 190–191.

Brown, H.D. & Kosslyn, S.M. (1995) Hemispheric differences in visual object processing: Structural versus allocation theories. In: Davidson, R.J. & Hugdahl, K. (Eds.), *Brain Asymmetry.* Cambridge, MA: MIT Press, pp. 77–97.

Brun, A. (1987) Frontal lobe degeneration of non-Alzheimer type. I. Neuropathology. *Archives of Gerontology and Geriatrics, 6,* 193–208.

Brun, A. & Gustafson, L. (1999) Clinical and pathological aspects of frontotemporal dementia. In: Miller, B.L. & Cummings, J.L. (Eds.), *The Human Frontal Lobes.* New York: Guilford Press, pp. 349–369.

Brust, J.C. (1996) Lesions of the supplementary motor area. In: Luders, H.O. (Ed.), *Advances in Neurology, Vol. 70: Supplementary Sensorimotor Area.* Philadelphia: Lippincott-Raven, pp. 237–248.

Brutkowski, S. (1965) Functions of prefrontal cortex in animals. *Physiological Review, 45,* 721–746.

Bryden, M.P. (1982) *Laterality: Functional asymmetry in the intact brain.* New York: Academic Press.

Buckner, R.L., Petersen, S.E., Ojemann, J.G., Miezin, F.M., Squire, L.R. & Raichle, M.E. (1995) Functional anatomical studies of explicit and implicit memory retrieval tasks. *Journal of Neuroscience, 15,* 12–39.

Butters, N. & Barton, M. (1970) Effect of parietal lobe damage on the performance of reversible operations in space. *Neuropsychologia, 8,* 205–214.

Butters, N. & Cermak, L. (1980) *Alcoholic Korsakoff's syndrome: An information processing approach to amnesia.* New York: Academic Press.

Butters, N, Granholm, E., Salmon, D.P, & Grant, I. (1987) Episodic and semantic memory: A comparison of amnesic and demented patients. *Journal of Clinical and Experimental Neuropsychology, 9,* 479–497.

Campbell, J.J., Duffy, J.D. & Salloway, S.P. (1994) Treatment strategies for patients with dysexecutive syndromes. *Journal of Neuropsychiatry and Clinical Neurosciences, 6,* 411–418.

Carmon, A. & Bechtoldt, H.P. (1969) Dominance of the right cerebral hemisphere for stereopsis. *Neuropsychologia, 7,* 29–39.

Carpenter, M.B. & Sutin, J. (1983) *Human Neuroanatomy.* Baltimore: Williams & Wilkins.

Cavada, C. & Goldman-Rakic, P.S. (1989) Posterior parietal cortex in rhesus monkeys: II Evidence for segregated corticocortical networks linking sensory and limbic areas with the frontal lobe. *Journal of Comparative Neurology, 287,* 422–445.

Chao, L.L. & Knight, R.T. (1995) Human prefrontal lesions increase distractibility to irrelevant sensory inputs. *NeuroReport, 6,* 1605–1610.

Chatterjee, A., & Southwood, M.H. (1995) Cortical blindness and visual imagery. *Neurology, 45,* 2189–2195.

Chauvel, P.Y., Rey, M., Buser, P. & Bancaud, J. (1996) What stimulation of the supplementary motor area in humans tells about its functional organization. In: Luders, H.O. (Ed.), *Advances in Neurology, Vol. 70: Supplementary Sensorimotor* Area. Philadelphia: Lippincott-Raven, pp. 199–209.

Chelune, G.J. & Bornstein, R.A. (1988) WMS-R patterns among patients with unilateral brain lesions. *The Clinical Neuropsychologist, 2,* 121–132.

Christensen, A-L. (1975) *Luria's Neuropsychological Investigation.* New York: Spectrum.

Christianson, S-A., Saisa, J, & Silfvenius, H. (1990) Hemispheric memory differences in sodium amytal testing of epileptic patients. *Journal of Clinical and Experimental Neuropsychology, 12,* 681–694.

Chui, H.C. & Damasio, A.R. (1980) Human cerebral asymmetries evaluated by computed tomography. *Journal of Neurology, Neurosurgery and Psychiatry, 43,* 873–878.

Clark, A.F., & Davidson, K. (1987) Mania following head injury. *British Journal of Psychiatry, 150,* 841–844.

Clarke, E. & O'Malley, C.D. (1968) *The Human Brain and Spinal Cord.* Los Angeles: University of California Press.

Cofer, C.N. & Appley, M.H. (1964) *Motivation: Theory and research.* New York: John Wiley & Sons.

Cohen, M. (1992) Auditory/verbal and visual/spatial memory in children with complex partial epilepsy of temporal lobe origin. *Brain and Cognition, 20,* 315–326.

Cohn, R. (1961) Dyscalculia. *Archives of Neurology, 4,* 301–307.

Corballis, M.C. (1994) Can commissurotomized subjects compare digits between the visual fields? *Neuropsychologia, 32,* 1475–1486.

Corkin, S. (1965) Tactually guided maze learning in man: Effects of unilateral cortical excisions and bilateral hippocampal lesions. *Neuropsychologia, 3,* 339–351.

Corkin, S. (1978) The role of different cerebral structures in somesthetic perception. In: Carterette, C.E. & Friedman, M.P. (Eds.), *Handbook of Perception, Vol. VI B.* New York: Academic Press, pp. 105–155.

Corkin, S. (1979) Hidden Figures Test performance: Lasting effects of unilateral penetrating head injury and transient effects of bilateral cingulotomy. *Neuropsychologia, 17,* 585–605.

Coslett, H.B., Bowers, D., Fitzpatrick, E., Haws, B, & Heilman, K.M. (1990) Directional hypokinesia and hemispatial inattention in neglect. *Brain, 113*, 475–486.

Crichton-Browne, J. (1880) On the weight of the brain and its component parts in the insane. *Brain, 2*, 467.

Critchley, M. (1969) *The Parietal Lobes.* NewYork: Hafner Publishing Co.

Critchley, M. (1966) The enigma of Gerstmann's syndrome. *Brain, 89*, 183–198.

Cronin-Golomb, A. (1986) Subcortical transfer of cognitive information in subjects with complete forebrain commissurotomy. *Cortex, 22*, 499–519.

Crowne, D.P. (1983) The frontal eye fields and attention. *Psychological Bulletin, 93*, 232–260.

Cummings, J.L. (1990) **Subcortical Dementia.** New York: Oxford Press.

Cummings, J.L. (1995) Anatomic and behavioral aspects of frontal-subcortical circuits. *Annals of the New York Academy of Sciences, 769*, 1–13.

Cummings, J.L. & Benson, D.F. (1992) *Dementia: A clinical approach, 2nd Edition.* Boston: Butterworth-Heinemann.

Cummings, J.L. & Coffey, C.E. (1994) Neurobiological basis of behavior. In: Coffey, C.E., Cummings, J.L., Lovell, M.R., & Pearlson, G.D. (Eds.), *The American Psychiatric Press of Geriatric Neuropsychiatry. Textbook* Washington, DC: American Psychiatric Press, pp. 71–96.

Cunningham, D.J. (1892) *Contribution to the Surface Anatomy of the Cerebral Hemispheres. Cunningham Memoirs, No VII.* Dublin: Royal Irish Academy.

Cutting, J. (1990) *The Right Cerebral Hemisphere and Psychiatric Disorders.* New York: Oxford University Press.

Cytowic, R.E. (1996) *The Neurological Side of Neuropsychology.* Cambridge, MA: MIT Press.

Dax, M. (1865) Lesion de la moitie gauche de l'encephale coincidant avec l'oubli des signes de la pensee. *Gazette Hebdomidaire de Medecine et de Chirurgie, 2*, 259–262.

Dahmen, W., Hartje, W., Bussing, A., & Strum, W. (1982) Disorders of calculation in aphasic patients—spatial and verbal components. *Neuropsychologia, 20*, 145–153.

Damasio, A.R. (1985) Disorders of complex visual processing: Agnosias, achromatopsia, Balint's syndrome, and related difficulties of orientation and construction. *In: Mesulam, M.M. (Ed.), Principles of Behavioral Neurology.* Philadelphia: F.A. Davis, pp. 259–288.

Damasio, A.R. & Anderson, S.W. (1993) The frontal lobes. In: Heilman, K.M. & Valenstein, E. (Eds.), *Clinical Neuropsychology.* New York: Oxford University Press, pp. 409–460.

Damasio, A.R., & Damasio, H. (1983) The anatomic basis of pure alexia. *Neurology, 33*, 1573–1583.

Damasio, A.R., & Damasio, H. (2000) Aphasia and the neural basis of language. In: *Principles of Behavioral and Cognitive Neurology, 2nd Edition.* New York: Oxford University Press, pp. 294–315.

Damasio, A.R., Damasio, H, & Chui, H.C. (1980) Neglect following damage to frontal lobe or basal ganglia. *Neuropsychologia, 18*, 123–132.

Damasio, A.R., Graff-Radford, N.R., Eslinger, P.J., Damasio, H., & Kassell, N. (1985) Amnesia following basal forebrain lesions. *Archives of Neurology, 42*, 263–271.

Damasio, A.R., Tranel, D., & Damasio, H.C. (1991) Somatic markers and the guidance of behavior: Theory and preliminary testing. In: Levin, H.S., Eisenberg, H.M., & Benton A.L. (Eds.), *Frontal Lobe Function and Dysfunction.* New York: Oxford University Press, pp. 217–229.

Damasio, A.R., Tranel, D., & Damasio, H. (1990) Individuals with sociopathic behavior caused by frontal damage fail to respond autonomically to social stimuli. *Behavioral Brain Research, 41*, 81–94.

Damasio, A.R., Tranel, D. & Rizzo, M. (2000) Disorders of complex visual processing. In Mesulam, M-M (Ed) *Principles of Behavioral and Cognitive Neurology.* New York: Oxford University Press, p. 332–372.

Damasio, A.R. & Van Hoesen, G.W. (1980) Structure and function of the supplementary motor area. Neurology, 30, 359.

Damasio, A.R., Yamada, T., Damasio, H., Corbett, J., & McKee, J. (1980) Central achromatopsia: behavioral, anatomic, and physiologic aspects. *Neurology, 30*, 1064–1071.

Damasio, H. (1991) Neuroanatomy of frontal lobe in vivo: A comment on methodology. In: Levin, H.S., Eisenberg, H.M., & Benton A.L. (Eds.), *Frontal Lobe Function and Dysfunction.* New York: Oxford University Press, pp. 92–121.

Damasio, H., Grabowski, T., Frank, R., Galaburda, A.M., & Damasio, A.R. (1994) The return of Phineas Gage: Clues about the brain from the skull of a famous patient. *Science, 264*, 1102–1105.

Damasio, H., Grabowski, T.J., Tranel, D., Hichwa, & Damasio, A. (1996) A neural basis for lexical retrieval. *Nature, 380,* 499–505.

Davidson, R.J. (1995) Cerebral asymmetry, emotion, and affective style. In: R.J. Davidson & K. Hugdahl (Eds.), *Brain Asymmetry.* Cambridge, MA: MIT Press, pp. 361–387.

Davidson, R.J. & K. Hugdahl, K. (1995) (Eds.) *Brain Asymmetry.* Cambridge, MA: MIT Press.

Davidson, R.J., Putnam, K.M. & Larson, C.L. (2000) Dysfunction in the neural circuitry of emotion regulation–A possible prelude to violence. Science, 289, 591–594.

DeAjuriaguerra, J. & Tissot, R. (1969) The apraxias. In: Vinkin, P.J. & Bruyn, G.W. (Eds.), *Handbook of Clinical Neurology, Vol. 4.* Amsterdam: North Holland.

Dee, H.L. (1970) Visuoconstructive and visuoperceptive deficits in patients with unilateral cerebral lesions. *Neuropsychologia, 8,* 305–314.

Deecke, L. (1987) Bereitschaftspotential as an indicator of movement preparation in supplementary motor areas and motor cortex. In: *Ciba Foundation Symposium, Motor areas of the cerebral cortex.* New York: John Wiley & Sons, pp. 231–250.

Dehaene, S. & Cohen, L. (1991) Two mental calculation systems: A case study of severe acalculia with preserved approximation. *Neuropsychologia, 29,* 1045–1054.

Delis, D.C., Kaplan, E. & Kramer, J.H. (2001) *D-KEFS:Executive Function System.* San Antonio: Psychological Corporation.

Delis, D.C., Kiefner, M. & Fridlund, A.J. (1988) Visuospatial dysfunction following unilateral brain damage: Dissociations in hierarchical and hemispatial analysis. *Journal of Clinical and Experimental Neuropsychology,* 10 421–431.

Delis, D.C., Robertson, L.C., & Efron, R. (1986) Hemispheric specialization of memory for visual hierarchical stimuli. *Neuropsychologia, 24,* 205–214.

De Luca, J., & Diamond, B.J. (1995) Aneurysm of the anterior communicating artery: A review of neuroanatomical and neuropsychological deficits. *Journal of Clinical and Experimental Neuropsychology, 17,* 100–121.

Denes, G. (1989) Disorders of body awareness and body knowledge. In: Boller, F. & Grafman, J. (Eds.), *Handbook of Neuropsychology, Vol. 2.* Amsterdam: Elsevier, pp. 207–228.

DeRenzi, E. (1997a) Visuospatial and constructional disorders. In: Feinberg, T. & Farah, M.J. (Eds.), *Behavioral Neurology and Neuropsychology.* New York: McGraw-Hill, pp. 297–307.

DeRenzi, E. (1997b) Prosopagnosia. In: Feinberg, T. & Farah, M.J. (Eds.), *Behavioral Neurology and Neuropsychology.* New York: McGraw-Hill, pp. 245–255.

DeRenzi, E. & Faglioni, P. (1967) The relationship between visuospatial impairment and constructional apraxia. *Cortex, 3,* 327–342.

DeRenzi, E., Faglioni, P. & Scotti, G. (1971) Judgment of spatial orientation in patients with focal brain lesions. *Journal of Neurology, Neurosurgery, and Psychiatry, 34,* 489–495.

DeRenzi, E., Perani, D., Carlesimo, G.A., Silveri, M.C. & Fazio, F. (1994) Prosopagnosia can be associated with damage confined to the right hemisphere: An MRI and PET study and a review of the literature. *Neuropsychologia, 32,* 893–902.

DeRenzi, E. & Scotti, G. (1969) The influence of spatial disorders in impairing tactual discrimination of shapes. *Cortex, 5,* 53–62.

DeRenzi, E. & Spinnler, H. (1967) Impaired performance on color tasks in patients with hemispheric damage. *Cortex, 3,* 194–216.

DeRenzi, E. & Spinnler, H. (1966) Visual recognition in patients with unilateral cerebral disease. *Journal of Nervous and Mental Disease, 142,* 515–525.

Diamond, B.J., Valentine, T., Mayes, A.R., Sandel, M.E. (1994) Evidence of covert recognition in a prosopagnosic patient. *Cortex, 30,* 377–393.

Dick, J.P.R., Benecke, R., Rothwell, J.C., Day, B.L., & Marsden, C.D. (1986) Simple and complex movements in a patient with infarction of the right supplementary motor cortex. *Movement Disorders, 1,* 255–266.

Diggs, C. & Basili, A.G. (1987) Verbal expression of right cerebrovascular accident patients: Convergent and divergent language. *Brain and Language, 30,* 130–146.

Dobbs, A.R. & Rule, B.G. (1987) Prospective memory and self-reports of memory abilities. *Canadian Journal of Psychology, 41,* 209–222.

Downer, J.L. (1961) Changes in visual gnostic functions and emotional behavior following unilateral temporal pole damage in the split-brain monkey. *Nature, 191,* 50–51.

Duffy, C.J. (1984) The legacy of association cortex. *Neurology, 34,* 192–197.

Duffy, J.D. & Campbell, J.J (1994) The regional prefrontal syndromes: A theoretical and clinical overview. *Journal of Neuropsychiatry, 6,* 379–387.

Duke, L. & Kaszniak, A.W. (2000) Executive control functions in degenerative dementias: A comparative review. *Neuropsychology Review, 10,* 75–99.

Dum, R.P. & Strick, P.L. (1991) Premotor areas: Nodal points for parallel efferent systems involved in the central control of movement. In: Humphrey, D.R. & Freund, H-J. (Eds.), *Motor Control: Concepts and issues.* New York: John Wiley & Sons, pp. 383–397.

Duncan, J., Rudiger, J.S., Kolodny, J., Bor, D., Herzog, H., Ahmed, A., Newell, F.N. & Emslie, H. (2000) A neural basis for general intelligence. *Science, 289,* 457–460.

Eastwood, M.R., Rifat, S.L., Nobbs, H. & Ruderman, J. (1989) Mood disorder following cerebrovascular accident. *British Journal of Psychiatry, 154,* 195–200.

Eberstaller, O. (1890) *Das Stirhirn.* Wein and Leipzig: Urban and Schwarzenberg.

Edwards, B. (1979) *Drawing on the Right side of the Brain.* Los Angeles: J.P. Tarcher.

Eslinger, P.J. & Damasio, A.R. (1985) Severe disturbance of higher cognition after bilateral frontal lobe ablation. *Neurology, 35,* 1731–1741.

Faglioni, P., Scotti, G. & Spinnler, H. (1970) Colouring drawings impairment following unilateral brain damage. *Brain Research, 24,* 546.

Falzi, G., Perrone, P., & Vignolo, L., (1982) Right–left asymmetry in anterior speech region. *Archives of Neurology, 39,* 239–240.

Fasanaro, A.M., Valiani, R., Russo, G., Scarano, E., De Falco, R., & Profeta, G. (1989) Memory performances after anterior communicating artery aneurysm surgery. *Acta Neurologica, 11,* 272–278.

Feinberg. T. (1997) Anosognosia and confabulation. In: Feinberg, T.E. & Farah, M.J. (Eds.), *Behavioral Neurology and Neuropsychology.* New York: McGraw-Hill, pp. 369–390.

Filley, C.M. (1995) *Neurobehavioral Anatomy.* Niwot, CO: University Press of Colorado.

Filley, C.M. & Cullum, C.M. (1993) Early detection of fronto-temporal degeneration by clinical evaluation. *Archives of Clinical Neuropsychology, 8,* 359–367.

Fisher, R. (1951) Psychosurgery. In: Walker, A.E. (Ed.), *A History of Neurosurgery.* Baltimore: Williams & Wilkins, pp. 272–284.

Finger, S. (1994) *Origins of Neuroscience: A history of explorations into brain function.* New York: Oxford University Press.

Fischer, S.C. & Pellegrino, J.W. (1988) Hemispheric differences for components of mental rotation. Brain and Cognition, 7, 1–15.

Foldi, N.S. (1987) Appreciation of pragmatic interpretation of indirect commands: Comparison of right and left hemisphere brain-damaged patients. *Brain and Language, 31,* 88–108.

Foundas, A.L., Eure, K.F., Luevano, L.F., & Weinberger, D.R. (1998) MRI asymmetries of Broca's area: The pars triangularis and pars opercularis. *Brain and Language, 64,* 282–296.

Foundas, A.L., Hong, K., Leonard, C.M., & Heilman, K.M. (1998). Hand preference and MRI asymmetries of the central sulcus. *Neuropsychiatry, Neuropsychology, and Behavioral Neurology, 2,* 65–71.

Foundas, A.L., Leonard, C.M., Gilmore, M., Fennel, E., & Heilman, K.M. (1994) Planum temporale asymmetry and language dominance. *Neuropsychologia, 32,* 1225–1231.

Foundas, A.L., Leonard, C., Gilmore, R., Fennell, E., & Heilman, K. (1996) Pars triangularis asymmetry and language dominance. *Proceedings of the National Academy of Sciences, 93,* 719–722.

Foundas, A.L., Leonard, C.M. & Heilman, K.M. (1995) Morphologic cerebral asymmetries and handedness: The pars triangularis and planum temporale. *Archives of Neurology, 52,* 501–508.

Frackowiak, R.S.J., Friston, K.J., Firth, C.D., Dolan, R.J. & Mazziotta, J.C. (1997) *Human Brain Functions.* New York: Academic Press.

Frederiks, J.A.M. (1969) Disorders of body schema. In: Vinken, P.J., & Bruyn, G.W. (Eds.), *Handbook of Clinical Neurology.* Amsterdam: North Holland, Vol. 4, pp. 207–240.

Freedman, M., Alexander, M.P. & Naeser, M.A. (1984) The anatomical basis of transcortical motor aphasia. *Neurology, 34,* 409–417.

Freeman, W. & Watts, J.W. (1942) *Psychosurgery.* Springfield: Charles C. Thomas.

Freund, H. & Hummelsheim, H. (1985) Lesions of premotor cortex in man. *Brain, 108,* 697–733.

Freund, H-J. (1987) Differential effects of cortical lesions in humans. In: *Ciba Foundation Symposium, Motor areas of the cerebral cortex.* New York: John Wiley & Sons, pp. 269–281.

Freund, H-J (1991) What is the evidence for multiple motor areas in the human brain? In: Humphrey, D.R. & Freund, H-J. (Eds.), *Motor Control: Concepts and issues.* New York: John Wiley & Sons, pp. 399–411.

Freund, H-J (1996a) Historical overview. In: Luders, H.O. (Ed.), *Advances in Neurology, Vol. 70: Supplementary Sensorimotor Area.* Philadelphia: Lippincott-Raven, pp. 17–27.

Freund, H-J. (1996b) Functional organization of the human supplementary motor area and dorsolateral premotor cortex. In: Luders, H.O. (Ed.), *Advances in Neurology, Vol. 70: Supplementary Sensorimotor Area.* Philadelphia: Lippincott-Raven, pp. 263–269.

Frisk, V. & Milner, B. (1990) The relationship of working memory to the immediate recall of stories following unilateral temporal or frontal lobectomy. *Neuropsychologia, 28,* 121–135.

Fuster, J.M. (1997) *The Prefrontal Cortex.* New York: Raven Press.

Gade, A. (1982) Amnesia after operations on aneurysms of the anterior communicating artery. *Surgical Neurology, 18,* 46–49.

Gainotti, G. (1997) Emotional disorders in relation to unilateral brain damage. In: Feinberg, T.E. & Farah, M.J. (Eds.), *Behavioral Neurology and Neuropsychology.* New York: McGraw-Hill, pp. 691–698.

Gainotti, G. (1972) Emotional behavior and hemispheric side of lesion. *Cortex, 8,* 41–55.

Gainotti, G., Cappa, A., Perri, R., & Silveri, M.C. (1994) Disorders of verbal and pictorial memory in right and left brain-damaged patients. International *Journal of Neuroscience, 78,* 9–20

Gainotti, G., Cianchetti, C, & Tiacci, C. (1972) The influence of hemispheric side of lesion on non-verbal tests of finger localization. *Cortex, 8,* 364–381.

Gainotti, G. & Tiacci, C. (1970) Patterns of drawing disability in right and left hemispheric patients. *Neuropsychologia, 8,* 379–384.

Galaburda, A. (1984) Anatomical asymmetries. In: Geschwind, N. & Galaburda, A. (Eds.), *Cerebral Dominance.* Cambridge, MA: Harvard University Press, pp. 11–25.

Galaburda, A.M., Corsiglia, J., Rosen, G.D. & Sherman, G.F. (1987) Planum temporale asymmetry, reappraisal since Geschwind and Levitsky. *Neuropsychologia, 25,* 853–868.

Garner, H., Brownell, H.H., Wapner, W., & Michelow, D. (1983) Missing the point: The role of the right hemisphere in the processing of complex linguistic materials. In: Perceman, E. (Ed.), *Cognitive Processing in the Right Hemisphere.* New York: Academic Press, pp. 169–191.

Garner, H., Ling, P.K., Flamm, L. & Silverman, J. (1975) Comprehension and appreciation of humorous material following brain damage. *Brain, 98,* 399–412.

Gates, A. & Bradshaw, J.L. (1977a) The role of the cerebral hemispheres in music. *Brain and Language, 3,* 451–460.

Gates, A. & Bradshaw, J.L. (1977b) Music perception and cerebral asymmetries. *Cortex, 13,* 390–401.

Gazzaniga, M.S., Bogen, J.E., & Sperry, R.W. (1962) Some functional effects of sectioning the cerebral commissures in man. *Proceedings of the National Academy of Science, 48,* 1765–1769.

Gazzaniga, M.S., Bogen, J.E., & Sperry, R.W. (1965) Observations on visual perception after disconnection of the cerebral hemispheres in man. *Brain, 88,* 221–236.

Gazzaniga, M.S., Bogen, J.E., & Sperry, R.W. (1967) Dyspraxia following division of the cerebral commissures. *Archives of Neurology, 16,* 606–612.

Gerstmann, J. (1940) Syndrome of finger agnosia, disorientation for right and left, agraphia and acalculia. *Archives of Neurology and Psychiatry, 44,* 398–408.

Geschwind, N. (1965) Disconnexion syndromes in man and animals. *Brain, 88,* 237–294, 585–644.

Geschwind, N. (1975) The apraxias: Neural mechanisms of disorders of learned movement. *American Scientist. 63,* 188–195.

Geschwind, N. & Galaburda, A.M. (Eds) (1984) *Cerebral Dominance: The biological foundations.* Cambridge, MA: Harvard University Press.

Geschwind, N. & Galaburda, A.M. (1985) Cerebral lateralization. Biological mechanisms, associations, and pathology: I. A hypothesis and a program for research. *Archives of Neurology, 42,* 428–459.

Geschwind, N. & Galaburda, A.M. (1987) *Cerebral Lateralization: Biological mechanisms, associations, and pathology.* Cambridge, MA: MIT Press.

Geschwind, N. & Galaburda, A. (1985) Cerebral lateralization. *Archives of Neurology, 42,* 428–459; 521–552; 634–654.

Geschwind, N. & Levitsky, W. (1968) Left-right asymmetry in temporal speech region. *Science*, *161*, 186–187.

Geschwind, N. & Strub, R. (1975) Gerstmann syndrome without aphasia: rely to Poeck and Orgass. *Cortex*, *11*, 296–298.

Glick, S.D., Ross, D.A. & Hough, L.B. (1982) Lateral asymmetry of neurotransmitter in human brain. *Brain Research*, *234*, 53–63.

Goldberg, E., Podell, K. & Lowell, M. (1994) Lateralization of frontal lobe functions and cognitive novelty. *Journal of Neuropsychiatry and Clinical Neurosciences*, *6*, 371–378.

Goldenberg, Georg (1997) Disorders of body perception. In: Feinberg, T.E. & Farah, M.J. (Eds.),*Behavioral Neurology and Neuropsychology*. New York: McGraw-Hill, pp. 289–296.

Goldman-Rakic, P.S. (1987a) Circuitry of primate prefrontal cortex and regulation of behavior by representational memory. In: Plum, F. & Mountcastle, V. (Eds.), *Handbook of Physiology, the Nervous System: V*. Bethesda: American Physiological Society.

Goldman-Rakic, P.S. (1987b) Motor control function of the premotor cortex. In: *Ciba Foundations Symposium, Motor areas of the cerebral cortex*. New York: John Wiley & Sons, pp. 187–200.

Goldman-Rakic, P.S (1988) Topography of cognition: parallel distributed networks in primate association cortex. *Annual Review of Neuroscience*, *11*, 137–156.

Goldman-Rakic, P.S., Bates, J.F. & Chafee, M.V. (1992) The prefrontal cortex and internally generated motor acts. *Current Opinion in Neurobiology*, *2*, 830–835.

Goldman-Rakic, P.S., Selemon, L.D., & Schwartz, M.L. (1984) Dual pathways connecting the dorsolateral prefrontal cortex with the hippocampal formation and the parahippocampal cortex in the rhesus monkey. *Neuroscience*, *12*, 719–743.

Goldstein, F.C. & Green, R.C. (1995) Problem solving and executive functions. In: Mapou, R.L & Spector, J. (Eds.), *Clinical Neuropsychological Assessment: A cognitive approach*. New York: Plenum Press, pp. 49–81.

Goldstein, K. (1939) Clinical and theoretical aspects of lesions of the frontal lobes. *Archives of Neurology and Psychiatry*, *41*, 865–867.

Goldstein, K. (1944) Mental changes due to frontal lobe damage. *Journal of Psychology*, *17*, 187–208.

Goodglass, H. (1993) *Understanding Aphasia*. New York: Academic Press.

Goodglass, H. & Kaplan, E. (1963) Disturbance of gesture and pantomime in aphasia. *Brain*, *86*, 703–720.

Goodglass, H. & Wingfield, A. (1997) *Anomia: Neuroanatomical and cognitive correlates*. New York: Academic Press.

Gorelick, P.B. & Ross, E.D. (1987) The aprosodias: Further functional-anatomic evidence for the organization of affective language in the right hemisphere. *Journal of Neurology, Neurosurgery and Psychiatry*, *50*, 553–560.

Gorenstein, E.E. (1982) Frontal lobe functions in psychopaths. *Journal of Abnormal Psychology*, *91*, 368–379.

Grafmann, J., Holyoak, K.J., & Boller, F. (Eds) (1995) *Structure and Functions of the Human Prefrontal Cortex*. New York: New York Academy of Science.

Grafman, J., Jones, B., & Salazar, A. (1990) Wisconsin Card Sorting Test performance based on location and size of neuroanatomical lesion in Vietnam veterans with penetrating head injury. *Perceptual and Motor Skills*, *71*, 1120–1122.

Grafman, J., Passafiume, D., Faglioni, P. & Boller, F. Calculation disturbances in adults with focal hemispheric damage. *Cortex*, *18*, 37–49.

Grant, D.A., & Berg, E.A. (1948) A behavioral analysis of degree of reinforcement and ease of shifting to new responses in a Weigl-type card-sorting problem. *Journal of Experimental Psychology*, *38*, 404–411.

Greenblatt, M., Arnot, R. & Solomon, H. (1950) *Studies in Lobotomy*. New York: Grune and Stratton.

Greene, K.A., Marciano, F.F., Dickman, C.A., Coons, S.W., Johnson. P.C., Bailes, J.E., Spetzler, R.F. (1995) Anterior communicating artery aneurysm paraparesis syndrome: Clinical manifestations and pathologic correlates. *Neurology*, *45*, 45–50.

Guariglia, C., & Antonucci, G. (1992) Personal and extrapersonal space: A case of neglect dissociation. *Neuropsychologia*, *30*, 1001–1010.

Gustafson, L. (1987) Frontal lobe degeneration on non-Alzheimer type. II. Clinical picture and differential diagnosis. *Archives of Gerontology and Geriatrics*, *6*, 209–223.

Gustafson, L., Brun, A. & Risberg, J. (1990) Frontal lobe dementia of non-Alzheimer type. In: Wurtman, R.J., Corkin, S., Growdon, J. & Ritter-Walker, E. (Eds.), *Alzheimer's Disease. Advances in Neurology, Vol 51.* New York: Raven Press, pp. 65–71.

Halpren, D.L. (1991) *Sex differences in Cognitive Abilities.* Hillsdale: Lawrence Erlbaum, and Associates.

Halsband, U., Ito, N., Tanji, J., & Freund, H-J. (1993) The role of premotor cortex and the supplementary motor area in the temporal control of movement in man. *Brain, 116,* 243–266.

Halsband, U., Matsuzaka, Y. & Tanji, J. (1994) Neuronal activity in the primate supplementary, presupplementary, and premotor cortex during externally and internally instructed sequential movements. *Neuroscience Research, 20,* 149–155.

Hannay, J., Varney, N., & Benton, A. (1976) Visual localization in patients with unilateral brain disease. *Journal of Neurology, Neurosurgery, and Psychiatry, 39,* 307–313.

Harlow, J.M. (1868) Recovery from the passage of a an iron bar through the head. *Publications of the Massachusetts Medical Society, 2,* 329–346.

Harris, L.J. (1978) Sex differences in spatial abilities: Possible environmental, genetic, and neurological factors. In: Kinsbourne, M. (Ed.), *Asymmetrical Function of the Brain.* Cambridge, MA: Cambridge University Press.

Haymaker, W. & Schiller, F. (Eds) (1970) *The Founders of Neurology, 2nd Edition.* Springfield, IL: Charles C. Thomas.

Hebb, D.O. (1945) Man's frontal lobes: A critical review. *Archives of Neurology and Psychiatry, 44,* 421–438.

Hecaen, H. (1962) Clinical symptomatology in right and left hemispheric lesions. In: V.B. Mountcastle (Ed.), *Interhemispheric Relations and Cerebral Dominance.* Baltimore: John Hopkins Press.

Hecaen, H. (1964) Mental changes associated with tumors of the frontal lobes. In: Warren, J.M. & Akert, K. (Eds.), *The Frontal Granular Cortex and Behavior.* New York: MCGraw-Hill, pp. 335–352.

Hecaen, H. & Albert, M.L. (1978) *Human Neuropsychology.* New York: John Wiley & Sons.

Hecaen, H. & Angelergues, R. (1962) Agnosia for faces (prosopagnosia). *Archives of Neurology, 7,* 92–100.

Hecaen, H. & Assal, G.A. (1970) A comparison of constructive deficits following right and left hemispheric lesions. *Neuropsychologia, 8,* 289–303.

Heilman, K.M. (1991) Anosognosia: Possible neuropsychological mechanisms. In: Prigatano, G.P. & Schacter, D.L.(Eds.), *Awareness of Deficit after Brain Injury: Clinical and theoretical issues.* New York: Oxford, pp. 53–62.

Heilman, K.M. (1995) Attentional asymmetries. In: R.J. Davidson & K. Hugdahl (Eds.), *Brain Asymmetry.* Cambridge, MA: MIT Press, pp. 217–234.

Heilman, K.M., Bowers, D., Speedie, L. & Coslett, B. (1984) Comprehension of affective and nonaffective speech. *Neurology, 34,* 917–921.

Heilman, K., Bowers, D. & Valenstein, E. (1993) Emotional disorders associated with neurological diseases. In: Heilman, K. & Valenstein, E. (Eds.), *Clinical Neuropsychology.* New York: Oxford University Press, pp. 461–497.

Heilman, K.M., Watson, R.T., & Valenstein, E. (2003) Neglect and related disorders. In Heilman, K.M., & Valenstein, E. (Eds) *Clinical neuropsychology.* New York: Oxford University Press, p. 296–346.

Heilman, K. & Rothi, L.J. (1993) Apraxia. In: Heilman, K. & Valenstein, E. (Eds.), *Clinical Neuropsychology.* New York: Oxford University Press, pp. 141–163.

Heilman, K.M. & Satz, P. (1983) *Neuropsychology of Human Emotion.* Ne York: Guilford Press.

Heilman, K.M., Scholes, R, & Watson, R.T., (1975) Auditory affective agnosia: Disturbed comprehension of affective speech. *Journal of Neurology, Neurosurgery and Psychiatry, 38,* 69–72.

Heilman, K.M., Tucker, D.M., and Valenstein, E. (1976) A case of mixed transcortical aphasia with intact naming. *Brain, 99,* 415–426.

Heilman, K.M. & Valenstein, E. (1972) Frontal lobe neglect in man. *Neurology, 22,* 660–664.

Heilman, K.M. & Valenstein, E. (Eds.) (1993) *Clinical Neuropsychology.* New York: Oxford University Press.

Heilman, K.M. & Valenstein, E. (Eds) (2003) *Clinical Neuropsychology.* New York: Oxford University Press.

Heilman, K.M., Valenstein, E. & Watson, R.T. (1983) Localization of neglect. In: Kertesz, A. (Ed.), *Localization in Neurology.* New York: Academic Press, pp. 471–492.

Heilman, K.M. & Watson, R.T. (1991) Intentional motor disorders. In: Levin, H.S., Eisenberg, H.M., & Benton A.L. (Eds.), *Frontal Lobe Function and Dysfunction*. New York: Oxford University Press, pp. 199–213.

Heilman, K.M., Watson, R.T., & Valenstein, E. (1995) Neglect and related disorders. In: Heilman, K. & Valenstein, E. (Eds.), *Clinical Neuropsychology*. New York: Oxford University Press, pp. 279–336.

Heilman, K.M., Watson, R.T., & Rothi, L.G. (1997) Disorders of skilled movements: Limb apraxia. In: Feinberg, T.E. & Farah, M.J. (Eds.), *Behavioral Neurology and Neuropsychology*. New York: McGraw-Hill, pp. 227–235.

Hellige, J.B. (1995) Hemispheric asymmetry for components of visual information processing. In: Davidson, R.J. & Hugdahl, K.(Eds.), *Brain Asymmetry*. Cambridge, MA: MIT Press, pp. 99–121.

Hellige, J.B. (1993) *Hemispheric Asymmetry: What's right and what's left*. Cambrige, MA: Harvard University Press.

Henderson, V.W. (1986) Anatomy of posterior pathways in reading: A reassessment. *Brain and Language, 29*, 119–133.

Henson, R.A. (1985) Amusia. In: Frederiks, J.A.M. (Ed.), *Handbook of Clinical Neurology, Vol* 45. New York: Elsevier, pp. 483–490.

Herron, J. (Ed) (1980) *Neuropsychology of Left-handedness*. New York: Academic Press.

Hochberg, F.H. & LeMay, M. (1975) Arteriographic correlates of handedness. *Neurology, 25*, 218–222.

Holmes, G. (1931) Mental symptoms associated with brain tumors. *Proceedings of the Royal Society of Medicine, 24*, 65–76.

Holmes, G. & Horax, G. (1919) Disturbances of spatial orientation and visual attention, with loss of stereoscopic vision. *Archives of Neurology and Psychiatry, 1*, 385–407.

Hough, M.S. (1990) Narrative comprehension in adults with right and left hemisphere brain damage: Theme organization. *Brain and Language, 38*, 253–277.

Humphrey, D.R. & Tanji, J. (1991) What features of voluntary motor control are encoded in the neuronal discharge of different cortical motor areas? In: Humphrey, D.R. & Freund, H-J. (Eds.), *Motor Control: Concepts and issues*. New York: John Wiley & Sons, pp. 413–443.

Iverson, S.D., & Mishkin, M. (1970) Perseverative interference in monkeys following selective lesions of the inferior prefrontal convexity. *Experimental Brain Research, 11*, 476–486.

Jacobson, C.F. (1936) Studies of cerebral functions in primates: I. The functions of the frontal association areas in monkeys. *Comparative Psychology Monographs, 13*, 3–60.

Janowsky, J.S., Shimamura, A.P., & Squire, L.R. (1989) Source memory impairment in patients with frontal lobe lesions. *Neuropsychologia, 27*, 1043–1056.

Javoy-Agid, F., Scatton, B., Ruberg, M., L'Heureux, R., Cervera, P., Raisman, R., Matloteaux, J-M., Beck, H. & Agid, Y. (1989) Distribution of monoaminergic, cholinergic, and GABAergic markers in the human cerebral cortex. *Neuroscience, 39*, 251–259.

Jones, E.G. (1987) Ascending inputs to, and internal organization of, cortical motor areas. In: *Ciba Foundations Symposium, Motor areas of the cerebral cortex*. New York: John Wiley & Sons, pp. 21–40.

Jones, E.G., & Powell, T.P.S. (1969a) Connexions of the somatic sensory cortex of the rhesus monkey. I. Ipsilateral cortical connections. *Brain, 92*, 477–502.

Jones, E.G., & Powell, T.P.S. (1969b) Connexions of the somatic sensory cortex of the rhesus monkey. II. Contralateral cortical connections. *Brain, 92*, 477–502.

Jones-Gotman, M. & Milner, B. (1977) Design fluency: The invention of nonsense drawings after focal cortical lesions. *Neuropsychologia, 15*, 653–674.

Jonides, J., Smith, E.E., & Koeppe, R.A. (1993) Spatial working memory in humans as revealed by PET. *Nature, 363*, 623–625.

Joseph, R. (1988) The right cerebral hemisphere. *Journal of Clinical Psychology, 44*, 630–673.

Joseph, R. (1990) *Neuropsychology, Neuropsychiatry, and Behavioral Neurology*. New York: Plenum Press.

Jouandet, M. & Gazzaniga, M.S. (1979) The frontal lobes. In: Gazzaniga, M.S. (Ed.), *Handbook of Behavioral Neurobiology, Vol. 2. Neuropsychology*. New York: Plenum, pp. 25–59.

Kaas, J.H., (1983) What, if anything is SI? Organization of first somatosensory area of cortex. *Physiological Review, 63*, 206–231.

Kahn, H.J. & Whitaker, H.A. (1991) Acalculia: An historical review of localization. *Brain and Cognition*, *17*, 102–115.

Kandel, E. & Freed, D. (1989) Frontal-lobe dysfunction and antisocial behavior: A review. Journal of Clinical Psychology, *45*, 404–413.

Kaplan, E. (1990) The process approach to neuropsychological assessment in psychiatric patients. *Journal of Neuropsychiatry and Clinical Neurosciences*, *2*, 72–87.

Kaplan, E. (1988), A process approach to neuropsychological assessment. In: T. Boll & B.K. Bryant (Eds.), *Clinical Neuropsychology in Brain Function: Research, measurement and practice*. Washington, DC: American Psychological Association, pp. 125–167.

Kaplan, E, Fein, D., Morris, R. & Delis, D. (1991) *WAIS-R as a Neuropsychological Instrument*. San Antonio, TX: Psychological Corporation, pp. 87–97.

Kaplan, R.F., Meadows, M.E., Verfaelie, M., Kwan, E., Ehrenberg, B.L., Bromfield, E.B. & Cohen, R.A. (1994) Lateralization of memory for the visual attributes of objects: Evidence from the the posterior cerebral artery amobarbital test. Neurology, 44, 1069–1073.

Kapur, S., Craik, F.I.M., Jones, C., Brown, G.M., Houle, S. & Tulving, E. (1995) Functional role of the prefrontal cortex in retrieval of memories: A PET study. *NeuroReport, 6*, 1880–1884.

Kaufer, D.I. & Lewis, D.A. (1999) Frontal lobe anatomy and cortical connectivity. In: Miller, B.L. & Cummings, J.L. (Eds.), *The Human Frontal Lobes: Functions and disorders*. New York: Guilford Press, pp. 27–44.

Kertesz, A., Davidson, W., & Munoz, D.G. (1999) Clinical and pathological overlap between frontotemporal dementia, primary progressive aphasia, and corticobasal degeneration: The Pick complex. *Dementia and Geriatric Cognitive Disorders, 10* (suppl) 46–49.

Kertesz, A. & Geschwind, N. (1971) Patterns of pyramidal decussation and their relationship to handedness. *Archives of Neurology, 24*, 326–332.

Kimberg, D.Y., D'Esposito, M., & Farah, M.J. (1997) Frontal lobes: Cognitive neuropsychological aspects. In: Feinberg, T.E. & Farah, M.J.(Eds.), *Behavioral Neurology and Neuropsychology*. New York: McGraw-Hill, pp. 409–418.

Kimura, D. (1963) Right temporal-lobe damage: Perception of unfamiliar stimuli after damage. *Archives of Neurology, 8*, 264–271.

Kimura, D. (1982) Left-hemisphere control of oral and brachial movement and their relation to communication. *Philosophical Transactions of the Royal Society of London, 298*, 135–149.

King, M.C. & Snow, W.G. (1981) Problem-solving task performance in brain-damaged subjects. *Journal of Clinical Psychology, 37*, 400–404.

Kingstone, A. & Gazzaniga, M.S. (1995) Higer-order subcortical processing in the split-brain patient: More illusory than real? *Neuropsychologia, 9*, 321–328.

Kinsbourne, M., & Warrington, E.K. (1962a) A disorder of simultaneous form perception. *Brain, 85*, 461–486.

Kinsbourne, M. & Warrington, E.K., (1962b) A study of finger agnosia. *Brain, 85*, 47–66.

Kirshner, H.S. (1986) *Behavioral Neurology: A practical approach*. New York: Churchill Livingstone.

Kittler, P., Turkewitz, G. & Goldberg, E. (1989) Shifts in hemispheric advantage during familiarization with complex visual patterns. *Cortex, 25*, 27–32.

Kluver, H., & Bucy, P.C. (1939) Preliminary analysis of the temporal lobes in monkeys. *Archives of Neurology and Psychiatry, 42*, 979–1000.

Koff, E., Naeser, M.A., Pieniadz, J.M., Foundas, A.L. & Levine, H.L. (1986) Computed tomographic scan of hemispheric asymmetries in right-and left-handed male and female subjects. *Archives of Neurology, 43*, 487–491.

Kohs, S.C. (1919) *Kohs Block Designs Test*. Wood Dale, IL: Stoelting.

Kolb, B. & Whishaw, I.Q. (1990) *Fundamental of Human Neuropsychology*. New York: W.H. Freeman.

Kooistra, C.A. & Heilman, K.M. (1988) Motor dominance and lateral asymmetry of the globus pallidus. *Neurology, 38*, 388–390.

Kosslyn, S.M. (1988) Aspects of a cognitive neuroscience of mental imagery. Science, 240: 1621–1626.

Kulynych, J.J., Vladar, K., Jones, D.W. & Weinberger, D.R. (1994) Gender differences in the normal lateralization of the supratemporal cortex: MRI surface-rendering morphometry of Heschl's gyrus and the planum temporale. *Cerebral Cortex, 4*, 107–118.

Kuypers, H.G.J.M. (1987) Some aspects of the organization of the output of the motor cortex. In: *Ciba Foundations Symposium, Motor areas of the cerebral cortex.* New York: John Wiley & Sons, pp. 63–82.

Lapierre, D., Braun, C.M.J., Hodgins, S. (1995) Ventral frontal deficits in psychopathy: Neuropsychological test findings. *Neuropsychologia, 33,* 139–151.

LeMay, M. (1977) Asymmetries of the skull and handedness. *Journal of the Neurological Sciences, 32,* 243–253.

LeMay, M. (1976) Morphological cerebral asymmetries of modern man, fossil man, and non-human primate. *Annals of the New York Academy of Sciences. 280,* 349–366.

Levin, H.S., Benton, A.L., & Grossman, R.G. (1982) *Neurobehavioral Consequences of Closed Head Injury.* New York: Oxford University Press.

Levin, H.S., Eisenberg, H.M., & Benton A.L. (Eds.) (1991) *Frontal Lobe Function and Dysfunction.* New York: Oxford University Press.

Levin, H.S., Goldstein, F.C. & Spiers, P.A. (1993) Acalculia. In: Heilman, K. & Valenstein, E. (Eds.), *Clinical Neuropsychology.* New York: Oxford University Press, pp 91–122.

Levin,H.S., Goldstein, F.C., Williams, D.H., & Eisenberg, H.M. (1991) The contributions of frontal lobe lesions to the neurobehavioral outcome of closed head injuries. In: Levin, H.S., Eisenberg, H.M., & Benton A.L. (Eds.), *Frontal Lobe Function and Dysfunction.* New York: Oxford University Press, pp. 318–338.

Lezak, M.D. (1982) The problem of assessing executive functions. *International Journal of Psychology, 17,* 281–297.

Lezak, M. (2004) *Neuropsychological Assessment.* New York: Oxford Press.

Lhermitte, F. (1983) "Utilization behaviour" and its relationship to lesions of the frontal lobes. *Brain, 106,* 237–255.

Liederman, J. (1995) A reinterpretation of the split-brain syndrome: Implications for the function of corticocortical fibers. In: Davidson, R.J. & Hugdahl, K. (Eds.), *Brain Asymmetry.* Cambridge, MA: MIT Press, pp. 451–490.

Liotti, M. & Tucker, D.M. (1995) Emotion in asymmetric corticolimbic networks. In: R.J. Davidson & K. Hugdahl (Eds.), *Brain Asymmetry.* Cambridge, MA: MIT Press, pp. 389–423.

Lishman, W.A. (1968) Brain damage in relation to psychiatric disability after head injury. *British Journal of Psychiatry, 114,* 373–410.

Lishman, W.A. (1973) The psychiatric sequelae of head injury: A review. *Psychological Medicine, 3,* 304–318.

Litvan, I. (1999). In: Miller, B.L. & Cummings, J.L. (Eds.), *The Human Frontal Lobes.* New York: Guilford Press, pp. 402–421.

Loring, D.W., Lee, G.P. & Meador, K.J. (1988) Revising the Rey-Osterrieth: Rating right hemisphere recall. *Archives of Clinical Neuropsychology, 3,* 239–247.

Loring, D.W., Lee, G.P., Meador, K.J., Smith, R., Martin, R.C., Ackell, A.B. & Flanigin, H.F. (1991) Hippocampal contribution to verbal recent memory following dominant-hemisphere temporal lobectomy. *Journal of Clinical and Experimental Neuropsychology, 13,* 575–586.

Luders, H.O. (1996) The supplementary sensorimotor area. In: Luders, H.O. (Ed.), *Advances in Neurology, Vol. 70: Supplementary Sensorimotor Area.* Philadelphia: Lippincott-Raven, pp. 1–16.

Lueger, R.J. & Gill, K.J. (1990) Frontal-lobe cognitive dysfunction in conduct disorder adolescents. *Journal of Clinical Psychology, 46,* 696–706.

Luppino, G., Matelli, M., Camarda, R. & Rizzolatti, G. (1993) Corticocortical connections of area F3 (SMA-proper) and areas F6 (pre-SMA) in the macaque monkey. *Journal of Comparative Neurology, 338,* 114–140.

Luppino, G., Matelli, M., Camarda, R. & Rizzolatti, G. (1994) Corticospinal projections from mesial frontal and cingulate areas in the monkey. *Neuroreport, 5,* 2545–2548.

Luria, A.R. (1959) Disorders of "simultaneous perception" in a case of bilateral occipitoparietal brain injury. *Brain, 83,* 437–449.

Luria, A.R. (1965) Two kinds of motor perseveration in massive injury to the frontal lobes. *Brain, 88,* 1–10.

Luria, A.R. (1966) *Higher Cortical Functions in Man.* New York: Basic Books.

Luria, A.R. (1970) *Traumatic Aphasia.* Paris: Mouton

Luria, A.R. (1973) *The Working Brain.* New York: Basic Books.

Luria, A.R., Karpov, B.A., & Yarbuss, A.L. (1966) Disturbances of active visual perception with lesions of the frontal lobes. *Cortex, 2,* 202–212.

Luria, A.R. & Tsvetkova, L.S. (1964) The programming of constructive activity in local brain injuries. *Neuropsychologia, 2,* 489–495.

Luria, A.R. & Tsvetkova, L.S. (1968) The mechanisms of "dynamic aphasia." *Foundations of Language, 4,* 296–307.

Macmillan, M.B. (1986) A wonderful journey through skull and brains: The travels of Mr. Gage's tamping iron. *Brain and Cognition, 5,* 67–102.

Malloy, P. & Duffy, J. (1994) The frontal lobes in neuropsychiatric disorders. In: Boller, F. & Grafman, J. (Eds.), *Handbook of Neuropsychology.* Amsterdam: Elsevier Science, pp. 203–232.

Malloy, P.F. & Richardson, E.D. (1994) Assessment of frontal lobe functions. *Journal of Neuropsychiatry, 6,* 399–410.

Mapou, R.L. (1995) A cognitive framework for neuropsychological assessment. In: Mapou, R.L & Spector, J. (Eds.), *Clinical Neuropsychological Assessment: A cognitive approach.* New York: Plenum Press, pp. 295–337.

Marsden, C.D., Deecke, L., Freund, H-J., Hallett, M., Passingham, R.E., Shibasaki, H., Tanji, J. & Wiesendanger, M. (1996) The functions of the supplementary motor area. In: Luders, H.O. (Ed) *Advances in Neurology, Vol. 70: Supplementary Sensorimotor Area.* Philadelphia: Lippincott-Raven, pp. 477–487.

Matsuzaka, Y., Aizawa, H., & Tanji, J. (1992) A motor area rostral to the supplementary motor area (presupplementary motor area) in the monkey: Neuronal activity during a learned motor task. *Journal of Neurophysiology, 68,* 653–662.

McCloskey, M., Aliminosa, D. & Sokol, S. (1991) Facts, rules, and procedures in normal calculation: Evidence from multiple single-patient studies of impaired arithmetic fact retrieval. *Brain and Cognition, 17,* 154–203.

McFie, J. & Zangwill, O. (1960) Visual-constructive disabilities associated with lesions of the left cerebral hemisphere. *Brain, 83,* 243–260.

McGlone, J. (1977) Sex differences in the cerebral organization of verbal functions in patients with unilateral brain lesions. *Brain, 100,* 775–793.

McGlone, J. (1986) The neuropsychology of sex differences in human brain organization. In: Goldstein, G. & Tarter, R.E. (Eds.), *Advances in Clinical Neuropsychology, Vol 3.* New York: Plenum Press, pp. 1–30.

McGlynn, S.M. & Kasizniak, A.W. (1991) Unawareness of deficits in dementia and schizophrenia. In: Prigatano, G.P. & Schacter, D.L.(Eds.), *Awareness of Deficit after Brain Injury: Clinical and theoretical issues.* New York: Oxford, pp. 84–110.

McHenry, L.C. (1969) *Garrison's History of Neurology.* Springfield, IL: Charles C. Thomas.

McKeith, I.G. (1999) Lewy body disorders. In: Miller, B.L. & Cummings, J.L. (Eds.), *The Human Frontal Lobes.* New York: Guilford Press, pp. 422–435

McKeith, I.G., Galasko, D., Kosaka, K., Perry, E.K., Dickson, D.W., Hansen, L.A., Salmon, D.P., Lowe, J., Mirra, S.S., Byrne, E.J., Lennox, G., Quinn, N.P., Edwardson, J.A., Ince, P.G., Bergeron, C., Burns, A., Miller, B.L., Lovestone, S., Collerton, D., Jansen, E.N.H., Ballard, C., deVos, R.A.I., Wilcock, G.K., Jellinger, K.A., & Perry, R.H. (1996) Consensus guidelines for the clinical and pathological diagnosis of dementia with Lewy bodies (DLB): Report of the consortium on DLB international workshop. *Neurology, 47,* 1113–1124.

McRae, D.L. (1948) Focal epilepsy: Correlation of the pathological and radiological findings. *Radiology, 50,* 439.

McRae, D.L., Branch, C.L. & Milner, B. (1968) The occipital horns and cerebral dominance. *Neurology, 18,* 95–98.

Meadows, J.C. (1974) Disturbed perception of colours associated with localized brain lesions. *Brain, 97,* 615–632.

Meerwaldt, J.D. (1982) *The Rod Orientation Test in Patients with Right-hemisphere Infarction.* Alblasserdam, Netherlands: Roceddavids.

Mega, M.S. & Cummings, J.L. (1994) Frontal-subcortical circuits and neuropsychiatric disorders. *Journal of Neuropsychiatry, 6,* 358–370.

Mendoza, J.E., Moore, J, Crane, M., Henderson, D. (1989) Effects of clock size on clock-drawing in neurological patients. *American Psychological Association,* New Orleans (Poster)

Mendoza, J.E., Hendrickson, R., & Apostolos, G.T. (1993) A combined qualitative and quantitative scoring system for clock drawing. *National Academy of Neuropsychology.* Phoenix (Poster)

Mercer, B., Wapner, W., Gardner, H., & Benson, D.F. (1977) A study of confabulation. *Archives of Neurology, 34,* 429–433.

Mesulam, M-M, (Ed) (2000a) *Principles of Behavioral and Cognitive Neurology.* New York: Oxford University Press.

Mesulam, M-M, (2000b) Behavioral neuroanatomy: Large–scale networks, association cortex, frontal syndromes, the limbic system, and hemispheric specialization. In Mesulam, M-M (Ed) *Principles of Behavioral and Cognitive Neurology.* New York: Oxford University Press, p. 1–120.

Mesulam, M-M, (2000c) Attentional networks, confusional states, and neglect syndromes. In Mesulam, M-M (Ed) *Principles of Behavioral and Cognitive Neurology.* New York: Oxford University Press, p. 174–256.

Mesulam, M-M. (1988) Central cholinergic pathways: Neuroanatomy and some behavioral implications. In: Avoli, M., Reader, T.A., Dykes, R.W. & Gloor, P. (Eds.), *Neurotransmitters and Cortical Function: From molecules to mind.* New York: Plenum Press, pp. 237–260.

Mesulam, M. & Mufson, E.J. (1985) The insula of Reil in man and monkey: Architectonics, connectivity and function. In: Peters, A. & Jones, E.G. (Eds.), *Cerebral Cortex.* New York: Plenum Press, Volume 4, pp. 179–226.

Meyer, J. & Barron, D. (1960) Apraxia of gait: A clinicophysiological study. *Brain, 83,* 261–284.

Meyer, A. (1971) *Historical Aspects of Cerebral Anatomy.* New York: Oxford University Press.

Meyers, C.A., Berman, S.A., Scheibel, R.S., & Hayman, A. (1992) Case report: Acquired antisocial personality disorder associated with unilateral left orbital frontal lobe damage. *Journal Psychiatry and Neuroscience, 17,* 121–125.

Milberg, W.P., Hebben, N. & Kaplan, E. (1986) The Boston process approach to neuropsychological assessment. In: Grant, I. & Adams, K. (Eds.), *Neuropsychological Assessment of Neuropsychiatric Disorders.* New York: Oxford University Press, pp. 65–86.

Miller, B.L. & Cummings, J.L. (Eds.) (1999) *The Human Frontal Lobes.* New York: Guilford Press.

Miller, B.L., Cummings, J.L., Vilanueva-meyer, J., Boone, K., Mehringer, C.M., Lesser, I.M., & Mena, I. (1991) Frontal lobe degeneration: Clinical, neuropsychological, and SPECT characteristics. *Neurology, 41,* 1374–1382.

Milner, B. (1963) Effects of different brain lesions on card sorting: The role of the frontal lobes. *Archives of Neurology, 9,* 90–100.

Milner, B. (1965) Visually guided maze learning in man: Effects of bilateral hippocampal, bilateral frontal, and unilateral cerebral lesions. *Neuropsychologia, 3,* 317–338.

Milner, B. (1968) Visual recognition and recall after right temporal-lobe excision in man. *Neuropsychologia, 6,* 191–209.

Milner, B. (1971) Interhemispheric differences in the localization of psychological processes in man. *British Medical Bulletin, 27,* 272–277.

Milner, B. (1972) Disorders of learning and memory after temporal lobe lesions in man. *Clinical Neurosurgery, 19,* 421–446.

Milner, B. (1974) Hemispheric specialization: Scope and limits. In: Schmitt, F.O. & Worden, F.G. (Eds.), *The Neurosciences: Third study program.* Cambridge, MA: MIT Press, pp. 75–92.

Milner, B. (1982) Some cognitive effects of frontal lobe lesions in man. In: Broadbent, D.E. & Weiskrantz, L. (Eds.), *The Neuropsychology of Cognitive Function.* London: The Royal Society, pp. 211–226.

Milner, B. & Teuber, H-L (1968) Alteration of perception and memory in man: Reflections on methods. In: Weiskrantz, L. (Ed.), *Analysis of Behavioral Change.* New York: Harper & Row, pp. 268–375.

Milner, B.L., Cummings, J.L., Villanueva-Meyer, J., Boone, K., Mehringer, C.M., Lesser, I.M. & Mena, I. (1991) Frontal lobe degeneration: Clinical, neuropsychological, and SPECT characteristics. *Neurology, 41,* 1374–1382.

Mishkin, M. and Appenzeller, T. (1987) The anatomy of memory. *Scientific American, 256,*(6) 80–89.

Molfese, D.M. & Segalowitz, S.J. (1988) *Brain Lateralization in Children: Developmental implications.* New York: Guilford Press.

Morrow, L., Vrtunsky, P.B., Kim, Y. & Boller, F. (1981) Arousal responses to emotional stimuli and laterality of lesion. *Neuropsychologia, 19*, 65–71.

Moruzzi, G. & Magoun, H.W. (1949) Brain stem reticular formation and activation of the EEG. *Electroencephalography and Clinical Neurophysiology, 1*, 455–473.

Mountcastle, V.B. (1997) The columnar organization of the neocortex. *Brain, 120*, 701–722.

Mountcastle, V.B. (1979) An organizing principle for cerebral function: The unit module and the distributed system. In: Schmitt, F.O. & Worden, F.G. (Eds.), *The Neurosciences Fourth Study Program*. Cambridge, MA: MIT Press, pp. 21–42.

Mountcastle, V.B., Lynch, J.C., Georgopoulos, A., Sakata, H., Acuna, C. (1975) Posterior parietal association cortex of the monkey. Command functions for operations for extra-personal space. *Journal of Neurophysiology, 38*, 871–908.

Mushiake, H., Inase, M. & Tanji, J. (1991) Neuronal activity in the primate premotor, supplementary, and precentral motor cortex during visually guided and internally determined sequential movements. *Journal of Neurophysiology, 66*, 705–718.

Myers, P.S. (1993) Narrative expressive deficits associated with right-hemisphere damage. In: Joanette, Y. & Brownell, H.H. (Eds.), *Narrative Discourse in Neurologically Impaired and Normal Aging Adults*. San Diego, CA: Singular, pp. 279–296.

Myers, P. (1994) Communication disorders associated with right-hemisphere brain damage. In: Chapey, R. (Ed.), *Language Intervention Strategies in Adults*. Baltimore: Williams and Wilkins, pp. 513–532.

Nadeau, S.E. & Heilman, K.M. (1991) Gaze-dependent hemianopia without hemispatial neglect. *Neurology, 41*, 1244–1250.

Naqvi, N.H., Rudrauf, D., Damasio, H. and Bechara, A. (2007) Damage to the insula disrupts addiction to cigarette smoking. *Science, 315*, 531–534.

Naugle, R.I., Chelune, G.J., Cheek, R., Luders, H., & Awad, I.A. (1993) Detection of changes in material-specific memory following temporal lobectomy using the Wechsler Memory Scale-Revised. *Archives of Clinical Neuropsychology, 8*, 381–395.

Navon, D. (1977) Forest before trees: The precedence of global features in visual perception. *Cognitive Psychology, 9*, 353–383.

Neary, D., Snowden, J.S., Northen, B. & Goulding, P. (1988) Dementia of frontal lobe type. *Journal of Neurology, Neurosurgery, and Psychiatry, 51*, 353–361.

Neary, D., Snowden, J.S., Gustafson, L., Passant, U., Stuss, D., Black, S., Freedamn, M., Kertesz, A., Robert, P.H., Albert, M., Boone, K., Miller, B.L., Cummings, J. & Benson, D.F. (1998) Frontotemporal lobar degeneration: A consensus on clinical diagnostic criteria. *Neurology, 51*, 1546–1554.

Nieuwenhuys, R., Voogd, J., & Van Huijzen, C. (1988) *The Human Nervous System*. New York: Springer-Verlag.

Nikkuni, S., Yashima, Y., Ishige, K., Suzuki, S., Ohno, E., Kumashiro, H. Kobayashi, E, Awa, H., Mihara, T. & Asakura, T. (1981) Left-right hemisphere asymmetry of critical speech zones in Japanese brains. *Brain and Nerve, 33*, 77–84.

Nolte, J. (1993) *The Human Brain*. St. Louis: Mosby-Year Book, Inc.

Oscar-Berman, M., McNamara, P., & Freedman, M. (1991) Delayed response tasks: Parallels between experimental ablation studies and findings in patients with frontal lesions. In: Levin, H.S., Eisenberg, H.M., & Benton A.L. (Eds.), *Frontal Lobe Function and Dysfunction*. New York: Oxford University Press, pp. 230–255.

Ogden, J.A. (1985) Autotopagnosia: Occurrence in a patient without nominal aphasia and with an intact ability to point to parts of animals and objects. *Brain, 108*, 1009–1022.

Osmon, D.C. & Suchy, Y. (1996) Fractionating frontal lobe functions: Factors of the Milwaukee Card Sorting Test. *Archives of Clinical Neuropsychology, 11*, 541–552.

Otto, M.W. (1992) Normal and abnormal information processing: A neuropsychological perspective on obsessive-compulsive disorder. *Psychiatric Clinics of North America, 15*, 825–848.

Partridge, M. (1950) *Prefrontal Leucotomy. A survey of 300 cases personally followed over 1 and 1/2–3 years*. Oxford: Blackwell.

Passingham, R.E. (1993) *The Frontal Lobes and Voluntary Action*. New York: Oxford University Press.

Passingham, R.E. (1996) Functional specialization of the supplementary motor area in monkeys and humans. In: Luders, H.O. (Ed.), *Advances in Neurology, Vol. 70: Supplementary Sensorimotor Area*. Philadelphia: Lippincott-Raven, pp. 105–116.

Paterson, A. & Zangwill, O.L., (1944) Disorders of space perception associated with lesions of the right cerebral hemisphere. *Brain, 67*, 331–358.

Penfield, W. & Jasper, H. (1954) *Epilepsy and the Functional Anatomy of the Human Brain.* Boston: Little, Brown Publications.

Penfield, W. & Welch, K. (1951) The supplementary motor area of the cerebral cortex: A clinical and experimental study. *Archives of Neurology and Psychiatry, 66*, 289–317.

Petrides, M. (1985) Deficits on conditional associative-learning tasks after frontal and temporal lobe lesions in man. *Neuropsychologia, 23*, 601–614.

Petrides, M. (1990) Nonspatial conditional learning impaired in patients with unilateral frontal—but not unilateral temporal—lobe excisions. *Neuropsychologia, 28*, 137–149.

Petrie, A. (1952) *Personality and the Frontal Lobes.* New York: Blakiston.

Pfeifer, R.A. (1936) Pathologie der Horstrahlung und der corticalen Horsphare. In: Bumke, O. & Forster, O. (Eds.), *Handbuch der Neurologie, Vol. 6.* Berlin: Springer, pp. 533–626.

Picard, N. & Strick, P.L. (1996) Motor areas of the medial wall: A review of their location and functional activation. *Cerebral Cortex, 6*, 342–353.

Pieniadz, J.M., Naeser, M.A., Koff, E., & Levine, H.L. (1983) CT scan hemispheric asymmetry measurements in stroke cases with global aphasia: Atypical asymmetries associated with improved recovery. *Cortex, 19*, 371–391.

Pieniadz, J.M. & Naeser, M.A. (1984) Computed tomagraphic scan asymmetries and morphologic brain asymmetries: Correlations in same cases post mortem. *Archives of Neurology, 41*, 403–409.

Piercy, M., Hacaen, H. & DeAjuriaguerra, J. (1960) Constructional apraxia associated with unilateral cerebral lesions—left and right sided cases compared. *Brain, 83*, 225–242.

Poeck, K. & Orgass, B. (1966) Gerstmann's syndrome and aphasia. *Cortex, 2*, 421–437.

Price, B.H., Daffner, K.R., Stowe, R.M. & Mesulam, M.M. (1990) The compartmental learning disabilities of early frontal lobe damage. *Brain, 113*, 1383–1393.

Prigatano, G.P. (1991) The relationship of frontal lobe damage to diminished awareness: Studies in rehabilitation. In: Levin, H.S., Eisenberg, H.M., & Benton A.L. (Eds.), *Frontal Lobe Function and Dysfunction.* New York: Oxford University Press, pp. 381–397.

Prigatano, G.P. and Schacter, D.L. (Eds.) (1991) *Awareness of Deficit after Brain Injury.* New York: Oxford.

Prigatano, G.P. & Schacter, D.L.(1991) *Awareness of Deficit after Brain Injury: Clinical and theoretical issues.* New York: Oxford, pp. 240–257.

Psychological Corporation (1997) *WAIS-III, WMS-III: Technical manual.* San Antonio, TX: Harcourt Brace & Co., pp. 201–202.

Raade, A.S., Rothi, L.J.G., & Heilman, K.M. (1991) The relationship between buccofacial and limb apraxia. Brain and Cognition, *16*, 130–146.

Rafal, R.D. (1997a) Balint syndrome. In, Feinberg, T. & Farah, M.J. *Behavioral Neurology and Neuropsychology.* New York: McGraw-Hill, pp. 337–356.

Rafal, R.D. (1997b) Hemispatial neglect: Cognitive neuropsychological aspects. In: Feinberg, T. & Farah, M.J. (Eds.), *Behavioral Neurology and Neuropsychology.* New York: McGraw-Hill, pp. 319–335.

Randolph, C., Braun, A.R., Goldberg, T.E., & Chase, T.N. (1993) Semantic fluency in Alzheimer's, Parkinson's, and Huntington's disease: Dissociation of storage and retrieval failures. *Neuropsychology, 7*, 82–88.

Ratcliff, G. (1979) Spatial thought, mental rotation and the right cerebral hemisphere. *Neuropsychologia, 17*, 49–54.

Ratcliff, G. & Newcombe, F. (1973) Spatial orientation in man: Effects of left, right and bilateral posterior lesions. *Journal of Neurology, Neurosurgery and Psychiatry, 36*, 448–454.

Reed, C.L., & Caselli, R.J. (1994) The nature of tactile agnosia: A case study. *Neuropsychologia, 32*, 527–539.

Risse, G.L., Gates, J., Lund, G., Maxwell, R., & Rubens, A. (1989) Interhemispheric transfer in patients with incomplete section of the corpus callosum. *Archives of Neurology, 46*, 437–443.

Rizzolatti, G. (1987) Functional organization of inferior area 6. In: *Ciba Foundations Symposium, Motor areas of the cerebral cortex.* New York: John Wiley & Sons, pp. 171–186.

Rizzolatti, G., Luppino, G. & Matelli, M. (1996) The classic supplementary motor area is formed by two independent areas. In: Luders, H.O. (Ed.), *Advances in Neurology, Vol. 70: Supplementary Sensorimotor Area.* Philadelphia: Lippincott-Raven, pp. 45–56.

Robin, A & Macdonald, D. (1975) *Lessons of Leucotomy*. London: Henry Kimpton.

Robinson, A.L., Heaton, R.K., Lehman, R.A.W., & Stilson, D.W. (1980) The utility of the Wisconsin Card Sorting Test in detecting and localizing frontal lobe lesions. *Journal of Consulting and Clinical Psychology, 48*, 605–614.

Robinson, R.G. & Benson, D.F. (1981) Depression in aphasic patient: Frequency, severity, and clinical-pathological correlations. *Brain and Language, 14*, 282–291.

Robinson, R.G., Kubos, K.L., Starr, L.B., Rao, K. & Price, T.R. (1984) Mood disorders in stroke patients: Importance of location of lesion. *Brain, 107*, 81–93.

Robinson, R.G. & Price, T.R. (1982) Post-stroke depressive disorders: A follow-up study of 103 patients. *Stroke, 13*, 635–641.

Robinson, R.G. & Szetela, B. (1981) Mood changes following left hemisphere brain injury. *Annals of Neurology, 9*, 447–453.

Roeltgen, D.P. (1993) Agraphia. In: Heilman, K. & Valenstein, E. (Eds.),*Clinical Neuropsychology*. New York: Oxford University Press, pp 63–89.

Roeltgen, D.P. & Heilman, K.M. 1983) Apractic agraphia in a patient with normal praxis. *Brain and Language, 18*, 35–46.

Roeltgen, D.P., Sevush, S. & Heilman, K.M. (1983) Pure Gerstmann's syndrome from a focal lesion. *Archives of Neurology, 40*, 46–47.

Roland, P.E. (1987) Metabolic mapping of sensorimotor integration in the human brain. In: *Ciba Foundation Symposium: Motor areas of the cerebral cortex*. New York: John Wiley & Sons, pp. 251–268.

Ross, E.D. (1981) The aprosodias: Functional-anatomical organization of the affective components of language in the right hemisphere. *Archives of Neurology, 38*, 561–569.

Ross, E.D. (1985) Modulation of affect and nonverbal communication by the right hemisphere. In: Mesulam, M. (Ed.), *Principles of Behavioral Neurology*. Philadelphia: F.A. Davis, pp. 239–257.

Ross, E.D. (1997) The aprosodias. In: Feinberg, T.E. & Farah, M.J. (Eds.), *Behavioral Neurology and Neuropsychology*. New York: McGraw-Hill, pp. 699–709.

Ross, E.D. & Rush, A.J. (1981) Diagnosis and neuroanatomical correlates of depression in brain damaged patients. *Archives of General Psychiatry, 38*, 1344–1354.

Rosselli, M. & Ardila, A. (1989) Calculation deficits in patients with right and left hemispheric damage. *Neuropsychologia, 27*, 607–617.

Rossi,G.F. & Rosadini, G. (1967) Experimental analysis of cerebral dominance in man. In: Darley, F.L. (Ed.), *Brain Mechanisms Underlying Speech and Language*. New York: Grune and Stratton, pp. 167–184.

Rubens, A.B., Mahowald, M.W. & Hutton, J.T. (1976) Asymmetry of lateral (Sylvian) fissures in man. *Neurology, 26*, 620–624.

Ruff, R.M., Allen, C.C., Farrow, C.E., Niemann, H., & Wylie, T. (1994) Figural fluency: Differential impairment in patients with left versus right frontal lobe lesions. *Archives of Clinical Neuropsychology, 9*, 41–55.

Rylander, G. (1948) Personality analysis before and after frontal lobotomy. *Research Publication of the Association of Nervous and Mental Disease, 27*, 591–700.

Sackeim, H.A., Greenberg, M.S., Weiman, A.L., Gur, R.C., Hungerbuhler, J.P. & Geschwind, N. (1982) Hemispheric asymmetry in the expression of positive and negative emotion. *Archives of Neurology, 39*, 210–218.

Saint-Cyr, J.A. & Taylor, A.E. (1992) The mobilization of procedural learning: The "key signature" of the basal ganglia. In: Squire, L.R. & Butters, N. (Eds.), *Neuropsychology of Memory*. New York: Guilford Press, pp. 188–202.

Saling, M.M., Berkovic, S.F., O'Shea, M.F., Kalnins, R.M., Darby, D.G., & Bladin, P.F. (1993) Lateralization of verbal memory and unilateral hippocampal sclerosis: Evidence of task-specific effects. *Journal of Clinical and Experimental Neuropsychology, 15*, 608–618.

Sandson, J. & Albert, M.L. (1987) Perseveration in behavioral neurology. *Neurology, 37*, 1736–1741.

Satz, P., Achenbach, K. & Fennel, P. (1967) Correlations between assessed manual laterality and predicted speech laterality in a normal population. *Neuropsychologia, 5*, 295–310.

Sauguet, J., Benton, A.L., & Hecaen, H. (1971) Disturbances of body schema in relation to language impairment and hemispheric locus of lesion. *Journal of Neurology, Neurosurgery, and Psychiatry, 34*, 496–501.

Schacter, D.L. (1991) Unawareness of deficit and unawareness of knowledge in patients with memory disorders. In: Prigatano, G.P. & Schacter, D.L. (Eds.), *Awareness of Deficit after Brain Injury: Clinical and theoretical issues.* New York: Oxford, pp. 125–151.

Schlanger, B.B., Schlanger, P. & Gerstmann, L.J. (1976) The perception of emotionally toned sentences by right-hemisphere damaged and aphasic subjects. *Brain and Language, 3,* 396–403.

Semmes, J. (1968) Hemispheric specialization: A possible clue to mechanism. *Neuropsychologia, 6,* 11–26.

Sergent, J. (1987) A new look at the human split brain. *Brain, 110,* 1375–1392.

Sergent, J. (1990) The neuropsychology of visual image generation: Data, method and theory. *Brain and Cognition, 13,* 98–129.

Sergent, J. (1995) Hemispheric contribution to face processing: Patterns of convergence and divergence. In: R.J. Davidson & K. Hugdahl (Eds.), *Brain Asymmetry.* Cambridge, MA: MIT Press, pp. 157–177.

Sergent, J. & Poncet, M. (1990) From covert to overt recognition of faces in a prosopagnosic patient. *Brain, 113,* 989–1004.

Sergent, J., Zuck., E., Terriah, S. & McDonald, B. (1992) Distributed network underlying musical sight-reading and keyboard performance. *Science, 257,* 106–109.

Shallice, T. (1982) Specific impairments of planning. *Philosophical Transactions of the Royal Society of London, 298,* 199–209.

Shallice, T. & Burgess, P. (1991) Higher-order cognitive impairments and frontal lobe lesions in man. In, Levin, H.S., Eisenberg, H.M., & Benton A.L. (Eds), Frontal Lobe Function and Dysfunction. N.Y.: Oxford Univ. Press, pp. 125–138.

Shannon, B. (1984) Asymmetries in musical aesthetic judgments. *Cortex, 20,* 567–573.

Shapiro, B.E., Alexander, M.P., Gardner, H., & Mercer, B. (1981) Mechanisms of confabulation. *Neurology, 31,* 1070–1076.

Sharpless, S. & Jasper, H. (1956) Habituation of the arousal reaction. *Brain, 79,* 655–680.

Shaywitz, B.A., Shaywitz, S.E., Pugh, K.R., Constable, R.T., Skudlarski, P., Fulbright, R.K., Bronen, R.A., et al., (1995) Sex differences in the functional organization of the brain. *Nature, 373,* 607–609.

Shima, K., Mushiake, H., Saito, N. & Tanji, J. (1996) Role for cells in the presupplementary motor area in updating motor plans. *Proceedings of the National Academy of Sciences* (USA), *93,* 8694–8698.

Shimamura, A.P., Janowsky, J.S., & Squire, L.R. (1991) What is the role of frontal lobe damage in memory disorders? In: Levin, H.S., Eisenberg, H.M., & Benton A.L. (Eds.), *Frontal Lobe Function and Dysfunction.* New York: Oxford University Press, pp. 173–195

Singer, C. (1957) *A Short History of Anatomy from the Greeks to Harvey.* New York: Dover.

Sinyour, D. Jacques, P., Kaloupek, D.G., Becker, R., Goldenberg, M. & Coopersmith, H.M. (1986) Post-stroke depression and lesion location: An attempted replication. *Brain, 109,* 537–546.

Smith, A. (1960) Changes in Porteus Maze scores of brain-operated schizophrenics after an eight-year interval. *Journal of Mental Science, 106,* 967–978.

Smith, A. (1966) Speech and other functions after left (dominant) hemispherectomy. *Journal of Neurology, Neurosurgery and Psychiatry, 29,* 467–471.

Smith, B.D., Meyers, M.B., & Kline, R. (1989) For better or for worse: Left-handedness, pathology and talent. *Journal of Clinical and Experimental Neuropsychology, 11,* 944–958.

Sperry, R.W. (1968) Mental unity following surgical disconnection of the cerebral hemispheres. *Harvard Lectures, 62,* 293–323.

Squire, L. (1987) *Memory and the Brain.* New York: Oxford University Press.

Squire, L. & Butters, N. (Eds.) (1984) *Neuropsychology of Memory.* New York: Guilford Press.

Starkstein, S.E., Pearlson, G.E., Boston, J., & Robinson, R.G. (1987) Mania after brain injury. *Archives of Neurology, 44,* 1069–1073.

Starkstein, S.E. & Robinson, R.G. (1989) Affective disorders and cerebral vascular disease. *British Journal of Psychiatry, 154,* 170–182.

Starkstein, S.E., Robinson, R.G., Berthier, M.L., Parikh, R.M. & Price, T.R. (1988) Differential mood changes following basal ganglia vs thalamic lesions. *Archives of Neurology, 45,* 725–730.

Starkstein, S.E., Robinson, R.G, & Price, T.R. (1987) Comparison of cortical and subcortical lesions in the production of poststroke mood disorders. *Brain, 110,* 1045–1059.

Stengel, E. (1944) Loss of spatial orientation, constructional apraxia and Gerstmann's syndrome. *Journal of Mental Science, 90,* 753–760.

Steinmetz, H., Rademacher, J., Haung, Y., Hefter, H., Zilles, K., Thron, A. & Freund, H. (1989) Cerebral asymmetry: MR planimetry of the human planum temporale. *Journal of Computer Assisted Tomography*, *13*, 996–1005.

Steinmetz, H., Volkmann, J., Jancke, L. & Freund, H. (1991) Anatomical left-right asymmetry of language-related temporal cortex is different in left and right handers. *Annuals of Neurology*, *29*, 315–319.

Strange, P.G. (1992) *Brain Biochemistry and Brain Disorders*. New York: Oxford University Press.

Strub, R.L., Black, F.W. & Leventhal, B. (1979) The clinical utility of reproduction drawing tests with low IQ patients. *Journal of Clinical Psychiatry*, *40*, 386–388.

Strub, R. & Geschwind, N. (1974) Gerstmann's syndrome without aphasia. *Cortex*, *10*, 378–387.

Strum-Olsen, J. (1946) Discussion on prefrontal leucotomy with reference to indication and results. *Proceedings of the Royal Society of Medicine*, *39*, 443–444.

Stuss, D.T. (1991) Interference effects on memory functions in postleukotomy patients: An attentional perspective. In: Levin, H.S., Eisenberg, H.M., & Benton A.L. (Eds.), *Frontal Lobe Function and Dysfunction*. New York: Oxford University Press, pp. 157–172.

Stuss, D.T. & Benson, D.F. (1984) Neuropsychological studies of the frontal lobes. *Psychological Bulletin*, *95*, 3–28.

Stuss, D.T., & Benson, D.F. (1986) *The Frontal Lobes*. New York: Raven Press.

Subirana, A. (1964) The relationship between handedness and language function. *International Journal of Neurology*, *5*, 215–234.

Szentagothai, J. (1979) Local circuits of the neocortex. In: Scmmitt, F.O. & Worden, F.G. (Eds.), *The Neurosciences: Fourth study program*. Cambridge, MA: MIT Press, pp. 399–415.

Szentagothai, J. & Arbib, M. (1974) Conceptual models of neural organization. *Neurosciences Research Program Bulletin*, *12*, 307–510.

Szentagothai, J. & Arbib, M.A. (1975) The module concept in cerebral cortex architecture. *Brain Research*, *95*, 475–496.

Talland, G.A., Sweet, W. & Ballantine, T. (1967) Amnestic syndrome with anterior communicating artery aneurysm. *Journal of Nervous and Mental Disease*, *145*, 179–192.

Tallis, F. (1997) The neuropsychology of obsessive-compulsive disorder: A review and consideration of clinical implications. *British Journal of Clinical Psychology*, *36*, 3–20.

Tanji, J. (1987) Neuronal activity in the primate non-primary cortex is different from that in the primary motor cortex. In: *Ciba Foundations Symposium, Motor areas of the cerebral cortex*. New York: John Wiley & Sons, pp. 142–150.

Tanji, J. (1994) The supplementary motor area in the cerebral cortex. *Neuroscience Research*, *19*, 251–268.

Tanji, J. & Mushiake, H. (1996) Comparison of neuronal activity in the supplementary motor area and primary motor cortex. *Cognitive Brain Research*, *3*, 143–150.

Tanji, J. & Shima, K. (1994) Role for supplementary motor area cells in planning several movements ahead. *Nature*, *371*, 413–416.

Tanji, J. & Shima, K. (1996) Contrast of neuronal activity between the supplementary motor areas and other cortical motor areas. In: Luders, H.O. (Ed.), *Advances in Neurology, Vol. 70: Supplementary Sensorimotor Area*. Philadelphia: Lippincott-Raven, pp. 95–103.

Teuber, H-L. (1964) The riddle of frontal lobe function in man. In: Warren, J.M., & Akert, K. (Eds.), *The Frontal Granular Cortex and Behavior*. New York: McGraw-Hill, pp. 410–444.

Tezner, D., Tzavaras, A., Gruner, J. & Hecaen, H. (1972) L'asymmetrie droite-gauche du lanum temporale: A propos de l'etude anatomique de 100 cerveaux. *Revue Neurologique*, *126*, 444–449.

Tow, P.M. (1955) *Personality Changes Following Frontal Leucotomy*. New York: Oxford University Press.

Tranel, D. (1997) Disorders of color processing (perception, imagery, recognition, and naming). In: Feinberg, T.E. & Farah, M.J. (Eds.), *Behavioral Neurology and Neuropsychology*. New York: McGraw-Hill, pp. 257–265.

Tucker, D.M., Watson, R.T. & Heilman, K.M. (1977) Affective discrimination and evocation in patients with right parietal disease. *Neurology*, *17*, 947–950.

Tusa, R.J., & Ungerleider, L.G. (1985) The inferior longitudinal fasciculus: A reexamination in humans and monkeys. *Annals of Neurology*, *18*, 583–591.

Ungerleider, L.G. & Mishkin, M. (1982) Two cortical visual systems. In: Ingle, D.J., Mansfield, R.J.W. & Goodale, M.D. (Eds.), *Analysis of Visual Behavior*. Cambridge, MA: MIT Press, pp 549–586.

Valenstein, E.S., (Ed) (1980) *The Psychosurgery Debate*. San Francisco: H.W. Freeman.

Van Essen, D.C. & Maunsell, J.H.R. (1983) Hierarchical organization and functional streams in the visual cortex. *Trends in Neuroscience, 6*, 370–375.

Varney, N.R. (1984) Gerstmann's syndrome without aphasia: A longitudinal study. *Brain and Cognition, 3*, 1–9.

Victor, M., Adams, R.D., & Collins, G.H. (1971) *The Wernicke–Korsakoff Syndrome*. Philadelphia: F.A. Davis.

Vilkki, J. (1989) Perseveration in memory for figures after frontal lobe lesion. *Neuropsychologia, 27*, 1101–1104.

von Economo, C. & Horn, L. (1930) Uber Windungsrelief Masse und Rindenarchitktonik der Supertemporalflache, ihre individullen und ihre Seitenunterschiede. *Zeitschrift fur ges. Neurol. Psychiat., 130*, 678–757.

Wada, J. & Rasmussen, T. (1960) Intracarotid injection of sodium amytal for the lateralization of cerebral speech dominance. Experimental and clinical observations. *Journal of Neurosurgery, 17*, 266–282.

Wada, J.A., Clarke, R. & Hamm, A. (1975) Cerebral hemispheric asymmetry in humans. *Archives of Neurology, 32*, 239–246.

Walsh, K. (1994) *Neuropsychology: A clinical approach, 3rd Edition*. New York: Churchill Livingstone.

Wang, P.L. (1987) Concept formation and frontal lobe function: The search for a clinical frontal lobe test. In: Perceman, E. (Ed.), *The Frontal Lobes Revisited*. New York: IRBN Press, pp. 189–205.

Wapner, W., Hamby, S. & Gardner, H. (1981) The role of the right hemisphere in the appreciation of complex linguistic materials. *Brain and Language, 14*, 15–33.

Warren, S., Yezierski, R.P., & Capra, N.F. (1997) The somatosensory system I: Discriminative touch and position sense. In: Haines,D.E. (Ed.), *Fundamental Neuroscience*. New York: Churchill Livingstone, pp. 219–235.

Warrington, E.K. (1969) Constructional apraxia. In: Vinken, P.J. & Bruyn, G.W. (Eds.), *Handbook of Clinical Neurology. Vol. 4*. Amsterdam: North Holland.

Warrington, E.K., James, M. & Kinsbourne, M. (1966) Drawing disability in relation to laterality of cerebral lesion. *Brain, 89*, 53–82.

Warrington, E.K. & Rabin, P. Perceptual matching in patients with cerebral lesions. *Neuropsychologia, 8*, 475–487.

Warrington, E.K. & Taylor, A.M. (1973) Contribution of the right parietal lobe to object recognition. Cortex, 9, 152–163.

Watson, R.T., Fleet, W.S., Rothi, L.J.G., & Heilman, K.M. (1986) Apraxia and the supplementary motor area. *Archives of Neurology, 43*, 787–792.

Watts, F.N., MacLeod, A.K., & Morris, L. (1988) Associations between phenomenal and objective aspects of concentration problems in depressed patients. *British Journal of Psychology, 79*, 241–250.

Wechsler, A.F., Verity, M.A., Rosenschein, S., Fried, I. & Scheibel, A.B. (1982) Pick's disease: A clinical, computed tomographic, and histological study with Golgi impregnation observations. *Archives of Neurology, 39*, 287–290.

Wechsler, D. (1981) *Wechsler Adult Intelligence Scale*. San Antonio, TX: Psychological Corporation.

Weinberger, D.R., Berman, K.F. & Zec, R.F. (1986) Physiological dysfunction of dorsolateral prefrontal cortex in schizophrenia. I: Regional blood flow (rCBF) evidence. Archives of General Psychiatry, 43, 114–125.

Weinstein, E.A. (1991) Anosognosia and denial of illness. In: Prigatano, G.P. & Schacter, D.L. (Eds.), *Awareness of Deficit after Brain Injury: Clinical and theoretical issues*. New York: Oxford, pp. 240–257.

Weintraub, S., Mesulam, M.M. & Kramer, L. (1981) Disturbances in prosody. *Archives of Neurology, 38*, 742–744.

Wernicke, C. (1874) *Der Aphasische Symptomcomplex*. Breslau: Cohn & Weigart.

White, L.E., Lucas, G., Richards, A., & Purves, D. (1994) Cerebral asymmetry and handedness. *Nature, 368*, 197–198.

Wiesendanger, M., Hummelsheim, H., Bianchetti, M., Chen, D.F., Hyland, B., Maier, V. & Wiesendanger, R. (1987) Input and output organization of the supplementary motor area. In: *Ciba Foundation Symposium, Motor areas of the cerebral cortex*. New York: John Wiley & Sons, pp. 40–62.

Wilson, S.A.K. (1924) Some problems in neurology. II: Pathological laughing and crying. *Journal of Neurological Psychopathology, 16*, 299–333.

Wise, S.P. (1996) Corticospinal efferents of the supplementary sensorimotor area in relation to the primary motor area. In: Luders, H.O. (Ed.), *Advances in Neurology, Vol. 70: Supplementary Sensorimotor Area*. Philadelphia: Lippincott-Raven, pp. 57–69.

Wise, S.P., Boussaoud, D., Johnson, P.B. & Caminiti, R. (1997) Premotor and parietal cortex: Connectivity and combinatorial computations. *Annual Review of Neuroscience, 20*, 25–42.

Wise, S.P., Fried, I., Olivier, A., Paus, T., Rizzolatti, G. & Zilles, K. (1996) Workshop on the anatomic definition and boundaries of the supplementary sensorimotor area. In: Luders, H.O. (Ed.), *Advances in Neurology, Vol. 70: Supplementary Sensorimotor Area*. Philadelphia: Lippincott-Raven, pp. 489–495.

Witelson, S.F. & Kigar, D.L. (1992) Sylvian fissure morphology and asymmetry in men and women: Bilateral differences in relation to handedness in men. *Journal of Comparative Neurology, 323*, 326–340.

Witelson, S.F. & Kigar, D.L. (1988) Asymmetry in brain function follows asymmetry in anatomical form: Gross, microscopic, postmortem and imaging studies. In: Boller, F. & Grafman, J. (Eds.), *Handbook of Neuropsychology, Vol. 1*. Amsterdam: Elsevier Science Publishers, pp. 111–142.

Witelson, S.F. & Pallie, W. (1973) Left hemisphere specialization for language in the newborn. Neuroanatomical evidence of asymmetry. *Brain, 96*, 641–643.

Wolf, S. (1973) Difficulties in right-left orientation in a normal population. *Archives of Neurology, 29*, 128–129.

Woods, D.L. & Knight, R.T. (1986) Electrophysiological evidence of increased distractibility after dorsolateral prefrontal lesions. *Neurology, 36*, 212–216.

Yamadori, A., Osumi, U., Mashuara, S. & Okuto, M. (1977) Preservation of singing in Broca's aphasia. *Journal of Neurology, Neurosurgery and Psychiatry, 40*, 221–224.

Zaidel, E. (1985) Language in the right hemisphere. In: Benson, D.F. & Zaidel, E. (Eds.), *Dual Brain: Hemispheric specialization in humans*. New York: Guilford Press, pp 205–231.

Zaidel, E. (1995) Interhemispheric transfer in the split-brain: Long-term status following complete cerebral commissurotomy. In: Davidson, R.J. & Hugdahl, K. (Eds.), *Brain Asymmetry*. Cambridge, MA: MIT Press, pp. 491–532.

Zaidel, E., Clarke, J., & Suyenobu, B. (1990) Hemispheric independence: A paradigm case for cognitive neuroscience. In: Scheibel, A.B. & Wechsler, A.F. (Eds.), *Neurobiology of Higher Cognitive Function*. New York: Guilford Press, pp. 297–362.

Zilles, K., Schlaug, G., Geyer, S., Luppino, G., Matelli, M., Qu, M., Schleicher, A. & Schormann, T. (1996) Anatomy and transmitter receptors of the supplementary motor areas in the human and nonhuman primate brain. In: Luders, H.O. (Ed.), *Advances in Neurology, Vol. 70: Supplementary Sensorimotor Area*. Philadelphia: Lippincott-Raven, pp. 29–43.

Zola, S. (1997) Amnesia: Neuroanatomic and clinical aspects. In: Feinberg, T.E. & Farah, M.J. (Eds.), *Behavioral Neurology and Neuropsychology*. New York: McGraw-Hill, pp. 447–461.

10   THE CEREBRAL VASCULAR SYSTEM

CHAPTER OVERVIEW

Cerebrovascular disorders represent one of the major causes of morbidity and mortality in adult populations by directly impacting the functional integrity of the CNS. This chapter will focus on three major aspects of the cerebrovascular system: **anatomy**, **pathology**, and **neurobehavioral syndromes**. As one cannot fully discuss the functional anatomy of the central nervous system without reviewing the vascular system that supplies it, the first priority will be to review the sources and patterns of distribution of the arterial blood supply to the brain. While it is not imperative for most clinicians to be able to name all the vessels, it is important to at least come away with a mental map of the spheres of influence for each of the major arterial groups. As the venous system was covered in the preceding chapter in conjunction with the discussion of the meninges, it will not be discussed in the same detail here.

Being under substantially greater pressure than venous blood and not having the benefit of the "filtering" action of the capillaries, the arterial system is prone to a wider range of pathology compared to the venous system. Also being responsible for the constant nourishment and oxygenation of the brain, any disruption of the system will have almost immediate and potentially catastrophic consequences. It is important for the clinician to have a general appreciation of the more common pathological conditions that can affect the cerebral arteries, including their premorbid risk factors, typical clinical presentation, effects on nervous tissue, and expected course over time. To this end, the major types of ischemic deficits, hemorrhagic events, and structural anomalies associated with cerebrovascular disease will be discussed.

Although there is obviously a significant interaction between the nature and severity of the specific pathological condition and the particular arterial system involved (anatomical locus or vascular distribution), the latter is clearly a major determinant of the nature of the neurobehavioral deficits or changes that will likely be manifested. As suggested earlier in this text, with the advances in neuroimaging over the past quarter century, the physical localization of lesions, including vascular, has become much less of a challenge. But as clinical neuroscientists our interest transcends the physical localization of the lesion. We want to understand the potential or probable behavioral consequences of such lesions in order to most effectively manage and care for our patients. Thus, the latter part of this chapter will be devoted to a review of the signs and symptoms or neurobehavioral syndromes commonly associated with lesions of the major cerebral arterial systems.

ANATOMY OF THE ARTERIAL SYSTEMS OF THE BRAIN

There are four ascending arteries that contribute to the cerebral circulation; two **carotid arteries** and two **vertebral arteries**. Although there is occasionally some variation in these vessels, Figure 10–1 represents the most common arrangement. The **left common carotid**

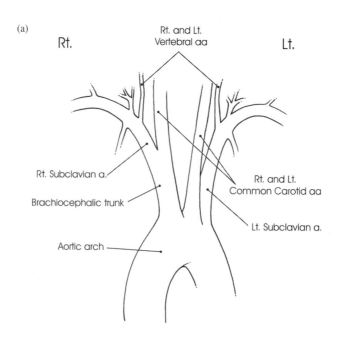

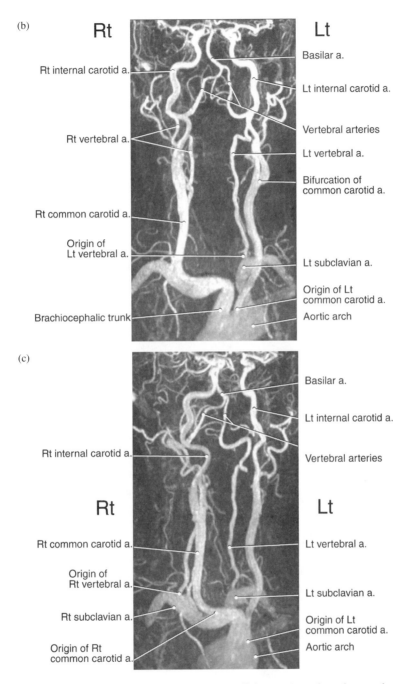

Figure 10–1. (a) Carotid and vertebral arteries coming off the aortic arch and secondary vessels. (b,c) MRAs of carotid and vertebral arteries, showing slightly rotated views from same individual.

artery emanates directly from the **aortic arch**, while the **right common carotid** derives from the **brachiocephalic** artery coming off the aortic arch. Slightly distal to the branching of the right common carotid artery, the brachiocephalic artery also gives rise to the **right vertebral** artery. The third major artery to branch off the aortic arch is the left subclavian artery. The **left vertebral** artery branches off the left subclavian shortly after it exits from the aortic arch.

VERTEBRAL SYSTEM

The vertebral arteries proceed in their cephalic course along the ventral surface of the cervical spine and actually becoming encased in its bony processes (the *transverse foramen*). After entering the foramen magnum, the vertebral arteries are found to lie adjacent to the ventral surface of the medulla in the brainstem. These two vertebral arteries then join to form the singular **basilar** artery at the level of the pontine–medullary junction (Figures 10–2 and 10–3). The basilar artery itself eventually will bifurcate just above the pons (at the level of the midbrain), giving rise to the two **posterior cerebral** arteries (PCA). As can be seen in Figures 10–3 and 10–4, there are several prominent cortical branches of the PCA: the **parietooccipital** artery, the **calcarine** artery (which supplies Brodmann's area 17, the primary visual cortex), and the **anterior** and **posterior** (not shown) **temporal branches** of the PCA (supplying the ventral and medial surfaces of the temporal lobes, including parts of the hippocampus).

In addition to the posterior cerebral arteries, the vertebral and basilar arteries give rise to multiple branches throughout their course along the brainstem. These branches, as seen in Figure 10–3, represent the major source of blood supply both to the brainstem and the

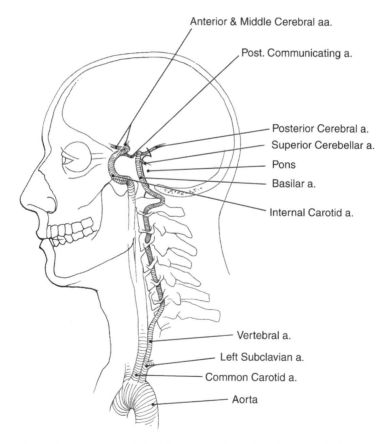

Figure 10–2.   Lateral view of the vertebral–basilar system showing its relationship to the internal carotid artery. Figure also illustrates the distal portion of the vertebral artery passing through the transverse processes of the cervical vertebrae before entering the foramen magnum and forming the basilar artery.

cerebellum, as well as to the cervical portion of the spinal cord.[1] The three main branches of the vertebral arteries are (1) the **anterior spinal** artery, (2) the **posterior spinal** arteries, and (3) the **posterior inferior cerebellar** arteries. After the vertebral arteries converge to form the basilar artery, the latter gives rise to two other major vessels supplying the cerebellum and brainstem: the **anterior inferior cerebellar** and the **superior cerebellar arteries**. The anterior inferior cerebellar artery supplies the anterior and inferior portions of the cerebellum and the caudal pons. The superior cerebellar artery supplies the superior aspect of the cerebellum, the rostral pons, and portions of the midbrain. Other smaller arteries (pontine branches) emanate from the basilar artery and serve to nourish brainstem structures, particularly the pons. The areas supplied by the posterior cerebal artery will be discussed below.

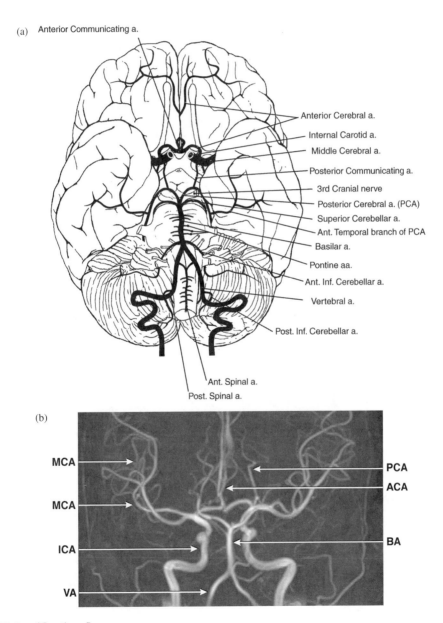

Figure 10–3. (*Continued*)

(c)

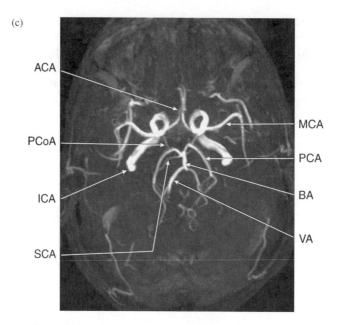

Figure 10–3. (a) Ventral view of vertebral–basilar system and its relationship to the brainstem and carotid circulation. Also illustrated is the "circle of Willis" creating anastomoses between the posterior and carotid circulations. MRA (magnetic resonance angiograms) showing same vessels from (b) A-P and (c) axial perspectives. Abbreviations: ACA, anterior cerebral artery; BA, basilar artery; ICA, internal carotid artery; MCA, middle cerebral artery; PCA, posterior cerebral artery; PCoA, posterior communicating artery; SCA, superior cerebellar artery; VA, vertebral artery.

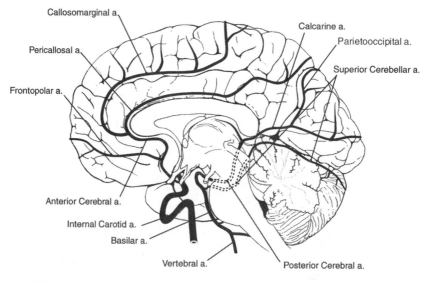

Figure 10–4. Midsagittal view of the cerebral hemisphere showing the general distribution of the main branches of the anterior and posterior cerebral arteries.

CAROTID SYSTEM

The common carotid arteries also proceed in a cephalic direction in a more anterior or ventral position in the neck. These are the arteries one feels as the fingers are placed on either side of the larynx. At approximately the level of the fourth cervical vertebra, the common carotids bifurcate, producing the **internal** and **external carotid** arteries. The latter go on to supply the extracranial tissues of the face and the scalp and most of the dura. The internal carotid arteries continue into the cranial vault where they eventually give rise to the four remaining cerebral vessels, the right and left **anterior** (ACA) and **middle cerebral** (MCA) arteries.

Just prior to entering the circle of Willis (see below) where the internal carotid and vertebral systems anastomose via "communicating" or connecting vessels, the internal carotids give rise to two major vessels, the **ophthalmic** artery and the **anterior choroidal** artery. In addition to supplying other structures in and around the eye and the anterior portion of the dura, one branch of the ophthalmic artery enters the eye along with the optic nerve and supplies the retina. The anterior choroidal artery will be reviewed in greater detail later. As will be seen, as its name implies, it is important in supplying the choroid plexus (lateral ventricles), but also supplies parts of the visual and motor systems and the temporal poles.

Circle of Willis

The internal carotid and vertebral vascular systems interconnect at the base of the brain, anterior to the brainstem and surrounding the optic chiasm (Figure 10–3). This interconnection or anastomosis is known as the *circle of Willis* after the 17th century anatomist (Thomas Willis, 1621–1675). The circle of Willis provides a potential diversion for collateral blood supply following the occlusion of one the major cranial arteries feeding into it. However, as we shall see, this potential collateral system is not uniform across individuals and can be influenced by several factors. This "circle" is completed by the presence of "communicating" arteries that connect the right and left internal carotids with the vertebral circulation. Just rostral to the third cranial nerves and slightly posterior to the mammillary bodies, the basilar artery bifurcates, forming the right and left posterior cerebral arteries (PCA). Shortly after their formation, each posterior artery sends off an anterior branch that connects to the ipsilateral internal carotids. These connecting vessels are the **posterior communicating** arteries. The internal carotids in turn divide into the middle and anterior cerebral arteries. The middle cerebral artery (MCA), which basically is the primary extension of the internal carotid, proceeds dorsal-laterally up through the lateral fissure between the temporal and frontal cortices. The anterior cerebral artery (ACA) initially remains more medial as it proceeds anteriorly toward the frontal pole. Just anterior to the optic chiasm, a small branching artery—the anterior communicating artery—connects the two anterior cerebral arteries, thus completing the circle.

The above description of the circle of Willis again represents the more typical pattern, but some individual variations may be noted. Despite the presence of a completed "circle" in the majority of individuals, there may not be much blood flowing around the circle, that is, there is little "communication" between the right and left internal carotids via these communicating arteries. This relative lack of flow around the circle of Willis seems primarily to be a function of the fact that:

1. The communicating arteries themselves are often relatively small.
2. There normally is relatively equal hemodynamic pressure from one arterial system to the other, thus not encouraging flow between the systems.

However, if over time one of the major feeder arteries (eg., one of the internal carotids) becomes gradually stenosed, a pressure gradient develops that encourages the shunting of blood from one side to the other. As a result of this shunting of blood through the posterior and/or anterior communicating arteries, the communicating arteries gradually enlarge, creating a larger lumen, thus facilitating more shunting of blood. Therefore, it is not unheard of to find a fairly complete thrombotic occlusion (see below) of one of the internal carotid arteries with little if any clinical manifestations of compromise of cerebral blood flow. However, if such an occlusion were to take place more acutely, for example, as a result of an embolus, the communicating vessels would not have time to adapt and a major stroke is likely to ensue.

Anterior Cerebral Artery

As noted above, the anterior cerebral artery (ACA) originates at the bifurcation of the internal carotid into the middle and anterior cerebral arteries. The ACA, typically smaller than the middle cerebral artery (MCA), proceeds rostrally (frontally) along the base of the frontal lobe. Shortly after their point of origin, the anterior communicating artery connects the two ACAs. As we will see, aneurysms have a tendency to develop at sites where proximal branching of the arteries occurs, and the ACA is a common site for such aneurysms. Distal to the anterior communicating artery, the main branch of the ACA proceeds anteriorly and dorsally through the interhemispheric fissure. It then curves around the genu of the corpus callosum and follows the corpus callosum posteriorly along its dorsal surface in the callosal sulcus, which lies between the corpus callosum and the cingulate gyrus (Figure 10–4). This portion of the ACA is known as the **pericallosal** artery. Typically, a second more dorsally positioned branching of the ACA follows the cingulate sulcus. This second branch is the **callosomarginal** artery. The anterior cerebral artery also sends off secondary arteries that supply the orbital (**orbital branch**) and polar (**frontopolar branch**) frontal cortices and the medial frontal and parietal cortices (including most if not all of the cingulate gyrus). Branches of this artery also supply the genu and body (more or less the anterior two thirds) of the corpus callosum, as well as parts of the anteroventral striatum and the anterior limb of the internal capsule (the **recurrent artery of Heubner**). As will be seen in Figure 10–5, the distal branches of this anterior system also tend to overlap slightly onto the dorsal–lateral surface of the frontal and parietal lobes.

Middle Cerebral Artery

The larger middle cerebral artery is a more direct continuation of the internal carotid artery. This, combined with its larger lumen, and hence, greater hemodynamic flow, increases the probability that emboli emanating from the heart or carotid vessels will affect the distribution of the MCA rather than going up through the ACA. After separating from the terminal end of the internal carotid, the MCA proceeds dorsolaterally into the inferior aspect of the lateral fissure. Within the frontal–parietal operculum in the region of the insular cortex, the MCA divides into various cortical branches (varying to some extent from one individual to another) that exit from the superior surface of the lateral fissure. As shown in Figure 10–5, these MCA branches (which include the **orbitofrontal, prefrontal, central, postcentral, anterior** and **posterior parietal, angular,** and the **posterior, middle, anterior,** and **polar temporal** arteries) supply almost the entire lateral convexities of the frontal, parietal, and temporal lobes. This amounts to most of the lateral surface of the brain. The MCA also supplies the cortex of the insula and the claustrum. Smaller, penetrating arteries (**medial** and **lateral lenticulostriate** arteries), coming off the MCA, supply other internal subcortical structures. These will be reviewed separately below. The terminal branches of both the anterior and posterior cerebral arteries, which primarily supply the medial cortical surfaces, extend slightly onto the dorsolateral and ventrolateral surface of all four lobes where they overlap with the terminal branches of the MCA. This region of overlap is referred to as the *watershed* or *borderzone* areas, and as we shall see may become important in certain hypotensive states or in some cases infarction of the internal carotid artery and hypertension.

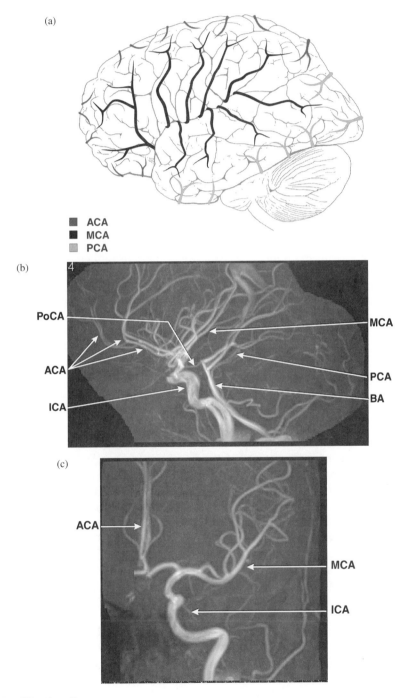

Figure 10–5. (*Continued*)

(d)

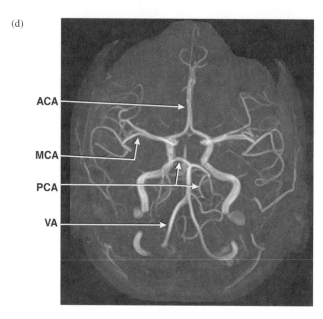

ACA

MCA

PCA

VA

Figure 10–5. (a) Lateral view illustrating the distribution of the middle cerebral artery. The main trunk (M-1) of the MCA comes off the internal carotid and reaches the surface of the hemispheres by passing through the lateral (Sylvian) fissure. As shown in Figure 10–4, the anterior and posterior circulations are most prominent on the medial surface of the cerebral hemispheres, but also extend slightly onto the lateral surface of the hemispheres where they anastomose with the terminal branches of the MCA. The MRAs in figures (b), (c), and (d), respectively, show the lateral, A-P, and oblique axial views of these major vessels. Abbreviations: ACA, anterior cerebral artery; BA, basilar artery; ICA, internal carotid artery; MCA, middle cerebral artery; PCA, posterior cerebral artery; PcoA, posterior communicating artery; VA, vertebral artery.

POSTERIOR CEREBRAL ARTERY

As noted earlier, the posterior cerebral arteries are part of the vertebral system and are formed from the bifurcation of the basilar artery. After their origination at the level of the midbrain, they curve posteriorly around the midbrain with the main trunk remaining on the medial surface of the occipital–temporal cortices (see Figure 10–4). Branches of the PCA supply the inferior and medial portions of the temporal lobe, except for the temporal pole. This includes at least part of the hippocampal gyrus and the parahippocampal and occipitotemporal (*fusiform*) gyri (part of the hippocampus is supplied by the anterior choroidal artery). As can be seen in Figure 10–4, occipital branches of the PCA supply the medial portions of the occipital lobe, the lingual gyrus and cuneus (which include the primary visual cortex), and parts of the medial superior parietal lobule. The splenium of the corpus callosum also is supplied by the PCA system. Again, the terminal branches of the PCA tend to overlap and anastomose with the terminal branches of the ACA and the MCA both on the margins of the lateral convexities as well as on the medial surfaces of the hemispheres.

CENTRAL PERFORATING ARTERIES

The ACA, MCA, PCA, and the posterior communicating artery also give off smaller perforating vessels that supply core brain structures. One group of these vessels derives from the area of the ACA. These vessels supply the hypothalamus, the optic chiasm, and the medial structures dorsal to it, including the preoptic area, the septum pellucidum, and the rostrum of the corpus callosum. The genu and body of the corpus callosum are supplied by the pericallosal branch of the ACA itself. Other more anterior perforating arteries supply the rostral portions of the head of the caudate nucleus, part of the putamen, and the anterior limb of the internal capsule.

The middle cerebral artery is the source for the **medial** and **lateral lenticulostriate** arteries (Figure 10–6). They emanate from the MCA shortly after its origination as it begins to penetrate the inferior portion of the lateral sulcus. These small vessels supply the "lenticular" nucleus (putamen and globus pallidus), the more posterior portions of both the caudate nucleus (except for the tail), and the anterior limb of the internal capsule, as well as the dorsal portion of its posterior limb. The more ventral aspects of the posterior limb of the internal capsule are supplied by the anterior choroid arteries. As we shall discuss later, the lenticulostriate arteries appear to be particularly vulnerable to hypertensive disease and are often involved in subcortical lacunar infarctions.

The PCA and the posterior communicating arteries also are the source of numerous small penetrating arteries. One portion of the larger **posteromedial group** proceeds rostrally to supply the mammillary bodies and hypothalamic structures, while other portions of this (the *thalamoperforating* and *inferior thalamic* branches) supply the anterior and medial thalamus, as well as other midbrain structures such as the red nucleus. A **posterolateral group**, arising either from the PCA or posterior communicating arteries, supplies the remainder of the thalamus, including the lateral and posterior nuclear groups and the geniculates. These latter arteries are known as the *thalamogeniculate* branches.

The **anterior choroidal** artery, which derives from the posterior surface of the internal carotid just distal to the branching of the posterior communicating artery, already was mentioned. In addition to supplying the choroid plexus of the lateral ventricles, it also may supply a number of other fiber pathways and nuclear groups. These include the optic tract, the tail of the caudate, the amygdala and portions of the hippocampal formation, and posterior portions of the internal capsule. It also may supply parts of the lateral

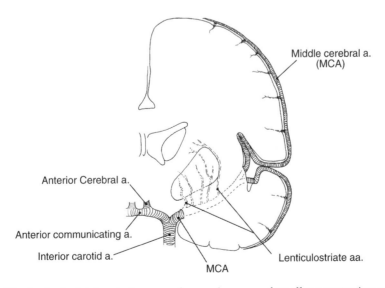

Figure 10–6.   The lenticulostriate arteries, one of several groups of smaller penetrating arteries, branch off the MCA within the fold of the lateral fissure to supply portions of the basal ganglia and internal capsule.

geniculates, globus pallidus, and posterior putamen, and other subthalamic nuclei, including the substantia nigra.

The **posterior choroidal** artery, which arises from the PCA, supplies the choroidal plexus of the third ventricle, the tectum or inferior and superior colliculi, and probably contributes to the dorsal medial nucleus of the thalamus.

SUMMARY OF THE ARTERIAL BLOOD SUPPLY TO THE BRAIN

The **brainstem** (midbrain, pons, and medulla) and **cerebellum** as well as the upper part of the **spinal cord** primarily are supplied by vessels arising from the vertebral and basilar arteries. Certain midbrain structures also are supplied by the anterior (e.g., substantia nigra) and posterior (e.g., tectum) choroidal arteries

The **lateral convexity of the cerebral hemispheres** is generally supplied by the branches of the MCA, which along with the ACA is derived from the bifurcation of the internal carotid arteries. However, both the ACA and PCA overlap the distribution of the MCA on the margins of the lateral surface of the hemispheres.

The **medial** and **inferior surfaces of the hemispheres** are supplied by the ACA and PCA. The ACA supplies the orbital and medial surface of the frontal lobes, including the sensorimotor regions for the lower limbs and most of the remaining medial parietal cortices, the cingulate gyrus, and most of the anterior two thirds of the corpus callosum. The PCA supplies the medial surface of the occipital cortex (includes the primary visual areas), parts of the medial superior parietal lobule, most of the medial and inferior temporal lobe, and the posterior third (splenium) of the corpus callosum.

The **basal ganglia** primarily are supplied by the lenticulostriate arteries that emanate from the proximal portion of the MCA. However, parts of these structures also are supplied by the anterior choroidal artery and small vessels coming off the posterior communicating artery.

The **thalamus** largely is supplied by the smaller vessels that branch off the PCA and posterior communicating arteries, including the thalamoperforating, inferior thalamic, thalamogeniculate, and posterior choroidal arteries. The anterior choroidal arteries also may make a minor contribution. The **anterior hypothalamus** largely is supplied by the small penetrating (anteromedial) vessels originating from the ACA and/or anterior communicating artery in the vicinity of the circle of Willis, while the more posterior portions of the hypothalamus are supplied by the posterior penetrating arteries.

Most of the anterior and posterior limb of the **internal capsule** also is nourished by the lenticulostriate arteries. Small arteries that originate directly from the internal carotid may supply the genu. The anterior choroidal artery generally supplies some of the more ventral portions of the posterior limb, as well as the retrolenticular portions.

The **visual system** is supplied by multiple vessels. The retina of each eye separately is supplied by its corresponding ophthalmic artery from the internal carotid. The optic nerves, optic chiasm, and the initial segment of the optic tracts largely are supplied by the small penetrating arteries derived from the ACA and the anterior communicating artery. Small vessels from the internal carotid or middle arteries primarily supply the more anterior portions of the optic tract. The more posterior portions of the tract, along with parts of the lateral geniculates, are supplied by the anterior choroidal arteries. The lateral geniculates also are supplied by the posterolateral penetrating arteries of the posterior system. The superior optic radiations are supplied by the posterior cortical branches of the middle cerebral artery, while the inferior radiations are nourished by the posterior cerebral artery (PCA). The primary visual cortex and parts of the secondary visual cortex are supplied by the PCA.

The **motor system** also is subserved by a variety of vessels. The disruption of any of these vessels can produce a weakness or paralysis. The primary motor cortex that mediates the face, hands, and trunk is served by the MCA, whereas the legs (especially the lower legs and feet) are represented on the medial surface of the hemispheres, and thus are supplied by the ACA. The MCA–ACA watershed territory supplies the motor cortex that mediates control of the proximal arm and proximal leg. The basal ganglia and internal capsule primarily are supplied by the lenticulostriate arteries (branches of the MCA), although some input is derived from the anterior choroidal and posterior penetrating arteries. The corticospinal tracts are supplied by various small vessels deriving both from the internal carotid and posterior cerebral artery systems at the level of the midbrain and by branches of the basilar and vertebral arteries at the level of the pons and medulla. The cerebellum, the disruption of which can lead to difficulties with balance, coordination, and weakness, is supplied by the superior cerebellar artery, the anterior and posterior cerebellar arteries that derive from the vertebral system.

PATHOLOGY OF THE VASCULAR SYSTEM

Thus far, the focus of this text has been primarily on functional neuroanatomy rather than on neuropathology. However, with the vascular system, it seems prudent to review, at least in general terms, the types of vascular problems that most frequently affect the brain and some of the more common syndromes that can result from specific vascular lesions. A detailed discussion of the pathophysiology of strokes will not be presented here.[2]

For present purposes, cerebrovascular disorders can be classified into three general types: (1) **ischemic lesions**, which can be either hypoxic or occlusive in origin, (2) **hemorrhagic lesions**, and (3) **blood vessel anomalies**. It should be noted, however, that these disorders are not mutually exclusive. For example, an ischemic infarct or a vascular anomaly, such as an aneurysm, may eventually hemorrhage, producing a hemorrhagic lesion. Ischemic lesions are the most common causes of stroke, accounting for about 80% of all strokes. Hemorrhagic lesions account for the remainder of all strokes, and most often are produced by ruptured berry aneurysms or hypertensive hemorrhages.

Ischemic Disease

Occlusive Lesions

Ischemic lesions can result from a (1) blockage of the vessel itself, (2) lack of sufficient oxygen in the blood, or (3) systemic circulatory problem that results in poor perfusion. Occlusive vascular disease, which results from a blockage (*infarction*) in a particular vessel, interrupts blood flow to the brain. Occlusive lesions may involve either arterial vessels or the venous system, although the former is much more common and will be the focus of the discussion. Three etiological subtypes of ischemic arterial occlusive disorders generally identified are: **thrombotic, thromboembolic,** and **embolic. Thrombotic disease** produces a gradual, progressive narrowing of the arterial lumen or *arterial stenosis* that inhibits or restricts the flow of blood. This condition most commonly results from atherosclerosis, which in turn results from a buildup of fibrous tissue within the interior walls of the vessel. This atherosclerotic condition is compounded by the adherence of fatty plaques and blood platelets to these fibrous plaques. Although atherosclerosis occurs with aging, certain conditions appear to accelerate this process. These include hypertension, diabetes, elevated serum lipoproteins, smoking, and genetic predisposition.

The large vessels of the neck (e.g., the internal carotid system) and the areas of bifurcation or branching in the proximal portions of the cerebral arteries are sites that offer particularly high risks as sources of thrombotic strokes. Not all thrombotic occlusions result in a stroke (i.e., persistent, focal neurological deficit associated with a disturbance of the circulatory system). If the arterial stenosis occurs very gradually, particularly if the site is proximal to the circle of Willis, compensatory collateral circulation may develop via other arterial systems (e.g., the opposite internal carotid or posterior vessels). The circle of Willis is not the only potential source of collateral blood supply. In certain circumstances, branches from the external carotid artery may shunt blood via anastomoses with the ophthalmic artery into the carotid system. While the areas in which the distal distributions of the ACA, MCA, and PCA overlap (watershed territories) also may form anastomoses providing potential sources of collateral interchanges between two arterial distributions, if present these generally have limited clinical significance.

In addition to thrombosis, another cause for occlusive vascular disease is emboli. An **embolus** is a bit of organic material or foreign matter that travels from one part of the body (or vascular system) through the arterial system where it eventually lodges, occluding the vessel. Common sources of emboli are bits of thrombotic plaques from the heart or cardiocephalic arteries, bacterial endocarditis, air or nitrogen bubbles, and fat. The heart is the most common source of embolic stroke (i.e., cardioembolic stroke). Cardioembolic stroke is associated with several cardiac conditions, including arrhythmias, acute myocardial infarction (MI), cardiac wall motion abnormalities due to an old MI or cardiomyopathy, valvular disease, and some congenital cardiac diseases. Cardiac arrhythmias (particularly atrial fibrillation) greatly increase the risk for the development of cardiac emboli in susceptible individuals, for example, those with history of rheumatic fever, myocardial infarction, or other conditions that lead to scarring or plaque development with the walls of the heart or its valves.

Fat emboli most commonly results following fractures of the long bones of the leg. Fat globules are carried by the venous system through the heart to the lungs where most are filtered out of the circulating blood. However, it is possible for some of the smaller fat cells to enter into the arterial system where they can be transported to the brain (or other organs) causing embolic infarctions. Because the lungs already have filtered out most of the larger fat emboli, the resulting cerebral infractions tend to be small, but may be multiple.

Distant emboli, again with the heart being the most common source, account for about 20% of ischemic strokes. Emboli that originate from thrombotic plaques in the neck or travel up the carotid system from the heart are considerably more likely to enter the middle as opposed to the anterior cerebral circulation. Emboli that originate from extracranial or large cerebral vessels that are stenosed often are referred to as thromboembolic strokes, as these occlusive strokes are produced by an embolus breaking off of a thrombosed vessel.

While there are no hard and fast rules, there are several factors that may differentiate thrombotic infarction from embolic infarctions. First, in general, embolic infarctions are somewhat more likely to occur during periods of activity rather than at rest (as is slightly more typical of thrombotic infarctions). Thrombotic infarctions are more likely to have been preceded by transient ischemic attacks (TIAs) than are embolic strokes. Embolic infarctions are more likely to evidence secondary hemorrhage (hemorrhagic infarctions) than are those caused by thrombosis. Thrombotic infarctions more often than not are likely to occur while sleeping or shortly after awakening in the morning. This is probably related to changes in hemodynamics and/or blood chemistries that can occur during periods of prolonged quiescence.

Although there are many ways to conceptualize stroke subtypes, one method is based on the vascular anatomy and pathophysiology. In addition to the MCA, ACA, and PCA distributions, another common dichotomy is *large* versus *small* vessels. Large cerebral vessels generally include the main surface vessels of the ACA, MCA, and PCA that supply the lateral, medial, and inferior aspects of the cerebral cortex. Small cerebral vessels, on the other hand, generally are defined as those central perforating vessels (such as the lenticulostriate arteries) that, as we saw earlier, tend to supply internal cortical structures (basal ganglia, internal capsule, and thalamus). Therefore, stroke syndromes often are subtyped as large vessel or small vessel strokes. The mechanisms and deficits associated with each of these subtypes differ. Large vessel strokes usually are caused by a localized thrombus to a large vessel or a branch of a large vessel (like a branch of the MCA) or by an embolus propagated from an area of atherosclerosis to a more distal branch. An example might be an atherosclerotic plaque at the internal carotid artery bifurcation throwing off an embolus to the MCA).

Deficits corresponding to the vascular territory that is occluded often are discernable with a comprehensive neurological and mental status examination, although neuroimaging generally will better define its anatomical boundaries. The more proximal the occlusion in the vascular tree, the larger the area of infarction, and hence, the greater the deficits that are likely to occur. By contrast, more distally located occlusions will produce smaller areas of infarction and, as a general rule, more restricted deficits (see Figure 10–7). The nature and prominence of the symptoms also will depend on the locus of the lesion. For example, a stroke produced by a more proximal occlusion of the internal carotid or MCA of the dominant hemisphere typically would result in massive motor, sensory, language (global aphasia), and other cognitive deficits, while an infarction that is limited to a particular branch of the MCA would result in an incomplete MCA syndrome with the exact symptoms dependent upon the vessel(s) involved.[3] Although less common, large-vessel strokes can be limited to the ACA distribution or the PCA (the latter being derived from the vertebrobasilar system). Again, specific complaints from the patient's history and signs and symptoms obtained from an examination of the patient often can identify the vascular territory most likely involved. Many of the specific findings and complaints commonly associated with particular arterial territories will be reviewed later in this chapter.

While hypertension frequently is a factor in the development of generalized, large-vessel atherosclerotic disease, a history of chronic hypertension almost universally is present in

(a)

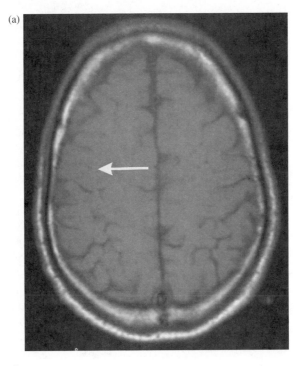

Figure 10–7. (*Continued*)

(b)

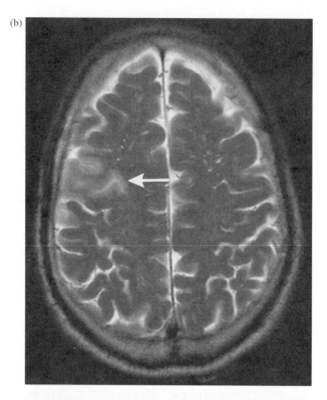

(c)

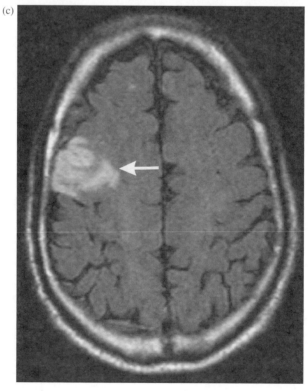

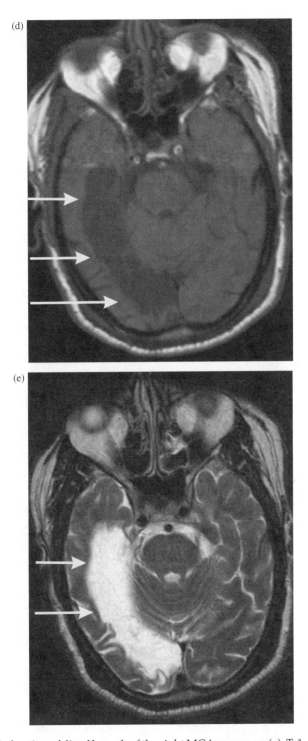

Figure 10–7. Acute infarction of distal branch of the right MCA as seen on (a), T-1, (b), T-2, and (c) FLAIR images. Compare the acute lesion as seen on (a) T-1 image with an old infarction of more proximal portions of the right PCA as seen on (d) T-1 and (e) T-2 weighted images. **Note:** Infarctions are less conspicuous in T-1 weighted images in their acute stage, but show up more prominently in latter stages as the infarcted tissue is displaced by water molecules. Nonetheless, careful observation of (a) reveals slight sulcal effacement in the affected area, a subtle or indirect sign of edema following acute infarction.

small-vessel disease, whether of occlusive or hemorrhagic origin. The mechanism behind hypertensive infarcts involves localized thrombosis of the small, penetrating vessels due to lipohyalinoid degeneration that leads to occlusion of the vessel.[4] These types of small-vessel strokes produce **lacunar infarcts**, which by definition are lesions of 1.0 to 1.5 centimeters in size (Figure 10–8). Because of the vascular anatomy and vulnerability of the deep penetrating small vessels that supply subcortical and brainstem areas, there are specific stroke syndromes

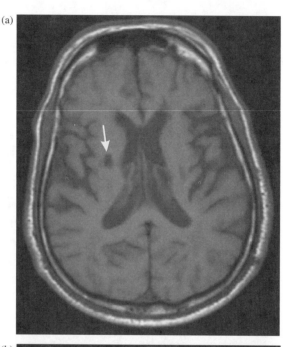

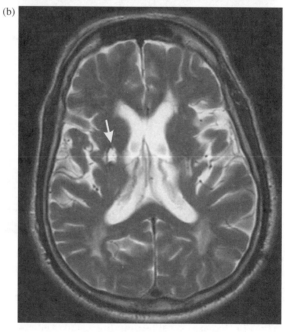

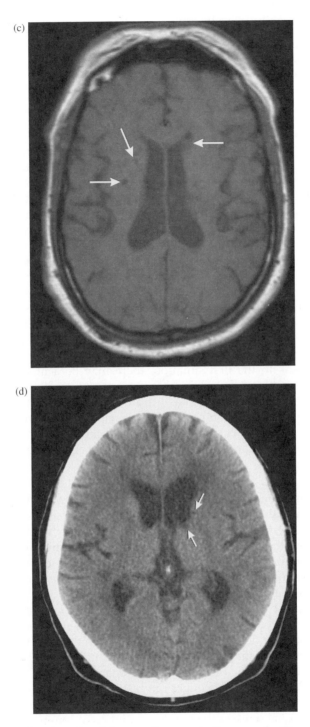

Figure 10–8. Lacunar infarctions involving smaller penetrating vessels. Figures (a) and (b) are the same lesions as seen on T-1 and T-2 MRI imaging, respectively. Multiple slightly smaller infarcts can be seen on (c) (T-1 weighted image), while (d) shows comparable small infarcts on a CT scan.

due to lacunar infarcts that most commonly involve the basal ganglia, thalamus, and internal capsule. These lacunar syndromes include:

1. Pure motor strokes
2. Pure sensory strokes
3. Mixed sensory-motor strokes
4. Ataxic hemiparesis
5. Clumsy hand, with dysarthria

Before proceeding to the hypoxic or insufficiency syndromes, it should be noted that while atherosclerotic disease and cardiac emboli are the most common causes of occlusive cerebrovascular disease (stroke), they are not the only causative factors. Other causes of strokes include trauma to the vessels themselves (especially in the neck resulting in carotid artery dissection), inflammatory or infectious processes directly involving the proximal or cerebral vessels (vasculitis), amyloid angiopathy, right-to-left shunt in the heart due to patent foramen ovale producing a paradoxical emboli, hematological disorders (e.g., sickle cell disease, thrombocytopenia), intravenous drug abuse, and pregnancy.

In addition to occlusive vascular disease that results in a total and extended shutting off of the blood supply of blood to a cerebral vessel and necrosis in its area of distribution (infarction), there are other conditions that produce more temporary or incomplete circulatory restrictions. The etiologies are similar to those for other forms of occlusive disorders: diseases affecting the vessels themselves (e.g., atherosclerosis), blood factors including emboli that contribute to the restriction of blood flow, or conditions that reduce cerebral perfusion (e.g., steal syndromes or reduced arterial pressure, especially in the presence of preexisting vascular disease). In the first situation, there may be a more or less complete but temporary restriction of blood flow through an artery as a result of either a thrombotic or embolic process. Commonly referred to as a **transient ischemic attack** (TIA) (see below), the individual may experience a temporary loss or impairment of functioning, particularly if a primary sensory, motor, or language region of the cortex is affected. Again, the specific symptoms will depend on the artery and branches affected and the hemisphere involved.

Hypoxic Lesions

Ischemic strokes result from lack of sufficient oxygenation of brain tissue. In addition to an inadequate profusion of the brain secondary to an occlusion of vessels (either as a result of a thrombus or embolus), there are other conditions in which the blood vessels themselves may remain patent but the brain is still oxygen-deprived. Commonly referred to as **hypoxic syndromes**, these conditions may result from either a lack (or displacement) of oxygen in the blood (*primary hypoxia*) or a systemic circulatory failure (*secondary hypoxia*), both of which result in failures to adequately replenish oxygen to the brain.

Less common than occlusive vascular disease, hypoxia or lack of sufficient oxygenation of the brain can produce either temporary or permanent neurological deficits. The nature, pattern, or severity of these deficits in part will depend on the severity, length, and etiology of the deprivation. Hypoxia may result from either chronic or acute conditions. Causes of chronic hypoxia include chronic obstructive pulmonary disease (COPD), congestive heart failure, or severe anemia. Subacute changes can result from unaccustomed exposure to high altitudes ("mountain sickness"). Chronic or subacute conditions are not usually associated with "strokes" in the usual sense of the word or with loss of consciousness. Chronic hypoxic syndromes frequently will present with a picture of more subtle cognitive or behavioral changes, although more dramatic effects, such as delirium, might occur particularly in the elderly who may have less cognitive reserve. Acute oxygen deprivation is more likely to be associated with loss of consciousness and more serious and permanent consequences. The latter can result from such events as airway obstructions (e.g., choking), cardiac arrest, carbon monoxide poisoning, near drowning, "partially successful" suicide attempts by hanging, respiratory paralysis (e.g., Guillain–Barré), or severe hypotension.

In more acute cases of anoxia, it is not uncommon to find more permanent, focal-type deficits, particularly if consciousness has been lost. In these cases, selective deficits may be related to the specific oxygen needs of the brain tissue itself (Adams & Graham, 1988). Certain brain regions are believed to be more vulnerable to oxygen depletion than others. The areas of greatest vulnerability are CA1 of the hippocampus, the basal ganglia, the cerebellum, and laminar necrosis to cortical layers 3 and 5. It is hypothesized that the brain tissue in these areas (i.e., the metabolism that is necessary to sustain function) is more sensitive to oxygen deprivation, and hence, more readily affected by oxygen deprivation. Thus, changes in memory and motor functions are fairly common in such cases. Certainly the brain as a whole has a very high rate of metabolism and uses vast amounts of oxygen relative to its size. Under normal circumstances, oxygen deprivation for more than 5 to 6 minutes can be expected to result in neuronal damage or death. Comas lasting more than 48 hours following an acute anoxic episode, especially if attended by absence of brainstem responses, generally are associated with severe, permanent brain damage. Conditions that tend to slow down brain metabolism and as a result its need for oxygen consumption, such as being immersed in frigid water, can lengthen the time of oxygen deprivation before permanent damage may result. Using this principle, it is common practice to lower the body temperature of patients undergoing certain types of heart surgery.

Conditions that result in reduced arterial pressure also may create states of hypoperfusion leading to either temporary or permanent changes in brain function. Two general types of conditions likely are to be responsible for brain lesions secondary to hypoprofusion. One is a systemic lowering of blood pressure. Some common causes include postural (orthostatic) hypotension, cardiac insufficiency or myocardial infarction, or severe systemic hemorrhaging. Postural hypotension (a sudden drop in blood pressure upon rising to a vertical position) can result in dizziness or a syncopal episode, without any permanent sequelae. If the hypotension is severe and long enough, all cerebral vessels can be affected and permanent deficits may ensue. However, arteries that are nearly completely occluded from atherosclerotic disease might be differentially affected, resulting in more focal deficits. In instances of decreased blood pressure or markedly reduced cardiac output, the areas maximally affected may be those where the distribution of the MCA, PCA, and ACA overlap, producing what is known as a *watershed* or *borderzone* syndrome (Figure 10–9).

(a)

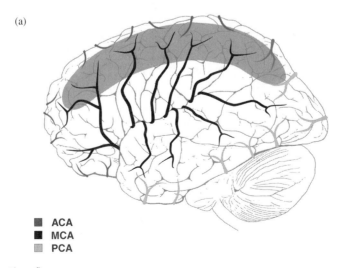

■ ACA
■ MCA
▫ PCA

Figure 10–9. (*Continued*)

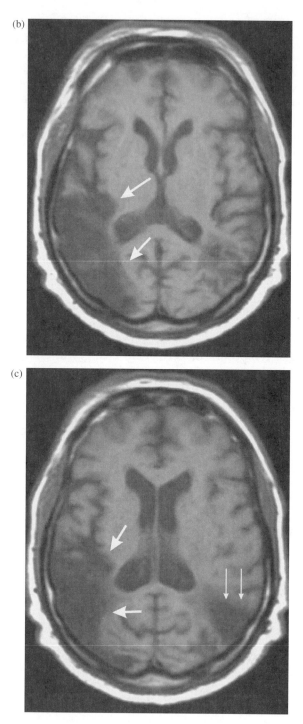

Figure 10–9. (a) Schematic representation of "borderzone" infarction involving the terminal branches of the anterior, middle, and posterior cerebral artery distributions. Figures (b) and (c) show right-hemisphere lesion (large arrows) thought to represent possible borderzone infarction involving the area of the MCA and PCA distributions, although direct occlusion of one or more of these vessels cannot be ruled out. (c) Also seen is an older, smaller infarction in the left posterior parietal area likely involving the MCA.

This syndrome will vary depending on the particular hemisphere and arterial distributions affected; however, the symptoms often will result in complex cognitive, behavioral disturbances (e.g., agnosias, aphasias, apraxias, frontal lobe-type syndromes).

A second situation that may produce secondary hypoxia is where arterial hypotension is limited to the arterial distribution distal to an area of infarction. A relatively common example might involve the occlusion of an internal carotid artery. While collateral circulation via the opposite carotid artery or via the vertebral system may allow for some circulation through the affected system, the pressure will be reduced. In this situation, the more proximal portions of the arterial distribution may be adequately perfused, but the more distal portions might suffer.

Transient Ischemic Attacks

By definition, if a deficit resulting from an ischemic infarction resolves (at least for most clinical purposes) within an hour, the episode is referred to as a *transient ischemic attack* (TIA). If it lasts longer but seems to completely clear within a relatively short time (e.g., within a few days), it may be referred to as a *reversible ischemic neurological deficit* (RIND).[5] The symptoms of a completed stroke generally come on rather suddenly, although they may evolve over time (typically a matter of minutes, but occasionally over hours or even days). An individual may have a TIA without ever experiencing a completed stroke, but the presence of multiple TIAs affecting the same vascular distribution normally is associated with increased risk for a permanent infarction. A single such episode may be either embolic or thrombotic in origin, while multiple episodes (again, within the same distribution) suggest arteriosclerotic disease. It should be noted that while by definition a TIA or RIND is completely clinically resolved within a short period of time, in what initially might appear to be a TIA the possibility of more subtle, residual deficits on occasion, can be detected if sufficiently sensitive measures are employed.[6]

Transient Global Amnesia

Transient global amnesia (TGA) is a syndrome in which the patient experiences a sudden loss of recent memory as well as an inability to lay down new memories (anterograde amnesia). Like TIAs, TGA is characterized by an acute onset, is generally limited in scope (the patient may evidence mild confusion secondary to the memory loss but no deterioration of global cognitive functions), and typically resolves within a few hours. While the capacity to lay down new memories returns, the patient typically remains amnestic for events surrounding the episode itself. Also like TIAs, transient global amnesia is thought to most likely represent a vascular type of event; however, unlike a TIA, it is not associated with increased risk of a subsequent stroke. If indeed this syndrome is of vascular origin (the exact etiology is not clear, but some type of vascular spasm is possible), the temporal branches of the PCA and/or some of the posterior penetrating arteries are most likely involved.

Recovery from Ischemic Strokes

Following occlusive or hypoxic syndromes, the patient often recovers significant function over time. What are the mechanisms that allow for this? Certainly one mechanism is that other areas of the brain "learn" to compensate or take over for the damaged area. The patient also may "relearn" how to do certain things using alternative pathways or behavioral strategies. However, a large amount of recovery probably takes place independent of what the patient learns or experiences poststroke and may be explained at a more mechanical level. When an infarct occurs, several things disrupt neuronal function, some of which appear to be reversible. First, it is commonly accepted that if neurons supplied by a vessel are deprived of oxygen for a sufficient period, they die, and once dead they do not regenerate. However, as one goes out from the center of the lesion (the *umbra*), there may be brain tissue that is not completely shut off from the supply of blood (perhaps as a result of partial collateral or overlapping circulation or vasospasms), which becomes hypoprofused. In this area (the *penumbra*), the cells become dysfunctional as a result of lack of sufficient oxygen, but still are viable and may recover over time as additional collateral circulation develops and/or the vasospasms diminish—the *idling neuron* hypothesis. Additionally, within the area of the infarct, some cytotoxic edema is invariably present.[7] As the edema subsides, those neurons that may have been adversely affected by the edema may recover. Some minor hemorrhaging around the site of an infarction into the surrounding brain tissue is not uncommon following an ischemic stroke. The direct imposition of red blood cells on neural cells is not compatible with their normal functioning. (i.e., the blood appears to have a toxic effect on the nerve cells). Once this blood breaks down and is absorbed, some of these affected cells may return to normal or near normal functioning. While there are likely additional explanations for partial recovery of function on a cellular level, the mechanisms discussed above are among the most commonly suggested.

Hemorrhagic Vascular Disease

Following occlusive (ischemic) disorders, hemorrhages are the next most common cause of stroke. Hemorrhages involve the rupture of a blood vessel, most commonly an artery. The major differences in the types of hemorrhagic stroke to be discussed below simply reflect the site of the bleed. Certainly a second major consideration is the size or extent of the resulting hemorrhage, which is largely a function of the size and site of the ruptured vessel. Intracranial hemorrhages tend to have increased mortality in the acute phase compared to occlusive strokes. However, if the patient survives the acute stage, the prognosis for improvement typically is somewhat better than that for occlusive infarction. While this may be an oversimplification of the situation, at one level this readily makes sense. As noted above, in a completed occlusive stroke or infarction, the blood supply is shut off from the cells in the arterial distribution distal to the occlusion. Without a supply of blood (oxygen), the cells die, and once dead there is no chance of regeneration.

Bleeds, by contrast, potentially are very destructive for other reasons. First, like occlusive infarcts they can disrupt the blood supply to adjacent nervous tissue or distal to the site of the hemorrhage. In addition, secondary vasospasms (particularly in subarachnoid bleeding) may cause additional expansion of the ischemic infarction. The extravasation (leakage) of free blood into the extracellular spaces will disrupt the normal functioning of the cells due to its direct toxic effect on neural tissue. These effects obviously will be maximal where the free or extravascular blood directly interfaces with the nervous tissue and the clinical signs and symptoms will depend on the location and size of the hemorrhage. Hemorrhagic strokes also can be associated with significant increases in intracranial pressure. As with infarcts, cytotoxic edema results from damage to the surrounding cells, with the degree of edema associated with the size of the bleed. However, in the case of hemorrhages, there potentially is another source of increased intracranial pressure, compounding the cytotoxic edema. The blood itself can act like a space-occupying lesion, causing compression of the brain tissue and possibly leading to herniation and death if the bleed is sufficiently large. Small-vessel or venous bleeds generally pose considerably less risk than a rupture of larger arterial vessels as might be found with aneurysms or epidural hematomas (see below). While these pressure effects may be disruptive to normal functioning of the cells, in many cases these effects are limited and transient. In the case of intraparenchymal hemorrhages, once the blood is reabsorbed, many cells might return to normal functioning. If the bleed is outside the brain tissue itself (e.g., subdural or epidural hematomas), the major destructive damage will be from pressure effects, which again might be overcome in the long run if the patient survives.

As suggested above, the classification of the sites of intracranial hemorrhages falls into four general categories:

1. Within the brain tissue itself (**intracerebral** or **intraparenchymal** hemorrhages).
2. In the subarachnoid space (**subarachnoid** hemorrhage).
3. In the potential space between the dura and the arachnoid (**subdural** hematoma).
4. Outside the dura (**epidural** hematoma).

As we shall see, the typical cause of such bleeds tends to vary depending on their sites.

Epidural hematomas most commonly result from skull fractures where the meningeal arteries are ruptured. Because the arterial blood is under greater pressure than venous blood, the bleeding is frequently profuse and can result in a rapid increase in intracranial pressure (Figure 10–10). Consciousness usually is impaired as the pressure increases, and unless the blood is surgically evacuated and the pressure relieved, death may ensue as a result of mass effect and brainstem herniation. Because the dura is attached to the skull along its suture lines, the lateral extent of the bleeding usually is somewhat restricted, resulting in a "lens"-shaped hematoma on imaging studies.

Subdural hematomas are produced when bleeding occurs into the "potential space" between the dura and the arachnoid (Figure 10–11). Subdural hematomas are also associated with head trauma but, unlike an epidural bleed, are more often associated with a closed head injury (i.e., without skull fractures). Subdural hematomas most frequently are due to rupture of the bridging veins into the dural sinuses as a result of the shearing forces, as in a deceleration-type injury. If the formation of the hematoma is acute and

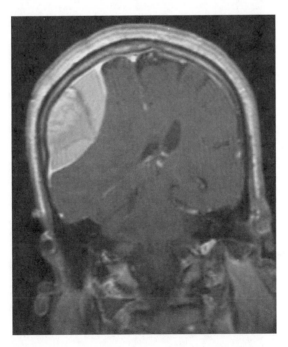

Figure 10–10. Example of an epidural hematoma. Due to the rapid increase in intracranial pressure, significant mass effect is seen. Unless evacuated soon after the injury, such lesions often prove fatal.

extensive, it also can produce increased intracranial pressure and brainstem compression. In such cases, surgery generally is indicated, especially if there are signs of rapidly increasing intracranial pressure. In many cases, the subdural bleeding is much more controlled or protracted (i.e., subacute). While signs of focal pressure may be present, the danger of herniation is much less. Particularly in chronic alcoholics, it is not uncommon

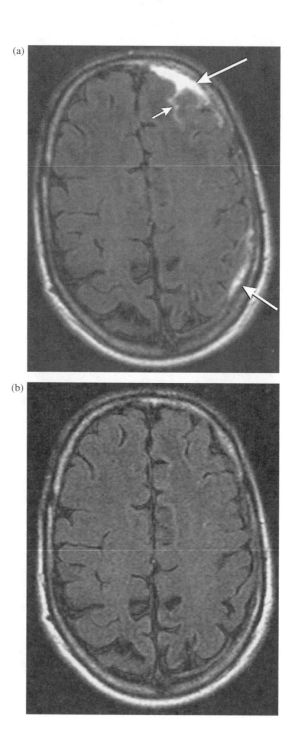

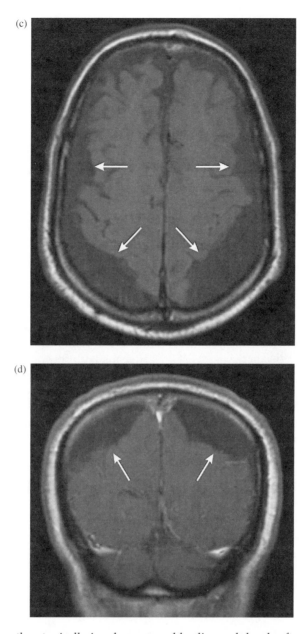

Figure 10–11. Because they typically involve venous bleeding, subdurals often develop more slowly and may be diffused over the surface of the hemisphere, and thus often associated with more minimal clinical effects. Figure 10–11a shows two more limited subdurals (large arrows), combined with area of subarachnoid bleeding (small arrow) as seen on FLAIR images (a CT scan on this same patient was read as "normal"). A second FLAIR image 2 months later (b) shows the same patient after the blood had been reabsorbed. Images (c) and (d) show residuals of large chronic bilateral subdural hematomas that have become filled with fluid. At this stage these fluid-filled pockets are often referred to as "hygromas."

to find evidence of chronic or old subdural hematomas on autopsy or subsequent MRI imaging that clinically had gone undetected at the time.

Subarachnoid hemorrhages, along with intraparenchymal bleeds, represent one of the most common causes of hemorrhagic strokes. Bleeding into the subarachnoid space may result from various events, but the most common are rupture of an aneurysm, leakage from an arteriovenous malformation (see below), extension of a primary intracerebral hemorrhage into the subarachnoid space, and trauma. At times the bleeding may be slow or subacute and very limited, and the symptoms, which usually include a headache, may go unheeded by the patient at the time. More commonly, the onset is quite acute and rather dramatic. Frequently occurring during strenuous physical activity, patients often experience the sudden onset of headache, which is typically described as the "worst of their life." There also may be some disturbance or loss of consciousness, confusion, nausea, dizziness, or vomiting. Seizures are not uncommon. Pain or stiffness of the neck commonly is present, and when present is strongly suggestive of a subarachnoid hemorrhage. Except in severe cases in which there may be a rapid rise in intracranial pressure, focal neurological signs, such as sensorimotor disturbances, often either are absent or transitory. Imaging studies and/or a spinal tap typically will confirm the diagnosis. In rare cases of small bleeds that may not show up on imaging, a tap will reveal grossly bloody or xanthochromic cerebral spinal fluid.

Additional complications of subarachnoid hemorrhage include the possibility of rebleeding (especially if a ruptured aneurysm was the initial cause of the bleed), bleeding into the parenchyma (brain tissue) itself, vasospasms with secondary ischemic infarctions, edema, and increased intracranial pressure (that may be exacerbated by the vasospasms). An acute hydrocephalus may result as the flow of CSF is obstructed. A communicating-type hydrocephalus (e.g., "normal pressure hydrocephalus") may develop at a later date due to the reduced absorption capacity of the villi in the subarachnoid space as a result of the original bleed. This latter syndrome typically presents with disturbances of gait, urinary urgency or incontinence, and changes in mental status.

Intracerebral or **intraparenchymal hemorrhages** are those that invade the brain tissue itself (Figure 10–12). The most common cause of intracerebral hemorrhage is hypertension, which produces changes in the walls of the small, central penetrating vessels. Chronic hypertension weakens the vessel wall and can result in a ballooning of the vessel wall, producing what is termed **pseudoaneurysms**.[8] These pseudoaneurysms may rupture with extravasation of blood into adjacent brain regions. The regions most vulnerable to small-vessel, hypertensive hemorrhagic strokes are the putamen, thalamus, cerebellum, and pons, but may occur in any subcortical structure supplied by these small, penetrating vessels (e.g., subcortical grey and white matter and brainstem). Although hypertension and its related arteriolosclerotic changes in the small cerebral vessels are the most common causes of spontaneous intracerebral hemorrhages, a variety of other conditions also are associated with increased risk for intracerebral hemorrhage. Among the more common are AVMs, aneurysms, amyloid angiopathy, blood dyscrasias, drugs (particularly anticoagulant therapy, amphetamines, and cocaine), primary brain tumors, and embolic infarctions. Trauma, especially any type of penetrating head wound, is another common cause of hemorrhaging within the brain.

The signs and symptoms of intracerebral hemorrhage, as well as its prognosis or morbidity, depend on a variety of factors including the size of the bleed, its progression, and its location. With regard to size and progression, two general scenarios may be identified. In the first, the bleeding may be slow and/or relatively circumscribed. Since the brain tissue

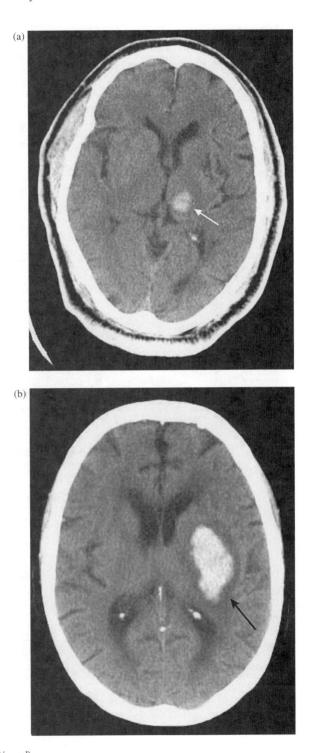

Figure 10–12. (*Continued*)

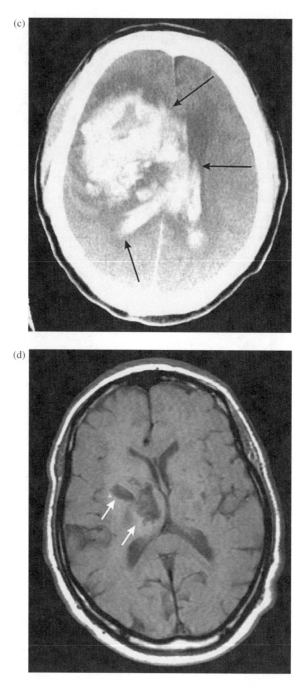

Figure 10–12. (a) Small thalamic, (b) larger putamenal, and (c) massive intraparenchymal hemorrhages. (d) Smaller hemorrhages may eventually (c) resolve with minimal to moderate clinical effects, although here signs of mass effect and gliosis are still present. Larger hemorrhages, as seen in (c) with intraventricular extension are typically incompatible with survival due to a rapid and massive increase in intracranial pressure and herniation.

itself as opposed to the meninges has no pain receptors, the patient may experience little or no headache. If headache does occur, it is most likely related to increased intracranial pressure. In addition to focal findings that are dependent on the site of the bleed, some patients may experience general confusion, nausea and vomiting, seizures, and/or changes in the level of consciousness. If the hemorrhage is relatively small, as in many small-vessel, hypertensive hemorrhages, the patient may present with focal findings and it may be difficult to differentiate clinically from an occlusive ischemic stroke, except on CT scan. In the second scenario, if the bleeding is more rapid and/or continues to expand to a critical size, the initial onset may be heralded by focal symptoms, as well as by headaches and nausea and vomiting (signs of increased intracranial pressure). Very quickly the patient will generally lapse into coma as a result of this rapidly increasing pressure and brain herniation. The latter picture is much more likely to be associated with mortality or increased morbidity.

Both the symptoms and the rate of survival from intracerebral hemorrhages also depend on the locus of the lesion. As will be discussed in greater detail in the next section, the clinical syndrome in large part will reflect the areas of the brain and/or pathways most directly impacted by the hemorrhage. Unlike occlusive (ischemic) infarcts, which deprive a large area of the brain cells of life-supporting oxygen, bleeding into the parenchyma affects the brain in a different fashion. While occlusive or ischemic lesions have their main effects in the regions of their distribution distal to the site of the blockage, hemorrhagic lesions primarily impact the area surrounding the bleed itself. First, intracerebral hemorrhage interrupts normal brain function by its direct mass effect. The hemorrhagic lesion puts increased pressure on the cells, axonal pathways, and capillary vessels, particularly those that are at the boundaries or in the immediate vicinity of the clot. In addition to this focal pressure, there also are the adverse effects due to a more generalized increase in intracranial pressure. These effects are enhanced by the secondary edema that occurs in addition to the space-occupying presence of the clot. Free blood in the extracellular spaces also seems to have a toxic or detrimental effect on the functioning of the cells that it directly impacts. Secondary ischemia may result from the various changes in blood flow and the effects on local vessels. Nonetheless, as noted earlier, if the organism survives the initial shock of an intracerebral hemorrhage, the possibility of recovery of function seems to be greater than if a comparable area was totally infarcted by occlusive cerebrovascular disease.

As might be expected, lesions that are confined to the area of the basal ganglia are likely to produce contralateral motor deficits (dyskinesias). However, somatosensory symptoms also may result. Thalamic bleeds may evoke similar symptoms except that, in the earlier stages especially, somatosensory deficits might predominate over motor deficits. The internal capsule commonly is involved following both thalamic and basal ganglia hemorrhages, frequently resulting in hemiplegia or hemiparesis. In capsular lesions, the upper face typically is spared because of its bilateral innervation, but the lower face, along with the leg and the arm, equally may be affected. In either site, but particularly with primary thalamic bleeds, a dysphasic syndrome (subcortical aphasia) may accompany left hemispheric hemorrhages. Other cognitive deficits, such as hemispatial neglect, also may follow thalamic lesions. Disturbances of oculomotor function, as well as visual field defects (hemianopia), are not uncommon with either basal ganglia or thalamic bleeds as a result of encroachment onto various pathways (e.g., from frontal eye fields, subthalamic connections, or optic tracts or radiations).

Intracerebral hemorrhages that occur in one of the lobes of the brain (lobar hemorrhage) will produce symptoms normally associated with that area of the brain. For example, contralateral motor findings are common with frontal lesions and contralateral sensory losses or neglect with parietal damage. Visual field loss may occur following occipital hemorrhages, although partial visual field cuts also can result from temporal (*superior quadrantanopia*) or parietal (*inferior quadrantanopia*) lesions, as a function of the interruption of the underlying optic radiations. Disturbances of higher cortical functions also would expect to be affected, such as language, visual spatial constructions, or selective memory. Often these cortical syndromes will be both subtler but at the same time more inclusive than those caused by occlusive disease. With lobar hemorrhages, the patient may experience focal headaches, (again due to pressure effects), especially with frontal or occipital involvement, and focal or generalized seizures are not uncommon.

Another relatively common site of hemorrhage is in the brainstem, particularly in the pons. Rapid loss of consciousness is typical and the mortality rate is high. Neurologically, the patient frequently is quadriplegic and may show decerebrate rigidity. If the signs of corticospinal involvement are unilateral, opposite-sided cranial nerve findings are likely. Oculomotor deviations and loss of pupillary responses are common in brainstem lesions involving the pons or midbrain. Breathing also is commonly affected.

Cerebellar hemorrhages typically are associated with nausea, vertigo, vomiting, and posterior headaches. Neurological examination typically reveals marked disturbances of gait and truncal ataxia, upper limb dysmetria, and dysarthric speech. Hemiparesis or visual field defects are not present as part of the cerebellar picture, but additional signs of brainstem compression may develop, including oculomotor findings, extensor plantar responses, and loss of consciousness, or more commonly obtundation if the hemorrhage reaches sufficient magnitude.

Vascular Anomalies

Aneurysms

Aneurysms represent the ballooning of the wall of a vessel, typically an artery (Figure 10–13. They can be extremely small, as in the case of the *microaneurysms* that may develop on the smaller penetrating arteries as a result of chronic hypertension. Microaneurysms likely represent one cause of lacunar infarctions or hemorrhages deep within the brain. In contrast, other aneurysms may enlarge to greater than 1 centimeter in diameter (*giant aneurysms*). Giant aneurysms usually are derived from the more proximal vessels, and if of sufficient size can act like a space-occupying lesion, exerting pressure on adjacent neurons or fiber

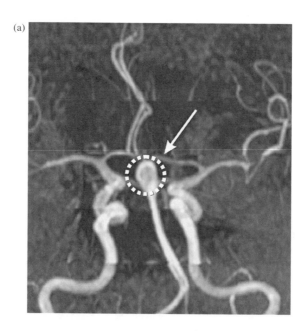

(a)

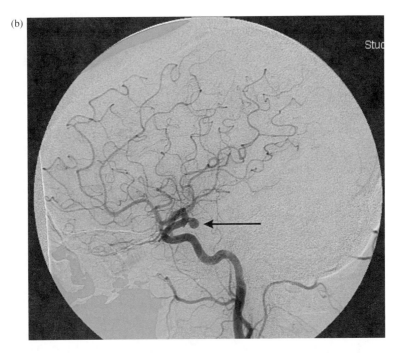

Figure 10–13. (a) Aneurysms (circled) at the tip of the basilar artery and (b) at the point of origin of the posterior communicating artery. The former is a standard MRA, while the latter represents a DSA (digital subtraction or conventional angiogram.

tracts. The majority of aneurysms appear to develop from a congenital weakness in the wall of the vessel, typically at the site of the bifurcation or branching of the artery or possibly at the site of a vestigial artery. These are known as *saccular* or *berry* aneurysms, the former representing a broad-based bulge in the artery, while the latter has a clearly defined stalk or "neck." Both of these are most commonly found in the carotid system, in or around the circle of Willis. Aneurysms also may be formed as a result of bacterial infection (*mycotic aneurysms*) or arteriosclerotic disease (*fusiform aneurysms*). The mycotic aneurysms often resemble berry-type aneurysms and more commonly are found in the carotid distribution (e.g., ACA or MCA). Fusiform aneurysms are more equally distributed in the carotid and vertebral systems and tend to be of the saccular type.

Many if not most aneurysms go undetected until they leak or rupture. Except for microaneurysms that may present as small infarcts, most ruptured aneurysms result in subarachnoid hemorrhages. The presenting symptoms are similar to that of other forms of subarachnoid hemorrhage discussed previously. Occasionally, a ruptured aneurysm may force blood directly into the parenchyma itself.

Special attention should be given to the rupture of aneurysms involving the area of the anterior communicating artery, a frequent site of cerebral aneurysms. Because the rupture of anterior communicating artery (ACoA) aneurysms directly affects the hypothalamus (including the mammillary bodies), the basal frontal regions, and medial temporal structures, significant behavioral symptoms typically are associated with this syndrome. Although ACoA aneurysms initially may present like most subarachnoid hemorrhages with severe headache and loss of consciousness, if and when the patient recovers, various personality, behavioral, and memory changes frequently are noted (Alexander & Freedman, 1984; Damasio et al., 1985; De Luca & Diamond, 1995; Fasanaro et al., 1989: Gade, 1982; Greene et al., 1995; Okawa et al., 1980; Talland et. al., 1967; Vilkki, 1985). These patients commonly demonstrate marked disorientation and major amnestic deficits, involving both retrograde and particularly severe anterograde losses. Elaborate confabulations are relatively frequent, although they typically will subside over time (as do most confabulatory tendencies). The memory disturbances also generally will improve with time, but the extent of final recovery will vary among patients. In addition to memory, behavioral and affective changes also are quite common. While some variation is found, these patients typically demonstrate a number of symptoms associated with frontal pathology, including apathy, disinhibition, poor judgment, euphoria, dysphoria, or increased irritability. Cognitive or intellectual impairment may occur, but the memory and behavioral changes usually are more dramatic. As with memory loss, many patients will evidence an improvement in behavioral disturbances over time.

Even without rupturing, aneurysms still may present symptomatically because of their mass effect. This is particularly true around the circle of Willis and brainstem where they may impinge on sensitive structures, such as cranial nerves. Thus, unilateral anosmia, visual field cuts, and oculomotor disturbances (diplopia, ptosis, gaze palsy, unequal pupillary responses) are not uncommon.

Arteriovenous Malformations

Arteriovenous malformations (AVMs) represent another type of developmental vascular malformation. There are multiple types of AVMs, but perhaps most typically they represent a lack of development of the normal capillary network between a group of arteries and veins. As a result, there is a more or less direct shunting of blood from the arterial to venous system in the affected vessels. Over time these vessels tend to enlarge into a massive network of engorged artery–venous anastomoses. The AVMs can be composed primarily of surface vessels or can be embedded deep within the hemisphere. Size of AVMs may vary between relatively small or focal vascular malformations to quite massive anomalies, sometimes encompassing huge portions of a hemisphere, as seen in Figure 10–14.

The effects of vascular malformations can be multiple. First and foremost, AVMs and other vascular anomalies are subject to leakage or rupture, producing either a subarachnoid or an intracerebral hemorrhage. Because there is a direct shunting of blood from the arteries to the veins, lowering of arterial blood pressure and focal ischemia may occur. Because of their size, which increases over time, AVMs also may create a mass effect; however, since their growth is extremely slow, this effect usually is minimal. It is not uncommon for patients to go into their third or fourth decade or beyond before the AVM is discovered, and even

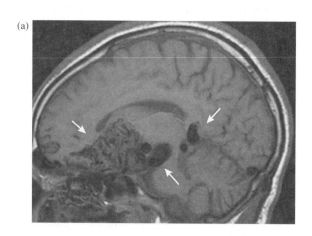
(a)

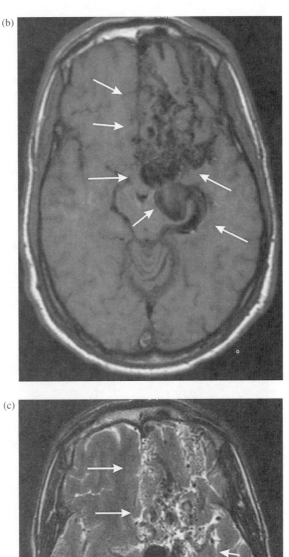

Figure 10–14. (*Continued*)

(d)

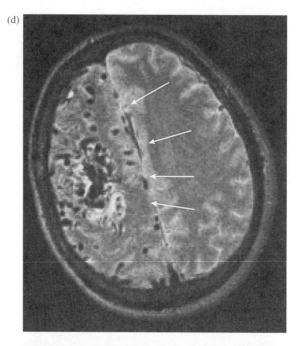

(e)

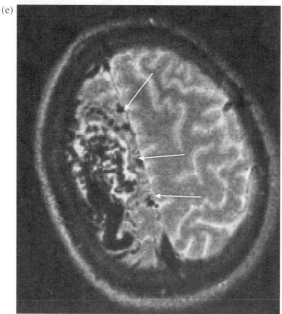

then usually only after some hemorrhaging occurs. Late-onset seizures perhaps are the most common presenting symptom of non-hemorrhagic AVMs, especially when accompanied by persistent focal vascular headaches. Over time, more neurological or behavioral symptoms may develop, as the AVM continues to enlarge and greater ischemic effects occur.

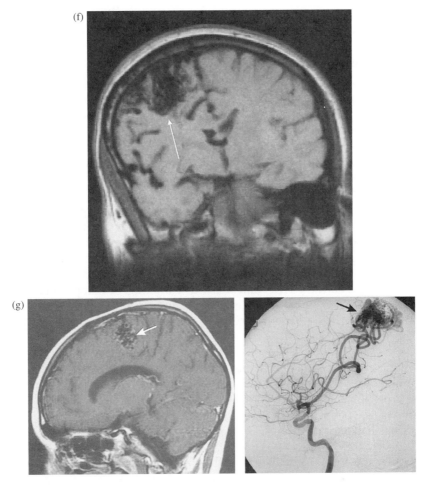

Figure 10–14. Large AVMs of (a–c) left and (d–f) right cerebral hemispheres. Because of slow growth, such lesions often are asymptomatic until they bleed and/or result in seizures. First patient (a–c), in addition to seizures, presented only with a mild right-sided weakness. The second patient (d–f) also had a history of seizures, but no other clear neurobehavioral or cognitive symptoms. Last image (g) compares the results of a MRI and DSA in the same patient. (MRIs shown in a, b, f, and g are T-1 weighted images; c, d, and e are T-2 weighted). (g) Courtesy of Dr. Jose Suros.

SPECIFIC VASCULAR SYNDROMES

As previously discussed, a number of variables play a role in determining the exact effects of vascular lesions. Hemorrhagic lesions produce very different effects from occlusive, ischemic lesions. The more proximal the lesion in a particular arterial distribution, the more devastating the impact and different patterns of deficits will emerge depending on the particular distal branches affected. Occlusive disease secondary to atherosclerotic processes, which develop slowly, allow for the possibility of greater collateral circulation development, thus possibly mitigating the effects of the stroke. Multiple strokes have more than a simple "additive" effect. While the overall vascular patterns generally are similar from one individual to the next, important variations may be common. The bottom line is that it is difficult to completely recreate or compare the particular stroke suffered by one patient with the stroke of another. However, it is possible to discuss, at least in general terms, the syndromes produced by the blockage of certain arteries or arterial systems.[9]

Internal Carotid Artery Syndrome

As noted, a major factor in internal carotid disease symptomatology is the period of time over which the stenosis takes place. If it happens slowly enough, sufficient collateral circulation may develop so that the patient may experience relatively mild, if any, overt neurological deficits. Often, however, transient ischemic attacks (TIAs) may signal the presence of internal carotid artery stenosis. Amaurosis fugax (temporary blindness or dimming of vision in one eye as a result of the involvement of the ophthalmic artery) is a relatively common warning sign of carotid disease. Other indications might include transient episodes of weakness, somatosensory disturbances, or impairments of speech or language. The presence of a bruit (abnormal swishing sound) in auscultation of the neck is further suggestive of carotid artery stenosis and might indicate either Doppler studies or angiography for a more definitive diagnosis. Then, depending on the nature and severity of the disease process, differential treatment options might be considered. The possibility of complete internal carotid artery occlusion potentially is catastrophic, especially in the absence of adequate collateral circulation. Not only might both the ACA and MCA distributions be compromised (resulting in massive functional deficits), but also the resulting edema from such a large area of infarction can be life-threatening. If the patient survives such a stroke, among the impairments that might be expected are:

1. Contralateral hemiplegia and hemisomatosensory disturbances.
2. Homonymous hemianopia (secondary to involvement of the optic radiations).
3. Marked or global aphasia (if dominant hemisphere is involved).
4. Severe visual–spatial and other "higher-cognitive" deficits.
5. Emotional–behavioral changes.

More commonly, the loss of circulation will be somewhat more restricted, especially in embolic disease, typically affecting the MCA more than the ACA. Some variants of the typical vascular pattern may occur in which, for example, both ACAs as well as one of the PCAs may be supplied by one internal carotid. In these cases it is possible, although not common, to develop a bilateral frontal syndrome or a PCA syndrome with or without additional findings from occlusions of this single carotid system. A third possible outcome of stenosis of the internal carotid artery is the development of a borderzone or watershed-type infarction. As previously noted, this syndrome also can result from systemic hypotension, rapid blood loss, or hypoxia. In this situation the patient typically will manifest various cognitive or behavioral deficits consistent with the hemisphere involved that usually will not be restricted to the symptom pattern seen with focal cerebral infarctions. Sensorimotor losses may be minimal compared to the cognitive disturbances following watershed infarctions involving the lateral cortices.

Anterior Cerebral Artery Syndrome

Focal infarctions of the ACA are considerably less common than infarctions of the MCA, especially those of embolic origin. Some of the expected findings following infarction of the ACA are contralateral weakness, greater in the lower extremity and possibly proximal portions of the upper extremity. Some cortical sensory disturbances (e.g., two-point discrimination) may be present in the same distribution due to involvement of the medial–dorsal parietal cortex. Involvement of the cingulate gyrus may lead to urinary urgency or incontinence, at least temporarily. This latter problem is likely to be more severe and more permanent if previous comparable lesions in the opposite hemisphere also are present. Since the ACA supplies the anterior portions of the corpus callosum, a partial disconnection syndrome may be detected, particularly ideomotor apraxia involving the nondominant hand. Again, while the effect is likely to be much more dramatic if bilateral lesions are present, frontal lobe symptomatology (e.g., decreased initiative or spontaneity, emotional lability or apathy, disinhibition, perseveration, frontal release signs) may be observed. In addition to contralateral leg weakness, gait ataxia may be present, along with an initial gaze preference toward the side of the lesion. If the dominant hemisphere is involved, elements of transcortical motor aphasia also may be seen.

Anterior Choroidal Artery Syndrome

Because this artery supplies blood to the posterior limb of the internal capsule, as well as to the lateral geniculates, infarction may lead to the combination of contralateral hemiparesis (or hemiplegia) and a contralateral homonymous hemianopia or quadrantanopia. Mild somatosensory deficits also may be found on the involved side.

Middle Cerebral Artery Syndrome

The anatomy of the MCA makes this vessel a prime candidate for embolic infarction from the carotid arteries, the heart, or lungs. Since the MCA is the largest cerebral artery, a proximal occlusion is quite devastating, producing a syndrome similar to that described for the internal carotid artery. Complete occlusion can result in:

1. Total contralateral hemiplegia (sparing the upper portion of the face), hemisensory impairment and hemianopia.
2. Unilateral neglect, global aprosodia, and/or marked visual spatial deficits (more common with right hemisphere lesions).
3. Global aphasia and ideomotor apraxia (left hemisphere lesions).
4. Memory and perceptual deficits (the exact nature of which will depend on the hemisphere involved).
5. Frontal executive deficits.[10]

Since the ACA is spared, initial lower extremity weakness is likely to improve faster than that of the upper extremity. It is not uncommon for more distal, individual branches of the MCA to be selectively occluded. In these cases the symptoms will depend on the specific territories and hemisphere involved, as well as the efficiency of collateral circulation.

Occlusion of one or more anterior branches of the MCA (e.g., the prefrontal or orbitofrontal branch) may produce frontal lobe symptoms (these often can be subtle with unilateral disease: see Chapter 9). Motor weakness or paralysis and/or cortical somatosensory symptoms with the upper limb generally being more affected than the lower can occur following an occlusion of the middle branches of the MCA (e.g., precentral, central branches, or postcentral branches). Syndromes of nonfluent aphasia (dominant hemisphere) or expressive aprosodia (nondominant hemisphere) also may occur following infarction of the anterior branches of the MCA. Occlusion of the more posterior parietal branches of the MCA (e.g., anterior, posterior, and/or angular parietal branches) may produce visual spatial problems (right or left hemisphere) or language problems (particularly visual-based processes such as reading and writing) if the left hemisphere is affected. An inferior quadrantanopia may result from an extension of the lesion to the underlying optic radiations. If the angular gyrus in the dominant hemisphere is affected, most if not all of the symptoms of Gerstmann's syndrome may be present. Contralateral neglect or anosognosia is common, especially in the acute stages of nondominant (right) hemispheric lesions.

If the posterior temporal branches of the MCA are involved, fluent aphasia (left hemisphere) or receptive aprosodia (right hemisphere) may be observed, with little or no motor weakness. Interruption of the optic radiations in the temporal region (Meyer's loop) may produce a superior quadrantanopia. Selective memory deficits may occur, probably secondary to perceptual or encoding difficulties. With selective involvement of the lenticulostriate arteries that supply the basal ganglia and parts of the anterior and posterior limbs of the internal capsule, dyskinesias or a contralateral hemiparesis or hemiplegia with dysarthria may result, in the relative absence of "higher-cortical" deficits.

Posterior Cerebral Artery Syndrome

The symptom most commonly associated with posterior cerebral artery disease is contralateral hemianopia due to lesions affecting the primary visual (striate) cortex (calcarine artery). If the calcarine branches of both PCAs are simultaneously affected, not only will the patient be cortically blind, but he or she may develop anosognosia for visual input (*Anton's syndrome*), characterized by lack of awareness or denial of his or her blindness) (Redlich & Dorsey, 1945). If both the left calcarine cortex, as well as the splenium of the corpus callosum are affected, the patient may manifest the syndrome of *alexia without agraphia* (Damasio & Damasio, 1983). However, depending on the particular branch of the PCA that is involved, visual field cuts may or may not be present. Despite the fact that there may not be a demonstrable field cut, other visual disturbances may be present, such as *apperceptive agnosias, color recognition* or *color-naming deficits, simultanagnosia* or a more complete *Balint's syndrome*, and *prosopagnosia* (difficulty recognizing faces). These syndromes usually occur following parietal–occipital or temporal–occipital lesions that may result from borderzone lesions affecting the PCA and MCA distributions (for review, see: Bauer, 1993; Damasio, 1985; DeRenzi, 1997; Farah, 1997; Tranel, 1997). Occasionally, visual hallucinations may result from occlusions of the PCA, but these are more commonly found with irritative or toxic lesions.

In addition to supplying the area around the calcarine fissure, the anterior and posterior temporal branches of the PCA also supply the inferior and medial temporal regions. Infarction of this area can produce severe impairment in the ability to encode new information (anterograde memory), especially if both hemispheres are involved. An acute state of confusion or delirium also may result. As noted above, the syndrome of transient global amnesia is thought to perhaps reflect a type of transient compromise of the temporal branches of the posterior cerebral arteries.

The posterior cerebral artery system also is responsible for supplying blood to parts of the upper portions of the brainstem and represents the main source of blood to the thalamus via the posteromedial and posterolateral penetrating or central arteries. Occlusion of these vessels can produce a variety of midbrain or thalamic syndromes. The resulting brainstem syndromes may include ipsilateral oculomotor (third nerve) findings, contralateral motor deficits (cerebral peduncle), tremor, hemiballismus or other extrapyramidal symptoms (red nucleus), and ataxia (superior cerebellar fibers). The combination of ipsilateral third nerve palsy and contralateral hemiplegia as a result of unilateral infarction of the midbrain is known as **Weber's syndrome**. Lesions of the thalamus can produce a combination of hemiparesis and hemisomatosensory loss, combined with hyperesthesia or thalamic pain syndrome. Disturbances of higher cortical functions, for example, aphasia or neglect, also may be observed.

Infarctions involving branches of the vertebral or basilar arteries will result in either brainstem and/or cerebellar lesions. In addition to cerebellar symptomatology (dyskinesias), infarctions of the pons (most frequent) or other areas of the brainstem can produce any of a large variety of symptoms associated with a disruption of the corticospinal and/or spinothalamic tracts, combined with cranial nerve findings. See Chapter 4 for a review of several common brainstem syndromes.

NEUROPSYCHIATRIC SYNDROMES ASSOCIATED WITH VASCULAR LESIONS

When examining patients for the sequelae of cerebral vascular accidents, there is a natural tendency to focus primarily on signs and symptoms of motor, sensory, language, and other cognitive dysfunctions. However, as has been previously noted, particularly in the preceding chapter, lesions in various parts of the brain often are associated with significant "behavioral disturbances." Recall, for example, problems of disinhibition and agitation that can be associated with orbital–frontal lesions, the paranoia that occasionally may accompany fluent aphasic syndromes, the emotional flatness (*expressive aprosodia*) that can result from frontal–temporal lesions in the right hemispheric, or conversely the emotional incontinence (pathological laughter or crying) that is most commonly associated with deep bilateral frontal lesions. Another important concept to keep in mind is the possibility of neurologically based depression and the differentiation between depression versus apathy. If an individual suffers a stroke (with associated physical or cognitive deficits), it is easy to attribute his or her sadness to this subjective sense of "loss." While such "reactive" depressions indeed may occur, the increased frequency of clinically depressive syndromes following left anterior CVAs strongly suggests the possibility of a neurological substrate in many of these cases. Finally, it also is important to distinguish between depressive syndromes, which are more likely to be associated with strokes affecting the MCA territories (especially the left frontal dorsolateral cortex) versus apathy that is more likely to result from strokes affecting the midline frontal cortices (ACA distribution).[11]

Finally, diffuse cerebrovascular disease that results in multiple large- or small-vessel strokes may present as a progressive dementia. There is some controversy regarding the subtypes of vascular dementias (for a brief review, see Cummings & Benson, 1992). Although the majority of dementias may be the result of degenerative rather than vascular changes, vascular disease also is recognized as a common cause of dementia. In fact, it now is suspected that in many cases both degenerative and vascular components are contributory factors. What types of vascular diseases or changes are most likely to produce a dementia? Certainly multiple, large cortical infarcts can result in significant, generalized deterioration of cognitive and behavioral function consistent with a dementia. Multiple and clearly demonstrable lacunar infarctions also may produce a comparable condition. Bilateral borderzone infarctions, although not common, will produce a clinical picture of dementia. Infarctions that appear predominately in the periventricular white matter, the basal ganglia, and thalamus (*Binswanger's disease*) also have been associated with dementias.[12] Etiological factors that may facilitate these conditions include chronic hypertension, amyloid angiopathy, and atherosclerosis.

Endnotes

1. The major parts of the thoracic and lower portions of the cord are supplied by the radicular arteries, which enter the cord at the various segmental levels.
2. For a more complete discussion of this topic, the following resources are suggested: Biller (1990); Bornstein and Brown (1991); Caplan and Bogousslavsky (2001); Millikan,

McDowell, and Easton (1987); Toole (1990); for exceptionally comprehensive treatment of this subject, see Welch et al. (1997) and Mohr et al. (2004).

3. While neuroimaging techniques (e.g., MRIs and CT scans) provide more definitive evidence of the locus and extent of an infarction, false-negative results may be obtained, especially in the early stages of a stroke. Regardless of the presence or absence of such radiographic evidence, knowledge of functional neuroanatomy is still essential when attempting to map out the functional deficits resulting from the stroke.

4. These changes also can lead to the formation of microaneurysms in walls of the small vessels that may subsequently hemorrhage.

5. The meaning and use of these terms are subject to change. TIAs used to be defined as any deficit that appeared to resolve within 24 hours and an RIND as a deficit that cleared after 24 hours. Now TIAs commonly are defined as deficits that last from minutes to an hour (more or less), while the term "RIND" is rarely used.

6. The author (JEM) found that both finger tapping and a simple bedside coin rotation task often revealed mild residual impairments days after what was reported to be a TIA supposedly had cleared. (Mendoza, et al, 2007).

7. Edema is simply a collection of fluid within or surrounding certain tissues. Three types of edema generally are identified with regard to the brain. **Cytotoxic** or cellular edema results from the disruption of the normal metabolic processes within the cell, particularly the disruption of the "sodium–potassium pump" (see Chapter 11). As a result of this disruption, sodium accumulates within the cell, which in turn attracts water molecules, thus producing the edema. This type of edema is most commonly associated with vascular disease. **Vasogenic** edema, on the other hand, results from a breakdown of the blood–brain barrier through a disruption of the walls of the capillary vessels allowing plasma fluids to leak into the interstitial spaces. This type of edema, which tends to be more commonly associated with tumors, is often characterized by its frondlike projections within the white matter surrounding the neoplasm. **Hydrocephalic** edema typically is associated with obstructive hydrocephalus. As a result of the increased pressure within the ventricles from the buildup of cerebrospinal fluid, there is leakage through the ependymal cell walls of the ventricle into the surrounding white matter.

8. These are also commonly referred to as "micro-aneurysms" or Charcot-Bouchard aneurysms, but the term pseudoaneurysms may be preferable, as they do not appear to be true aneurysms, as classically defined. Like true aneurysms, however, they do represent a weakness in the walls of the vessel.

9. The following will focus primarily on arterial syndromes affecting the cerebral hemispheres. See respective chapters for lesions affecting other regions of the CNS.

10. See Chapter 9 for a more detailed listing of functions associated with right and left hemispheric lesions, as well as descriptions of "frontal executive functions."

11. For a more detailed review of the neuropsychiatric sequelae of strokes, the reader is referred to Robinson (1997, 1998).

12. Not infrequently, CT scan or MRI reports will note the presence of periventricular changes in the absence of any evidence of mental status change. Hence, caution is advised against making a diagnosis of subcortical leukoencephalopathy or Binswanger's disease based on neuroimaging techniques alone.

REFERENCES

Adams, J.H. & Graham, D.I. (1988) *An Introduction to Neuropathology.* New York: Churchill Livingstone.

Alexander, M. & Freedman, M. (1984) Amnesia after anterior communicating artery rupture. *Neurology,* 34, 752–757.

Bauer, R.M. (1993) Agnosia. In: Heilman, K. & Valenstein, E. (Eds.), *Clinical Neuropsychology.* New York: Oxford University Press, pp. 215–278.

Biller, J. (1990) Vascular syndromes of the cerebrum. In: Brazis, P.W., Msadeu, J.C., & Biller, J. (Eds.), *Localization in Clinical Neurology.* Boston: Little, Brown, and Co., pp.

Bornstein, R. & Brown, G. (1991) *Cerebrovascular Disease.* New York: Oxford University Press.

Caplan, L.R. & Bogousslavsky, J. (2001) *Stroke syndromes and uncommon causes of stroke.* Cambridge: Cambridge University Press.

Cummings, J.L. & Benson, D.F. (1992) *Dementia: A clinical approach. 2nd Edition.* Boston: Butterworth-Heinemann.

Damasio, A.R. (1985) Disorders of complex visual processing: Agnosias, achromatopsia, Baliant's syndrome, and related difficulties of orientation and construction. In: Mesulam, M-M (Ed.), *Principles of Behavioral Neurology.* Philadelphia: F.A. Davis, pp. 259–288.

Damasio, A.R., & Damasio, H. (1983) The anatomic basis of pure alexia. *Neurology, 33,* 1573–1583.

Damasio, A.R., Graff-Radford, N.R., Eslinger, P.J., Damasio, H., & Kassell, N. (1985) Amnesia following basal forebrain lesions. *Archives of Neurology, 42,* 263–271.

De Luca, J., & Diamond, B.J. (1995) Aneurysm of the anterior communicating artery: A review of neuroanatomical and neuropsychological deficits. *Journal of Clinical and Experimental Neuropsychology, 17,* 100–121.

DeRenzi, E. (1997) Prosopagnosia. In: Feinberg, T. & Farah, M.J. (Eds.), *Behavioral Neurology and Neuropsychology.* New York: McGraw-Hill, pp. 245–255.

Farah, M.J. (1997) Visual object agnosia. In: Feinberg, T. & Farah, M.J. (Eds.), *Behavioral Neurology and Neuropsychology.* New York: McGraw-Hill, pp. 239–244.

Fasanaro, A.M., Valiani, R., Russo, G., Scarano, E., De Falco, R., & Profeta, G. (1989) Memory performances after anterior communicating artery aneurysm surgery. *Acta Neurologica, 11,* 272–278.

Gade, A. (1982) Amnesia after operations on aneurysms of the anterior communicating artery. *Surgical Neurology, 18,* 46–49.

Greene, K.A., Marciano, F.F., Dickman, C.A., Coons, S.W., Johnson. P.C., Bailes, J.E., Spetzler, R.F. (1995) Anterior communicating artery aneurysm paraparesis syndrome: Clinical manifestations and pathologic correlates. *Neurology, 45,* 45–50.

Millikan, C.H., McDowell,F., & Easton, J.D. (1987) *Stroke.* Philadelphia: Lea & Febiger.

Mendoza, J.E., Pladdy, B.H., Apostolos, G.T. & Hendrickson, R.K. (2007) *The coin rotation task: A bedside measure of motor dexterity.* Unpublished manuscript.

Mohr, J.P, Choi, D., Grotta, J. & Wolf, P. (2004) *Stroke: Pathophysiology, Diagnosis, and Management.* New York: Churchill Livingstone.

Okawa, M., Maeda, S., Nukui, H., & Kawafuchi, J. (1980) Psychiatric symptoms in ruptured anterior communicating artery aneurysms: Social prognosis. *Acta Psychiatrica Scandinavia, 61,* 306–312.

Redlich, F.C., & Dorsey, J.F. (1945) Denial of blindness by patients with cerebral disease. *Archives of Neurology and Psychiatry, 53,* 407–417.

Robinson, R.G. (1997) Neuropsychiatric consequences of stroke. *Annual Review of Medicine, 48,* 217–229.

Robinson, R.G. (1998) *The Clinical Neuropsychiatry of Stroke.* Cambridge University Press.

Talland, G.A., Sweet, W. & Ballantine, T. (1967) Amnestic syndrome with anterior communicating artery aneurysm. *Journal of Nervous and Mental Disease, 145,* 179–192.

Toole, J.F. (1990) *Cerebrovascular Disorders.* Raven Press. New York.

Tranel, D. (1997) Disorders of color processing (perception, imagery, recognition, and naming). In: Feinberg, T. & Farah, M.J.(Eds.), *Behavioral Neurology and Neuropsychology.* New York: McGraw-Hill, pp. 257–265.

Vilkki, J. (1985) Amnesic syndromes after surgery of anterior communicating artery aneurysms. *Cortex, 21,* 431–444.

Welch, K.M.A., Caplan, L.R., Reis, D.J., Siesjo, B.K., & Weir, B. (1997) *Primer on Cerebrovascular Diseases.* New York: Academic Press.

11   NEUROCHEMICAL TRANSMISSION

CHAPTER OVERVIEW

Thus far, this book has focused primarily on structural pathways and presumed hardwired connections among various regions of the central and to a lesser extent peripheral nervous system. However, another process needs to occur in order for these structural "connections" to become functional. The latter is a biochemical process, carried out by a variety of *neurotransmitter* and *neuromodulator* substances. The diffusion of these substances across the synaptic clefts that separate individual neurons is what permits one individual neuron to be able to communicate with the next. In turn, these biochemical links allow for communications among diverse regions of the nervous system necessary for the integration of different functional units required in complex behaviors. Thus, neurotransmitters and neuromodulators comprise an integral part of behavioral neurology and neuropsychology. This chapter initially will provide a basic review of the putative action of neurochemical transmitters in general within the neuron and in the synaptic cleft. Special attention will be given to various types of receptors and their suspected role in behavior. The main part of the chapter will focus on what are thought to be the major neurotransmitters in the central nervous system: acetylcholine, glutamate, gamma-aminobutyric acid, norepinephrine, dopamine, and serotonin. For each of these substances, the manner in which they are synthesized, broken down, the nature of their receptors, and their general distribution in the CNS briefly will be reviewed. Next, the suspected major contributions to the behavior of each of these CNS transmitters in turn, as well as their suspected clinical and pharmacological implications, will be discussed in greater detail. Brief mention of neuropeptides and their probable contribution to neurochemical transmission will conclude the chapter. For those who might be interested, an appendix to this chapter will review the basic physiology of nerve conduction.

At the conclusion of this chapter the reader should have at least a rudimentary knowledge of neurochemical transmission within the CNS and the implications of the major transmitters in various neurological and psychiatric conditions. Perhaps of equal importance, for those readers who already might not be familiar with these processes, one should also derive a basis for the understanding of at least some of the concepts underlying clinical psychopharmacology.

INTRODUCTION

Nervous transmission, and subsequently behavior, can be disrupted by various mechanisms. In discussing neuropathology, perhaps one's initial inclination is to think of those conditions that primarily cause structural damage to or disrupt neural networks. Common examples might include vascular accidents (as reviewed in Chapter 10), brain tumors, head traumas, and degenerative, infectious, or metabolic diseases. Such processes can damage cell bodies and/or neuronal pathways. However, these or other, often vague or unidentified conditions, including genetic or even psychological factors, also can disrupt basic biochemical processes in the brain. This could result from changes in the ability of cells to synthesize, transport, utilize, or break down neurochemical transmitters or neuromodulators. As noted in Chapter 8, changes in the relative availability of these biochemical substrates or integrity of these systems are the putative source of numerous neurological and psychiatric disorders, from Parkinson's disease to panic disorders. Many drugs (both prescribed and illicit) exert their effects through their interactions with this system.

This chapter will attempt to provide an overview of the better-known neurotransmitters, their general distribution and probable functional significance within the nervous system, and their relevance to both normal and abnormal behavior. While the role of certain drugs in the treatment of pathological conditions associated with anomalies of neurotransmitter systems will be discussed, this chapter should in no way be construed as a comprehensive review of pharmacotherapy for neurobehavioral or neuropsychiatric disorders. For his purpose, the reader is referred to any number of texts that provide a more in-depth treatment of this topic.

GENERAL PRINCIPLES

Before discussing the specific neurotransmitters, it may be useful to review a few key concepts. As most readers are probably aware, the propagation of a nervous impulse ("action potential") along a neuron is the result of a progressive, sequential depolarization along the axon. This depolarization is characterized by an influx of sodium (NA^+) ions and the subsequent efflux of potassium (K^+) ions across the cell membrane, primarily via the opening of voltage-gated ion channels.[1] For those who are interested, a more detailed

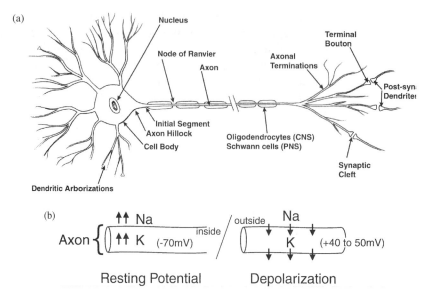

Figure 11–1. General model of the neuron. (a) The basic components of a neuron, including the cell body with its nucleus and dendritic arborizations on the left, the axon, and the axonal terminations on the right. Illustrated is a myelinated axon in which the axon is covered with either oligodendrocytes (in the central nervous system) or Schwann cells (in the peripheral nervous system) that facilitate (speed) conduction of the action potential along the axon. Not all neurons are so myelinated (see Appendix). (b) The major changes that take place in the axon during an axon potential. In the "resting stage" there is a disproportionate amount of sodium ions (Na^+) outside the neuronal membrane and a disproportion of potassium ions (K^+) intracellularly. During depolarization of the neuron, sodium influxes into the intracellular space, while potassium ions tend to efflux into the extracellular space. These changes result in a relative voltage change from negative to positive between the intracellular and extracellular spaces. A more detailed discussion of the mechanics of the action potential and depolarization process will be found in the Appendix.

discussion of nerve conduction and action potentials can be found in the Appendix. For now, Figure 11–1 offers a basic representation of a neuron and the chemical interchange that takes place during depolarization. Recall that the activation of ligand-gated ion channels and the resulting graded (partial) depolarization of the cell ultimately precede the opening of the voltage-gated Na^+ and K^+ ion channels in the action potential. As will be discussed shortly, these ligand-gated ion channels are directly or indirectly activated by the action of neurochemical transmitters at the chemical synapse.[2] The chemical synapse thus not only represents the site where neurotransmitter abnormalities, and hence, many clinical disorders, are expressed (at least physiologically), but also is the place where many drugs produce their behavioral effects. Because of their importance in these clinical processes, certain key features of synaptic neurochemical transmission briefly will be reviewed.

Production and Storage of Neurotransmitters

Most of the neurotransmitters are synthesized by enzymes within the **presynaptic terminals** and stored in **presynaptic vesicles** where they remain safe from enzymatic degradation until they are ready to be used. Other neurotransmitters, such as acetylcholine, are synthesized elsewhere in the cell and subsequently transported into their vesicles. When the action potential of the neuron reaches the **presynaptic terminal**, it causes an influx of calcium (Ca^{2+}) ions into the terminal, depolarizing it. This results in vesicles being transported to the

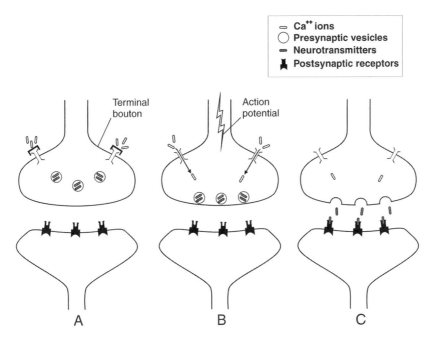

Figure 11–2. Migration of the synaptic vesicles. Neurotransmitters are stored or encapsulated in synaptic vesicles in the terminal boutons awaiting release into the synaptic cleft by an action potential, thereby stimulating the next neuron in the chain via their postsynaptic receptors. The arrival of the action potential at the terminal bouton causes a release of calcium (Ca^{2+}) into the terminal bouton. This influx of calcium causes the vesicle to migrate to the surface of the bouton and release its contents into the synaptic cleft.

presynaptic membrane where they discharge their contents (i.e., neurochemical transmitters) into the **synaptic cleft** (Figure 11–2). It is important to note that a given presynaptic terminal may synthesize, store, and release multiple neurotransmitters. The precursors and particular enzymes responsible for the production of the different neurotransmitters will be reviewed as they are individually discussed.

Neurotransmitter Receptors

Neurotransmitters act by binding to a particular receptor after they are released into the synaptic cleft (Figure 11–3). As a result of this binding, the receptor initiates a secondary response that leads to any one of several cellular events, which in turn leads to some behavioral change. Before discussing some of these potential events, it should be noted that each neurotransmitter has its own unique set of receptors into which it fits, not unlike a lock and a key. However, just as a master key may open several different locks, each of which may be serving a different purpose, each neurotransmitter generally has multiple receptor subtypes to which it can bind. Often various receptor subtypes may be found at a single synapse or different receptor subtypes may predominate at different anatomical sites. For example, cholinergic nicotinic receptors typically are associated with preganglionic neurons of both the sympathetic and parasympathetic nervous system as well as somatic muscle innervation, while muscarinic receptors, also responsive to acetylcholine, are found at the end organs innervated by postganglionic parasympathetic fibers. Like the locks in the analogy above, each of these different receptors, even though stimulated by the same neurotransmitter, may result in a very different behavioral response, both clinically and at

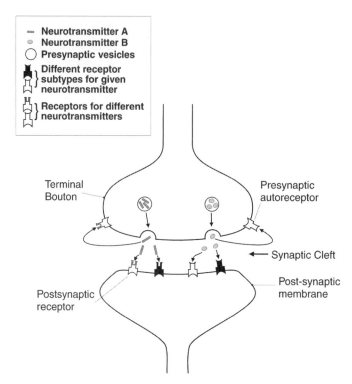

Figure 11–3. Chemical synapse. The chemical synapse provides a means for information to be communicated from one neuron to the next. Its major elements are the presynaptic membrane, the synaptic cleft, postsynaptic membrane, neurotransmitter(s), and receptors. Here the pre- and postsynaptic membranes are represented by a terminal bouton of an axon synapsing on a dendrite of a second neuron, although it is possible for an axon to synapse directly on cell body, axon, or terminal bouton of another neuron. Once neurotransmitters (a single neuron may have more than one) are released into the synaptic cleft from the presynaptic terminal, they bind to their respective receptors on the postsynaptic membrane, triggering some response in the second neuron. They also may diffuse to autoreceptors located on the presynaptic membrane.

the cellular level. An example of the latter is the response to an adrenergic antagonist drug like *propranolol*. Used to treat cardiovascular problems as well as specific psychiatric (e.g., social phobia or *stage fright*) or neurological (e.g., neuroleptic-induced akathisia) conditions, propranolol is a nonselective beta-blocker.[3] Being nonselective, it blocks the action of both beta$_1$ and beta$_2$ receptors. While blocking of the former may be desirable in the individual who gets anxious when having to give a public speech, it may cause problems if the person also suffers from asthma (as a result of bronchoconstriction due to beta$_2$ blockage). As will be noted later in greater detail, it is their action on multiple but unwanted receptors that accounts for the undesirable side effects of medications.

Allosteric Modulation

Just as individual neurons are capable of producing more than one type of neurotransmitter, so too each synapse may have receptors that are responsive to more than one type of neurotransmitter. Thus, a given postsynaptic membrane may contain both noradrenergic and serotonergic receptors. However, the situation is even more complex. In addition to receptors from multiple neurotransmitters possibly occupying a given postsynaptic membrane, even single receptors may have multiple binding sites that are responsive to different chemical molecules. For example, a primary receptor for neurotransmitter A might have sites that also are responsive to neurotransmitter B. These secondary neurotransmitters or "modulators" (e.g., neurotransmitter B) are unable to effect a response independent of the primary neurotransmitter (i.e., they cannot effect a response in this particular receptor outside the presence of the primary neurotransmitter). However, they can alter significantly the response of the receptor once the primary neurotransmitter adheres to its binding site. This process is known as *allosteric modulation* (Figure 11–4).

Allosteric modulation may be either positive or negative. If positive, the effect of binding by these secondary transmitter substances will be to enhance or augment the effect produced by the primary neurochemical transmitter at that particular receptor. Negative allosteric modulation, by contrast, would reduce or possibly even reverse the normal response elicited by the primary neurotransmitter. The receptor for **gamma-aminobutyric acid** (GABA), a ligand-gated receptor (see below), provides a good example of this phenomenon. The GABA receptor (specifically, the GABA-A receptor) is responsible for controlling the flow of chloride (Cl^-) ions into the intracellular space. When GABA binds to these receptors, it facilitates the opening of these chloride channels, thus allowing for an increased influx of chloride ions down their concentration gradient. Recall that the interior of the membrane already is negative in the resting state. Thus, this increased infusion of negative chloride ions serves to **hyperpolarize** the nerve, making it less likely to fire.[4] However, the GABA-A receptor also is suspected of having several positive allosteric binding sites, including those for benzodiazepines, barbiturates, and alcohol. When benzodiazepines or barbiturates bind to the GABA-A receptor in the presence of GABA, they respectively increase the frequency and the length of time the channel remains open (compared with that produced by the binding of GABA alone), permitting an even greater influx of Cl^- ions. The result is yet a greater hyperpolarization of the neuron (i.e., greater inhibition).

Allosteric Modulation

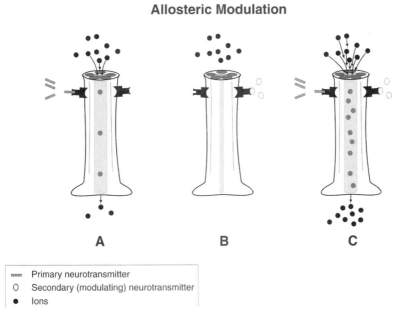

— Primary neurotransmitter
O Secondary (modulating) neurotransmitter
• Ions

Figure 11–4. Allosteric modulation. (a) The primary neurotransmitter produces a response, in this case the opening of an ion channel. (b) Binding by the secondary neurotransmitter itself has no impact. (c) However, when combined with primary neurotransmitter, the effect of the latter is enhanced.

In addition to causing a more dramatic change in the diameter of the ion channel, it also is suspected that allosteric modulation may influence the relative polarization of the neuron by affecting (1) the frequency with which an ion channel will open, (2) the duration for which the channel will remain open, or (3) the ease or probability that the channel will open given the presence of an agonist (e.g., despite the binding of a ligand, the opening of an ion channel may be greatly affected by the resting potential of the membrane at the time). These mechanisms would help explain how substances such as the benzodiazepines and barbiturates might exert their anxiolytic, anticonvulsant, and depressive effects on the central nervous system. Conversely, *picrotoxin*, a proconvulsant drug, appears to bind to the GABA-A receptor but in a negative allosteric fashion. Instead of increasing the opening of the chloride channel, when bound to the receptor it tends to have the opposite effect, closing the channel relative to its normal resting state. This results in reduced chloride infusion into the cell, resulting in a relative depolarization, making the discharge of an action potential more likely. Thus, allosteric modulation allows for multimodal influences on cells (receptors), beyond that which might be afforded by the primary neurotransmitters themselves.

Autoreceptors

Not all receptors for a given transmitter are located on the postsynaptic membrane of the receiving neuron or target organ. Receptors sensitive to the released transmitter commonly are found either on the dendrites (or cell body itself) or at the same presynaptic terminal from which the neurotransmitter was released (Figure 11–5). These may be (1) an integral part of the reuptake transport process (see below) or (2) serve as feedback receptors for neurochemical transmitters within a particular cell. Insofar as they mediate the latter function, they are termed *autoreceptors*. As will be discussed later, either too much or too little release of a neurotransmitter at the synapse may be associated with behavioral dysfunction. For optimal performance, the release of neurotransmitter substances must be kept within a certain range. Autoreceptors on the presynaptic membrane provide a mechanism whereby the synthesis and/or release of a specific neurotransmitter may be modified through their

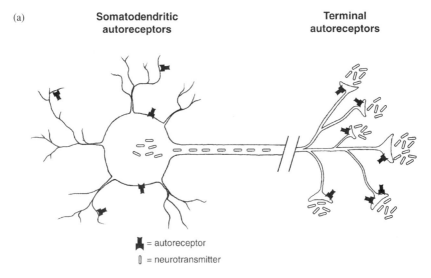

(a)   **Somatodendritic autoreceptors**   **Terminal autoreceptors**

= autoreceptor
= neurotransmitter

"Uninhibited" Flow of Neurotransmitter from Soma and from Terminal Boutons

Figure 11–5. (*Continued*)

(b)   **Somatodendritic**   **Terminal**
                 autoreceptors   **autoreceptors**

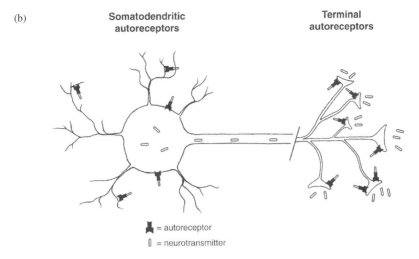

= autoreceptor

= neurotransmitter

Reduced Flow (Release) of Neurotransmitter Following Occupation of Autoreceptors

Figure 11–5.   Autoreceptors. (a) In the absence of autoreceptor stimulation, the production and release of neurotransmitters by the neuron is "uninhibited." (b) With stimulation of the somatodendritic autoreceptors, the production of the neurotransmitter is retarded. Similarly, with stimulation of the terminal autoreceptors, there is a reduction of neurotransmitter release into the synaptic cleft.

interaction with organelles within the cell. Thus, the cell may synthesize or release increased amounts of the neurotransmitters when the autoreceptors are insufficiently stimulated or by slowing down these processes when there is an oversupply of the neurotransmitter in the synaptic cleft.

In addition to the regulation of the synthesis and release of the neurotransmitter at the neuronal terminal, autoreceptors also may be found at the dendritic (receiving) end of certain neurons. These also serve an important regulatory function. For example, certain neurons (e.g., within the raphe nuclei) are responsible for the synthesis and eventual distribution of serotonin to other parts of the CNS. Again, since there is strong reason to suspect that maintaining serotonin availability within certain limits at these distant sites is critical for normal functioning, some form of feedback mechanism needs to be in place. These dendritic autoreceptors (also known as *somatodendritic receptors* because of their proximity to the soma or cell body) likely control this process either by causing a hyperpolarization of the cell (thereby reducing its firing rate) or by sending a message to the DNA in the nucleus of the cell to slow down production.

Heteroreceptors

A neurotransmitter not only may affect its own release via autoreceptors, but may diffuse to its own receptors located on other neurons, where it either inhibit or facilitate release of a second neurotransmitter. Because of their capacity to exert a heteromodal influence (i.e., over different neurotransmitters), such receptors are known as *heteroreceptors*. One well-known example in psychopharmacology is norepinephrine receptors. **Alpha-1** receptors located on the cell bodies of serotonergic neurons (e.g., in the raphe nuclei) when stimulated *facilitate* the release of 5-hydroxytryptamine (5-HT) (serotonin). The opposite occurs when norepinephrine stimulates the α-**2** receptors located on the terminals of 5-HT neurons. In this case, the release of serotonin is *inhibited* (Figure 11–6). Serotonin, on the other hand, via $5-HT_{2A}$ heteroreceptors located both on terminal and somatodendritic and portions of

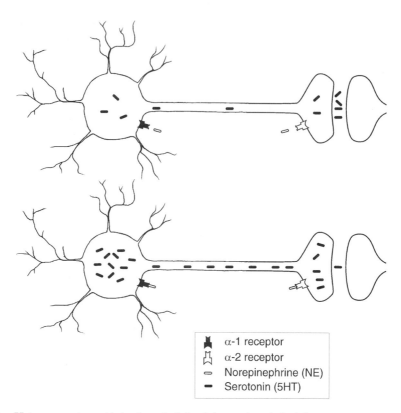

Figure 11–6. Heteroreceptors. Alpha-1 and alpha-2 (norepinephrine) heteroreceptors are seen at the cell body and terminal boutons, respectively, on a serotonergic neuron. In the top figure, without stimulation of the alpha heteroreceptors, there is minimal impetus for production of serotonin at the level of the cell body, but relatively uninhibited release of serotonin at the terminals. However, occupancy of the alpha-1 heteroreceptors by norepinephrine at the soma (lower figure) stimulates increased production of serotonin by the neuron. Conversely, occupancy of the alpha-2 terminal receptor would tend to inhibit the release of serotonin.

dopaminergic neurons can inhibit the release of dopamine. As will be seen later, drugs that have the capacity to block these 5-HT$_{2A}$ heteroreceptors (e.g., the atypical antipsychotics), and thereby increase the release of dopamine in certain areas of the brain, can decrease the side effects normally associated with traditional antipsychotics [e.g., by increasing the release of dopamine (DA) in the nigrostriatal system]. Additionally, these drugs are thought to improve the "negative" symptoms of schizophrenia (possibly by increasing DA in the frontal regions) (Stahl, 2000). Finally, dopamine itself is noted to have inhibitory effects on the release of acetylcholine, an observation that as we shall see underlies the use of anticholinergics to help control the extrapyramidal side effects of traditional antipsychotic drugs.

G-Protein-Linked and Ligand-Gated Ion Receptors

In addition to receptor subtypes for a particular neurotransmitter, there is what has been termed *receptor superfamilies*. These superfamilies represent more generic classes of receptors that cut across individual neurotransmitters. Two major subtypes are defined: **G-protein-linked receptors**, and **ligand-gated receptors** (Figure 11–7). These two types of receptors differ both structurally and functionally. Recall that receptors are composed of strings of proteins (amino acids) manufactured within each cell. These *strings* of protein weave in and out of the cell membrane, creating a circular arrangement. While weaving in and out of the cell membrane, these amino acids in effect form extracellular, intracellular, and transmembrane (i.e., within the multilayered cell membrane itself) *domains*. While the G-protein-linked receptors have seven such transmembrane sections, the ligand-gated ion channel receptors are composed of four or five subunits, with each subunit generally containing four transmembrane segments or domains.

Ligand-gated ion channel (also known as *ionotropic*) receptors operate in a relatively direct manner as compared with G-protein-linked receptors. When stimulated by an agonist or neurotransmitter they alter (increase) the flow of ions through the channel.[5] Different receptors may permit the passage of various ions (e.g., Na^+, K^+, Ca^{2+}, Cl^-), although each receptor may be activated selectively by a particular neurotransmitter. Thus, glutamate generally is considered an excitatory neurotransmitter and can lead to the opening of Na^+ channels, which in turn have a depolarizing effect on the neuron. However, in addition to Na^+, glutamate receptors at times also can serve as channels for K^+ and Ca^{2+}, depending on respective concentration gradients and the presence of specific subunits in the receptor itself. Conversely, GABA, an inhibitory neurotransmitter, opens Cl^- channels, thereby increasing the negative valence of the intracellular milieu relative to the outside of the cell, making it more difficult for the action potential to be initiated (hyperpolarization).

In contrast to ligand-gated receptors, **G-protein-linked** (or *metabotropic*) receptors have a more diversified impact on the cell and produce their effects via different mechanisms. While glutamate and GABA have some metabotropic binding sites, these G-protein receptors are most commonly associated with other neurotransmitters, particularly the amines [e.g., dopamine (DA), norepinephrine (NE), and serotonin (5-HT)]. One major distinction of the metabotropic receptors is that they do not have a direct impact on ion channels like

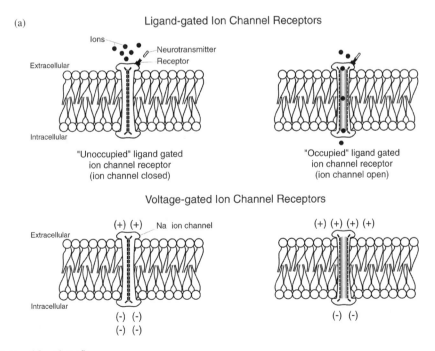

Figure 11–7. (*Continued*)

(b) **Unbound G-Protein receptor**

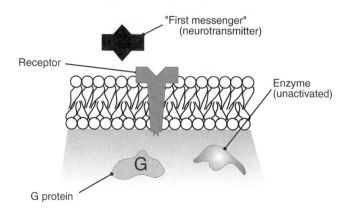

Unbound G-Protein receptor

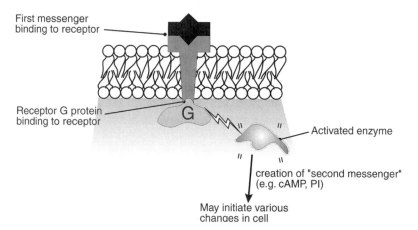

Figure 11–7. Ligand-gated, voltage-gated, and G-protein receptors. (a) The binding of an appropriate neurotransmitter to a ligand-gated ion channel receptor (top) might cause it to open, allowing the passage of ions through the membrane, whereas changes in voltage potentials across the membrane (bottom) may trigger the opening of a voltage-gated ion receptor. (b) In G-protein receptors, the process is more complicated. Binding of the neurotransmitter ("first messenger") to the receptor (bottom figure) causes a conformational change in the receptor that enables it to bind with a G protein inside the cell membrane. This leads to the activation of an intracellular enzyme that in turn produces a "second messenger." This second-messenger molecule that has been produced might effect any number of changes within the cell or may impact directly on other receptors.

ligand-gated receptors. Rather, the binding of the neurotransmitter to the G-protein-linked receptor results in a conformational change in the receptor that then enables it to interact with another protein (known as a *G-protein*) in the transmembrane space. After this interaction occurs, it allows the receptor/G-protein complex to bind with a specific intracellular enzyme. Once linked to the receptor/G-protein complex, this enzyme is capable of synthesizing another molecule [e.g., **cyclic adenosine monophosphate** (cAMP) or **phosphatidyl inositol** (PI)]. These secondarily produced molecules are known as *second messengers* (the original neurotransmitter being the *first messenger*). Once produced, these second messengers can effect a variety of changes within the cell, including:

1. Activating other enzymes.
2. Impacting on the DNA of the cell, which in turn may set in motion the synthesis or degradation of other enzymes and proteins (including receptors).
3. Modulation or "opening" (through phosphorylation) of ion channels.

Since these G-protein-linked receptors involve this multistep process, their effects tend to be carried out significantly more slowly than those mediated by ligand-gated receptors. Hence, the amino acid transmitters glutamate and GABA, which typically are associated with ligand-gated receptors, are known as *fast-acting* neurotransmitters, while the monoamine transmitters (DA, NE, 5-HT), which selectively bind to G-protein receptors, are referred to as *slow acting*.

Receptor Regulation

One of the effects of G-protein-linked receptors or the *second messenger* system is the capacity to influence DNA within the nucleus of the neuron. DNA in turn through messenger RMA(mRNA) directs all functions of the cell, including protein synthesis. Recall that the receptors are nothing more than strings of specially designed proteins manufactured within the endoplasmic reticulum and Golgi apparatus of the neuron. Thus, the binding of DA, NE, or 5-HT to G-protein-linked receptors directly can impact (via "second messenger systems") the formation of new receptors. Conversely, under certain conditions they can also trigger the destruction of previously formed receptors by the lysosomes.

Why might the regulation of the synthesis of receptors be important? Suppose that for whatever reason there is a relative deficiency of a particular neurotransmitter at a given synapse. How can the cell try and adapt to this situation in order to maintain an optimal level of response? One way might be for the cell to produce more receptors, thus helping to ensure that whatever molecules of the neurotransmitter that are available in the synaptic cleft will find a receptor to which they might bind (and hence, effect a response postsynaptically). Conversely, if there is too much of a transmitter available, the postsynaptic neuron may want to reduce the number of its receptors to avoid becoming overly stimulated by the excess of neurotransmitter molecules.

In effect, this is what seems to happen. When a relative surplus of a particular neuro-transmitter is present, the excessive stimulation of the G-protein-linked receptors for that neurotransmitter sends a signal for the cell to slow down the production of new receptors and/or to speed up the process of destroying old receptors. Although these changes take time, eventually there will be fewer receptors in the postsynaptic membrane. This process is referred to as **down-regulation** of receptors (Figure 11–8). The opposite is true when there is a relative deficiency of a specific neurotransmitter. In this case, the cell may be instructed through second messengers to speed up the production of that particular receptor, creating a larger number of receptors (**up-regulation**). As we shall see shortly, this change in the availability or number of receptors tentatively has been linked to changes in clinical behavior and response to drug treatment.

Spectrum of Drug–Receptor Interactions

Although the body's natural neurotransmitters such as GABA, dopamine, acetylcholine (ACh), norepinephrine, and others are the normal stimulants for receptors, they are not the only substances that can bind to receptor sites. Clearly, many drugs, various toxic

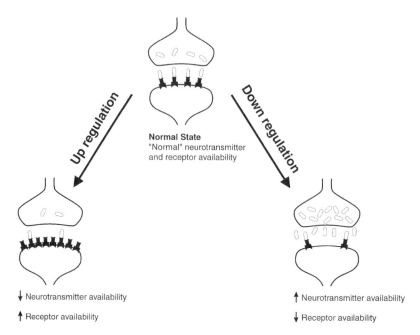

Figure 11–8. Receptor regulation. The nucleus of the neuron through its DNA and mRNA can call either for the production or degradation of receptors. The number of receptors generally would appear to be inversely proportional to the long-term availability of their respective neurotransmitters. As seen in the left of the figure, if a given neurotransmitter is in relatively short supply, there is a compensatory increase in the number of receptors ("up-regulation"), perhaps as a means of ensuring that a response will occur. Conversely, given a relative abundance of a neurotransmitter (right figure), the cell may guard against overstimulation by reducing the number of available receptors ("down-regulation").

agents, and a host of other psychoactive substances produce their effects through their capacity to occupy specific receptors. The clinical effects of many of these substances were discovered by accident, well before there was any understanding of their impact on neurotransmitter systems and before the synapse or receptors themselves were discovered. Now, however, drugs specifically are being designed to impact in highly specific ways on particular types of receptors or other intracellular processes. These drugs can impact on receptors in a variety of ways. The most commonly known mechanisms are as **agonists** and **antagonists**.

   Agonists are any substance that activate the receptor. Thus, if the normal response of a receptor when bound to its natural neurotransmitter is to open an ion channel, then any drug that when bound to the receptor produces a similar response would be termed an *agonist*. An **antagonist**, on the other hand, simply blocks the receptor site, preventing an agonist (or other substance) from binding at that site (Figure 11–9). **Note:** In and of themselves, antagonists have no effect on the receptor (i.e., in the case of a ligand-gated receptor, they would neither open nor close an ion channel). However, they will prevent another potentially activating drug from binding to that site.[6] Another way of expressing the distinction between agonists and antagonists is that while the former has both affinity and exerts full intrinsic activity, the latter, while having affinity, exerts no intrinsic activity within the receptor. Many drugs produce their effects by antagonistic mechanisms. One class of drugs mentioned earlier, beta-blockers (e.g., **propranolol** and **atenolol**) fall into this category. They exert their effect by selectively blocking adrenergic beta (sympathetic)

Agonists, Antagonists, Partial Agonists

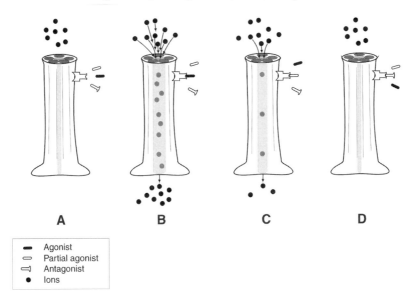

A   B   C   D

- Agonist
- Partial agonist
- Antagonist
- Ions

Figure 11–9. Agonists, antagonists, and partial agonists. (a) The receptor is not being stimulated and the ion channel is closed (in some, the "resting" condition could be *opened*). (b) With stimulation by an agonist, the channel may fully open (i.e., responds in a manner similar to its normal neurotransmitter). (c) When the receptor is occupied by a partial agonist, the response is greater than with no stimulation, but less than to a (full) agonist. (d) Occupation by an antagonist produces no response on its own. It simply blocks the receptor from being stimulated by any of the various types of agonists or neurotransmitters. In certain instances, however, this blockage can be overcome.

receptors, thus allowing for the unopposed action of the parasympathetic system to slow down the heart rate.

At rest, ion channels may be opened or closed. If closed, for example, an agonist for a ligand-gated receptor may cause it to open. What, if anything, might serve to actively prevent the ion channel from opening? This might accomplished by either an antagonist an *inverse agonist*. Inverse agonists produce effects opposite those of agonists.[7] Although some compounds have been discovered that act as benzodiazepine (BZD) inverse agonists, because they tend to be anxiogenic and convulsant (effects opposite those of BZD agonists), they are used for research rather than clinically (Sarter, Nutt, & Lister, 1995).

Partial Agonists

To complete the spectrum, there are two other ways in which drugs theoretically might impact on a receptor. They may function either as a **partial agonist** or as a **partial inverse agonist** (see Figure 11–9). In these instances, the drug would have an intermediate effect on the receptor. Again, with regard to a ligand-gated ion channel, a partial agonist would tend to open the ion channel relative to its resting state but not open it as much as would be accomplished by the binding of a "full" agonist. In the language used above, a partial agonist might be said to have affinity, but has less-than-full intrinsic activity. **Buspirone** (BuSpar), **clonidine** (Catapres), and **aripiprazole** (Abilify) are examples of partial agonists. Similarly, a partial inverse agonist would reduce the flow of ions through this same channel relative to its resting state, but not restrict the flow to the same extent as a "full" inverse agonist. Like "full" agonists and inverse agonists, these "partial" agonists (and inverse agonists) can be blocked by antagonists or they may replace a full agonist or full inverse agonist at a given receptor site. As noted by Stahl (2000), what is interesting about these partial agonists (or partial inverse agonists) is that their effect can vary depending on what substances they may replace. For example, in the absence of a full agonist, the partial agonist might "partially open" the ion channel (thus serving as a *net agonist* in relation to the channel's previous state). By contrast, in the presence of a full agonist, which might fully open an ion channel, the binding of a partial agonist now would have the effect of only "partially opening" (i.e., partially closing) the ion channel (thus having a *net antagonistic* effect on the ion channel).

Enzymatic Actions

Finally, antagonist drugs block not only pre- and postsynaptic membrane receptors but also block may the actions of enzymes. An enzymes typically is activated when a specific substrate binds to it, thus enabling the enzyme to carry out its assigned task. This may be the creation of a new substance (*product*) or the breakdown of an old one. As we shall see shortly, this process is critical to chemical neurotransmission. For now, it should be noted that different drugs might act as either *reversible* or *nonreversible* inhibitors of enzymes. If a drug forms an irreversible bond to an enzyme, that particular enzyme is permanently disabled. In order to carry out the normal activity of the enzyme, the cell must synthesize new enzymes, a process that takes a certain amount of time. In contrast, a reversible enzyme inhibitor may be successfully challenged by the substrate under certain conditions, thus permitting the enzyme to carry out its normal function. As will be discussed, this would have important implications for certain antidepressant medications.

Neurotransmitter Degradation/Reuptake

Once a neurotransmitter has been released into the synaptic cleft and carried out its intended activity, it needs to be cleared from the synapse. This prevents the possibility of tonic stimulation of the postsynaptic membrane or presynaptic autoreceptors and prepares the synapse for the next stimulus generated by the presynaptic neuron. There are three basic possibilities as to how this might be accomplished:

1. The neurotransmitter might be degraded by enzymatic action while still in the synaptic cleft.
2. It might be "transported" back into the presynaptic terminal where it is either broken down by enzyme action or simply returned to the presynaptic vesicles where it is again stored until needed.
3. It may simply diffuse away from the synaptic cleft.

Degradation

Different enzymes are involved in the breakdown of different neurotransmitters. **Catechol-0-methyltransferase** (COMT), which is found in the synaptic cleft, is important in the degradation of NE and DA. **Acetylcholinesterase** (AChE), which is present in the synaptic cleft, breaks acetylcholine down into choline and acetate, thus inactivating it. **Monoamine oxidase** (MAO) is instrumental in the destruction of NE, DA, and 5-HT once they reenter the presynaptic terminal. Similarly, GABA is broken down in the presynaptic terminal by **gamma-aminobutyric acid transaminase** (GABA-T). Glutamate is an exception among the major neurotransmitters. Rather than being destroyed by enzymatic mechanisms, it is simply transported back into either the presynaptic terminal or the neighboring glial cells. In certain disease states that are believed to be associated with neurotransmitter deficiencies, such as Alzheimer's disease (acetylcholine) and depression (monoamines), drug treatments may be designed to inhibit these destructive enzymes, thus increasing the supply of the neurotransmitter. **Donepezil** (Aricept) and the MAO inhibitors [e.g., **phenelzine** (Nardil], **tranylcypromine** (Parnate)] used for the management of Alzheimer's and depression, respectively, are classic examples of these approaches (see Figure 11–10).

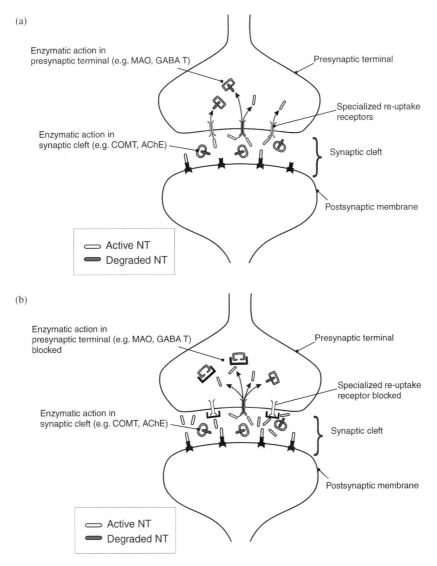

Figure 11–10. Degradation of neurotransmitters. (a) Two of the ways in which the actions of neurotransmitters may be terminated once they have been released into the synaptic cleft. They may be broken down by enzymatic action while still in the synaptic cleft or may be transported back into the presynaptic membrane where they may be degraded by yet other enzymes. They also simply may diffuse away from the synaptic cleft or in the case of acetylcholine be picked up by nearby glial cells where it will be degraded by butyrycholinesterase (BuChE) (not shown). (b) Degradation may be blocked either by blocking reuptake receptors (allowing increased accumulation in the synaptic cleft) or inhibiting the action of enzymes in the presynaptic terminals. Both mechanisms are utilized in antidepressant medications.

As noted above and illustrated in Figure 11–10a, a second way to terminate the activity of a neurotransmitter in the synaptic cleft (other than subjecting it to a catabolizing enzyme in the cleft itself) is to shepherd it back into the synaptic terminal from which it came. Here it also could be destroyed by enzymes or be funneled back into the presynaptic vesicles for storage and reuse. This is accomplished by yet another specialized chain of proteins known as *transport pumps*, *reuptake proteins*, or *transport carriers*. As the presynaptic cell membrane basically is designed to keep intracellular components separated from extracellular substances, these transport mechanisms typically require energy to carry out this work. The energy is supplied by **sodium–potassium adenosine triphosphatase** (ATPase), an energy-providing enzyme. These transport mechanisms, which are unique for each neurotransmitter, can be seen as specialized, presynaptic receptors to which an energy molecule is attached. However, rather than opening an ion channel (like the ligand-gated receptors) or creating a second messenger (like the G-protein-linked receptors), the sole purpose of these transport pumps is to transfer neurotransmitter molecules from outside the presynaptic membrane to the inside of the presynaptic terminal.

Reuptake Inhibition

These transport carriers share some of the same characteristics of other receptors, namely, the capacity to bind with other substances. One very practical result of this is that when certain molecules bind to these protein strings (i.e., the transport pumps), they effect a change in the configuration of the protein molecules that in effect precludes the binding of the neurotransmitter. Since they cannot bind to the carrier, they cannot be transported into the cell. If they cannot be transported back into the presynaptic terminal, they must remain in the synaptic cleft, either slowly diffusing away from the synapse or awaiting enzymatic catabolization in the synaptic cleft itself (e.g., by COMT). In the meantime, this results in an increased supply of neurotransmitter substance in the synapse. This basically is the principle by which the **tricyclic antidepressants** and **selective serotonin reuptake inhibitors** (SSRIs) currently are thought to operate (see Figure 11–10b). By binding to and thus blocking the ability of their respective neurotransmitters to bind to the transport pumps, they sharply increase the availability of various monoamines within the synapse (recall that depression is thought to be associated with a relative depletion of monoamines, particularly NE and 5-HT, in the brain).

However, the effectiveness of tricyclic and SSRI medications in the treatment of depression raises another critical consideration, that is, *association does not necessarily establish causality*. Consider the following:

1. Antidepressants result in an increase in the availability of monoamines at the synaptic cleft.
2. While this effect takes place immediately (within a couple of days at most), the full clinical effects usually take much longer before they are seen (weeks to months).
3. When supplies of the neurotransmitters were low, the postsynaptic neuron attempted to compensate by generating an increased number of receptors (up-regulation). But when the availability of the neurotransmitters was increased, the number of receptors diminished (down-regulation).
4. The time period for this down-regulation to stabilize was approximately 6 to 8 weeks, corresponding to the time frame frequently seen for clinical improvement.
5. Thus, one hypothesis is that the clinical response to antidepressants may be due to the down-regulation of receptors rather than to increases in availability of NE or 5-HT per se.[8]

However, even here, although the time frames are at least comparable, one is still working with correlations, not necessarily cause and effect. While the change in the relative number of receptors indeed may be the critical factor in the clinical response, why is this the case? Will another, yet to be discovered factor prove to be more critical? Future research may provide yet other insights into the mechanism of action behind these antidepressant drugs.

Side Effects and Receptors

As everyone who has opened the PDR (*Physician's Desk Reference*) or simply taken aspirin is well aware that most drugs have potentially significant, if not actually harmful, side effects. Such side effects generally result from the fact that a given medication either acts on:

1.   Receptors other than the one(s) primarily targeted,
2.   Receptors (of the targeted transmitter) that may be present in organs or functional systems other than the one(s) intended.

In the first instance, unlike endogenous neurotransmitters that tend to be specific to individual classes of receptors, drugs often have the capacity to bind to various classes of receptors. It is this feature that accounts for the side effects in drugs such as the tertiary amine tricyclic antidepressants. These drugs, in addition to blocking monoamine (primarily NE and 5-HT) reuptake at the synapse, also block M_1 (cholinergic), α_1 (adrenergic), and H_1 (histaminic) receptors. It is the blockading of these latter receptors that accounts for the problems with blurred vision, constipation, dry mouth, and (in certain vulnerable individual, such as the elderly) impaired cognition (M_1 block), orthostatic hypotension, dizziness (α_1 block), weight gain (H_1 block), and drowsiness (H_1 and, to a lesser extent, α_1 and M_1 blocks) that are frequently associated with these medications.

In the second instance, in addition to interacting with different classes of receptors, some drugs, while exerting their effect on the intended class of receptors, do so at unwanted sites. Another category of antidepressant medications, the MAO inhibitors, falls within this group. In addition to preventing the catabolization of monoamines in the brain associated with depression (NE, DA, 5-HT), MAO also is important in breaking down certain exogenous amines in the liver and small intestine that are found in certain foods and drinks. When MAO (type A) enzymes are inhibited, this results in a buildup of these amines, stimulating a release of endogenous catecholamines, which in turn stimulates sympathetic receptors, thus causing a potentially dangerous increase in blood pressure. This is where a "reversible" MAO inhibitor, such as **moclobemide** (currently not available in the United States) would be advantageous. While the MAO enzyme would be blocked under normal circumstances, when the patient eats food containing significant amounts of tyramine (or other problematic amines), the excess of amines displaces the reversible MAO inhibitor, freeing the enzyme to catabolize these substrates.

Many neuroleptic medications have similar, undesirable side effects, primarily because of their penchant for blocking dopaminergic receptors in the nigrostriatal, tuberoinfundibular, and mesocortical pathways, in addition to those of primary interest in the mesolimbic pathways.[9] The blocking of dopamine receptors in the nigrostriatal system may lead to the parkinsonian and other motor symptoms often associated with neuroleptics, particularly with the higher potency drugs, such as haloperidol.[10] In general, the search for newer drugs to treat disorders like depression and psychosis in large part has been motivated by the desire to find drugs that more specifically target receptors of primary interest, thus reducing some or all of the side effects associated with the older generation medications.

Rebound Effect

Another potential adverse effect of medication is associated with changes in receptor sensitivity. Recall that an up-regulation (increase) of receptors normally occurs with the relative depletion of a given neurotransmitter. A similar phenomenon is thought to take place when drugs block or have an antagonistic effect on postsynaptic receptors. Take, for example, neuroleptic drugs. In addition to their primary effect of blocking dopamine receptors, some also have potent anticholinergic properties (secondary effects). Because of this blocking there tends to be a respective up-regulation (increase) of dopaminergic and muscarinic receptors, creating a potential supersensitivity to dopamine and acetylcholine. As long as the drug dosages are maintained, two things might be expected to happen. First, the psychotic symptoms should be reduced due to the blocking of D_2 receptors. Second, anticholinergic side effects (e.g., constipation, dry mouth, or drowsiness) likely are due to blocking of muscarinic receptors. What happens, however, if the drug is either abruptly withdrawn or replaced by another antipsychotic with minimal or weak anticholinergic effects? If simply stopped, the original psychotic symptoms not only might return (relapse), but also, at least temporarily, they become exacerbated due to supersensitivity of the D_2 receptors (rebound effect). Additionally, because of the prior up-regulation of muscarinic receptors (while on the drug), a cholinergic rebound effect could occur as acetylcholine now is no longer being blocked. These effects might include nausea, vomiting, excessive sweating, restlessness, and insomnia. If replaced by another antipsychotic with minimal anticholinergic effects, no exacerbation of psychotic symptoms would be expected (due to the antidopaminergic effects of the second drug); however, an enhanced cholinergic effect still could occur (Luchins, Freed, & Wyatt, 1980; Stahl, 2000). In the latter instance, this effect could be reduced or eliminated by a gradual tapering of the first drug or adding an anticholinergic to the second.

It might be noted that tardive dyskinesia is thought to be related, at least in part, to such an up-regulation of dopaminergic receptors in the basal ganglia. This might explain why these symptoms commonly become manifest on either reducing or discontinuing the neuroleptic medication and may improve by increasing the dosage (Keltner & Folks, 2005).

THE CHEMICAL NEUROTRANSMITTERS

Whether as a result of the accidental fermentation of grapes or the sampling of mushrooms in the search for sustenance, mankind likely has been aware of the capacity for certain (chemical) substances to have a profound effect on behavior even prior to recorded history. However, it was not until the 20th century that the nature of these interactions began to be understood. The nervous system was presumed to be a continuous structure until 1889 when Ramon y Cajal, using the recently developed Golgi staining technique, was the first to discover the presence of synapses. In doing so, he established that neurons were noncontinuous, individual units, but how these individual neurons communicated with one another still was somewhat of a mystery. The assumption seemed to be that it was likely electrical in nature. Around this time, there also was increased scientific interest in the realization that certain chemical substances consistently could induce specific (autonomic) behavioral responses. This eventually led to a hypothesis by T.R. Elliott in 1905 that the nerves themselves might release certain chemicals in producing their behavioral effects. However, it was not until 1921 that Loewi began to produce experimental data that eventually established the existence of the release of a behaviorally active neurochemical substance following electrical stimulation of the heart by the vagus nerve. This substance subsequently was identified as acetylcholine, yet it was not until the second half of the 20th century that all of the currently recognized major neurotransmitters were identified. The recognition that still other substances (e.g., neuropeptides) also may serve as neurotransmitters was an even later, and still ongoing, development.

Currently, several different classes of neurotransmitter substances have been identified. First, this includes those relatively small molecular amines and amino acids with which we are most familiar (see Table 11–1). These commonly are referred to as the "classic neurotransmitters." Other small molecular nucleotides (or nucleosides) and even one gas [nitric oxide (NO)] also have been found to have neurotransmitter properties in certain circumstances. Larger molecular substances such as peptides and hormones more recently have been identified as putative neurotransmitters.[11]

Other ways of classifying neurotransmitters had been to identify them as either being *fast* or *slow* transmitters or as *neurotransmitters* versus *neuromodulators*. Recall that earlier we spoke of certain neurotransmitters that activated ligand-gated ion channels, most notably glutamate and GABA. Because of the relatively simple and direct action of the ligand-gated receptors, the effect of the binding by the neurotransmitter occurs quite rapidly (within a few milliseconds). Thus, glutamate and GABA are referred to as *fast transmitters*. The action of acetylcholine (ACh) also is relatively fast at its nicotinic receptors (e.g., neuromuscular

Table 11–1. List of Neurochemical Transmitters[a]

Amines	*Amino Acids*
Acetylcholine (ACh)	Excitatory
	Glutamate
Monoamines	Aspartate[d]
	Inhibitory
Catecholamines	Glycine[e]
Norepinephrine (NE)	Gamma-aminobutyric acid (GABA)
Epinephrine[b]	
Dopamine (DA)	
Indolamines	*Peptides/Hormones*[f]
Serotonin (5HT)	
Histamine[c]	Endorphin
	Enkephalin
	Kassinin
	Neurokinin (A and B)
	Neurotensin
Other	Oxytocin and Vasopressin
	Somatostatin
Adenosine	Substance P
Adenosine triphosphate (ATP)	Thyrotropin-releasing
Nitric oxide (gas)	hormone

[a]Not all of these will discussed in detail as relatively little is currently known regarding their role(s) in neurochemical transmission.

[b]Epinephrine, which has important implications for cardiac function, appears to have minimal effects on the CNS compared to NE.

[c]Histamine, not considered one of the major neurotransmitters, most often is associated with immunologic, cardiovascular, and smooth muscle responses in the gastrointestinal and pulmonary systems. However, it is not without CNS effects. Blockage of H-1 receptors (probably in the ventral, posterior hypothalamus) by certain drugs may produce sedative effects (e.g., tricyclic antidepressants, phenothiazines, and first generation antihistamines like *hydroxyzine* (Atarax) and *diphenhydramine* (Benadryl)).

[d&e]While aspartate and glycine have excitatory and inhibitory effects, respectively, on the CNS, their role would appear to be relatively minor compared to the more abundant glutamate and GABA.

[f] This is only a partial list of the various neuropeptides or hormones that may have neurotransmitter functions (for more complete listing, see Rockhold, 1997).

junctions). Conversely, the monoamines (e.g., NE, DA, 5-HT) bind to G-protein-linked receptors. Because of the intermediary processes involved (see above), the elapsed response time is much greater (perhaps 100 milliseconds or more); hence, their designation as *slow transmitters*. Also, as a result of these slower times, these slower neurotransmitters have more time to interact with, and hence modulate, other synaptic transmissions. As a result, these monoamines or G-protein-coupled transmitters are sometimes referred to as *neuromodulators*. As we shall see later on, having multiple transmitters, acting at different speeds, and having the capacity to modulate or regulate the activity of other transmitter substances (or receptors) provides the nervous system with far greater flexibility and range of responses, beginning at the synaptic level. Most of the neuropeptides and hormones, which often work in association with other neuromodulators, also fall within this category. Certainly some hormones can have a protracted effect on the nervous system. While the term *neurotransmitter* by default may be reserved for those fast acting transmitters, it typically is applied to the entire range of substances that are capable of transferring information from one neuron to the next (i.e., including the "neuromodulators").

At this time the more well-known neurotransmitters (and neuromodulators) will be reviewed, focusing on their origins and modes of action, distribution within the CNS, putative role(s) in behavior, and the clinical effects resulting from disturbances or imbalances of these substances in the brain. Where appropriate, a brief description of the use of psychoactive drugs in relation to these neurotransmitters will also be discussed. Included here is either how drugs may disrupt certain chemical systems or how they may be used in an attempt to correct "suspected" imbalances.

Acetylcholine

As previously noted, acetylcholine (ACh) was the first neurotransmitter to be identified. It is relatively prominent in both the peripheral and central nervous systems. Acetylcholine is produced in the cytoplasm of the neuron by combining choline with acetyl CoA, through the intervention of the enzyme *choline acetyltransferase*. Once released into the synaptic cleft, it is quickly degraded by the action of the enzyme, *acetylcholinesterase*. Unlike some other neurotransmitters, ACh is not taken back up into the presynaptic membrane by high-affinity transport systems, however, the choline resulting from its breakdown in the synaptic cleft is recycled back into the cholinergic neuron through such systems.

Receptors

Two general classes of cholinergic receptors have been identified: **muscarinic** and **nicotinic**. The **muscarinic receptors**, of which several subtypes have been identified, appear to be largely G-protein-type receptors (i.e., slower acting) and are found mostly in the postganglionic parasympathetic nervous system, as well as in the CNS. Because of their association with the parasympathetic system cholinergic stimulation tends to be associated with increased gastric secretions, bronchoconstriction, decreased rate and inotropic (force) contractility of the heart, relaxation of gastrointestinal and genitourinary sphincters, pupillary constriction, and penile erection. In the brain, there is some evidence that ACh is important for both cognitive as well as motor functions (see below).[12] The effect of ACh on muscarinic receptors may be either excitatory or inhibitory, depending on the receptor subtypes and their location (e.g., M_1 and M_3 receptors are more likely to be excitatory, whereas M_2 are more likely to be inhibitory, as in the case of cardiac innervation).

Nicotinic receptors (so designated because they were found to respond to nicotine, as opposed to muscarine) primarily are found in the preganglionic cells of both the sympathetic and parasympathetic nervous system. They also are found at the neuromuscular junctions of striated muscles (those that are responsible for carrying out voluntary motor activity). Like the muscarinic receptors, they also are found in the CNS, but tend to be more concentrated in the spinal cord, in contrast to the greater predominance of muscarinic receptors in the brain. Recent studies, however, have suggested that nicotinic receptors in the brain also may play an important role in cognition, perhaps both directly as well as indirectly through their ability to influence other receptors and transmitters (Levin & Simon, 1998). Whereas muscarinic receptors are of the G-protein-coupled type, nicotinic receptors appear to be largely of the excitatory ion channel type. Among other things, this means that their response times tend to be quicker, as might be expected for innervation of striate muscles.

Distribution in the CNS

As noted above, the preganglionic fibers of both the sympathetic and parasympathetic pathways are cholinergic. As might be expected because of the role of ACh at the neuromuscular junctions, the motor cells in the ventral horn of the cord as well as the motor nuclei of the brain stem also are cholinergic, as are the Renshaw cells in the cord. These latter interneuron cells in the cord are important in providing feedback to associated muscle groups in the wake of voluntary or reflex motor activity. With regard to higher-order functions, there would appear to be two general cholinergic pathways that are of major importance. One originates in the **pedunculopontine nucleus** of the midbrain at the level of the inferior colliculus and projects rostrally to the thalamus and globus pallidus as part of the reticular system. The second and more extensive pathway, as can be seen in Figure 11–11, appears to have its origin in the basal forebrain, especially in the area of the **medial septal nuclei** and more ventrally in the **nucleus basalis of Meynert.** The latter is represented by distinct groups of cells in the area referred to as the *substantia innominata,* which encompasses the gray matter lying below the anterior commissure. Cholinergic neurons emanating from these latter areas project to diffuse areas of the neocortex, to the hippocampus (probably via the fornix), and to the amygdala. Finally, in addition to these longer projection pathways, there are local or intrinsic cholinergic neurons that originate in and whose projections are confined to the striatum (caudate and putamen). Although apparently confined to the striatum, these latter cholinergic neurons nevertheless may influence outside structures as a result of their action on GABA or the monoamine transmitters (DA, 5-HT, NE) (Hasey, 1996).

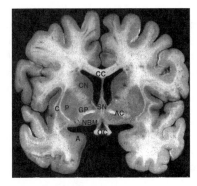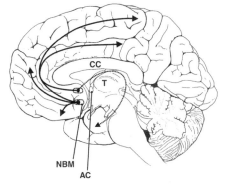

Figure 11–11.   Distribution of acetylcholine in the CNS. Cholinergic fibers emanate from the area of the nucleus basalis of Meynert (NBM) and from around the septal nuclei (SN) and project to the cortex. Abbreviations: A, amygdala; AC, anterior commissure; C, claustrum; CC, corpus callosum; CN, caudate nucleus; GP, globus pallidus; OC, optic chiasm; P, putamen T, thalamus. Brain image (left) was adapted from the *Interactive Brain Atlas* (1994), courtesy of the University of Washington.

Probable Role(s) of Acetylcholine in Behavior

In general, ACh is known to affect motor systems, cognition, and most likely general arousal. We already have alluded to its role in motor systems:

1. The primary neurotransmitter at the neuromuscular junctions involved in voluntary movement.
2. Affects smooth muscles (via the postganglionic neurons of the parasympathetic nervous system).
3. Works in conjunction with dopamine and other neurotransmitter substances in the basal ganglia.

Because of acetylcholine projections to the thalamus from the midbrain reticular system, it also is thought to be important in arousal or sleep/wake cycles. Finally, because of its diffuse projection to the neocortex and areas in the medial temporal lobes, ACh probably is important for memory and cognition. This latter assumption has been supported primarily by two lines of evidence. The first was experimental data showing that the administration of **physostigmine** (a cholinergic agonist) tended to enhance memory performance, while **scopolamine** (an anticholinergic drug) was shown to result in confusion and decreased memory (Davis et al., 1978; Bartus et al., 1985; Beatty, Butters, & Janowsky, 1986; Kopelman & Corn, 1988; Thal, 1992).[13] The second is evidence for degeneration of cholinergic neurons in the basal forebrain in Alzheimer's disease and the negative correlation between choline acetyltransferase (required in the synthesis of ACh) and cognitive abilities (Perry et al., 1978; Whitehouse et al., 1981; Coyle, Price, & DeLong, 1983; Dekker, Conner, & Thal, 1991; Bierer et al., 1995; Mattson & Pedersen, 1998).

Clinical and Drug Effects Associated with Cholinergic Mechanisms

As we shall see is true of many neurotransmitters, either relative depletions or excess amounts of ACh in selective pathways of the CNS can result in behavioral anomalies. As noted above, degeneration of cholinergic neurons in the basal forebrain has been linked to Alzheimer's disease and possibly to other dementing and neuropsychiatric illnesses (Chozick, 1987; Berger-Sweeney, 1998; Sarter & Bruno, 1998; Sarter, Bruno, & Turchi, 1999). While depletion of ACh is not necessarily the only or perhaps even the primary factor in this or other dementias, at this time it is assumed to play a critical role. It is this assumption that led to the use of **tacrine** (Cognex) and later **donepezil** (Aricept), **rivastigmine** (Exelon), and **galantamine** (Reminyl) for the partial management of the cognitive and behavioral deficits associated with Alzheimer's disease. All of these are centrally acting acetylcholinesterase inhibitors, thereby increasing the availability of synaptic ACh. While the effects are not always dramatic, some benefits have been established, especially when given in the earlier stages of the disease process (Rogers & Friedhoff, 1996; Keltner & Folks, 2005).

At this point, it is important to emphasize that although it is common practice to speak of a particular behavioral disorder as being related to a deficiency or excess of a particular neurotransmitter, any given disorder or its behavioral effects generally are not believed to be exclusively related to the disturbance of any single chemical abnormality. Single synapses may utilize multiple neurotransmitters. Even more, we expect that complex behaviors (that likely involve millions of synapses) are governed by the interactions of multiple neurotransmitters. As we have seen, one neurotransmitter may have a modulating influence on other transmitters, thus changes in the concentration of one may affect the availability of others. Furthermore, there is evidence that neurotransmitters counterbalance one another in many behavioral systems, so that a relative depletion of one may throw the system off balance. Such would appear to be the case with acetylcholine in **Parkinson's disease** and other dystonic reactions.

As previously discussed, ACh is known to be involved in the local circuitry of the basal ganglia, along with dopamine, glutamate, GABA, and various neuropeptides. Although the hallmark of Parkinson's typically is characterized by the relative depletion of dopaminergic neurons (see below for additional discussion), this situation would appear to create at least one other problem, namely an imbalance between dopaminergic and cholinergic systems. This situation likely is exacerbated by the fact that dopamine and acetylcholine reciprocally are related in the nigrostriatal pathways, so with either the blockage of or reduction in dopaminergic systems one might expect a concomitant increase in ACh (Stahl, 2000). This would explain why before the development of DA replacement therapy, anticholinergics such as **benztropine** (Cogentin) and **trihexyphenidyl** (Artane) were used to treat Parkinson's.[14] A related phenomena can be seen with the use of antipsychotic drugs, particularly with the high-potency, "typical" neuroleptics, such as **haloperidol** (Haldol) and **fluphenazine** (Prolixin), which tend to be associated with "extrapyramidal" and dystonic reactions. In this case, the effect is drug-induced, resulting from the blockage of dopamine receptors in the nigrostriatal pathways. Again, the use of a second anticholinergic drug such as Cogentin or Artane often is prescribed prophylactically to correct this potential imbalance between ACh and DA.[15]

While most of the recent emphasis with regard to the use of AChE inhibitors has focused on CNS disorders, there is at least one condition, **myasthenia gravis**, where such drugs act at the periphery. Myasthenia gravis is an autoimmune disorder characterized by depletions of acetylcholine receptors at the neuromuscular junction, resulting in progressive muscular weakness. Among other treatments, AChE inhibitors (e.g., **pyridostigmine**) have been used to increase the availability of ACh at the neuromuscular junction, thereby facilitating muscular contraction (i.e., strength).

We just have seen that anticholinergic drugs (e.g., Artane, Cogentin) have been effectively used in situations where there is a suspected imbalance between cholinergic and dopaminergic activity in the CNS. We also have seen where just as too much ACh can be disruptive to normal function, so can too little. Unfortunately, many drugs that are used to treat a variety of medical and neuropsychiatric conditions also have anticholinergic properties among their side effects.[16] These effects generally can be subsumed under one of two broad categories: central and peripheral. The "peripheral" effects, which are thought to be due to the blocking of muscarinic receptors, include dry mouth, blurred near vision (primarily resulting from sluggish accommodation by the lens), urinary retention, constipation, and possible tachycardia at higher dosages (low-dose anticholinergics may result in a relative bradycardia). As might be surmised, these side effects result from the blockage of parasympathetic (muscarinic) receptors at the end organs. Because of their capacity to also block muscarinic receptors in the CNS (i.e., the brain), drugs possessing anticholinergic properties also can result in confusion, memory disturbances, and sedation, although the latter often primarily is due to the concomitant blocking of histaminic receptors by some of these same drugs. Both the peripheral and central anticholinergic effects generally pose greater difficulties for the elderly or, as noted earlier, for those whose central cholinergic systems already may be compromised (e.g., patients suffering from Alzheimer's disease).

Glutamate

Glutamate is considered the primary excitatory neurotransmitter in the CNS (Hicks & Conti, 1996). **Aspartate** is another amino acid, excitatory neurotransmitter, but with more limited effects. Glutamate, like most other amino acids, is involved in protein synthesis and other metabolic activities within the neuron. Its role as a potential mammalian neurotransmitter was not appreciated until the 1950s, when it was noted that L-glutamate applied directly to the surface of the brain resulted in convulsions (Hyashi, 1954). Because its effects on nervous tissue were found to be so diffuse and apparently nonspecific, it was years later before it was recognized as a unique neurotransmitter. It is thought that glutamate (L-glutamate) is synthesized from glutamine by the actions of two enzymes found in the presynaptic terminal: glutaminase and aspartic acid aminotransferase. Glutamate then is stored in presynaptic vesicles where it awaits release. As noted earlier, unlike other neurotransmitters, once released glutamate is not broken down by enzymatic action in the synaptic cleft. Rather it is transported intact either back into the presynaptic terminal or into nearby glial cells. Within these glial cells, glutamate is transformed into glutamine (via the enzyme **glutamine synthetase**). This glutamine then is transported back into the neuron where it once again can be converted into glutamate.

Ionotropic Receptors

Glutamate receptors are divided into two broad categories—ionotropic (ligand-gated ion channel) and metabotropic (G-protein-linked) receptors—each having multiple identified subtypes. As their primary purpose, ionotropic receptors have to permit an influx of ions through the cell membrane. In contrast to GABA receptors that allow the passage of negatively charged chloride ions and create inhibitory postsynaptic potentials (IPSPs) (see below), glutamatergic ionotropic receptors permit the flow of catons[17] [e.g., calcium (Ca^{2+}), sodium (NA^+), and probably to a lesser extent, potassium (K^+)]. The influx of positively charged ions, primarily Na^+, into the postsynaptic neuron have a depolarizing effect on the cell, setting up excitatory postsynaptic potentials (EPSPs) (see Appendix).

The three major subtypes of glutamatergic ionotropic receptors that have been identified are the **NMDA** (N—methyl-D-aspartate), **AMPA** (alpha-amino-3-hydroxy-5-methyl-4-isoxazolepropionic acid), and **kainate** receptors. The NMDA is primarily a voltage-dependent calcium channel receptor.[18] In addition to being the primary conduit for Ca^{2+} ions, the NMDA receptor has a number of other unique features among the glutamatergic ionotropic receptors. First, being voltage-dependent, even when bound by glutamate, the NMDA channel does not open until the postsynaptic membrane has been partially depolarized (to about –30mV), typically as a result of AMPA or kainate receptor activation. This is thought to be the result of a blockade by magnesium (Mg^{2+}) ions, which are released when the cell becomes less polarized. As was noted earlier to be the case with the GABA receptors (see also below), the NMDA receptors have sites for several allosteric modulators, including the aforementioned magnesium, glycine, zinc, polyamines, and PCP (phencyclidine). The presence of glycine appears necessary for the opening of the NMDA channel, and along with zinc and the polyamines appears to enhance its response. Phenyl-cyclidine, like magnesium, can cause a voltage-dependent blockade of the NMDA receptor. Finally, the NMDA receptor has a slower and longer-lasting response than do the AMPA and kainate receptors. This latter characteristic of the NMDA receptor likely contributes to long-term potentiation effects (prolonged excitatory postsynaptic potentials) seen particularly in the hippocampus and thought to be associated with learning and memory (Bliss & Collinridge, 1993). On the downside, the NMDA receptors likely play a key role in the development of **excitotoxicity** (see below).

In contrast to the NMDA receptors, the AMPA and kainate receptors primarily are monovalent cation receptors. Although serving as a primary channel for sodium (Na^+) to enter the cell, depending on the presence of specific subunits the AMPA receptor, for example, may admit Ca^{2+} ions. Also, as opposed to NMDA ion channels, both AMPA and kainate receptors are chemical-dependent or ligand-gated (i.e., they can be activated, when bound by glutamate, regardless of the resting potential of the membrane). Finally, both have much shorter response and recovery times compared with the NMDA receptors, particularly the AMPA ion channels. As a result, the AMPA and kainate receptors are thought to mediate fast, highly specific responses, in addition to creating an environment (EPSP) conducive to NMDA conductance of calcium ions.

Metabotropic Receptors

The second broad class of glutamate receptors is designated as metabotropic receptors (mGluRs). These latter receptors, again of which there are multiple subtypes, are G-protein-linked receptors that utilize **PI** (phosphatidyl inositol) and **cAMP** (cyclic adenosine monophosphate) as **second messengers**.[19] Recall from the earlier discussion that rather than effecting a simple, direct change in the electrical potential of the cell membrane by opening ion channels, metabotropic (G-protein-linked) receptors generally produce more complex intracellular changes via second-messenger enzymes. The functional significance of these receptors still is unclear at this time. However, it is believed that they are important in the more general, tonic regulation of both pre- and postsynaptic excitability. If this indeed is the case, then these metabotropic receptors likely also play a critical role in such processes as long-term potentiation and neuronal (synaptic) plasticity, as well as serving a neuropro-tective function (i.e., prevention of excitotoxicity) (MacDonald, Wojtowicz, & Baskys, 1996).[20]

Distribution in the CNS

Although the areas to which they project may be quite diffuse, the cell bodies for ACh and monoamine[21] neurons are localized in discrete, typically subcortical regions of the CNS. They rely on axonal transport to convey the enzymes and/or precursors necessary for the synthesis (and degradation) of the specific neurotransmitter at the nerve terminals. In the case of the amino acids, such as glutamate, there is a more ubiquitous distribution of these neurons throughout the CNS, both at cortical and subcortical levels. This is particularly true of the NMDA, AMPA, and metabotropic receptors. The kainate receptors, which may have a somewhat more restricted distribution, are noted to be particularly plentiful in the hippocampus and other subcortical regions. Specific subtypes of each of these receptors may show even greater selectivity, although their functional specificity is not always clear (Cotman et al., 1995; MacDonald et al., 1996).

Probable Role(s) of Glutamate in Behavior

As suggested earlier, glutamate acts as an amino acid in cell metabolism, including serving as a precursor in the synthesis of GABA, as well as functioning as a neurotransmitter. As a neurotransmitter, its primary purpose appears to be to depolarize (activate) postsynaptic neurons in the brain and central nervous system. For example, glutamate appears to have an excitatory effect throughout the brain, including cortical commissural, association, and projection pathways; thalamocortical and limbic connections; and cerebellar afferents and granule cells. As a neurotransmitter, glutamate's effects probably are not limited to the CNS as it also may be employed as an excitatory transmitter by peripheral sensory fibers, as well as at some neuromuscular junctions (Cooper, Bloom, & Roth, 1991; Penny, 1996). However, not all of its actions may be excitatory; some presynaptic metabotropic receptors would appear to exert an inhibitory influence on postsynaptic selective fields (Schoepp & Conn, 1993). Additionally, as was seen in discussions of the basal ganglia (see Chapter 8), glutamate may synapse on and thus activate inhibitory (GABAergic) neurons. By exciting these inhibitory neurons, glutamate can exert an inhibitory influence on the neuronal fields to which these GABAergic neurons project.[22]

As previously noted, glutamatergic activation of both NMDA and metabotropic receptors have been associated with long-term potentiation or prolonged tonic changes in postsynaptic potentials (Bortolotto & Collinridge, 1992; Rison & Stanton, 1995). Behaviorally, these latter phenomena have been associated with learning and memory or *synaptic plasticity*. While the number and general arrangement of neurons may be more or less genetically determined, the nature and complexity of their dendritic connections is thought to be determined largely by learning or experience. The excitation of neuronal networks that occur as a result of these experiences is thought to promote the establishment of new or more elaborate dendritic arborizations (synapses). This process is believed to be critical during early development, where it may help explain the effects of enriched versus deprived environments in early studies of neuronal plasticity (LoTurco, Blanton, & Kriegstein, 1991; Katz & Shatz, 1996; Kriegstein, Flint, & LoTurco, 1996; McAllister, Katz, & Lo, 1999). Such neuronal excitation probably also is important not only during adulthood (learning) but even in the elderly. It has been suggested, for example, that continued mental and sensory stimulation may help preserve the functional integrity of the brain in the elderly (Katz, 1999). Finally, as will be seen in the next section, just as under certain circumstances excessive stimulation may be associated with a variety of brain diseases or pathological conditions, it also has been suggested that the reverse may occur. That is, lack of stimulation in the adult brain may lead to loss of dendritic connections (i.e., a decline in mental ability). In summary, glutamate and its excitatory impact on neuronal tissue not only are critical for sensory, motor, and higher cognitive activities within the brain, they likely also serve to expand the potentials of the developing brain, enable one to expand their knowledge throughout the life span, and may help ward off the ravages of age in the elderly.

Clinical and Drug Effects Associated with Glutamatergic Mechanisms

The importance of calcium in the skeletal and muscular systems is well known. Actually calcium is critical for most organ systems, including as we have just seen the neurological system. The firing of all neurons is made possible in part by the depolarization resulting from an influx of calcium. However, like chocolates and cream cheese, there can be too much of a good thing. If too much calcium is permitted to enter the neuron, it can have a toxic or even deadly effect. When this happens, the phenomena are referred to as *excitotoxicity*. First described by Olney et al. (1971a,b), the primary candidates that might contribute to the development of this condition are either (1) an increase in extracellular glutamate, or (2) toxic or chemical imbalances that affect the modulation of the NMDA receptor. The increase in extracellular glutamate may result from a breakdown in the presynaptic membrane or a failure of glutamate reuptake transport mechanisms, both of which might result from injury or disease. Whether due to an increase in extracellular glutamate or a toxic mechanism, the end result is the same, namely, an increased infusion of Ca^{2+} into the cell, which is potentially destructive (Choi & Rothman, 1990; Beal, 1992a; Choi, 1992; Rosenberg, Amin, & Leitner, 1992; Lipton & Rosenberg, 1994; Martin, Lloyd, & Cowan, 1994; Paschen, 1996; Fornai et al., 1997; Tapia, Medina-Ceja, & Pena, 1999).

Excitotoxicity has been implicated in either the development or subsequent exacerbation of various neurological disorders, including **epilepsy**, **AIDS dementia complex**, **ischemia** or **hypoxia**, **amyotrophic lateral sclerosis**, **olivopontocerebellar atrophy**, and other primary **neurodegenerative disorders**, such as Alzheimer's, Parkinson's, and Huntington's diseases (DiFiglia, 1990; Beal, 1992b, 1998; Lipton & Rosenberg, 1994; Choi, 1995; Blandini et al., 1996; Good et al., 1996; Paschen, 1996). As a result of these observations, there are current attempts to study the potential utility of NMDA receptor antagonists in the prevention of neuronal loss in these various conditions (Blandini et al., 1996; Schousboe, Belhage, & Frandsen, 1997; Fornai et al., 1997; Marsden & Olanow, 1998; Doble, 1999; Simon & Standaert, 1999).

One drug, memantine (Namenda), a NMDA receptor antagonist, recently was approved for use in the United States for the treatment of Alzheimer's disease. Although targeting multiple neurochemical receptors, its effectiveness is thought to be due to its ability to block NMDA receptors in the brain. During "resting" phases, magnesium serves to transiently block NMDA receptors, preventing transport of excessive calcium into the cell, but temporarily is displaced by normal glutamatergic activity. An example of the latter might be during attempted information recall or other cognitive activity. However, as noted above, in certain disease states (like dementia) there can be excessive amounts of glutamate leaking into the extracellular spaces. This excess tends to displace the magnesium ions, thus leading to an increase of intracellular CA^{2+}. These effects on the cell are twofold. First, it renders the neuron less sensitive to normal transmission due to an excitotoxic state (or what has been termed excessive "background noise"). Second, as was seen, excessive influx of CA^{2+} ions eventually can lead to cell death. Memantine acts like magnesium in that it also blocks the NMDA receptor, but its antagonism is stronger than that of magnesium. Thus it continues to block the pathologically chronic low levels of glutamate, while still allowing the cell to respond to normal glutamatergic stimulation. As a result, there is commonly some more or less immediate improvement in routine information processing (at therapeutic levels). In addition, NMDA antagonism might be expected to reduce the long-term effects of excitotoxicity (i.e., cellular degeneration (Danysz et al., 2000).

For some time now there has been strong reason to suspect monoamine (particularly, DA and 5-HT) imbalances as an etiological mechanism in the development of schizophrenic disorders. This hypothesis will be reviewed in greater detail later in this chapter. For now, however, it also should be noted that even recently a relative depletion of glutamate has been the focus of attention as a possible contributing factor in schizophrenic illness (Bunney, Bunney, & Carlsson, 1995). The possible association between glutamate and schizophrenia initially was based on the observation that PCP (phencyclidine) when ingested by "normal" individuals could produce a schizophreniclike psychosis. This hypothesis was strengthened when it was discovered that other NMDA antagonists (e.g., ketamine, an anesthetic agent) could produce similar effects (Lodge et al., 1987). Although there has been some variation with regard to the specific nature of the lesion and/or behavioral pathology involved, including excitotoxicity, the underlying assumption is one of a hyporeactivity of NMDA receptors (Coyle, 1996; Hirsch et al., 1997; Farber et al., 1998; Tamminga, 1998, 1999a; Tsai et al., 1998).

If a relative depletion of glutamate indeed is one of the etiological factors in schizophrenic process, should this not prevent rather than contribute to excitotoxicity? Recall that GABAergic neurons may be stimulated by glutamatergic receptors. If there is a depletion of glutamate in certain GABAergic pathways, this can result in a disinhibition of other "downstream" pathways, resulting in enhanced rather than diminished neural activity. It has been postulated that this selective disinhibition of certain frontolimbic pathways during early development could result in excitotoxic lesions that lay the groundwork for the development of a schizophrenic reaction later in life (Deakin & Simpson, 1997). A similar mechanism, that is disinhibition of the thalamus with resulting failure to selectively activate the cortex, also has been postulated to help account for the difficulty schizophrenic patients often have in screening out irrelevant stimuli (Carlsson et al., 1999). Relative depletions of glutamate in the frontal cortices also have been used to explain some of the negative symptoms of schizophrenia (Coyle, 1996).[23] Thus, it seems likely that glutamatergic mechanisms may be involved not only in early developmental changes in the brain that establish a lifelong vulnerability to schizophrenia and contribute to its expression later in life, but also may provide clues to new treatment approaches (e.g., use of selective glutamate agonists or conversely use of GABA antagonists). These various elements are interwoven into a rather comprehensive theory concerning the role of glutamate in the development and expression of schizophrenia by Olney and Farber (1995).[24]

Gamma Aminobutyric Acid

GABA, like glutamate, is an amino acid that is found throughout the central nervous system. It has been estimated that approximately 40% of all synaptic terminals within the brain are GABAergic (Bloom & Iversen, 1971). In contrast to glutamate, GABA receptors exert an inhibitory influence on postsynaptic neurons. Although overly simplified, even casual consideration suggests the importance of inhibition in the central nervous system. We already have noted (Chapters 3 and 8) the role of inhibition in the coordination and execution of motor activities and in preventing unwanted spontaneous motor actions (e.g., tremors, ballismus). On the input side, it would be pure chaos if our brain attempted to give equal weight to all of the stimuli to which the organism is subjected at any given moment in time. Those that are deemed irrelevant are "filtered" from conscious attention. Similarly, thought processes, if they are to follow predicable paths and arrive at logical conclusions, must remain goal-directed, suppressing divergent or unproductive tracks. Inhibitory mechanisms are ideally suited to these tasks. In fact, as the organism becomes more complex, with a greater range of potential cognitive and behavioral responses, the need for more elaborate inhibitory capacity would be expected to increase accordingly. This may underlie the observation reported by Hicks and Conti (1996) that the density of GABA receptors appears roughly correlated with phylogenetic development.

Interestingly, GABA, the primarily inhibitory neurotransmitter in the CNS, is synthesized from glutamate, the primary excitatory transmitter. This results from the removal of a carboxyl group[25] from the glutamate molecule by the action of the enzyme glutamic acid decarboxylase. Like other neurotransmitters, GABA is stored in synaptic vesicles until it is released (by depolarization of the presynaptic terminal as a result of an action potential). Following its release into the synaptic cleft, where it may bind to either pre- or postsynaptic receptors, it is transported back into the presynaptic terminal (or into nearby glia cells) where it is catabolized by the enzyme GABA transaminase.

GABA_A Receptors

Two major types of GABA receptors have been identified. These are the $GABA_A$ and the $GABA_B$ receptors,[26] and they may be found in either the pre- or postsynaptic membranes.[27] $GABA_A$ are among the more complex receptors known. A number of drugs with significant neuropsychiatric implications, including anxiolytic, muscle relaxant, anticonvulsant, and sedative effects, operate at these sites. $GABA_A$ receptors are ligand-gated ion channel (ionotropic) receptors. In contrast to glutamate, which permit the flow of positive ions (Ca^{2+}, Na^+) into the cell, creating an EPSP, GABA receptors allow the passage of additional chloride ions (Cl^-) through the cell membrane (by opening additional Cl^- channels). Recalling that the resting potential of the neuron already is negative, the influx of Cl^- effectively hyperpolarizes the postsynaptic membrane, creating an **inhibitory postsynaptic potential** (IPSP). This, in turn, increases the threshold for the initiation of an action potential, which is why GABA is classified as an inhibitory neurotransmitter.

While this action might appear relatively straightforward, the $GABA_A$ receptor itself is quite complex, being composed of various combinations of multiple subunits (designated as *alpha*, *beta*, *gamma*, *delta*, and *rho*). These subunits make up the ion channel itself, as well as create binding sites, not only for GABA but also allosteric binding sites for several other molecules. Among the latter are sites that respond selectively to benzodiazepines (BZD), barbiturates, alcohol, and certain anesthetics. While GABA itself must be present for the ion channel to open, in the presence of one or more of these other drugs, the GABAergic response is enhanced. This positive allosteric response results in increased hyperpolarization of the postsynaptic membrane.[28] These allosteric sites, in addition to enhancing the effect of GABA, also are responsible for some of the additional therapeutic and/or side effects associated with these drugs. Thus a BZD, which may be chosen for its anxiolytic effects, also may produce muscle relaxation (potentially a positive side effect) or sedation, ataxia, and decreased memory (possible negative side effects).[29] Whether a given GABA receptor will respond to one or another of these drugs (e.g., BZDs versus barbiturates versus ETOH) depends on the particular configuration of the receptor molecule itself (which subunits are present to configure the specific binding site). In turn, the response profile of some of these drugs (e.g., BZD versus barbiturates versus ETOH) will depend on the specific localization (within the CNS) and characteristics of the binding site itself. Not only are there variations in the combinations of subunits making up the receptor complex, but also there are multiple variations within the subunits themselves. This helps explain why the different types of BZD or barbiturates have different response and side effect profiles. Capitalizing on these differences, drug companies can attempt to develop designer drugs that will have the desired therapeutic effect, while eliminating undesirable side effects by targeting select GABA/BZD receptors.

Not all GABA-philic drugs act to enhance the GABAergic response. **Bicuculline** is a competitive antagonist that works at the GABA binding site. Because it tends to block the effect of GABA, it effectively decreases the frequency with which the Cl^- channel is opened, as well as its duration. **Picrotoxin** is classified as a noncompetitive GABA receptor antagonist. Rather than competing with GABA for its binding site, it appears to act by blocking the chloride channel, effectively blocking GABA's inhibitory influence. Because it acts at a site other than that which normally is occupied by the primary agonist (in this case, GABA) and because its action is opposite that of GABA, picrotoxin would be considered to be an example of **negative allosteric modulation**. Some compounds may work as inverse agonists. Recall that an inverse agonist has a behavioral effect opposite that which an agonist would have at the same receptor site. While such compounds might be expected to have potential beneficial effects on memory, unfortunately they also tend to produce seizures and are anxiogenic. Because both bicuculline and picrotoxin tend to interfere with the normal inhibitory influence of GABA on neurons, they have the capacity to cause seizures. Consequently, these drugs are used only for experimental purposes (Sarter, Nutt, & Lister, 1995). On the other hand, **flumazenil** (Romazicon), a BZD antagonist (which means that it simply blocks the enhancing effects of benzodiazepines on GABA), is used therapeutically to reverse the sedation and psychomotor effects of midazolam following surgical procedures or in cases of suspected benzodiazepine overdose.[30]

GABA_B Receptors

$GABA_B$ receptors belong to the G-protein-linked (metabotropic) class of receptors that are found both pre- and postsynaptically. In contrast to the $GABA_A$ receptors, they are not associated with chloride ion channels. Rather, working through second-messenger systems, they (1) increase the permeability of the ion channels that permit K^+ to exit the cell, (2) reduce the influx of NA^+ and/or Ca^{2+} into the cell, and (3) inhibit the formation of cyclic AMP by inhibiting adenylyl cyclase, the enzyme responsible for converting ATP into cyclic AMP. All these effects are inhibitory in nature, for example:

1. Since K^+ is concentrated within the neuron during its resting state, by making it easier for K^+ to flow down its concentration gradient (i.e., out of the cell), this tends to hyperpolarize the interior of the cell.
2. Similarly, by making it more difficult for NA^+ and Ca^{2+} to diffuse into the cell (a depolarizing influence), this decreases the likelihood that the cell will fire.

3. A reduction of cyclic AMP (cAMP) within the postsynaptic membrane may set up enzymatic changes that could affect ion channels, but its probable effect on presynaptic membranes is clearer. $GABA_B$ receptors often are found on the presynaptic terminals of neurons that are responsible for the release of other neurotransmitters (heteroreceptors). As cAMP is critical to the synthesis of neurotransmitters within the cell, reducing cAMP in effect will inhibit the release of these other transmitters.

While generally much less is known about the $GABA_B$ receptors as compared to $GABA_A$, it is known that the $GABA_B$ receptors have a particular affinity for **baclofen**, a compound with antispasmotic properties. Also, because $GABA_B$ receptors are metabotropic, they tend to have a slower onset of response and a longer duration. Because of this slower, longer response profile and because they can inhibit the synthesis of other transmitters, their behavioral effects are more tonic in nature than those of $GABA_A$ receptors.

Distribution in the CNS

Like glutamate, GABA and its receptors are diffusely scattered throughout the CNS but are relatively limited in the peripheral nervous system. GABAergic neurons are especially prominent in the cerebrum and the cerebellum, whereas glycine, which serves a comparable inhibitory function, is found in greater proportions in the brain stem and spinal cord. Most GABA neurons within the CNS would appear to be short interneurons involved in local circuitry. Here they predominately have an inhibitory influence on postsynaptic neurons or inhibit the synthesis or release of GABA or other neurotransmitters through the mediation of presynaptic GABA receptors.[31] In addition to its influence on local circuits, projection neurons also may utilize GABA. Recall from Chapters 3 and 6 that the primary output of the cerebellar Purkinje cells and the caudate, putamen, and the globus pallidus largely are GABAergic. Diversity in the composition of individual GABA receptors (i.e., the presence or absence of specific subunits) in different parts of the CNS provides the structural basis for diversity in function and selective sensitivity to certain drugs (Wisden & Seeberg, 1992; Wisden et al., 1992).

Probable Role(s) of GABA in Behavior

As a neurotransmitter, first and foremost, GABA's primary role appears to be inhibition. As noted above, $GABA_A$ receptors contribute to the hyperpolarization of neurons by permitting an influx of Cl^- ions when stimulated. While $GABA_B$ receptors may have the same impact, as we just saw, they do so through different mechanisms. Found primarily in short interneurons within the CNS, GABA thus would appear to be important in screening out irrelevant and/or competing stimuli, associations, or responses at a very elementary level. By acting in opposition to the various excitatory mechanisms in the brain, GABA can be seen as providing one source of checks and balance, keeping the brain from "running amok," so to speak. Although not specifically linked to GABA depletion, as we shall see, seizures and panic-type disorders might be considered clinical examples of the latter phenomenon, and in fact these conditions do respond to GABA agonists.

In addition to inhibiting the discharge of neurons, GABA receptors, particularly $GABA_B$ receptors, likely influence the release not only of other neurotransmitters (e.g., the monoamines) but apparently also have an effect on the endocrine system, modulating the release of various hormones and hormone releasing factors (see Davies, 1996; Bowery, 1993). However, contrary to what might be expected, it has been suggested that in some instances GABA may actually facilitate rather than inhibit the release of monoamines. Thus, depression, for example, which is typically associated with relative deficiencies in monoamine transmitters, also might be associated with GABAergic depletions in local circuits (Petty, 1995).

If summarized from a strictly clinical perspective (particularly as judged from the pharmacological effects of GABA agonists), GABA receptors can be said to play a major role in (1) anxiolysis, (2) sedation, (3) seizure control, and (4) muscle relaxation. These effects will be discussed in greater detail in the next section.

Clinical and Drug Effects Associated with GABAergic Mechanisms

As noted earlier, multiple allosteric receptor sites have been identified on the GABA receptor. From a clinical perspective, among the more important ones appear to be those for (1) benzodiazepines (BZDs), (2) barbiturates, (3) certain neurosteroids, and (4) alcohol (ETOH). All of these can be found on the $GABA_A$ receptor.[32]

Barbiturates. One of the first drugs discovered to act at the $GABA_A$ receptors (aside from ETOH) are the barbiturates. Known for their sedating, hypnotic, and anticonvulsant properties, barbiturates still are used for certain types of epilepsy and for anesthesia and sedation during medical procedures. Although in the past barbiturates were frequently are used to treat anxiety and sleep disorders on an outpatient basis, they largely have been replaced by BZDs for this purpose because of the greater safety profile of the benzodiazepines. Nonetheless, barbiturates occasionally are still prescribed for this purpose. Further capitalizing on their ability to inhibit cortical neurons, shorter-acting barbiturates such as **amobarbital** (Amytal) or **thiopental** (Pentothal) are used in WADA testing to assess functional hemispheric asymmetries (see Chapter 9), and to elicit "suppressed" memories or information. The latter effect most likely results from a selective disinhibition of frontal systems. While WADA testing still remains a widely used clinical and experimental tool, there has been a general decline in the use of the *sodium pentothal* (or *sodium amytal*) interview, although occasional recent references still can be found (Cornell et al., 1996; Russo et al., 1997; Fackler, Anfinson, & Rand, 1997).

The current more cautious approach to the use of barbiturates stems not only from their greater propensity for inducing tolerance, dependence, and hepatic enzyme induction, but also from their greater danger in the event of an overdose, including the possibility of respiratory suppression. Their effectiveness (e.g., sedative potential) apparently derives from the fact that at lower dosages they act like BZDs in that they increase the influx of Cl^- (in this case, by prolonging the opening of the chloride channels). On the other hand, the greater risk they impose stems from the fact that at higher dosages the barbiturates can open the chloride channels independently of GABA. Thus, the upper limits of their inhibitory potential are determined primarily by the concentration (dosage) of the drug itself.

Benzodiazepines. Although acting in a similar manner to the barbiturates (i.e., augmenting the opening of the chloride ion channel) and having similar response profiles (i.e., anxiolysis, sedation/hypnosis, and anticonvulsant action), the benzodiazepines are much safer drugs. Unlike the barbiturates, the BZDs cannot open the chloride channels independently of GABA, regardless of dose (Teboul & Chouinard, 1990). One of the more widely prescribed psychotropic medications, the benzodiazepines have multiple clinical uses, all apparently stemming from their ability to bind to the GABA receptor. In general, all benzodiazepines potentially can produce all the effects mentioned above, although some effects (e.g., sedation) may be more prominent in some individual drugs than others. Selection frequently is based on considerations of potency, response latency, half-life of the drug, and available modes of administration. Among the conditions treated by BZDs are acute situational stress, generalized anxiety and panic disorders,[33] social phobias,[34] sleep disturbances,[35] muscle spasms, seizures, sedation in acute, psychotic agitation,[36] and preoperative sedation (Hyman, Arana, & Rosenbaum, 1995; Chapter 6; Levy, 1996; Nelson & Chouinard, 1996).

As noted above, ETOH, which has binding sites on GABA receptors, also tends to increase its inhibitory effects. This helps account for behaviors commonly associated with acute intoxication (e.g., motor incoordination and slurred speech, cognitive impairments, behavioral disinhibition, and sedation). With sustained ETOH use, however, there is a down-regulation of GABA receptors (Gilman et al., 1996; Brailowsky & Garcia, 1999). This chronic down-regulation of GABA receptors (and consequent diminished inhibitory capabilities) could explain some of the clinical effects seen during withdrawal states (e.g., muscle tremors, nervousness, sleep disturbances, dilirium tremens, seizures). Conversely, this also helps explain how the administration of a benzodiazepine (e.g., diazepam or lorazepam), which acts on these same receptors to maximize GABAergic responses, serves to attenuate these withdrawal effects. However, just as the chronic use of ETOH can result in increased tolerance, dependence, and withdrawal upon discontinuation of use, similar phenomena can occur following long-term, *indiscriminate* use of benzodiazepines.[37] While most if not all benzodiazepines have the potential for abuse and dependency, the greater danger lies in the long-term use of the higher potency, shorter-acting drugs, such as **lorazepam** (Ativan) and **alprazolam** (Xanax).

Although when used within prescribed guidelines, BZDs generally are safe, effective drugs, they can have unwanted side effects. This would be particularly true if used in excess or by individuals who have trouble metabolizing the drug (e.g., the elderly or those with hepatic disease).[38] As might be expected given the inhibitory effects of these drugs, among the more common side effects encountered with BZDs are fatigue, lethargy, impaired coordination, and reduced cognitive and memory functions (Ghoneim & Merwaldt, 1990; Woods, Katz, & Winger, 1992). However, as noted earlier, in certain cases these "side effects" can be employed to one's advantage, for example, the use of **midazolam** in certain medical procedures.[39] Finally, the onset of depressive symptoms, or worsening of preexisting depressive symptoms also can occur with the use of benzodiazepines.

Anticonvulsants and Mood Stabilizers. For many mood stabilizers and anticonvulsants, the mode of action is not clearly understood. There seems to be a general presumption that many of these drugs likely enhance GABA receptors, although this may not be their only effect. For some drugs with anticonvulsant and/or mood stabilizing properties, the GABA connection is either well known (e.g., **phenobarbital, diazepam,** and other BZDs) or is strongly suspected [e.g., **carbamazepine** (Tegretol), **divalproex** (Depakote), **clonazepam** (Klonopin), **tiagabine** (Gabitril), **topiramate** (Topamax), and **lithium**]. Interestingly, **gabapentin** (Neurontin), which structurally is very similar to GABA, does not appear to act at GABA receptors, although it is thought to increase intracellular concentrations of GABA (Fisher, 1998).

Clinical Syndromes

The role of GABA in the treatment, if not in the etiology, of some types of epilepsy and anxiety disorders already has been suggested. However, unlike acetylcholine deficiencies or excitotoxicity resulting from excesses of glutamate that have been associated with dementia processes, as we shall see shortly, of the various neuropsychiatric disturbances associated with aberrations of monoamine systems, specific disruptions of GABA transmission are less frequently associated with particular syndromes. However, it has been suggested (Mann & Kupfer, 1993; Shiah & Yatham, 1998) that GABA deficiencies may contribute to depression or other mood disorders.

Although the current trend is to link schizophrenic symptomatology primarily to disturbances of dopaminergic and serotonergic systems (see below), GABA interneurons are seen as important not only for neurodevelopment, but also in maintaining a homeo-static balance among the various neurochemical systems. Disturbances in this neurochemical network resulting from disruptions of GABAergic mechanisms are believed to potentially contribute both to one's predisposition to schizophrenia as well as to expression of its symptoms (Benes, 1997; Keverne, 1999). Following this logic, it has been proposed that GABA agonists may have an adjunctive role in the treatment of schizophrenia. While the inclusion of BZDs in treatment of this disorder has yielded positive results in some cases, their effect apparently tends to be minimal (except in calming acute agitation, where they can be very effective). However, in some instances they actually may exacerbate symptoms (Wolkowitz & Pickar, 1991; Hyman, Arana, & Rosenbaum, 1995; Levy, 1996; Stimmel, 1996).

The one neuropsychiatric disorder that long has been associated with relative depletions of GABA in the basal ganglia is **Huntington's disease** (Bird et al., 1973; Perry, Hanson, & Kloster, 1973). In Huntington's disease, there is a significant loss of neurons in the striatum (putamen and caudate), particularly of GABAergic medium spiny neurons, although these reductions in GABA neurons alone cannot fully explain the resulting symptoms. More recent investigators have focused on disruptions of the balance between GABA, dopamine, and other neurotransmitters, both within the basal ganglia as a whole and in the globus pallidus in particular (Albin, Young, & Penney, 1989; Pearson et al., 1990; Storey & Beal, 1993; Jankovic, 1998).[40] While losses of GABA routinely are found in the basal ganglia, including the neostriatum (putamen and caudate), GABA would appear to be relatively well preserved in the neocortex in patients with Huntington's (Bonilla, Prasad, & Arrieta, 1988; Reynolds, Pearson, & Heathfield, 1990; Storey et al., 1992). Although various attempts have been made to improve symptoms using some form of GABA-enhancing drugs, these attempts have met with relatively little if any success (Barr et al., 1978; Shoulson et al., 1978; Perry et al., 1982).

THE MONOAMINES

As previously shown in Table 11–1, the monoamine transmitters include the centrally active **catecholamines**, *norepinephrine* (NE) and *dopamine* (DA), and the **indolamine**, *serotonin* (5-HT). As a group, these neurotransmitters have considerable importance from a neurobe-havioral perspective. Monoamine transmitters have been associated with a wide range of neuropsychiatric disorders from obsessive–compulsive, anxiety, mood, and thought disorders to dementias and movement disorders. These transmitters continue to represent a major target of pharmacological research and development. While each will be discussed separately, the reader should keep in mind that as is true of acetylcholine and the amino acids just discussed most if not all functional systems and neuropsychiatric syndromes are multiply determined. Thus, rather than a single transmitter being the key, these systems require a certain "balance" of interaction among the various transmitters, peptides, hormones, and other chemically active substances in the brain in order to function properly.

Norephnephrine

Norepinephrine (NE) or noradrenaline perhaps is best known for its association with the body's response to stress or the "fight–flight" phenomena associated with activation of the sympathetic branch of the autonomic nervous system. While epinephrine (*adrenaline*) also plays a critical part in the body's response to stress, its major effects are peripheral (see below). As a central neurotransmitter, epinephrine has a much more limited role.[41] Norepinephrine is synthesized from tyrosine, an amino acid that is actively transported into the neuron from the blood in a three-step enzymatic process. As can be seen in Figure 11–12, (1) tyrosine hydroxylase converts the tyrosine into 3,4-dihydroxphenylalanine (DOPA) (2) Next, DOPA decarboxylase converts DOPA into dopamine (DA). (3) Finally, another enzyme—dopamine beta hydroxylase—which is present in noradrenergic but obviously not in dopaminergic neurons transforms dopamine into norepinephrine (NE). Like other transmitters, NE is stored in vesicles until it is ready to be released. Once released into the synaptic cleft by a nerve impulse at the presynaptic terminal,[42] NE then must be deactivated. This is accomplished by one of three methods: (1) NE may be destroyed directly while in the synaptic cleft by catechol-0-methyltransferase (COMT), (2) NE may be transported

Synthesis/Degradation
of Norepinephrine (NE)

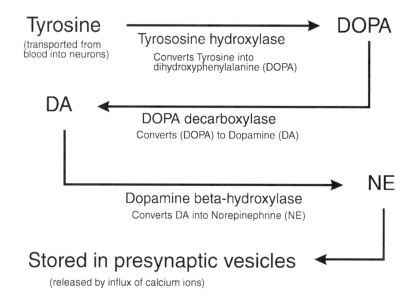

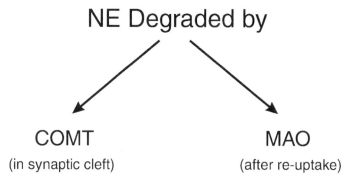

Figure 11–12. Synthesis/degradation of norepinephrine (NE). Figure illustrates several key steps in this process.

back into the presynaptic terminal where it can be stored simply for future use, or (3) once transported back into the presynaptic terminal, NE may be broken down by monoamine oxidase (MAO).

Receptors

Although various subtypes have been identified, for present purposes, four basic types of adrenergic receptors are known to be present in the CNS: α-1, α-2, β-1, and β-2. As noted earlier, these receptors also can be found in many peripheral organ systems and are part of the body's sympathetic response to stress (see below). All adrenergic receptors are of the G-protein-linked type, meaning that all use second messenger systems. While all four types can be found postsynaptically, it is the β-1 that appears to be most sensitive to norepinephrine within the CNS (Cooper, Bloom, & Roth, 1991, p. 253). Finally, α-2 receptors also are found presynaptically, both in the brain and in the sympathetic nervous system. Here they serve as autoreceptors, regulating the release of NE from the presynaptic terminals by inhibiting adenylyl cyclase.[43]

Distribution in the CNS

The primary source of norepinephrine in the CNS is the **locus ceruleus**, a group of adrenergic neurons situated along the lateral edges of the periaqueductal gray in the rostral pons.[44] Fibers from the locus ceruleus are widely distributed in the CNS. Some fibers enter the cord, others the cerebellum via the superior cerebellar peduncles, while still others are distributed to other parts of the brain stem.[45] However, the majority of the projections from the locus ceruleus are directed rostrally, primarily via the **medial forebrain bundle**. The noradrenergic neurons that make up these ascending pathways then are distributed to the hypothalamus, thalamus, limbic structures (including the hippocampal formation, septal nuclei, amygdala, and cingulate gyrus), as well as the cerebral cortex (see Figure 11–13).

Although not as well organized as the locus ceruleus, other adrenergic neurons are found scattered in the **medullary tegmentum**. Together with fibers from the subceruleus, these neurons appear to project primarily to the spinal cord and brain stem, but some may find their way to the basal forebrain and limbic structures (Cooper, Bloom, & Roth, 1991).

Note should be made of two other major sources of norepinephrine. One already has been mentioned, namely the postganglionic neurons of the sympathetic nervous system. Ultimately under the control of the hypothalamus, these peripheral adrenergic neurons

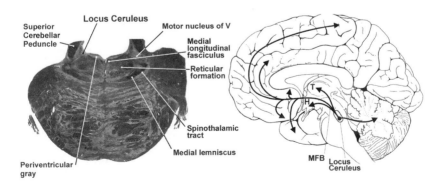

Figure 11–13. Distribution of norepinephrine in the CNS. Noradrenergic pathways in the brain primarily have their origin in the locus ceruleus that lies in the dorsal lateral tegmentum of the pons and caudal midbrain. Abbreviations: H, hypothalamus; T, thalamus. Brain image (left) was adapted from the *Interactive Brain Atlas* (1994), courtesy of the University of Washington.

basically prepare the body to take either aggressive or defensive action in response to threat, stress, or other high-energy states. Finally, both epinephrine and norepinephrine are synthesized by the adrenal medulla. In fact, the adrenals are the primary source of epinephrine. Because these substances are released directly into the blood by the adrenal glands, strictly speaking they act like hormones rather than neurotransmitters. However, they nonetheless can effect a sympathetic response(s). Because the catecholamines do not easily cross the blood–brain barrier, it is thought that their effects may be primarily peripheral. However, even if this is indeed the case, if one suddenly and unexpectedly begins to experience palpitations, sweating or flushing, and headaches (as might happen with a pheochromocytoma), secondary anxiety (a CNS reaction) would not be an unexpected response.

Probable Role(s) of Norepinephrine in Behavior

Although this chapter primarily is concerned with the central effects of neurotransmitters, we have discussed the autonomic nervous system throughout this text and autonomic responses are an integral part of the brain's response to stress. As previously noted, both NE and ACh play a role in the modulation of various organ systems depending on external and/or internal events. Table 11–2 summarizes the behavioral correlates of these transmitters via the autonomic nervous system.

The role of NE in the sympathetic nervous system is reasonably clear; however, such is not the case when it comes to its central mechanisms. Some of the suspected clinical and behavioral correlates of norepinephrine in the CNS will be explored in greater detail in the following section. However, it is important to reiterate that most if not all of the behaviors addressed below are exceedingly complex biochemically and multiply determined. Thus, it is difficult to establish a specific role for any given neurotransmitter or neuromodulator. However, for the purpose of this discussion, it can be stated that central noradrenergic systems are thought to play a significant role in **arousal**, **attention/vigilance**, **affect**, **stress/threat**, and **memory**,[46] a list that has remained fairly constant at least over the past 20 years (see Moore & Bloom, 1979; Charney, Nestler, & Bremner, 1999).

Table 11–2. Autonomic Responses Associated with Norepinephrine and Acetylcholine

Organ System	*Sympathetic Response[a]*	*Parasympathetic Response[b]*
Eyes	Dilates pupil	Constricts pupil
Lungs	Dilates bronchioles	Constricts Bronchioles
Heart	Increases heart rate	Decreases heart rate
	Increases contractility	Decreases contractility
Blood Vessels		
(Surface)	Constricts (α_1)	
(Striated muscle)	Dilates (β_2)	
(Blood pressure)	Increases	
Energy	Increases glycogenolysis	
Systems	Increases release of stored glucagon	
	Increases lipolysis	
Gastrointestinal	Inhibits GI processes	Promotes GI processes
GI/GU Sphincters	Constricts	Relaxes
Skin	Sweating[b]	
	Piloerection	

[a] Adrenergic
[b] Cholinergic

Clinical and Drug Effects Associated with Noradrenergic Mechanisms

Anxiety-Related Disorders. The adrenergic response of the peripheral sympathetic nervous system (i.e., "fright–flight–fight") has been well documented. It is believed that "central" adrenergic responses likely serve related, if not similar goals (Aston-Jones, Chiang, & Alexinsky, 1991; McCormick, Pape, & Williamson, 1991; Robbins & Everitt, 1995). Noradrenergic systems in the brain appear to respond not only to perceived threats or stress, but also likely are involved in maintaining vigilance (i.e., an increased state of "arousal") against potential or imminent threats. The latter state likely would involve being "on guard" both psychologically and physiologically in the event some form of defensive (or offensive) action is required. If this indeed is the situation, problems theoretically arise if such sympathetic-like arousal occurs either (1) in the absence of a real or impending threat, (2) to inappropriate stimuli, or (3) continues to occur even when the threat is passed. These latter conditions might be seen as reflecting the clinical conditions of (1) "uncued" panic attacks or generalized anxiety disorders (GAD), (2) "cued" panic attacks (e.g., panic disorder with agoraphobia) or specific phobic disorders, and (3) posttraumatic stress disorders (PTSD), respectively. In turn, most if not all of these anxiety-related disorders are associated with an increased sensitization and/or responsivity of the locus ceruleus or the central noradrenergic system to one degree or another (Chrousos & Gold, 1992; Charney et al., 1993, 1995; Valentino & Aston-Jones, 1995; Brawman-Mintzer & Lydiard, 1997; Southwick et al., 1997b; Bourin, Baker, & Bradwejn, 1998; Charney, Nestler, & Bremner, 1999).[47]

Adrenergic Drug Effects and Anxiety-Related Disorders. One of the bases for demonstrating the relationship between increased NE and anxiety-related disorders has been the ability of **yohimbine** (Yocon) to elicit or exacerbate symptoms of anxiety, particularly in individuals already prone to these disorders (Charney et al., 1987b; Albus, Zahn, & Breier, 1992; Bremner et al., 1997). Since yohimbine is an α-2 antagonist, its anxiogenic effect is thought to result from its ability to block the autoreceptors of the neurons in the locus ceruleus. By blocking this feedback mechanism, an excess of NE is released by these neurons. Conversely, by stimulating these adrenergic somatodendritic and/or terminal autoreceptors, **clonidine** (Catapres), an α-2 agonist, should diminish the release of NE and thus reduce anxiety. Although some anxiolytic and autonomic effects have been reported with clonidine, particularly in opiate withdrawal, it apparently has a limited impact on "psychological" symptoms. On the other hand, drugs such as alcohol, morphine, or the BZDs, which exert an inhibitory influence on NE, do have anxiolytic properties. If anxiety is related to an excess of NE, one also might expect that drugs that block postsynaptic adrenergic receptors also should have anxiolytic effects. The most well-known class of such drugs is the "beta-blockers." While such drugs such as **propranolol** (Inderal) or **atenolol** (Tenormin) have proven quite useful for "social anxiety states" (e.g., public speaking), they have not proven particularly effective for some of the major anxiety disorders such as panic disorders, GAD, or PTSD.

As noted in the previous discussion of GABA, the BZDs[48] often can be useful in the management of anxiety-related disorders (see Table 11–3). However, a common feature of many of the more chronic anxiety-related disorders is that they also are more likely than not to respond positively to various antidepressant medications. This would include monoamine oxidase inhibitors (MAOIs), classic tricyclics (TCAs), selective serotonin reuptake inhibitors (SSRIs), and other newer antidepressants such as venlafaxine, mirtazapine, and nefazodone (Davidson, Tupler, & Potts, 1994; Hyman, Arana, & Rosenbaum, 1995; Katz, Fleisher, Kjernisted, & Milanese, 1996; Stahl, 1996, 2000; Davidson, 1997, 1998; Pine, Grun, & Gorman, 1998; Taylor, 1998; Goddard, Coplan, Gorman & Charney, 1999; Keltner & Folks, 2005). The fact that these disorders respond to the drugs raises some interesting questions. Certainly all the antidepressants mentioned above, including the SSRIs, have

Table 11–3. Drugs Commonly Used to Treat Anxiety-related Disorders[a]

Panic Disorders
  SSRIs[b]
  e.g., - sertraline (Zoloft)
        - paroxetine (Paxil)
        - fluoxetine (Prozac)
  TCAs
  e.g., - imipramine (Tofranil)
        - clomipramine (Anafranil)
        - desipramine (Norpramin)
  MAOIs
  BZDs[c,d]
  e.g., - alprazolam (Xanax)
        - clonazepam (Klonopin)

Generalized Anxiety Disorders
  BZDs[h]

  Buspirone (BuSpar)

  TCAs
  e.g., imipramine (Tofranil)

  SSRIs

  Venlafaxine (Effexor)[i]

  MAOIs[j]

Obsessive–Compulsive Disorders
  TCAs
    clomipramine (Anafranil)

  SSRIs[e,f]

  MAOIs[g]

Posttraumatic Stress Disorder
  SSRIs, MAOIs, TCAs

  BZDs[k]

  Beta-blockers[l]

  Anticonvulsants[m]

Social Phobias
  Beta-blockers
  BZDs
  MAOIs
  SSRIs

---

[a]Cognitive and/or behavioral therapy commonly recommended in addition to chemotherapy for these disorders.

[b]Usually start with low dosages because may cause anxiety initially.

[c]If used, high potency BZDs generally are recommended, especially clonazepam due to its longer half-life.

[d]May be used in combination with antidepressant in resistant cases.

[e]Usually require higher dosages than used when treating depression and may take longer to respond.

[f]May require combining with other drugs such as *buspirone* (BuSpar) or serotonin-2 antagonist/reuptake inhibitors (SARIs) such as *trazodone* (Desyrel) or *nefazodone* (Serzone) to increase 5-HT effects, or a BZD such as clonazepam (Stahl, 1996).

[g]May be useful when there is concomitant panic or GAD.

[h]High-potency, short-acting BZDs may produce more immediate relief, but as this condition is generally chronic, lower-potency, longer-acting BZDs present less immediate risk of dependency (for review, see Thompson, 1996).

[i]see Davidson (1998).

[j]Consider when there are associated panic attacks.

[k]For control of anxiety components.

[l]For reducing autonomic overreactivity.

[m]To assist with mood stabilization.

varying degrees of adrenergic activity in that they block the reuptake and/or destruction of NE (Richelson, 1996, 2001). But recall that we just have pointed out the probable relationship between excess activation of the locus ceruleus (i.e., increased noradrenergic activity) and anxiety. If this is the case, should not blocking the reuptake or destruction of NE then lead to greater not less anxiety? One possible explanation may lie in the complexity and highly interactive nature of the catecholamines both within themselves and with other neurotransmitter and neuropeptide systems (e.g., see Petty, Davis, Kabel & Kramer, 1996; Goddard, Coplan, Gorman & Charney, 1999). In the case of the antidepressants (and their capacity to inhibit reuptake of NE), the effects of the excess NE on the somatodendritic and/or terminal autoreceptors of the adrenergic neurons of the locus ceruleus (thereby limiting their firing and/or release of NE) may override any anxiogenic effects they may have at the postsynaptic terminals. This then might ensure a more modulated response of the locus ceruleus to anxiety-provoking stimuli.[49]

Norepinephrine and Depression. As previously discussed, the early antidepressants, specifically the TCAs and the MAOIs,[50] were noted to enhance the available supply of norepinephrine (along with dopamine and serotonin) in the synaptic cleft by either interfering with their reuptake (TCAs) or their enzymatic breakdown (MAOIs). It also was noted that **reserpine**, a drug with both hypotensive and tranquilizing effects (apparently resulting from its ability to lower catecholamine levels), had a tendency to induce depression in some individuals. Conversely, amphetamines, which were found to enhance NE, earlier had been found to have antidepressant effects. These observations have led to what is now termed the *monoamine hypothesis* of depression (Schildkraut, 1965; Coppen, 1967). Essentially, this hypothesis suggests that depression is associated with a relative depletion of the monoamines, especially NE and 5-HT.

Indeed, the various antidepressant medications all act to enhance one or more of the monoamine transmitters (see Table 11–4). However, the issue of cause and effect is still far

Table 11–4. Antidepressants: Major Effects on Monoamine and Other Receptors

Drug	NE	DA	5HT	5HT2$_A$	H-1[a]	M-1[b]	α-1[c]
MAOIs[d]	+++	+	+++		?	?	?
TCAs[e]	+++	+	+++	+	+++	++	+++
SNRIs[f]	+++	+	+++				
SSRIs	+		+++		+/-		
Bupropion[g]	+	++					
Mirtazapine[h]	+++		+++	+++	+++	+	+
Nefazodone[i]	+		++	+++			+

Abbreviations: Strong (+++), moderate (++), weak (+) effect
[a] Associated with drowsiness, weight gain;
[b] may result in dry mouth, constipation, blurred vision, urinary retention, and mild confusion (especially in the elderly);
[c] can produce dizziness, orthostatic hypotension;
[d] side effects suggest possible H-1, M-1, and α-1 antagonism;
[e] effects on NE, 5-HT, and other receptors can vary considerably depending on specific drug (e.g., desipramine and clomipramine have greater respective effects in blocking NE and 5-HT reuptake);
[f] e.g., venlafaxine, duloxetine;
[g] weakly inhibits NE and DA; active metabolite inhibits NE and DA reuptake;
[h] blockade of NE reuptake is more pronounced as dose levels increase;
[i] trazodone, a related compound, has considerably weaker antidepressant effects, but increased H-1 effects make it an effective sedative.

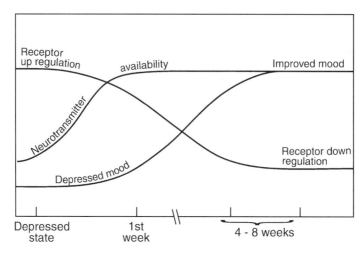

Figure 11–14. Correlation of antidepressant effects and receptor down-regulation. The apparent temporal correlation between down-regulation of postsynaptic monoamine receptors and clinical improvement, in contrast to the much earlier increase in monoamine availability following treatment with antidepressants. Adapted from Stahl (2000).

from clear. As can be seen in Table 11–4, while most of the classic antidepressants target NE, 5-HT, and to a lesser extent DA, many of the newer ones selectively act on serotonin. As discussed earlier in this chapter (see under Neurotransmitter Reuptake/Degradation), an even more confounding observation is the fact that while monoamine levels apparently return to normal baselines fairly quickly upon the initiation of treatment, clinically the effects may be delayed for weeks or more. One potential explanation for these findings has to do with changes in receptor sensitivity. Recall that receptors may be either in their "normal state," up-regulated, or down-regulated, depending on the relative availability of a given neurotransmitter. Thus, when there is a relative depletion of a transmitter(s), as appears to be the case in depression, the receptors are "up-regulated" (increased), ostensibly to take maximum advantage of whatever molecules of the transmitter are present in the synaptic cleft (see Figure 11–8). Once the availability of the neurotransmitter(s) in the synaptic cleft increases (e.g., as a result of reuptake inhibition), the receptors tend to "down-regulate" (undergo a reduction in number). What is intriguing is the fact that the clinical response time of the drugs is approximately the same as the time it takes for the receptors to down-regulate (Figure 11–14). However, how and why such changes should lead to antidepressant effects is still a mystery.[51] Finally, it also has been suggested (see Stahl, 2000) that depression may not so much result from an anomaly of monoamine transmitters or their receptors per se, but rather either from a breakdown of second-messenger systems in their capacity to produce changes within the cell (such as the expression of certain genes) or from a dysregulation of certain neuropeptides.

Catecholamines and Attention-Deficit/Hyperactivity Disorder. The precise etiological and neuropathological bases of **attention-deficit hyperactivity disorder** (ADHD) are still pretty much of an enigma (Faraone & Biederman, 1999). Neurochemically, however, there is evidence that a catecholaminergic dysregulation may be involved. This conclusion in large part is based on the observation that drugs that affect these chemical systems often have been shown to reduce many of the symptoms associated with ADHD. **Dextroamphetamines** (Dexedrine) and other more recently developed stimulant-type medications, such as **methylphenidate** (Ritalin), **pemoline** (Cylert), and Adderall, an amphetamine mix, are among the drugs that have been used to treat ADHD. What all these stimulants have in common is that they tend to increase DA and NE (pemoline has a greater relative effect on DA than methylphenidate) by blocking the reuptake of these transmitters (Greenhill, 1995). This also may help explain their antidepressant effects. Other drugs that also tend to increase catecholamine levels, such as **selegiline** (Eldepryl), **bupropion** (Wellbutrin), and the tricyclics all have been shown to be efficacious in the management of ADHD in both children and adults (Riddle et al., 1988; Green, 1995; Wilens, Biederman, Spencer, & Prince, 1995; Conners et al., 1996; Spencer et al., 1996; Ernst et al., 1997).

However, as was true of the other system disorders previously discussed, currently ADHD cannot be reduced to any single etiological factor or any one neurotransmitter system (Pliszka, McCracken, & Maas, 1996; Ernst et al., 1997). As seems clear from Faraone and Biederman's (1999) review, in ADHD one undoubtedly is dealing with complex neuro-chemical, neurobehavioral, neuroanatomical, and perhaps even environmental phenomena. Reflecting the neurochemical complexity involved is the observation that patients with ADHD often respond favorably to clonidine (Hunt et al., 1990, 1995). Recall that **clonidine** (Catapres), along with guanfacine, which also is used in treating this disorder (Hunt et al., 1995), basically is an α-2 agonist. Because of its capacity to stimulate autoreceptors, we would expect that clonidine would reduce the release of norepinephrine at the synapse, just the opposite of what appears to be accomplished by the stimulants bupropion and the TCAs. While the reason for clonidine's effectiveness is not clear, two potential explanations come to mind. One is that since clonidine is described as a "partial agonist" at the α-2 receptor (Katzung, 1998), it may have a reduced inhibitory effect on the release of NE as compared with the presence of a "full agonist" (i.e., NE). Second, while α-2 receptors are found presynaptically (autoreceptors), it has been suggested that NE and other α-2 agonists (such as clonidine) may have a positive impact on ADHD via their capacity to bind to additional α-2 subtype receptors postsynaptically (Arnsten, Steere, & Hunt, 1996).

Dopamine

Like norepinephrine, **dopamine** (DA) is a centrally acting catecholamine. As will be discussed in the following sections, dopaminergic neurons have extensive cortical and subcortical distributions in the brain and are thought to play a major role in multiple neurobehavioral processes. Among the better publicized of these has been its role in diseases of the basal ganglia, such as parkinsonism, as well as its apparent implications for the expression and treatment of primary psychiatric disorders (e.g., schizophrenia). As is the case with norepinephrine, DA is synthesized from tyrosine. The process is the same as that described for norepinephrine, except for the final step that converts DA into NE (see Figure 11–8). In summary, tyrosine hydroxylase converts the tyrosine into 3,4-dihydroxphenylalanine (DOPA), and then DOPA decarboxylase converts DOPA into DA. Once formed, DA is stored in presynaptic vesicles until released by an action potential. Once released, again like NE, DA either may be inactivated by COMT in the synaptic cleft or may be taken back up into the presynaptic terminal by its own high-affinity transport carrier where it either can be stored in presynaptic vesicles or metabolized into **homovanillic acid**. Drugs, whose effect is to enhance the available supply of DA, often work by either inhibiting this reuptake process (e.g., TCAs, cocaine, amphetamines, and other stimulant-type drugs) or interfering with their metabolic breakdown (e.g., MAOIs).

Receptors

Dopamine receptors initially were classified as either one of two types: D_1 or D_2 (Spano, Govoni, & Trabucchi, 1978), both being G-protein-linked receptors. Subsequent studies have identified many different subtypes (Seeman, 1995; Baskys & Remington, 1996; Jaber, Robinson, Missale, & Caron, 1996; Kulkarni & Ninan, 1996; Missale et al., 1998). For most current purposes, however, they are usually were designated as D_1, D_2, D_3, D_4, and D_5. These, in turn, typically are divided into two general categories: D_1-like, which includes the original D_1 and the D_5 receptors, and the D_2-like, which includes the original D_2 as well as the D_3 and D_4 receptors. The D_1-type receptors are associated with the stimulation of cyclic AMP, while the D_2-type generally inhibits cyclic AMP. The D_2 receptors function both as postsynaptic receptors and as autoreceptors at both the somatodendritic portion of the neuron (where they inhibit the firing of the neuron) and on the presynaptic terminals (where they inhibit the synthesis and release of dopamine) (Wolf & Roth, 1990).[52] As will be seen, postsynaptic D_2 receptors in the mesolimbic system likely are the primary site of action of most neuroleptics (with the exception of **clozapine,** which selectively targets D_4 receptors), where they have an antagonistic effect. On the other hand, the antagonism of D_2 receptors in the striatum seems to be responsible for the extrapyramidal side effects of these medications, while D_2 agonists such as **bromocriptine** may benefit parkinsonian-type symptoms, which are thought to result in part from relative depletions in DA.

Distribution in the CNS

The various subtypes of DA receptors are not equally distributed within the CNS (Baskys & Remington, 1996; Jaber, Robinson, Missale, & Caron, 1996; Mansour et al., 1998; Missale et al., 1998). The D_1 and D_2 receptors, the latter of which has an extremely high affinity for many psychotropic drugs, are the most widespread, with both being extensively represented in the caudate and putamen, limbic-related structures (especially the nucleus accumbens), as well as in the cerebral cortex. D_3 also has been associated in particular with limbic structures, including the hypothalamus and olfactory tubercle, but is apparently relatively sparse in the motor portions of the basal ganglia. Like the D_2 receptor, D_3 also may serve as an autoreceptor presynaptically. D_4, which would appear to have important implications for the pharmacological treatment of psychiatric disorders (see below), is found primarily in the frontal and temporal cortices (at least in primates), as well as in the amygdala and hypothalamus. However, D_4 receptors also have been found in other areas of the brain, including motor areas where they are thought to modulate GABA systems. D_5 receptors, which might be recalled are similar to D_1, also appear to have both cortical and subcortical distributions, but their particular significance has not yet been established.

A substantial majority of the central dopaminergic neurons originate in the mesencephalon, particularly in the **pars compacta of the substantia nigra** (SNc), and the medially adjacent **ventral tegmental area** (VTA). Another smaller group of dopaminergic cells are located in the nearby **mesencephalic reticular formation**, referred to as the *retrorubral area.* Finally, additional sources of dopaminergic cells are found in the hypothalamic region.[53] Although there is some overlap, a definite pattern can be seen in terms of the rostral projections of these various populations of dopamine neurons. In general, the more ventral portions of the SNc project to those areas of the striatum that mediate sensorimotor functions. The VTA and the adjacent dorsal portions of the SNc contribute primarily to those portions of the striatum (e.g., anteroventral striatum, caudate nucleus, nucleus accumbens) that are linked to the limbic system and association cortices, as well as to portions of the association cortices themselves (especially to the mesial portions of the frontal lobe. The dopaminergic neurons in the hypothalamic region remain fairly local, projecting to hypothalamic nuclei and to the pituitary gland (Fallon, 1988; Cooper, Bloom, & Roth, 1991; Haber & Fudge, 1997a).

Based in part on these projections, anywhere from three to five (depending on the particular investigator) "dopamine pathways" commonly have been identified (see Figure 11–15). For current purposes, the pathways outlined by Stahl (1996) would seem most

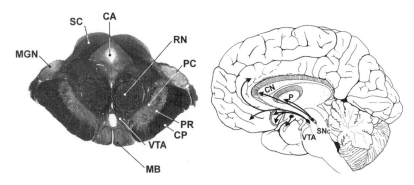

Figure 11–15. Dopamine pathways in the CNS. The various pathways described in the text are shown, with the mesolimbic and mesocortical pathways deriving primarily from dopaminergic cells in the ventral tegmental area (VTA), while those projecting to the basal ganglia and hypothalamic region coming from the substantia nigra, pars compacta (SNc) and diencephalons, respectively. Abbreviations: CA, cerebral aqueduct; CN, caudate nucleus; CP, cerebral peduncles; MB, mammillary body; MGN, medial geniculate nucleus; P, putamen; PC, substantia nigra, pars compacta; PR, substantia nigra, pars reticulata; RN, red nucleus; SC, superior colliculus. Brain image (left) was adapted from the *Interactive Brain Atlas* (1994), courtesy of the University of Washington.

useful. These are the (1) **nigrostriatal**, (2) **mesolimbic**, (3) **mesocortical**, and (4) **tuberoinfundibular pathways**.[54] The **nigrostriatal system** is typically associated with "motor" functions, meaning that part of the striatum (primarily putamen) receives projections from the sensorimotor cortices, supplemental motor areas, and the frontal eye fields. It receives its dopaminergic input primarily from the SNc. The **mesolimbic system**, which receives the majority of its dopamine from the VTA (and the dorsal tier of the SNc), generally is believed to encompass the nucleus accumbens, olfactory tubercle, bed nucleus of the stria terminalis, septal nuclei, and possibly the amygdala. The **mesocortical pathway** similarly derives its input from the VTA and dorsal SNc. This latter pathway usually is described as consisting of the cingulate gyrus (particularly the anterior cingulate), the mesial portions of the prefrontal cortex, and the entorhinal and pyriform cortices of the temporal lobe.[55] The **tuberoinfundibular pathway**, as noted above, derives its input from dopaminergic neurons in the diencephalon itself and targets cells in hypothalamic–pituitary axis. While such classifications are useful, it should be noted that such a schema once again likely represents a major oversimplification of what appears to be a highly complex system (Roth & Elsworth, 1995). The possible clinical significance of these receptors and pathways will be addressed in the following sections.

Probable Role(s) of Dopamine in Behavior

As has been repeatedly emphasized, neurotransmitter systems are exceedingly complex and multidimensional. Although for the sake of providing a general overview, attempts were made to ascribe specific functions to the transmitters previously discussed; this was done with the awareness that any such characterizations almost assuredly would understate their actual role(s). This is certainly the case with dopamine. As will be discussed in the following section, we know that increases or decreases in dopamine are suspected of being associated with various dramatic behavioral aberrations. However, these purported associations, even if true, do not necessarily reveal the specific role played by dopamine in these processes. Perhaps the first clue to dopamine's complex role might come from the earlier observation that D_1-type receptors facilitate adenylyl cyclase activity, while D_2-type receptors inhibit it. Again, recall that adenylyl cyclase is critical in the formation of cAMP, which in turn is part of the second-messenger system in neurons. Thus, indirectly, dopamine can effect numerous processes within the cell, including the synthesis and release of other neurotransmitters and other functions mediated by mRNA. Additionally, dopamine can affect the concentrations of charged ions within the cell (e.g., Ca^{2+}, K^+, Na^+), which affects the excitability of neurons (Missale et al., 1998).

As a result of its rather extensive anatomical distribution and behavioral associations, dopamine has been implicated in such diverse processes as locomotion, neuroendocrine functions, behavioral inhibition and other frontal executive and/or cognitive functions, as well as learning and memory (Roth & Elsworth, 1995; LeMoal, 1995; Baskys & Remington, 1996; Haber & Fudge, 1997b; Missale et al., 1998). Dopaminergic stimulation of the mesolimbic areas of the brain (particularly the nucleus accumbens) also has been associated with the reward and reinforcement, including the reinforcing effects of stimulant drugs such as cocaine, methamphetamine, and nicotine (Koob, 1992; Wise, 1996, 1999). Our current understanding of the role of DA may have been best characterized by LeMoal and Simon (1991).[56] In summarizing their conclusions, LeMoal states:

The conclusions [of LeMoal and Simon] can be summarized briefly as follows: (i) DA neurons do not have specific functions; (ii) they regulate and enable integrative functions in the neuronal systems onto which they project; (iii) a lesion of their terminals induces neuropsychological deficits that are characteristic of the functions of the neuronal systems they regulate; and (iv) the deficits observed depend on the behavioral situations or tasks used to explore them. (LeMoal, 1995, p. 283)

Clinical and Drug Effects Associated with Dopaminergic Mechanisms
Psychotic Disorders, Neuroleptics and Their Side Effects:

The Dopamine Hypothesis of Schizophrenia. Chlorpromazine (Thorazine), the first truly effective antipsychotic medication, initially was known for its antiemetic and antihistaminic effects. In the early 1950s, a French surgeon, Laborit, found that when administered preoperatively chlorpromazine had calming, sedating effects on surgical patients without disrupting consciousness. Shortly thereafter, it also was discovered to have similar effects on psychotic patients, making them considerably more tractable. Following the success of Thorazine, additional drugs were developed that also tended to reduce psychotic symptoms, particularly in schizophrenic populations. They continued to be used for this purpose despite the fact that their mechanism of action was not known. Eventually, it was discovered that (1) these drugs not only had an affinity for dopamine receptors where they exerted an antagonistic effect, but (2) their clinical effectiveness and/or potency seemed to be directly correlated with their capacity to bind to these receptors (van Rossum, 1966; Seeman et al., 1975, 1976; Creese, Burt, & Snyder, 1976). Additionally, it had been noted that drugs such as methamphetamine and cocaine, which have a dopaminergic agonist effect either by increasing the release or inhibiting the reuptake of DA, tend to produce psychosis (see Jaber et al., 1996). These discoveries led to what has been termed the **dopamine hypothesis** of schizophrenia (Matthyse, 1973; Snyder, 1976). In essence, this theory suggested that schizophrenia and perhaps most psychotic type reactions resulted from an excess stimulation (or excessive sensitivity) of D_2 receptors, primarily in the mesolimbic regions. Although some refinement of this hypothesis has been required to meet the demands of subsequent data and clinical insights, at its core the dopamine hypothesis still remains a major force in the search for improved therapeutic models of schizophrenia.

The need to revise the original dopamine hypothesis initially stemmed from the realization that not all patients responded to neuroleptics, and even those for whom these drugs were helpful not all of their symptoms equally benefited from treatment. In fact, in many cases, some of their symptoms actually seemed to get worse. If the symptoms of schizophrenia were broken down into the **positive** and **negative symptoms** of schizophrenia

Table 11–5. "Positive" and "Negative" Symptoms of Schizophrenia

Positive Symptoms[a]	Negative Symptoms[b]
Delusions	Blunted, shallow affect
Hallucinations	Anhedonia
Disorganization of speech (loosening of associations)	Cognitive impairments (e.g., concrete thinking, poor
Language disturbances (e.g., neologisms; clang assoc.)	abstractive capacity and problem-solving skills, bradyphrenia,
Disorganization of behavior (e.g., social impropriety)	attention/concentration deficits)
Catatonia	Abulia
Agitation, lability of affect	Decreased verbal fluency
	Social withdrawal/apathy

[a] Usually associated with increased dopamine in mesolimbic region.
[b] Usually associated with decreased dopamine in mesocortical/prefrontal region(s).

(see Table 11–5), it was noted that the classic neuroleptics generally were effective in treating the "positive" symptoms, but often appeared to exacerbate the "negative" symptoms. This suggested that different biochemical and/or anatomical substrates were likely responsible (Davis et al., 1991; Kahn & Davis, 1995). It also was noted that, like antidepressants, the full effectiveness of drug treatment typically was delayed well beyond what might have been expected if the problem were simply an excess of dopamine in certain regions of the brain (Meltzer, 1989; Hyman, Arana, & Rosenbaum, 1995; Baldessarini, 1996a; van Veelen & Kahn, 1999).[57] Finally, it also was becoming clear that other neurotransmitters besides dopamine were involved. As noted earlier, it was hypothesized that glutamate likely plays a role both in the etiology and in the symptomatology of schizophrenia. As will be discussed in greater detail later, serotonin also has been implicated in the etiology and treatment of schizophrenia (Meltzer, 1991a,b; Roth & Elsworth, 1995; van Veelen & Kahn, 1999).

As suggested above, in "revising" the "dopamine hypothesis" of schizophrenia, first and foremost, it was necessary to account for, and distinguish between, the positive and negative symptoms of schizophrenia (as listed in Table 11–5) and their biological correlates. The **revised dopamine hypothesis** suggested that the *positive* symptoms of schizophrenia were likely due to increases in dopamine activity in the mesolimbic areas of the brain, particularly in the nucleus accumbens (most likely involving D_2 and D_3 receptors), while the *negative* symptoms were thought to be associated with a decrease in dopamine in the mesocortical region and/or prefrontal cortices (PFC), possibly linked more to D_1 and/or D_4 receptors (Kahn & Davis, 1995; Kulkarni & Ninan, 1996; Josselyn, Miller, & Beninger, 1997; Knable & Weinberger, 1997; Willner, 1997; Lidow, Williams, & Goldman-Rakic, 1998). Furthermore, it was observed that due to apparent feedback mechanisms operating between cortical and VTA neurons, there was a reciprocal relationship between prefrontal dopaminergic systems and the striatal and mesolimbic pathways, such that diminished PFC dopamine activity would lead to increased dopaminergic activity in the nucleus accumbens and other mesolimbic areas (Deutch, 1991; Kahn & Davis, 1995; Knable & Weinberger, 1997; Willner, 1997). The implications of this change in the perception of the role of dopamine in the treatment of schizophrenia will be addressed in the following section.

The Advent of "Atypical" Antipsychotics: Clozapine. The dopamine hypothesis of schizophrenia was proffered, in large part, because all of the neuroleptics that effectively reduced many of its core symptoms were known to block dopamine, particularly in the mesolimbic system. Despite their great boon to the field of psychiatry, all of the early compounds, which included the phenothiazines (e.g., Thorazine, Mellaril, Stelazine), butyrophenones (e.g., Haldol), and thioxanthenes (e.g., Navane), these drugs had two major drawbacks. First, when used over time, they had a tendency to produce motor disturbances [e.g., "extrapyramidal" side effects (EPS) and tardive dyskinesia], presumably as a result of dopamine depletion in the nigrostriatal system. Second, they tended to have little impact on the "negative" symptoms of schizophrenia, at times even making them worse. The advent of **clozapine** (Clozaril), however, ushered in a new era in the treatment of schizophrenia. This drug not only was found to be effective in patients who had previously failed to benefit from the older, "classic" neuroleptics, but clozapine also seemed to offer something that was not typical of the older drugs. It appeared to be better at alleviating the negative symptoms of schizophrenia and at the same time had fewer propensities to cause motor or "extrapyramidal" side effects. Clozapine, however, did have one major drawback: it had a tendency to precipitate agranulocytosis, a potentially fatal blood disorder.[58] However, Clozaril's clinical efficacy prompted continued research into its mechanisms of action and how they might elucidate our understanding of the biology of schizophrenia. Clozapine eventually became the prototypical example of the class of drugs that are now referred to as *atypical* antipsychotics. In addition to clozapine (Clozaril), other "atypicals" currently approved for clinical use in the United States are **risperidone** (Risperdal), **olanzapine** (Zyprexa), **quetiapine** (Seroquel), **ziprasidone** (Geodon), and **aripiprazole** (Abilify). All these newer drugs share several common features, namely, they:

1. Have less or little tendency to produce EPS or tardive dyskinesia.
2. Are generally more efficacious in treating refractory cases of schizophrenia.
3. Have a more favorable impact on the negative symptoms of schizophrenia.
4. As a group, they have less tendency to increase prolactin levels.

Additionally, clozapine and other "atypical" antipsychotics (with the exception of aripiprazole: see below) have two other major properties that set them apart from the older neuroleptics, they have a relatively:

5. Lower affinity for D_2 receptors.
6. Greater affinity for $5-HT_{2A}$ receptors, where they also have an antagonistic function (i.e., they show greater affinity for $5-HT_2$, as compared to D_2, receptors)

As we shall see below and later when discussing the suspected role of serotonin in schizophrenia, these latter two differences are thought to account both for the improved response profile of these drugs, as well as for their reduced side effects [59,60] (Barnes & McPhillips, 1998; Casey, 1998; Stahl 1999a; Remington & Kapur, 2000 (1-4); Farde et al., 1989; Meltzer, Matsubari & Lee, 1989a; Meltzer, 1989, 1991b, 1995; Deutch et al., 1991; Kinon & Lieberman, 1996; O'Donnell & Grace, 1996; Kapur, Zipursky, & Remington, 1999; Remington & Kapur, 1999; Stahl, 1999a) (5&6).

Atypical Antipsychotics: Proposed Mechanisms of Action. How then do these newer serotonin/dopamine antagonists impact on the symptoms of schizophrenia? First, consider the **negative** symptoms for which the atypical antipsychotics appear to provide more effective treatment. Recall that these symptoms are thought to be related to a relative *decrease* in dopaminergic output in the mesocortical pathways. In order to appreciate their possible therapeutic effect, also recall that the $5-HT_{2A}$ receptor serves as a heteroreceptor on selective dopaminergic neurons and through these receptors serotonin typically exerts an inhibitory influence on dopaminergic neurons (effectively reducing the amount of dopamine released into the synaptic cleft; see Figure 11–16). As a result of the atypicals blocking $5-HT_{2A}$

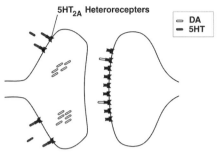

5HT normally inhibits release of DA

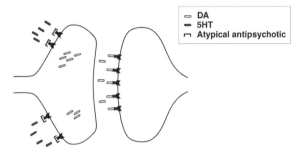

Atypicals block 5HT$_{2A}$ receptors in
mesocortical and nigrostriatal pathways

Figure 11–16. Serotonin-dopamine interaction. By blocking 5HT$_{2A}$ receptors in the mesocortical and nigrostriatal pathways (bottom figure), the atypicals overcome the normal inhibiting influence of serotonin on dopaminergic neurons in these areas, thus theoretically reducing the cognitive and motor disturbances typically associated with older neuroleptics.

receptors, dopaminergic neurons in this pathway are disinhibited, thus at least partially making up for the dopamine deficit.

What about the **positive** symptoms of schizophrenia, which ostensibly are related to a surplus of dopamine in the mesolimbic system (see Bymaster et al., 1996; Kapur & Remington, 1996; Meltzer, 1999; Stahl, 1999a; van Veelen & Kahn, 1999)? If the same disinhibitory mechanisms were to hold true in the mesolimbic and mesocortical pathways, should this not exacerbate the positive symptoms (by facilitating the release of even more dopamine)? Among the possible explanations are:

1. There may be far fewer 5-HT$_{2A}$ receptors in the mesolimbic compared with the other pathways (Stahl, 2000).
2. While these drugs have a relatively lower affinity for D$_2$ receptors, they would appear to have a special affinity for the mesolimbic pathways (with less disinhibitory effect from 5-HT$_{2A}$ receptor binding) (Farde et al., 1989).
3. Their clinical effectiveness may be mediated, at least in part, by D$_3$ and D$_4$ receptors for which these drugs do have a strong affinity (Kinon & Lieberman, 1996; Owens & Risch, 1998).[61]

Thus, the blockade of dopamine receptors in the mesolimbic pathway was hypothesized to be instrumental in reducing the more florid, "positive" symptoms of schizophrenia. On the other hand, the indiscriminate blocking of these receptors in the mesocortical pathway by the classic antipsychotics was thought to have contributed to the blunting of affect, diminished spontaneity, cognitive impairments, and other "negative" symptoms often associated with these older drugs. Recall that these latter symptoms are thought to be associated with *reduced* dopamine activity in the mesocortical areas. See accompanying Box for summary of suspected mechanisms behind clinical effects of the atypicals.

Box 11–1. Possible Explanations for Reduced Extrapyramidal Symptoms (EPS) and Reduction of Positive and Negative Symptoms with the Atypicals

Dopamine (DA) usually has an inhibitory influence on acetylcholine (ACh). Thus if DA is reduced, this may lead to excess of ACh (relative to DA) in nigrostriatal system. This imbalance is thought to contribute to EPS. Serotonin (5-HT) normally has an inhibitory influence on DA. This inhibitory influence is thought to be mediated by the presence of 5-HT$_2$ receptors on the presynaptic terminal of the DA neurons. When stimulated by the presence of 5-HT, the 5-HT$_2$ receptors inhibit the release of DA from the presynaptic terminals. However, the presence of these 5-HT$_2$ receptors on DA neurons seems to be most prominent in the nigrostriatal and mesocortical (frontal) systems and minimal in the mesolimbic system. The newer, atypical antipsychotics that are serotonin–dopamine antagonists cause less EPS as well as treat the positive and negative symptoms of schizophrenia. Possible explanations are as follows:

- **Less EPS.** Blocking of DA in nigrostriatal system normally would cause EPS. However, since 5-HT also is blocked, the normal inhibitory influence of 5-HT on the DA neurons (via the 5-HT$_2$ receptors) is absent. This causes a relative disinhibition of the release of DA, helping to negate the effects of DA antagonism in the nigrostriatal system. Clozaril, in particular, has minimal affinity for D$_2$ receptors, which are an integral part of the nigrostriatal system. Since the D$_2$ receptors in this system are not blocked, this likely explains its greater efficaciousness with regard to EPS effects.
- **Treatment of negative symptoms.** Since the same mechanisms apply to the mesocortical or frontal systems, again the end result is a relative disinhibition of DA, and increased DA is associated with less negative symptoms (greater activation).
- **Treatment of positive symptoms.** Since there are suspected fewer 5-HT$_2$ receptors on the DA neurons in the mesolimbic system, the antagonism of DA in this system (which is the key to treating the positive symptoms of schizophrenia) is NOT negated by the blocking of 5-HT$_2$ receptors (since there are so few of them in the mesolimbic system), thus the overall effect in the mesolimbic system is to diminish DA activity only.

Other Modifications of the "Atypical" Approach. Aripiprazole represents a somewhat different atypical antipsychotic. Instead of a relatively low affinity for and antagonist effect on D$_2$ receptors, this compound actually has a relatively high affinity for the D$_2$ receptor. In addition, rather than having an antagonistic effect on these receptors, it acts as a partial agonist at both the D$_2$ receptor and at the 5-HT$_{1A}$ somatodendritic autoreceptors along with its antagonism at 5-HT2A receptors (similar to the other atypicals) (Source: manufacture's product information). If you will recall the action of partial agonists (see pages 559), one might expect that this drug would tend to have a net agonist effect in the mesocortical area (where dopaminergic output supposedly is diminished), while having a net antagonistic effect in the mesolimbic region (where an excess of dopamine is thought to occur).

The current assumptions regarding the possible neurochemistry underlying schizophrenia continue to prompt research into other compounds that would meet the following requirements: (1) the need to increase dopamine in the mesocortical and prefrontal cortex (to reduce the negative symptoms), (2) decrease dopamine in the mesolimbic pathways (to treat the positive symptoms), while (3) maintaining dopamine levels in the nigrostriatal pathway (to prevent EPS). In addition to the recently developed partial dopamine agonists, these might include drugs that specifically target mesolimbic and/or mesocortical pathways and more powerful 5-HT$_{2A}$ antagonists (Willner, 1997; Belanoff & Glick, 1998).

Another interesting compound that has been approved in Europe is **amisulpride**. At lower dosages, it selectively blocks presynaptic dopamine autoreceptors (thus increasing the release of dopamine necessary to treat negative symptoms), while at higher dosages it tends to selectively block mesolimbic D$_2$ and D$_3$ receptors (reducing positive symptoms) (Moller et al., 1997; Loo et al., 1997; Scatton et al., 1997; Rein & Turjanski, 1997; Puech, Fluerot, & Rein, 1998; Danion, Rein, & Fleurot, 1999). These, like all antipsychotic drugs, continue to be predicated on the assumption that in order to be effective it is essential to alter mesocorticolimbic dopamine levels, either directly or indirectly. While no one is likely to argue about the relevance of dopamine in psychosis at the present time, it should be noted that most of these compounds interact at multiple receptor sites, and viewing schizophrenia as merely a dysregulation of dopamine in these pathways is again an oversimplification of a complex syndrome in an even more complex and interactive nervous system (Kinon & Lieberman, 1996; Willner, 1997; Knable, Kleinman, & Weinberger, 1998; Stahl, 1999a). However, even though Remington and Kapur (1999) also are quick to point out that any single (e.g., DA) or dual (DA and 5-HT) hypothesis of schizophrenia is likely to be overly simplistic, they emphasize the importance of understanding the mechanism of action for these drugs, including their side effects.

Side Effects of Neuroleptic Medications In addition to their varying capacity to block histaminic, adrenergic, and cholinergic receptors (which accounted for their differing side effect profiles) essentially all the conventional or "typical" antipsychotics were demonstrated to have a high affinity for D$_2$ receptors. As we just have seen, it was this capacity to block D$_2$ receptors in the mesolimbic system that was thought to be the basis for their clinical effect in the treatment of psychotic symptoms. Unfortunately, the impact of this D$_2$ blockade was not limited to the mesolimbic region. The conventional neuroleptics also blocked D$_2$ receptors in the nigrostriatal and tuberoinfundibular pathways. It is this blockade of D$_2$ receptors in these latter areas that is thought to contribute to (1) EPS and tardive dyskinesia,[62] and (2) increase in prolactin levels, respectively.

Extrapyramidal Side Effects. The decreased affinity for D$_2$ receptors alone might help explain the diminished side effects of the atypical antipsychotics, as there would be less antagonism of D$_2$ receptors in the nigrostriatal (as well as in the tuberoinfundibular) pathways. This easily might appear to be the case with clozapine and quetiapine, in which even following extremely high dosages D$_2$ blockade remains relatively low as measured during steady states of these drugs (Remington & Kapur, 1999; Kapur et al., 2000). However, additional explanations for the relative lack of EPS with the atypical antipsychotics have been offered. Recall that one of the main differential characteristics of the atypical antipsychotics is their affinity for 5-HT$_{2A}$ receptors. It is this affinity that has been thought to account for the relative lack of EPS with these drugs. Consider the nigrostriatal pathway. Because of the antagonism of D$_2$ receptors, even with the atypicals some degree of dopamine blocking can be expected.[63] However, since the atypicals also block 5-HT$_{2A}$ heteroreceptors, this results in a disinhibition of the dopamine neurons, causing an increase in the release of dopamine in the striatum, which helps compensate for the D_2 blockade (see Figure 11–16). This combined effect helps maintain dopamine at acceptable levels, thus reducing the probability of EPS (Stahl, 1999, 2000).[64]

Tardive Dyskinesia. Tardive dyskinesia (TD) also is thought to be related to dopamine depletion, but other mechanisms have been proposed (Casey, 1995; Egan, Apud, & Wyatt, 1997). While the exact cause remains unknown, these authors suggest TD may represent a hypersensitivity to dopamine as a result of long-term up-regulation of DA receptors. Although tardive dyskinesia typically is listed as another type of "extrapyramidal syndrome," it has several distinctive features. First, as previously noted, it has a more delayed onset than parkinsonism, acute dystonia, or akathisia (from 3 months to years after the initiation of neuroleptic treatment). Second, whereas, most other EPS symptoms often rapidly improve with a dosage reduction of the neuroleptic drug, often such a reduction will make the symptoms of TD worse, at least initially. In fact, contrary to other types of EPS, TD often can be improved by *increasing* the neuroleptic dosage (Gardos & Cole, 1995; Egan, Apud, & Wyatt, 1997). Similarly, the use of antiparkinsonian drugs (e.g., anticholinergics and dopamine agonists) may aggravate the symptoms of TD. While the continued use of conventional neuroleptics has not been shown to result in an inexorable exacerbation of TD (in some cases, there even may be a tapering of symptoms despite continuing with the same dose), there is some evidence that switching to one of the newer neuroleptics may help reduce the symptoms of preexisting TD (Lieberman et al., 1991; Lamberti & Bellnier, 1993; Remington, 1993; Chouinard, 1995; Egan, Apud, & Wyatt, 1997: Barnes & McPhillips, 1998; Stanilla & Simpson, 1998).

Increased Prolactin Levels. Dopamine normally inhibits the release of prolactin. With the D_2 blockade of the lactotroph cells of the pituitary gland, the resulting disinhibition leads to an increase in the release of prolactin. Chronic increases in prolactin in turn can result in amenorrhea, galactorrhea (inappropriate lactation), or gynecomastia (breast enhancement) (Pies, 1998). Since antipsychotic medications, particularly the conventional neuroleptics, tend to block D_2 receptors in both the nigrostriatal and tuberoinfundibular pathways, a situation similar to that observed with EPS might account for the changes seen in prolactin levels with the newer antipsychotics (e.g., $5\text{-}HT_2$ blockage facilitating the release of DA).[65] However, in this instance, somewhat different mechanisms are thought to be at work. While dopamine does inhibit the release of prolactin, serotonin facilitates it. Thus, with the antagonism of $5\text{-}HT_{2A}$ receptors by the atypicals, in addition to possibly disinhibiting DA, the blockade of the $5\text{-}HT_{2A}$ receptors themselves tends to inhibit the release of prolactin (thus negating or balancing out the effects of the D_2 blockade) (Stahl, 2000).

Transient Occupancy Hypothesis. Kapur and others offer an alternate tentative explanation for the reduced side effects (both EPS and increased prolactin) found in the atypical antipsychotics. Recall that, when measured during "steady states" (usually after 2 to 3 weeks and approximately 12 hours after the last administration of the drug), the percentage of D_2 receptors showing blockage is less than that evidenced with the conventional antipsychotics. What these researchers were able to demonstrate is that within the first 2 to 3 hours following the administration of the drug, the dopamine occupancy along with prolactin levels are relatively high (perhaps comparable to that seen with typical neuroleptics). It is only when measured later (e.g., 12 hours), that a reduction in D_2 occupancy is noted. They account for this phenomena by suggesting that while all antipsychotics have similar binding affinities for D_2 receptors (i.e., they all bind easily and quickly), the atypicals are bound more "loosely," and thus, more transiently (Seeman & Tallerico, 1998; Kapur & Seeman, 2000; Kapur et al., 2000). Referred to as *the transient occupancy hypothesis,* it suggests that (1) transient D_2 occupancy may be sufficient to manage psychotic symptoms, (2) prolactin levels readily fluctuate depending on D_2 blockage (with only sustained increases causing problems), but (3) EPS may follow only both high and sustained D_2 occupancy (Kapur, 2000).

Dopamine and Affective Disorder.   Earlier it was noted that most major depressive disorders are believed to be associated with relative monoamine deficiencies. In addition, as has been noted the mechanism of action of most if not all antidepressants in current use (e.g., MAOIs, TCAs, SSRIs) is thought to be their ability to increase the availability of these neurotransmitters (including DA) at the synaptic cleft by blocking either their reuptake or degradation. Since the preservation of norepinephrine and serotonin appears to be the major target of most of these drugs, the assumption is that these latter two transmitters are of primary importance in most cases of depression (Baldessarini, 1996b; Kamil, 1996; Stahl, 1997; Musselman et al, 1998; Schatzberg & Nemeroff, 1998, particularly Chapters 10–13). However, just as it is unlikely that dopamine alone is responsible for the schizophrenic syndrome, so too is it unlikely that NE and 5-HT alone are responsible for depression. Dopamine is thought to also play a significant role in perhaps most if not all psychiatric syndromes, particularly depression (Randrup, Munkvad, & Fog, 1975; Schatzberg et al., 1985; Kapur & Mann, 1996; Willner, 1983, 1995).

Additional evidence pointing to dopamine's potential role in depression is the fact that dopaminergic agonists like **d-amphetamines** and **methylphenidate** (Ritalin) were used to treat depressive symptoms prior to the development of the tricyclics (albeit not as success-fully as the TCAs and SSRIs). These drugs still are used, especially in certain organic or elderly cases, to treat depression or to augment the effect of modern antidepressants (Ayd & Zohar, 1987; Hyman, Arana, & Rosenbaum, 1995; Willner, 1995; Fogel, 1996; Kelsey & Nemeroff, 1998; Pies, 1998).[66] Finally, bupropion (Wellbutrin), an atypical antidepressant, is (via its metabolites) a potent reuptake inhibitor of dopamine.[67] Musselman et al. (1998) also report that a selective dopamine reuptake inhibitor (nomifensine) had proved to be an effective antidepressant before it was withdrawn from the market due to adverse neuro-logical reactions. For a more extensive review of this topic, the reader is referred to Kapur and Mann (1996) and Willner (1995).

Finally, depression frequently is associated with Parkinson's disease (Cummings, 1992; Dooneief et al., 1992) and its diagnosis can be dissociated from the motor manifestations of this disease (Starkstein, Preziosi, Forrester, & Robinson, 1990). These findings might suggest dopaminergic mechanisms may be responsible. The reasoning is, *if DA deficiencies in the basal ganglia lead to motor symptoms, then do similar deficiencies in the mesolimbic pathways lead to depression?* However, it has been suggested that this conclusion might be overly simplistic, and that while DA may be at least partially responsible for mood swings in patients with Parkinson's disease, serotonin rather than dopamine deficiency may be the primary cause of clinical depression in these patients (Mayeux, Stern, & Williams, 1986; Sano, Marder, & Dooneief, 1996).

Dopamine and Movement Disorders.   In contrast to the evidence linking dopamine and depression, the association between DA and movement disorders is quite robust. Among the more common and well-documented examples of this putative association is Parkinson's disease and those symptoms associated with neuroleptic medications, specifically their "extrapyramidal" side effects (EPS) and tardive dyskinesia (TD). Parkinson's disease, first described in 1817, perhaps is primarily characterized pathologically by a degen-eration of dopaminergic pathways and symptomatically by hypokinesis. Although the substantia nigra, pars compacta (A9) commonly is identified as a site of extensive degeneration, A8 (retrorubral group) and A10 (ventral tegmental area) also are affected

(Hornykiewicz & Kish, 1987; Hirsch, Graybiel, & Agid, 1988). Because of its projections to the mesocorticolimbic areas, changes in the latter group in part may be responsible for some of the cognitive or behavioral symptoms often associated with Parkinson's.[68] However, it is the loss of cells in the substantia nigra and the impact of this loss on striatal connections (particularly those in the putamen) that appear to have the major impact on motor functions.

The basic motor pathways thought to be disrupted in Parkinson's disease and its clinical symptomatology has been outlined in Chapter 6. To briefly review, dopaminergic input to the neostriatum from the substantia nigra, pars compacta (SNc) can be either facilitatory or inhibitory.[69] As can be seen in Figures 6–3 and 11–17, the connections to the *direct system* (which itself facilitates motor activity) are believed to be *facilitatory*, while those to the *indirect pathway* (thought to normally inhibit motor responses) are *inhibitory* (Albin, Young, & Penney, 1989; DeLong, 1990; Mink & Thach, 1993). As outlined in Figure 11–17, depletion of dopaminergic input to both the "direct" and "indirect" pathways would have similar effects, namely, reduced thalamocortical activation. It is this reduction of cortical activation through

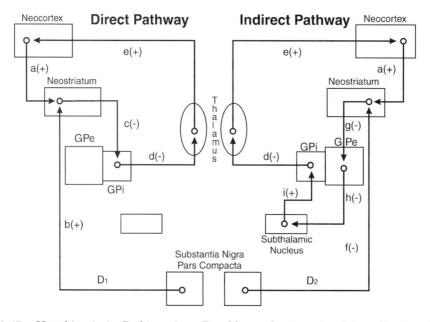

Figure 11–17. Hypokinesia in Parkinsonism: Possible mechanisms involving direct and indirect pathways. DIRECT PATHWAYS (normally facilitatory): (1) The normal facilitatory influence on the neostriatum (a) and (b) is reduced as a result of diminished dopaminergic input (b) from SNc;. (2). resulting in a relative disinhibition of GPi neurons by (c) striatopallidal pathway; and subsequent (3). increased inhibition of thalamus via the pallidothalamic inputs (d), which, in turn, (4). reduces (e) cortical activation, thus contributing to the hypokinetic state. Indirect Pathways (normally inhibitory): (1) Normal modulating influence on neostriatum is reduced as a result of (f) diminished dopaminergic input from SNc, which (2) effectively increases the inhibitory influences on GPe via the (g) neostriatal-GPe pathway, (3) resulting in a relative disinhibition of the subthalamic nuclei by way of the (h) GPe-subthalamic inputs, which in turn (4) result in (i) increased stimulation of Gpi, and subsequent (5) increased inhibition of thalamus via the (d), pallidothalamic inputs, thus leading to (6) (e) reduced cortical activation further contributing to the hypokinesis. Abbreviations: Gpe, globus pallidus externa; Gpi, globus pallidus, interna; NS, neostriatum-putamen and caudate n.; SNc, substantia nigra, pars compacta; STN, subthalamic nuclei; T, thalamus; "+", glutamate pathway; "−", GABA pathway.

the basal ganglia/thalamic feedback loop that is thought to underlie the hypokinesis seen in parkinsonism.

Although Parkinson's disease is the most common naturally occurring manifestation of striatal dopamine deficiency, other neurological conditions can manifest related findings. **Progressive supranuclear palsy** and **striatonigral degeneration** (including the *Shy-Drager syndrome*) are among the conditions that also appear to be characterized by relative depletions of dopamine in the striatal pathways (Adams, Victor, & Ropper, 1997). However, among the more frequent dopamine-related motor disorders are those with iatrogenic etiologies, typically resulting from the use of antipsychotics. These disorders may occur more acutely, with symptoms of EPS generally occurring within a few weeks of initiating treatment or may surface only after years of taking the drugs, as typically is the case with TD (*tardive* meaning *late*). Estimates of the patients who suffer from EPS as a result of taking the classic neuroleptics have been placed as high as 90% and TD, as a result of the more chronic use of these drugs, at 20% or more (Kane, 1995; Swartz et al., 1997; Casey, 1998).[70]

Specific, motor-related symptoms associated with the use of neuroleptics, their onset, severity, and response to treatment may vary considerably from one individual to the next. Among the more common manifestations of acute EPS are (1) **acute dystonias**, (2) **akathisia**, and (3) **parkinsonism**. A rarer, but considerably more dangerous, acute reaction is the **neuroleptic malignant syndrome**. Tardive dyskinesia, as already has been noted, is the most common and often the most troublesome late-onset consequence of neuroleptic medication. **Perioral tremor** (also known as *rabbit syndrome*) is another less frequent syndrome with a tardive onset (Yassor & Lal, 1986). The major elements of these syndromes are outlined in Table 11–6. As can be seen from Table 11–6, the exact mechanism behind some of these syndromes still is uncertain, However, two generalizations are possible. First, it seems likely that changes in dopamine, particularly in the nigrostriatal pathways, are a key factor in most if not all of these syndromes. Second, not all neuroleptics are equally likely to result in these syndromes. For reasons that will be explained below, the lower-potency typical neuroleptics with greater anticholinergic effects are less likely to cause problems than are the higher potency drugs that commonly have lower anticholinergic effects.[71] Also, the newer atypical antipsychotics also appear to have a much lower incidence of both EPS and TD compared with the older drugs, but probably for different reasons (e.g., lower affinity for D_2 receptors and $5HT_2$ receptor antagonism).[72]

Cognitive Disturbances in Parkinson's Disease. As noted in Chapter 6, cognitive changes are not uncommon in Parkinson's disease. Such changes may range from a more benign general slowing of cognitive processes (*bradyphrenia*) (Pate & Margolin, 1994) to more severe "dementia" (Biggins et al., 1992). While the former usually is believed to be associated with "frontal–subcortical" mechanisms resulting in decreased cortical arousal/activation, the latter may involve (1) a more advanced stage of a similar process, (2) a "cortical"-type dementia of the Alzheimer type, or (3) a combination of the two.[73] As mentioned earlier, to further complicate matters in terms of trying to link cognitive deficits with changes in dopamine levels in Parkinson's is the fact that disruptions in multiple transmitter systems have been associated with this disease (see, for example, Hornykiewicz & Kish, 1987; Cummings, 1988).

Dopaminergic Drugs and the Treatment of Movement Disorders

As noted above, most of the movement disorders under present consideration, whether idiopathic (e.g., Parkinson's) or drug-induced (e.g., EPS, TD), are suspected of being related to (1) the lowering of dopamine in the nigrostriatal pathways, and/or (2) other neurotransmitter imbalances as a result of these changes. Thus, in Parkinson's disease, the goal of drug

Table 11–6.   Motoric Side Effects of Neuroleptic Drugs[a]

Side Effect	Major Symptoms	Potential Treatment[b]
Acute dystonia	Facial grimacing Oculogyric crises Opisthotonus Torticollis Dysphagia Respiratory distress	Anticholinergics (Cogentin) (Benadryl) Benzodiazepines
Akathisia[c]	Restlessness Need to walk May appear agitated	Beta-blockers Anticholinergics Benzodiazepines
Parkinsonism	Bradykinesia Rigidity Masked facies Gait disturbances Resting tremor	Anticholinergics (Cogentin) (Artane)
Neuroleptic malignant syndrome[d]	Rigidity Fever Autonomic lability Altered consciousness Confusion/delirium Elevated liver enzymes	Medical management required as condition is associated with high mortality rate (10–20%)
Tardive dyskinesia[e]	Involuntary movements: face / mouth / lips / tongue / eyelids / jaw / extremities	No consensus[f]
Perioral tremor (rabbit syndrome)	Late onset, perioral, parkinsonian-like movements	Anticholinergics

[a] Adapted from Baldessarini (1996a)

[b] In most cases, the first response to acute reactions is to temporarily discontinue (or markedly reduce) the dosage of the neuroleptic until the crisis has passed, restarting at a lower dose, gradually increasing, if necessary, as patients frequently accommodate. A current strategy is to consider restarting with an "atypical" antipsychotic.

[c] It has been suggested that akathisia may be related more to mesocortical, rather than the nigrostriatal pathways (Marsden & Jenner, 1980).

[d] For additional reviews, see Addonizio, (1991); Caroff etal. (1991); Mann, Caroff, & Lazarus, (1991).

[e] The pathophysiology of TD is controversial, but hypersensitivity to DA, probably secondary to receptor up-regulation, appears to be most widely accepted hypothesis at present (Casey, 1995)

[f] Increasing dosage of neuroleptic may provide temporary improvement in symptoms. General recommendations, however, are to discontinue the drug, lower the dose, or switch to atypical (Kane etal, 1992; Gardos & Cole, 1995; Raja, 1996; Egan, Apud, & Wyatt, 1997; Stanilla & Simpson, 1998; Casey, 1999; Gardos, 1999).

management is to either (1) increase the supply of DA diminished by nigral degeneration, or (2) restore or prevent the imbalance between DA and other neurotransmitters, especially ACh. Table 11–7 lists some of the more common approaches and types of drugs used in the management of this disorder that will be reviewed below.

Dopamine Replacement.   The introduction of **levodopa** (ʟ-dopa), a precursor of dopamine that is able to cross the blood–brain barrier, transformed the treatment of Parkinson's. One

Table 11–7.   Drugs Used in the Treatment of Parkinson's Disease and EPS

Dopamine replacement:	Levodopa/carbidopa (Sinemet, Atamet)
Dopamine agonists:	Bromocriptine (Parlodel), pergolide (Permax), ropinirole (Requip), pramipexole (Mirapex)
Anticholinergics:	Trihexyphenidyl (Artane), benzotropine mesylate (Cogentin), diphenhydramine hydrochloride (Benedryl), ethopropazine (Parsidol), biperiden (Akineton)
MAO-B inhibitors:	Selegiline (deprenyl) (Eldepryl, Carbex)
COMT inhibitors:	Tolcapone (Tasmar)
Amantadine:	(Symmetrel) (facilitates release of DA)

drawback was that if given alone L-dopa is metabolized peripherally as well as in the brain, which increases its toxic effects. However, **carbidopa** and **benserazide** were found to interfere with the decarboxylation of L-dopa to dopamine. Since these latter compounds do not cross the blood–brain barrier when given in combination with levodopa, they allow more of the L-dopa to reach the brain, thus reducing the required dose of L-dopa, and, as a result, the peripheral side effects (e.g., nausea, hypotension).[74]

Dopamine Agonists.   Despite revolutionizing the management of Parkinson's disease, L-dopa treatment is not without potential complications [possible neurotoxic effects (Marsden & Olanow, 1998), "on/off" phenomena, nausea, hypotension], all of which appear to be dose related. In order to minimize these adverse effects, dopamine replacement therapy frequently is augmented with dopamine agonists, particularly in more advanced cases of Parkinson's disease (Standaert & Young, 1996). These drugs exert their effect by directly stimulating striatal dopamine receptors. Although not without some of the same side effects as L-dopa, the DA agonists serve to potentiate the effects of L-dopa, thereby reducing the required dose, hence reducing the potential complications of long-term use of levodopa, including possible neurotoxic effects (Koller, Silver, & Lieberman, 1994; Olanow, Jenner, & Brooks, 1998).

Anticholinergics.   Another mainstay of treatment for parkinsonism has been anticholinergic drugs. Dopamine normally exerts an inhibitory influence on ACh. When the supply of dopamine is reduced, its inhibitory influence also is diminished, which may result in a relative excess of ACh in the basal ganglia. As noted earlier (see section Clinical and Drug Effects Associated with Cholinergic Mechanisms), this imbalance between ACh and DA is thought to contribute to movement disorders associated with Parkinson's disease or antipsychotics, especially those with minimal anticholinergic effects (e.g., **haloperidol**). Because the anticholinergics used in the treatment of Parkinson's, especially **ethopropazine** (Parsidol), often are most effective in alleviating tremors (Gilroy, 1990; Koller, Silver, & Lieberman, 1994; Standaert & Young, 1996), it may be reasonable to speculate that ACh/DA imbalance likely is important in the development of this latter symptom.[75] These anticholinergics probably act primarily on striatal interneurons. However, since they do not increase DA, they are not likely to produce or exacerbate psychotic symptoms (as can be the case with levodopa or DA agonists). However, they may produce the usual anticholinergic side effects, including dry mouth, constipation, urinary retention, and in vulnerable patients (e.g., elderly, demented) confusion or other cognitive problems.

MAO-B inhibitors.   When faced with diminished supplies of DA, another potential solution is to preserve as much as possible whatever supply of DA is present by inhibiting its breakdown. As we saw earlier, the MAO enzymes are involved in the breakdown of the catecholamines. However, since MAO-B is the enzyme that is primarily at work in the striatum, if that enzyme alone could be targeted, one could avoid the problematic side effects of the nonselective MAO inhibitors used in the treatment of depression. It just so happens that **selegiline** (Deprenyl), in low to moderate dosages, is a selective inhibitor of MAO-B that can slow down the metabolism of DA in striatum. Because of its limited effect in the face of nigral degeneration, it often is used in conjunction with levodopa to reduce on–off phenomena and better stabilize the amount of DA available (Standaert & Young, 1996). In a related strategy, COMT (synaptic cleft enzyme) inhibitors also are used.

Amantadine.   Amantadine, an antiviral agent, was serendipitously discovered to benefit Parkinson's disease patients. While the mechanisms behind its efficacy are not clear, it is thought to affect the release or reuptake of DA (it also has a mild anticholinergic effect). It frequently is used in patients with mild to moderate symptoms in the early stages of treatment (Koller, Silver, & Lieberman, 1994).

Parkinsonism and Psychosis.   In terms of treatment, although apparently caused by essentially the same mechanisms responsible for Parkinson's disease (i.e., depletion of DA in the nigros-triatal pathways), the range of treatment options is more limited in neuroleptic-induced parkinsonism or other EPS symptoms. The reason is relatively simple. In Parkinson's disease, L-dopa and DA agonists are used to increase the supply of DA in the striatum. However, the effects of these drugs are not limited to the nigrostriatal pathway. They also tend to increase DA in the mesolimbic pathways. Since reduction of DA in the mesolimbic theoret-ically was behind the use of the neuroleptic in first place, use of these drugs likely would exacerbate the initial pathology (i.e., the psychosis). Consequently, the first drugs of choice in the treatment of the parkinsonian type symptoms of neuroleptic-induced EPS, as well as acute dystonic reactions, are anticholinergics (Stanilla & Simpson, 1998). As in Parkinson's, the primary mechanism of action of these drugs appears to be the restoration of DA/ACh balance in the system.[76]

Serotonin

The last of the monoamine neurotransmitters to be discussed is serotonin, or *5-hydroxytryptamine* (5-HT). In contrast to NE and DA, 5-HT (along with melatonin) is classified as an **indolamine**, being derived from tryptophan rather than tyrosine. Serotonin also has several other distinctions that set it apart from NE and DA: (1) It has many more receptor subtypes than any other neurotransmitter, (2) through its different receptors, it can exert either a facilitatory or inhibitory influence on cAMP activity, and (3) it utilizes both ligand-gated ion channel and G-protein linked receptors. First identified as important in smooth muscle contraction, it subsequently was discovered to be widely dispersed throughout the brain, where it is thought to serve both *neurotransmitter* and *neuromodulator* functions (Sanders-Bush & Mayer, 1996).[77] Finally, as we already have witnessed, 5-HT is thought to have a profound influence on a wide variety of neurobehavioral, especially neuropsychiatric functions. As we shall see, most of the more recent antidepressant and antipsychotic drugs are known for their capacity to affect serotonin receptors. We also shall see where many of these drugs with selective serotonin affinities are being used to treat an ever expanding range of psychopathological conditions.

Serotonin is synthesized in the brain (primarily in raphe nuclei) from tryptophan, an amino acid derived from dietary proteins. In this process, tryptophan is first converted into 5-hydroxytryptophan by the enzyme tryptophan hydroxylase. The next and final step is the conversion of 5-hydroxytryptophan into 5-hydroxytryptamine (5-HT, or serotonin) by the enzyme aromatic L-amino acid decarboxylase (Baskys & Remington, 1996). Like the other monoamines, the newly synthesized 5-HT is stored in vesicles in the presynaptic terminals until it is ready to be released into the synaptic cleft.[78]

After being released into the synaptic cleft, most of the serotonin is taken back up into the presynaptic terminal by specific transporter mechanisms. Once there, serotonin is converted into an inactive metabolite by MAO, primarily MAO-A. As is the case with NE and DA, the **MAO inhibitors** tend to increase the availability of 5-HT by interfering with the degradation process, whereas the TCAs and SSRIs and other newer antidepressants apparently act by blocking the 5-HT reuptake transporters. While this action alone would tend to increase the available supply of 5-HT in the synaptic cleft, other secondary reactions also may be involved. According to Blier and his associates (Blier et al., 1987, 1990; de Montigny, Chaput, & Blier, 1990), this blockage of reuptake mechanisms (with the subsequent increased availability of 5HT) causes a desensitization (down-regulation) of 5-HT$_{1A}$ somatodendritic and terminal autoreceptors that results in increased firing of serotonergic neurons, as well as increased release of 5-HT at the axon terminals.

Receptors

As in the case of dopamine, multiple subtypes of serotonin receptors have been identified (Glennon & Dukat, 1995; Baskys & Remington, 1996; Sanders-Bush & Mayer, 1996). These are divided into seven different classes, designated as 5-HT$_1$ through 5-HT$_7$, and several have been further divided into various subtypes (e.g., 5-HT$_{1A}$, 5-HT$_{1B}$, etc). As noted above, most are G-protein-linked receptors, meaning that they utilize second messengers, may have multiple effects on other transmitters and intracellular processes, and relatively speaking have slower response time. The one identified exception is the 5-HT$_3$ receptor, which is a ligand-gated, ion channel receptor. The various subtypes of 5-HT G-protein-linked receptors may exert either excitatory or inhibitory influences, depending on their effects on cAMP or phospholipase C (see Glennon & Dukat, 1995; Baskys & Remington, 1996; Sanders-Bush & Mayer, 1996, for brief reviews).

Most but not all serotonin receptors can be found in the CNS. Just as serotonin receptors can help regulate the release of other neurotransmitters (e.g., the inhibition of DA release, most likely by 5-HT$_2$ receptors), serotonin itself is regulated by a combination of autoreceptors and heteroreceptors. For example, 5-HT$_{1A}$ is thought to function as a somatodendritic autoreceptor, inhibiting the generation of impulses in serotonergic neurons when stimulated. On the other hand, 5-HT$_{1D}$ appears to act as a terminal autoreceptor, inhibiting the further release of serotonin from the presynaptic vesicles in the presence of excess 5-HT in the synaptic cleft. Norepinephrine also appears to play a role in 5-HT activity. The stimulation (by NE) of α_2 presynaptic (terminal) heteroreceptors has an inhibitory effect on the release of 5-HT, whereas stimulation of postsynaptic α_1 heteroreceptors has an excitatory effect on serotonergic neurons (Stahl, 1996). Finally, although the functional or clinical significance of many of these receptors still are under investigation, most are thought to play a role in neuropsychiatric disorders (see section Role(s) of Serotonin in Behavior).

Distribution in the CNS

The primary source of serotonin in the CNS is the **raphe nuclei**. Centered within the general area of the reticular formation, these series of midline nuclei that lie ventral (anterior) to the cerebral aqueduct and fourth ventricle, extend from the midbrain, through the pons, and into the medulla (Carpenter & Sutin, 1983; Azmitia & Whitaker-Azmitia, 1995). In general, the more rostral of these nuclei tend to project rostrally to the brain, while the more caudal of the nuclei primarily project to the brainstem and spinal cord. The former ascend primarily through the medial forebrain bundle. These pathways supply most parts of the brain,

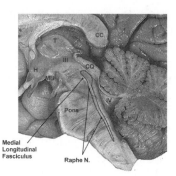 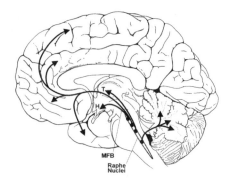

Figure 11–18. Distribution of serotonin in the CNS. The raphe nuclei provide the major source of serotonin to the brain, primarily via the medial forebrain bundle (MFB). Abbreviations: III, 3rd ventricle; IV, 4th ventricle; CC, corpus callosum; CQ, corpora quadrigemina; H, hypothalamus; MB, mammillary body; T, thalamus. Brain image (left) was adapted from the Interactive Brain Atlas (1994), courtesy of the University of Washington.

including the diencephalon, basal ganglia, limbic structures, and neocortex (see Figure 11–18) (Jacobs & Azmitia, 1992). However, these serotonergic projections are not equally distributed throughout the brain, nor are the individual receptors (Glennon & Dukat, 1995; Penny, 1996). Limbic and sensory areas of the cortex appear particularly well supplied with 5-HT receptors (Azmitia & Whitaker-Azmitia, 1995; Baskys & Remington, 1996).

Probable Role(s) of Serotonin in Behavior

Initially found to affect smooth muscle tone in the arterial vessels and internal organs, serotonin's presence also soon was discovered in the CNS (Brodie & Shore, 1957; Page, 1976). Given the multiplicity of serotonin's receptors, its capacity to interact with other neurotransmitters, and its diverse distribution in the CNS, it is easy to see how serotonin might be involved in a broad range of behaviors, albeit most likely through a complex interaction with multiple neuronal systems and neurotransmitters. Thus, in discussing the role of serotonin in behavior, Jacobs and Fornal (1995) observe that "[serotonin] is at once implicated in virtually everything, but responsible for nothing" (p. 461). Although serotonin has been implicated in numerous neuropsychiatric and neurobehavioral disorders, as well as with other body systems, particularly the cardiovascular and gastrointestinal, in keeping with the purpose of this book we will focus primarily on those behaviors associated with the nervous system. The apparent implications of serotonin in the management of schizophrenia and other psychoses, affective disorders, and extrapyramidal motor symptoms were introduced in preceding discussions. Table 11–8 lists a number of symptoms or conditions that have been associated with serotonin. Some of the more common disorders will be discussed in greater detail in the following section.

Clinical and Drug Effects Associated with Serotonergic Mechanisms

Depressive Disorders. As discussed earlier, the monoamine hypothesis of depression essentially was derived from the fact that this condition was found to respond to drugs that increased the availability of these chemical neurotransmitters, particularly NE and 5-HT. Both the MAOI and the TCA antidepressants led to an increase in NE and 5-HT at the synaptic cleft, along with DA (see Table 11–4). Although the importance of serotonin itself in depression was suggested much earlier (Coppen, 1967), it was following the success of **fluoxetine** (Prozac), which was released in the late 1980s, that serotonin's role in the

Table 11–8.   **Symptoms   and   Disorders   Associated   with   Serotonin**

Neuropsychiatric	*Neurobehavioral*
Anxiety/panic disorders	Sleep disturbances
Affective disorders	Eating disorders
Obsessive–compulsive disorders	Sexual dysfunction
Psychoses (esp. schizophrenia)	Headaches
Aggression and impulse control	Motor disturbances
Attention deficit disorder	Cognitive disorders
Phobic disorders	Tourette's syndrome
Substance abuse	Prader-Willi syndrome
Posttraumatic stress disorders	Pain
Autism	Emesis

References: Heninger (1995); Enna & Coyle (1998); Charney, Nestler, & Bunney (1999); Davidson, Putnam, & Larson (2000).

treatment of depression became firmly established (Stahl, 1994; Kamil, 1996; Musselman et al, 1998; Garlow, Musselman, & Nemeroff, 1999). Fluoxetine and the other selective serotonin reuptake inhibitors (SSRIs) that followed [e.g., **sertraline** (Zoloft), **paroxetine** (Paxil), **fluvoxamine** (Luvox), **citalopram** (Celexa), and **escitalopram** (Lexapro)] differed from the other antidepressants in several important ways: (1) they primarily blocked the reuptake of serotonin, (2) they had none of the antimuscarinic (anticholinergic), antihistaminic, or antiadrenergic effects of the TCAs, or any of the adverse effects of the MAOIs on dietary amines, and (3) they lacked the cardiotoxic or lethal overdose potential of the TCAs. Because these newer drugs were found to be frequently as efficacious as the TCAs and the MAOIs, while lacking the same adverse side effects of the latter, they have rapidly become the first-line choice for the treatment of depression (Kamil, 1996; Tollefson & Rosenbaum, 1998).[79] Figure 11–19 illustrates a proposed mechanism of action for the SSRIs, which also helps explain the delayed onset of drug effects. However, it should be reemphasized that the biochemical basis of depression (as well as the other psychiatric disorders to be discussed below) is, in all probability, not limited to any single neurotransmitter. Keep in mind the SSRIs alone are not effective in all cases of depression, and that serotonin exerts a modulating influence on numerous other transmitters. Following Stahl (2000), Table 11–9 summarizes the various classes of antidepressants

Anxiety-Related Disorders.   Not only have the SSRIs proven to be safe and effective for the treatment of depression, but they also have been found to be helpful in managing a variety of anxiety-related disorders (Lydiard, Brawman-Mintzer & Ballenger, 1996: Kent, Coplan, & Gorman, 1998). As we shall see, this is particularly true of obsessive–compulsive disorders, where SSRIs have become the drug of choice for many clinicians. What makes this situation particularly fascinating is the fact that, whereas a relative depletion of serotonin is thought to contribute to depression, it has been suggested that a relative excess of serotonin may be one of the factors contributing to anxiety states (Iversen, 1984; Eison, 1990; Sheehan, Raj, & Knapp, 1993; Kahn, Westenberg, & Moore, 1995; Stahl, 1996; Leonard, 1997). The fact that 5-HT agonists (e.g., *m*-chlorophenylpiperazine) have been shown to precipitate or enhance both generalized anxiety and panic disorders would certainly appear to support such an hypothesis (Charney, Woods, & Goodman, 1987; Germine et al., 1992).[80] However, if this is indeed the case, why then should drugs such as the SSRIs and **buspirone** (which can

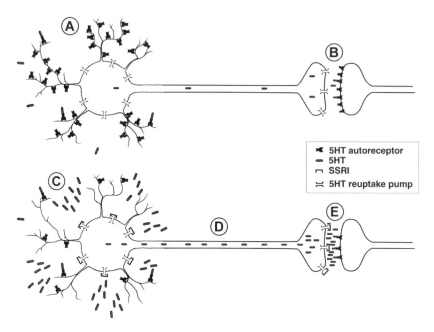

Figure 11–19. Mechanism of action for SSRIs. In a depressed state there is thought to be a relative depletion of 5-HT both in the somatodendritic area of the 5-HT neuron in the (a) raphe nuclei, (b) as well as in its terminal branches. The initial critical effect may be to block the somatodendritic reuptake pumps causing an increased concentration of 5-HT in the somatodendritic areas. As a consequence, there is a gradual down-regulation of (c) somatodendritic autoreceptors. This down-regulation of somatodendritic autoreceptors is thought to eventually result in a disinhibition of the 5-HT neuron, causing (d) increased production/release of 5-HT, which in turn results in an (e) increase of 5-HT at the terminal branches and subsequent down-regulation of postsynaptic receptors. After Stahl (2000).

increase 5-HT) be used as anxiolytics? Before addressing this question, it might useful to briefly review the role(s) of serotonergic drugs in treating anxiety related disorders.

Among anxiety driven disorders, probably none is so closely linked to serotonin as is **obsessive–compulsive disorder** (OCD). For at least the past 30 years, it has been recognized that among the TCAs **clomipramine** (Anafranil) alone was effective in the treatment of OCD (Renynghe de Voxurie, 1968). What sets clomipramine apart from the other TCAs is the fact that, while not specific to it is a particularly potent serotonin reuptake inhibitor (Goodman et al., 1990; Janicak, Davis, Preskorn & Ayd, 1997). With the advent of the SSRIs (which primarily target 5-HT), along with clomipramine they have become the mainstay for the pharmacological management of this disorder (McDougle, Goodman, Leckman, & Price, 1993; Goodman & Murphy, 1998).[81] In spite of the frequent treatment success with serotonergic enhancing drugs, the biological and neurochemical basis of OCD remains largely unknown.

Despite the continuing uncertainty regarding the exact biochemical mechanisms underlying OCD, there does appear to be widespread agreement that anatomically the orbitofrontal and/or cingulate pathways through the basal ganglia probably are critically involved (Stahl, 1988; Baxter et al., 1992, 1996; Insel, 1992; Swerdlow, 1995; Schwartz, 1997; Baxter, 1999). It has been hypothesized that obsessive compulsive symptoms may reflect a dopamine/serotonin imbalance in these circuits (Goodman et al., 1990; McDougle, Goodman, Leckman, & Price, 1993), and/or an imbalance between the direct and indirect striatal–pallidal pathways (Saxena, Brody, Schwartz, & Baxter, 1998). Additional support

Table 11–9. Classes of Antidepressants

SSRIs:	Selective Serotonin Reuptake Inhibitors. Examples: citalopram, escitalopram, fluoxetine, fluvoxamine, paroxetine, sertraline.
SARIs:	Serotonin 5-HT$_{2A}$ Antagonist and Reuptake Inhibitors. Potent antagonist effect on postsynaptic 5-HT$_{2A}$ receptors and less potently block the reuptake of 5-HT. Stimulation of the postsynaptic 5-HT$_{2A}$ receptors tends to mitigate the effects of postsynaptic stimulation of 5-HT$_{2A}$ receptors. Blocking of the 5-HT$_{2A}$ receptors enhances the effects of 5-HT$_{1A}$ stimulation (an antidepressant effect). Examples: nefazodone, trazodone.
SNRIs:	Serotonin and Norepinephrine Reuptake Inhibitors. Examples: duloxetine and venlafaxine.
NDRIs:	Norepinephrine and Dopamine Reuptake Inhibitor (tend to be more activating). Example: bupropion.
NRIs:	Norepinephrine Reuptake Inhibitors. While potentially more "activating," no NRI (such as Roboxetine) has yet been approved for depression in the U.S., although desipramine and protriptyline have strong NRI effects.)
NaSSA:	Norepinephrine and Specific Serotonergic Antidepressant. Blocking of alpha-2 presynaptic NE autoreceptors and alpha-2 heteroreceptors on 5-HT neurons results in increased release of NE and 5-HT, respectively. Increased NE also impacts on alpha-1 postsynaptic somatodendritic receptors in the raphe nuclei, results in increased firing of 5HT neurons. Blocking of H1 and other 5-HT receptors reduces nausea and sexual dysfunction but can cause sedation and weight gain. Example: mirtazapine.
TCAs:	Tricyclic Antidepressant. They primarily block reuptake of 5-HT and NE. Increased side effects result from antagonism of H1 (histaminic), alpha-1 (adrenergic), and M1 (muscarinic) receptors. Examples: amitriptyline, amoxapine, desipramine, doxepin, imipramine, nortriptyline.
MAOIs:	Monoamine Oxidase Inhibitors. They inhibit breakdown of NE, DA, and 5-HT in presynaptic terminals. Examples: tranylcypromine, phenelzine.

for these speculations have been derived from the fact that (1) a large percentage of patients suffering from **Tourette's** (a basal ganglia disease thought to be associated with excessive dopamine activity) manifest obsessive–compulsive symptomatology (Stahl, 2000; Goodman & Murphy, 1998), and (2) that **cingulotomies** (Jenike et al., 1991; Baer et al., 1995) have proven effective in relieving obsessive–compulsive symptoms. Whether chemical, surgical, or perhaps even behavioral treatment approaches are employed, one model suggests that these interventions help to disrupt what has been referred to as pathologically ritualized *worry circuits*, which in turn are thought to be mediated in part by orbitofrontal–limbic–striatal loops (see Swerdlow, 1995).

Although serotonergic drugs also have proven effective in the management of other anxiety-related disorders, they do not enjoy the same exclusive prominence in the treatment of these other conditions as they do with OCD. While **buspirone** (BuSpar), a partial serotonergic agonist, frequently is used for the more chronic management **of generalized anxiety disorder** (GAD) (in large part because of its safer side effect profile), the lower-potency benzodiazepines (BZDs) also are commonly used and probably have somewhat greater overall effectiveness, particularly if used during acute episodes (Hyman, Arana, & Rosenbaum, 1995; Janicak, Davis, Preskorn & Ayd, 1997; Lader, 1998). For the more long-term management of GAD, however, the demonstrated efficacy of various antidepressants have led to their increased use in the treatment of this disorder (Rocca et al., 1997; Casacalenda & Boulenger, 1998; Conner & Davidson, 1998; Davidson, DuPont, Hedges, & Haskins, 1999; Goodnick, Puig, DeVane, & Freund, 1999).

With **panic disorder**, it is even more critical to differentiate between acute attacks versus long-range management when considering pharmacological management. Once again, the benzodiazepines, in this case, a short-acting, high potency BZD such as **alprazolam** (Xanax), **clonazepam** (Klonopin), or **lorazepam** (Ativan) are the preferred first-line choice in quelling acute attacks. Although BZDs also are used for the long-term maintenance of panic disorder, antidepressants are becoming increasingly popular for long-term management (Boyer, 1995; Raj & Sheehan, 1995; Sheehan & Harnett-Sheehan, 1996; Janicak, Davis, Preskorn & Ayd, 1997; Goddard & Charney, 1998; Kent, Coplan, & Gorman, 1998; Pine, Grun, & Gorman, 1998; Goddard, Coplan, Gorman & Charney, 1999). Among available antidepressants, the SSRIs are often the drugs of choice because of their combined effectiveness and more benign side effects.[82]

Two other anxiety-related conditions that are known to respond to SSRIs are **social phobias** and **posttraumatic stress disorders** (**PTSD**). **Social phobia** is defined as a fear of certain types of social situations, especially, although not necessarily exclusively, where one might be expected to perform, and thus where there is the perceived potential to suffer social embarrassment. Unlike other specific phobias (e.g., fear of snakes) that with few exceptions (Balon, 1999) respond better to cognitive/behavioral therapies, *social phobias are more amenable to psychotropic interventions*. In addition to the use of BZDs and beta-blockers for acute, episodic relief of social phobia, various antidepressants, particularly SSRIs, MAOIs, and other atypical antidepressants such as **venlafaxine** (Effexor) have been found to be reasonably effective in treating this disorder (Keck & McElroy, 1997; Ballenger et al., 1998; Bouwer & Stein, 1998; Lyiard, 1998; Allgulander, 1999; Altamura, Pioli, Vitto, & Mannu, 1999; Pollack, 1999; Stein, Fyer, Davidson, et al., 1999).

PTSD is a complex disorder, typically encompassing elements of a variety of psychological disorders, including:

1. Generalized anxiety (*hyperreactivity*).
2. Phobic (avoidance) and/or paranoid reactions (*hypervigilance*).
3. OCD (intrusive, disturbing thoughts or recollections).
4. Depression (anhedonia, blunted affect).
5. Sleep disturbances, including nightmares.
6. Impulse control problems (anger, irritability).

Because of the complexity of this syndrome, it is not surprising that multiple transmitter and other biological systems have been linked to this disorder (van der Kolk, 1997). While multiple psychopharmacological strategies are employed in the management of this disorder, including anxiolytics, neuroleptics, anticonvulsants, and mood stabilizers, the mainstay of most treatment paradigms are antidepressants, especially the SSRIs (Davidson, 1997, 1999; Davidson & Connor, 1999; Davis et al., 1997; Dow & Kline, 1997; Meek & Kablinger, 1998; Clark et al., 1999).[83] However, even with a polypharmacy approach, response to treatment in PTSD may be limited (Shalev, Bonne, & Eth, 1996; Connor, Davidson, Weisler, & Ahearn, 1999; Hidalgo et al., 1999), which perhaps should not be too surprising given the complexity of the disorder. To make matters even more complicated, Southwick and colleagues (1997) suggest there may be at least two major subgroups of PTSD patients: one that might respond better to noradrenergic drugs and the other to drugs that maximally affect the serotonergic system.

5HT Agonists and SSRIs as Both Anxiolytics and Antidepressants.   At the beginning of the preceding section, it was noted that while depression was thought to be related to a relative monoamine deficiency, an excess of serotonin was believed to be linked to anxiety disorders.

Although, as noted by Stahl (1996), this clearly represents an oversimplification of both disorders,[84] even if partially correct this raises some interesting questions as to why and how the same medications might be used to treat both conditions. Before briefly considering the potential roles of two different classes of serotonergic drugs (SSRIs, azapirones) on anxiety and depression, it should be noted that as in depression the neurobiological substrates of the anxiety disorders still are largely speculative (Conner & Davidson, 1998; Goddard & Charney, 1998). However, since the same drugs (the SSRIs, in particular) are effective in treating both anxiety and depression, there is likely to be some common linkages (Eison, 1990; Casacalenda & Boulenger, 1998).

Consider the case of **buspirone** (BuSpar), an anxiolytic with mild antidepressant potential. Buspirone and related azapirones (e.g., **ipsapirone** and **gepirone**) are considered partial 5-HT agonists. That means, recalling our earlier discussion of partial agonists, these drugs may have either a *net agonist* or a *net antagonist* effect depending on whether they are competing with a full agonist or with an antagonist. A similar phenomenon is hypothesized to occur in states of relative excess (anxiety states) or deficiency (depression) of serotonin. We should note that despite the fact that it may seem counterintuitive, studies have suggested that increases in serotonin resulting from the chronic administration of an SSRI result in a down-regulation of both somatodendritic and terminal 5-HT_{1A} autoreceptors (see Blair et al., 1990; de Montigny et al., 1990). Similarly, Stahl (1996) has postulated a comparable desensitization of 5-HT_{1A} autoreceptors in states of anxiety (excess serotonin).

One theory is that buspirone acts as a full agonist at presynaptic 5-HT_{1A} autoreceptors, while acting as a partial agonist at postsynaptic serotonergic receptors (Eison, 1990; Casacalenda & Boulenger, 1998). By acting as a full agonist at the presynaptic autoreceptors, this drug would tend to reduce firing of serotonergic neurons, as well as inhibit the release of additional 5-HT at the presynaptic terminals. By acting as a partial agonist postsynaptically, it would be expected to have a "net antagonistic" effect on the postsynaptic receptors (where there already is n hypothesized excess of 5-HT).[85] In this case, the agonist effects on the presynaptic autoreceptors might be expected to result in their down-regulation, leading to an eventual increase in serotonergic activity.

Again, the situation seems paradoxical. If anxiety is associated with an excess of 5-HT, would not this further increase in 5-HT activity exacerbate the anxiety? Two points should be considered. First, if the reduction in 5-HT activity is indeed anxiolytic, then the response to buspirone should be immediate, but it is not. Like the antidepressants, BuSpar's anxiolytic effect is delayed, which could be explained by the time it takes for the autoreceptors to down-regulate. Second, the SSRIs, which also have been shown to have anxiolytic effects (see below), also traditionally have been associated with increases in serotonergic activity.

Although buspirone seems to have minimal effects on clinical depression (it is occasionally used to augment other antidepressants), we briefly can consider how that might work. In the case of depression (decreased serotonergic output), it is speculated that there is a relative proliferation of somatodendritic receptors. In this case, buspirone's initial agonist effect (i.e., stimulation of the 5-HT_{1A} receptors) would tend to inhibit the firing of the serotonergic neurons, thus allowing them time to replenish their depleted stores of 5-HT. At the same time, buspirone would be expected to have a net agonist effect on the minimally stimulated postsynaptic 5-HT receptors. Meanwhile, the chronic agonist effects of buspirone on the autoreceptors eventually would be expected to result in a desensitization of 5-HT_{1A} somatodendritic autoreceptors, thus facilitating the activity of these serotonergic neurons, which results in an increased supply of 5-HT at the postsynaptic terminals of the corticolimbic pathways.

Now let us consider the possible actions of the SSRIs and the role of serotonin in anxiety-related disorders. As previously noted, the mechanism of action of the SSRIs in depressive illness is thought to be related, at least in part, to an increase in serotonergic activity and the gradual desensitization (down-regulation) of both the presynaptic and postsynaptic 5-HT receptors. The question is, how do the SSRIs assist in the alleviation of anxiety-related syndromes that, as has been suggested, may be characterized by an excess of serotonin? First, as noted in Kent, Coplan, and Gorman (1998), it is *unlikely* that this effect is related to "a paradoxical reduction in 5-HT neurotransmission." Recall that 5-HT was thought to serve both neurotransmitter and neuromodulator functions. For example, 5-HT is known to modulate other neurochemicals, including NE and cholecystokinin, and is involved in various neuroanatomical structures, such as the amygdala, hypothalamus, and related pathways, all of which have been associated with the expression of anxiety (Conner & Davidson, 1998; Goddard & Charney, 1998; Kent, Coplan, & Gorman, 1998). It is suspected that despite the previous evidence associating increased 5-HT with anxiety (mostly from animal models), in humans at least selective increases in 5-HT likely can have antidepressant as well as anxiolytic effects.[86],[87]

Schizophrenia. The role of serotonin in psychosis, particularly schizophrenic illness, was first suggested in the 1950s when it was noted that lysergic acid diethylamide (LSD), which is chemically related to serotonin, evoked psychotic type symptoms (Wooley & Shaw, 1954). Clozapine (Clozaril), a potent $5-HT_{2A}$ antagonist, was discovered to have antipsychotic effects in the mid-1960s, but was withdrawn at least from use in the United States after it became associated with the development of agranulocytosis (Hippius, 1989). It later was approved for use in this country in 1990. Since then, various other "atypical" antipsychotics with similar $5-HT_{2A}$ antagonistic properties not only have been introduced, but generally have replaced the older generation neuroleptics as the "first-line" treatment choice for schizophrenia, both because of their clinical efficacy and their more favorable side effect profile (Owens & Risch, 1998; Markowitz, Brown, & Moore, 1999; Meltzer, 1999; Meltzer & McGurk, 1999; Stahl, 1999; Remington & Kapur, 2000).[88] Nonetheless, the conventional, older neuroleptics still are seen as having a significant role in the management of these disorders (Feltus & Gardner, 1999; Remington & Chong, 1999).

As noted earlier, among the major advantages cited for these atypical ("novel") drugs are their (1) improved response rates in previously treatment-resistant patients, (2) greater improvement in "negative" symptoms and improved cognition, along with good response to "positive" symptoms, in addition to (3) possible therapeutic impact on mood disorders, and (4) markedly reduced EPS and TD. These effects would appear to be largely although probably not exclusively due to the differential affinities to the $5-HT_{2A}$ and D_2 receptors and the apparent relative affinity of these drugs for D_2 receptors in the mesolimbic cortex (Kapur & Remington, 1996; Arnt, 1998; Owens & Risch, 1998).[89]

The mechanisms of action of these "novel" antipsychotics were introduced earlier under athe section Dopaminergic Drugs and the Treatment of Schizophrenia. For the sake of brevity, several key points made earlier will be reviewed (and occasionally expand upon) here. These are:

1. Serotonin, via $5-HT_{2A}$ receptors, exerts an inhibitory influence on dopaminergic neurons.
2. While clozapine (Clozaril), along with other newer, "atypical" neuroleptics, have an antagonistic effect on both $5-HT_{2A}$ and D_2 receptors, they have much higher affinity for the $5-HT_{2A}$ than for the D_2 receptors.
3. The *positive* symptoms of schizophrenia are thought to be related to *increased* dopamine activity in the mesolimbic region.
4. *Negative* symptoms are thought to be associated with *decreased* dopaminergic activity in the prefrontal or mesocortical areas.

5. EPS and tardive dyskinesia (TD) have been linked to the excessive blockade of D_2 receptors in the nigrostriatal system.
6. Because of their capacity to block 5-HT_{2A} presynaptic terminal receptors in the striatum, as well as blocking 5-HT_{2A} somatodendritic heteroreceptors on the primary DA neurons in the brain stem (e.g., SN and VTA nuclei), these newer antipsychotics result in a disinhibition of dopaminergic neurons. This combined with their relative weak binding to D_2 receptors in the striatum prevents the relative depletions of striatal DA that has been linked to EPS and TD with the conventional neuroleptics.[90]
7. A similar process is thought to occur in the mesocortical (prefrontal) regions that again are thought to result in a disinhibition of dopaminergic neurons. The resulting relative increase in DA in these frontal cortical regions is thought to be instrumental in helping alleviate the "negative" symptoms of schizophrenia and improving cognition.[91]
8. If the same pattern of receptor interactions were to occur in the mesolimbic region as described for the prefrontal and striatal areas, the resulting increase in DA activity theoretically would exacerbate the "positive" symptoms of schizophrenia. Clearly this does not happen with the novel antipsychotics. The explanation is twofold. First, although these drugs have a relatively weak affinity for D_2 receptors (where they have an antagonistic function), they do show strong affinities for mesolimbic neurons in general (80 to 90% according to Farde et al., 1989); hence, they selectively exert an inhibitory influence on the DA neurons in this area. Second, there appears to be a relative paucity of 5-HT_{2A} receptors in the mesolimbic region, so that the antagonistic action on the D_2 receptors is relatively unopposed by the potentially disinhibitory effects of 5HT blockade.
9. While the novel antipsychotics bind to D_2 receptors, their affinity for these receptors is weaker, and thus more "transient," than is the case with the older neuroleptics. This more transient binding, while sufficient to impact on the positive symptoms of schizophrenia, tends not to produce EPS or elevated prolactin levels.

Again, following Stahl (2000), some of the above relationships can be seen graphically in Table 11–10.

Aggression. As in the case of most complex behaviors, aggression and impulse control problems are multiply determined by such factors as genetics, early environmental or learning experiences, immediate environmental circumstances, and current psychological state of the organism. Also, as in the case of the other behaviors just discussed, neurochemical activity also can influence the expression of aggression. Among the various neurotransmitters, serotonin has been most closely linked to aggressive behavior across multiple species, from fish to man (Enserink, 2000; Nelson & Chiavegatto, 2001; Walsh & Dinan, 2001). More specifically, it has been suggested that relative depletions or dysregulation of serotonin, particularly in the frontal systems, may constitute a contributing factor to pathological aggression in man (Davidson, Putnam, & Larson, 2000; Filley et al., 2001; Lee & Coccaro, 2001). Consequently, it also has been suggested that drugs that enhance serotonin levels may be useful in managing some forms of aggressive behavior (Cherek et al., 2002; Kim, Moles, & Hawley, 2001; Mintzer 2001; Ryan, 2000).[92]

Side Effects Associated with Centrally Acting Serotonergic Drugs

As a general rule, the centrally acting serotonergic drugs discussed thus far (SSRIs, "atypical" antipsychotics, and buspirone) have a more favorable side-effect profile as compared to the TCAs, conventional neuroleptics, and BZDs. Because of the greater purity of the SSRIs, it might be tempting to attribute their side effects to the effects of increased 5-HT. It should

Table 11–10. Suspected Mechanisms of Action for Antipsychotics

Sites of Origins of Neurotransmitters	Flow	Principle Sites of Action of Antipsychotics	Effects of "Conventional" Drugs	Effects of "Atypicals"
Hypothalamus **Raphe nuclei**	DA→ 5-HT→	**Pituitary Gland** Prolactin release is normally inhibited by dopamine and facilitated by serotonin	Blocks D_2 receptors, resulting in increased prolactin levels	Blocking 5-HT$_{2A}$ receptors inhibits release of prolactin, thus compensating for any D_2 blockade in maintaining normal prolactin levels[a]
Substantia nigra (pars compacta) **Nucleus basalis of Meynert** **Raphe nuclei**	DA→ ACh→ 5-HT→	**Striatum** Motor control requires delicate balance of neurotransmitters	BlocksD_2receptors, thus increasing probability of EPS by disrupting DA/ACh balance	Blocking of 5-HT$_{2A}$ receptors, along with decreased affinity for D_2 receptors, helps maintain dopamine levels[b]
Ventral tegmental area / SN (pc) **Raphe nuclei**	DA→ 5-HT→	**Mesolimbic Cortices** Excess of dopamine thought to underlie the "positive" symptoms of schizophrenia	Blocking of D_2 receptors is thought to reduce the "positive" symptoms of schizophrenia	Blocking of D_2 receptors, along with decreased presence of 5-HT$_{2A}$ receptors, help lower dopamine levels[c]
Ventral tegmental area / SN (pc) **Raphe nuclei**	DA→ 5-HT→	**Mesocortical Cortices** Dopamine deficit thought to underlie the "negative" symptoms of schizophrenia	Blocking of D_2 receptors thought to possibly exacerbate the "negative" symptoms of schizophrenia	Blocking of 5-HT$_{2A}$ receptors thought to disinhibit the release of DA, overcoming the theoretical deficit

[a]Blocking of 5-HT$_{2A}$ receptors also possibly could have a disinhibiting effect on dopamine.
[b]More transient blocking action of the atypicals may also help reduce EPS.
[c]Aripiprazole's effect influenced by the fact that it is a partial agonist.

be recalled, however, that serotonin may directly or indirectly affect other transmitter systems. Thus, the presence or absence of particular side effects is not necessarily the result of agonist or antagonist effects of these drugs on 5-HT receptors, but rather the effects these particular compounds may have on other receptors. For example, one of the major advantage of SSRIs is their failure to block cholinergic and histaminic receptors, while some of the positively expressed side effects of the serotonin–dopamine antagonists (atypical antipsychotics), particularly clozapine, may be attributable to the numerous receptors to which they actively bind.[93] The major side effects of the SSRIs that are associated with stimulation of 5-HT$_{2A}$ receptors are agitation, anxiety, panic, insomnia, akathisia, and sexual dysfunction (usually both decreased libido and anorgasmia), whereas nausea and vomiting, gastrointestinal distress, and headaches are thought to be associated with excitation of 5-HT$_3$ receptors (Stahl, 1997).[94]

Although relatively rare, one potentially serious (deadly) adverse effect of SSRIs is what has been termed the *serotonin syndrome*. This syndrome, which results from an excess of

5-HT, is characterized by changes in mental status (delirium, affective, behavioral disturbances), motor (restlessness, myoclonus, hyerreflexia, tremor), and autonomic activity (fever,
sweating, tachycardia, increased blood pressure) (Sternbach, 1991; Lane & Baldwin, 1997).
This syndrome is most likely to occur when two or more serotonin-enhancing drugs are
given simultaneously or in too close proximity of one another (e.g. an SSRI and a MAOI)
(Tollefson & Rosenbaum, 1998).

NEUROPEPTIDES

Throughout this chapter, as attention was focused on specific neurotransmitters, care was
repeatedly taken to emphasize the complex and interactive nature of these chemicals. It was
noted that behaviors, although often linked to a particular transmitter (e.g., dopamine with
Parkinson's disease and schizophrenia), in fact, were multiply determined, that is, multiple
transmitter substances are likely involved. The advancements in the study of neuropeptides
in the latter part of the 20th century only have added to this complexity. While it is certainly
well beyond the intended scope of this chapter to attempt a discussion of neuropeptides, for
the sake of completeness, at least a few general observations are in order. For more detailed
review, see: de Wied, 1997a; Strand, 1999; Sandman, et al., 1999.

Peptides are composed of sequences of amino acids. Initially associated with the
endocrine glands and gastrointestinal functions, by the 1970s a number of peptides
were noted to impact directly on the nervous system. These specialized peptides, which
subsequently were found to be present even in one-celled organisms, became known as
neuropeptides. Synthesized within the ribosomes of nerve cells, as well as elsewhere in the
body, neuropeptides provide an additional mechanism for chemical communication both
within the CNS and between neurons and other organ systems. Many of the neuropeptides are closely linked to the hypothalamus and pituitary gland (including, **adreno-
corticotropin, oxytocin, vasopressin, somatostatin, enkephalins, endorphins, cholecystokinin, luetinizing, growth, gonadotropin,** and **thyroid stimulating hormones**). These are
important in maintaining homeostasis and controlling other basic physiological functions,
such as eating, drinking, blood pressure, temperature control, sexual development and
sexual function, pH balance, and respiration. However, neuropeptides also have a more
direct impact on general behavior, sensorimotor functions, cognition, including learning and
memory, and affective and emotional responses. They also are believed to play an important
role in certain disorders or disease states (see also Kovacs & de Wied 1994; Bennett, Ballard,
Watson & Fone 1997; de Wied 1997b; Schulkin, 1999).

One of the more fascinating aspects of neuropeptides is their functional flexibility. The
same neuropeptide alternately may assume the role of a neurotransmitter, a hormone, or a
neuromodulator and it can have excitatory or inhibitory effects depending on the particular
site or conditions under which it acts. Here we briefly will consider the role of neuropeptides
as neuromodulators.

Neuromodulators, unlike neurotransmitters, do not have the capacity to initiate an action
potential in a nerve cell in and of themselves, despite the fact that they are considerably
more potent than the latter. However, they do have the capacity to alter some of the
basic physiology of the cell, such as the synthesis or degradation of enzymes or receptors,
thus influencing the reaction of the neuron to neurotransmitters. In fact, neuromodulators
typically work in conjunction with neurotransmitters, frequently being released from the
same presynaptic terminal similar to a *co-transmitter*. However, in addition to being unable
to independently initiate an action potential, neuropeptides as neuromodulators differ from
the neurotransmitters discussed thus far in several other important ways. First, they appear

to act exclusively on G-protein-linked receptors, thus their actions are slower and often more prolonged than those transmitters that directly affect ion channels. Second, their release may be more prolonged (apparently dependent, in part, on such factors as the pattern and intensity of stimulation) and their distribution (hence, their influence) is more widespread than that of the "classic" neurotransmitters. Thus, in general, these neuromodulators seem to provide a more prolonged or relatively long-term change in the "tone" or "predisposition" of the postsynaptic neurons or neuronal systems. One other difference that is important to note is that being produced in the nucleus of the nerve cell and being dependent on axonal transport to reach the terminal boutons (as well as not being "recycled" back into the presynaptic terminal like the transmitters), it is likely that available supplies of neuropeptides may be influenced by certain tonic behavioral states. Although still under intense investigation, it is clear that the presence of neuropeptides as neuromodulators adds not only more complexity to an already complex system, but at the same time offers considerably more behavioral flexibility as well as a challenge to understand their contribution to this most remarkably sophisticated and mysterious entity: the central nervous system.

Endnotes

1. At rest, NA^+ and chloride (Cl^-) are concentrated outside the cell membrane, while K^+ and mostly negatively charged, large-molecule amino acids and proteins (which do not diffuse through the membrane) are concentrated intracellularly. This results in a resting potential in which the intracellular space is negative (approximately $-70\,mV$) compared to the extracellular spaces).

2. Electrical synapses have been found in some organisms, but at least in man, they appear to be quite rare.

3. Norepinephrine, which is associated with sympathetic arousal, utilizes both alpha and beta receptors, each of which is further subdivided into two types. Alpha$_1$ receptors primarily are associated with vasoconstriction, while α_2 receptors are *autoreceptors* (see below) associated with feedback mechanisms in adrenergic neurons. Beta (as well as alpha) receptors are found both in the brain and in peripheral organs, particularly the heart (β_1), lungs, and blood vessels (β_2). Under conditions of stress, norepinephrine (noradrenalin), by stimulating the beta receptors, will result in increased heart rate and force of contractility, as well as dilation of the bronchioles and blood vessels.

4. This is why GABA is considered an inhibitory neurotransmitter.

5. Different drugs or molecules other than the primary agonist may bind to these receptors, either augmenting, diminishing, blocking, or reversing the action of the agonist (see below).

6. This is the case with *irreversible* or *noncompetitive antagonists*. Some drugs, however, are classified as *competitive antagonists*. This means that while they block the receptor, preventing the binding of agonists, their effect is not insurmountable. Given a sufficient amount of an agonist, this blockade can be overcome. In the case of drugs, this would mean that higher levels of an agonist would be required in the presence of a competitive antagonist to produce an effect. For example, the beta-blocker, atenolol, which acts as a competitive antagonist in the heart, reduces one's resting heart rate by blocking the natural agonist, norepinephrine. However, as one begins to exercise, the subsequent sympathetic arousal leads to an increase in the availability of norepinephrine. This increase in NE overcomes the effects of the beta-blocker, allowing the heart rate to increase to meet the demands of increased physical activity. As will be seen, a similar phenomenon occurs with the blocking of enzymes.

7. Thus, an inverse agonist, rather than an antagonist, is the opposite of an agonist.

8. See also section, *Norepinephrine and Depression*, p. 585.

9. Many different areas of the CNS have similar receptor subtypes (e.g., D_2 receptors), thus limiting the specificity of many drugs, including the neuroleptics. Thus, in the latter case, adverse effects may result from blocking D_2 receptors in the tuberoinfundibular and mesocortical pathways. These will be reviewed when the dopamine pathways are discussed in greater detail.

10. The lower potency neuroleptics, in contrast (such as chlorpromazine), tend to have less effect on the nigrostriatal pathways, but tend to have enhanced anticholinergic side effects, similar to the tricyclics.

11. The criteria for establishing a substance as a neurotransmitter has varied somewhat over time, as well as from author to author. However, the following represent fairly commonly agreed upon criteria:

    • The substance should be produced (synthesized) within and released by the neuron, usually but not necessarily at the presynaptic terminal.

    • The release takes place as a result of the electrical activation of the presynaptic neuron.

    • There is some means of terminating the action of the substance once it has produced its effect (e.g., reabsorption into the presynaptic membrane or being destroyed by enzymes within the synaptic cleft).

    • Application of the identified substance directly to the postsynaptic membrane should produce an effect comparable to its natural release by the presynaptic neuron.

12. ACh also has been associated with disturbances of other behavioral phenomena such as mood, affect ,and sleep, although overall the data seem less robust than for its effect on cognitive functions (Janowsky et al., 1972, 1974; Yoemans et al., 1984; Hasey & Hanin, 1991: Hasey, 1996; Devinsky, Kernan, & Bear, 1992). One bit of evidence that has been forwarded with regard to this hypothesis is the behavioral changes that are associated with smoking and its withdrawal. However, it also is unclear to what extent the observed effects of ACh receptor stimulation are due directly to ACh versus its impact on other neurochemical systems (Levin & Simon, 1998).

13. It has been argued, however, that lesions of the basal forebrain may affect memory in a more indirect manner by disrupting attention (Voytko et al., 1994).

14. While partially effective, these drugs are less commonly used now in large part because of their undesirable side effects (see below).

15. Extrapyramidal side effects are at a lower risk for occurrence with the use of the lower-potency, typical neuroleptics, such as *chlorpromazine* (Thorazine) and *thioridazine* (Mellaril), because these drugs have their own intrinsic anticholinergic properties. In contrast, one of the advantages of the newer, atypical antipsychotic medications, such as risperidone or olanzapine, is that while most have minimal anticholinergic effects, like the high-potency, typical neuroleptics, they still manage to have minimal extrapyramidal effects. Most likely, this results from their ability to block serotonin (specifically, $5\text{-}HT_{2A}$ receptors) in the nigrostriatal system whose effect is normally to inhibit the release of DA (see under section Dopamine for more detail).

16. This includes certain antiarrythmics, antihistamines, antipsychotics (especially the phenothiazines), and the tricyclic antidepressants.

17. Positively charged ions.

18. It may also allow passage to some monovalent ions (e.g., NA^+), but it is primarily known as a CA^{2+} channel.

19. For the purpose of this chapter, the various subtypes of mGluRs need not be discussed. For brief reviews of the subtypes of metabotropic receptors, see MacDonald, Wojtowicz, and Baskys (1996) and Cotman et al. (1995).

20. Calcium, when released into the cell by NMDA receptors, itself can function as a second messenger, initiating other intracellular changes. For example, it can facilitate the production of nitric oxide (a neurotransmitter in its own right), which also is believed to be involved in long-term potentiation and memory. Through its ability in turn to facilitate the synthesis of cyclic GMP that relaxes smooth muscles in the arterioles, nitric oxide is critical for penile erections.

21. Such as NE, DA, and 5-HT, which will be discussed later.

22. Recall that many neurotransmitters, including glutamate, GABA, and the monoamines, frequently interact with one another either in a facilitatory or inhibitory manner, thus adding to the diversity and breadth of their potential effects.

23. The reader should keep in mind that when one speaks of depletions of glutamate or glutamate receptor hypofunction, this does not imply that **all** glutamatergic pathways or receptors are necessarily involved. Subtle differences in the subunits making up receptors (even within a major subtype), as well as their patterns of interconnections with other neurotransmitters, can affect their behavioral expression.

24. Other neuropsychiatric conditions, at least in part, may be related to glutamatergic factors. For example, Tsai, Gastfriend, and Coyle (1995) and Tsai and Coyle (1998) present cogent arguments relating many of the effects of chronic alcohol abuse to changes in glutamate activity, including (1) acute intoxication (NMDA inhibition), (2) alcoholic withdrawal and seizures (up-regulation of NMDA receptors), (3) blackouts (impaired long-term potentiation), (4) Wernicke–Korsakoff syndromes and neuronal atrophy (excitotoxicity secondary to NMDA supersensitivity), and (5) fetal alcohol syndrome (diminished NMDA receptors with resulting selective impairment of neuronal development). Posttraumatic stress disorder (PTSD) would appear to be another likely candidate for excitotoxic changes due to prolonged or severe states of arousal (stress); however, a review of the recent literature failed to document such findings.

25. A carboxyl group is a carbonyl group (a carbon and oxygen atom attached by a double bond) to which a hydroxyl group (OH) is attached (to the carbon atom).

26. A third type, GABA$_C$, reportedly has been identified, but its clinical relevance is not clear (see Davies, 1996).

27. Recall that when a transmitter binds to its own presynaptic (terminal) autoreceptors, its primary function is to regulate the release (or synthesis) of the transmitter (in this case, GABA) from the presynaptic terminal.

28. Different drugs effect this response in different ways. For example, BZD increases the frequency with which the ion channel can be opened, whereas barbiturates tend to prolong the duration of the opening (at lower dosages), while increasing the size of the channel opening at higher dosages. However, both BZDs and barbiturates can increase the affinity of the GABA receptor to GABA (see Paul, 1995).

29. **Midazolam** (Versed) is a BZD used for certain medical procedures, specifically because of its sedative and amnestic propensities.

30. A number of psychotropic medications (e.g., phenothiazines, tricyclic antidepressants) have the potential for exacerbating or causing seizures, especially if abused or given to patients with preexisting seizure disorders (Malatynska et al., 1998; Zorumski & Isenberg, 1991). Among some of the newer generation drugs, bupropion (Wellbutrin) and clozapine (Clozaril) are well known for their capacity to induce seizures in certain individuals. Although the mechanisms behind these effects in these latter two drugs

apparently have not been clearly established, it is likely that $GABA_A$ receptors may be involved (Squires & Saederup, 1997). Similarly, alterations in the GABAergic response, as well as that of other neurotransmitters (e.g., norepinephrine), may help explain the anxiogenic effects of some medications.

31. It is thought that some GABA neurons in the spinal cord may have a depolarizing effect (Cooper, Bloom, & Roth, 1991, p. 161).

32. The functional implications of $GABA_B$ receptors are less well known at the present time, and hence, generally have not been targeted as extensively in clinical drug research. Baclofen, which has a specific affinity for $GABA_B$ receptors, shares muscle relaxation properties with the BZDs, but likely accomplishes this effect through a separate mechanism (i.e., reduction of Ca^{2+} conductance) (Cooper, Bloom, & Roth, 1991, p. 149; Bowery, 1993). There also are data to suggest that baclofen also may have a role in mediating certain types of pain (Green & Selman, 1991).

33. These conditions likely involve multiple neurotransmitter and chemical interactions, including NE, 5-HT, and cholecystokinin (Brawman-Mintzer & Lydiard, 1997; Bourin, Baker, & Bradwejn, 1998). BZDs, while providing symptomatic relief in generalized anxiety disorders (GAD), generally are not effective in producing long-term amelioration of the syndrome (Stahl, 1996). Pointing to the diverse mechanisms involved in these disorders, it might be noted that **buspirone** (BuSpar) often is beneficial in treating GAD (although apparently impacting on 5-HT rather than GABA receptors) and that antidepressant drugs, particularly the SSRIs and **clomipramine** (Anafranil) in addition to high-potency BZDs such as **alprazolam** (Xanax), **clonazepam** (Klonopin), or **lorazepam** (Ativan) frequently are used to treat panic disorders.

34. Beta-blockers such as **propranolol** (Inderal) also frequently are used to treat this condition.

35. **Zolpidem** (Ambien), one of the newer $GABA_A$ agonists that act at the BZD site, currently is being used extensively for sleep disorders because of its narrower response profile (primarily sedation) when compared to other BZDs, as well as having a relatively short half-life.

36. Usually a high-potency BZD given I.M. in combination with a neuroleptic such as *haloperidol* (Haldol) for severe agitation.

37. At this point it may be useful to review the meaning of a few terms. *Tolerance* refers to the fact that with continued use of a particular substance, the body adapts so that it takes more of the substance to produce the same clinical effect. *Dependence* refers to the fact that on discontinuation of certain drugs, the individual will experience new, unwelcome physical and/or psychological (withdrawal) symptoms. If, on discontinuation of the drug, the original symptoms recur but in a more intense fashion than were present prior to the initiation of the drug, this is known as *rebound*. Finally, *addiction* refers to the situation where a person continues to compulsively engage in drug-seeking (taking) behaviors, despite significant adverse effects to his or her physical, psychological, financial, and/or social well-being as a result of the continued use of the substance.

38. Because **lorazepam** (Ativan) and **oxazepam** (Serax) are metabolized without producing any active metabolites, they are safer to use in these at-risk populations.

39. As noted earlier, there is considerable variation both within $GABA_A$ receptors themselves and their distribution within the CNS, as well as in their response to specific ligands. For example, *type 1* BZD receptors are thought to mediate anxiolytic effects, while *type 2* receptors have been associated with sedation and anticonvulsant effects (Penny, 1996). In addition, the anxiolytic and anticonvulsant effects of BZDs often are present at lower dosages than their muscle relaxant and sedative effects. Thus a goal of

pharmaceutical current research is to develop "designer drugs" that will target a specific symptom (e.g., anxiety) without producing other unwanted side effects. Effective partial BZD agonists may be one solution to this problem (Haefely, 1990; Haefely, Martin, & Schoch, 1990).

40. The loss of the spiny cells from the striatum is thought to result in an increased inhibition of the subthalamus in the "indirect, inhibitory pathway" (see Chapter 6). This in turn eventually results in thalamic disinhibition that may explain the extraneous (choreiform) movements.

41. Apparently there are minor adrenergic connections between the lower brain stem and the periaqueductal gray of the midbrain via the central tegmental tract. Epinephrine also may bind to autoreceptors in the locus ceruleus, where it exerts an inhibitory influence on these noradrenergic neurons (Gilman & Newman, 1996).

42. Action potentials reaching the nerve terminal cause an influx of extracellular CA^{2+} ions, which in turn causes the transmitter vesicles to attach themselves to the wall of the presynaptic membrane. Once attached, they then release their contents into the synaptic cleft.

43. In order for biochemical processes to occur within the cell, energy is required. The primary source of energy within the cell is adenosine triphosphate (ATP), which is derived from glucose. Adenylyl cyclase, an enzyme, is needed to convert ATP into cAMP, which serves as a second messenger within the cell. Presynaptically, cAMP is essential in the synthesis and/or release of norepinephrine. Thus, as in the case of $GABA_B$, inhibition of adenylyl cyclase by α_2 autoreceptors exerts an inhibitory influence on NE presynaptically.

44. The locus ceruleus itself apparently extends slightly into the caudal midbrain. A locus subceruleus, a group of less-well-demarcated adrenergic cells lying ventrolateral to the locus ceruleus, has also been identified.

45. It has been suggested that while the sensory and association nuclei of the brain stem likely receive their adrenergic inputs from the locus ceruleus, the motor and visceral nuclei may derive their inputs from adrenergic neurons originating in the medulla (see below) (Carpenter & Sutin, 1983).

46. There also is some suggestion that NE may be involved in the modulation of pain at the spinal level through its inhibitory action on substance P (Kuraishi et al., 1985).

47. Again, it should be emphasized that, while disturbances in the noradrenergic system have been repeatedly associated with many anxiety related disorders, it is not the only brain chemical associated with these disorders. Endocrinological disturbances associated with the dysregulation of the hypothalamic–pituitary–adrenal (HPA) axis, cholecystokinin (CCK), 5-HT (all of which may impact on the adrenergic system), and even CO^2 also have been associated with these disorders (Yehuda, Boisoneau, Lowy, & Giller, 1995; Brawman-Mintzer & Lydiard, 1997; Southwick et al., 1997a; Friedman, 1997; van der Kolk, 1997; Bourin, Baker, & Bradwejn, 1998; Stein & Uhde, 1998; Sullivan, Coplan, & Gorman, 1998; Charney & Bremner, 1999).

48. Recall that $GABA_B$ heteroreceptors can exert an inhibitory influence on catecholamine neurons.

49. As will be seen in the discussions to follow, even though the TCAs and MAOIs have been in use for decades and additional novel antidepressants have been developed since the clinical introduction of the SSRIs in 1987, the exact mechanism(s) by which they exert their therapeutic effect still remains unknown.

50. In the 1950s, an MAOI (iproniazid) used in the treatment of tuberculosis was found to have antidepressant effects when given to these patients. Shortly thereafter, imipramine,

a drug being investigated for potential antipsychotic qualities, likewise was found to be effective in treating depression.

51. For additional discussion of the biochemistry of depression, see Mayberg, (2004); Musselman et al, (1998); Duman, (1999); and Garlow, Musselman, & Nemeroff, (1999).

52. Some D_2 autoreceptors are not directly linked to cyclic AMP, but rather appear to increase the flow outflow of potassium, thus hyperpolarizing the neuron, leading to an inhibition of DA and ACh in the striatum and nucleus accumbens (Roth & Elsworth, 1995).

53. Other designations occasionally are applied to these monoamine cell groups. Thus, SNc may be referred to as "A9" neurons, VTA as "A10," the retrorubral group as "A8," while the various hypothalamic dopamine cell groups have been designated as "A11" through "A15" (Anden et al., 1964; Dahlstrom & Fuxe, 1964; Pearson et al., 1990; Moore & Lookingland, 1995).

54. The reason for the difference in the number of categorizations is that some authors collapse the mesolimbic and mesocortical into one mesocorticolimbic pathway and others distinguish two separate hypothalamic pathways: the tuberoinfundibular and incertohypothalamic (Moore & Lookingland, 1995; Mansour et al., 1998)

55. Not all authors agree or are explicit on which structures constitute the mesolimbic and mesocortical pathways. For example, Baldessarini and Tarazi (1996) include the entorhinal cortex in the mesolimbic pathway and appear to limit the mesocortical to the neocortex and underlying white matter.

56. It should be noted that most of their findings primarily were based on studies using rodents, which not only are behaviorally less complex than man but anatomically different with regard to some of their dopaminergic projections, specifically with regard to mesocorticolimbic pathways.

57. Consequently, it is likely that, as with depression, relief of some symptoms may be more related to slow changes in receptor or other metabolic activities of the cell (e.g., possible increase in dopamine receptors in mesolimbic areas) than to the more immediate changes in the levels of dopamine activity.

58. Originally available in Europe the mid-1960s, clozapine was withdrawn from that market in the mid-1970s because of this side effect. In the late 1980s it became available in the United States as a treatment option for refractory cases of schizophrenia, with the mandate that weekly blood counts be obtained (Hippius, 1989).

59. Akathisia may occur in a small percentage of patients (Beasley et al., 1996; Kopala, Good, & Honer, 1997; Stahl, 1999a).

60. It also been has noted that while typical antipsychotics affect both A9 (SN) and A10 (VTA) neurons, the atypicals tend to selectively affect only VTA (for review, see Kinon & Lieberman, 1996; Owens & Risch, 1998). Because A9 neurons project primarily to the neostriatum, while A10 projects to the mesocorticolimbic areas, it has been suggested that this could help account for the relative lack of EPS in these newer drugs. But (as pointed out by both reviewers) the clinical significance of these findings are still unresolved. Other hypotheses will be reviewed below.

61. All neuroleptics apparently bind to both D_1 and D_2 type receptors, which include both D_3 and D_4 receptors. Because clozapine has a relatively higher affinity for both D_3 and D_4 receptors when compared to the older antipsychotics and since D_3 receptors predominately are found in the mesolimbic region, it has been speculated that these latter two receptors may have special implications for schizophrenia. However, thus far neither any clear link between D_3 or D_4 receptors nor any special efficacy for drugs specifically targeting these receptors has been established. Owens & Risch (1998) provide a brief review of this topic.

62. So wedded were these notions, that drugs that didn't produce these extrapyramidal side effects were once thought to likely lack any antipsychotic effect (Owens & Risch, 1998).

63. It is estimated that somewhere between 60 and 80% reduction of postsynaptic dopaminergic stimulation (D_2 receptor blockade) is required before reduction of clinical symptoms become apparent, with the risk of EPS being significantly elevated with occupancy of greater than 80% (Fardeet et al., 1992; Hirsch & Herrero 1997; Korczyn, 1995; Remington & Kapur, 1999).

64. Among the typical neuroleptics, the **lower potency drugs** [e.g., *chlorpromazine* (Thorazine), *thioridazine* (Mellaril)] generally are less problematic than the **higher potency drugs** [e.g., *haloperidol* (Haldol), *fluphenazine* (Prolixin)]. This generally is thought to be because of the greater anticholinergic effects of the former that helps to maintain a better DA/ACh balance. To reduce the likelihood of precipitating motor problems, anticholinergics often are given in conjunction with the higher potency, typical neuroleptic. However, over the long haul, both the low- and high-potency drugs are likely to produce EPS (Collaborative Working Group on Clinical Trial Evaluations, 1998). At higher dosages, even risperidone (and, to a lesser extent, olanzapine) has the capacity for sufficient blockade of D_2 receptors to produce EPS (Kapur, Zipursky, & Remington, 1999; Kapur et al., 1998).

65. Clozapine and quetiapine have not been associated with chronic elevations of prolactin (Remington & Kapur, 2000). Olanzapine occasionally has been associated with mild, transient prolactin elevations in adults (Beasley et al., 1997; Crawford, Beasley, & Tollefson, 1997), with possibly more problematic elevations in children or adolescents (Wudarsky et al., 1999). Risperidone, however, has been found to more consistently produce significant elevations in prolactin (in addition to an increased incidence of EPS) compared to the other atypicals (Lavalaye et al., 1999; Conley, 2000; Stanniland & Taylor, 2000).

66. Baldessarini (1996b) suggests that the *stimulant* effects of these drugs should be differentiated from their *antidepressant* effects.

67. Although bupropion also is reported to block the reuptake of norepinephrine, this effect is probably minimal (Richelson, 1996, 1999b).

68. As is the case with other disease states, Parkinson's disease is associated with changes in other neurotransmitter systems in addition to dopamine, particularly, NE, ACh, and 5-HT (Hornykiewicz & Kish, 1987; Cummings, 1988; Korczyn, 1995; Miyawaki, Meah, & Koller, 1997).

69. Recall that it is the putamen that primarily is linked to the sensorimotor cortices, and hence motor activity, while the caudate nucleus is more associated with the prefrontal and more posterior association cortices and cognitive functions.

70. In fact, the term *neuroleptics* came to be applied to this class of drugs because of their dampening effect on behavior, which included neurological type manifestations (Baldessarini, 1996a; Stahl, 1996).

71. An example of a lower potency, high anticholinergic drug is *thioridazine* (Mellaril), while *haloperidol* (Haldol), *thiothixene* (Navane), and *fluphenazine* (Prolixin) are examples of high potency, low anticholinergic antipsychotics.

72. **Clozapine** (Clozaril) is the prototypic example of the new generation antipsychotic medications that have become known for their reduced tendency to produce EPS or TD. **Olanzapine** (Zyprexa), **quetiapine** (Seroquel), and **risperidone** (Risperdal), in lower dosages, appear to have similar margins of safety with regard to motor difficulties.

73. Parkinson's disease is known to differentially affect the basal ganglia, including the caudate. Recall from Chapter 6 that the head of the caudate is intimately connected to frontal systems, while the remainder of the caudate receives and sends projections (via

the thalamus) back to other cortical association cortices. This might help account for the "frontal" type syndrome often associated with the so-called "subcortical dementias." For a more detailed review of the cognitive syndromes associated with Parkinson's, see (Freedman 1990).

74. While levodopa frequently is quite successful in reducing symptoms when given during the early stages of Parkinson's disease, as the disease progresses, more and more dopaminergic neurons and terminals are lost. As a result, two things occur: (1) patients require increased dosages of the drug, and (2) its effects wear off sooner, possibly as a result of the loss of neuron terminals to store and metabolize DA (Korczyn, 1995; Standaert & Young, 1996). This can result in what has been referred to as the *on* and *off* phenomena. During the "on" period (usually shortly after ingesting the drug) the excess of DA from the higher dosages results in the development of dyskinesias. However, the effects of the drugs also dissipate more rapidly, resulting in an exacerbation or return of the hypokinesis ("off" phenomenon).

75. Miyawaki, Meah, and Koller (1997) suggest that a "dysregulation" of serotonin systems in the basal ganglia also may play a key role in the expression of tremors and other forms of dyskinesia. In this regard, it also is important to note that **clozapine**, which is known to block 5-HT_{2A} receptors, has been shown to be effective in treating some cases of parkinsonian tremor (Fisher, Bass, & Hefner, 1990; McCarthy, 1994).

76. Other treatment options include reducing the neuroleptic dosage or switching to either a lower potency neuroleptic or one of the newer, "atypical" drugs. Occasionally, other drugs also may be tried, including amantadine, benzodiazepines, or beta-blockers. The latter often is effective in the management of akathisia.

77. Basically, neuromodulators are considered to be substances (e.g., steroids and neuropeptides) that are released at sites other than the synapse itself, and that as a result have the capacity to exercise a more diverse influence on other neurotransmitters and more distal synapses.

78. In addition to being released directly into the synaptic cleft from the presynaptic terminals, some 5-HT also is thought to be released from sites along the axon itself known as *varicosities* (Descarries et.-al., 1990). This apparent, more diffused release of 5-HT outside of the synapse is offered as evidence of serotonin's potential additional role as a neuromodulator (Sanders-Bush & Mayer, 1996).

79. Although on the whole the SSRIs have comparable response profiles to the older drugs, not all patients respond to them even after dose adjustments. In such instances, either the addition of another agent (e.g., thyroid supplement, stimulant, mood stabilizer, anticonvulsant, buspirone, atypical antipsychotic, estrogen), or either adding or switching to another antidepressant with a different pharmacological profile (e.g., with NRI and/or DRI properties) is usually recommended (Keltner & Folks, 2005; Stahl, 2000; Gelenberg & Delgado, 1997; Janicak, Davis, Preskorn & Ayd, 1997; Schatzberg, 1998). Also, the SSRIs, although indeed a clinically safer drug, are not without their own side effects (see below).

80. While serotonin certainly is believed to play a role in anxiety states, numerous other potential contributors have been identified. Among these are norepinephrine, GABA, lactate, carbon dioxide, yohimbine, caffeine, and cholecystokinin (Asnis & van Praag, 1995; Brawman-Mintzer & Lydiard, 1997; Goddard & Charney, 1997; Bourin, Baker, & Bradwejn, 1998; Stein & Uhde, 1998).

81. Although these serotonin reuptake inhibitors, when given in adequate dosages (which are usually higher than those for treating depression), are the most effective medications to date for the relief of the symptoms of OCD, their success rate is still limited. Meaningful clinical improvement has been estimated to occur in only 40–60% of patients

treated (Goodman & Murphy, 1998; McDougle, 1999). Again, this points to the neuro-chemical complexity of neurobehavioral syndromes.

82. Other classes of antidepressants (i.e., the TCAs and MAOIs) also have been shown to be effective in reducing the frequency and severity of panic attacks, particularly *imipramine* (Tofranil) and *clomipramine* (Anafranil), but it has been suggested that it is the serotonergic effects of these drugs that in large part account for their effectiveness (Leonard, 1997).

83. Although PTSD often has a depressive component, it should be clear by now that most of the anxiety-related disorders discussed thus far have been found to respond to antidepressant drugs, especially those that have a strong serotonergic component. These findings have led to a rethinking of the classic distinctions that were made between antidepressants and anxiolytics as these differences have begun to blur. Just as the anxiolytic potential of antidepressants has begun to be appreciated, so has the antide-pressant potential of the anxiolytics (Casacalenda & Boulenger, 1998; Stahl, 1999b,c).

84. For example, 5-HT is known to interact with NE and GABA, as well as other neuromod-ulators (see Eison, 1990; Kahn, Westenberg, & Moore, 1995; Goddard & Charney, 1998; Kent, Coplan, & Gorman, 1998), and we already have noted that various other chemicals, including neuropeptides, have been associated with anxiety states (see Endtnote 82).

85. Stahl (1996) suggests that buspirone also acts as a partial agonist at the presynaptic autoreceptors. In either case, the expected results would be similar. If this is true, then the displacement of 5-HT by the 5-HT_{1A} partial agonist (with its subsequent weaker agonist effect) would lead to an eventual increase (up-regulation) in the number of the 5-HT_{1A} somatodendritic autoreceptors. The increase in these presynaptic autoreceptors then would exert a net increased inhibitory effect on the firing of the neuron, thus reducing serotonergic output (anxiolytic effect). Because buspirone's mechanism of action is thought to involve the synthesis of new receptors, this would also explain its delayed effect as compared to the action of the benzodiazepines.

86. For example, 5-HT and NE would appear to have a reciprocal relationship, in that 5-HT tends to exert inhibitory effects on the noradrenergic neurons in the locus ceruleus.

87. These observations can offer only very tentative, partial "explanations" for what are obviously exceedingly complex and as of yet poorly understood phenomena.

88. Again, these include risperidone (Risperdal), olanzapine (Zyprexa), quetiapine (Seroquel), and ziprasidone (Geodon), as well as others that are currently pending approval, such as sertindole.

89. Many of these newer drugs bind to several other 5-HT and DA receptors and likely impact on several other neurotransmitters (e.g., glutamate, ACh, NE) through second messenger systems involved with the various 5-HT receptors (see Roth & Meltzer, 1995; Owens & Risch, 1998; King, 1998; Stahl, 1999).

90. If enhanced stimulation of 5-HT_{2A} receptors in the nigrostriatal pathway has the potential to inhibit the release of dopamine, then might one expect that drugs that increase 5-HT (e.g., the SSRIs) would have a tendency to produce EPS. While such symptoms have been reported with the use of these drugs (Arya & Szabadi, 1993; Arya, 1994), the incidence of such reactions is relatively rare compared to conventional neuroleptics. One possible explanation is that an increase in 5-HT also might tend to stimulate 5-HT_{1A} somatodendritic autoreceptors that would tend to reduce the output of 5-HT neurons in the striatum. Also, as previously noted, in order to produce parkin-sonian symptoms, it has been estimated that a 75 to 80% blockage of DA may be required. It is likely that such a level of DA depletion with SSRIs may occur only in those individuals who already are predisposed to Parkinson's disease subclinically or those who are already on other dopamine-depleting drugs (Kapur & Remington, 1996).

91. It also has been proposed that 5-HT also may exert a direct inhibitory effect on prefrontal neurons. If this is indeed the case, then the antagonistic actions of the novel antipsychotics on serotonergic neurons also may have a direct effect in improving the responsiveness of this area (see Kapur & Remington, 1996, p. 469).

92. Reduced irritability is often one of the earliest changes noted by patients started on SSRIs for depression.

93. Some of the more common side effects associated with these drugs are weight gain, sedation, and orthostatic hypotension. As noted earlier, an infrequent but potentially serious side effect associated with clozapine is agranulocytosis. For a more complete review of the major side effects associated with these individual compounds, see the *Physician's Desk Reference*, or Stahl (1999), Janicak (1999), Tamminga (1999b).

94. *Cyproheptadine*, a 5-HT$_2$ antagonist, may help reverse this syndrome (Cohen, 1992; Graudins, Stearman, & Chan, 1998; Woodrum & Brown, 1998).

REFERENCES

Adams, R.D., Victor, M. & Ropper, A.H. (1997) *Principles of Neurology.* New York: McGraw-Hill.

Addonizio, G. (1991) The pharmacologic basis of neuroleptic malignant syndrome. *Psychiatric Annals, 21*, 152–156.

Aghajanian, G.K. (1995) Electrophysiology of serotonin receptor subtypes and signal transduction pathways. In: Bloom, F.E. & Kupfer, D.J. (Eds.) *Psychopharmacology: The fourth generation of progress.* New York: Raven Press, pp. 451–460.

Albin, R.L., Young, A.B. & Penney, J.B. (1989) The functional anatomy of basal ganglia disorders. *Trends in Neuroscience, 12*, 366–375.

Albus, M. Zahn, T.P. & Breier, A. (1992) Anxiogenic properties of yohimbine: Behavioral, physiological and biochemical measures. *European Archives of Psychiatry and Clinical Neuroscience, 241*, 337–344.

Allgulander, C. (1999) Paroxetine in social anxiety disorder: A randomized placebo-controlled study. *Acta Psychiatrica Scandinavica, 100*, 193–198.

Altamura, A.C., Pioli, R., Vitto, M. & Mannu, P. (1999) Venlafaxine in social phobia: A study in selective serotonin reuptake inhibitor non-responders. *International Clinical Psychopharmacology, 14*, 239–245.

Anden, N.E., Carlsson, A., Dahlstrom, A., Fuxe, K., Hillarp, N.A. & Larsson, K. (1964) Demonstration and mapping out of nigroneostriatal dopamine neurons. *Life Sciences, 3*, 523–530.

Arnsten, A.F.T., Steere, J.C. & Hunt, R.D. (1996) The contribution of [alpha-2] noradrenergic mechanisms to prefrontal cortical cognitive function. *Archives of General Psychiatry, 53*, 448–455.

Arnt, J. (1998) Pharmacological differentiation of classical and novel antipsychotics. *International Clinical Psychopharmacology, 13* (Suppl 3), S7-S14.

Arya, D.K. (1994) Extrapyramidal symptoms with selective serotonin reuptake inhibitors. *British Journal of Psychiatry, 165*, 728–733.

Arya, D.K. & Szabadi, E. (1993) Dyskinesia associated with fluvoxamine. *Journal of Clinical Psychopharmacology, 13*, 365–366 (letter).

Asnis, G.M. & van Praag, H.M. (Eds.) (1995) *Panic Disorder: Clinical, biological, and treatment aspects.* New York: John Wiley & Sons.

Aston-Jones, G., Chiang, C. & Alexinsky, T (1991) Discharge of noradrenergic locus ceruleus neurons in behaving rats and monkeys suggests a role in vigilance. *Progress in Brain Research, 88*, 501–520.

Ayd, F.J. & Zohar, J. (1987) Psychostimulant (amphetamine or methylphenidate) therapy for chronic and treatment resistant depression. In: Zohar, J. & Belmaker, R.H. (Eds.) *Treating Resistant Depression.* New York: PMA Corp., pp. 343–355.

Azmitia, E.C. & Whitaker-Azmitia, P.M. (1995) Anatomy, cell biology, and plasticity of the serotonergic system. In: Bloom, F.E. & Kupfer, D.J. (Eds.) *Psychopharmacology: The fourth generation of progress.* New York: Raven Press, pp. 443–449.

Baer, L., Rauch, S.L., Ballantine, H.T., Martuza, R., Cosgrove, R., Cassem, E., Giriunas, I., Manzo, P.A., Dimino, C. & Jenike, M.A. (1995) Cingulotomy for intractable obsessive-compulsive disorder. Prospective long-term follow-up of 18 patients. *Archives of General Psychiatry, 52,* 384–392.

Baldessarini, R.J. (1996a) Drugs and the treatment of psychiatric disorders: Psychosis and anxiety. In: Molinoff, P.B. & Ruddon, R.W. (Eds.) *Goodman & Gilman's the Pharmacological Basis of Therapeutics, Ninth Edition.* New York: McGraw-Hill, pp. 399–430.

Baldessarini, R.J. (1996b) Drugs and the treatment of psychiatric disorders: Depression and mania. In: Molinoff, P.B. & Ruddon, R.W. (Eds.) *Goodman & Gilman's the Pharmacological Basis of Therapeutics, Ninth Edition.* New York: McGraw-Hill, pp. 399–430.

Baldessarini, R.J. & Tarazi, F.I. (1996) Brain dopamine receptors: A primer on their current status, basic and clinical. *Harvard Review of Psychiatry, 3,* 301–325.

Ballenger, J.C., Davidson, J.R., Lecrubier, Y., Nutt, D.J., Bobes, J., Beidel, D.C., Ono, Y. & Westenberg, H.G. (1998) Consensus statement on social anxiety disorder from the International Consensus Group On Depression and Anxiety. *Journal of Clinical Psychiatry, 59* (Suppl 17), 54–60.

Balon, R. (1999) Fluvoxamine for phobia of storms. *Acta Psychiatrica Scandinavica, 100,* 244–245.

Barker, E.L. & Blakely, R.D. (1995) Norepinephrine and serotonin transporters. In: Bloom, F.E. & Kupfer, D.J. (Eds.) *Psychopharmacology: The fourth generation of progress.* New York: Raven Press, pp. 321–333.

Barnes, T.R.E. & McPhillips, M.A. (1998) Novel antipsychotics, extrapyramidal side effects, and tardive dyskinesia. *International Clinical Psychopharmacology, 13* (Suppl 3), S49-S57.

Barr, A., Heinze, W., Mendoza, J.E., Perlik, S. (1978) Long-term treatment of Huntington's disease with L-glutamate and pyridoxine. *Neurology, 28,* 1280–1282.

Barr, L.C., Goodman, W.K. & Price, L.H. (1993) The serotonin hypothesis of obsessive–compulsive disorder. *International Clinical Psychopharmacology, 8* (Suppl 2), 79–82.

Barr, L.C., Goodman, W.K., Price, L.H., McDougle, C.J. & Charney, D.S. (1992) The serotonin hypothesis of obsessive–compulsive disorder: Implications of pharmacologic challenge studies. *Journal of Clinical Psychiatry, 53* (Suppl 4), 29–37.

Bartus, R.T., Dean, R.L., Pontecorvo, M.J., & Flicker, C. (1985) The cholinergic hypothesis: a historic review, current perspectives, and future directions. *Annals of the New York Academy of Science, 444,* 332–358.

Baskys, A. & Remington, G. (1996) Serotonin and dopamine as neuroreceptors. In: Baskys, A. & Remington, G. (Eds.) *Brain Mechanisms and Psychotropic Drugs.* Boca Raton, FL: CRC Press, pp. 55–71.

Baxter, L.R. (1999) Functional imaging of brain systems mediating obsessive–compulsive disorder. In: Charney, D.S., Nestler, E.J. & Bunney, B.S. (Eds.) *Neurobiology of Mental Illness,* New York: Oxford University Press, pp. 534–547.

Baxter, L.R., Saxena, S., Brody, A.L., Ackermann, R.F., Colgan, M., Schwartz, J.M., Allen-Martinez, Z., Fuster, J.M. & Phelps, M.E. (1996) Brain mediation of obsessive–compulsive disorder symptoms: Evidence from functional brain imaging studies in the human and nonhuman primate. *Seminars in Clinical Neuropsychiatry, 1,* 32–47.

Baxter, L.R., Schwartz, J.M., Bergman, K.S., Szuba, M.P., Guze, B.H., Mazziotta, J.C., Alazraki, A., Selin, C.E., Ferng, H-K., Munford, P. & Phelps, M.E. (1992) Caudate glucose metabolic rate changes with both drug and behavior therapy for obsessive–compulsive disorder. *Archives of General Psychiatry, 49,* 681–689.

Beal, M.F. (1992a) Mechanisms of excitotoxicity in neurologic diseases. *FASEB Journal, 6,* 3338–3344.

Beal, M.F. (1992b) Role of excitotoxicity in human neurological disease. *Current Opinion in Neurobiology, 2,* 657–662.

Beal, M.F. (1998) Excitotoxicity and nitric oxide in Parkinson's disease pathogenesis. *Annals of Neurology, 44* (Suppl) S110-S114.

Beasley, C.M.Jr, Hamilton, S.H., Crawford, A.M., Dellva, M.A., Tollefson, G.D., Tran, P.V., Blin, O. & Beuzen, J.N. (1997) Olanzapine versus haloperidol: Acute phase results of the international double-blind olanzapine trial. *European Neuropsychopharmacology, 7,* 125–137.

Beasley, C.M., Tollefson, G.D., Tran, P.V., Satterlee, W., Sanger, T. & Hamilton, S. (1996) Olanzapine versus placebo and haloperidol: Acute phase results of the North American double-blind olanzapine trial. *Neuropsychopharmacology, 14,* 111–123.

Beatty, W.W., Butters, N. & Janowsky, D.S. (1986) Patterns of memory failure after scopolomine treatment: implications for cholinergic hypothesis of dementia. *Behavioral and Neural Biology*, 45, 196–211.

Belanoff, J.K. & Glick, I.D. (1998) New psychotropic drugs for Axis I disorders: Recently arrived, in development, and never arrived. In: Schatzberg, A.F. & Nemeroff, C.B. (Eds.) *Textbook of Psychopharmacology*. Washington, DC: American Psychiatric Press, pp. 1015–1027.

Benes, F.M. (1997) The role of stress and dopamine-GABA interactions in the vulnerability for schizophrenia. *Journal of Psychiatric Research*, 31, 257–275.

Bennett, G.W., Ballard, T.M., Watson, C.D., & Fone, K.C. (1997) Effects of neuropeptides on cognition function. *Experimental Gerontology*, 32, 451–469.

Berger-Sweeney, J. (1998) The effects of neonatal basal forebrain lesions on cognition: Towards understanding the developmental role of the cholinergic basal forebrain. *International Journal of Developmental Neuroscience*, 16, 603–612.

Bierer, L.M., Haroutunian, V., Gabriel, S., Knott, P.J., Carlin, L.S., Purohit, D.P., Perl, D.P., Schmeidler, J., Kanof, P. & Davis, K.L. (1995) Neurochemical correlates of dementia severity in Alzheimer's disease: Relative importance of the cholinergic deficits. *Journal of Neurochemistry*, 64, 749–760.

Biggins, C.A., Boyd, J.L., Harrop, F.A., Maddeley, P., Mindham, R.H., Randall, J.I. & Spokes, E.G. (1992) A controlled, longitudinal study of dementia in Parkinson's disease. *Journal of Neurology, Neurosurgery and Psychiatry*, 55, 566–571.

Bird, E.D., Mackay, A.V.P., Rayner, C.N. & Iversen, L.L. (1973) Reduced glutamic-acid-decarboxylase activity of post-mortem brain in Huntington's chorea. *Lancet*, 1, 1090–1092.

Birtwistle, J. & Baldwin, D. (1998) Role of dopamine in schizophrenia and Parkinson's disease. *British Journal of Nursing*, 7, 838–841.

Blandini, F., Porter, R.H.P. & Greenamyre, J.T. (1996) Glutamate and Parkinson's disease. *Molecular Neurobiology*, 12, 73–94.

Blier, P., de Montigny, C. & Chaput, Y. (1987) Modifications of the serotonin system by antidepressant treatments: Implications for the therapeutic response in major depression. *Journal of Clinical Psychopharmacology*, 7, 24S-35S.

Blier, P., de Montigny, C. & Chaput, Y. (1990) A role for the serotonin system in the mechanism of action of antidepressant treatments: Preclinical evidence. *Journal of Clinical Psychiatry*, 51 (Suppl), 14–20.

Blin, O. (1999) A comparative review of new antipsychotics. *Canadian Journal of Psychiatry*, 44, 235–244.

Bliss, T.V. & Collinridge, G.L. (1993) A synaptic model of memory: Long term potentiation in the hippocampus. *Nature*, 361, 31–39.

Bloom, F.E. (1996) Neurotransmission and the central nervous system. In: Molinoff, P.B. & Ruddon, R.W. (Eds.) *Goodman & Gilman's the Pharmacological Basis of Therapeutics, Ninth Edition*. New York: McGraw-Hill, pp. 267–293.

Bloom, F.E. & Iversen, L.L. (1971) Localizing 3H-GABA in nerve terminals of rat cerebral cortex by electron microscopic autoradiography. *Nature*, 229, 628–630.

Bloom, F.E. & Kupfer, D.J. (Eds.) *Psychopharmacology: The fourth generation of progress*. New York: Raven Press.

Bonilla, E., Prasad, A.L. & Arrieta, A. (1988) Huntington's disease: Studies on brain free amino acids. *Life Sciences*, 42, 1153–1158.

Bortolotto, Z.A. & Collinridge, G.L. (1992) Activation of glutamate metabotropic receptors induces long-term potentiation. *European Journal of Pharmacology*, 214, 297–298.

Bourin, M., Baker, G.B. & Bradwejn, J. (1998) Neurobiology of panic disorder. *Journal of Psychosomatic Research*, 44, 163–180.

Bouwer, C. & Stein, D.J. (1998) Use of the selective serotonin reuptake inhibitor citalopram in the treatment of generalized social phobia. *Journal of Affective Disorders*, 49, 79–82.

Bowery, N.G. (1993) GABAB receptor pharmacology. *Annual Review of Pharmacological Toxicology*, 33, 109–147.

Boyer, W. (1995) Serotonin uptake inhibitors are superior to imipramine and alprazolam in alleviating panic attacks. *International Clinical Psychopharmacology*, 10, 45–49.

Brailowsky, S. & Garcia, O. (1999) Ethanol, GABA and epilepsy. *Archives of Medical Research*, 30, 3–9.

Brawman-Mintzer, O. & Lydiard, R.B. (1997) Biological basis of generalized anxiety disorder. *Journal of Clinical Psychiatry*, 58 (Suppl 3), 16–25.

Bremner, J.D., Innis, R.B., Innis, C.K., Staib, L.H., Salmin, R.M., et.al. (1997) Positron emission tomography measurement of cerebral; metabolic correlates of yohimbine administration in combat-related posttraumatic stress disorder. *Archives of General Psychiatry*, 54, 246–254.

Brodi.e., B.B. & Shore, P.A (1957) A concept for a role of serotonin and norepinephrine as chemical mediators in the brain. *Annals of the New York Academy of Sciences*, 66, 631–642.

Bunney, B.G., Bunney, W.E. & Carlsson, A. (1995) Schizophrenia and glutamate. In: Bloom, F.E. & Kupfer, D.J. (Eds.) *Psychopharmacology: The fourth generation of progress.* New York: Raven Press, pp. 1205–1214.

Bymaster, F.P., Calligaro, D.O., Falcone, J.F., Marsh, R.D., Moore, N.A., Tye, N.C., Seeman, P. & Wong, D.T. Radioreceptor binding profile of the atypical antipsychotic olanzapine. *Neuropsychopharmacology*, 14, 87–96.

Caroff, S.N., Mann, S.C., Lazarus, A., Sullivan, K. & Macfadden, W. (1991) Neuroleptic malignant syndrome: Diagnostic issues. *Psychiatric Annals*, 21, 130–147.

Carlsson, A., Hansson, L.O., Waters, N. & Carlsson, M.L. (1999) A glutamatergic deficiency model of schizophrenia. *British Journal of Psychiatry*, 37 (Suppl), 2–6.

Carpenter, M.B. & Sutin, J. (1983) *Human Neuroanatomy.* Baltimore: Williams & Wilkins.

Casacalenda, N. & Boulenger, J.P. (1998) Pharmacologic treatments effective in both generalized anxiety disorder and major depressive disorder: clinical and theoretical implications. *Canadian Journal of Psychiatry*, 43, 722–730.

Casey, D.E. (1995) Tardive dyskinesia: Pathophysiology. In: Bloom, F.E. & Kupfer, D.J. (Eds.) *Psychopharmacology: The fourth generation of progress.* New York: Raven Press, pp. 1497–1502.

Casey, D.E. (1998) Effects of clozapine therapy in schizophrenic individuals at risk for tardive dyskinesia. *Journal of Clinical Psychiatry*, 59 (Suppl) 31–37.

Casey, D.E. (1999) Tardive dyskinesia and atypical antipsychotic drugs. *Schizophrenia Research*, 35 (Suppl) S61-S66.

Charney, D.S., Bremner, J.D. & Redmond, D.E. (1995) Noradrenergic neural substrates for anxiety and fear: Clinical associations based on preclinical research. In: Bloom, F.E. & Kupfer, D.J. (Eds.) *Psychopharmacology: The fourth generation of progress.* New York: Raven Press, pp. 387–395.

Charney, D.S. & Bremner, J.D. (1999) The neurobiology of anxiety disorders. In: Charney, D.S., Nestler, E.J. & Bunney, B.S. (Eds.) *Neurobiology of Mental Illness.* New York: Oxford University Press, pp. 494–517.

Charney, D.S., Deutch, A.Y., Krystal, J.H., Southwick, S.M., & Davis, M. (1993) Psychobiologic mechanisms of posttraumatic stress disorder. *Archives of General Psychiatry*, 50, 295–305.

Charney, D.S., Nestler, E.J. & Bunney, B.S. (Eds.) *Neurobiology of Mental Illness.* New York: Oxford University Press.

Charney, D., Woods, S. & Goodman, W. (1987) Serotonin function in anxiety: II. Effects of the serotonin agonist mCPP in panic disorder and healthy subjects. *Psychopharmacology*, 92, 14–24.

Charney, D.S., Woods, S.W., Goodman, W.K. & Heninger, G.R. (1987) Neurobiological mechanisms of panic anxiety: Biochemical and behavioral correlates of yohimbine-induced panic attacks. *American Journal of Psychiatry*, 144, 1030–1036.

Cherek, D.R., Lane, S.D., Pietras, C.J, & Steinberg, J.L. (2002) Effects of chronic paroxetine administration on measures of aggressive and impulsive responses of adult males with a history of conduct disorder. *Psychopharmacology*, 159(3), 266–274.

Choi, D.W. (1992) Excitotoxic cell death. *Journal of Neurobiology*, 23, 1261–1276.

Choi, D.W. (1995) Calcium: Still center-stage in hypoxic-ischemic neuronal death. *Trends in Neuroscience*, 18, 58–60.

Choi, D.W. & Rothman, S.M. (1990) The role of glutamate neurotoxicity in hypoxic-ischemic neuronal death. *Annual Review of Neuroscience*, 13, 171–182.

Choo, V. (1993) Paroxetine and extrapyramidal reactions (letter). *Lancet*, 341, 624.

Chouinard, G. (1995) Effects of risperidone in tardive dyskinesia: An analysis of the Canadian Multicenter Risperidone Study. *Journal of Clinical Psychopharmacology*, 15 (Suppl), 36S-44S.

Chozick, B. (1987) The nucleus basalis of Meynert in neurological dementing disease: A review. *International Journal of Neuroscience*, 37, 31–48.

Chrousos, G.P. & Gold, P.W. (1992) The concepts of stress and stress system disorders: Overview of physical and behavioral homeostasis. *Journal of the American Medical Association, 267,* 1244–1252.

Clark, R.D., Canive, J.M., Calais, L.A., Qualls, C.R. & Tuason, V.B. (1999) Divalproex in posttraumatic stress disorder: An open-label clinical trial. *Journal of Traumatic Stress, 12,* 395–401.

Cohen, A.J. (1992) Fluoxetine-induced yawning and anorgasmia reversed by cyproheptadine. *Journal of Clinical Psychiatry, 53,* 174.

Collaborative Working Group on Clinical Trial Evaluations. (1998) Assessment of EPS and tardive dyskinesia in clinical trials. *Journal of Clinical Psychiatry, 59* (Suppl 12), 23–27.

Conley, R.R. (2000) Risperidone side effects. *Journal of Clinical Psychiatry, 61* (Suppl 8), 20–23.

Conners, C.K., Casat, C.D., Gualtieri, C.T., Weller, E., Reader, M., Reiss, A., Weller, R.A., Khayrallah, M. & Ascher, J. (1996) Bupropion hydrochloride in attention deficit disorder with hyperactivity. *Journal of the American Academy of Child and Adolescent Psychiatry, 35,* 1314–1321.

Conner, K.M. & Davidson, J.R. (1998) Generalized anxiety disorder: Neurobiological and pharmacotherapeutic perspectives. *Biological Psychiatry, 44,* 1286–1294.

Connor, K.M., Davidson, J.R., Weisler, R.H. & Ahearn, E. (1999) A pilot study of mirtazapine in post-traumatic stress disorder. *International Clinical Psychopharmacology, 14,* 29–31.

Cooper, J.R., Bloom, F.E. & Roth, R.H. (1991) *The Biochemical Basis of Neuropharmacology,* New York: Oxford University Press.

Coppen, A. (1967) The biochemistry of affective disorders. *British Journal of Psychiatry, 113,* 1237–1264.

Cornell, M.S., Ahktar, T., Slater, R.N.S. & Walker, C.J. (1996) The pentothal test in the management of orthopaedic patients with hysterical illness. *Journal of the Royal Society of Medicine, 89,* 37P-38P.

Cotman, C.W., Kahle, J.S., Miller, S.E., Ulas, J. & Bridges, R.J. (1995) Excitatory amino acid neurotransmission. In: Bloom, F.E. & Kupfer, D.J. (Eds.) *Psychopharmacology: The fourth generation of progress.* New York: Raven Press, pp. 75–85.

Coyle, J.T. (1996) The glutamatergic dysfunction hypothesis for schizophrenia. *Harvard Review of Psychiatry, 3,* 241–253.

Coyle, J.T., Price, D.L., & DeLong, M.R. (1983) Alzheimer's disease: A disorder of cortical cholinergic innervation. *Science, 219,* 1184–1190.

Crawford, A.M., Beasley, C.M. Jr., & Tollefson, G.D. (1997) The acute and long-term effect of olanzapine compared with placebo and haloperidol on serum prolactin concentrations. *Schizophrenia Research, 26,* 41–54.

Creese, I., Burt, D.R. & Snyder, S.H. (1976) DA receptor binding predicts clinical and pharmacological potencies of antipsychotic drugs. *Science, 192,* 481–483.

Cummings, J.L. (1988) Intellectual impairment in Parkinson's disease: Clinical, pathologic, and biochemical correlates. *Journal of Geriatric Psychiatry and Neurology, 1,* 24–36.

Cummings, J.L. (1992) Depression and Parkinson's disease: A review. *American Journal of Psychiatry, 149,* 443–454.

Dahlstrom, A. & Fuxe, K. (1964) Evidence for the existence of monoamine containing neurons in the central nervous system. I. Demonstration of monoamines in the cell bodies of brainstem neurons. *Acta Physiologica Scandinavica* (Suppl), *62,* 1–55.

Danion, J.M., Rein, W. & Fleurot, O. (1999) Improvement in schizophrenic patients with primary negative symptoms treated with amisulpride. American Journal of Psychiatry, *156,* 610–616.

Danysz, W., Parsons, C.G., Mobius, H.J., Stoffler, A. & Quack, G. (2000) Neuroprotective and symptomatological action of memantine relevant for Alzheimer's disease—a unified glutamatergic hypothesis on the mechanism of action. *Neurotoxicity Research, 2,* 85–98.

Davidson, J.R.T. (1997) Biological therapies for posttraumatic stress disorder: An overview. *Clinical Psychiatry,* (Suppl). *58,* 29–32.

Davidson, J.R.T. (1998) What's new in treating GAD? *18th Annual Meeting of the Anxiety Disorders Association of America,* Boston.

Davidson, J.R.T. (1999) Managing posttraumatic stress disorder: Treatment options. Teleconference sponsored by Postgraduate Institute for Medicine, November, 1999.

Davidson, J.R. & Connor, K.M. (1999) Management of posttraumatic stress disorder: Diagnostic and therapeutic issues. *Journal of Clinical Psychiatry, 60* (Suppl 18), 33–38.

Davidson, J.R., DuPont, R.L., Hedges, D. & Haskins, J.T. (1999) Efficacy, safety, and tolerability of venlafaxine extended release and buspirone in outpatients with generalized anxiety disorder. *Journal of Clinical Psychiatry*, 60, 528–535.

Davidson, J.R.T, Tupler, L.A. & Potts, N.L. (1994) Treatment of social phobia with benzodiazepines. *Journal of Clinical Psychiatry*, (Suppl), 5, 28–32.

Davidson, R.J., Putnam, K.M. & Larson, C.L. (2000) Dysfunction in the neural circuitry of emotion regulation—A possible prelude to violence. *Science*, 289, 591–594.

Davies, M.F. (1996) The pharmacology of the gamma-aminobutyric acid system. In: Baskys, A. & Remington, G. (Eds.) *Brain Mechanisms and Psychotropic Drugs*. Boca Raton, FL: CRC Press, pp. 101–116.

Davis, K.L., Kahn, R.S., Ko, G. & Davidson, M. (1991) Dopamine and schizophrenia: A review and reconceptualization. *American Journal of Psychiatry*, 25, 1474–1486.

Davis, K.L., Mohs, R.C., Tinklenberg, J.R., Pfefferbaum, A., Hollister, L.E. & Kopell, B.S. (1978) Physostigmine: Improvement in long-term memory processes in humans. *Science*, 201, 272–274.

Davis, L.L., Suris, A., Lambert, M.T., Heimberg, C. & Petty, F. (1997) Post-traumatic stress disorder and serotonin: New directions for research and treatment. *Journal of Psychiatry and Neuroscience*, 22, 318–326.

Deakin, J.F. & Simpson, M.D. (1997) A two-process theory of schizophrenia: Evidence from studies in post-mortem brain. *Journal of Psychiatric Research*, 31, 277–295.

Dekker, A.J., Conner, D.J. & Thal, L.J. (1991) The role of cholinergic projections from the nucleus basalis in memory. *Neuroscience and Biobehavioral Reviews*, 15, 299–317.

DeLong, M.R. (1990) Primate models of movement disorders of basal ganglia origin. *Trends in Neuroscience*, 13, 281–289.

de Montigny, C., Chaput, Y. & Blier, P. (1990) Modification of serotonergic neuron properties by long-term treatment with serotonin reuptake blockers. *Journal of Clinical Psychiatry*, 51 (Suppl B), 4–8.

de Wied, D. (1997a) The neuropeptide story. *Frontiers in Neuroendocrinology*, 18, 101–113.

de Wied, D. (1997b) Neuropeptides in learning and memory processes. *Behavioural Brain Research*, 83, 83–90.

Descarries, L., Audet, M.A., Doucet, G., Garcia, S., Oleskevich, S., Seguela, P., Soghomonian, J.J. & Watkins, J.C. (1990) Morphology of central serotonin neurons: Brief review of quantified aspects of their distribution and ultrastructural relationships. *Annals of the New York Academy of Sciences*, 600, 81–92.

Deutch, A.Y. (1991) The regulation of subcortical dopamine systems by the prefrontal cortex: Interactions of central dopamine systems and the pathogenesis of schizophrenia. *Journal of Neural Transmission*, (Suppl), 36, 61–89.

Deutch, A.Y., Moghaddam, B., Innis, R.B., Krystal, J.H., Aghajanian, G.K., Bunney, B.S. & Charney, D.S. (1991) Mechanisms of action of atypical antipsychotic Drugs. Implications for novel therapeutic strategies for schizophrenia. *Schizophrenia Research*, 4, 121–156.

Devinsky, O., Kernan, J. & Bear, D.M. (1992) Aggressive behavior following exposure to cholinesterase inhibitors. *Journal of Neuropsychiatry and Clinical Neurosciences*, 4, 189–194.

DiFiglia, M. (1990) Excitotoxic injury of the neostriatum: A model for Huntington's disease. *Trends in Neurosciences*, 13, 286–289.

Dinan, T.G. (1996) Noradrenergic and serotonergic abnormalities in depression: Stress-induced dysfunction? *Journal of Clinical Psychiatry*, 57 (Suppl 4), 14–18.

Doble, A. (1999) The role of excitotoxicity in neurodegenerative disease: Implications for therapy. *Pharmacology and Therapeutics*, 81, 163–221.

Dooneief, G., Mirabello, E., Bell, K., Marder, K., Stern, Y. & Mayeux, R. (1992) An estimate of the incidence of depression in idiopathic Parkinson's disease. *Archives of Neurology*, 49, 305–307.

Dow, B. & Kline, N. (1997) Antidepressant treatment of post-traumatic stress disorder and major depression in veterans. *Annals of Clinical Psychiatry*, 9, 1–5.

Duman, R.S. (1999) The neurochemistry of mood disorders: Preclinical studies. In: Charney, D.S., Nestler, E.J. & Bunney, B.S. (Eds.) *Neurobiology of Mental Illness*. New York: Oxford University Press, pp. 333–347.

Egan, M.F., Apud, J. & Wyatt, R.J. (1997) Treatment of tardive dyskinesia. *Schizophrenia Bulletin*, 23, 583–609.

Eison, M.S. (1990) Serotonin: A common neurobiologic substrate in anxiety and depression. *Journal of Clinical Psychopharmacology, 10* (Suppl), 26–30.

Enna, S.J. & Coyle, J.T. (1998) *Pharmacological Management of Neurological and Psychiatric Disorders.* New York: McGraw-Hill.

Enserink, M. (2000) Searching for the mark of Cain. *Science, 289*(July), 575–579.

Ernst, M., Liebenauer, L.L., Tebeka, D., Jons, P.H., Eisenhofer, G., Murphy, D.L. & Zametkin, A.J. (1997) Selegiline in ADHD adults: Plasma monoamines and monoamine metabolites. *Neuropsychopharmacology, 16,* 276–284.

Fackler, S.M., Anfinson, T.J. & Rand, J.A. (1997) Serial sodium Amytal interviews in the clinical setting. *Psychosomatics, 38,* 558–564.

Fallon, J.H. (1988) Topographical organization of ascending dopaminergic projections. *Annals of the New York Academy of Sciences, 537,* 1–9.

Faraone, S.V. & Biederman, J. (1999) The neurobiology of attention deficit hyperactivity disorder. In: Charney, D.S., Nestler, E.J. & Bunney, B.S. (Eds.) *Neurobiology of Mental Illness.* New York: Oxford University Press, pp. 788–801.

Farber, N.B., Newcomer, J.W. & Olney, J.W. (1998) The glutamate synapse in neuropsychiatric disorders. Focus on schizophrenia and Alzheimer's disease. *Progress in Brain Research, 116,* 421–437.

Farde, L., Nordstrom, A.L. & Weisel, F.A. & Sedvall, G. (1989) D1 and D2 dopamine occupancy during treatment with conventional and atypical neuroleptics. *Psychopharmacology* (Suppl), *99,* S28-S31.

Farde, L., Nordstrom, A.L., Wiesel, F.A., Pauli, S., Halldin, C & Sedvall, G. (1992) Positron emission tomographic analysis of central D1 and D2 dopamine receptor occupancy in patients treated with classical neuroleptics and clozapine: Relation to extrapyramidal side effects. *Archives of General Psychiatry, 49,* 538–544.

Feltus, M.S. & Gardner, D.M. (1999) Second generation antipsychotics for schizophrenia. *Canadian Journal of Clinical Pharmacology, 6,* 187–195.

Filley, C.M., Price, B.H., Nell, V., Antionette, T., Morgan, A.S., Bresnahan, J.F., Pincus, J.H., Gelbort, M.M., Weissberg, M, & Kelly, J.P. (2001) Toward and understanding of violence: Neurobehavioral aspects of unwarranted physical aggression. *Neuropsychiatry, Neuropsychology, Behavioral Neurology, 14,* 1–14.

Fisher, R.S. (1998) Epilepsy. In: Enna, S.J. & Coyle, J.T. (Eds.) *Pharmacological Management of Neurological and Psychiatric Disorders.* New York: McGraw-Hill, pp. 459–503.

Fisher, P.A., Bass, H. & Hefner, R. (1990) Treatment of parkinsonian tremor with clozapine. *Journal of Neural Transmission, 2,* 233–238.

Fogel, B.S. (1996) Drug therapy in neuropsychiatry. In: Fogel, B.S., Schiffer, R.B. & Rao, S.M. (Eds.) *Neuropsychiatry.* Baltimore: Williams & Wilkins, pp. 223–256.

Fornai, F., Vaglini, F., Maggio, R., Bonuccelli, U. & Corsini, G.U. (1997) Species differences in the role of excitatory amino acids in experimental parkinsonism. *Neuroscience and Biobehavioral Reviews, 21,* 401–415.

Freedamn, M. (1990) Parkinson's disease. In: Cummings, J.L. (Ed.) *Subcortical Dementia.* New York: Oxford Press, pp.108–122.

Friedman, M.J. (1997) Posttraumatic stress disorder. Journal of *Clinical Psychiatry, 58* (Suppl 9), 33–36.

Gardos, G. (1999) Managing antipsychotic-induced tardive dyskinesia. *Drug Safety, 20,* 187–193.

Gardos, G. & Cole, J.O. (1995) The treatment of tardive dyskinesias. In: Bloom, F.E. & Kupfer, D.J. (Eds.) *Psychopharmacology: The fourth generation of progress.* New York: Raven Press, pp. 1503–1511.

Garlow, S.J., Musselman, D.L. & Nemeroff, C.B. (1999) The neurochemistry of mood disorders: Clinical studies. In: Charney, D.S., Nestler, E.J. & Bunney, B.S. (Eds.) *Neurobiology of Mental Illness.* New York: Oxford University Press, pp.348–364.

Gelenberg, A.J. & Bassuk, E.L. (Eds.) *The Practitioners Guide to Psychotropic Drugs,Fourth Edition.* New York: Plenum Press.

Gelenberg, A.J. & Delgado, P.L. (1997) Depression. In: Gelenberg, A.J. & Bassuk, E.L. (Eds.) *The Practitioners Guide to Psychotropic Drugs, Fourth Edition.*New York: Plenum Press, pp. 19–97.

Germine, M., Goddard, A.W., Woods, S.W., Charney, D.S. & Heninger, G.R. (1992) Anger and anxiety responses to *m*-chlorophenylpiperzine in generalized anxiety disorder. *Biological Psychiatry, 32,* 457–461.

Gerson, S.C. & Baldesssarini, R.J. (1980) Motor effect of serotonin in the central nervous system. *Life Sciences, 27,* 1435–1451.

Ghoneim, M.M. & Merwaldt, S.P. (1990) Benzodiazepines and human memory: A review. *Anesthesiology*, 72, 926–938.

Gilman, S., Koeppe, R.A., Adams, K., Johnson-Greene, D., Junck, L., Kluin, K.J., Motorello, S. & Lohman, M. (1996) Positron emission tomographic studies of cerebral benzodiazepine receptor binding in chronic alcoholics. *Annals of Neurology*, 40, 163–171.

Gilman, G. & Newman, S.W. (1996) *Essentials of Neuroanatomy and Neurophysiology, Ninth Edition*. Philadelphia: F.A. Davis.

Gilroy, J. (1990) *Basic Neurology*. New York: Pergamon Press.

Glennon, R.A. & Dukat, M. (1995) Serotonin receptor subtypes. In: Bloom, F.E. & Kupfer, D.J. (Eds.) *Psychopharmacology: The fourth generation of progress*. New York: Raven Press, pp. 415–429.

Goddard, A.W. & Charney, D.S. (1997) Toward an integrated neurobiology of panic disorder. *Journal of Clinical Psychiatry*, 58 (Suppl 2), 4–11.

Goddard, A.W. & Charney, D.S. (1998) SSRIs in the treatment of panic disorder. Depression and Anxiety, 8 (Suppl. 1), 114–120.

Goddard, A.W., Coplan, J.D., Gorman, J.M. & Charney, D.S. (1999) Principles of the pharmacology of anxiety disorders. In: Charney, D.S., Nestler, E.J. & Bunney, B.S. (Eds.) *Neurobiology of Mental Illness*. New York: Oxford University Press, pp. 548–563.

Good, P.F., Werner, P., Hsu, A., Olanow, C.W. & Perl, D.P. (1996) Evidence for neuronal oxidative damage in Alzheimer's disease. *American Journal of Pathology*, 149, 21–28.

Goodman, W.K., McDougle, C.J., Price, L.H., Riddle, M.A., Pauls, D.L. & Leckman, J.F. (1990) Beyond the serotonin hypothesis: A role for dopamine in some forms of obsessive-compulsive disorders. *Journal of Clinical Psychiatry*, 51, 36–43.

Goodman, W.K. & Murphy, T. (1998) Obsessive-compulsive disorder and Tourette's syndrome. In: Enna, S.J. & Coyle, J.T. (Eds.) *Pharmacological Management of Neurological and Psychiatric Disorders*. New York: McGraw-Hill, pp. 177–211.

Goodman, W.K., Price, L.H., Delgado, P.L., Palumbo, J., Krystal, S.H., Nagy, L.M., Rasmussen, S.A., Heninger, G.R. & Charney, D.S. (1990) Specificity of serotonin reuptake inhibitors in the treatment of obsessive-compulsive disorder: Comparison of fluvoxamine and desipramine. *Archives of General Psychiatry*, 47, 577–585.

Goodnick, P.J., Puig, A., DeVane, C.L. & Freund, B.V. (1999) Mirtazapine in major depression with co-morbid generalized anxiety disorder. *Journal of Clinical Psychiatry*, 60, 446–448.

Graudins, A., Stearman, A. & Chan, B. (1998) Treatment of the serotonin syndrome with cyproheptadine. *Journal of Emergency Medicine*, 16, 615–616.

Green, H.W. (1995) The treatment of attention-deficit disorder with non-stimulant medications. In: Riddle, M. (Ed.) *Pediatric Psychopharmacology*. Philadelphia: W.B. Saunders; *Child and Adolescent Clinics of North America*, 4, 169–195.

Green, M.W. & Selman, J.E. (1991) Review article, the medical management of trigeminal neuralgia. *Headache*, 31, 588–592.

Greenhill, L.L. (1995) Attention-deficit hyperactivity disorder. In: Riddle, M. (Ed.) *Pediatric Psychopharmacology*. Philadelphia: W.B. Saunders; *Child and Adolescent Clinics of North America*, 4, 123–168.

Haber, S.N. & Fudge, J.L. (1997a) The primate substantia nigra and VTA: Integrative circuitry and function. *Critical Reviews in Neurobiology*, 11, 323–342.

Haber, S.N. & Fudge, J.L. (1997b) The interface between dopamine neurons and the amygdala: Implications for schizophrenia. *Schizophrenia Bulletin*, 23, 471–482.

Haefely, W.E. (1990) The GABA-benzodiazepine reaction fifteen years later. *Neurochemical Research*, 15, 169–174.

Haefely, W., Martin, J.R. & Schoch, P. (1990) Novel anxiolytics that act as partial agonists at benzodiazepine receptors. *Trends in Pharmacological Sciences*, 11, 452–456.

Hasey, G.M. (1996) Acetylcholine. In, Baskys, A. & Remington, G. (Eds.) *Brain Mechanisms and Psychotropic Drugs*. Boca Raton, FL: CRC Press, pp. 73–100.

Hasey, G. & Hanin, I. (1991) The cholinergic-adrenergic hypothesis of depression reexamined using clonidine, metoprolol, and physostigmine in an animal model. *Biological Psychiatry*, 29, 127–138.

Heninger, G.R. (1995) Indolamines: The role of serotonin in clinical disorders. In: Bloom, F.E. & Kupfer, D.J. (Eds.) *Psychopharmacology: The fourth generation of progress*. New York: Raven Press, pp. 471–482.

Hicks, T.P. & Conti, F. (1996) Amino acids as the source of considerable excitation in cerebral cortex. *Canadian Journal of Physiology and Pharmacology, 74,* 341–361.

Hidalgo, R., Hertzberg, M.A., Mellman, T., Petty, F., Tucker, P., Weisler, R., Zisook, S., Chen, S., Churchill, E. & Davidson, J. (1999) *International Clinical Psychopharmacology, 14,* 61–68.

Hippius, H. (1989) The history of clozapine. *Psychopharmacology, 99,* 53–55.

Hirsch, E.C., Graybiel, A.M. & Agid, Y. (1988) Melanized dopaminergic neurons are differentially affected in Parkinson's disease. *Nature, 334,* 345–348.

Hirsch, E.C. & Herrero, M.T. (1997) Neurochemical correlates of parkinsonism: Role of dopaminergic lesions. In: Obeso, J.A., DeLong, M.R., Ohye, C. & Marsden, C.D. (Eds.) *The Basal Ganglia and New surgical Approaches for Parkinson's Disease. Advances in Neurology,* Vol. 74. Philadelphia: Lippincott-Raven Publishers, pp. 119–126.

Hirsch, S.R., Das, I., Garey, L.J. & de Belleroche, J. (1997) A pivotal role for glutamate in the pathogenesis of schizophrenia, and its cognitive dysfunction. *Pharmacology, Biochemistry, and Behavior, 56,* 797–802.

Hornykiewicz, O. & Kish, S. (1987) Biochemical pathophysiology of Parkinson's disease. *Advances in Neurology, 45,* 19–34.

Hunt, R.D., Arnsten, A.F.T. & Asbell, M.D. (1995) An open trial of guanfacine in the treatment of attention-deficit hyperactivity disorder. *Journal of the American Academy of Child and Adolescent Psychiatry, 34,* 50–54.

Hunt, R.D., Capper, S. & O'Connell, P. (1990) Clonidine in child and adolescent psychiatry. *Journal of Child and Adolescent Psychopharmacology, 1,* 87–102.

Hyashi, T. (1954) Effects of sodium glutamate on the nervous system. *Keio Journal of Medicine, 3,* 183–???

Hyman, S.E., Arana, G.W. & Rosenbaum, J.F. (1995) *Handbook of Psychiatric Drug Therapy.* Boston: Little, Brown and Company.

Insel, T. (1992) Toward a neuroanatomy of obsessive-compulsive disorder. *Archives of General Psychiatry, 49,* 739–744.

Iversen, S.D. (1984) 5-HT and anxiety. *Neuropharmacology, 23,* 1353–1360.

Jaber, M., Robinson, S.W., Missale, C. & Caron, M.G. (1996) Dopamine receptors and brain function. *Neuropharmacology, 35,* 1503–1519.

Jacobs, B.L. & Azmitia, E.C. (1992) Structure and function of the brain serotonin system. *Physiological Reviews, 72,* 165–229.

Jacobs B.L. & Fornal, C.A. (1995) Serotonin and behavior. In: Bloom, F.E. & Kupfer, D.J. (Eds.) *Psychopharmacology: The fourth generation of progress.* New York: Raven Press, pp. 461–469.

Janicak, P.G. (1999) *Handbook of Psychopharmacology.* Philadelphia: Lippincott Williams & Wilkins.

Janicak, P.G., Davis, J.M., Preskorn, S.H. & Ayd, F.J. (1997) *Principles and Practice of Psychopharmacotherapy, Second Edition.* Baltimore: Williams & Wilkins.

Jankovic, J. (1998) Movement disorders. In: Joynt, R.J. & Griggs, R.C. (Eds.) *Clinical Neurology, Volume 3.* Philadelphia: Lippincott, Williams & Wilkins, pp. 1–106.

Janowsky, D.S., El-Yousef, M.K., Davis, J.M. & Sckerke, H.J. (1972) A cholinergic-adrenergic hypothesis of mania and depression. *Lancet, 2,* 632–635.

Janowsky, D.S., El-Yousef, M.K. & Davis, J.M. (1974) Acetylcholine and depression. *Psychosomatic Medicine, 36,* 248–257.

Jenike, M.A., Baer, L., Ballantine, T., Martuza, R.L., Tynes, S., Giriunas, I., Buttolph, M.L. & Cassem, N.H. (1991) Cingulotomy for refractory obsessive-compulsive disorder: A long-term follow-up of 33 patients. *Archives of General Psychiatry, 48,* 548–555.

Josselyn, S.A., Miller, R. & Beninger, R.J. (1997) Behavioral effects of clozapine and dopamine receptor subtypes. *Neuroscience and Biobehavioral Reviews, 21,* 531–558.

Kahn, R.S. & Davis, K.L. (1995) New developments in dopamine and schizophrenia. In: Bloom, F.E. & Kupfer, D.J. (Eds.) *Psychopharmacology: The fourth generation of progress.* New York: Raven Press, pp. 1193–1203.

Kahn, R.S., Westenberg, H.G.M. & Moore, C. (1995) Increased serotonin function and panic disorder. In: Asnis, G.M. & van Praag, H.M. (Eds.) *Panic Disorder: Clinical, biological, and treatment aspects.* New York: John Wiley & Sons, pp. 151–180.

Kamil, R. (1996) Antidepressants. In: Baskys, A. & Remington, G. (Eds.) *Brain Mechanisms and Psychotropic Drugs.* Boca Raton, FL: CRC Press, pp. 153–180.

Kane, J.M. (1995) Tardive dyskinesia: Epidemiological and clinical presentation. In: Bloom, F.E. & Kupfer, D.J. (Eds.) *Psychopharmacology: The fourth generation of progress.* New York: Raven Press, pp. 1485–1495.

Kane, J.M., Jeste, D.V., Barnes, T.R.E., et.al. (1992) Treatment of tardive dyskinesia. In: *American Psychiatric Association Task Force on Tardive Dyskinesia.* Washington, DC: American Psychiatric Association, pp. 103–120.

Kapur, S. (2000) The clinical implications of neuroreceptor occupancy in the management of psychosis. *PsychLink* (EDA# 550–0520). New Orleans, LA.: Interactive Medical Networks.

Kapur, S. & Mann, J.J. (1996) Role of the dopaminergic system in depression. *Biological Psychiatry, 32,* 1–17.

Kapur, S. & Remington, G. (1996) Serotonin-dopamine interaction and its relevance to schizophrenia. *American Journal of Psychiatry, 153,* 466–476.

Kapur, S. & Seeman, P. (2000) Antipsychotic agents differ in how fast they come off the dopamine D2 receptors. Implications for atypical antipsychotic's action. *Journal of Psychiatry and Neuroscience, 25,* 161–166.

Kapur, S., Zipursky, R.B. & Remington, G. (1999) Clinical and theoretical implications of 5-HT$_2$and D$_2$ receptor occupancy of clozapine, risperidone, and olanzapine in schizophrenia. *American Journal of Psychiatry, 156,* 286–293.

Kapur, S., Zipursky, R., Jones, C., Shammi, C.S., Remington, G. & Seeman, P. (2000) A positron emission tomography study of quetiapine in schizophrenia: A preliminary finding of an antipsychotic effect with only transiently high dopamine D2 receptor occupancy. *Archives of General Psychiatry, 57,* 553–559.

Kapur, S., Zipursky, R., Remington, G., Jones, C., DaSilva, J., Wilson, A.A. & Houle, S. (1998) 5-HT2 and D2 receptor occupancy of olanzapine in schizophrenia: A PET investigation. *American Journal of Psychiatry, 155,* 921–928.

Katz, L. (1999) *Keep Your Brain Alive.* New York: Workman Publishing.

Katz, L., Fleisher, W., Kjernisted, K. & Milanese, P. (1996) A review of the psychobiology and pharmacotherapy of posttraumatic stress disorder. *Canadian Journal of Psychiatry, 41,* 233–238.

Katz, L.C. & Shatz, C.J. (1996) Synaptic activity and the construction of cortical circuits. *Science, 274,* 1133–1138.

Katzung, B.G. (1998) *Basic and Clinical Pharmacology.* Stamford,CT:: Appleton & Lange.

Keck, P.E. Jr. & McElroy, S.L. (1997) New uses for antidepressants: Social phobia. *Journal of Clinical Psychiatry, 58* (Suppl 14), 32–36.

Kelsey, J.E. & Nemeroff, C.B. (1998) Affective disorders. In: Enna, S.J. & Coyle, J.T. (Eds.) *Pharmacological Management of Neurological and Psychiatric Disorders.* New York: McGraw-Hill, pp. 95–136.

Keltner, N.L. & Folks, D.G. (2005) *Psychotropic Drugs.* St. Louis: Mosby.

Kent, J.M., Coplan, J.D. & Gorman, J.M. (1998) Clinical utility of the selective serotonin reuptake inhibitors in the spectrum of anxiety. *Biological Psychiatry, 44,* 812–824.

Keverne, E.B. (1999) GABA-ergic neurons and the neurobiology of schizophrenia and other psychoses. Brain Research Bulletin, 48, 467–473.

Khan, A.U. (1998) *Neurochemistry of Schizophrenia and Depression.* Clovis, CA: A.J. Publisher.

Kim, K.Y., Moles, J.K., & Hawley, J.M. (2001) Selective serotonin reuptake inhibitors for aggressive behaviors in patients with dementia after head injury. *Pharmacotherapy, 21*(4), 498–501.

King, D.J. (1998) Drug treatment of the negative symptoms of schizophrenia. *European Neuropsychopharmacology, 8,* 33–42.

Kinon, B.J. & Lieberman, J.A. (1996) Mechanisms of action of antipsychotic drugs: A critical analysis. *Psychopharmacology, 124,* 2–34.

Knable, M.B., Kleinman, J.E. & Weinberger, D.R. (1998) Neurobiology of schizophrenia. In: Schatzberg, A.F. & Nemeroff, C.B. (Eds.)*Textbook of Psychopharmacology.* Washington, DC: American Psychiatric Press, pp. 589–607.

Knable, M.B. & Weinberger, D.R. (1997) Dopamine, the prefrontal cortex and schizophrenia. *Journal of Psychopharmacology, 11,* 123–131.

Koller, W.C., Silver, D.E. & Lieberman, A. (1994) An algorithm for the management of Parkinson's disease. *Neurology, 44* (Suppl. 10), S1-S52.

Koob, G.F. (1992) Drugs of abuse: Anatomy, pharmacology, and function of reward pathways. *Trends in Pharmacological Science, 13,* 177–184.

Kopala, L.C., Good, K.P. & Honer, W.G. (1997) Extrapyramidal signs and clinical symptoms in first-episode schizophrenia: Response to low-dose risperidone. *Journal of Clinical Psychopharmacology, 17,* 308–313.

Kopelman, M.D. & Corn, T.H. (1988) Cholinergic "blockade" as a model for cholinergic depletion. *Brain, 111,* 1079–1110.

Korczyn, A.D. (1995) Parkinson's disease. In: Bloom, F.E. & Kupfer, D.J. (Eds.) *Psychopharmacology: The fourth generation of progress.* New York: Raven Press, pp. 1479–1484.

Kovacs, G.L. & de Wied, D. (1994) Peptidergic modulation of learning and memory processes. *Pharmacological Reviews, 46,* 269–91.

Kriegstein, A.R., Flint, A.C. & LoTurco, J.J. (1996) The role of excitatory amino acids in cortical development. In: Conti, F. & Hicks, T.P. (Eds.) *Excitatory Amino Acids and the Cerebral Cortex.* Cambridge, MA: MIT Press, pp. 201–214.

Kulkarni, S.K. & Ninan, I. (1996) Current concepts in the molecular diversity and pharmacology of dopamine receptors. *Methods and Findings in Experimental and Clinical Pharmacology, 18,* 599–613.

Kuraishi, Y., Hirota, N. Sato, Y., Kaneko, S., Satoh, M. & Takagi, H. (1985) Noradrenergic inhibition of the release of substance P from the primary afferents in the rabbit spinal dorsal horn. *Brain Research, 359,* 177–182.

Lader, M.H. (1998) The nature and duration of treatment for GAD. *Acta Psychiatrica Scandinavica, 393* (Suppl), 109–117.

Lamberti, J.S. & Bellnier, T. (1993) Clozapine and tardive dyskinesia. *Journal of Nervous and Mental Disorders, 181,* 137–138.

Lane, R. & Baldwin, D. (1997) Selective serotonin reuptake inhibitor-induced serotonin syndrome: Review. *Journal of Clinical Psychopharmacology, 17,* 208–221.

Lavalaye, J., Linszen, D.H., Booij, J., Reneman, L., Gersons, B.P. & van Royen, E.A. (1999) Dopamine D2 receptor occupancy by olanzapine or risperidone in young patients with schizophrenia. *Psychiatry Research, 92,* 33–44.

Lee, R. & Coccaro, E. (2001) The neuropsychopharmacology of criminality and aggression. *Canadian Journal of Psychiatry, 46,* 35–44.

Le Moal, M. (1995) Mesocorticolimbic dopaminergic neurons: Functional and regulatory roles. In: Bloom, F.E. & Kupfer, D.J. (Eds.) *Psychopharmacology: The fourth generation of progress.* New York: Raven Press, pp. 283–294.

Le Moal, M. & Simon, H. (1991) Mesocorticolimbic dopaminergic network: Functional and regulatory roles. *Physiological Review, 71,* 155–234.

Leonard, B.E. (1997) *Fundamentals of Psychopharmacology.* New York: John Wiley & Sons.

Levin, E.D. & Simon, B.B. (1998) Nicotinic acetylcholine involvement in cognitive functions in animals. *Psychopharmacology, 138,* 217–230.

Levy, R.H. (1996) Sedation in acute and chronic agitation. *Pharmacotherapy* (Suppl), *16,* 152S-159S.

Lidow, M.S., Williams, G.V. & Goldman-Rakic, P.S. (1998) The cerebral cortex: A case for a common site of action of antipsychotics. *Trends in Pharmacological Sciences, 19,* 136–140.

Lieberman, J.A., Saltz, B.L., Johns, C.A., Pollack, S., Borenstein, M. & Kane, J. (1991) The effects of clozapine on tardive dyskinesia. *British Journal of Psychiatry, 158,* 503–510.

Lipton, S.A. & Rosenberg, P.A. (1994) Excitatory amino acids as a final common pathway for neurological disorders. *New England Journal of Medicine, 330,* 613–622.

Littrell, R.A. & Schneiderhan, M. (1996) The neurobiology of schizophrenia. *Pharmacotherapy, 16,* 143S-147S.

Lodge, D., Aram, J.A., Church, J., Davies, S.N., Martin, D, O'Shaughnessey, C.T. & Zeman, S. (1987). Excitatory amino acids and phencyclidine-like drugs. In: Hicks, T.P., Lodge, D. & McLennan, H. (Eds.) *Excitatory Amino Acid Transmission.* New York: Alan R. Liss, pp. 83–90.

Loo, H., Poirier-Littre, M.F., Theron, M., Rein, W. & Fleurot, O. (1997) Amisulpride versus placebo in the medium-term treatment of the negative symptoms of schizophrenia. *British Journal of Psychiatry, 170,* 18–22.

LoTurco, J.J., Blanton, M.G. & Kriegstein, A.R. (1991) Initial expression and endogenous activation of NMDA channels in early neocortical development. *Journal of Neuroscience, 11,* 792–799.

Luchins, D.J., Freed, W.J., Wyatt, R.J (1980) The role of cholinergic supersensitivity in the medical symptoms associated with withdrawal of antipsychotic drugs. *American Journal of Psychiatry, 137,* 1395–1398.

Lydiard, R.B. (1998) The role of drug therapy in social phobia. *Journal of Affective Disorders, 50* (Suppl.1), S35-S39.

Lydiard, R.B., Brawman-Mintzer, O. & Ballenger, J.C. (1996) Recent developments in the psychopharmacology of anxiety disorders. *Journal of Consulting and Clinical Psychology, 64,* 660–668.

MacDonald, J.F., Wojtowicz, J.M. & Baskys, A. (1996) Glutamate receptors. In: Baskys, A. & Remington, G. (Eds.) *Brain Mechanisms and Psychotropic Drugs.* Boca Raton, FL: CRC Press, pp. 117–130.

Malatynska, E., Knapp, R.J., Ikeda, & Yamamura, H.I. (1988) Antidepressants and seizure interactions at the GABAA receptor chloride-ionophore complex. *Life Sciences, 43,* 303–307.

Mann, S.C., Caroff, S.N. & Lazarus, A. (1991) Pathogenesis of neuroleptic malignant syndrome. *Psychiatric Annals, 21,* 175–180.

Mann, J.J. & Kupfer, D.J. (Eds) (1993) *GABA and Depression.* New York: Plenum Press.

Mansour, A., Meador-Woodruff, J.H., Lopez, J. & Watson, S.J.Jr. (1998) Biochemical anatomy: Insights into cell biology and pharmacology of the dopamine and serotonin systems in the brain. In: Schartzberg, A.F. & Nemeroff, C.B. (Eds.) *The American Psychiatric Press: Textbook of psychopharmacology.* Washington, DC: American Psychiatric Press, pp. 55–73.

Marder, S.R. (1998) Antipsychotic medications. In: Schatzberg, A.F. & Nemeroff, C.B. (Eds.) *Textbook of Psychopharmacology.* Washington, DC: American Psychiatric Press, pp. 309–321.

Markowitz, J.S., Brown, C.S. & Moore, T.R. (1999) Atypical antipsychotics. Part I. Pharmacology, pharmacokinetics, and efficacy. *Annals of Pharmacotherapy, 33,* 73–85.

Marsden, C.D. & Jenner, P. (1980) The pathophysiology of extrapyramidal side-effects of neuroleptic drugs. *Psychological Medicine, 10,* 55–72.

Marsden C.D. & Olanow, C.W. (1998) The causes of Parkinson's disease are being unraveled and rational neuroprotective therapy is close to reality. *Annals of Neurology, 44* (Suppl) S189-S196.

Martin, R.L., Lloyd, H.G.E. & Cowan, A.I. (1994) The early events of oxygen and glucose deprivation: Setting the scene for neuronal death? *Trends in Neurosciences, 17,* 251–257.

Matthyse, S. (1973) Antipsychotic drug actions: A clue to the neuropathology of schizophrenia. *Federation Proceedings, 32,* 200–205.

Mattson, M.P. & Pedersen, W.A. (1998) Effects of amyloid precursor derivatives and oxidative stress on basal forebrain cholinergic systems in Alzheimer's disease. *International Journal of Developmental Neuroscience, 16,* 737–753.

Mayberg, H. (2004) Depression: A neuropsychiatric perspective. In, Panksepp, J. (Ed) *Textbook of biological psychiatry.* Hoboken, N.J.: Wiley-Liss, pp. 197–230.

Mayeux, R., Stern, Y. & Williams, J.B.W. (1986) Clinical and biochemical features of depression in Parkinson's disease. *American Journal of Psychiatry, 143,* 756–759.

McAllister, A.K., Katz, L.C. & Lo, D.C. (1999) Neurotropins and synaptic plasticity. *Annual Review of Neuroscience, 22,* 295–318.

McCarthy, R.H. (1994) Clozapine reduces essential tremor independent of its antipsychotic effect: A case report. *Journal of Clinical Psychopharmacology, 14,* 212–213.

McCormick, D.A., Pape, H.C. & Williamson, A. (1991) Actions of norepinephrine in the cerebral cortex and thalamus: Implications for function of the central norepinephrine system. *Progress in Brain Research, 88,* 293–305.

McDougle, C.J. (1999) The neurobiology and treatment of obsessive-compulsive disorder. In: Charney, D.S., Nestler, E.J. & Bunney, B.S. (Eds.) *Neurobiology of Mental Illness.* New York: Oxford University Press, pp. 518–533.

McDougle, C.J., Goodman, W.K., Leckman, J.F. & Price, L.H. (1993) The psychopharmacology of obsessive-compulsive disorder. Implications for treatment and pathogenesis. *Psychiatric Clinics of North America, 16,* 749–766.

Meek, J.K. & Kablinger, A. (1998) Antidepressants and post-traumatic stress disorder. *Journal of the Louisiana State Medical Society, 150,* 487–489.

Meltzer, H.Y. (1989) Duration of a clozapine trial in neuroleptic-resistant schizophrenia. *Archives of General Psychiatry, 46,* 672.

Meltzer, H.Y. (1991a) The significance of serotonin in neuropsychiatric disorders. *Journal of Clinical Psychiatry* (Suppl), *52*, 70–72.

Meltzer, H.Y. (1991b) The mechanism of action of novel antipsychotic drugs. *Schizophrenia Bulletin, 17*, 263–287.

Meltzer, H.Y. (1995) Role of serotonin in the action of atypical antipsychotic drugs. *Clinical Neuroscience, 3*, 64–75.

Meltzer, H.Y. (1999) The role of serotonin in antipsychotic drug action. Neuropsychophar*macology, 21* (Suppl) 106S-115S.

Meltzer, H.Y., Matsubari, F. & Lee, J.C. (1989a) The ratios of serotonin 2 and dopamine 2 affinities differentiate atypical and typical antipsychotic drugs. *Psychopharmacology Bulletin, 25*, 390–392.

Meltzer, H.Y., Matsubari, F. & Lee, J.C. (1989b) Classification of typical and atypical antipsychotic drugs on the basis of dopamine D-1, D-2 and serotonin-2 pKi values. *Journal of Pharmacology and Experimental Therapeutics, 251*, 238–246.

Meltzer, H.Y. & McGurk, S.R. (1999) The effects of clozapine, risperidone, and olanzapine on cognitive function in schizophrenia. *Schizophrenia Bulletin, 25*, 233–255.

Mink, J.W. & Thach, W.T. (1993) Basal ganglia intrinsic circuits and their role in behavior. *Current Opinions in Neurobiology, 3*, 950–957.

Mintzer, J.E. (2001) Underlying mechanisms of psychosis and aggression in patients with Alzheimers disease. *Journal of Clinical Psychiatry, 62* (Suppl 21), 23–25.

Missale, C, Nash, S.R., Robinson, S.W., Jaber, M. & Caron, M.G. (1998) Dopamine receptors from structure to function. *Physiological Reviews, 78*, 189–225.

Miyawaki, E., Meah, Y. & Koller, W.C. (1997) Serotonin, dopamine and motor effects in Parkinson's disease. *Clinical Neuropharmacology, 20*, 300–310.

Moller, H.J., Boyer, P., Fleurot, O. & Rein, W. (1997) Improvement of acute exacerbations of schizophrenia with amisulpride: A comparison with haloperidol. *Psychopharmacology, 132*, 396–401.

Moore, K.E. & Lookingland, K.J. (1995) Dopaminergic neuronal systems in the hypothalamus. In: Bloom, F.E. & Kupfer, D.J. (Eds.) *Psychopharmacology: The fourth generation of progress.* Ne wYork: Raven Press, pp. 245–256.

Moore, R.Y. & Bloom, F.E. (1979) Central catecholamine neuron systems: Anatomy and physiology of the norepinephrine and epinephrine systems. *Annual Review of Neuroscience, 2*, 113–168.

Musselman, D.L., DeBattista, C., Nathan, K.I., Kilts, C.D., Schatzberg, A.F. & Nemeroff, C.B. (1998) Biology of mood disorders. In: Schartzberg, A.F. & Nemeroff, C.B. (Eds.) *Textbook of Psychopharmacology.* Washington, DC: American Psychiatric Press, pp. 549–588.

Nelson, R.J. & Chiavegatto, S. (2001) Molecular basis of aggression. *Trends in Neuroscience, 24*, 713–719.

Nelson, J. & Chouinard, G. (1996) Benzodiazepines: Mechanisms of action and clinical indications. In: Baskys, A. & Remington, G. (Eds.) *Brain Mechanisms and Psychotropic Drugs.* Boca Raton, FL: CRC Press, pp. 213–238.

O'Donnell, P.O. & Grace, A.A. (1996) Basic neurophysiology of antipsychotic drug action. In: Csernansky, J.C. (Ed.) *Handbook of Experimental Pharmacology: Antipsychotics.* New York: Springer-Verlag, pp. 163–202.

Olney, J.W. & Farber, N.B. (1995) Glutamate receptor dysfunction and schizophrenia. *Archives of General Psychiatry, 52*, 998–1007.

Olney, J.W., Ho, O.L. & Rhee, V. (1971) Cytotoxic effects of acidic and sulfur containing amino acids on the infant mouse central nervous system. *Experimental Brain Research, 14*, 61–76.

Olanow, C.W., Jenner, P. & Brooks, D. (1998) Dopamine agonists and neuroprotection in Parkinson's disease. Annals of Neurology, *44* (Suppl), S167-S174.

Olney, J.W., Sharpe, L.G. & Feigin, R.D. (1971) Glutamate-induced brain damage in infant primates. *Journal of Neuropathology and Experimental Neurology, 31*, 464–488.

Owens, M.J. & Risch, S.C. (1998) Atypical antipsychotics. In: Schatzberg, A.F. & Nemeroff, C.B. (Eds.) *Textbook of Psychopharmacology,* Washington, DC: American Psychiatric Press, pp. 323–348.

Page, I.H. (1976) The discovery of serotonin. *Perspectives in Biology-and Medicine, 20*, 1–8.

Paschen, W. (1996) Glutamate excitotoxicity in transient global cerebral ischemia. *Acta Neurobiologiae Experimentalis, 56*, 313–322.

Pate, D.S. & Margolin, D.I. (1994) Cognitive slowing in Parkinson's and Alzheimer's patients: Distinguishing bradyphrenia from dementia. *Neurology, 44*, 673–674.

Paul, S.M. (1995) GABA and glycine. In: Bloom, F.E. & Kupfer, D.J. (Eds.) *Psychopharmacology: The fourth generation of progress*. New York: Raven Press, pp. 87–94.

Pearson, J., Halliday, G., Sakamoto, N., et.al. (1990) Catacholeminergic neurons. In: Paxinos, G. (Ed.) *The Human Nervous System*. San Diego, CA: Academic Press, pp. 1023–1050.

Pearson, S.J., Heathfield, K.W. & Reynolds, G.P. (1990) Pallidal GABA and chorea in Huntington's disease. *Journal of Neural Transmission, 81*, 241–246.

Penny, J.B. (1996) Neurochemical neuroanatomy. In: Fogel, B.S., Shiffer, R.B. & Rao, S.M. (Eds.) *Neuropsychiatry*. Baltimore: Williams & Wilkins, pp. 145–171.

Penny, J.B. & Young, A.B. (1998) Extrapyramidal disorders. In: Enna, S.J. & Coyle, J.T. (Eds.) *Pharmacological Management of Neurological and Psychiatric Disorders*. New York: McGraw-Hill, pp. 351–376.

Perry, E.K., Tomlinson, B.E., Blessed, G., Bergmann, K., Gibson, P.H., Perry, R.H. (1978) Correlation of cholinergic abnormalities with senile plagues and mental test scores in senile dementia. *British Medical Journal, 2*, 1457–1459.

Perry, T.L., Hanson, S. & Kloster, M. (1973) Huntington's chorea. Deficiency of gamma aminobutyric acid in brain. *New England Journal of Medicine, 288*, 337–342.

Perry, T.L., Wright, J.M., Hansen, S., Thomas, S.M., Allan, B.M., Baird, P.A. & Diewold, P.A. (1982) A double-blind clinical trial of isoniazid in Huntington's disease. *Neurology, 32*, 354–358.

Petty, F. (1995) GABA and mood disorders: A brief review and hypothesis. *Journal of Affective Disorders, 24*, 275–281.

Petty, F., Davis, L.L., Kabel, D., & Kramer, G.L. (1996) Serotonin dysfunction disorders: A behavioral neurochemistry perspective. *Journal of Clinical Psychiatry, 57*, (Suppl 8), 11–16.

Pickar, D. (1986) Neuroleptics, dopamine, and schizophrenia. *Psychiatric Clinics of North America, 9*, 35–48.

Pies, R.W. (1998) *Handbook of Essential Psychopharmacology*. Washington, DC: American Psychiatric Press.

Pine, D.S., Grun, J. & Gorman, J.M. (1998) Anxiety disorders. In: Enna, S.J. & Coyle, J.T. (Eds.) *Pharmacological Management of Neurological and Psychiatric Disorders*. New York: McGraw-Hill, pp. 53–94.

Pliszka, S.R., McCracken, J.T. & Maas, J.W. (1996) Catecholamines in attention-deficit hyperactivity disorder: Current perspectives. *Journal of the American Academy of Child and Adolescent Psychiatry, 35*, 264–272.

Pollack, M.H. (1999) Social anxiety disorder: Designing a pharmacologic treatment strategy. *Journal of Clinical Psychiatry, 60* (Suppl.9), 20–26.

Puech, A., Fluerot, O. & Rein, W. (1998) Amisulpride and atypical antipsychotic, in the treatment of acute episodes of schizophrenia: A dose-ranging study vs. haloperidol. Acta Psychiatrica Scandinavica, *98*, 65–72.

Raja, M. (1996) The treatment of tardive dyskinesia. *Schweizer Archiv fur Neurologie und Psychiatrie, 147*, 13–18.

Raj, B.A. & Sheehan, D.V. (1995) Somatic treatment strategies in panic disorder. In: Asnis, G.M. & van Praag, H.M. (Eds.) *Panic Disorder: Clinical, biological, and treatment aspects*. New York: John Wiley & Sons, pp. 279–313.

Randrup, A., Munkvad, I. & Fog, R. (1975) Mania, depression, and brain dopamine. In: Essman, W.B. & Vaizelli, L. (Eds.) *Currents Developments in Psychopharmacology*. New York: Spectrum, pp. 207–229.

Rein, W. & Turjanski, S. (1997) Clinical update on amisulpride in deficit schizophrenia. *International Clinical Psychopharmacology, 12* (Suppl), S19-S27.

Remington, G.J. (1993) Clinical consideration in the use of risperidone. *Canadian Journal of Psychiatry, 38* (Suppl 3), S96-S100.

Remington, G. (1996) Neuroleptics. In: Baskys, A. & Remington, G. (Eds.) *Brain Mechanisms and Psychotropic Drugs*. Boca Raton, FL: CRC Press, pp. 193–211.

Remington, G. & Chong, S.A. (1999) Conventional versus novel antipsychotics: Changing concepts and clinical implications: *Journal of Psychiatry and Neuroscience, 24*, 431–441.

Remington, G. & Kapur, S. (1999) D_2 and 5-HT$_2$Receptor effects of antipsychotics: Bridging basic and clinical findings using PET. *Journal of Clinical Psychiatry, 60*, 15–19.

Remington, G. & Kapur, S. (2000) Atypical antipsychotics: Are some more atypical than others? Psychopharmacology, *148*, 3–15.

Renynghe de Voxuri.e., G.E. (1968) Anafranil (G34586) in obsessive neurosis. *Acta Neurologia Belgica*, *68*, 787–792.

Reynolds, G.P., Pearson, S.J. & Heathfield, K.W. (1990) Dementia in Huntington's disease is associated with neurochemical deficits in the caudate nucleus, not the cerebral cortex. *Neuroscience Letters, 113*, 95–100.

Richelson, E. (1996) Synaptic effects of antidepressants. *J. Clinical Psychopharmacology, 16*, (Suppl 2), 1S–9S.

Richelson, E. (1999a) Receptor pharmacology of neuroleptics: Relation to clinical effects. *Journal of Clinical Psychiatry, 60* (Suppl 10), 5–14.

Richelson, E. (1999b) Basic neuropharmacology of antidepressants relevant to the pharmacotherapy of depression. *Clinical Cornerstone, 1*, 17–30.

Richelson, E. (2001) Pharmacology of antidepressants. *Mayo Clinic Proceedings, 76*, 511–527.

Riddle, M.A., Hardin, M.T., Cho, S.C., etal. (1988) Desipramine treatment of boys with attention-deficit hyperactivity disorder and tics: Preliminary clinical experience. *Journal of the American Academy of Child and Adolescent Psychiatry. 27*, 811–814.

Rison, R.A. & Stanton, P.K. (1995) Long-term potentiation and N-methyl-D-aspartate receptors: Foundations of memory and neurological disease? *Neuroscience Biobehavioral Review, 19*, 533–552.

Robbins, T.W. & Everitt, B.J. (1995) Central norepinephrine neurons and behavior. In: Bloom, F.E. & Kupfer, D.J. (Eds.) *Psychopharmacology: The fourth generation of progress.* New York: Raven Press, pp. 363–372.

Rocca, P., Fonzo, V., Scotta, M., Zanalda, E., Ravizza, L. (1997) Paroxetine efficacy in the treatment of generalized anxiety disorder. *Acta Psychiatrica Scandinavica, 95*, 444–450.

Rockhold, R.W. (1997) The chemical basis for neuronal communication. In: Haines, D.E. (Ed.) *Fundamental Neuroscience.* New York: Churchill Livingstone, pp. 51–63.

Rodriguez, M.C., Obeso, J.A. & Olanow, C.W. (1998) Subthalamic nucleus-mediated excitotoxicity in Parkinson's disease: A target for neuroprotection. *Annals of Neurology, 44* (Suppl), S175–S188.

Rogers, S.L. & Friedhoff, L.T. (1996) The efficacy and safety of donepezil in patients with Alzheimer's disease: Results of a U.S. multicenter, randomized, double-blind, placebo controlled trial. *Dementia, 7*, 293–303.

Rosenberg, P.A., Amin, S. & Leitner, M. (1992) Glutamate uptake disguises neurotoxic potency of glutamate agonists in cerebral cortex in dissociated cell culture. *Journal of Neuroscience, 12*, 56–61.

Roth, R.H. & Elsworth, J.D. (1995) Biochemical pharmacology of midbrain dopamine neurons. In:Bloom, F.E. & Kupfer, D.J. (Eds.) *Psychopharmacology: The fourth generation of progress.* New York: Raven Press, pp. 227–243.

Roth, B.L. & Meltzer, H.Y. (1995) The role of serotonin in schizophrenia. In: Bloom, F.E. & Kupfer, D.J. (Eds.) *Psychopharmacology: The fourth generation of progress.* New York: Raven Press, pp. 1215–1227.

Russo, M.B., Brooks, F.R., Fontenot, J.P., Dopler, B.M., Neely, E.T. & Halliday, A.W. (1997) Sodium pentothal hypnosis: A procedure for evaluating medical patients with suspected psychiatric co-morbidity. *Military Medicine, 162*, 215–218.

Ryan, J.M. (2000) Pharmacologic approach to aggression in neuropsychiatric disorders. *Seminars in Clinical Neuropsychiatry, 5(4)*, 238–249.

Sanders-Bush, E. & Canton, H. (1995) Serotonin receptors. In: Bloom, F.E. & Kupfer, D.J. (Eds.) *Psychopharmacology: The fourth generation of progress.* New York: Raven Press, pp. 431–441.

Sanders-Bush, E. & Mayer, S.E. (1996) 5-hydroxytryptamine (serotonin) receptor agonists and antagonists. In: Molinoff, P.B. & Ruddon, R.W. (Eds.) *Goodman & Gilman's the Pharmacological Basis of Therapeutics, Ninth Edition.* New York: McGraw-Hill, pp. 249–263.

Sandman, C.A., Strand, F.L., Beckwith, B. (1999) Neuropeptides: Structure and function in biology and behavior. *Annals of New York Academy of Sciences*, (v. 897).

Sano, M., Marder, K. & Dooneief, G. (1996) Basal ganglia diseases. In: Fogel, S., Schiffer, R.B. & Rao, S.M. (Eds.) *Neuropsychiatry.* Baltimore: Williams & Wilkins, pp. 805–825.

Sarter, M. & Bruno, J.P. (1998) Cortical acetylcholine, reality distortion, schizophrenia, and Lewy-Body dementia: Too much or too little cortical acetylcholine? *Brain and Cognition, 38*, 297–316.

Sarter, M., Bruno, J.P., & Turchi, J. (1999) Basal forebrain afferent projections modulating cortical acetylcholine, attention, and implications for neuropsychiatric disorders. *Annals of the New York Academy of Sciences, 877*, 368–382.

Sarter, M., Nutt, D. & Lister, R.G. (1995) *Benzodiazepine Receptor inverse Agonists*. New York: Wiley-Liss.

Saxena, S., Brody, A.L., Schwartz, J.M. & Baxter, L.R. (1998) Neuroimaging and frontal-subcortical circuitry in obsessive-compulsive disorder. *British Journal of Psychiatry*, 173 (Suppl 35), 26–37.

Scatton, B., Claustre, Y., Cudennec, A., Oblin, A., Perrault, G., Sanger, D.J. & Schoemaker, H. (1997) Amisulpride: From animal pharmacology to therapeutic action. *International Clinical Psychopharmacology*, 12 (Suppl), S29-S36.

Schatzberg, A.F. (1998) Noradenergic versus serotonergic antidepressants: Predictors of treatment response. *Journal of Clinical Psychiatry*, 59 (Suppl 14), 15–18.

Schatzberg, A.F. & Nemeroff, C.B. (Eds) (1998) *Textbook of Psychopharmacology*. Washington, DC: American Psychiatric Press.

Schatzberg, A.F., Rothschild, A.J., Langlais, P.J., Bird, E.D. & Cole, J.O. (1985) A corticosteroid/dopamine hypothesis for psychotic depression and related states. *Journal of Psychiatric Research*, 19, 57–64.

Schildkraut, J.J. (1965) The catecholamine hypothesis of affective disorders: A review of supporting evidence. *American Journal of Psychiatry*, 122, 509–522.

Schoepp, D.D. & Conn, P.J. (1993). Metabotropic receptors in brain function and pathology. *Trends in Pharmacological Sciences*, 14, 13–20.

Schotte, A., Janssen, P.F.M., Gommeren, W., Luyten, W.H.M.I., Van Gompel, P., Lesage, A.S., DeLoore, K. & Leyse, J.E. (1996) Risperidone compared with new and reference antipsychotic drugs: In vitro and in vivo receptor binding. *Psychopharmacology*, 124, 57–73.

Schousboe, A., Belhage, B. & Frandsen, A. (1997) Role of Ca^{2+} and other second messengers in excitatory amino acid receptor mediated neurodegeneration: Clinical perspectives. *Clinical Neuroscience*, 4, 194–198.

Schulkin, J. (1999) *The Neuroendocrine Regulation of Behavior*. NewYork: Cambridge University Press.

Schwartz, J.M. (1997) Obsessive-compulsive disorder. *Science and Medicine*, 4, 14–23.

Seeman, P. (1995) Dopamine receptors: Clinical correlates. In: Bloom, F.E. & Kupfer, D.J. (Eds.) *Psychopharmacology: The fourth generation of progress*. New York: Raven Press, pp. 295–302.

Seeman, P., Chau-Wong, M., Tedesco, J. & Wong, K. (1975) Brain receptors for antipsychotic drugs and dopamine: Direct binding assays. *Procedures of the National Academy of Science*, 72, 4376–4380.

Seeman, P., Lee, T., Chau-Wong, M., & Wong, K. (1976) Antipsychotic drug doses and neuroleptic/dopamine receptors. *Nature*, 261, 717–719.

Seeman, P. & Tallerico, T. (1998) Antipsychotic drugs which elicit little or no parkinsonism bind more loosely than dopamine to brain D2 receptors, yet occupy high levels of these receptors. *Molecular Psychiatry*, 3, 123–134.

Shalev, A.Y., Bonne, O. & Eth, S. (1996) Treatment of posttraumatic stress disorder: A review. *Psychosomatic Medicine*, 58, 165–182.

Sheehan, D.V. & Harnett-Sheehan, K. (1996) The role of SSRIs in panic disorder. *Journal of Clinical Psychiatry*, 57 (Suppl), 51–58.

Sheehan, B.A., Raj, R.R. & Knapp, E.L. (1993) Serotonin in panic disorder and social phobia. *International Clinical Psychopharmacology*, 8, 63–77.

Shiah, I-S & Yatham, L.N. (1998) GABA function in mood disorders: An update and critical review. *Life Sciences*, 63, 1289–1303.

Shiloh, R., Zemishlany, Z., Aizenberg, D., Radwan, M., Schwartz, B., Dorfman-Etrog, P., Modai, I., Khaikan, M. & Weizman, A. (1997) Sulpiride augmentation in people with schizophrenia partially responsive to clozapine: A double-blind, placebo controlled study. *British Journal of Psychiatry*, 171, 569–573.

Shoulson, I., Goldblatt, D., Carlton, M. & Joynt, R.J. (1978) Huntington's disease: Treatment with muscimol, a GABA-mimetic drug. *Annals of Neurology*, 4, 279–284.

Simon, D.K. & Standaert, D.G. (1999) Neuroprotective therapies. *Medical Clinics of North America*, 83, 509–523.

Snyder, S.H. (1976) The dopamine hypothesis of schizophrenia: Focus on the dopamine receptor. *American Journal of Psychiatry*, 133, 197–202.

Southwick, S.M. Krystal, J.H., Bremner, J.D., Morgan, C.A., Nicolaou, A.L., Nagy, L.M., Johnson, D.R., Heninger, G.R. & Charney, D.S. (1997) Noradrenergic and serotonergic function in posttraumatic stress disorder. *Archives of General Psychiatry*, 54, 749–758.

Southwick, S.M., Morgan, C.A., Grillon, C.G., Krystal, J.H., Nagy, L.M. & Charney, D.S. (1997) Noradrenergic alterations in posttraumatic stress disorder. *Annals of the New York Academy of Sciences, 821,* 125–141.

Spano, P.F., Govoni, S. & Trabucchi, M. (1978) Studies on the pharmacological properties of dopamine receptors in various areas of the central nervous system. *Advances in Biochemical Psychopharmacology, 19,* 155–165.

Speller, J.C., Barnes, T.R., Curson, D.A., Pantelis, C. & Alberts, J.L. (1997) One-year, low dose neuroleptic study of in patients with chronic schizophrenia characterized by persistent negative symptoms. Amisulpride vs. haloperidol. *British Journal of Psychiatry, 171,* 564–568.

Spencer, T., Biederman, J., Wilens, T., Harding, M., O'Donnell, D. & Griffin, S. (1996) Pharmacotherapy of attention-deficit hyperactivity disorder across the life cycle. *Journal of the American Academy of Child and Adolescent Psychiatry, 35,* 409–432.

Squires, R.F. & Saederup, E. (1997) Clozapine and some other antipsychotic drugs may preferentially block the same subset of GABA(A) receptors. *Neurochemical Research, 22,* 151–162.

Stahl, S.M. (1988) Basal ganglia neuropharmacology and obsessive-compulsive disorder: The obsessive-compulsive disorder hypothesis of basal ganglia dysfunction. *Psychopharmacology Bulletin, 24,* 370–374.

Stahl, S.M. (1992) Serotonin neuroscience discoveries usher in a new era of novel drug therapies for psychiatry. *Psychopharmacology Bulletin, 28,* 3–9.

Stahl, S. (1994) Is serotonin receptor down-regulation linked to the mechanism of action of antidepressant drugs? *Psychopharmacology Bulletin, 30,* 39–43.

Stahl, S.M. (1996) *Essential Psychopharmacology.* Cambridge, UK: Cambridge University Press.

Stahl, S.M. (1997) *Psychopharmacology of Depression.* London: Martin Dunitz.

Stahl, S.M. (1999a) *Psychopharmacology of Antipsychotics.* London: Martin Dunitz Ltd.

Stahl, S.M. (1999b) Mergers and acquisitions among psychotropics: Antidepressant takeover of anxiety may be complete. *Journal of Clinical Psychiatry, 60,* 282–283.

Stahl, S.M. (1999c) Antidepressants: the blue-chip psychotropic for the modern treatment of anxiety disorder. *Journal of Clinical Psychiatry, 60,* 356–357.

Stahl, S.M. (2000) *Essential Psychopharmacology.* Cambridge, UK: Cambridge University Press.

Standaert, D.G. & Young, A.B. (1996) Treatment of central nervous system degenerative disorders. In: Molinoff, P.B. & Ruddon, R.W. (Eds.) *Goodman & Gilman's the Pharmacological Basis of Therapeutics, Ninth Edition.* New York: McGraw-Hill, pp. 399–430.

Stanilla, J.K. & Simpson, G.M. (1998) Treatment of extrapyramidal side effects. In: Schatzberg, A.F. & Nemeroff, C.B. (Eds.) *Textbook of Psychopharmacology.* Washington, DC: American Psychiatric Press, pp. 349–375.

Stanniland, C. & Taylor, D. (2000) Tolerability of atypical antipsychotics. *Drug Safety, 22,* 195–214.

Starkstein, S.E., Preziosi, T.J., Forrester, A.W. and Robinson, R.B. (1990) Specificity of affective and autonomic symptoms in depression and Parkinson's disease. *Journal of Neurology, Neurosurgery and Psychiatry, 53,* 869–973.

Stein, M.B., Fyer, A.J., Davidson, J.R., Pollack, M.H. & Wiita, B. (1999) Fluvoxamine treatment of social phobia. *American Journal of Psychiatry, 156,* 756–760.

Stein, M.B. & Uhde, T.W. (1998) Biology of anxiety disorders. In: Schartzberg, A.F. & Nemeroff, C.B. (Eds.) *The American Psychiatric Press: Textbook of psychopharmacology.* Washington, DC: American Psychiatric Press, pp. 609–628).

Sternbach, H. (1991) The serotonin syndrome. *American Journal of Psychiatry, 148,* 705–713.

Stimmel, G.L. (1996) Benzodiazepines in schizophrenia. *Pharmacotherapy,* (Suppl), *16,* 148S–151S.

Storey, E. & Beal, M.F. (1993) Neurochemical substrates of rigidity and chorea in Huntington's disease. *Brain, 116,* 1201–1222.

Storey, E. Kowall, N.W., Finn, S.F., Mazurek, M.F. & Beal, M.F. (1992) The cortical lesion of Huntington's disease: Further neurochemical characterization, and reproduction of some of the histological and neurochemical features by N-methyl-D-aspartate lesions of the rat cortex. *Annals of Neurology, 32,* 526–534.

Strand, F.L. (1999) *Neuropeptides: Regulators of physiological Processes.* Cambridge, MA: MIT Press.

Sullivan, G.M., Coplan, J.D. & Gorman, J.M. (1998) Psychoneuroendocrinology of anxiety disorders. *Psychiatric Clinics of North America, 21,* 397–412.

Swartz, J.R., Burgoyne, K., Smith, M., Gadasally, R., Ananth, J. & Ananth, K. (1997) Tardive dyskinesia and ethnicity: Review of the literature. *Annals of Clinical Psychiatry*, *9*, 53–59.

Swerdlow, N.R. (1995) Serotonin, obsessive compulsive disorder and the basal ganglia. *International Review of Psychiatry*, *7*, 115–129.

Tamminga, C. (1998) Schizophrenia and glutamatergic transmission. *Critical Reviews in Neurobiology*, *12*, 21–36.

Tamminga, C. (1999a) Glutamatergic aspects of schizophrenia. *British Journal of Psychiatry*, *37* (Suppl), 12–15.

Tamminga, C.A. (1999b) Principles of pharmacotherapy of schizophrenia. In: Charney, D.S., Nestler, E.J. & Bunney, B.S. (Eds.) *Neurobiology of Mental Illness*. New York: Oxford University Press, pp. 272–289.

Tapia, R., Medina-Ceja, L. & Pena, F. (1999) On the relationship between extracellular glutamate, hyperexcitation, and neurodegeneration, in vivo. *Neurochemistry International*, *34*, 23–31.

Tassin, J.P. (1992) NE/DA interactions in prefrontal cortex and their possible roles as neuromodulators in schizophrenia. *Journal of Neural Transmission*, *36* (Suppl), 135–162.

Taylor, C.B. (1998) Treatment of anxiety disorders. In: Schartzberg, A.F. & Nemeroff, C.B. (Eds.) *Textbook of Psychopharmacology*. Washington, DC: American Psychiatric Press, pp. 775–789

Teboul, E. & Chouinard, G. (1990) A guide to benzodiazepine selection. Part I. Pharmacological aspects. *Canadian Journal of Psychiatry*, *35*, 700–710.

Thal, L.J. (1992) Cholinomimetic therapy in Alzheimer's disease. In: Squire, L.R., &Butters, N. (Eds.) *Neuropsychology of Memory*. New York: Guilford Press, pp. 277–284.

Thompson, P.M. (1996) Generalized anxiety disorder treatment algorithm. Psychiatric Annals, *26*, 227–232.

Tollefson, G.D. & Rosenbaum, J.F. (1998) Selective serotonin reuptake inhibitors. In: Schatzberg, A.F. & Nemeroff, C.B. (Eds.) *Textbook of Psychopharmacology*. Washington, DC: American Psychiatric Press, pp. 219–237.

Trichard, C., Paillere-Martinot, M-L., Attar-Levy, D., Recassens, C., Monnet, F. & Martinot, J-L. (1998) Binding of antipsychotic drugs to cortical 5-HT2A Receptors: A PET study of chlorpromazine, clozapine, and amisulpride in schizophrenic patients. *American Journal of Psychiatry*, *155*, 505–508.

Tsai, G. & Coyle, J.T. (1998) The role of glutamatergic neurotransmission in the pathophysiology of alcoholism. *Annual Review of Medicine*, *49*, 173–184.

Tsai, G., Gastfriend, D.R. & Coyle, J.T. (1995) The glutamatergic basis of human alcoholism. *American Journal of Psychiatry*, *152*, 332–340.

Tsai, G., van Kammry, D.P., Chen, S., Kelley, M.E., Grier, A. & Coyle, J.T. (1998) Glutamatergic neurotransmission involves structural and clinical deficits in schizophrenia. *Biological Psychiatry*, *44*, 667–674.

Valentino, R.J. & Aston-Jones, G.S. (1995) Physiological and anatomical determinants of locus ceruleus discharge: Behavioral and clinical implications. In: Bloom, F.E. & Kupfer, D.J. (Eds.) *Psychopharmacology: The fourth generation of progress*. New York: Raven Press, pp. 373–385.

van der Kolk, B.A. (1997) The psychobiology of posttraumatic stress disorder. *Journal of Clinical Psychiatry*, *58* (Suppl 9), 16–24.

van Rossum, J.M. (1966) The significance of dopamine receptor blockade for the mechanism of action of neuroleptic drugs. *Archives Internationales de Pharmacodynamie et de Therapie*, *160*, 492–494.

van Veelen, N.M.J. & Kahn, R.S. (1999) Dopamine, serotonin, and schizophrenia. In: Stern, G.M. (Ed.) *Parkinson's Disease. Advances in Neurology*, Vol. *80*. Philadelphia: Lippincott Williams & Wilkins, pp. 425–429.

Voytko, M., Olton, D., Richardson, R., Gorman, L., Tobin, J. & Price, D. (1994) Basal forebrain lesions in monkeys disrupt attention, but not learning and memory. *Journal of Neuroscience*, *14*, 167–186.

Walsh, M.T. & Dinan, T.G. (2001) Selective serotonin inhibitors and violence: A review of the available evidence. *Acta Psychiatrica Scandinavica*, *104*, 84–91.

Walters, V.L., Tognolini, R.Z., Rueda, H.M., Rueda, R.M. & Torres, R.G. (1997) New strategies for old problems: Tardive dyskinesia (TD). *Schizophrenia Research*, *28*, 231–246.

Whitehouse, P.J., Price, D.L., Clark, A.W., Coyle, J.T., & DeLong, M.R. (1981) Alzheimer's disease: evidence for selective loss of cholinergic neurons in the nucleus basalis. *Annals of Neurology*, *10*, 122–128.

Wickelgren, I. (1998) A new route to treating schizophrenia. *Science, 281*, 1264–1265.

Wilens, T.E., Biederman, J., Spencer, T.J. & Prince, J. (1995) Pharmacotherapy of adult attention deficit/hyperactivity disorder: A review. *Journal of Clinical Psychopharmacology, 15*, 270–279.

Willner, P. (1983) Dopamine and depression: A review of recent evidence. Brain Research. *Brain Research Reviews, 6*, 211–246.

Willner, P. (1995) Dopaminergic mechanisms in depression and mania. In: Bloom, F.E. & Kupfer, D.J. (Eds.) *Psychopharmacology: The fourth generation of progress.* New York: Raven Press, pp. 921–931.

Willner, P. (1997) The dopamine hypothesis of schizophrenia: Current status, future prospects. *International Clinical Psychopharmacology, 12*, 297–308.

Wisden, W. & Seeberg, P.H. (1992) GABAA receptor channels: Subunits to functional entities. *Current Opinion in Neurobiology, 2*, 263–269.

Wisden, W., Lauri.e., D.J., Monyer, H. & Seeberg, P.H. (1992) The distribution of 13 GABAA receptor subunit mRNAs in the rat brain. I. Telencephalon, diencephalon, mesencephalon. *Journal of Neuroscience, 12*, 1040–1062.

Wise, R.A. (1999) Animal models of addiction. In: Charney, D.S., Nestler, E.J. & Bunney, B.S. (Eds.) *Neurobiology of Mental Illness.* New York: Oxford University Press, pp. 569–577.

Wise, R.A. (1996) Neurobiology of addiction. *Current Opinions in Neurobiology, 6*, 243–251.

Wolf, M.E. & Roth, R.H. (1990) Autoreceptor regulation of dopamine synthesis. *Annals of the New York Academy of Sciences, 604*, 323–343.

Wolkowitz, O.M. & Pickar, D. (1991) benzodiazepines in the treatment of schizophrenia: A review and reappraisal. *American Journal of Psychiatry, 148*, 714–726.

Woodrum, S.T. & Brown, C.S. (1998) Management of SSRI-induced sexual dysfunction. *Annals of Pharmacotherapy, 32*, 1209–1215.

Woods, J.H., Katz, J.L. & Winger, G. (1992) Benzodiazepines: Use, abuse, and consequences. IV. Adverse behavioral consequences of benzodiazepine use. *Pharmacological Reviews, 44*, 151–347.

Wooley, D.W. & Shaw, E. (1954) A biochemical and pharmacological suggestion about certain mental disorders. *Proceedings of the National Academy of Sciences (USA), 40*, 228–231.

Wooten, G.F. (1997) Functional anatomical and behavioral consequences of dopamine receptor stimulation. *Annals of the New York Academy of Sciences, 835*, 153–156.

Wudarsky, M., Nicolson, R., Hamburger, S.D., Spechler, L., Gochman, P., Bedwell, J., Lenane, M.C. & Rapoport, J.L. (1999) Elevated prolactin in pediatric patients on typical and atypical antipsychotics. *Journal of Child and Adolescent Psychopharmacology, 9*, 239–245.

Yassor, R. & Lal, S. (1986) Prevalence of the rabbit syndrome. *American Journal of Psychiatry, 143*, 656–657.

Yehuda, R., Boisoneau, D., Lowy, M.T. & Giller, E.L. (1995) Dose-response changes in plasma cortisol and lymphocyte glucocorticoid receptors following dexamethasone administration in combat veterans with and without posttraumatic stress disorder. *Archives of General Psychiatry, 52*, 583–593.

Yeomans, J.S., Kofman, O. & McFarlane, V. (1984) Cholinergic involvement in lateral hypothalamic rewarding brain stimulation. *Brain Research, 329*, 19–26.

Zorumski, C.F. & Isenberg, K.E. (1991) Insights into the structure and function of GABA-benzodiazepine receptors: Ion channels and psychiatry. *American Journal of Psychiatry, 148*, 162–173.

CELL MEMBRAINES AND ION GRADIENTS

The cell membrane consists of a fluid bilayer of **phosphoglyceride molecules** whose *hydrophilic (water-attracting)* phosphoric acid *heads* are arranged on the inner and outer edges of the membrane, with their *hydrophobic* glyceride *tails* oriented toward each other in the center of the membrane (see Figure A–1). This bilayered membrane establishes a partial barrier between what lies within and what lies outside the cell. But this barrier is not absolute. Certain substances, which are vital to the basic metabolism of the cell such as oxygen and carbon dioxide (which are lipid soluble), easily can diffuse through these lipid barriers which constitute the cell membrane. However, other chemicals, such as sodium (Na^+), potassium (K^+), chloride (Cl^-), and calcium (Ca^{2+}), which are essential to signal conduction, cannot diffuse through the membrane at will. Special mechanisms are required to permit the passage of such substances in and out of the cell. Also there needs to be mechanisms in place that will allow neurotransmitters to communicate between neurons and between neurons and muscles. Both are accomplished by the means of large protein molecules that "float" within this bilayered membrane. These long strings of amino acids (proteins), which often transect both sides of the membrane, form *receptors* which provide the *gateways* or points of communication between the inside and outside of the cell.

There are basically two types of receptors with which we are concerned. One is what is known as a **ligand-gated ion channel receptor**; the other is a **G-protein-linked receptor** (see Figure 11–7). For the moment, we will focus on the ligand-gated receptor, which also is referred to as an *ion channel receptor*, since its primary function is to permit the exchange or movement of ions (particularly, Cl^-, Na^+, K^+, and Ca^{2+}) between the inside and outside of the cell. These ions, along with other larger charged particles, are unequally distributed on either side of the cell membrane (Figure A–2; see also Figure 11–1b). While this differential distribution of ions is true of all cells, here we are concerned only with neurons. When at rest (i.e., in its normal resting state), there is a greater concentration of Na^+ and Cl^- outside the cell membrane, while K^+ and larger protein or **amino acid molecules**, which primarily carry a negative charge, are concentrated inside the cell membrane (see Table A–1). Calcium (Ca^{2+}) also is differentially distributed, with extracellular Ca^{2+} being approximately 10,000 times greater than cytosolic Ca^{2+}. Since these various ions and amino acid molecules all carry electrical charges, this differential distribution sets up the possibility of unequal electrical charges on either side of the cell membrane. In the case of the neuron, this electrical difference (approximately $-70\,mV$ on the inside of the membrane as compared with the outside) acts like a capacitor, creating the substrate for the action potential of the neuron. But before reviewing this action potential in greater detail, it will be necessary to review the role of the various transport mechanisms and the forces that act to maintain the concentration gradients.

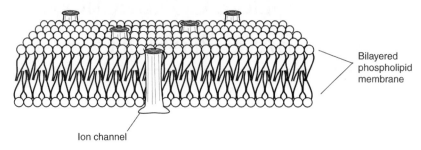

Figure A–1. Cell membrane with ion channel receptors. Schematic representation of ion channel receptors embedded in cell membrane.

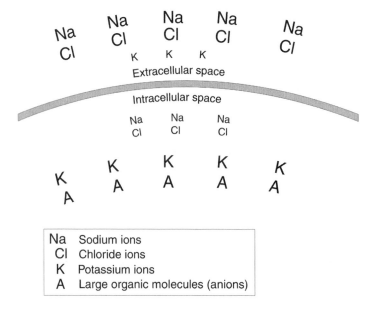

Figure A–2. Differential distribution of ions across cell membrane. Distribution of selected ions across cell membranes. Larger lettering represents greater concentrations.

Table A–1. Distribution of Approximate Na^+, K^+, and Cl^- Ions Across the Neuronal Membrane[a]

	Extracellular	Intracellular
Sodium ions[a]	150.0 mM/liter	15.0
Potassium ions	5.0	100.0
Chloride ions	150.0	13.0

From Bear, Connors, and Paradiso (2001), p. 65.
[a] Numbers may vary with different sources.

DIFFUSION AND THE MAINTENANC OF ION GRADIENTS

Ion channel receptors are key to maintaining the **resting potential**, the **action potential** *depolarization*), and eventually the return of the neuron back to its **resting state** (*repolarization*). Ion channels might be thought of as sleeves that penetrate the lipid layers of the cell membrane, creating a potential opening (channel) through which various ions may pass into or be extruded from the cytoplasm of the cell. Made up of four or five subunits (each consisting of four transmembrane segments), each ion channel receptor is highly selective as to under what conditions, to what degree, and what type of ion it will permit to pass through the cell membrane. First, ion channels basically have two states: opened or closed. When ion channels are closed (by what has been characterized as a "gatelike" mechanism that is part of the receptor itself), ions are unable to flow through them (Figure A–3). The probability of a given ion channel opening (as well as the frequency, duration, and possibly even the degree of such openings) can be affected by a variety of factors that differentially affect ion flow. Additionally, for each ion group, there may be multiple types of ion channels, each operating through different mechanisms and having different resting states and/or patterns of response. It would appear, for example, that at least some ion channels for most if not all ion groups remain open at any given point in time, permitting the "leakage" of ions across the membrane. The percentage of channels that remain *open* during normal "resting" states vary among the various ion groups, so that some ions (e.g., potassium) tend to flow more freely than others (e.g., sodium) during this stage. As conditions or stimuli change, more ion channels may open, permitting an even greater exchange of ions between the inner and outer cellular compartments.

There generally are two mechanisms by which ion channels may open (or close). One is **chemical** and the other **electrical** (see Figure 11–7). The first, known as *chemical gating*, results from a particular chemical (e.g., a neurotransmitter or other related agonist) binding to the receptor itself. In the second instance, the ion channel will open or close as a result of a change in the background voltage within the cell itself (*voltage gating*). In either case, what happens is that through these mechanisms, the flow of ions (and hence, membrane potentials) can be dramatically altered. This flow of ions across the cell membrane is referred to as **diffusion**. As we shall see, it is through this process that electrical potentials are established, changed (increased or decreased), and nerve conduction is made possible. Several factors or "forces"

Ligand-gated ion channel receptor

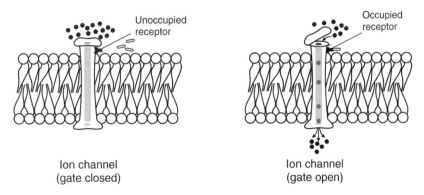

Figure A–3. Ligand-gated ion channel receptors. Occupation of receptor by appropriate neurotransmitter or chemical will cause "gate" to open, allowing passage of ions.

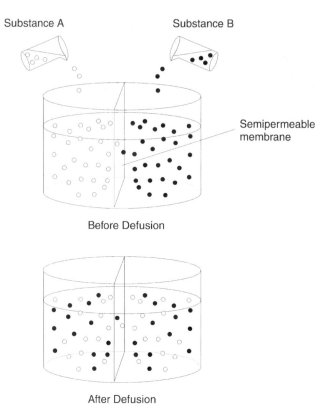

Figure A–4.   Diffusion. Simple diffusion down concentration gradient in the presence of a semiper-meable membrane.

are constantly at work that influences diffusion across the cell membrane. One of these is **osmolarity**. To briefly explain, the various ions or molecules (solute) present in any solution are constantly in motion. This motion of the molecules exerts a certain amount of pressure against a membrane which contains or separates that solution. At a given temperature (which also affects the movement of molecules, hence pressure), the total pressure created by the solution varies with the absolute number of molecules per unit of volume (i.e., by the *density* of the solution). When a semipermeable membrane separates two volumes of aqueous solutions, water will flow from the side that contains a lower density of the solute (lower pressure) to the side of *greater* solute density (higher pressure). This diffusion of water molecules across membranes as a result of pressure differences between solutions (Figure A–4) is a common biological phenomenon, referred to as **osmosis**. Thus, *osmolarity is a means of expressing the osmolar concentration per liter of a given solution, where an osmole equals one gram molecular weight of a given solute*. This factor is obviously important in maintaining optimal fluid ratios between cells and extracellular fluids throughout the body, including neurons, and in the function of the kidneys in particular.

CHEMICAL AND ELECTRICAL GRADIENTS

The two factors that have a more direct impact on nerve conduction per se are **concentration gradients** and **electrical gradients**. Concentration gradients are similar to the osmolarity factor discussed above. Essentially what this means is, if you have more of one substance (e.g., Na^+ or Cl^- ions) on one side of a membrane than the other, there is a natural tendency for the ions to diffuse across the membrane until the concentration of that particular ion is the same on both sides of the membrane. Electrical gradients, on the other hand, derive from the fact that ions carry an electrical charge (either positive or negative). Consider, for example, the following situation. You have a membrane separating two fluids in which various ions are suspended. Both positive and negative ions are present on both sides and while (at least for the moment) the electrical charges (i.e., total number of positive and negative ions) are balanced with respect to either side of the membrane, the concentration gradients are not. Suppose there are many more (positively charged) sodium ions on the outside of the membrane and more potassium ions (also positively charged) on the inside. Furthermore, assume that the membrane is much more permeable to the potassium ions than it is to either the sodium ions or the negatively charged ions. What then might be expected to happen is that, given the greater opportunity (i.e., increased permeability to K^+ ions), the potassium ions will tend to diffuse across the membrane (down their concentration gradient) at a much greater pace than the sodium ions can diffuse to the inside of the membrane. The result will be a net loss of positive ions on the inside of the membrane, making the inside electrically "negative" relative to the outside. An *electrical gradient* thus is established (Figure A–5). This is essentially what happens in the neuron.

What is it that maintains these gradients? Why do these imbalances not simply work themselves out? Would not the slower influx of sodium ions eventually balance out the electrical gradient, or would the fact that the inside of the neuron is now "negative" tend to attract the positively charged potassium ions back inside the membrane? In addition to the previously mentioned differences in permeability with regard to the various ions, there are two other major factors that seem to prevent this. First, the electrical and concentration gradients are not independent of one another. Suppose, for example, there is a higher concentration of Cl^- outside of the cell than on the inside. Because of the concentration gradients, Cl^- would have a natural tendency to diffuse to the inside. However, given that at rest the inside of the cell already is negatively charged with respect to the outside (in this case, as a function of large, negatively charged amino acids that cannot diffuse through the ion channels) and since similar electrical charges tend to repel one another, this sets up a force that opposes the reduction of the Cl^- concentration gradient. Similarly, while the negative charge on the inside of the cell membrane might tend to attract the K^+ ions back into the cell, such a movement would tend to be opposed by the K^+ concentration gradient. When these two opposing forces (i.e., the concentration and electrical gradients) tend to balance each other out, a state of equilibrium is established for that particular ion. The difference in the electrical potential at which that state occurs is known as the **equilibrium potential** for that ion. Without going into detail, it should be known that it is the equilibrium potential of potassium that primarily contributes to the resting potential of the neuron.

The second factor that contributes significantly to maintaining the electrical and concentration gradients of the nerve cell is the "**sodium/potassium pump**." This will be discussed in greater detail in the following section.

TYPES OF DIFFUSION

Passive Diffusion

While electrical gradients and resulting nerve (*action*) potentials are dependent on the diffusion or transfer of various ions from one side of the cell membrane to the other, there are multiple ways in which this transfer can take place. The easiest way is through what might be referred to as **passive diffusion**. In this case, although governed by forces imposed

Forces Affecting Diffusion

Concentration Gradients

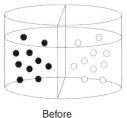 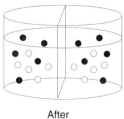

Before   After

Electrical Gradients

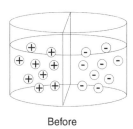 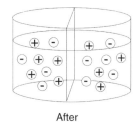

Before   After

Pressure Gradients

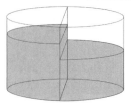 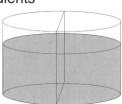

Figure A–5. Forces affecting diffusion. Diffusion tends to occur in the presence of unequal concentration gradients, differences in electrical potentials (i.e., positive and negative charges attracting one another), and pressure differences. Multiple forces may act in combination or competitively. Thus, the tendency to "even out" a concentration difference across the cell membrane may be partially offset by the electrical gradients. This is what takes place in the neuron and these combined forces are referred to as the *electrochemical gradient*.

by concentration and electrical gradients as discussed above, diffusion occurs though open ion channels (see Figure A–3) or possibly though permeable pores in the cell membrane itself. Known as *simple diffusion*, the rate of diffusion will be modulated by such factors as the degree of concentration or electrical gradients present (i.e., the "force" driving the diffusion), as well as the number, frequency, and duration of ion channel openings.

Mediated Diffusion

Another type of diffusion is termed **mediated diffusion**. Simple diffusion may be thought of as a strictly "open door policy" in which ions can more or less enter or leave the cell at will as long as the ion channel is "open." Mediated diffusion, on the other hand, is akin to being permitted access or egress; however, in this case each person (ion) must be individually accompanied by an armed guard or escort as they pass through. With ion transfer this essentially means that in order to pass through the ion channel, the ion (or molecule) must be "bound" to another protein in order for the ion channel to permit it to pass (Figure A–6). This process effectively puts an upper limit on the rate and/or conditions under which ion transfer may take place. There are essentially two types of mediated diffusion: **facilitated diffusion** and **active transport**. In facilitated diffusion, the ion must be bound to another protein, which causes a conformational change in the receptor or transport mechanism, permitting the ion to enter the cell. Glucose absorption by cells is an example of facilitated diffusion.[1]

Both simple and facilitated diffusion are "driven" by natural pressures to alleviate either concentration or electrical gradients, and as such no additional "force" is necessary to implement these actions. Simple or facilitated diffusions cannot go against the **electrochemical gradient** (i.e., the combined forces of the electrical and concentration gradients). However, in some instances, the functioning of the cell (neuron) demands that ions (or any other solute) be transferred "against" existing concentration and/or electrical gradients. This requires another type of mediated transport referred to as **active transport**. Rather than diffusion following the natural, passive dictates of the electrochemical gradients, the cell must actively transport these ions against these gradients. In order for this to happen, the cell must utilize some type of energy to diffuse (transport) the ions against these force fields.[2] The energy in this case is derived from the cell itself, typically from adenosine triphosphate (**ATP**), which is termed **primary active transport**, or by the use of the potential energy derived from differential ion concentrations (**secondary active transport**). In active transport, in addition to the ion or solute (to be transferred) binding to a site on the transport protein, an ATP molecule or high concentration ion (usually sodium) also must bind to the transporter. As illustrated in Figure A–7, the energy supplied by the latter causes a conformational change in the protein which carries ("transports") the solute into (or out of) the cell against the electrochemical gradients.

The Sodium–Potassium Pump

One of the better-known examples of active transport is what has been termed the **sodium–potassium pump,** associated with the repolarization of neurons. As will be seen in the

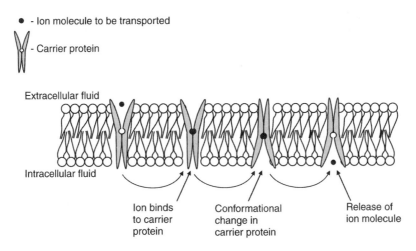

Figure A–6. Facilitated diffusion. In some situations the diffusion of ion or molecule across membrane requires the intervention of a protein to which it is temporarily bound.

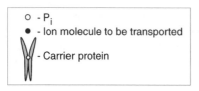

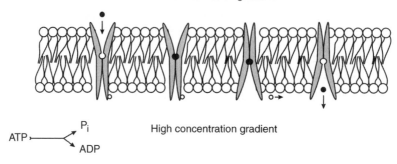

Figure A–7. Active transport. If an ion or molecule must, for example, be transported against an electrochemical gradient, energy must be expended. The breakdown of adenosine triphosphate (ATP) into adenosine diphosphate (ADP) resulting in the attachment of phosphate group (P_i) to the transporter protein provides the energy for this process. (Adapted from Vander, Sherman, and Luciano, 2001)

next section, during the course of depolarization associated with the action potential of the neuron, sodium enters the cell while potassium is extruded from the cell. In order to help restore the initial polarization of the neuron, as well as the concentration gradients, it is necessary for sodium to be "pumped" back out of the cell (neuron), while potassium is "pumped" back into the cell. The term *pump* is used to imply that this is an active, energy-consuming process, necessary in this case since sodium, in being extruded from the cell, will be going against both the concentration gradient (Na^+ is more abundant outside the cell) and the electrical gradient. Potassium, on the other hand, which is already more plentiful within the cell membrane, will be fighting the concentration gradient.

As its name might suggest, the sodium–potassium pump actually serves a twofold function: for every three Na^+ ions it pumps out of the cell, it simultaneously carries two K^+ ions into the cell. As represented in Figure A–8, this is accomplished by a specialized protein (receptor) within the membrane of the cell. Basically this protein complex has three high affinity binding sites for Na^+ on the inside of the neuron and two comparable binding sites for K^+ on the outside of the cell. Again, the energy for this process is supplied by ATP. As each molecule of ATP is hydrolyzed into **ADP** (adenosine diphosphate) and a phosphate group (P_i), the energy released by this process causes conformational changes in the transport protein. Because for every three positive ions (Na^+) extruded from the cell only two positive ions (K^+) are brought back into the cell, the inner negativity of the cell is thus restored and maintained (see below).[3]

GRADED POTENTIALS

To briefly review, recall that there is a differential distribution of ions on each side of the cell membrane, with Na^+ and Cl^- being concentrated outside the membrane, while K^+ and large, negatively charged amino acids (proteins) are more abundant inside the cell. This ion

Sodium-Potassium Pump

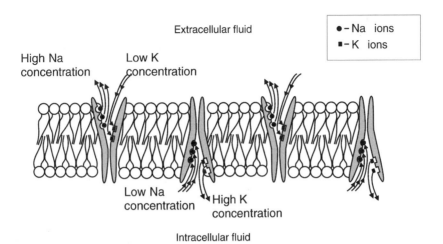

Figure A–8. Sodium/potassium pump. The sodium/potassium pump provides an example of an active transport mechanism. Because both sodium (Na^+) and potassium (K^+) are being moved against their concentration gradients, energy is required. As in Figure A-7, the breakdown of ATP into ADP and P_i provides the energy source. As noted in the text, three Na^+ ions are extruded from the intracellular space while two K^+ ions are pumped into the cell.

imbalance creates a situation where an electrical potential exists across the cell membrane, with the interior of the cell being negative (anywhere from approximately -50 to -90 mV, depending on the size of the nerve fiber) with respect to the outside. This is referred to as the *resting potential* of the neuron. It is this electrical potential that establishes the basis for the nerve impulse (the "firing" of the neuron).

The cell body (*soma*) of each neuron has multiple dendritic processes. In turn, each dendrite has multiple receptor sites that can be stimulated by various neurotransmitters. When these dendritic receptors are stimulated, basically one of two things can happen, depending on the specific receptor/transmitter interface being stimulated. The stimulus can have either an **excitatory** (*depolarizing*) or an **inhibitory** (*hyperpolarizing*) effect on the cell via the dendritic receptor. An excitatory signal (e.g., the release of certain transmitters, such as glutamate or acetylcholine, into the synaptic cleft) will cause the opening of additional sodium channels. The influx of Na^+ (or Ca^{2+})[4] ions into the cell causes a decrease in its relative negativity (depolarization), thus facilitating the initiation of an action potential. Such stimulation also is referred to as an *excitatory postsynaptic potential* or **EPSP** (Figure A–9).

Conversely, inhibitory dendritic connections tend to affect chloride and/or potassium channels. While some chloride channels are thought to remain open when the neuron is "at rest," it appears that many remain closed.[5] However, as a result of an inhibitory stimulus (e.g., release of GABA on postsynaptic receptors), additional Cl^- channels may be opened, permitting an increased influx of Cl^- ions as a result of the large concentration gradient (more outside the cell than inside). Similar inhibitory stimuli also may trigger the opening of additional potassium channels (this may occur along with or independent of the opening of additional chloride channels) (Thompson, 1993). Although many K^+ channels apparently remain constantly open, the relative negativity inside the cell tends to inhibit any massive efflux of potassium ions to the outside of the cell. However, with the opening of additional potassium channels there does tend to be some diffusion of K^+ ions down the concentration

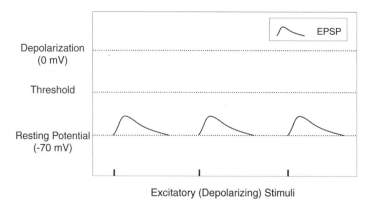

Figure A–9. Excitatory postsynaptic potentials. This figure shows a series of non-overlapping excitatory hypopolarizing) potentials, none of which is sufficient (meets the threshold) for triggering an action potential.

gradient to the outside. This influx of negative (Cl⁻) ions along with the loss of positive (K⁺) ions results in a state of increased internal negativity (*hyperpolarization*) of the neuron. This phenomenon is known as an *inhibitory postsynaptic potential* or **IPSP** (see Figure A–10).

In order for a full depolarization (action potential) to occur, the internal negativity of the cell must drop below a certain threshold.[6] It appears unlikely that any one dendritic connection (synapse) typically would result in such a depolarization of the neuron as a whole. However, at any given time, the cell body (via its dendrites or synapses directly on the soma itself) is being bombarded by multiple inputs, some inhibitory, some excitatory. It is the **summation** of these inputs that determines whether hypopolarizing influences significantly outweigh the hyperpolarizing inputs so as to trigger a full depolarization of the neuron (i.e., cause it to fire). This summation effect may be either temporal (several excitatory impulses coming in immediate succession) or spatial (similar impulses arrive simultaneously in close spatial approximation to one another) (Figure A–11). Should inhibitory (e.g., GABAergic) inputs prevail, it would take even stronger or more frequent excitatory inputs in order to trigger an action potential.

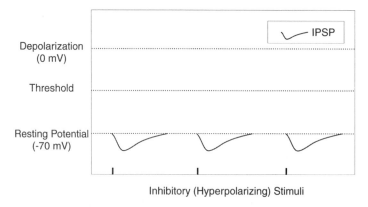

Figure A–10. Inhibitory postsynaptic potentials. In this instance there is a series of inhibitory (hyperpolarizing) potentials, creating the situation where an action potential (depolarization of the neuron) somewhat more difficult.

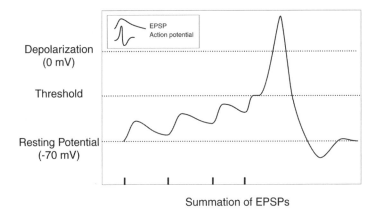

Summation of EPSPs

Figure A–11. Spatial/temporal summation of postsynaptic potentials. This figure illustrates the effect of multiple excitatory postsynaptic potentials being temporally summated. Each may augment the next until the firing threshold is reached resulting in depolarization and an action potential being propagated down the axon.

ACTION POTENTIALS

In the situation discussed above, unless sufficient summation of excitatory potentials is achieved, full depolarization of the neuron will not occur and the response to the stimulus will remain local and graded. However, once a certain depolarization threshold is reached, very different phenomena occur. At this point, the full depolarization of the neuron takes place (an *all-or-nothing*–type response), initiating an action potential that now travels without decrement the full length of the neuronal axon to the presynaptic terminals. The major steps in this process are outlined below and illustrated in Figure A–12.

1. Once the depolarization threshold is passed, an increased number of voltage-sensitive Na^+ channels are opened.[7] Because of the marked difference in the concentration gradient, there is a major influx of sodium ions into the cell. This influx of positive sodium ions not only erases the preexisting negative potential on the inside of the cell, it actually creates a slightly positive potential inside the neuron when compared with the extracellular fluid. During this stage, the neuron is incapable of firing again, a state that has been termed the *absolute refractory period*.
2. In addition to opening Na^+ channels, the depolarization process also leads to the opening of additional K^+ channels, which prompts a greater efflux of potassium ions outside of the cell (down potassium's concentration gradient, and initially at least down the electrical gradient) than normally occurs during the resting stage.[8]
3. Although initially the influx of Na^+ ions overwhelms the outpouring of K^+ ions (resulting in a reversal of the polarity of the cell), once the inside of the cell becomes positive, the internal voltage-sensitive "inactivation" gates of the sodium close, while the newly opened potassium gates remain open for a while longer. This allows a continued efflux of K^+ without a counterbalancing influx of Na^+. The result is a return to a relative negativity inside the cell.
4. As the additional K^+ channels remain open, like with sodium there tends to be an overshooting response, such that the interior of the cell actually becomes slightly more negative than is the case during the normal resting potential. During this stage of hyperpolarization, the neuron can fire but it would require a stronger than normal stimulus. This stage is known as the *relative refractory period*.

Stages of Action Potential

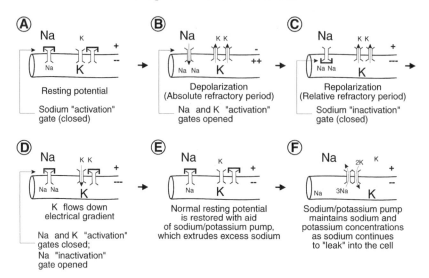

Figure A–12. Stages of the action potential. Some of the key proposed mechanisms involved in maintaining the "normal" resting potential of the neuron and how that resting potential is restored following depolarization and propagation of the action potential.

5. This reversion to the normal polarity results in a closing of the additional potassium channels (the normally opened K^+ channels remaining open) and an opening of the "inactivation" gates of the sodium channels (since the "activation" gates of the Na^+ channels remain closed, there is minimal diffusion of sodium across the membrane). With this closing of the additional potassium channels, there is a gradual return of the membrane to its normal resting potential (due to the diffusion of some of the K^+ ions back into the cell, down the electrical gradient).

6. The restoration of the resting potential is facilitated by the *sodium/potassium pump,* which extrudes some of the excess sodium brought into the membrane during depolarization. At the same time, it brings back into the cell potassium ions lost during the same process. As noted earlier, the Na^+/K^+ pump, in addition to restoring the resting potential, also serves to maintain it. Although more K^+ than Na^+ channels remain open during the resting stage, more Na^+ tends to "leak" into the cell (due to the electrical gradient) than does K^+ (down its concentration gradient). Thus, by exchanging three Na^+ ions for every two K^+ ions, the chemical equilibrium within the cell is preserved (Vander, Sherman, & Luciano, 2001, p. 187).

7. Once this depolarization process has begun, the excitation spreads to adjacent sections of the neuron, causing successive sections of the axonal process to also undergo non-decremental depolarization. With the larger axons, which are covered by myelin sheaths, this depolarization occurs only at the regular breaks in this sheath (called **nodes of Ranvier**). This latter phenomenon, called **saltatory conduction,** actually serves to speed up the process by which nerve impulses are transmitted down the axon (see Figure A–13).

8. Once this action potential reaches the axon terminal, it opens voltage-gated calcium ions channels. The resulting influx of Ca^{2+} into the terminal boutons of the axon terminals causes the vesicles that store neurotransmitters to migrate to the presynaptic membrane and subsequently discharge their contents into the synaptic cleft. Through

Saltatory Conduction

Propagation of nerve conduction ⟶

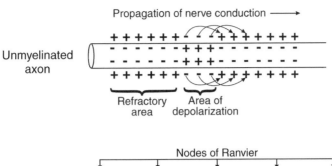

Unmyelinated axon

Refractory area   Area of depolarization

Nodes of Ranvier

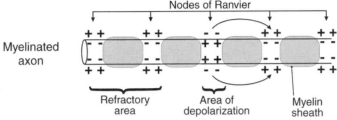

Myelinated axon

Refractory area   Area of depolarization   Myelin sheath

Figure A–13. Saltatory conduction. By "jumping" from one node to the next, myelinated fibers are able to transmit nerve impulses faster than unmyelinated fibers.

the chemical-gating that now can take place at the postsynaptic membrane, the nerve impulse that began in one neuron now can influence the next neuron (or muscle fiber) in the chain (see Figure 11–2).

REFERENCES

Bear, M.F., Conners, B.W. & Paradiso, M.A. (2001) *Neuroscience—Exploring the brain.* Baltimore: Lippincott, Williams & Wilkins.

Guyton, A.C. & Hall, J.E. 2000 *Textbook of Medical Physiology.* Philadelphia: W.B. Saunders.

Thompson, R.F. 1993 *The Brain: A neuroscience primer.* New York: W.H. Freeman and Company.

Vander, S., Sherman, J. & Luciano, D. (2001) *Human Physiology—The mechanisms of body function.* Boston: McGraw Hill

Endnotes

1. Insulin "facilitates" the transfer of glucose into the cells. This is why, in the case of insulin deficiency (e.g., in diabetes) there is so much glucose (sugar) in the blood. It is there because, lacking the proper supply of insulin, the glucose cannot be adequately diffused out of the blood and into the cells. Thus, the cells are "starving" in the midst of plenty.

2. Passive diffusion may be compared to water falling from a dam. Once the gates are open, the water, following the force of gravity, falls passively to the river below. On the other hand, if one wishes to withdraw water from a well (which goes against the force of gravity), some type of force needs to be applied to actively pump the water from the well. This would be active transport.

3. The diffusion (reuptake) of neurotransmitters back into the presynaptic cleft after they have been released from the postsynaptic receptors is another example of active transport at work.

4. Calcium ions are also present in greater abundance extracellularly. In addition to or essentially in place of sodium, they can precipitate an action potential through their influx into cell. The latter appears to be particularly true in the case of smooth or cardiac muscles (Guyton & Hall, 2000))

5. There seems to be a relative paucity of discussion of the state of chloride channels in neurons, and authors seem divided on whether most remain open or closed when the neuron is at rest.

6. Different authors cite different resting potentials for the typical neuron (from -70 to -90 mV), as well as different thresholds for initiating an action potential (from -50 to -70 mV). What is important here is simply to understand the basic process.

7. The sodium channels actually have two gates, one near the outside of the membrane and one near the inside. During the normal resting stage, the outer gate (the "activation" gate) normally is closed, while the inner ("inactivation") gate normally is open. During the initial stage of the action potential, the "activation" gate also is opened, thus allowing sodium to enter the cell. After both are opened for only a few milliseconds, the "inactivation" gate (which also is voltage sensitive) is closed and remains closed until the neuron once again approaches its normal resting potential.

8. Although chloride also is differentially distributed across the membrane, like potassium many of its channels remain open, and it apparently contributes little in comparison to Na^+ (or Ca^{2+}) and K^+ to the action potential. As noted above, the opening of additional Cl^- channels is important in establishing IPSPs.

GLOSSARY

Abdominal reflex: Elicited by scratching the abdomen with a sharp object in a supine patient, the umbilicus should deviate toward the source of the stimulation. Absence of reflex is suggestive of upper motor neuron lesion in the lower thoracic region or above.

Abducens nerve (CN VI): Controls the lateral rectus muscle that enables to eye to look laterally or "abduct"; i.e., allows the right eye to look to the right and the left to the left.

Abducens nucleus: Pontine nucleus for cranial nerve VI (abducens) controlling the lateral rectus muscle of the eye.

Abduction: To pull or draw away from the midline, outwardly.

Abulia: Lack of will or capacity to formulate and/or follow through with an intention. Most commonly seen in large or deep frontal lesions.

Acalculia: An acquired deficit in the ability to carry out mathematical operations, typically as a result of the loss of the symbolic meaning of the numbers or arithmetical symbols, or of the operations themselves.

Accessory cuneate nucleus: Medullary nucleus that gives rise to the cuneocerebellar tract providing sensory feedback from upper extremities to the cerebellum.

Accessory nerve: See: Spinal Accessory nerve.

Accessory nucleus: See: Spinal Accessory nucleus.

Accommodation: Adjustment of the lens of the eyes when focusing on a near object, normally accompanied by convergence and pupillary constriction.

Accumbens nucleus: That region of the striatum rostral to the plane of the anterior commissure and ventral to the internal capsule, where the caudate and putamen are conjoined. Functionally, this area seems to serve as the limbic equivalent to the cortical-striatal feedback loops of the caudate nucleus and putamen.

Acetylcholinesterase: Enzyme present in the synaptic cleft of cholinergic neurons that breaks down ACh. Certain cognitive-enhancing drugs such as donepezil (Aricept) act by inhibiting this enzyme.

Achromatopsia: A condition in which the ability to perceive colors or hues is either lost or diminished (dyschromatopsia), such that colors may be seen as "dull" or "faded." Typically associated with lesions of the inferior occipitotemporal cortex, either unilaterally (hemiachromatopsia) or bilaterally.

Action tremor: Form of intention tremor induced when carrying out a dynamic action, as opposed to a "resting" tremor or simply holding a position against gravity ("sustention" tremor). Usually associated with cerebellar disease.

Adduction: To draw or pull toward the midline, inwardly.

Adenohypophysis: Anterior lobe of pituitary gland.

Affective prosody: Meaning added to verbal communication (speech) through the use of emotional inflections or tone, such as happy, angry, or sad (see **Aprosodia and Linguistic Prosody**).

Afferent: While commonly associated with the notion of sensory input, the term actually refers to any type of fiber connections or information coming into a nerve cell or body of cells, regardless of their nature (see **Efferent**).

Afferent paresis: Difficulty executing smooth, coordinated, fine movements as a result of loss of afferent (somatosensory) feedback from the muscles engaged in the motor activity.

657

Agonist: Drug that binds to and stimulates a receptor in the same manner as the receptor's natural neurotransmitter (which is also an agonist for that receptor).

Agnosia: The inability to recognize an object or stimulus as a result of an acquired brain injury, despite the relative preservation of elementary sensory input.

Agranular cortex: Cortex characterized by the relative absence of granule cells (layers II and IV) and the prominence of layers III and V in which an abundance of pyramidal cells are found. It is found in the primary motor and motor association cortices.

Agraphia: Loss of writing ability despite adequate motor functions.

Ahylognosia: Refers to the inability to determine its substance, i.e., such features as its texture, density, resistance to pressure, or "thermal conductivity," which would give a clue as to the material(s) from which it is made.

Akathisia: A manifestation of an "extrapyramidal syndrome" associated with the use of classic neuroleptics, it is characterized by both an externally observable increase in nervous activity (fidgeting, often especially noticeable in the lower extremities), as well as a subjectively experienced "need to move."

Alexia: Acquired loss of reading ability despite sufficient visual acuity.

Alexia without agraphia: Syndrome usually associated with a lesion in the territory of the left posterior cerebral artery affecting the left occipital cortex and the splenium of the corpus callosum in which reading is impaired, while writing and visual–object recognition is generally intact.

Alien-hand syndrome: Relatively rare phenomenon in which the affected hand/arm will engage in quasi-meaningful activities without being consciously willed or directed by the individual. Possibly associated with lesions of the contralateral supplementary motor cortex.

Allesthesia: Misperception of the location of tactile stimulation. Typically used to describe the displacement of a stimulus applied to the "affected" side of the body (contralateral to lesion) to the "unaffected" side (ipsilateral to the lesion).

Allocortex: The most primitive type cortex, consisting of only three layers and represented by the primary olfactory cortices (e.g., lateral olfactory gyrus and anterior perforated substance) or paleocortex, and the hippocampal formation proper or archicortex. **Note:** Some authors use the term *allocortex* more broadly to include the juxta-allocortex as well.

Allokinesia: Moving the limb ipsilateral to a cortical lesion, when movement of the contralateral limb is requested.

Allosteric modulation: Alteration of the effect (e.g., enhancing or attenuating) of a neurotransmitter via the binding of a second compound at a different site on the same receptor (e.g., the binding of a benzodiazepine to a GABA receptor magnifying the response to GABA).

Alpha motor neuron: Large diameter neuron that supplies the extrafusal (body) of striated muscles causing contraction (see **Gamma motor neuron**).

Alveus: Thin sheet of axonal fibers covering the lateral or ventricular surface of the hippocampus that consists of efferent (and some afferent) fibers that eventually will coalesce to form the fimbria or foot of the fornix.

Amorphognosis: Refers to the inability to appreciate the external form of the object, i.e., its size, shape, and/or idiosyncratic contours that set its shape off from other objects.

Amusia: Inability to appreciate pitch, tone, or other musical qualities as a result of a brain lesion, more commonly in the region of the right temporal lobe.

Amygdala: Relatively large, heterogeneous nuclear grouping underlying the uncus (the external surface feature) anterior to the hippocampus in the parahippocampal gyrus. Divided into two major components, the corticomedial and basolateral groups, the former has extensive connections with the olfactory system, while the latter is closely associated with the cortex, basal ganglia, and other "limbic" structures. Functionally, it seems to play a major role in the experience and/or expression of emotion.

Anarithmetria: Deficit in performing calculations (dyscalculia) that is not the result of other primary deficits such as the recognition of numbers or difficulty handling spatial relationships.

Aneurysm: A ballooning of a blood vessel, most commonly at a site of bifurcation, as a result of a weakening in the vessel wall.

Angular gyrus: Gyrus in inferior parietal lobule that envelops the distal portion of the superior temporal sulcus (Brodmann's area 39). Lesions to this area in the "dominant" hemisphere are often associated with "Gerstmann's syndrome."

Anomalous dominance: Any pattern of cortical dominance that is different from the left hemisphere being the controlling hemisphere for language, the dominant hand, and praxis.

Anosmia: Loss of sense of smell.

Anosognosia: Lack of awareness or appreciation of a neurological deficit (e.g., sensory or motor loss) as a result of a destructive lesion to the brain.

Anosodiaphoria: Although not denying its presence, the failure to fully appreciate the significance or implications of a neurological deficit as a result of a brain lesion.

Ansa lenticularis: Ventral fibers of passage from the globus pallidus to the area of the zona incerta.

Antagonist: Drug that binds to a receptor, thus preventing (noncompetitive antagonist) or interfering with (competitive antagonist) another drug or neurotransmitter from binding to and stimulating the receptor. The antagonist has no effect of its own; it simply prevents other drugs or neurotransmitters from exerting their effect.

Anterior cerebral artery: Branch of the internal carotid artery that supplies the orbital frontal area and the anterior two thirds of the medial surface of the cerebral hemisphere.

Anterior choroidal artery: Branching off the internal carotid at the juncture of the middle cerebral, supplies numerous subcortical structures as well as the choroid plexus of the lateral ventricles.

Anterior commissure: Lying above the region of the optic chiasm along the lamina terminalis, this pathway connects olfactory areas between the two hemispheres, although the majority of its fibers interconnect the more anterior portions of the contralateral temporal lobes, especially the middle and inferior gyri, which probably includes the amygdala.

Anterior communicating artery: Connects the two anterior cerebral arteries and, along with the posterior communicating arteries, form the circle of Willis.

Anterior nucleus (of thalamus): Principal nucleus of the anterior division, it receives much of its input from the mammillary bodies and projects to the cingulate gyrus.

Anterior perforated substance: Region just posterior to the splitting of the olfactory tract and anterior to the optic chiasm marked by the entrance of multiple tiny blood vessels "perforating" this tissue. May help process olfactory input.

Anton's syndrome: Cortical blindness, although the patient denies any visual problems. Usually the result of an acute, vascular lesion involving either the occipital cortex or the optic radiations bilaterally.

Aphasia: An acquired deficit in the ability to use or understand language as a result of brain injury. Although it may be associated with simple motor speech impairments, it primarily is a language disorder.

Apperceptive visual agnosia: Inability to visually identify an object as a result of impaired perceptual abilities (for contrast, see **Associative visual agnosia**).

Apraxia: Difficulty carrying out a motor program (goal-directed movement or series of movements) that is unrelated to any primary motor or sensory deficits.

Apraxia of speech: see **Dysarthria**.

Apraxic agraphia: Specific difficulty forming letters when writing. Although theoretically independent of other language and visual–motor difficulties, since it usually is associated with dominant parietal lesions, ideomotor apraxia, constructional deficits, as well as disturbances of language (anomia, spelling, and reading difficulties) may be present.

Aprosodia: An acquired deficit in the ability to either endow one's speech with appropriate, meaningful affective intonations (expressive aprosodia) and/or to correctly perceive or interpret the emotional nuances (meaning) in the speech of others (receptive aprosodia).

Arachnoid: Middle meningeal layer that is contiguous to and follows the contours of the dura. In the space between the arachnoid and the pia are the larger cerebral blood vessels and cerebral spinal fluid. It is through the arachnoid granulations that this CSF is passed into the venous system.

Arcuate fasciculus: see **Superior longitudinal fasciculus**.

Argyll-Robertson pupil: Pupil that fails to react to light but does react (constricts) to accommodation; more commonly seen with syphilis.

Arteriosclerosis: Also commonly refered to as "hardening of the arteries," in this general disease process the arteries become less elastic as a result of the buildup of calcium within the walls of the vessel. Usually associated with this process is a narrowing of the lumen of the vessel, potentially restricting the flow of blood (see **Atherosclerosis**).

Arteriolosclerosis: "Anteriosclerosis" of small vessels.

Arteriovenous malformation: A congenital abnormality resulting in a proliferation and enlargement of blood vessels over time. Often not apparent until the third or fourth decade, it can become symptomatic as a result of mass effects or hemorrhage.

Asomatognosia: A disturbance of body schema following a cortical insult. It may involve neglect of one side of the body (contralateral to the lesion) or difficulty identifying specific parts of the body (usually bilaterally). The former is more likely to be associated with right hemisphere lesions, the latter with left-sided lesions.

Association pathway: Fiber bundles that connect one part of the cerebral cortex with another cortical area within the same hemisphere (e.g., occipito-frontal pathways).

Associative visual agnosia: Inability to verbally identify or categorize an object despite adequate visual perception (e.g., can draw it) as a result of disconnection between the secondary visual and the heteromodal association areas (for contrast, see **Apperceptive visual agnosia**).

Astereognosis: Inability to adequately describe the physical features of an object through the sense of touch, despite the relative preservation of more elementary tactile sensations. Implies lesions affecting the medial lemniscal system, possibly from the dorsal columns of the spinal cord to the cortex.

Asynergy: The breakdown of the normal smooth flow of movement into disconnected or "jerky" components.

Ataxia: Inability to regulate or control the direction, speed, or intensity of movement. Most commonly applied to disturbances of gait, but may apply to any movements.

Atherosclerosis: A common subtype of arteriosclerosis, this condition refers to the buildup of fatty deposits (plaques) within the lumen of an arteriosclerotic blood vessel. In addition to potentially either reducing or completely blocking the flow of blood, bits of these fatty deposits can dislodge (creating an embolus) and subsequently block a more distal portion of the artery.

Athetosis: Motor disorder characterized by irregular, involuntary, writhing type movements. May appear similar to choreiform but tends to be somewhat slower and more likely restricted to the distal portions of the extremities. The term "choreoathetosis" reflects the combination or lack of ability to differentiate between the two.

Atrium (of lateral ventricles): Area of widening of the body of the lateral ventricles (posterior) that represents the confluence of the posterior and temporal horns.

Autoreceptors: Receptors that modulate the activity of a neuron when stimulated by the same transmitter (or an agonist) that is released by that same neuron. Located near the cell body (somatodendritic autoreceptor), along the axon, or at the presynaptic terminal (terminal autoreceptor), these feedback mechanisms can influence the rate of neuronal firing, the synthesis, and/or release of the neurotransmitter.

Autotopagnosia: Loss of the ability to recognize or identify parts of one's body. Most frequently manifested as finger agnosia or right–left disorientation.

Babinski: Abnormal reflex in adults characterized primarily by extension of the great toe (upgoing toe) when the sole of the foot is appropriately stimulated. Suggestive of a upper motor neuron lesion.

Balint's syndrome: Typically the result of bilateral parieto-occipital lesions, this syndrome is characterized by (1) difficulty fixating on (interpreting) more than one feature of a visual stimulus at a time (simultanagnosia); (2) difficulty with eye–hand coordination (e.g., pointing to or localizing objects in space); and (3) difficulty with visual tracking or other signs of disturbances of eye movement.

Ballismus: A sudden, violent, involuntary, non-purposeful movement, typically involving one or more extremities and resulting from a lesion affecting the subthalamus. If the condition is associated with only one side of the body, it is referred to as hemiballismus.

Bands of Baillarger: Horizontal (transverse) fibers located with the cellular layers of the cortex, primarily those in layers IV and V. These fibers represent either terminal, horizontal branches of ascending thalamocortical inputs, or horizontal branches or axons of cortical cells. One functional significance of these fibers is to interconnect local vertical cortical columns. In the striate cortex, these bands are visible to the naked eye and are known as the line of Gennari.

Basal ganglia: Collective term for various subcortical structures, a common, but probably not exclusive, function of which are to modulate motor tone and activities. The structures most frequently included under this rubric are the caudate nucleus, putamen, and globus pallidus; however, depending on the context, other nuclear groups, including the claustrum, amygdala, subthalamus, some of the thalamic nuclei, and the substantia nigra, may be encompassed by this term.

Basis pedunculi: see **Cerebral peduncles**.

Benedikt's syndrome: Combination of unilateral third nerve palsy with contralateral sensory changes and contralateral ataxia and tremor, typically resulting from lesions of midbrain tegmentum ipsilateral to the eye findings.

Bilateral simultaneous stimulation: see **Double simultaneous stimulation**.

Blepharospasm: Involuntary, forceful closing of the eyelids. A form of focal dystonia.

Borderzone infarctions: see **Watershed syndrome**.

Brachium conjunctivum: see **Superior cerebellar peduncle**.

Brachium pontis: see **Middle cerebellar peduncle**.

Bradykinesia: Motor disorder characterized by slowness in the initiation or execution of voluntary movement or a paucity of movement. Frequently seen with basal ganglia disease.

Broca's aphasia: Disturbance of speech and language characterized by effortful, dysarthric, nonfluent speech, in which comprehension is relatively spared.

Brown-Sequard syndrome: Results from hemisection of the spinal cord and is characterized by spastic paralysis and loss of position sense and stereognosis below level of the lesion ipsilaterally, with a contralateral loss of pain and temperature.

Calcarine fissure: Found on the medial surface of the occipital lobe, separates the lingual gyrus (ventrally) from the cuneus (gyrus) and the visual cortex representing the superior and inferior visual fields, respectively.

cAMP: Cyclic adenosine-3′,5′-monophosphate. Acts as a "second messenger," affecting numerous processes in the cytoplasma of the neuron (e.g., receptor, neurotransmitter synthesis, opening or closing of ion channels).

Caudal: From the Latin for "tail"; refers to the direction away from the head or brain. In the brain itself, it most commonly means toward the brainstem. Less frequently (and more confusingly) it may imply toward the occipital pole.

Caudate nucleus: Large, subcortical nuclear mass whose enlarged "head" lies between the internal capsule and the anterior portion the lateral ventricle and whose "tail" follows the inferior curvature of the lateral ventricles to end near the amygdala. Its major input would appear to be from the frontal association cortex (to its "head"), but receives input from other association cortices. Part of the basal ganglia, it is likely involved in modulating motor activity, along with other higher-level functions.

Central sulcus: That which separates the primary motor (area 4) rostrally from the primary somatosensory cortex (areas 3,1,2) caudally or posteriorly. Also known as *Rolandic* sulcus.

Central tegmental tract: Primarily visualized in the pontine tegmentum, this tract carries an assortment of ascending and descending motor associated fibers, especially between the red nucleus and the inferior olivary nucleus and to and from the reticular nuclei (e.g., corticoreticular and reticulospinal fibers).

Cerebellar hemispheres: The largest and most conspicuous parts of the cerebellum that lie on each side of the vermis primarily are responsible for controlling discrete movements of the limbs, as well as possibly being involved in other cognitive and/or affective processes.

Cerebral aqueduct: Slender passageway in the midbrain that allows the flow of CSF between the third and fourth ventricles.

Cerebral dominance: Concept that one hemisphere is predominately responsible (i.e., is the "leading" hemisphere) for certain behavioral functions, e.g., that in most right-handers the left hemisphere is "dominant" for (controls) propositional speech and language comprehension.

Cerebral peduncles: Term is commonly used interchangeably with "crus cerebri," which consists of those corticofugal fibers descending in the anterior or basal portion of the midbrain and which are targeted to terminate in the brainstem (corticobulbar), cerebellum (corticopontocerebellar), and spinal cord (corticospinal). However, strictly speaking, the term also encompasses the tegmental portion of the midbrain.

Chorea: Movement disorder associated with basal ganglia disease and characterized by involuntary, spastic, non-goal directed movements of the limbs and/or trunk.

Choreiform: see **Chorea**.

Choroid plexus: Formed from the ependymal cells lining the ventricles, these glandular-like structures produce the cerebral spinal fluid (CSF). They are found in all four ventricles and are typically the source of the calcification seen in adult CT scans.

Cingulate gyrus: Juxt-allocortex seen lying above the corpus callosum on a midsagittal section of the brain. Often connected with the "limbic system," it has extensive connections both with the overlying neocortex as well as with "limbic" structures, including the septal region, amygdala, hippocampus, and anterior nuclei of the thalamus. In humans, it seems to play a prominent role in modulating affect or drive states.

Cingulotomy: Surgical lesion involving the anterior cingulate gyrus and the cingulum.

Cingulum: Association pathway most easily seen as the "white matter" within the folds of the cingulate gyrus on a coronal section of the brain. This fiber system carries information between the hippocampal gyrus and many other parts of the forebrain, including the frontal, parietal, and cingulate gyri and the septal area.

Cingulumotomy: Lesioning of the anterior cingulum, typically for the treatment of psychiatric disorders such as OCD and intractable pain.

Circle of Willis: Pattern of vessels lying on the ventral surface of the hypothalamus interconnecting the right and left internal carotid system and the posterior cerebral arteries creating the potential for the shunting of blood from one system to another in the case of blockage.

Cisterns: Larger spaces in the subarachnoid region, primarily around the base of the brain, where greater collections of cerebral spinal fluid can be visualized.

Claustrum: A narrow band of gray matter underlying the insular cortex between the extreme and external capsule. While it appears to have reciprocal connections with the cortex, particularly the sensory cortices, its function is unclear.

Climbing fibers: Fibers entering the cerebellum from the inferior olivary nuclei.

Clonus: Repetitive jerking of a limb (typically the hand or foot) when the flexor tendons are subjected to sudden and sustained stretching. Suggestive of an upper motor neuron lesion.

Collateral sulcus: Sulcus separating the parahippocampal from the fusiform gyrus on the ventral surface of the brain.

Color agnosia: A controversial concept and, if it exists in a pure form, is probably quite rare. It must be differentiated from achromatopsia (perceptual deficit), pure color anomia (visual-verbal dissociation), or simply color imagery deficit. Theoretically, this might be characterized by preserved ability to match colors, inability match colors to objects, inability to name or point to colors presented, but preserved ability to verbally to indicate that "apples are red" and "grass is green" (verbal-verbal association).

Color anomia: Inability to name colors, despite preserved ability to sort or match them. Usually associated with left occipitotemporal lesions and a right homonymous hemianopia.

Color imagery: The capacity to "picture" or conjure up the color of things (e.g., inside, outside, and seeds of a watermelon). Deficits in this ability are often associated with achromatopsia.

Commissure: A crossing of fibers from one side of the neural axis to homologous structures on the opposite side, e.g., the corpus callosum.

Commissurotomy: The surgical cutting of the fibers of a commissure, most commonly those of the corpus callosum, usually as a treatment for intractable seizures.

COMT (Catechol-*O*-methyltransferase): Typically found in the synaptic cleft, this enzyme helps to break down monoamine transmitters after they have been released from the presynaptic terminal.

Conduction aphasia: Typically associated with a deep, "middle zone" lesion that impinges on the superior longitudinal (arcuate) fasciculus and characterized by marked deficits in repetition, despite relatively intact comprehension and articulate (although paraphasic) output.

Confabulation: Presenting as definitive or factual information, a memory of something that has no basis in reality. Often seen in the early stages of Korsakoff's. The more improbable (fantastic) the confabulation, the more likely that extensive disruption of frontal systems are involved (loss of self-monitoring ability).

Conjugate gaze: Coordination of the movements of the two eyes so that each is looking in the same direction at the same time.

Constructional disability: An acquired deficiency in copying (drawing) or reconstructing (using objects or forms) two- or three-dimensional geometric designs or other models.

Contralateral: On the opposite side, as opposed to ipsilateral or same side.

Convergence: The two eyes "converging" toward the midline (simultaneous contraction of the medial rectus muscles of both eyes) when looking at a near object. Typically accompanied by accommodation (lens adjustment) and pupillary constriction.

Convolution: The process of enfolding of the cortex into gyri and sulci that allows for expansion of the cortical area without necessitating excessive expansion of the cranial vault.

Coronal radiations: Fanlike array of ascending projection (corticopetal) fibers from the thalamus to the cerebral cortex and descending (corticofugal) cortical fibers converging in the internal capsule.

Corpora quadrigemina: The superior and inferior colliculi.

Corpus callosum: Main fiber pathway that interconnects the cerebral hemispheres. From it's anterior to posterior extent, it is divided into the rostrum, genu, body, and splenium.

Corpus striatum: Collective term for the caudate nucleus, putamen, and globus pallidus.

Corticobulbar tract: Projection fibers from the cortex (whose origin are similar to those of the corticospinal tract) to the cranial nerve nuclei responsible to movements of the face and head.

Corticofugal: Traveling away from the cortex (e.g., to brainstem or spinal cord).

Corticopetal: Traveling toward the cortex from subcortical structures.

Corticopontine tract: Consists of cortical fibers, primarily from the sensorimotor regions of the brain, which descend through the internal capsule and then the crus cerebri to terminate on the (basilar) pontine nuclei. From there, most secondary fibers enter the cerebellum via the middle cerebellar peduncles.

Corticospinal tract: Pathway that begins in the cortex (primarily in the sensorimotor cortex surrounding the central sulcus) and terminates with the spinal cord. See **Lateral and Ventral corticospinal tract**.

Corticotectal tract: Fibers from various areas of the cortex (particularly from the occipitoparietal cortices) to the superior colliculi. These along with secondary connections help mediate visual tracking and reflex movements triggered by visual stimuli.

Cremasteric reflex: In males, stimulation by stroking or scratching the inner surface of the thigh causes the ipsilateral testicle to elevate. Suppressed by upper motor neuron lesions.

Cribriform plate: That part of the ethmoid bone lying at the base of the skull through which the axons of the olfactory neurons in the nasal cavity pass to make connections with the secondary neurons in the olfactory bulbs.

Crus cerebri: see Cerebral peduncles.

Cuneocerebellar tract: Carries sensory information ipsilaterally from the upper extremities via the accessory cuneate nucleus to the cerebellum (see **Rostral spinocerebellar tract**).

Cuneus: Cortical gyrus in occipital cortex which lies on the medial surface just above the calcarine fissure. Mediates visual information coming from the inferior visual fields.

Cytoarchitecture: The pattern and relative development of the cellular layers making up cortical tissue. For the cerebral neocortex, this consists of six layers: the molecular, external granular, external pyramidal, internal granular, internal pyramidal, and the fusiform layers.

Cytotoxic edema: Edema associated with disruption of adenosine triphosphate's (ATP) capacity to drive the "sodium-potassium pump" resulting in a buildup of intracellular sodium (NA+) and subsequent increase in intracellular fluid. Frequently resulting from strokes. (Contrast with **Vasogenic edema**)

Decerebrate rigidity: Extension of all four extremities, with internal rotation of the wrists following midbrain lesion (at mid-collicular level or below). Can present unilaterally.

Decorticate posturing: Extension of the lower extremities, with flexion of the upper extremities. Results from lesion transecting the most rostral portion of the midbrain. Can result from herniation.

Decussation: A crossing of fibers from one side of the neural axis to nonhomologous areas of the other, e.g., the decussation of the corticospinal, spinothalamic, and medial lemniscal fibers.

Dejerine-Roussy syndrome: Contralateral loss of, or diminished, somatosensory sensation typically resulting from vascular lesion of the thalamus and usually involving the ventral posterior nucleus (VPL & VPM). Most notable, however, is the concomitant presence of diffuse, lingering pain, often occasioned by a relatively minor stimulus.

Dentate gyrus: Three-layered allocortex lying between the subiculum and the hippocampus proper, the three of which constitute the hippocampal formation.

Dentate nuclei: The largest and most lateral of the cerebellar nuclei that receive most of their input from the cerebellar hemispheres and project back to the cortex via the thalamus and to various brainstem nuclei.

Diagonal band of Broca: Band of fibers in the posterior region of the anterior perforated substance connecting the basolateral amygdala with the septal nuclei.

Disconnection syndrome: A consistent pattern of behavioral deficits that results from lesions affecting the fiber pathways connecting two or more areas of the brain (e.g., ideomotor apraxia, alexia without agraphia).

Disinhibition: The tendency to respond impulsively according to how one is inclined at the moment without consideration of social appropriateness or longer-term consequences. Can result from multiple causes including immaturity, inebriation, generalized dementia, and frontal lobe pathology.

Distributed system: The idea that most behavioral functions are highly complex and their successful execution requires the integration of multiple elementary functions, which in turn are widely "distributed" throughout the brain. In turn, multiple behaviors may utilize the same, common elementary functions.

Doll's eyes: see **Oculocephalic reflex**.

Dominance: Most commonly refers to the combined localization of propositional language and preferred handedness to one side of the brain or another. Less frequently, may refer to other, non-linguistic functions (e.g., dominance for certain visual–spatial skills).

Dorsal cochlear nuclei: Along with the ventral cochlear nuclei, they represent the first relay for auditory fibers in the brainstem (upper medulla). In turn, they send fibers across the midline via the dorsal acoustic stria where most join the contralateral lateral lemniscus to the inferior colliculi.

Dorsal longitudinal fasciculus: Fiber pathway linking the hypothalamus with tegmental and tectal nuclei of the midbrain. Another of the pathways that likely mediate autonomic responses.

Dorsal motor nucleus of X: Situated in the dorsal aspect of the rostral medulla, this nucleus is the source of the majority of the preganglionic parasympathetic fibers of the vagus nerve that will eventually regulate thoracic and abdominal organs.

Dorsal nucleus of Clarke: Column of neurons found in the thoracic and upper lumbar regions of the spinal cord (lamina VII) which is the origin of the dorsal spinocerebellar tracts.

Dorsal spinocerebellar tract: Uncrossed tract providing sensory input from joints and muscles of the lower extremities to the cerebellum via the dorsal nucleus of Clarke (see **Ventral spinocerebellar tract**).

Dorsal trigeminal lemniscus: Along with the ventral trigeminal lemniscus, conveys somatosensory information from the face to the VPM nucleus of the thalamus.

Dorsomedial nucleus (of thalamus): Primary nucleus of the medial division of the thalamus, its primary connections are with the frontal association cortex.

Double simultaneous stimulation: Act of applying similar stimulation (visual, tactile, or auditory) to both sides of the body at the same time to test for suppression or incomplete neglect. (see **Extinction**)

Down-regulation: Reduction in the synthesis (hence, reduction in the absolute number) of either pre- or postsynaptic receptors following an increase in the availability of a particular neurotransmitter. Also referred to as "desensitization" of receptors.

Dressing apraxia: The inability to dress one's self, despite adequate motor skills, usually associated with nondominant parietal lesions. However, one must distinguish between errors resulting from (1) failure to recognize an article of clothing (possible agnosia resulting from a left hemisphere lesion), (2) unilateral neglect or inattention, or (3) difficulty orienting and/or manipulating the clothes in a visual–spatial environment.

Dura mater: An extremely tough, fibrous tissue that forms the outermost meningeal layer. Attached to the inner surface of the skull, it also forms sinuses through which venous blood is channeled into the jugular veins for its return to the heart and lungs. That portion of the dura that extends vertically between the hemispheres is called the falx cerebri, while that the portion that separates the occipital lobe from the cerebellum is called the tentorium.

Dynamic aphasia: see **Transcortical aphasia—motor**.

Dysarthria: A disturbance of motor speech as opposed to a disturbance of language (aphasia), although both may be present at the same time. Can result from either (1) lesions of motor association areas, especially area 44 ("cortical dysarthria" or "apraxia of speech"), or (2) subcortical lesions affecting motor neurons that serve the organs of speech ("subcortical dysarthria").

Dyscalculia: Acquired difficulty in performing arithmetical operations. May be due to primary loss of mathematical ability (anarithmetria), difficulty maintaining the spatial relationships of the numbers (spatial dyscalculia), or other more fundamental perceptual/behavioral deficits.

Dysdiadochokinesia: A manifestation of asynergy following cerebellar lesions characterized by difficulty carrying out rapid, alternating movements of a distal extremity, e.g., rapid pronation and supination of the hand.

Dyskinesia: Any of several types of involuntary or abnormal movements such as tremors, choreiform, ballistic, or athetoid movements typically associated with lesions of the basal ganglia.

Dysmetria: Failure to properly judge distance when reaching for an object, usually associated with cerebellar lesions.

Dystonia: Motor disorder characterized by slow, sustained, involuntary muscle contractions, usually involving the trunk or more proximal extremities, although it also may be seen in the more distal aspects of the limb.

Edinger-Westphal nucleus: Parasympathetic component of CN III which mediates constriction of the pupil and accommodation of the lens.

Efferent: Refers to neuronal processes (e.g., axons, fiber tracts) that exit a nerve cell or collection of cells, the output of a nerve cell. While often associated with "motor" pathways, the term applies to any type of information being conveyed away from the cell.

Emboliform nuclei: One of the pairs of deep cerebellar nuclei.

Embolism: A piece of foreign matter (air, fat, fungus, tissue) that travels in an artery to a spot distal to its origin and occludes the vessel.

Engram: More or less permanent change in the cell or pattern of synaptic connections among groups of cells, which is thought to provide the physiological basis for memory.

Entorhinal cortex: Seen on the medial surface of the temporal lobe, this area represents the more anterior portion of the parahippocampal gyrus (Brodmann's area 28) and serves as the major channel of multisensory input from the cortex into the hippocampus via the perforant and alvear pathways.

Epidural hematoma: Bleeding that occurs between the dura and the skull, usually as a result of a skull fracture.

Epithalamus: The habenular nuclei and the pineal body.

Equipotentially: Belief that any part of the brain is capable of assuming the function of any other part.

Excitatory postsynaptic potential (EPSP): A decrease in the electrical gradient of the neuronal cell membrane resulting from a stimulus which increases the permeability of sodium (Na^+) ion channels. While not sufficient in and of itself to produce an action potential (complete depolarization), it increases the probability of such an event given the spatial or temporal summation of similar events.

Excitotoxicity: Destruction of a cell resulting from an excessive influx of calcium as a result of abnormal excitation or a breakdown in the cell membrane.

Executive functions: Those functions thought to be mediated largely by the frontal cortices (especially the prefrontal regions) which reflect the "conscious" decision to act or to modulate ongoing actions based on consideration of internal and external circumstances (feedback), plans or intentions, and anticipated consequences.

Extended amygdala: Includes the medial extension of the centromedial portion of the amygdala to the bed nucleus of the stria terminalis.

External medullary lamina (of the thalamus): Thin band of fibers lying on the lateral aspect of the thalamus separating this main nuclear mass from the reticular nucleus (of the thalamus).

Extinction: Refers to the "fading out" of a stimulus despite being continuously applied. However, this term also commonly is used synonymously with "suppression" or "neglect" with double simultaneous stimulation or the failure to perceive a stimulus when another stimulus is present (typically in the same modality but on the contralateral side). Suggests a lesion in the contralateral hemisphere.

Extrafusal muscle fibers: Those fibers that constitute the main body of the muscle and innervated by alpha and beta fibers.

Extrapyramidal: General term used to define descending motor pathways other than the corticobulbar or corticospinal tracts, traditionally those effects thought to be mediated by the basal ganglia. However, it would now appear that these distinctions are not so clear-cut.

Facial nerve (CN VII): Controls the muscles of facial expression and carries information for taste (anterior two-thirds of tongue).

Facial nucleus: Central pontine nucleus that mediates control over muscles of expression and closing the eyelid via CN VII.

Falx cerebri: That portion of the dura which extends down between the two cerebral hemispheres.

Fasciculus: Large collection of fibers (pathway or tract) within the brain or spinal cord.

Fasciculus cuneatus: Somatosensory pathway in the dorsal funiculus of the upper thoracic and cervical portions of the spinal cord which carries information regarding proprioception and stereognosis from the upper extremities to the cortex via the medial lemniscus and thalamocortical projections.

Fasciculus gracilis: Somatosensory pathway in the dorsal funiculus of the spinal cord which carries information regarding proprioception and stereognosis from the lower extremities to the cortex via the medial lemniscus and thalamocortical projections.

Fasciculus mammillaris princeps: Name of fiber tract exiting the mammillary bodies before it divides into the mammillothalamic and mammillotegmental tracts.

Fasciculus retroflexus (habenulointerpeduncular tract): As implied by its alternate nomenclature, this tract proceeds from the habenular nuclei (of the epithalamus) to the interpeduncular ("between the cerebral peduncles") nuclei of the midbrain. Because of the sources of input to the former, including the septal nuclei, the hypothalamus, and the raphe nuclei (serotonin) in the midbrain, this would appear to be one of the pathways important for modulating visceral or autonomic functions as a result of emotional arousal.

Fastigial nuclei: Innermost pair of the deep cerebellar nuclei which relay efferent fibers from the flocculonodular lobe to the vestibular and reticular nuclei.

Fibrillation: Minute, rapid, continuous contractions of a muscle detectable only via EMG. Seen in lower motor neuron disease. Distinguish from fasciculations that are grosser movements and observable with the naked eye. Suggestive of lower motor neuron disease.

Fimbria: Collection of hippocampal and subicular fibers from the alveus on the dorsomedial aspect of the hippocampus which forms the foot of the fornix. From the fimbria, the efferent fibers form the crus of the fornix as they proceed anteriorly through the body and columns (of the fornix).

Finger agnosia: A loss of the ability to name or otherwise identify or differentiate individual fingers, despite relatively intact elementary sensory feedback, e.g., vision or somesthetic sensation.

Flocculonodular lobe: Oldest and smallest division of the cerebellum that lies adjacent to the inferior aspect of the fourth ventricle and consists of the nodulus (in the midline) and the two flocculi on each side. Strongly associated with the vestibular nuclei, it is important for balance, posture, and eye movements in response to postural changes.

Fluent aphasia: Typically characterized by effortless, but paraphasic speech and comprehension difficulties (e.g., Wernicke's aphasia). Sometimes also referred to as "sensory," "posterior," or "receptive" aphasia. Lesions are commonly in the area of the superior temporal gyrus.

Foramen magnum: Opening at base of skull at medulla-spinal cord junction.

Foramen of Magendie: An opening in the roof of the fourth ventricle below the pons that, along with the foramina of Luschka, allows the CSF to circulate outside the brain in the subarachnoid space from where it is reabsorbed.

Foramen of Monro: Also known as the interventricular foramen, these paired openings in the anterior horns of the lateral ventricles allow the flow of CSF between the lateral and third ventricles.

Foramina of Luschka: Paired lateral openings in the fourth ventricle at the level of the pons that provide an access, along with the Foramen of Magendie, for the CSF to the subarachnoid space.

Forel's field H: Area in the subthalamus that primarily is characterized by fibers of passage, although some cells also are present. Three distinct portions commonly are identified: tegmental field H of Forel (prerubral area), located dorsolaterally to the red nucleus, consists of fibers coming from the dentate nucleus of the cerebellum and red nucleus on their way to the thalamus, as well as pallidothalamic fibers; H1, which consists of pallidofugal fibers (ansa lenticularis) that make their way around the posterior limb of the internal capsule to the area of the zona incerta and are joined by fibers from the dentate nucleus of the cerebellum on their way to the thalamus (thalamic fasciculus); and H2, which consists of fibers (lenticular fasciculus) from the globus pallidus traversing the posterior limb of the internal capsule on their way directly to the subthalamus.

Fornix: Prominent fiber tract that has its origin in the hippocampal formation and curving over and around the dorsal thalamus projects to various diencephalic, basal forebrain, and midbrain. Some of its major connections are with the mammillary bodies, anterior thalamic and septal nuclei. Its main components, beginning at the hippocampal formation, are the alveus and the fimbria (which represent the initial collection and congregation of hippocampal fibers), the crus (or crura) which then converge to form the body, which in turn splits again anteriorly to form the descending columns of the fornix.

Frontal ("magnetic") **ataxia:** Difficulty walking secondary to problems initiating leg movements, as if the foot is "stuck" to the floor. Usually the result of bilateral frontal involvement, as in normal pressure hydrocephalus.

Frontal eye fields: Represented by Brodmann's area 8 on the lateral surface of the frontal lobes, this area is responsible for controlling voluntary eye movements, such as in active visual searching.

Function: Theoretically, the smallest, meaningful and identifiable unit of behavior that contributes to an integrated response of the organism.

Funiculi: Fiber tracts that run in a rostral–caudal direction within the spinal cord. Each half of the spinal cord contains the dorsal, lateral, and ventral funiculi, which are themselves composed of discrete pathways running to (sensory) or from (motor) the brain, brainstem, or cerebellum.

Fusiform gyrus: Also known as the occipitotemporal gyrus, it is seen on the ventral view of the brain lying between the parahippocampal gyrus medially and the inferior temporal gyrus laterally. It is separated from the former by the collateral sulcus and, from the latter by the inferior temporal sulcus.

Gamma-aminobutyric acid transaminase (GABA$_T$): Enzyme that breaks down GABA once it has reentered the presynaptic terminal.

Gamma motor neuron: Small diameter neuron that innervates the intrafusal fibers of muscle spindles within striated muscles, which, in turn, are important for kinesthetic feedback and muscular reflexes.

Geniculocalarine tract: Also known as "optic radiations," visual fiber tract from the lateral geniculates to the calcarine (visual) cortex.

Genu: Means "knee" and refers to the anterodorsal portion of the corpus callosum or the central portion (horizontal view) of the internal capsule.

Gerstmann's syndrome: Typically resulting from a lesion involving the left angular gyrus, this syndrome consists of finger agnosia, right–left disorientation, agraphia, and acalculia. Although not an original part of the syndrome, constructional deficits also commonly are seen as part of the clinical picture.

Global versus local processing: The tendency to perceive either the whole (global) or the part (local), as in the outline of a square (global unit) generated by stringing together many small circles (local units). Right hemisphere is supposedly superior with "global," and the left with "local" processing.

Globose nuclei: One of the pairs of deep cerebellar nuclei.

Globus pallidus: Part of the basal ganglia lying between the putamen and internal capsule, it receives input from both "above" (caudate and putamen) and "below" (subthalamus and substantia nigra). In turn, it projects back to the subthalamus and to the thalamic nuclei. (The latter completes the loop by projecting back to the cortex, thus exerting a modulating influence on certain cortical activities.)

Glossopharyngeal nerve (CN IX): Subserves the sense of taste on the posterior aspect of the tongue, assists in elevating the pharynx, and helps control the salivary glands.

G-protein-linked receptors: Receptor that utilizes G-proteins and secondary enzymes ("second messengers") to open ion channels or effect other metabolic changes within the cell.

Graded potential: A local change in the potential of a membrane that dissipates as it moves away from the initial point of change. The direction and degree of change can vary depending on the stimulus or agent of change.

Granular cortex: Cortex in which there are few, if any, pyramidal cells and layer IV is quite prominent (primarily found in primary sensory projection areas. Also known as koniocortex.

Gyrus: The ridge or elevation between the sulci or grooves on the surface of the cortex.

Hematoma: A collection of blood within tissue as a result of a hemorrhage or leakage from a vessel.

Hemiakinesia: Reduction in spontaneous motor activity contralateral to the side of the lesion. Cannot be explained by motor weakness, which may be mild or grossly absent. Most likely seen when lesions involve the supplementary motor system.

Hemianopia: Vertical visual field cut involving one half of the visual field.

Hemiballismus: see **Ballismus**.

Hemiplegia: Paralysis on one side of the body.

Hemiparesis: Weakness on one side of the body (monoplegia = weakness of one limb).

Hemispatial neglect: The tendency to ignore or inadequately attend to stimuli emanating from one half of extrapersonal space as a result of a lesion in the contralateral hemisphere.

Hemispheric specialization: see **Lateralization**

Heschl's gyrus: Located in the temporal operculum of the superior temporal gyrus, it is the primary projection site for auditory fibers from the medial geniculates.

Heteromodal association cortex: see **Tertiary cortex**.

Heteroreceptors: Receptor that by binding to one neurotransmitter alters the response of a second neurotransmitter (e.g., serotonergic heteroreceptors on dopaminergic neurons exerting an inhibitory influence on the latter).

Heterotypical cortex: Defines those cortical regions where it is difficult to clearly differentiate all six cellular layers. The two representative subtypes are the agranular (primary motor) and granular (primary sensory) cortices (see **Primary cortex** and **Idiotypic cortex**)

Hippocampal formation: Lying deep along the medial surface of the temporal lobe this three-layered allocortex appears to be critical for the laying down (encoding) of new memories. Consists of the hippocampus proper, the dentate gyrus and the subiculum.

Hippocampus proper: Three-layered allocortex that represents the innermost portion of the hippocampal formation and lies adjacent to the temporal horn of the lateral ventricle. Also known as Ammon's horn.

Hoffmann's sign: The adduction (flexion) of the thumb when the distal joint of the middle finger is briskly "flicked" by the examiner. Associated with hypertonicity or spasticity, it may be suggestive of an upper motor neuron lesion, especially if asymmetrical in its presentation.

Homonymous hemianopia: Vertical visual field cut (one half of field) that is comparable in both eyes, i.e., the right half or left half in each eye.

Homonymous quadrantanopia: Visual loss restricted to one quadrant of the visual field that is comparable in both eyes. Typically results from lesions affecting the optic radiations.

Homotypical cortex: Describes those cortical regions where the six individual types of cellular layers are clearly distinguished. These are represented by the frontal, parietal and polar types.

Horner's syndrome: Resulting from an interruption of the sympathetic fibers to the eye, this syndrome typically includes a constricted pupil in the ipsilateral eye, a drooping of the eyelid (ptosis), and a warm, dry face on the affected side.

Hydrocephalic edema: Periventricular edema associated with transependymal migration of fluids, commonly seen in the presence of obstructive hydrocephalus.

Hydrocephalus: Condition reflecting increased collection of CSF in the brain. May be either communicating or noncommunicating. The former reflects a problem of reabsorption, the latter a blockage of the foramina or the cerebral aqueduct.

Hyperreflexia: Increased reflexes. Typically associated with increased muscle tone as a result of an upper motor neuron lesion.

Hypertonia: More or less chronic or continuous state of increased excitation of a muscle or muscle groups. May result from diminished cortical inhibitory influences. Suggestive of upper motor neuron lesion.

Hypoglossal nerve (CN XII): Supplies the muscles of the tongue.

Hypoglossal nuclei: Located in the rostral medulla, these nuclei (CN XII) are responsible for movements of the tongue.

Hypophysis: The pituitary gland, including the adenohypophysis and the posterior lobe.

Hyporeflexia: Decreased reflexes. Typically associated with decreased muscle tone as a result of a lower motor neuron lesion.

Hypothalamus: Part of the diencephalon, this collection of individual nuclei lie at the base of the brain on either side of the third ventricle, beginning anteriorly at the level of the optic chiasm and extending to the level of the mammillary bodies posteriorly. In general, these structures appear to be largely responsible for monitoring and controlling, primarily through its influence on the pituitary (endocrine system) internal homeostasis as well as facilitating (or, in some instances, inhibiting) behaviors that are important for the survival of the individual as well as of the species (e.g., feeding and sexual behaviors).

Hypotonia: More or less chronic or continuous state of hypoarousal of muscle or muscle groups. May result from disturbance of either lower motor neuron lesion or disruption of sensory feedback loop.

Hypoxia: State of reduced oxygenation of tissue that may result from lack of sufficient cardiac output, oxygen deprivation, or inability of the blood to transport oxygen.

Ideomotor apraxia: The inability to carry out a symbolic gesture (e.g., a salute) or a learned, transitive action (especially without benefit of the object) to command, despite relatively intact sensorimotor and language abilities.

Ideational apraxia: The inability to carry out a concrete action or related series of actions (e.g., folding a letter and placing it envelope) when provided with the actual objects. Commonly associated with perceptual and/or generalized cognitive deficiencies.

Idiotypic cortex: see **Primary cortex** and **Heterotypical cortex**.

Impersistence (motor): Inability to sustain a motor action on command. Typically assessed by asking the patient to maintain an unnatural posture (at least for the situation), e.g., protrusion of the tongue or closing of the eyes.

Indusium griseum: Band of gray matter running between the corpus callosum and the cingulate gyrus. Contains fiber tracts (the medial and lateral longitudinal stria of Lancisi) that connect the hippocampal region with the cingulate gyrus and septal nuclei.

Inferior brachium: Auditory fibers from inferior colliculi to medial geniculates.

Inferior cerebellar peduncle: Chief final common pathway for all somatosensory information entering the cerebellum except for the ventral spinocerebellar tract (which enters through the superior cerebellar peduncle)

Inferior colliculi: Nuclear group on the posterior aspect (tectum) of the midbrain. Relay for auditory stimuli and probably involved in auditory reflexes.

Inferior longitudinal fasciculus: Less well-defined pathway underlying the temporal cortex. May interconnect more anterior temporal regions with the occipital lobes.

Inferior occipitofrontal fasciculus: Fiber pathway lying below the insula which connects the frontal cortex with the inferior temporal and occipital cortices.

Inferior oblique: Eye muscle that facilitates looking up when the eye is turned inwardly (adducted), mediated by CN III.

Inferior olivary nucleus: Large nucleus in the anterior portion of the upper medulla which is part of a feedback loop involving cerebral cortex, red nuclei, inferior olivary nuclei, and the cerebellum, along with proprioceptive input from the spinal cord.

Inferior rectus: Eye muscle that facilitates looking down, especially when eye is deviated laterally, mediated by CN III.

Inferior sagittal sinus: Deep, dural sinus running along the top of the corpus callosum which collects venous blood from the cortical surfaces within the longitudinal fissure.

Infundibulum (infundibular stem): Stalk connecting the hypothalamus with the pituitary gland, specifically that portion of the stem which contains neural elements.

Inhibitory postsynaptic potential: An increase in the electrical gradient of the neuronal cell membrane resulting from a stimulus that increases the permeability of chloride (Cl–) channels. By "hyperpolarizing" the cell membrane, it makes it more difficult to initiate an action potential.

Insula: Area of cortex "buried' in the lateral fissure with extensive connections to both sensory cortices and limbic structures. Also known as the "island of Reil."

Intention tremor: Tremor that is present when attempting to carry out a motor action rather than when the limb is at rest. Tends to be more characteristic of cerebellar disease.

Internal capsule: Large sheath of myelinated fibers lying between the thalamus and head of the caudate nucleus medially and the lentiform nuclei laterally. Carrying both descending cortical fibers and thalamocortical projections, it is divided into an anterior and posterior limb on either side of its central curvature (the genu) and its most posterior and ventral aspects, the retrolenticular and sublenticular portions.

Internal medullary lamina (of thalamus): A "Y"-shaped band of myelinated fibers that transverse the interior of the thalamus dividing it into its anterior, medial, and lateral nuclear groups.

Internuclear ophthalmoplegia: Failure of one eye to adduct (move toward the midline) on attempted lateral gaze to the opposite side as a result of a brainstem lesion affecting the ascending medial longitudinal fasciculus (MFL) between the abducens and oculomotor nuclei.

Intrafusal muscle fibers: Fibers that lie within or constitute the muscle spindles, which in turn are contained within the larger muscle body (or extrafusal fibers). Intrafusal fibers ("nuclear bag" or "nuclear chain") are innervated by both beta and gamma neurons and provide feedback about the state of the muscle's contraction.

Intralaminar nuclei (of thalamus): Nuclei (e.g., centromedian and parafascicular) that lie within the internal medullary lamina and that have extensive subcortical (e.g., basal ganglia, reticular formation, as well as other thalamic nuclei) and cortical interconnections.

Intraparenchymal hemorrhage: An intracerebral hemorrhage, i.e., bleeding into the brain tissue itself, most commonly associated with small vessel disease.

Intraparietal sulcus: Separates the superior parietal from the inferior parietal (angular and supramarginal gyrus) lobules.

Inverse agonist: Drug or compound whose behavioral response is exactly opposite that of an agonist. For example, if an agonist at a GABA receptor has anxiolytic and anticonvulsant effects, an inverse agonist at that same site would be anxiogenic and would tend to produce seizures.

Ionotropic receptors: Also known as "ion channel receptors," respond directly (and rapidly) to ligands by either opening or closing ion channels.

Ipsilateral: Same side, as opposed to contralateral or opposite side.

Ischemia: Tissue damage as a result of insufficient blood flow.

Ischemic infarction: Stroke resulting from a blockage of a blood vessel in contrast to a bleed or hemorrhage.

Island of Reil: see **Insula**.

Isocortex: Six-layered cortex that forms the majority of all the cortical structures in man. Also known as homotypical or simply, more commonly, "neocortex."

Juxta-allocortex: Also known as "mesocortex," refers to those limbic areas that lie adjacent to and represent a transition from the simpler allocortex (e.g., parts of the parahippocampal gyrus).

Juxtarestiform body: Small, medial portion of the inferior cerebellar peduncle (restiform body) that carries afferent and efferent fibers between the cerebellum (particularly the flocculonodular lobe) and the vestibular nuclei.

Kinesthesia: The ability to perceive or judge, without the aid of vision, the direction and speed or intensity of either active or passive movement of the limbs. It is mediated by the posterior columns and, along with proprioception, enables one to develop and maintain a "position" sense.

Kluver-Bucy syndrome: First described in monkeys with bilateral lesions of the anterior temporal cortex (including the amygdala), it is characterized by "psychic blindness" or a form of visual agnosia that is manifested by a tendency to mouth small objects and to approach previously fear-provoking visual stimuli. There also tends to be a decrease in aggressiveness and a concomitant decrease in social dominance, as well as an increase in sexual behaviors. Elements can also occasionally be found in humans following similar lesions.

Koniocortex: see **Granular cortex**.

Korsakoff's syndrome: Pathology most commonly associated with a history of chronic alcohol abuse appears to result from thiamine deficiency and with lesions predominately in and around the thalamus (especially the dorsomedial nuclei) and hypothalamus. Characterized primarily by severe memory loss (anterograde > retrograde) and frequent early tendency to confabulate against a background of mild to moderate more generalized cognitive impairments.

Lamina terminalis: Thin sheet of tissue that once was the anterior limits of the neural tube and now represents the anterior or rostral limits of the third ventricle.

Lateral corticospinal tract: The largest component of the corticospinal tract that has decussated at the level of the lower medulla and travels caudally in the lateral funiculi of the spinal cord. Primarily responsible for mediating skilled, discrete movements of the extremities.

Lateral dorsal nucleus (of thalamus): Part of the lateral division, it has reciprocal connections with the posterior cingulate gyrus, as well as with the posterior medial cortices.

Lateral fissure: Also known as the *Sylvian fissure* and composed of five segments, the most prominent being the posterior horizontal ramus, which separates the temporal lobe ventrally from the frontal and anterior aspect of the parietal lobes. The other four segments are the anterior horizontal, anterior ascending, posterior ascending, and posterior descending rami.

Lateral geniculates: Posteroventral thalamic nuclei that serve as the primary thalamic relay and integration centers for visual information.

Lateral lemniscus: Brainstem tract beginning at the level of the pontomedullary junction consisting of fibers from the cochlear nuclei carrying auditory information to the inferior colliculi.

Lateral longitudinal stria (of Lancisi): Tiny strand of fibers (along with the medial longitudinal stria) that lie within the indusium griseum along the dorsal surface of the corpus callosum and interconnect the septal and hippocampal regions.

Lateral medullary lamina: Small white fiber tract that separates the lateral globus pallidus from the putamen.

Lateral olfactory stria: That portion of the olfactory tract that proceeds (laterally) to the region of the medial amygdala and prepyriform cortex.

Lateral posterior nucleus (of thalamus): Receives information from the superior colliculus and projects to the somatosensory association areas.

Lateral rectus: Eye muscle that enables the eye to look toward the outside (abduct), CN VI.

Lateral spinothalamic tract: Spinal tract whose inputs cross the midline of cord within a few segments of their point of entry and ascend contralaterally through the cord and brainstem to synapse primarily in the VPL nuclei of the thalamus. In conjunction with the ventral spinothalamic tract, it is largely responsible for carrying information relating to temperature, acute, sharply localized pain, and simple touch.

Lateralization (of function): Refers to the concept that different sides of the brain are responsible for or take the lead in processing certain classes of stimuli or in executing certain types of behaviors.

Lemniscal system: Refers to those posterior (spinal) column pathways (fasciculus gracilis and fasciculus cuneatus) which coalesce into the medial lemniscus in the medulla and carry information regarding stereognosis, proprioception, kinesthesia, and vibratory sense to the cortex via the thalamus.

Lemniscus: A generic term for a fiber tract that lies within the brainstem (see **Medial lemniscus** and **Lateral lemniscus**).

Lenticular fasciculus: Efferent fibers from the globus pallidus crossing the posterior limb of the internal capsule to enter the subthalamus (also known as Forel's field H2).

Lenticular (lentiform) nuclei: Collective term for the putamen and globus pallidus.

Lenticulostriate arteries: Small, deep penetrating arteries coming off the proximal portion of the middle cerebral artery that supply parts of the basal ganglia and internal capsule.

Leptomeninges: Term used to designate the combination of the pia and the arachnoid layers of the meninges.

Ligand: Substance (drug, neurotransmitter) that directly binds to a receptor.

Ligand-gated channels: Receptor composed of five transmembrane regions that form an ion channel, governing the flow of receptor-specific ions in and out of the cell.

Limbic lobe: Name given to a series of interconnected, older anatomical structures along the more medial aspects of the cerebral hemispheres that form an incomplete ring around the diencephalon. This includes the subcallosal gyrus (septal region) anteriorly, then continuing dorsally, posteriorly, and then looping forward and ventrolaterally encompassing the cingulate gyrus, the isthmus of the cingulate gyrus, and the parahippocampal gyrus (including the hippocampus proper).

Limbic system: Term derived from a combination of Broca's "limbic lobe" and Papez's description of a system of structures (including the limbic lobe, anterior nuclei of the thalamus, hypothalamus (mammillary bodies) and their interconnections (e.g., fornix, mammillothalamic tract) which was thought to play a significant role in the experience and expression of emotion. Since individually these structures likely reflect broad functional heterogeneity, the utility of the term "system" is widely disputed.

Line of Gennari: see **Bands of Baillarger**.

Lingual gyrus: Lying on the medial surface of the occipital lobe just below the calcarine fissure, this gyrus receives projections from the more temporal (inferior) optic radiations that carry information originating in the superior visual fields.

Linguistic prosody: Altering the meaning of spoken language by changing such features as the pitch or stress that is applied to a word(s) within a sentence or pauses between words (contrast with "affective prosody").

Locked-in syndrome: State where patient retains consciousness but is unable to make any type of response, except perhaps opening and closing of the eyelids and some eye movements. Typically results from lesions in the lower pons that disrupt the corticobulbar and corticospinal tract bilaterally.

Locus ceruleus: Pigmented pontine nuclei that gives rise to ascending and descending norepinephrine fibers.

Lower motor neuron lesion: Lesion affecting the anterior horn cells, ventral nerve roots, or peripheral nerves. Produces flaccid (decreased muscle tone) paralysis accompanied by significant muscle atrophy, hyporeflexia, and fibrillations.

Macula: The central portion of the eye that contains most of the color sensitive receptors (cones) and the fovea (site of sharpest vision).

Mammillary bodies: Small, rounded prominences visible on the ventral surface of the brain between the optic chiasm and the cerebral peduncles of the midbrain that represent the most posterior of the hypothalamic nuclei. Among their major connections are with the hippocampal formation via the fornix, with the anterior thalamus via the mammillothalamic tract and with the midbrain via the mammillotegmental tract.

Mammillotegmental tract: Arises from the mammillary bodies and projects to the dorsal and ventral tegmental nuclei. Along with the fasciculus retroflexus and dorsal longitudinal fasciculus, this tract is probably important in initiating visceral (autonomic) responses.

Mammillothalamic tract (tract of Vicq d'Ayzr): Arises from the medial portion of the mammillary bodies and projects to the anterior nuclei of the thalamus, which in turn projects to the cingulate gyrus.

Masked facies: Face characterized by lack or paucity of expression. Commonly seen in Parkinson's disease.

Mass action: Belief that essentially all behavior is a result of the entire brain functioning as a whole.

Medial forebrain bundle: A complex, bi-directional group of fiber tracts interconnecting basilar forebrain structures (e.g., septal nuclei), amygdala, hypothalamic nuclei, and brainstem nuclei. Also likely serves as a major pathway for monoaminergic (e.g., dopamine, serotonin, norepinephrine) inputs into the cortex.

Medial geniculates: Posterioventral thalamic nuclei that serve as the primary thalamic relay and integration centers for auditory information.

Medial lemniscus: Brainstem tract beginning with the crossing of the arcuate (sensory) fibers in the lower medulla and ascending to the VPL nucleus of the thalamus. Carries proprioceptive, tactile form (stereognosis), and vibratory sense information derived from the posterior columns.

Medial longitudinal fasciculus (MLF): A midline tract composed of various ascending and descending fibers seen from the midbrain through the cervical cord. At the upper levels, helps integrate eye movements with movements of the head, while at the cervical level it likely helps to maintain balance (equilibrium) with movements of the head.

Medial longitudinal stria (of Lancisi): see **Lateral longitudinal stria of Lancisi**.

Medial medullary lamina: Faint collection of white matter fibers dividing the globus pallidus into medial and lateral portions.

Medial rectus: Eye muscle that enables the eye to look toward the midline (adduct), CN III.

Medulla (medulla oblongata): The lower portion of the brainstem between the pons and the spinal cord. Some of its primary components include the pyramids and their decussation; the nucleus cuneatus and nucleus gracilis, the decussation of the lemniscal fibers and the beginning of the medial lemniscus; the olivary nuclei; the nucleus ambiguous (CN IX, X), hypoglossal nucleus (XII), solitary nucleus (VII, IX, X), dorsal motor nucleus (X); and the inferior cerebellar peduncles.

Meninges: Series of supportive and protective coverings for the brain and spinal cord which includes the dura mater, arachnoid, and pia mater.

Mesencephalic nucleus of V: Midbrain nucleus that receives proprioceptive and kinesthetic feedback from the muscles of mastication.

Mesencephalon: The upper part of the brainstem above the pons, the midbrain.

Mesocortex: Those cortical areas designated as "paralimbic," representing a transition between the more primitive allocortex (e.g., hippocampal formation) and the neocortex. The cingulate gyrus and entorhinal cortex of the parahippocampal gyrus are the two more prominent examples of this type of cortex.

Mesocortical pathway: Generally consisting of mesial frontal (probably including the anterior portion of the cingulate gyrus) and mesial temporal cortices, this region has been associated with dopamine depletion and resulting "negative" symptoms of schizophrenia.

Mesolimbic pathway: Generally consisting of subcortical nuclei in the mesial and orbitofrontal regions, this region has been associated with excessive dopaminergic activity in schizophrenic patients and is thought to contribute to the "positive" symptoms (e.g., hallucinations, delusions) of this disorder.

Metabotropic receptors: see **G-protein linked receptors**.

Metencephalon: The pons and the cerebellum.

Meyer's loop: That portion of the optic radiations that loop slightly forward through the temporal regions on their way to the occipital cortex. These fibers carry superior visual field information.

Midbrain: The upper portion of the brainstem lying between the diencephalon and the pons. Primary components include the cerebral peduncles, the decussation of the superior cerebellar peduncle, the superior and inferior colliculi, the red nucleus, the nuclei for the third and fourth cranial nerves, and the substantia nigra.

Middle cerebellar peduncle ("brachium pontis"): Reflects the continuation of the transverse corticopontocerebellar fibers visible on the anterior and lateral surface of the pons. Thus, it carries information from the contralateral cerebral hemisphere to the cerebellum.

Middle cerebral artery: Supplies most of the lateral surface of the cerebral hemisphere.

Midline nuclei (of thalamus): Small nuclear groups that are associated with the more primitive cortical areas (limbic and basal frontal regions).

Mossy fibers: Fibers entering the cerebellum, except those (climbing fibers) from the inferior olivary nuclei.

Muscarinic receptors: Cholinergic receptors, primarily G-protein-linked, which are prominent in the postganglionic parasympathetic nervous system, as well as in the brain.

Myelencephalon: The medulla oblongata (or, medulla).

Neurohypophysis: The neural or posterior portion of the pituitary gland and the infundibular stem.

Neuroleptic malignant syndrome: Potentially fatal side effect of neuroleptic medications. Symptoms can include muscular rigidity, elevated temperature, altered consciousness, low blood pressure, tachycardia, and sweating.

Neuromodulator: A compound (e.g., a neuropeptide) that, while not evoking an independent response on its own, has the capacity to alter (modulate) the behavioral response to the neurotransmitters with which it is associated.

Nicotinic receptors: Cholinergic receptors, primarily ion channel type, which are found in the ganglionic cells of both the sympathetic and parasympathetic nervous system, as well as being very prominent in neuromuscular junctions.

NMDA (N-methyl- D-aspartate) receptor: One of three glutaminergic (excitatory) ionotropic receptors (the other two being AMPA and kainate receptors), which is primarily known as a voltage-dependent, calcium channel receptor.

Nodulus: Most ventral or anterior portion of the vermis of the cerebellum–the central portion of the flocculonodular lobe.

Nonfluent aphasia: Typically characterized by effortful, dysarthric articulations; reduced, "telegraphic" output; with comprehension relatively preserved (e.g., Broca's aphasia). Sometimes also referred to as "motor," "anterior," or "expressive" aphasia. Lesions are usually in the area of the frontal operculum.

Nucleus accumbens: see **Accumbens nucleus**.

Nucleus ambiguus: A visually rather indistinct nucleus (hence, its name) in the central medulla which is the source of the lower motor neurons of CN IX and X supplying the muscles of the larynx, pharynx, and soft palate used in swallowing and phonation. Also the source of preganglionic, parasympathetic vagal fibers to the heart.

Nucleus cuneatus: Nucleus in lower medulla which, along with the nucleus gracilis, gives rise to the medial lemniscus which, in turn, transmits proprioceptive and stereognostic information to the cortex via the VPL nucleus of the thalamus. The nucleus cuneatus mediates input from the upper extremities.

Nucleus gracilis: Similar to nucleus except mediates input from lower extremities.

Nystagmus: An involuntary oscillation of the eyes most commonly seen on extreme lateral gaze and characterized by a slow component in one direction and a fast component in the other. Most often seen with lesions affecting the vestibular system or cerebellum.

Occipitotemporal gyrus: see **Fusiform gyrus**.

Oculocephalic reflex: If the head of a comatose patient is briskly moved from side to side, the eyes should move in the opposite direction (doll's eyes movement). Failure to show such a response in a comatose (but not in an awake patient) suggests brainstem dysfunction.

Oculovestibular reflex: Automatic adjustment of the eyes to focus on target as the head is moved voluntarily.

Oculomotor nerve (CN III): Controls superior, inferior and medial rectus, and inferior oblique muscles, elevates eyelid and constricts pupil.

Oculomotor nucleus: Midbrain nucleus that mediates elevation of eyelid and eye movements of CN III.

Operculum: Refers to that portion of the cerebral cortex (frontal, temporal, or parietal) that is enfolded into the lateral fissure.

Optic chiasm: Point at which the nasal (medial) fibers of each optic nerve cross the midline, while the lateral fibers remain ipsilateral. Marks the end of the optic nerves and the beginning of the optic tracts.

Optic nerve: Axonal fibers that arise from the ganglion cells of the eye. Although they are continuous fibers (do not synapse), after they leave the optic chiasm they are then referred to as the optic tract.

Optic radiations: Visual fiber pathways from the lateral geniculates of the thalamus to the occipital cortex (area 17).

Optic tract: Fibers representing a continuation of the axons of the ganglion cells of the eye from the optic chiasm to the lateral geniculates.

Pain asymbolia: Rare condition in which, although accurately able to describe painful stimuli, there is no affective response to it (i.e., no subjective sense of "hurt" or discomfort). Most commonly associated with lesions of parietal lobe and likely represents a disconnection between cortical and limbic systems. (Does not refer to individuals with congenital inability to experience pain).

Pallidum: Globus pallidus.

Papez's circuit: Interconnection of subcortical structures, including the hippocampal formation, fornix, mammillary bodies, mammillothalamic tract, anterior nucleus of the thalamus, and cingulate gyrus, that were thought to be important for the experience and expression of emotion.

Papilledema: Swelling of the optic disk (region where optic nerves exit and blood vessels enter the eye), typically associated with increased intracranial pressure, as from a mass lesion, hemorrhage, or infectious process.

Parahippocampal gyrus: Most medial of the "temporal" gyri when viewing the ventral surface of the brain. Contains the amygdala and is continuous with the hippocampal formation that lies buried beneath its surface.

Paralimbic: Those limbic structures (such as the cingulate gyrus, parts of the parahippocampal gyrus, and posterior orbital–frontal regions) that evidence increasingly complex cytoarchitecture (although still lack the six-layered features of neocortex) and lack direct connections with the hypothalamus.

Paramedian pontine reticular formation (PPRF): Cells in pontine portion of the reticular formation that facilitate lateral conjugate gaze.

Paraphasia: Aphasic symptom involving either the substitution of one phoneme or syllable for another (literal paraphasia), one word for another (verbal paraphasia); or the creation of a word with no recognizable referent (neologism).

Paraplegia: Paralysis of both lower extremities.

Parastriate cortex: Brodmann's area 18 (visual association area).

Parasubiculum: Part of parahippocampal gyrus interposed between the entorhinal cortex and the presubiculum.

Parasympathetic nervous system: Balances the action of the sympathetic system (maintains a more "normal" resting baseline). Cells of origin are located in the brainstem and sacral region of the spinal cord. Postsynaptic fibers are cholinergic.

Parenchyma: The tissue which is characteristic of a particular organ. In this context, the brain, or nervous tissue itself.

Parinaud's syndrome: Disturbance of upper conjugate gaze and pupils that react sluggishly to light, normally to accommodation, resulting from lesions of the dorsal midbrain.

Partial agonist: Drug (e.g., buspirone) that has the capacity to interact with (stimulate) a receptor but in a less powerful fashion than a full receptor. Because of their unique characteristics, partial agonists may exert an agonist-like effect when there is a relative shortage of a given receptor, while acting like a weak antagonist when there is an overabundance of a given transmitter.

Pars opercularis: Most posterior portion of the inferior frontal gyrus (Brodmann's area 44). Part of what is normally referred to as "Broca's area."

Pars orbitalis: Most ventral portion of the inferior frontal gyrus (Brodmann's area 47).

Pars triangularis: Central portion of the inferior frontal gyrus (Brodmann's area 45). Part of what is normally referred to as "Broca's area."

Pedunculopontine nucleus: Located in the midbrain tegmentum, it receives input both from the motor cortex and the cerebellum (via the superior cerebellar peduncle) and is interconnected with the basal ganglia via the subthalamus and substantia nigra.

Perforant pathway: Primary input into the hippocampal formation from the entorhinal cortex. Dentate gyrus is likely the primary recipient of these fibers.

Perirhinal: Area bordering either side of the collateral sulcus. This would include the parahippocampal and rostral portions of the fusiform gyrus.

Peristriate cortex: Refers to Brodmann's area 19, representing (along with area 18) the visual association cortices. Areas 18 and 19 are sometimes collectively referred to as the "prestriate" cortex.

Phospholipase C: An enzyme that is linked to another "second messenger" system in the neuron (see **cAMP**).

Pia mater: Very thin, innermost meningeal layer that contains fine blood vessels and follows the contour of the brain into all the sulci.

Planum temporale: That portion of the superior temporal gyrus lying enfolded within the temporal operculum (i.e., inside the lateral fissure) and posterior to Heschl's gyrus. Considered part of Wernicke's area in the left hemisphere, it typically tends to be larger on the left than on the right in the human brain.

Pons: The "middle" section of the brainstem between the midbrain and the medulla. Some of its unique components include the pontine nuclei and the pontocerebellar fibers making up the middle cerebellar peduncle, nuclei for CN VI, the motor nuclei for VII, and most of the nuclei for VIII (the nuclei for CN V extend throughout all three segments of the brainstem).

Pontine Nuclei: Multiple nuclei scattered in the basal portion of the pons serving as a relay for the crossed corticopontocerebellar fibers making up the middle cerebellar peduncle.

Posterior cerebral artery: Supplies the occipital pole, inferior, and medial temporal regions and medial aspect of posterior parietal and occipital cortices.

Posterior choroidal artery: Derived from the posterior cerebral supplies the choroid plexus of the third ventricle and the tectal region of the midbrain.

Posterior columns: Ascending sensory fibers in the dorsal funiculi of the spinal cord consisting of the fasciculus cuneatus (upper cord only) and fasciculus gracilis (see **Lemniscal system**).

Posterior commissure: Located at the upper limits of the brainstem, probably interconnects local nuclei that subserve visual reflexes (i.e., eye movements).

Posterior communicating arteries: Connect the posterior (vertebral) system with the carotid system in the circle of Willis

Posterior nucleus (of thalamus): Receiving more diffuse somatosensory input than the ventral posterior nucleus and projecting to the insular cortex, it is thought to play a role in the perception of pain.

Posterior perforated substance: Region of the hypothalamus lying between the mammillary bodies and the crus cerebri.

Praxis: The ability to carry out a skilled action, either an intransitive gesture (e.g., saluting) or a transitive response, whether by pantomime or with the actual object (e.g., using a hammer), as a result of having established the motor programs for that action by virtue of prior practice (see **Ideomotor apraxia**).

Prefrontal cortex: The more anterior portions of the frontal lobe, also known as the "frontal granular cortex."

Premotor area: Typically used to refer to that part of Brodmann's area 6 that lies on the lateral surface of the cortex. More broadly, it may include areas 8, 44, and 45. Apparently plays a major role in the organization, timing, and integration of fine or complex motor skills.

Presubiculum: Part of the parahippocampal gyrus that lies adjacent to the subiculum of the hippocampal formation.

Pretectal area: Lying at the diencephalic–midbrain junction just rostral to the superior colliculi, this region receives direct input from the optic tracts and appears to be important for light reflexes.

Primary cortex: Also known as heterotypic or idiotypic cortex, it either receives direct input from a single sensory modality (organ) via the sensory relay nuclei of the thalamus e.g., medial (auditory) or lateral (visual) geniculates or the ventral posterior nuclei (somatosensory) of the thalamus; or represents the final common cortical motor pathway (e.g., area 4). Unlike homotypic isocortex, there is less clear differentiation between pyramidal and stellate (granular) cell layers.

Primary motor cortex: Also referred to as the "precentral gyrus," or Brodmann's area 4, this highly topographically organized region represents the final common cortical pathway for carrying out all voluntary motor activity, except for eye movements.

Principal sensory nucleus of V: Located in the pons, this nucleus mediates the finer aspects of touch for the face (comparable to the medial lemniscal system for the extremities).

Projection fibers: Tracts that interconnect an area of the cortex with subcortical or non-cortical structures (e.g., thalamus, basal ganglia, or spinal cord).

Pronator drift: The tendency for the arm contralateral to a cortical lesion, particularly in the area surrounding the central sulcus, to "drift" from the horizontal position when the arms are extended with the palms facing upward and the eyes are closed.

Proprioception: Ability to perceive, through somatosensory feedback, the position of a part of the body in space (e.g., being aware of where one's hand is without having to look at it). It is mediated by posterior columns of the spinal cord and parietal association cortex.

Prosopagnosia: In its pure form, it reflects an inability to recognize familiar faces (or learn to recognize new ones) in the absence of otherwise significant disturbances of visual perception. Usually associated with bilateral occipital-temporal lesions.

Pseudobulbar palsy: A condition generally produced by bilateral lesions (e.g., lacunar infarcts) which affect descending frontal pathways. The effects can mimic brainstem lesions ("pseudobulbar") by producing difficulties in swallowing, motor speech, or other facial movements, as well as producing upper motor neuron type changes in the extremities. In addition to motor changes, there is an increase in emotional lability or emotional "incontinence" that tends to be quantitatively rather than qualitatively inappropriate.

PTO cortex: Cortex that is at the juncture of the parietal, temporal, and occipital lobes where clear differentiation is difficult.

Ptosis: Drooping of the eyelid may result from disruption of either sympathetic fibers (milder) or those of the third cranial nerve (generally more severe).

Pulvinar: The largest of the thalamic nuclei, it has extensive connections with the posterior association cortices.

Pure alexia: Inability to comprehend written language, except through letter-by-letter analysis of each word (see: **Alexia without agraphia**).

Pure word blindness: see **Alexia without agraphia**.

Pure word deafness: Inability to comprehend spoken language, while hearing and reading comprehension remain intact. Relatively rare disorder resulting from a functional disconnection between auditory input and Wernicke's area in the dominant hemisphere, usually associated with deep, bitemporal lesions of the middle portions of the superior temporal gyrus.

Purkinje cells: Very large cells in the cerebellum that project to the deep cerebellar nuclei and are the source of most of efferent output of the cerebellum.

Putamen: Part of the basal ganglia, the nuclear mass lies just lateral to the globus pallidus and medial to the external capsule. Receiving input from the cortex, particularly from the areas surrounding the central sulcus, as well as other subcortical structures and sending fibers to the globus pallidus and the substantia nigra, it likely is involved with modulating motor activity.

Pyramids: That remaining portion of the cerebral peduncles that exits from the lower pons and can be seen as an eminence along the ventral surface of the medulla and that eventually constitutes the lateral and ventral corticospinal tracts.

Pyriform cortex: Associated with the olfactory system, this area refers to the region surrounding the lateral olfactory stria in the anterior portion of the parahippocampal gyrus.

Quadrantanopia: Deficit that involves one quadrant of the visual field for a given eye. A right superior, homonymous quadrantanopia, for example, would describe a right upper field defect in each eye.

Raphe nuclei: Midline tegmental nuclei (part of reticular formation) that are a source of serotonin.

Rebound: Exacerbation of original (or de novo) symptoms resulting from supersensitive (up-regulated) receptors following the discontinuation of a medication. Rebound should be distinguished from *relapse* (simply a return of symptoms in an unresolved disorder) and from *withdrawal* (the effect of stopping a physiologically addicting drug).

Red nucleus: Large nucleus located in the lower midbrain with connections to the cerebral cortex, inferior olivary nucleus, cerebellum, and spinal cord, hence thought to be involved in coordination.

Restiform body: The major portion of the inferior cerebellar peduncle carrying all somatosensory information to the cerebellum, except for that conveyed by the ventral spinocerebellar tract.

Reticular activating system (RAS): As classically defined, refers to that functional component of the reticular system that is sensitive to polymodal sensory inputs and that has an "activating" influence on the thalamus, which in turn has arousal effects on the cortex. However, other neurochemical pathways outside the reticular formation also likely have implications for cortical "arousal."

Reticular formation: Collective term for a somewhat heterogeneous group of cells extending throughout the brainstem tegmentum that appear to receive inputs (of a somewhat diffuse nature) from sensory systems, the cortex, and the cerebellum. Its ascending fibers generally have been associated with cortical arousal ("reticular activating system"), but it also has a descending component that may play a role in motor and/or sensory gating.

Reticular nuclei (of thalamus): Thin sheet of cells along the lateral surface of the thalamus that receive diffuse input from the cortex and project back to the thalamus and to the reticular formation.

Reticulospinal tract: Consists of descending pathways (medial and lateral reticulospinal tracts) from the pontine and medullary components of the reticular system. Probably helps maintain balance, posture, and gross axial movements.

Right–left disorientation: Acquired inability to reliably differentiate right **versus** left either on one's self (personal) or on another (extrapersonal).

Rigidity: Stiffness of movement associated with increased muscle tone. May be of an intermittent (cogwheel) or steady (lead-pipe) variety. Frequently seen in Parkinson's.

Rinne test: An aid to differentiating between a conduction **versus** a neural hearing loss, this test consists of asking a patient to compare the relative intensity (loudness) of a tuning fork placed against the mastoid process **versus** simply being held close to the external ear. In a conduction loss, the sound is louder (and lasts longer) via the bone (mastoid) conduction; in a neurosensory loss, both bone and air conduction are diminished (also see **Weber test**).

Romberg test: Having the patient stand with feet together and eyes closed. If patient has trouble maintaining balance under these (but not other) conditions, the test is said to be "positive" and is suggestive of posterior column disturbance.

Rostral: From the Latin word for beak, it has to symbolize the foremost, front, or most anterior portion. In anatomy, it refers to the direction toward the head or in the case of the brain itself toward the frontal region.

Rostral spinocerebellar tract: Carries information from the upper extremities to the cerebellum (see **Cuneocerebellar tract**).

Rostrum: Refers to anterior descending portion of the corpus callosum.

Rubrospinal tract: Fiber tract connecting the red nucleus to spinal neurons. This tract travels in the lateral funiculus (adjacent and anterior to the lateral corticospinal tract). Because of the red nucleus' extensive interconnections with the cerebellum, it likely plays a role in coordination of the limbs.

Salivatory nucleus: Nuclear groups (inferior and superior) in the rostral medulla that contribute fibers to CN VII and IX and innervate salivary, lacrima,l and parotid glands.

Secondary cortex: Represents a higher stage or level of processing or integration of sensory input while remaining unimodal in character (e.g., visual or auditory percepts or symbols); or (in frontal cortices) responsible for the organization and sequencing of the discrete, individual aspects of a complex motor response or movement. Also known as modality specific or unimodal association cortex.

Second messengers: Complex protein produced within the cell through the interaction of G-proteins and enzymes (G-protein-linked receptors) which then can carry out multiple chores (e.g., open or close ion channels, create new enzymes, alter the synthesis of receptors).

Sella turcica: Indentation in base of skull in which the pituitary gland rests.

Sensory ataxia: Ataxia resulting from a disruption of somatosensory feedback.

Septal nuclei: Nuclei that lie below the rostrum of the corpus callosum and anterior to the anterior commissure. Major connections are with the amygdala (via the stria terminalis), the habenular nuclei of the thalamus (stria medullaris thalami), the hippocampus (fornix), and the basal forebrain, amygdala, hypothalamus, and brainstem (by way of the diagonal band of Broca and the medial forebrain bundle). Like the amygdala, these nuclei often are linked to emotional expression, although often in a manner opposite that of the amygdala.

Simultanagnosia: The inability to recognize a complex visual picture or correctly interpret the thematic content of a visual scene because of an inability of the patient to focus on, visually attend to, or recognize more than one feature or aspect of the visual image at a time. Distinctions are sometimes made between dorsal and ventral simultanagnosia. Dorsal simultanagnosia, which is associated with bilateral parieto-occipital lesions and Balint's syndrome, purportedly reflects an inability to attend to more than one object or feature of an object at a time. Patients with ventral simultanagnosia, reportedly resulting from left temporal–occipital lesions, are described as being better able to shift visual focus or attention but having greater difficulty recognizing and/or integrating complex visual elements or scenes, and reading is likely to be accomplished letter-by-letter.

Solitary nucleus: Medullary nucleus whose more rostral segment receives information regarding taste from CN VII, IX, and X, while its more caudal portion is dedicated to visceral feedback ranging from the oxygen content of the blood (CN IX) to the state of internal organs (CN X).

Somatodendritic receptors: Autoreceptors found on the cell bodies of neurons or adjacent dendrites that regulate the rate of firing of the neuron.

Somatosensory: Sensory system that mediates sensations arising from the skin, joints, or muscles such as touch, temperature, relative position, and movement of the limbs and pain.

Spatial dyscalculia: Difficulty in carrying out certain arithmetical operations (mentally or written) because of difficulties handling the spatial alignment of the numbers, or other "spatial" operations (e.g., carrying numbers). Associated with lesions of the non-dominant hemisphere.

Spinal accessory nerve (CN XI): Exiting along the lateral medulla, CN XI innervates trapezius and sternocleidomastoid muscles, which respectively elevate the shoulders and turn the head.

Spinal accessory nucleus: Lies in upper cord (C1-C5) and is origin of motor fibers innervating the trapezius and sternocleidomastoid muscles (CN XI).

Spinal nucleus of V: Nucleus that extends from pons into the cord and is thought to be primarily responsible for sensations of pain and temperature (and probably some simple touch as well) in the face.

Spinotectal tract: Provides somatosensory information to the superior colliculi. Unclear, but may serve role in visual orientation or visual reflexes.

Spinothalamic tract: see **Lateral spinothalamic tract**.

Splenium: The most posterior aspect of the corpus callosum.

Split-brain: Surgical procedure in which the cerebral commissures (typically only the corpus callosum) are cut, thus preventing the normal exchange of information between the hemispheres.

Stereognosis: Ability to recognize the form of an object, e.g., its shape, size, and texture. It is mediated by the posterior columns of the spinal cord and parietal association cortex.

Strabismus: Lack of conjugate gaze due to weakness of one or more extrinsic eye muscles. May result in double vision when looking in critical direction.

Straight sinus: Lying along the midline on the dorsal surface of the cerebellum, this vessel provides the final common pathway for the inferior sagittal sinus, the deep, internal, and basal cerebral veins to the transverse sinus.

Stria medullaris (thalami): Fiber tract that lies on the dorsomedial surface of the thalamus connects the septal nuclei with the habenular nuclei (of the epithalamus).

Stria terminalis: Another of the C-shaped fiber pathways, it lies along the ventromedial surface of the caudate nucleus and represents a major output of the amygdala. After completing almost a 360-degree loop around the dorsal thalamus, it establishes connections with the hypothalamus and portions of the basal forebrain (e.g., bed nuclei of the stria terminalis and likely the septal nuclei and medial forebrain bundle).

Striatum: Collective term for the caudate nucleus and the putamen (also known as the neostriatum).

Subarachnoid hemorrhage: Bleeding into the space within the arachnoid layer, most commonly resulting from a spontaneous hemorrhage (e.g., from an arteriovenous malformation or aneurysm, or hypertension).

Subdural hematoma: Bleeding that accumulates between the dura and the arachnoid, usually as a result of trauma.

Subiculum: Outermost portion of the hippocampal formation that represents a transition between the six-layered entorhinal cortex and the three-layered hippocampus proper. It is the major source of postcommissural fibers of the fornix, as well as a major source of other subcortical and cortical efferents from the hippocampal formation.

Substantia gelatinosa: Column of neurons in lamina II of the spinal cord that are largely involved in the cortical modulation of pain.

Substantia innominata: Literally, "unnamed substance," refers to the area that lies roughly between the anterior perforated substance and the anterior commissure in the basilar portion of the forebrain. Now known to incorporate such structures as the basal nucleus of Meynert, as well as portions of other nuclear groupings such as the extreme anteroventral aspect of the basal ganglia and the medial extension of the centromedial portion of the amygdala.

Substantia nigra: Dark band of cells in the midbrain tegmentum. Closely linked to the basal ganglia and the source of dopamine.

Subthalamic fasciculus: Two-way fiber bundle carrying GABAergic fibers from the external segment of the globus pallidus to the subthalamus and returning glutaminergic fibers to the internal segment.

Subthalamus: Collection of nuclear groupings between the thalamus proper and the substantia nigra. Is heavily involved in the feedback loop encompassing the corpus striatum, substantia nigra and the thalamus.

Sulcus: The grooves or valleys on the surface of the cortex, some of the larger of which are called fissures.

Superior cerebellar peduncle "*brachium conjunctivum*": it is the chief efferent pathway from the cerebellum back to the cortex and brainstem nuclei (except for those fibers from the fastigial nuclei which project to the vestibular and reticular nuclei via the juxtarestiform body). Also carries afferent fibers from the ventral spinocerebellar tract.

Superior colliculi: Nuclear grouping on the dorsal aspect (tectum) of the midbrain. Serves as a relay for visual information and is probably involved in involuntary visual tracking and other visual reflexes.

Superior longitudinal fasciculus: Lying above the insula and claustrum, this pathway (also known as arcuate fasciculus) provides one of the major interconnections between the frontal and posterior cortices.

Superior longitudinal fissure: see **Superior sagittal fissure**.

Superior oblique: Eye muscle that facilitates looking down when the eye is turned inwardly (adducted), mediated by CN IV.

Superior occipitofrontal fasciculus: Lying dorsolateral to the head of the caudate on a coronal section (also known as subcallosal bundle), along with the superior longitudinal fasciculus, links the frontal cortices to the posterior regions of the brain.

Superior olivary nucleus: Pontine nucleus that is a relay for many of the auditory fibers from the cochlear nuclei in route to the inferior colliculi via the lateral lemniscus.

Superior rectus: Eye muscle that facilitates looking up, especially when eye is deviated laterally, mediated by CN III.

Superior sagittal fissure: Also known as the *superior longitudinal fissure*, this large, deep sulcus or fissure separates the left and right cerebral hemispheres.

Superior sagittal sinus: Formed by the dura, it runs longitudinally along the inner vertex of the skull providing venous drainage for the vessels on the dorsolateral surface of the brain. Posteriorly, it joins with the straight and transverse sinuses.

Suppression: Tendency to ignore or failure to perceive stimuli on one side of the body when comparable stimuli are applied at the same time bilaterally (double simultaneous stimulation).

Supplementary motor area: Motor association cortex (area 6) that lies on the medial aspect of the frontal lobe. Divided into pre-SMA and SMA proper, these areas appear to be important in the transition between the intention to act and initiating or facilitating the execution of motor programs.

Supramarginal gyrus: Located anterior to the angular gyrus in the inferior parietal lobule, it surrounds the terminal portion of the posterior ascending ramus of the lateral fissure. Lesions to this region (Brodmann's area 40) in the dominant hemisphere have been associated with disturbances of writing, praxis, and body schema.

Supranuclear: Refers to pathways or processes between a nucleus and the cortex or other "higher" cortical centers.

Sustention tremor: Tremor induced when limb is held in fixed position against gravity. Suggestive of cerebellar dysfunction.

Sylvian fissure: see **Lateral fissure**.

Sympathetic nervous system: Primarily responsible for preparing the organism for a "fight or flight" response. Cells of origin are located in the thoracic and upper lumbar regions of the spinal cord (lateral horn). Postsynaptic fibers are adrenergic.

Symptom: An integrated behavioral response (or perhaps a lack thereof despite a normally sufficient stimulus) that suggests an underlying lesion or disease process.

Tactile agnosia: Inability to recognize (i.e., identify) an object either through naming or demonstrating its use by tactual means alone, despite the preserved ability to accurately describe its physical characteristics, such as size, shape, and texture.

Tactile asymboly: Older term for tactile agnosia.

Tectum: The dorsal-most part of the midbrain lying above (or behind) the cerebral aqueduct. Primarily made up by the inferior and superior colliculi.

Tegmentum: The "central" portion of the brainstem which contains the reticular system, all the cranial nerve nuclei, the substantia nigra and red nuclei (midbrain), and the olivary nuclei (medulla).

Tentorial notch: The fairly rigid opening in the dura at the base of the brain (at the level of the midbrain) through which the brainstem passes.

Tentorium: That part of the dura that lies over ("tents") the cerebellum and separates it from the occipital lobe.

Tertiary cortex: Those areas responsible for the highest level of multisensory integration or associations (e.g., being able to recall the smell, texture, and taste of a crisp, juicy red apple when seeing or just thinking about one), or (in the frontal cortices), the supraordinate planning, organization and ongoing monitoring of behavioral responses to the environment. Also known as heteromodal association cortex.

Thalamic fasciculus: Rostral continuation of the fibers of the ansa lenticularis joined by the lenticular fasciculus and fibers from the dentate nucleus of the cerebellum as both make their way to the thalamus (also known as Forel's field H1).

Thrombus: The occlusion (in whole or in part) of a vessel as a result of a buildup of tissue within its walls.

Tinnitus: A persistent ringing in the ear. May be a sign of inner ear or nerve damage or excessive use of aspirin.

Titubation: A tremor of the head. May be associated with cerebellar disease.

Tonsillar herniation: Compression of the cerebellar tonsils into the foramen magnum (opening at the base of the skull), usually as a result of a posterior fossa lesion. Initial symptoms may include a occipital headache and stiff neck. However due to compression of the medulla, respiratory arrest (and death) may occur.

Topographical disorientation: Specific difficulty in negotiating external space or creating topographical maps (e.g., floor plan of one's house). Primary deficit is the inability to utilize or generate internal, spatial maps. Typically the result of bilateral (occasionally, unilateral right) parietal-occipital lesions (see **Topographical memory**).

Topographical memory: Memory for geographical or topographical relationships. May be tested by having patient reconstruct the floor plan of their house, the layout of the major streets or buildings in the city, the location of major cities on a state or national map (assuming previous familiarity). In the absence of more primary deficits, would suggest bilateral or right posterior lesions.

Topographical representation: Pattern of neuronal arrangement in which the different parts of the body are mapped in a distinct and orderly fashion in a tract or on the cortex.

Tractus solitarius: (*"fasciculus solitarius"*)**:** Consists of gustatory and other visceral afferent fibers from CN VII, IX, and X, which enter the solitary nucleus (see **Solitary nucleus**).

Transcortical aphasia: Types of aphasia where the ability to repeat is relatively preserved in the face of either severe comprehension (transcortical sensory) or expressive (transcortical motor) deficits.

Transient global amnesia: A sudden and more or less total loss of episodic memory that typically resolves within a matter of hours. Most likely of vascular etiology, but dynamics are uncertain.

Transient ischemic attack (TIA): Typically a small, focal stroke in which, by definition, the symptoms, for most practical purposes, appear to resolve in less than an hour.

Transverse sinuses: Formed at the convergence of the superior sagittal and straight sinuses, the transverse sinuses proceed anteriorly around the dorsal curvature of the cerebellum to become the sigmoid sinuses and then the jugular veins.

Trapezoid body: Auditory pathway in the lower pontine tegmentum that represents ventral cochlear nucleus fibers (via the ventral acoustic stria) crossing the midline to enter the lateral lemniscus either directly or indirectly after synapsing in the ipsilateral or contralateral superior olivary nuclei.

Tremor: Rhythmical shaking of an extremity. May be present either at rest ("resting tremor"), as in Parkinson's disease, or more noticeable when attempting to carry out an action ("intention tremor"), as in cerebellar lesions. If the tremor involves the head, it is referred to as titubation.

Trigeminal nerve (CN V): Primary source of somatosensory feedback from face and anterior portion of the head and controls muscles of mastication.

Trigeminal neuralgia (tic douloureux): Sharp, paroxysmal facial pains secondary to irritations of one or more branches of CN V.

Trigeminothalamic tract: Major (crossed) pathway that carries information concerning pain, temperature, touch, and proprioception from the face to the ventral posterior medial nucleus of the thalamus, which in turn projects to the somatosensory cortex.

Trigone (of the ventricles): see **Atrium**

Trochlear nerve (CN IV): Controls the superior oblique muscle of the eye.

Trochlear nucleus: Located in the midbrain and serving CN IV, it is the only cranial nerve nucleus whose nerve roots exit from the dorsal aspect of the brainstem and completely decussate.

Uncinate fasciculus: That part of the more extensive inferior occipitofrontal fasciculus that interconnects the orbital frontal regions with the anterior temporal cortex.

Uncal herniation: Compression of the medial aspect of the temporal lobe (uncus) and upper brainstem against the tentorial notch as a result of increased intracranial pressure. Initial symptoms are usually unilateral weakness and a dilated, unresponsive pupil as a result of compression of the cerebral peduncle and third CN, respectively. If severe, symptoms can become bilateral and coma may ensue.

Uncus: Refers to the external prominence on the anteromedial surface of the parahippocampal gyrus created by the presence of the amygdala beneath it.

Unilateral neglect: The tendency to ignore or failure to properly or consistently attend to one side of one's body or to external stimuli or environmental cues impinging on that side. When the neglect is projected into space (i.e., ignoring visual or auditory cues to either the right or left of midline–more commonly the latter), the term *unilateral spatial neglect* is frequently used. Neglect may also include motor acts or responses. Typically the result of a lesion in the contralateral hemisphere.

Unimodal cortex: see **Secondary cortex**.

Upper motor neuron lesion: Lesion that affects the brain or corticospinal tracts (prior to synapse in the anterior horn cells) resulting in paresis or paralysis characterized by increased muscle tone (spasticity), hyperreflexia, and "primitive" reflexes (e.g., palmomental, Babinski).

Up-regulation: The relative increase of receptor sites in response to a relative depletion of a given neurotransmitter (also referred to as "sensitization").

Utilization behavior: The tendency for a patient to pick up and handle objects within reach, even when there is no reason to do so. Suggestive of frontal pathology.

Vagus: The **tenth** cranial nerve, which innervates the muscles of the soft palate, pharynx, and larynx (swallowing and talking) as well as the source of parasympathetic innervation for many of the internal organs. It also carries sensory information (e.g., tactile and visceral feedback) from these same target muscles or organs.

Vasogenic edema: Commonly associated with neoplastic lesions, it results from the breakdown of the blood–brain barrier at the capillary level. Seen as finger-like extensions on neuroimaging surrounding a central lesion. (Contrast with **Cytotoxic edema**)

Ventral amygdalofugal tract: Pathway connecting the amygdala with the hypothalamus, part of the medial forebrain bundle.

Ventral anterior nucleus (of thalamus): Part of a motor feedback loop as it projects to the motor association cortices.

Ventral cochlear nuclei: Along with the dorsal cochlear nuclei, they represent the first relay for auditory fibers in the brainstem (upper medulla). In turn, they send fibers to the contralateral (majority) or ipsilateral inferior colliculi either directly or via the superior olivary nuclei. The crossing fibers exiting the ventral cochlear nuclei via the ventral acoustic stria (along with the crossing fibers from the superior olivary nuclei) constitute the trapezoid body in the lower pons.

Ventral corticospinal tract: A smaller (approximately 10%) component of the corticospinal tract that does not decussate at the lower medulla and travels ipsilaterally in the ventral funiculi of the spinal cord. Probably involved in mediating axial as opposed to appendicular (limb) movements.

Ventral median fissure: Fissure that runs rostrally and caudally along the ventral or anterior aspect of the cord.

Ventral posterior nucleus (of thalamus): Divided into two parts, the ventral posterior medial (head and face) and the ventral posterior lateral (trunk and extremities)—this nucleus receives somatosensory input and projects to the secondary somatosensory cortices.

Ventral spinocerebellar tract: Doubly crossed tract that carries information from the trunk and lower extremities to the cerebellum. Only spinocerebellar tract entering via the superior cerebellar peduncle.

Ventral spinothalamic tract: see **Lateral spinothalamic tract**.

Ventral trigeminal lemniscus (ventral trigeminothalamic tract): Along with the dorsal trigeminal lemniscus, conveys somatosensory information from the face to the VPM nucleus of the thalamus.

Vermis: That portion of the cerebellum that lies along its midline and along with the flocculonodular lobe is important for axial movements, postural corrections, and balance.

Vertebral arteries: Arteries lying on the ventral-lateral surface of the medulla that unite to form the basilar artery on the ventral surface of the pons, which, in turn, gives rise to the posterior cerebral arteries.

Vertigo: A visual sensation of external space as spinning or as assuming an abnormal orientation.

Vestibular nuclei: Collection of nuclei in the upper medulla and lower pons consisting of the superior, inferior, medial, and lateral vestibular nuclei. Collectively, these nuclei have extensive connections with the motion and position detectors of the inner ear, cerebellum (especially the flocculonodular lobe), spinal cord, cerebral cortex, and other brainstem nuclei and help mediate balance, visual tracking, and awareness of one's position in space.

Vestibulocerebellum: That portion of the cerebellum (e.g., flocculonodular lobe and anterior portions of the vermis) that are intimately connected with the vestibular nuclei and play a large role in maintaining equilibrium or balance.

Vestibulocochlear nerve (CN VIII): Gives rise to two divisions: the cochlear division, which carries auditory information, and the vestibular division, which mediates orientation with respect to gravity and acceleration (deceleration).

Vestibulo-ocular reflex: see **Oculovestibular reflex**.

Vestibulospinal tract: Pathway that has its origin in the lateral (lateral vestibulospinal tract) and medial (medial vestibulospinal tract) vestibular nuclei of the brainstem. Collectively, these tracts likely assist in making postural adjustments during whole body movement and/or helping to maintain balance against the forces of gravity.

Voltage-gated channels: Ion channels that preferentially respond to changes in background voltage (e.g., hyper- or hypopolarization) rather than strictly to the binding of a ligand.

Wallenberg syndrome: Results from a lesion of the lateral or dorsolateral medulla, which affects the lateral and ventral spinothalamic tracts, the spinal nucleus and tract of CN V, inferior cerebellar peduncle, and the vestibular and ambiguus nuclei. This syndrome typically presents with contralateral loss of pain and temperature in the lower extremities with an ipsilateral loss of facial pain and temperature and Horner's syndrome, ipsilateral ataxia and nystagmus, and difficulty swallowing. Because the more ventral and medial aspects of the brainstem remain intact, motor movement (corticospinal tracts) and position sense and stereognosis (medial lemniscus) are preserved.

Watershed syndrome: Resulting from a reduction in blood flow from a proximal vessel (typically carotid or internal carotid), the distal portions of the vessels involved are poorly perfused, resulting in areas of relative infarction where their distributions normally overlap (watershed areas). Symptoms generally are of a more complex cognitive or behavioral variety, as opposed to major sensory or motor deficits.

Weber's syndrome: Resulting from a lesion affecting the basilar portion of the midbrain, it is characterized by contralateral hemiparesis (due to involvement of the cerebral peduncles) and accompanied by an ipsilateral third nerve palsy (dilated pupil, ptosis, and paralysis of medial and vertical eye movements.

Weber test: An aid to helping differentiate between unilateral neurosensory versus conduction hearing loss, this test consists of placing a vibrating tuning fork on the center of the skull. If there is a conduction deficit, the sound will be perceived as louder in that ear or in the opposite ear with a neural loss (see **Rinne test**).

Wernicke's aphasia: Syndrome characterized by fluent, paraphasic speech, word finding, repetition, and comprehension difficulties, typically associated with lesions involving the posterior portion of the superior temporal gyrus.

Witzelsucht: Given to impulsive, inappropriate, often hurtful or inconsiderate remarks, usually done in a humorous vein and typically associated with orbitofrontal lesions.

Yakovlev's circuit: Another circuit (in addition to the one proposed by Papez) that was thought to subserve the more visceral aspects of emotion. This circuit included the dorsomedial nucleus of the thalamus, orbitofrontal cortex, temporal pole and amygdala.

Zona incerta: Small nuclear group in the subthalamus lying between two fiber pathways (lenticular fasciculus and thalamic fasciculus) that may represent a rostral continuation of the midbrain reticular formation.

INDEX